Raghu Murtugudde

PALEOCLIMATOLOGY
Second Edition

This is Volume 68 in the
INTERNATIONAL GEOPHYSICS SERIES
A series of monographs and textbooks
Edited by RENATA DMOWSKA and JAMES R. HOLTON

PALEOCLIMATOLOGY

Reconstructing Climates of the Quaternary

Second Edition

Raymond S. Bradley

University of Massachusetts
Amherst, Massachusetts

Amsterdam   Boston   Heidelberg   London   New York   Oxford
Paris   San Diego   San Francisco   Singapore   Sydney   Tokyo

Front Cover Photograph:
Rock Art at Zalat el Hammad, Southeastern Sahara, Northwest Sudan. Rock art from Zalat el Hammad (17°50' N, 26°45' E), a circular group of fractured sandstone hills, in the Wadi Howar, southeastern Sahara. The Wadi (1100 km long and 10 km wide) is a now defunct watercourse which 10,000–2000 yr ago was the Nile's largest tributary from the Sahara. The engravings depict domestic cattle, giraffes, elephants, lions, antelopes, monkeys, ostriches, desert foxes, barbary sheep and other wildlife, as well as portrayals of round-headed humans with dogs. Based on their different stylistic characteristics, grades of patination, and the composition of species, the depictions must be from different epochs of the early and mid-Holocene wet phase in the Sahara. Bones of all the big game fauna depicted have been found in nearby excavations of paleo-lake sediments of early and mid-Holocene age; they suggest that the engravings were based on local observations, and not reproduced from memory of other regions. The absence of camels suggests that at least since the beginning of the camel period (about 2000 yr ago) no more engravings were rendered due to the worsening living conditions. Through their art, our prehistoric ancestors have provided a vivid record of the dramatic changes of climate that have occurred in what is today one of the most arid locations on earth. (Photograph courtesy of Stefan Kröpelin, University of Köln.)

REFERENCES
Kröpelin, S. (1993). Zur Rekonstruktion der spätquartären Umwelt am Unteren Wadi Howar (Südöstliche Sahara / NW-Sudan). *Berliner Geographische Abhandlungen* 54.
Rhotert, H. (1952). *Libysche Felsbilder*. Darmstadt: Wittich.

This book is printed on acid-free paper. ∞

Copyright © 1999, 1985 by Academic Press

Permissions may be sought directly from Elsevier's Science and Technology Rights Department in Oxford, UK. Phone: (44) 1865 843830, Fax: (44) 1865 853333, e-mail: permissions@elsevier.co.uk. You may also complete your request on-line via the Elsevier homepage: http://www.elsevier.com by selecting "Customer Support" and then "Obtaining Permissions".

ACADEMIC PRESS
An Imprint of Elsevier
525 B Street, Suite 1900, San Diego, CA 92191-4495, USA
http://www.apnet.com

Academic Press Limited
24–28 Oval Road, London NW1 7DX, UK
http://www.hbuk.co.uk/ap/

Academic Press
An Imprint of Elsevier
200 Wheeler Road, Burlington, MA 01803
http://www.harcourt-ap.com

Library of Congress Number: 98-83154
ISBN-13: 978-0-12-124010-3
ISBN-10: 0-12-124010-X

PRINTED IN THE UNITED STATES OF AMERICA
07 DSG 9 8 7 6 5

To Jane

CONTENTS

3 Dating Methods I

4 Dating Methods II

5 Ice Cores

6 Marine Sediments and Corals

7 Non-marine Geological Evidence

8 Non-marine Biological Evidence

9 Pollen Analysis

10 Dendroclimatology

11 Documentary Data

12 Paleoclimate Models

PREFACE TO THE SECOND EDITION

When I wrote the first edition of *Quaternary Paleoclimatology* in the early 1980s the field of paleoclimatology was still in its infancy. Since then there has been an explosion of interest and research on the subject. It is amazing to realize that in the early 1980s AMS radiocarbon dating was hardly being used, we knew nothing of Dansgaard-Oeschger cycles and their relationship to the North Atlantic thermohaline circulation, the significance of Heinrich events had not been fully recognized, the first preliminary carbon dioxide measurements on ice cores were just being made, and no long ice cores had been recovered from the Tropics. General circulation models were crude and paleoclimatic simulations were rare. By contrast, paleoclimatology today is a major field in earth systems research and of vital importance to concerns over future global changes. As a result, the literature on the subject has grown immensely and it is becoming increasingly difficult to stay on top of the entire field.

In this edition I provide a contemporary overview of the field, but inevitably there will be topics that I may not have adequately reviewed. There are certainly topics currently under debate on which I may not have represented the full range of perspectives. Some important topics are omitted, or have been given only a cursory introduction. Such are the dangers of trying to cover such a wide field. However, I believe there are advantages in having one lens through which this rapidly evolving field is viewed, rather than a spectrum of perspectives that an edited volume of specialists might present. I hope that those specialists who turn to their particular areas

xiii

of expertise will do so with the overall objectives of the book in mind; a comprehensive review of every subfield cannot be reconciled with an up-to-date overview of the rest of paleoclimatology. The final product is thus a compromise between completeness, expediency, and (eventually) exhaustion. Nevertheless, I hope I have done justice to most topics, and that the new references I have included will enable interested readers to access the important literature quickly. Certainly, there is no substitute for reading the original scientific papers.

Apart from being more up to date, I believe this edition is a much more comprehensive overview of paleoclimatology and the record of climatic changes during the Quaternary than the first edition. All sections have been revised and updated. Particularly noteworthy changes include new material on dating (calibration of the radiocarbon timescale, amino acid geochronology, thermoluminescence [TL], optically stimulated luminescence [OSL], and infrared stimulated luminescence [IRSL]), a completely new chapter on ice cores, a longer and extensively revised chapter on marine sediments and ocean circulation in the past, new sections on corals, alkenones, loess, lake sediments (albeit brief), and greatly revised chapters on pollen analysis, tree rings, and historical records. I have also included a new chapter on paleoclimate models, emphasizing the increasing use of general circulation models in paleoclimatology. Over 1100 new references have been added and there are approximately 200 new figures. My goal has been to enable nonspecialists in any one subfield of paleoclimatology to learn enough of the basics in other subfields to allow them to read and appreciate the literature they might not otherwise understand. This will facilitate better communication of ideas within paleoclimatology and beyond. I leave it to the reader to decide how well this goal has been achieved.

Ray Bradley
Amherst, Massachusetts

ACKNOWLEDGMENTS

As in the first edition, I have benefited greatly from discussions with colleagues who are often far more deeply immersed in a particular aspect of paleoclimatology than I am. In particular, I have had the great luxury of meeting frequently with some of the leaders of the field while I was a member of the Scientific Steering Committee of the IGBP PAGES Project. Discussions with members of this committee were always stimulating.

Special thanks go to those individuals who took the time to read early drafts of this extensively revised edition and give me advice on what needed to be corrected or rewritten. They include: Ed Cook; Jean Jouzel; Lloyd Keigwin; Michael Mann; Gerry McCormac; Bill McCoy; Jon Pilcher; Reed Scherer; Tom Webb; and Ann Wintle (especially for guidance in OSL & IRSL). I also thank the graduate students who were subjected to drafts of various chapters and who gave me plenty of useful feedback: Caspar Ammann; Lesleigh Anderson; Ruediger Dold; Tammy Rittenauer; and Anne Waple. Additionally, I have been helped at various times by Mark Besonen, Jane Bradley, Julie Brigham-Grette, Laurie Brown, Tom Davis, Dirk Enters, Jennifer Hwang, Frank Keimig, Mark Leckie, Lorna Stinchfield, and Bernd Zolitschka. Many thanks to them all. Whatever faults may remain, I know the book is a lot better as a result of their efforts and I appreciate their help very much. If there are still errors or lapses in coverage, this is of course my responsibility and probably means I did not do what they suggested! So, blame me, not them. . . . I hope they find the final product justifies their time.

PALEOCLIMATIC RECONSTRUCTION

1.1 INTRODUCTION

Paleoclimatology is the study of climate prior to the period of instrumental measurements. Instrumental records span only a tiny fraction ($<10^{-7}$) of the Earth's climatic history and so provide a totally inadequate perspective on climatic variation and the evolution of climate today. A longer perspective on climatic variability can be obtained by the study of natural phenomena which are climate-dependent, and which incorporate into their structure a measure of this dependency. Such phenomena provide a proxy record of climate and it is the study of proxy data that is the foundation of paleoclimatology. As a more detailed and reliable record of past climatic fluctuations is built up, the possibility of identifying causes and mechanisms of climatic variation is increased. Thus, paleoclimatic data provide the basis for testing hypotheses about the causes of climatic change. Only when the causes of past climatic fluctuations are understood will it be possible to fully anticipate or forecast climatic variations in the future (Bradley and Eddy, 1991).

Studies of past climates must begin with an understanding of the types of proxy data available and the methods used in their analysis. One must be aware of the difficulties associated with each method used and of the assumptions each entails. With such a background, it may then be possible to synthesize different lines of evidence into a comprehensive picture of former climatic fluctuations, and to test hypotheses about the causes of climatic change. This book deals with the different types of proxy data and how these have been used in paleoclimatic reconstructions. The organization is methodological, but through discussion of examples, selected from major contributions in each field, an overview of the climatic record during the late Quaternary (the last ~1 Ma) is also provided. The climate of earlier periods can be studied using some of the methods discussed here (particularly those in Chapters 6, 7, and 9) but the farther back in time one goes, the greater are the problems of dating, preservation, disturbance, and hence interpretation. For a thorough discussion of climate over a much longer period, the reader is referred to Frakes *et al.* (1992).

Although our perspective on the past is obviously somewhat myopic, the Quaternary was a period of major environmental changes that were possibly greater than at any other time in the last 60 million years (Fig. 1.1). Nevertheless, there is no doubt that an understanding of climatic variation and change during the Quaternary period is necessary not only to appreciate many features of the natural environment today, but also to comprehend fully our present climate. Different components of the climate system change and respond to external factors at different rates (see Section 2.2); in order to understand the role such components play in the evolution of climate it is necessary to have a record considerably longer than the time it takes for them to undergo significant changes. For example, the growth and decay of continental ice sheets may take tens of thousands of years; in order to understand the factors leading up to such events and the effects such events subsequently have on climate, it is necessary to have a record considerably longer than the cryospheric (snow and ice) changes which have taken place. Furthermore, as major periods of global ice build-up and decay appear to have occurred on a quasi-periodic basis during at least the late Quaternary, a much longer record than the mean duration of this period ($\sim 10^5$ yr) is necessary to determine the causative factors, and to appreciate how those factors play a role in climate today. A detailed paleoclimatic record, spanning at least the late Quaternary period, is therefore fundamental to comprehension of modern climate, and the causes of climatic variation and change (Kutzbach, 1976). Furthermore, unless the natural variability of climate is understood, it will be extremely difficult to identify with confidence any anthropogenic effects on climate.

Computer models can be used to estimate the spatial and temporal pattern of climate change as greenhouse gas concentrations increase in the atmosphere. This provides a "target" of expected change against which contemporary observations can be compared. If the climate system evolves towards such a target, one could then argue that anthropogenic effects have been detected on a global scale (Santer *et al.*, 1996). But natural variability, unless fully represented in model simulations, may confound such detection efforts. Whatever anthropogenic effects there are on climate, they will be superimposed on the underlying background of "natural" climate variability, which may be varying on all timescales in response to different forcing factors. Paleoclimatic research provides the essential understanding of climate system variability, and its relationship to both forcing mechanisms and feedbacks, which may amplify or reduce the direct consequences of particular forcings.

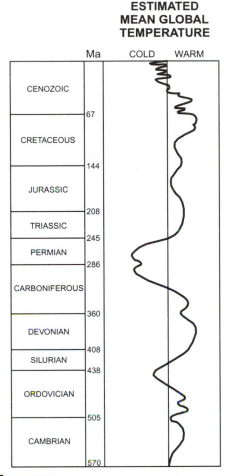

ESTIMATED MEAN GLOBAL TEMPERATURE

FIGURE 1.1 Generalized temperature history of the Earth plotted as relative departures from the present global mean (from Frakes *et al.,* 1992). Studying the proxy record of paleoclimate is rather like looking through a telescope held the wrong way around; for recent periods there is evidence of short-term climatic variations, but these cannot be resolved in earlier periods.

It is abundantly clear from the paleoclimate record that abrupt changes have occurred in the global climate system at certain times in the past. Nonlinear responses apparently have occurred as critical thresholds were passed. Our knowledge of what these thresholds are is completely inadequate; we cannot be certain that anthropogenic changes in the climate system will not lead us, inexorably, across such a threshold, beyond which may lie a dramatically different future climate state (Broecker, 1987). Only by careful attention to such episodes in the past can we hope to comprehend fully the potential danger of future global changes due to human-induced effects on the climate system.

1.2 SOURCES OF PALEOCLIMATIC INFORMATION

Many natural systems are dependent on climate; where evidence of such systems in the past still exists, it may be possible to derive paleoclimatic information from them. By definition, such proxy records of climate all contain a climatic signal, but that signal may be relatively weak, embedded in a great deal of extraneous "noise" arising from the effects of other (non-climatic) influences. The proxy material has acted as a filter, transforming climatic conditions at a point in time, or over a period, into a more or less permanent record, but the record is complex and incorporates other signals that may be irrelevant to the paleoclimatologist.

To extract the paleoclimatic signal from proxy data, the record must first be calibrated. Calibration involves using modern climatic records and proxy materials to understand how, and to what extent, proxy materials are climate-dependent. It is assumed that the modern relationships observed have operated, unchanged, throughout the period of interest (the principle of uniformitarianism). All paleoclimatic research, therefore, must build on studies of climate dependency in natural phenomena today. Dendroclimatic studies, for example, have benefited from a wealth of research into climate-tree growth relationships, which have enabled dendroclimatic models to be based on sound ecological principles (see Chapter 10). Significant advances have also been made in palynological research by improvements in our understanding of the relationships between modern climate and modern pollen rain (see Chapter 9). It is apparent, therefore, that an adequate modern data base and an understanding of contemporary processes in the climate system are important prerequisites for reliable paleoclimatic reconstructions. However, not all environmental conditions in the past are represented in the period of modern experience. Obviously, situations existed during glacial and early Postglacial times that defy characterization by modern analogs. One must therefore be aware of the possibility that erroneous paleoclimatic reconstructions may result from the use of modern climate-proxy data relationships when past conditions have no analog in the modern world (Sachs *et al.*, 1977). By the use of more than one calibration equation it may be possible to detect such periods and avoid the associated errors (Hutson, 1977; Bartlein and Whitlock, 1993; see also Section 6.4).

Major types of proxy climatic data available are listed in Table 1.1. Each line of evidence differs according to its spatial coverage, the period to which it pertains, and its ability to resolve events accurately in time. For example, ocean sediment cores are potentially available from 70% of the Earth's surface, and may provide continuous proxy records of climate spanning many millions of years. However, these records are difficult to date accurately; commonly there is a dating uncertainty of ± 1% of a sample's true age (the absolute magnitude of the uncertainty thus increasing with sample age). Mixing of sediments by marine organisms and generally low sedimentation rates also make it difficult to obtain samples from the open ocean that represent less than 500–1000-yr intervals (depending on depth in the core). This large minimum sampling interval means that the value of most marine sediment studies lies in low-frequency (long-term) paleoclimatic information (on the

TABLE 1.1   Principal Sources of Proxy Data for Paleoclimatic Reconstructions

(1) Glaciological (ice cores)

      (a) geochemistry (major ions and isotopes of oxygen and hydrogen)

      (b) gas content in air bubbles

      (c) trace element and microparticle concentrations

      (d) physical properties (e.g., ice fabric)

(2) Geological

    (A) Marine (ocean sediment cores)

      (i)   Biogenic sediments (planktonic and benthic fossils)

         (a) oxygen isotopic composition

         (b) faunal and floral abundance

         (c) morphological variations

         (d) alkenones (from diatoms)

      (ii) Inorganic sediments

         (a) terrestrial (aeolian) dust and ice-rafted debris

         (b) clay mineralogy

    (B) Terrestrial

      (a) glacial deposits and features of glacial erosion

      (b) periglacial features

      (c) shorelines (Eustatic and glacio-eustatic features)

      (d) aeolian deposits (loess and sand dunes)

      (e) lacustrine sediments, and erosional features (shorelines)

      (f) pedological features (relict soils)

      (g) speleothems (age and stable isotope composition)

(3) Biological

      (a) tree rings (width, density, stable isotope composition)

      (b) pollen (type, relative abundance, and/or absolute concentration)

      (c) plant macrofossils (age and distribution)

      (d) insects (assemblage characteristics)

      (e) corals (geochemistry)

      (f) diatoms, ostracods, and other biota in lake sediments (assemblages, abundance, and/or geochemistry)

      (g) modern population distribution (refugia and relict populations of plants and animals)

(4) Historical

      (a) written records of environmental indicators (parameteorological phenomena)

      (b) phenological records

order of 10^3–10^4 yr; see Chapter 6). However, areas with high sedimentation rates (that can provide higher resolution data) are now the focus of major coring efforts (IMAGES Planning Committee, 1994). Sediments from such areas can document changes on the ~10^2 yr timescale (e.g., Keigwin, 1996) or even at the decadal scale in exceptional circumstances (e.g., Hughen et al., 1996b). By contrast, tree rings from much of the (extratropical) continental land mass can be accurately dated to an individual year, and may provide continuous records of more than a thousand years duration. With a minimum sampling interval of one year, they provide primarily high-frequency (short-term) paleoclimatic information (see Chapter 10). Table 1.2 documents the main characteristics of these and other sources of paleoclimatic data. The value of proxy data to paleoclimatic reconstructions is very dependent on the minimum sampling interval and dating resolution, as it is this that primarily determines the degree of detail available from the record. Currently, annual and even seasonal resolution of climatic fluctuations in the timescale 10^1–10^3 yr is provided by ice-core, coral, varved sediment, and tree-ring studies (see Chapters 5–7 and 10). Detailed analyses of pollen in varves may provide annual data, but it is likely that the pollen itself is an integrated measure of the pollen rain over a number of prior years (Jacobson and Bradshaw, 1981). On the longer timescale (10^5–10^6 yr) ocean cores provide the best records at present, although resolution probably decreases to ± 10^4 years in the early Quaternary. Historical records have the potential of providing annual (or intra-annual) data for up to a thousand years in some areas, but this potential has been realized only for the last few centuries in a few areas (see Chapter 11).

Commonly, there is a frequency-dependence that precludes reconstruction of past climates over part of the spectrum because of inherent attributes of the archive itself. Marine sediments typically have a strong red noise spectrum with most of the variance at low frequencies due to low sedimentation rates and bioturbation. Tree rings, on the other hand, rarely provide information at very low frequencies (i.e., greater than a few hundred years); removal of the biological growth function (a necessary prerequisite to paleoclimatic analysis) essentially filters out such low frequency components from the raw data. All paleo records have some frequency-dependent bias, which must be understood to make sensible use of the data.

Not all paleoclimatic records are sensitive indicators of abrupt changes in climate; the climate-dependent phenomenon may lag behind the climatic perturbation so that abrupt changes appear as gradual transitions in the paleoclimatic record. Different proxy systems have different levels of inertia with respect to climate, such that some systems vary essentially in phase with climatic variations whereas others lag behind by as much as several centuries (Bryson and Wendland, 1967). This is not simply a question of dating accuracy but a fundamental attribute of the proxy system in question. Pollen, for example, derives from vegetation that might take up to a few hundred years to adjust to an abrupt change in climate. Even with interannual resolution in the pollen record, sharp changes in climate are unlikely to be reflected in pollen assemblages, as the vegetation affected may take many centuries to adjust to a new climatic state (though interannual values of total pollen influx may provide clues to rapid shifts in circulation patterns). By contrast,

■■■■ **TABLE 1.2   Characteristics of Natural Archives**

Archive	Minimum sampling interval	Temporal range (order: yr)	Potential information derived
Historical records	day/hr	$\sim 10^3$	T, P, B, V, M, L, S
Tree rings	yr/season	$\sim 10^4$	T, P, B, V, M, S
Lake sediments	yr (varves) to 20 yr	$\sim 10^4$–10^6	T, B, M, P, V, C_W
Corals	yr	$\sim 10^4$	C_W, L, T, P
Ice cores	yr	$\sim 5 \times 10^5$	T, P, C_a, B, V, M, S
Pollen	20 yr	$\sim 10^5$	T, P, B
Speleothems	100	$\sim 5 \times 10^5$	C_W, T, P
Paleosols	100 yr	$\sim 10^6$	T, P, B
Loess	100 yr	$\sim 10^6$	P, B, M
Geomorphic features	100 yr	$\sim 10^6$	T, P, V, L, P
Marine sediments	500 yr[a]	$\sim 10^7$	T, C_W, B, M, L, P

T = temperature

P = precipitation, humidity, or water balance (P-E)

C = chemical composition of air (C_a) or water (C_W)

B = information on biomass and vegetation patterns

V = volcanic eruptions

M = geomagnetic field variations

L = sea level

S = solar activity

After Bradley and Eddy (1991).

[a] In rare circumstances (varved sediments) ≤10 yr.

the record of fossil insects may point to short-term changes of climate (because insect populations are often highly mobile and sensitive to temperature fluctuations) that are not resolvable using pollen analysis alone (see Section 8.4). Thus, not all proxy data are readily comparable because of differences in response time to climatic variations.

In terms of the resolution provided by proxy data, it is also worth noting that not all data sources provide a continuous record. Certain phenomena provide discontinuous or episodic information; glacier advances, for example, may leave geomorphological evidence of their former extent (moraines, trim-lines, etc.) but these represent discrete events in time, resulting from the integration of climatic conditions prior to the ice advance (see Section 7.4). Such deposits say nothing about times of ice recession. Furthermore, major ice advances may obliterate evidence of previous, smaller advances, so the geomorphological record is likely to be not only discontinuous but also incomplete. Studies of *continuous* paleoclimatic records can help to place such episodic information in perspective and, for this reason, the continuous marine sedimentary records are commonly used as a chronological and paleoclimatic frame of reference for long-term climatic fluctuations recorded on land (Kukla, 1977). This does not mean that the growth and decay of ice sheets in different areas were globally synchronous; indeed there is much evidence that this was not the case.

So far the focus has been on paleorecords of past climatic change (i.e., the response of the climate system to some external or internal forcing). However, paleorecords can also provide critical information on the nature of past forcing factors. Ice cores, for example, register the occurrence of major explosive eruptions in the record of non sea-salt sulfate, resulting from acidic fallout after such events. [10]Be in ice also provides insight into past solar variability, and the dust content of ice records past atmospheric turbidity. Changes in radiatively important greenhouse gases (CO_2, CH_4, N_2O) are also recorded in air bubbles in the ice. Such records are extremely valuable in understanding what factors may have been important in bringing about changes in past climate, and in defining the significance of future environmental changes.

In all paleorecords, accurate dating is of critical importance. Without accurate dating it is impossible to determine if events occurred synchronously or if certain events led or lagged others. This is a fundamental requirement if we are to understand the nature of global changes of the past (see Chapters 3 and 4). Accurate dating is required in any assessment of the rate at which past environmental changes occurred, particularly when considering high frequency, short-term changes in climate. Indeed, the duration of such events may be shorter than the normal error associated with many dating methods.

1.3 LEVELS OF PALEOCLIMATIC ANALYSIS

Paleoclimatic reconstruction may be considered to proceed through a number of stages or levels of analysis. The first stage is that of data collection, generally involving fieldwork, followed by initial laboratory analyses and measurements. This results in primary or level 1 data (Hecht *et al.*, 1979; Peterson *et al.*, 1979). Measurements of tree-ring widths or the isotopic content of marine foraminifera from an ocean core are examples of primary data. At the next stage, the level 1 data are calibrated and converted to estimates of paleoclimate. The calibration may be entirely qualitative, involving a subjective assessment of what the primary data represent (e.g., "warmer," "wetter," "cooler" conditions, etc.) or may involve an explicit, reproducible procedure that provides quantitative estimates of paleoclimate. These derived or level 2 data provide a record of climatic variation through time at a particular location. For example, tree-ring widths from a site near the alpine or arctic treeline may be transformed into a paleotemperature record for that location, using a calibration equation derived from the relationship between modern climatic data and modern tree-ring widths (see Chapter 10).

Different level 2 data may also be mapped to provide a regional synthesis of paleoclimate at a particular time, the synthesis providing greater insight into former circulation patterns than any of the individual level 2 data sets could provide alone (Nicholson and Flohn, 1980). In some cases, three-dimensional (3D) arrays of level 2 data (i.e., spatial patterns of paleoclimatic estimates through time) have been transformed into objectively derived statistical summaries. For example, spatial patterns of drought in the eastern United States over the last 300 yr (based on level 1 tree-ring data) have been converted into a small number of principal components (eigenvectors) that account for most of the variance in the level 2 data set (Cook *et al.*, 1992b). The eigenvectors show that there are a small number of modes, or patterns of drought, which characterize the data. The statistics derived from such analyses constitute a third level of paleoclimatic data (level 3 data).

Most paleoclimatic research involves level 1 and level 2 data at individual sites, though regional syntheses are becoming more common (e.g., see individual chapters in Wright *et al.*, 1993). At the larger, hemispheric or global scale, there are few studies of the spatial dimensions of climate at particular periods in the past. Notable exceptions are the CLIMAP and COHMAP reconstructions of marine and continental conditions at 3000-yr intervals from 18,000 yr B.P. to the present (CLIMAP, 1976; COHMAP, 1988; Webb *et al.*, 1993a). Such syntheses provide rigorous tests of the ability of general circulation models to simulate climate under different boundary conditions and different forcing mechanisms (see Chapter 12).

1.4 MODELING IN PALEOCLIMATIC RESEARCH

In addition to studies of natural archives, paleoclimatic research also involves numerical models of the climate system. These models are necessarily based on studies of the contemporary environment, but are applied to those periods in the past when boundary conditions were different from those of today. This provides a test of the models' ability to simulate distinctly different environmental states, by providing a database of environmental conditions in the past that were quite different from those of today. For example, general circulation models have been used to produce global paleoclimatic reconstructions at 3 ka intervals, from 18 ka B.P. to the present, which can be verified (or nullified) by research on natural archival materials (Kutzbach *et al.*, 1993b). Models are also used to test hypotheses about the causes of past environmental changes, to quantify the relative importance of one factor compared to another, and to examine the sensitivity of the system to different forcing mechanisms. If the models prove to be reliable in such tests, more confidence can be placed in their ability to predict future climatic changes in response to anthropogenic forcing (Rind, 1993).

Models and field data are used interactively to stimulate new hypotheses about the nature and causes of environmental change, and to assess their validity. As an example of this evolving process of data collection and model-building, consider the long-held hypothesis that orbital variations were primarily responsible for the onset and cessation of the major glaciations of the late Quaternary. Modeling studies suggest that the changes in solar radiation produced by orbital variations were insufficient, by themselves, to produce the observed environmental changes. Other processes or feedbacks were evidently involved. Recent studies of polar ice cores have revealed significant changes in CO_2, atmospheric aerosols, and CH_4 from interglacial to glacial periods; CO_2 and CH_4 levels were much lower during the last glaciation and the tropospheric aerosol load was much greater. Models have been used to assess the relative importance of these factors in the growth and decay of continental ice sheets and to provide *quantitative* estimates of their significance. However, these results have raised new questions about the processes involved in such dramatic environmental changes. What changes occurred first? Was there a primary triggering factor? What caused the changes in atmospheric composition? Are there critical thresholds in earth systems that, once crossed, lead to new and different quasi-stable environmental states? Much new research has been fueled by such questions, and so the search for more information and better understanding continues. This is not an irrelevant academic exercise. Changes in atmospheric CO_2 and CH_4 levels recorded in ice cores were of the same magnitude as changes wrought by human activities over the last century (though anthropogenic changes have been more rapid). Together, paleoenvironmental data and modeling can help us to evaluate known changes in the past, and to comprehend the feedbacks and system responses that are of direct relevance to understanding the future impact of greenhouse gases.

2 CLIMATE AND CLIMATIC VARIATION

2.1 THE NATURE OF CLIMATE AND CLIMATIC VARIATION

Climate is the statistical expression of daily weather events; more simply, climate is the expected weather. Naturally, for a particular location, certain weather events will be common (or highly probable); these will lie close to the central tendency or mean of the distribution of weather events. Other types of weather will be more extreme and less frequent; the more extreme the event, the lower the probability of recurrence. Such events would appear at the margins of a distribution of weather events characterizing a particular climate. The overall distribution of climatic parameters defines the climatic variability of the place. If we were to measure temperature in the same location for a finite period of time, the statistical distribution of measured values would reflect the geographical situation of the site (in relation to solar radiation receipts and degree of continentality) as well as the relative frequency of synoptic weather patterns and the associated airflow over the region. Given a long enough period of observations, it would be possible to characterize the temperature of the site in terms of mean and variance. Similarly, observations of other meteorological parameters, such as precipitation, relative humidity, solar radiation, cloudiness, wind speed, and direction, would enable a more comprehensive understanding of the climate of the site to be obtained. However, implicit in such statistics is the element of time. For how long should observations be taken to obtain a reliable picture of the climate at a particular place? The World Meteorological Organization has recommended the adoption of standardized 30-yr periods to characterize climate (Mitchell *et al.,* 1966; Jagannathan *et al.,* 1967). Adoption of a standard reference period is necessary because the statistics that define climate in one area may vary over time so that climate, strictly speaking, should always be defined with reference to the period used in its calculation. Recent studies of global warming express global temperature changes relative to the 1961–1990 mean (Jones, 1994) but most paleoclimate studies rely on climatic data from earlier decades. This becomes important when attempting to compare the fairly subtle climatic variations of the recent past (or general circulation model simulations) with the climate of "today." Presumably, in such a context "today" means the most recent 30-yr mean, but in many areas the last 30 years have been significantly warmer than in previous decades; in fact, on a global scale, 1986–1995 was probably one of the warmest decades for many centuries. The problem is even more difficult in dealing with precipitation, where one 30-yr climatic average may be quite different from another (Bradley, 1991). There is no simple solution, so changes in climate should always be expressed relative to some defined time interval, to allow different reconstructions to be appropriately compared.

    Climate may vary in different ways. Some examples of climatic variation are shown in Fig. 2.1. Variations may be periodic (and hence predictable) quasi-periodic (predictable only in the very broadest terms) or non-periodic. Central tendencies

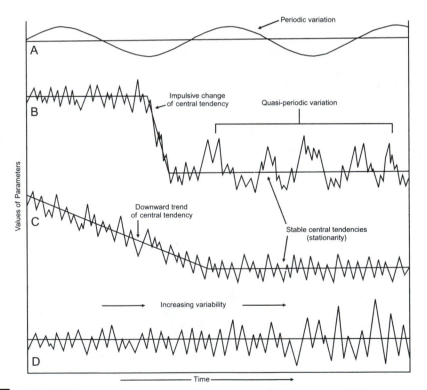

FIGURE 2.1   Examples of climatic variation and variability (from Hare, 1979).

(mean values) may remain more or less constant or exhibit trends or impulsive changes from one mean to another (Hare, 1979). Such occurrences may appear to be random in a time series but this does not necessarily mean they are not predictable. For example, a number of studies have shown that abrupt changes in climate generally result from large explosive volcanic eruptions (e.g., Bradley, 1988). Consequently, the climatic effects of similar eruptions can be anticipated. Hansen *et al.* (1996), for example, used a general circulation model to estimate the changes in temperature expected from the 1991 eruption of Mount Pinatubo (Philippines). Their estimates tracked very closely observed temperature changes in the years following the eruption. Such studies indicate that in some circumstances reliable climate predictions can be made, even though the eruptions themselves are non-periodic.

A very important aspect of variability in the climate system involves non-linear feedbacks, in which drastic changes may occur if some critical threshold is exceeded. One example of this is the oceanic thermohaline circulation, which may cease to operate if the salinity-density balance in near-surface waters of the North Atlantic Ocean is disturbed beyond a certain point. The circulation would then cease until salinity increased to the level where density-induced overturning of the water column could resume (see Section 6.9).

Finally, climatic variation may be characterized by an increase in variability without a change in central tendency, though commonly a change in variability accompanies a change in overall mean. Climatic variability is an extremely important characteristic of climate in our increasingly overstressed world. Every year, unexpected weather events (extremes in the climate spectrum) result in hundreds of thousands of deaths and untold economic and social hardships. If climatic variability increases, the unexpected becomes more probable and the strain on social and political systems increases. High resolution paleoclimatic data can shed light on this important aspect of climatic variation.

In the light of these discussions it is appropriate to consider the term *climatic change*. Clearly, climates may change on different scales of time and in different ways. In paleoclimatic studies, climatic changes are characterized by significant differences in the mean condition between one time period and another. Given enough detail and chronological control, the significance of the change may be calculated from statistics describing the time periods in question. Markedly different climatic conditions between two time periods imply an intervening period of climate characterized by an upward or downward trend, or by an impulsive change in central tendency (see Fig. 2.1). Many paleoclimatic records appear to provide evidence for there being distinct modes of climate, within which short-term variations are essentially stochastic (random). Brief periods of rapid, step-like, climatic change appear to separate these seemingly stable interludes (Bryson *et al.*, 1970). Analysis of several thousand ^{14}C dates on stratigraphic discontinuities (primarily in pollen records from western Europe, but including data from elsewhere) lends some support to this idea (Wendland and Bryson, 1974). Certain periods stand out as having been times of environmental change on a worldwide scale[1] (Fig. 2.2). Such widespread discontinuities imply abrupt, globally

---

[1] As discussed in Chapter 3, changes in the ^{14}C content of the atmosphere may result in periods of apparently rapid change, because events which were in reality separate in time appear near synchronous when ^{14}C dated. This effect may have influenced the pattern of change noted by Wendland and Bryson (1974).

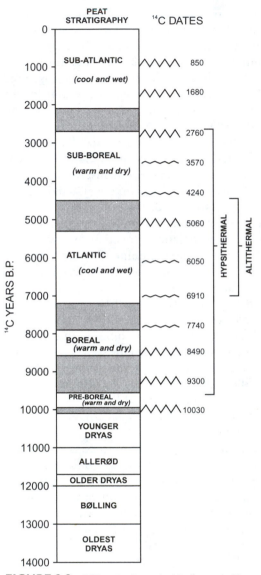

FIGURE 2.2 "Climatic discontinuities" revealed by analysis of over 800 ^{14}C dates on stratigraphic discontinuities in paleoenvironmental (primarily botanical) records (based on data in Wendland and Bryson, 1974). Major and minor discontinuities are shown by large and small jagged lines, respectively. Time limits of the Hypsithermal and the Altithermal are from Deevey and Flint (1957). The Blytt-Sernander scheme of Scandinavian peat stratigraphy was developed before ^{14}C dating techniques were available. It is based on changes in peat growth that were considered to be climate related. Radiocarbon dates now indicate that the boundaries are not precise, but vary over the ranges indicated (based on summaries by Godwin, 1956 and Deevey and Flint, 1957). Objective analyses of peat stratigraphy indicate that the "classic" stages of peat stratigraphy may not be of regional significance after all (except for the Sub-Boreal/Sub-Atlantic transition at about 2500 years B.P.) (Birks and Birks, 1981). Nevertheless, the descriptors (Atlantic, Sub-Atlantic, etc.) are still commonly used to refer to a particular time period, albeit vaguely defined, both climatically and chronologically. Late Glacial/Early Holocene chronozones are from Mangerud *et al.* (1974).

synchronous climatic changes, presumably brought about by some large-scale forcing. In particular the period 2760–2510 yr B.P. (the beginning of sub-Atlantic time) stands out in both palynological and archeological data as a period of major environmental and cultural change, the cause of which is not known. If such a disruption of the climate system were to recur today, the social, economic, and political consequences would be nothing short of catastrophic (Bryson and Murray, 1977).

If climate is considered from a mathematical viewpoint, it is theoretically possible that a particular set of boundary conditions (solar radiation receipts, Earth surface conditions, etc.) may not give rise to a unique climatic state. Two or more distinct sets of statistics ("climate") may result from a single set of controls on the atmospheric circulation (Lorenz, 1968, 1970, 1976). In a practical sense, this suggests that climate (taken here to mean a particular mode of the general circulation of the atmosphere) may be essentially stable until some external factor (e.g., a change in solar radiation output or in volcanic dust loading of the atmosphere) causes a perturbation in the system. This perturbation may be only a short-term phenomenon, after which boundary conditions return to their former state; however, the resultant climate may not be the same, even though the boundary conditions are nearly identical to those before the perturbation — one of the other "solutions" to the mathematical problem of climate may have been adopted. Such a system is said to be intransitive. If, on the other hand, there is only one unique climatic state corresponding to a given set of boundary conditions, the system is said to be transitive. If climate does operate as an intransitive system, this poses intractable problems for mathematical models of climate, and for attempts to use these for climate forecasting. There is another possibility that complicates matters further. With only a minor change in boundary conditions, it is theoretically possible for there to be two or more time-dependent solutions, each with different statistics (climate) when considered over a moderate time span (i.e., the system may appear to be intransitive). However, if the time span is made sufficiently long, the different statistics converge to an essentially stable state. This is referred to as an almost-intransitive system (Lorenz, 1968). In practice, this means that a single set of boundary conditions may result in different climate states over discrete time intervals. If we were observing these different states through the paleoclimatic record, they would appear to represent periods separated by a "climatic change" (implying an external causative factor) whereas they would be, in actuality, merely stages on the way to a long-term stable state.

2.2 THE CLIMATE SYSTEM

Although it is common to consider climate as simply a function of the atmospheric circulation over a period of time, to do so overlooks the complexity of factors that determine the climate of a particular region. Climate is the end-product of a multitude of interactions between several different subsystems — the atmosphere, oceans, biosphere, land surface, and cryosphere — which collectively make up the climate system. Each subsystem is coupled in some way to the others (Fig. 2.3) such that

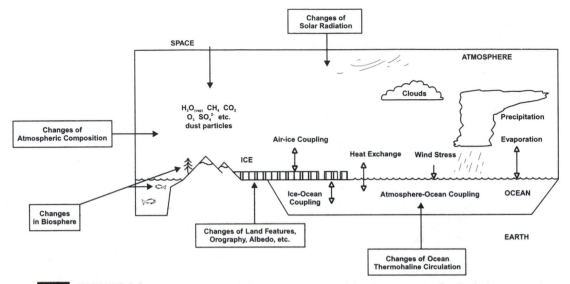

FIGURE 2.3 Schematic diagram of major components of the climatic system. Feedbacks between various components play an important role in climate variations.

changes in one subsystem may give rise to changes elsewhere (see Section 2.3). Of the five principal subsystems, the atmosphere is the most variable; it has a relatively low heat capacity (low specific heat) and responds most rapidly to external influences (on the order of 1 month or less). It is coupled to other components of the climate system through energy exchanges at the surface (the atmospheric boundary layer) as well as through chemical interactions that may affect atmospheric composition (Junge, 1972; Jaenicke, 1981; Bolin, 1981). Only recently has it been possible to assess variations in atmospheric composition and turbidity through time (Raynaud *et al.*, 1993; Zielinski, 1995). Such variations are of particular importance because they may be a fundamental cause of past climatic variations.

The oceans are a much more sluggish component of the climate system than the atmosphere. Surface layers of the ocean respond to external influences on a timescale of months to years, whereas changes in the deep oceans are much slower; it may take centuries for significant changes to occur at depth. Because water has a much higher heat capacity than air, the oceans store very large quantities of energy, and act as a buffer against large seasonal changes of temperature. On a large scale, this is reflected in the differences between seasonal temperature ranges of the Northern and Southern Hemispheres (Table 2.1). On a smaller scale, proximity to the ocean is a major factor affecting the climate of a region. Indeed, it is probably the single most important factor, after latitude and elevation.

At the present time, the oceans cover 71% (361×10^6 km²) of the Earth's surface and hence play an enormously important role in the energy balance of the Earth (see Section 2.4). The oceans are most extensive in the Southern Hemisphere, between 30 and 70° S, and least extensive in the zone 50–70° N and poleward of 70° S (Fig. 2.4). This distribution of land and sea is of great significance; it is largely

TABLE 2.1   Mean Temperatures (°C) and Temperature Differences

	Extreme months		Year
(a) Surface			
Northern Hemisphere	8.0 (January)	21.6 (July)	15.0
Southern Hemisphere	10.6 (July)	16.5 (January)	13.4
Entire globe	12.3 (January)	16.1 (July)	14.2
(b) Middle Troposphere (300–700-mb layer)			
Mean temperatures			
Equator	-8.6	-8.6	-8.6
North Pole	-41.5 (January)	-25.9 (July)	-35.9
South Pole	-52.7 (July)	-38.3 (January)	-47.7
Temperature differences			
Equator–North Pole	32.9 (January)	17.3 (July)	27.3
Equator–South Pole	29.7 (January)	44.1 (July)	39.1

After Flohn, 1978 and Van Loon *et al.,* 1972.

responsible for the differences in atmospheric circulation between the two hemi-spheres, and has important implications for glaciation of the Earth (Flohn, 1978). On a global scale, the relative proportions of land and sea have changed little during the Quaternary, in spite of sea-level changes due to the growth and decay of continental ice sheets. When sea level was 100 m below current levels ocean area decreased by only 3% (though this is equivalent to a 10% increase in land-surface area). Such changes undoubtedly had regional significance; in particular, sea-level changes may

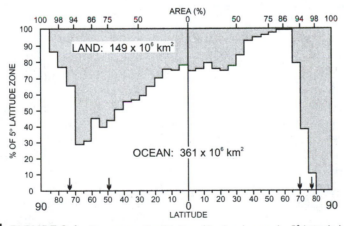

FIGURE 2.4   Percentage distribution of land and ocean by 5° latitude band. Land area shaded. Upper fig-ures give percentage of hemispheric surface area equatorward of latitudes shown. Arrows indicate mean latitu-dinal ranges of seasonal snow cover (see Table 2.3).

have had important effects on oceanic circulation and certainly must have influenced the degree of continentality of some areas (e.g., Barry, 1982; Nix and Kalma, 1972).

The oceans play a critical role in the chemical balance of the atmospheric system, particularly with respect to atmospheric carbon dioxide levels. Because the oceans contain very large quantities of CO_2 in solution, even a small change in the oceanic CO_2 balance may have profound consequences for the radiation balance of the atmosphere, and hence climate (Sundquist, 1985). The role of the oceans in global CO_2 exchanges is of particular importance, not only for an understanding of past climatic variations but also for insight into future CO_2 trends in the atmosphere (Baes, 1982; Bolin, 1992).

The land surface of the Earth interacts with other components of the climate system on all timescales. Over very long periods of time, continental plate movements (in relation to the Earth's rotational axis) have had major effects on world climate (Tarling, 1978; Frakes *et al.*, 1992). It is no coincidence that the frequency of continental glaciation increased as the plates moved to increasingly polar positions. Similarly, mountain-building episodes (orogenies) have had major effects on world climate. Apart from the dynamic effects on atmospheric circulation (Yoshino, 1981; Ruddiman and Kutzbach, 1989) the presence of elevated surfaces at relatively high latitudes, where snow can persist throughout the year, may be a prerequisite for the development of continental ice sheets (Ives *et al.*, 1975).

The latitudinal distribution of land and sea is of fundamental significance for both regional and global climate. In particular, the presence of highly reflective snow- and ice-covered regions at high latitudes strongly affects Equator–Pole temperature gradients (Table 2.1b). In the Southern Hemisphere, the presence of the high elevation Antarctic plateau south of ~75° S (Fig. 2.4) causes there to be a much stronger Equator-Pole temperature gradient than in the Northern Hemisphere. As a result, an intense westerly circulation pattern develops above the surface layers (60% stronger, on average, than westerlies in the Northern Hemisphere [Peixoto and Oort, 1992]). The stronger temperature gradient also results in the subtropical high pressure belt of the Southern Hemisphere being located closer to the Equator than in the Northern Hemisphere (29–35° S as compared with 33–41° N; Fig. 2.5). This difference, stemming primarily from the polar location of Antarctica and its associated low temperatures, gives rise to a basic asymmetry in the position of climatic zones in both hemispheres (Korff and Flohn, 1969; Flohn, 1978).

The cryosphere consists of mountain glaciers and continental ice sheets, seasonal snow and ice cover on land, and sea ice. Its importance in the climate system stems from the high albedo of snow- and ice-covered regions, which greatly affects global energy receipts (Kukla, 1978). At present, about 8% of the Earth's surface is permanently covered by snow and ice (Table 2.2) but seasonal expansion of the cryosphere causes this figure to double (Table 2.3). The hemispheric differences are particularly profound. In the Northern Hemisphere, 4% of the total area is permanently ice covered (mainly the Arctic Ocean [~3%] and Greenland). In winter months, sea-ice formation and snowfall on the continents results in a 6-fold increase in snow and ice cover. By midwinter, 24% of the Northern Hemisphere is generally covered by snow and ice. In the Southern Hemisphere, most of the permanent ice cover is land-based on the Antarctic continent, and seasonal changes are due almost

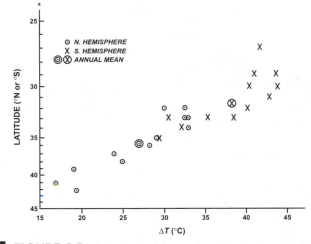

FIGURE 2.5 Relationship between latitude of main axis of subtropical anticyclones and hemispheric (Equator–Pole) temperature gradient in preceding month (after Korff and Flohn, 1969).

TABLE 2.2 Present Extent of Permanent Snow and Ice (Glaciers, Ice Caps, and Sea Ice)[a]

	Area (× 10⁶ km²)	Volume (× 10⁶ km³)	Sea-level equivalent (m)
Northern Hemisphere			
Greenland	1.73	3.0	7.5
Other locations	0.5	0.12	0.3
Total land-based snow and ice	2.23		
Sea ice	8.87		
Total for Northern Hemisphere	**11.0**		
Southern Hemisphere			
Antarctica	13.0	29.4	73.5
Other locations	0.032	<0.01	<0.02
Total land-based snow and ice	13.032		
Sea ice	4.2		
Total for Southern Hemisphere	**17.23**		
Entire Globe			
Total land-based snow and ice	~15.3		
Sea ice	~13.0		
Total for entire globe	~28.3		

[a] From Kukla (1978), Hughes *et al.* (1981), and Hollin and Schilling (1981).

TABLE 2.3   Seasonal Changes in Snow- and Ice-cover Area ($\times 10^6 \text{ km}^2$); Snow and Ice Extent Based on the Period 1967–74

	Maximum extent			Minimum extent		
	Month	Area	Percentage (%)	Month	Area	Percentage (%)
Northern Hemisphere	February	60.1	24[a]	August	11.0	4[a]
Southern Hemisphere	October	34.0	13[a]	February	17.2	7[a]
Entire globe	December	79.1	16[b]	August	42.3	8[b]

From Kukla (1978).
[a] Percentage of area of hemisphere.
[b] Percentage of area of entire globe.

entirely to an increase in sea-ice formation (Fig. 2.6). By midwinter, 13% of the Southern Hemisphere is generally covered by snow and ice. It is of particular interest that the cryosphere, considered on a global scale, doubles in area over a relatively short period—from August to December, on average. Given the variability in seasonal timing of snow- and ice-cover changes in both hemispheres, it is quite probable that very large area increases may occur over an even shorter period, and this has important implications for theories of climatic change (Kukla, 1975). Clearly, part of the cryosphere undergoes extremely large seasonal variations and hence has a very short response time. Glaciers and ice sheets, on the other hand, respond very slowly to external changes, on the timescale of decades to centuries; for large ice sheets, adjustment times may be measured in millennia.

The final component of the climate system is the biosphere, consisting of the plant and animal worlds, though vegetation cover and type are mainly of significance for climate. Vegetation not only affects the albedo, roughness, and evapotranspiration characteristics of a surface, but also influences atmospheric composition through the removal of carbon dioxide and the production of aerosols and oxygen. Absence of vegetation may result in significant increases in particulate loading of the atmosphere, at least locally, and this may of itself be a significant factor in altering climate (Charney *et al.*, 1975; Overpeck *et al.*, 1996). Vegetation type varies greatly from one region to another (Table 2.4). Forests and woodlands cover 34% of the continents and play a major role in the removal of atmospheric CO_2 (Woodwell *et al.*, 1978; Potter *et al.*, 1993; Ciais *et al.*, 1995). Deserts and desert scrublands occupy ~13% of the continents, and are the major sources of wind-blown dust (though cultivated lands are increasingly susceptible to wind erosion also). The response time of the biosphere varies widely, on the order of years for individual elements of the biosphere to centuries for entire vegetation communities. Carbon sequestration in terrestrial ecosystems has varied over glacial-interglacial cycles because of large-scale changes in the area of different ecosystem types. Thus, the area of forests during the Last Glacial Maximum (LGM) was reduced to less than one-third of the forest cover today, with a corrresponding reduction in carbon storage in forest ecosystems (Van Campo *et al.*, 1993; Peng *et al.*, 1998). Overall, carbon storage on land was 30% lower during the LGM than it is today.

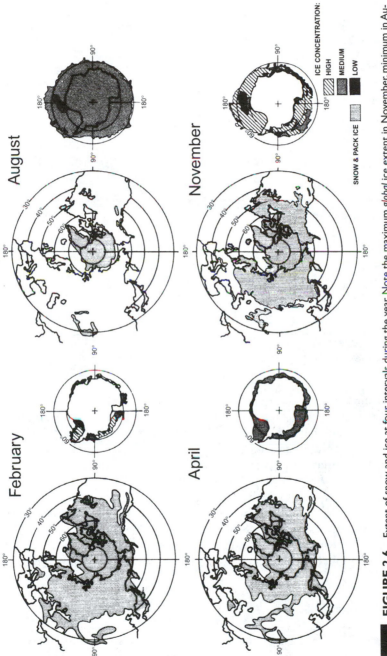

FIGURE 2.6 Extent of snow and ice at four intervals during the year. Note the maximum *global* ice extent in November, minimum in August, see Table 2.3 (from Kukla, 1978).

TABLE 2.4   Areas of Major Ecosystems of the World and Their Estimated Carbon Content and Albedo — Today and at the Last Glacial Maximum

Ecosystem	Modern area (10^6 km^2)	Modern carbon storage (Pg)	Albedo[a] (%)	LGM area (10^6 km^2)	LGM carbon storage (Pg)
Boreal forest	11.8	310.2	7–15	2.3	63.5
Temperate forest	13.0	343.9	13–17	3.9	109.2
Tropical forest	14.3	399.9	7–15	6.1	159.4
Xerophytic woodlands	11.3	147.0	15–20	19.0	249.2
All forests and woodlands	50.4	1201.0		31.3	581.3
Arctic and Alpine tundra	10.7	204.4	10–15	14.7	281.9
Steppes and mountain shrublands	30.8	337.9	15–20	41	444.7
All steppes and tundras	41.5	542.3		55.7	726.6
Cool and polar deserts	4.0	27.4	10–20	15.8	64.0
Hot deserts	14.5	21.8	25–44	19.7	29.6
All deserts	18.5	49.2		35.5	93.6
Cultivated lands	14.1	195.0	8–20		
Bogs	0.7	128.1			
TOTAL	125.2	2115.6		122.5	1401.5

From Van Campo *et al.* (1993)
[a] From Lieth (1975).

Human beings are, of course, part of the biosphere and human activities play an increasingly important role in the climate system. Increases in atmospheric CO_2 concentration, changes in natural vegetation, increases in particulate loading of the lower troposphere, and reductions in atmospheric ozone concentrations in the stratosphere may all be attributed to man's worldwide activities (see Chapter 4 in MacCracken *et al.*, 1990; Schimel *et al.*, 1996). The rate of such changes is rapid and the extent to which the climate system can adjust to them without drastic changes in climate or climatic variability remains uncertain. The only certainty is that mankind has become exceedingly vulnerable to any unexpected perturbations of climate. Common sense argues for action to limit those activities that may contribute to global-scale climatic effects (see Chapters 3 and 4 in Abrahamson, 1990).

2.3 FEEDBACK MECHANISMS

Interactions within the climate system often involve complex, nonlinear relationships. All components of the climate system are intimately linked or coupled with all other components, such that changes in one subsystem may involve compensatory changes throughout the entire climate system. These changes may amplify the initial disturbance (anomaly) or dampen it. Interactions that tend to amplify the disturbance are termed positive feedback mechanisms or processes; they operate in such a way that the system is increasingly destabilized. Interactions that tend to dampen the initial disturbance are termed negative feedback mechanisms or processes; they provide a stabilizing influence on the system, tending to preserve the status quo (Prentice and Sarnthein, 1993).

Growth of continental ice sheets provides an example of positive feedback mechanisms. Whatever the initial perturbation of the climate system that led to continental ice-sheet growth in the past (see Section 2.6) once snow and ice persisted year round, the higher continental albedo would have resulted in lower global radiation receipts, hence lower temperature, and a more favorable environment for ice-sheet growth. Clearly, at some point other factors (such as precipitation starvation and bedrock depression) must have come into play as the ice sheet grew in size to reverse this trend toward increasing glacierization of the planet (Budd and Smith, 1981).

Changes in atmospheric CO_2 concentration also may induce positive feedbacks. As CO_2 levels increase, there will be an increase in the absorption of longwave (infrared) terrestrial radiation by CO_2; concomitantly, there will be an increase in longwave absorption by water vapor, resulting from enhanced earth surface and atmospheric infra-red emissions. Lower tropospheric temperatures will thus increase (the "greenhouse effect") though the magnitude of this increase remains controversial (Schneider, 1993; Lindzen, 1993). As atmospheric temperatures increase, the temperature of the upper layers of the ocean may also increase, causing CO_2 in solution to be released to the atmosphere, thereby reinforcing the trend toward higher temperatures. This (rather simplistic) example of a physical-biochemical feedback is sometimes referred to as the "runaway greenhouse effect." That such an eventuality will occur due to the anthropogenic production of excess CO_2 is unlikely. It might be argued that as temperatures increase there would be more evaporation from the oceans, increased cloudiness (higher global albedo), and hence a decrease in energy to the system. In addition, higher temperatures at high latitudes, associated with increased poleward advection of moisture, might be accompanied by more snowfall, resulting in higher continental albedo (and/or a shorter snow-free period) and hence lower overall global energy receipts. Such mechanisms are examples of negative feedbacks, whereby the system tends to become stabilized after an initial perturbation.

Interactions between different parts of the climate system that are brought about by a process within the system are considered internal mechanisms of climatic variation. They involve initiation by an internal factor, such as the upwelling of cool deep-ocean water or an unusually persistent snow cover over an extensive area of the land surface, which may be amplified by other components of the climate system and eventually lead to an adjustment in the atmospheric circulation. These adjustments within the climate system may in turn alter, and perhaps eliminate, the original factor that initiated the climatic variation. Generally, such mechanisms are stochastic in nature, so that the climatic consequences are not predictable over timescales much longer than the timescale of the initiating process. By contrast, there are factors external to the climate system that may bring about ("force") adjustments in climate, but those changes have no influence on the initiating factor (Mitchell, 1976). Changes in solar output and/or spectral characteristics, changes in the Earth's orbital parameters, and changes in atmospheric turbidity due to explosive volcanic eruptions are examples of external factors that may cause changes in the climate system but are not affected by those changes (Robock, 1978). Some of these mechanisms of climate variation are deterministic (predictable) as they vary in a known way. This is particularly the case with the Earth's orbital variations, which have been calculated accurately both for periods back in time and into the future (Berger and Loutre, 1991; Berger *et al.,* 1991). There is therefore an element of predictability in the consequent climatic changes, though these may, in turn, depend on the particular internal conditions of the climate system prevailing at the time of the external forcing.

2.4 ENERGY BALANCE OF THE EARTH AND ITS ATMOSPHERE

As the Earth sweeps through space on its annual revolution around the Sun, it intercepts a minute fraction of the energy emitted by this all-important star. Because the Earth is (approximately) spherical and rotates on an axis inclined (at present) 23.4° to the plane across which it moves around the Sun, (the ecliptic) energy receipts vary greatly from one part of the globe to another. Furthermore, the pattern of energy receipts is constantly changing. These differential energy receipts are the fundamental driving force of the atmospheric circulation. If solar output is assumed to be invariant, the spatial and temporal patterns of energy receipts impinging on the outer atmosphere can be calculated (Fig. 2.7; Newell and Chiu, 1981). However, for conditions near the surface of the Earth, the role of the atmosphere must be considered because it greatly diminishes potential solar radiation receipts. A consideration of energy exchanges in the Earth-atmosphere system also provides some insight into the potentially important factors involved in climatic variations and variability. For the system as a whole, energy receipts at the outer limits of the atmosphere during the course of a year are 342 W m^{-2} (Fig. 2.8).

As radiation penetrates the atmosphere, as a global average 77 W m^{-2} (23%) is either reflected from cloud tops or scattered upward by molecules and particulate matter in the air. Because the Earth's surface is also reflective, another ~9% of incoming solar radiation is returned to space without heating the atmosphere or the Earth's surface. A further 67 W m^{-2} (20%) is absorbed by ozone, by water vapor and water droplets in clouds, and by particulates, thereby raising the temperature of the atmosphere.[2] Thus, only approximately half of the energy impinging on the outer atmosphere reaches the surface, where it is absorbed, increasing the surface temperature. Energy is re-radiated from the Earth's surface at longer wavelengths (terrestrial radiation), much of which is absorbed by water vapor and carbon dioxide in the atmosphere (the greenhouse effect). This is eventually re-radiated by the atmosphere and ultimately lost to space. Only ~39% of the energy absorbed by the Earth's surface (66 W m^{-2}) is lost by radiative emissions in this way. The balance, or net

[2] Because the atmosphere absorbs short-wave solar radiation as well as long-wave radiation from the Earth, it also emits long-wave radiation both upward and downward (counter-radiation). Overall, however, there is a net loss of long-wave radiation (66 W m^{-2}) from the Earth to space via the atmosphere.

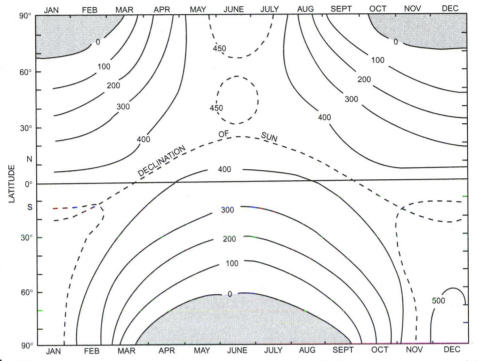

FIGURE 2.7 Distribution of solar radiation at top of atmosphere (in Watt hours per square meter). Apparent position of Sun overhead at noon (declination) is shown by dotted line.

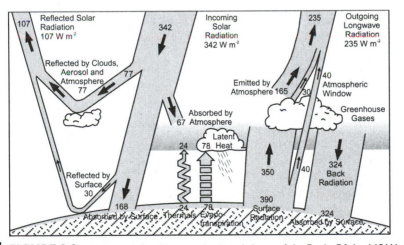

FIGURE 2.8 Mean annual radiation and energy balance of the Earth. Of the 168 W m^{-2} absorbed by the Earth's surface, there is a net radiative loss of 66 W m^{-2} (the *net loss* from long-wave emissions, less counter-radiation from the atmosphere to the surface). The balance of energy at the surface (the net radiation) is transferred by latent heat transfer (evaporation) and sensible heat transfer ("thermals" in this diagram) (from Kiehl and Trenberth, 1997).

radiation, is transferred to the atmosphere via sensible and latent heat transfers. Sensible heat flux (H) involves the transfer of heat directly from the surface to layers of air immediately adjacent to it by the processes of conduction and convection. Latent heat flux (LE) involves the transfer of heat from the surface via the evaporation of water; as water evaporates from the surface, latent heat is extracted, only to be released to the atmosphere later when the water condenses. This is the most important mechanism by which energy is transferred from the Earth to the atmosphere, accounting for ~46% (78 W m^{-2}) of the incoming energy absorbed by the Earth's surface (Fig. 2.8). The relative importance of sensible and latent heat mechanisms in the transfer of heat from the Earth's surface is sometimes characterized by the Bowen ratio (H/LE); high values (≥10) are typical of desert areas where values of latent heat flux are very low, whereas low Bowen ratios (≤1) are typical of oceanic areas where most energy is transferred through the evaporation of water.

The global mean values for the energy balance provide a basis for appreciating the importance of a number of parameters in the climate system. Consider, for example, the role of cloudiness in global energy receipts. On a global scale approximately one-fifth of all energy entering the atmosphere is reflected by cloud tops as a result of their extremely high albedo. Small variations in global cloud cover, or even of cloud type, may thus have very large consequences for global energy balance but we have no clues from the paleoclimatic record as to how cloudiness may have varied through time on a global scale (Bradley *et al.,* 1993). Albedo is of particular significance at the Earth's surface, and this is particularly apparent when zonal (latitude band) averages are considered (Fig. 2.9). The distribution of snow and ice

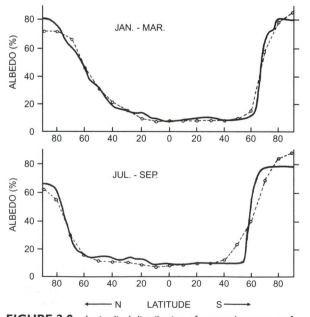

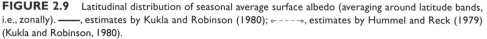

FIGURE 2.9 Latitudinal distribution of seasonal average surface albedo (averaging around latitude bands, i.e., zonally). ———, estimates by Kukla and Robinson (1980); ∘−−−−∘, estimates by Hummel and Reck (1979) (Kukla and Robinson, 1980).

dominates this pattern (see Fig. 2.4) and is largely responsible for the large energy deficits at high latitudes (i.e., higher radiative losses than gains, accommodated by energy transfers from low latitudes). Only during the last 25 years have satellites provided a global perspective on snow- and ice-cover variations, both seasonally and interannually. Although the records are quite short, it is clear that variations in snow and ice extent from year to year can alter area-weighted hemispheric surface albedo by 3–4% (compare the interannual troughs, or peaks, in Fig. 2.10), which may influence atmospheric circulation in subsequent seasons, providing a positive feedback to the system (Groisman *et al.*, 1994a,b). Over longer time periods, changes in surface albedo have been very large, and their effects on albedo must

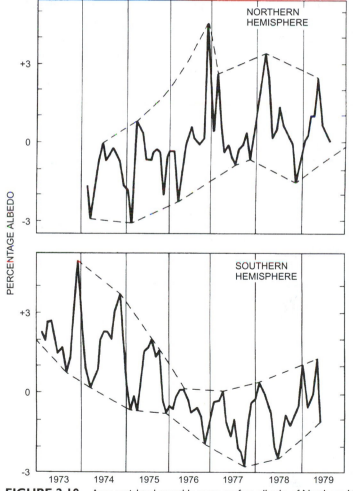

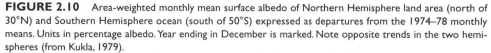

FIGURE 2.10   Area-weighted monthly mean surface albedo of Northern Hemisphere land area (north of 30°N) and Southern Hemisphere ocean (south of 50°S) expressed as departures from the 1974–78 monthly means. Units in percentage albedo. Year ending in December is marked. Note opposite trends in the two hemispheres (from Kukla, 1979).

have been profound. Not only did continental ice sheets and more extensive sea ice (Table 2.5) increase global albedo but the more extensive deserts and savanna grasslands at the time of glacial maxima would have accentuated this effect.

The significance of atmospheric CO_2 and water vapor is also apparent from Fig. 2.8; these gases play a vital role in global energy balance because of their relative opacity to terrestrial radiation. An increase in CO_2 would reinforce this energy exchange, increasing atmospheric temperatures. However, many other interactions and consequences would also ensue and it is this complexity that makes forecasts of the climatic impact of CO_2 increases so difficult (Dickinson *et al.*, 1996).

This thumbnail sketch of the radiation balance of the Earth-atmosphere system is very much a simplification of reality. Most importantly, there are large regional differences in values of net radiation and of latent and sensible heat flux due to the geography of the earth (distribution of continents and oceans, surface relief, vegetation, and snow cover) and the basic climatic differences from one region to another (principally variations in cloud cover and type) (Budyko, 1978). This is readily apparent from a consideration of annual energy balance components for the Earth's surface, shown as zonal averages in Table 2.6, and mapped in Figs. 2.11–2.13. Net radiation varies from near zero at high latitudes to >140 kcal cm^{-2}a^{-1} (186 W m^{-2}) over parts of the tropical and equatorial oceans (Fig. 2.11). On the continents, net radiation is lower than the zonal average due to higher albedo of the surface (e.g., in desert regions) or because of higher cloud amounts, which reduce surface radiation receipts (Table 2.6). For the Earth as a whole (Table 2.6, bottom line) 84% of net radiation is accounted for by latent heat expenditures (66 of 79 kcal cm^{-2} a^{-1}, or 88 of 105 W m^{-2}). If we just consider the oceans, however, 90% of net radiation is utilized in evaporation compared to only 54% (27 of 50 kcal cm^{-2} a^{-1}, or 36 of 66 W m^{-2}) on

TABLE 2.5   Maximum Extent of Land-Based Ice Sheets During the Pleistocene

	Area ($\times$ 10^6 km^2)
North America	16.22
Greenland	2.30
Europe	7.21
Asia	3.95
South America	0.87
Australasia	0.03
Antarctica	13.81

From Flint (1971) and Hollin and Schilling (1981).

Note that not all areas experienced maximum ice cover at the same time during the Pleistocene. It is therefore not appropriate to total these values. Also, seasonal snow cover and sea-ice extent are not included, so these figures represent minimum changes in the area of the overall cryosphere (see Tables 2.3 and 2.4).

TABLE 2.6 **Mean Latitudinal Values of the Heat Balance Components of the Earth's Surface (W m⁻²)**[a]

Latitude	Land			Ocean				Earth			
	R	LE	P	R	LE	P	F_0	R	LE	P	F_0
70–60°N	29	21	8	30	41	29	-40	29	27	15	-12
60–50	42	30	12	57	62	25	-31	49	44	17	-12
50–40	60	33	27	85	89	21	-25	72	60	24	-12
40–30	77	31	46	119	127	19	-27	101	86	31	-16
30–20	85	25	60	147	145	9	-7	125	100	28	-3
20–10	98	42	56	161	155	97	-4	145	126	21	-3
10–0	105	76	29	165	138	9	17	151	123	13	15
0–10°S	105	81	24	169	131	8	29	154	119	12	23
10–20	100	60	40	162	150	12	0	149	130	19	0
20–30	94	37	57	145	141	15	-11	133	117	24	-8
30–40	82	38	44	122	109	15	-1	117	101	19	-3
40–50	58	29	29	96	68	8	20	94	66	9	19
50–60	46	29	17	61	46	12	3	61	46	12	3
Earth as a whole[b]	66	36	30	121	109	12	0	105*	88*	17*	0

From Budyko (1978).

[a] R is the radiative flux of heat (radiation balance of the Earth's surface) equal to the difference of absorbed short-wave radiation and the net long-wave radiation outgoing from the Earth's surface; LE is the heat expenditure for evaporation (L is the latent heat of vaporization, E is the rate of evaporation); P is the turbulent flux of heat between the Earth's surface and the atmosphere; F_0 is the heat income resulting from heat exchange through the sides of the vertical column of a unit section going through the Earth's surface with the ambient layers.

[b] Values for the earth as a whole are slightly different from those given in Fig. 2.8, which is based on more recent satellite-derived data (Kiehl and Trenberth, 1997). These recent estimates give $R = 102$, $LE = 78$, and $P = 24$, so the zonal average values given here will no doubt require some revision. Nevertheless, the broad patterns depicted in this table will not change substantially, and the values are likely to be correct to within ± 10%.

the continents. In fact, in extremely arid areas, latent heat transfer may account for only 15–20% of the net radiation (see Figs. 2.11 and 2.12). In those areas, sensible heat flux is of primary importance (Fig. 2.13). For the continents as a whole, 46% of net radiation is utilized in sensible heat transfers. Over the oceans, sensible heat flux is only important at high northern latitudes where northward-flowing currents bring warm water into contact with cold polar air masses (see Fig. 2.13). Ocean currents themselves play a very important role in energy transport, as is clear from column 8 in Table 2.6. "Excess" heat is transferred from equatorial and tropical regions to higher latitudes where the energy thereby made available may even exceed net radiation at the surface (e.g., 60–70° N; see Figs. 2.11 and 2.13).

From this overview of the energy balance of different regions it is only a short step to consider how components of the energy balance of some areas may have varied in the past, and how human activities may affect the energy balance of some

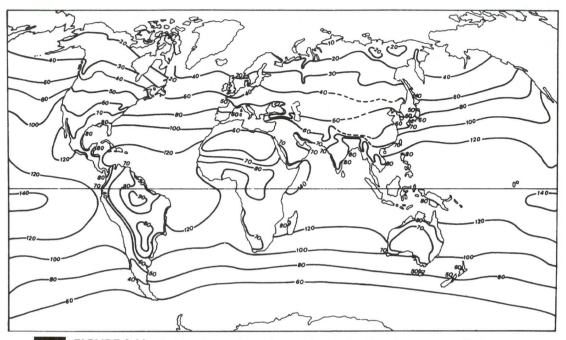

FIGURE 2.11 Radiative balance (net radiation, R_n) of the Earth's surface (in kcal cm^{-2} yr^{-1}). Note discontinuities at ocean/land boundaries (from Budyko, 1978).

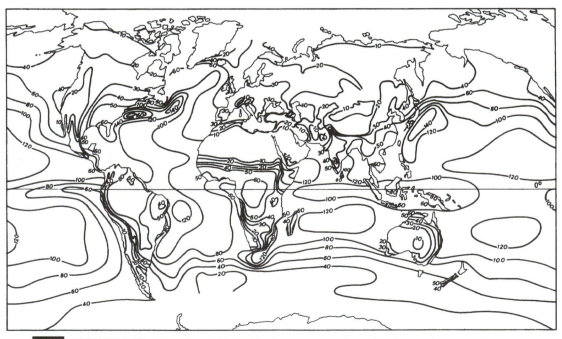

FIGURE 2.12 Expenditure of latent heat for evaporation (latent heat flux, L, in kcal cm^{-2} yr^{-1}) (from Budyko, 1978).

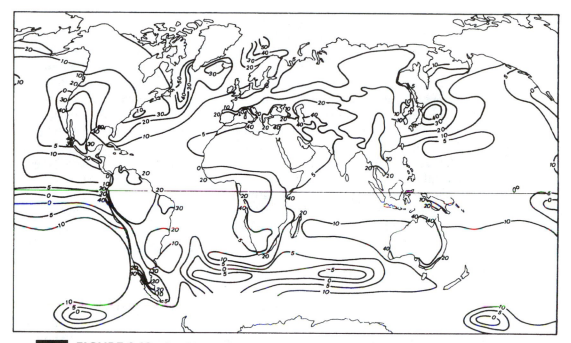

FIGURE 2.13 Sensible heat flux between Earth's surface and the atmosphere (in kcal cm^{-2} yr^{-1}) (from Budyko, 1978).

areas in the future. Of course, it will only be possible to do this in a crude way because the energy balance of any one site is a function of a great many variables, including parts of the climate system far from the site in question. Nevertheless, some general points can be made. Consider, for example, the vast Saharan Desert region. At the present time, net radiation in this area averages ~53 kcal cm^{-2} a^{-1} (70 W m^{-2}) with a Bowen ratio of ~8 (Table 2.7; Baumgartner, 1979). During the early to mid-Holocene, the area was wetter and supported a sparse grassland vegetation cover, increasing to savanna along the Sahelian margin to the south (Fabre and Petit-Maire, 1988; Lézine, 1989); if modern analogies are any guide, the area would have had a lower albedo, higher net radiation, and much lower Bowen ratio. Other desert regions also experienced similar changes in vegetation and hence in energy balance (though changes elsewhere were commonly greatest at the last glacial maximum). As

TABLE 2.7 Energy Balance for Different Surfaces (W m^{-2})

	R	L	H	a	H/L
Tropical rainforest	110	85	25	13	0.3
Savanna	65	40	25	33	0.6
Desert	70	8	62	46	8.0

From Baumgartner (1979).

deserts and semideserts today occupy more than 10% of the continental area, such changes had major consequences for the energy balance of the world as a whole. It also seems likely that overgrazing and desertification of marginal environments today, as well as the destruction of tropical forest ecosystems, will bring about marked changes in the energy balance of low latitudes, with possible global consequences.

The energy balance changes associated with alterations in natural vegetation do not, of course, provide information on why the environmental changes occurred in the first place. However, they do provide important baseline data for computer models of the general circulation at particular time periods in the past (e.g., Kutzbach *et al.*, 1993a) and point to potentially important feedbacks between the atmosphere and underlying surface once the vegetation changes have occurred. The development of a particular vegetation type may, in fact, bring about changes in the energy balance that would favor persistence of the "new" vegetation type (Charney *et al.*, 1975; Foley *et al.*, 1994).

2.5 TIMESCALES OF CLIMATIC VARIATION

Climate varies on all timescales and space scales, from interannual climatic variability to very long-period variations related to the evolution of the atmosphere and changes in the lithosphere. Examples of known climatic fluctuations are shown in Fig. 2.14. In this diagram, each row represents an expansion, by a factor of ten, of each interval on the row above it. Thus, one can envisage short-term (high-frequency) variations nested within long-term (lower-frequency) variations (Webb, 1991). However, in the paleoclimatic record, as we delve farther and farther back in time, it is increasingly difficult to resolve the higher-frequency variations. As climatic variations on the timescale of decades to centuries are of the utmost importance to modern society, increasing attention must be focused on paleoclimatic data pertinent to this problem (Bradley and Jones, 1992a).

Climatic fluctuations on different timescales may be brought about by internal or external mechanisms that operate at different frequencies (Fig. 2.15). Changes in the Earth's orbital parameters, for example, are likely candidates for climatic variations on the timescale of glacials and interglacials during the late Quaternary but cannot account for climatic variations that have occurred over the last thousand years. For fluctuations on that timescale, other factors such as volcanic dust loading of the atmosphere, solar variability or internal adjustments between different subsystems in the climate system, are more likely to be involved (Jones *et al.*, 1996). Of course, different forcing factors may have operated together to cause climatic fluctuations of varying magnitude at different times in the past, though individual factors may account for the variance of climate at a particular frequency. Mitchell (1976) pointed out that much of the variance of the climate record results from stochastic processes internal to the climate system. This includes short-period atmospheric processes (e.g., turbulence) with time constants on the scale of minutes or hours, to slower-acting processes or feedback mechanisms that add to climatic variance over longer timescales. However, these factors only contribute

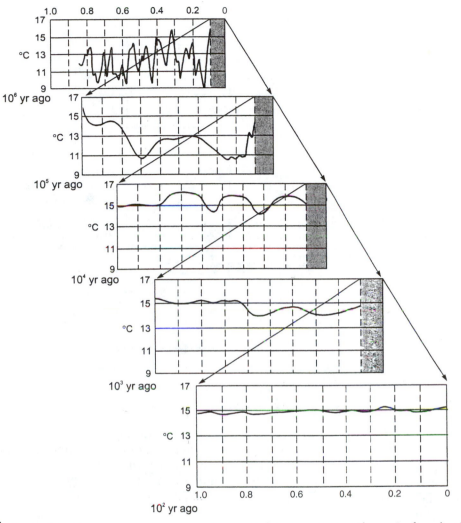

FIGURE 2.14 Schematic diagram illustrating climatic fluctuations at timescales ranging from decadal (the last 100 yr, lowest panel) to centennial (the last 1000 yr, second panel) to millennial (the last 10,000 yr), and so on, to the last million yr (top panel). Each successive panel, from the back to the front, is an expanded version (expanded by a factor of 10) of one-tenth of the previous column. Thus, higher-frequency climatic variations are "nested" within lower-frequency changes. Note that the temperature scale (representing global mean annual temperature) is the same on all panels. This demonstrates that temperature changes over the last 100 yr (lower panel) have been minor compared to changes over long periods of time. Such changes have occurred throughout history, but they are lost in the noise of the longer-term climatic record; only the larger amplitude changes are detectable as we look far back in time.

white noise to the climate spectrum on timescales longer than the timescale of the process in question (i.e., they contribute to the variance of climate in a random, unpredictable manner, with no effects concentrated at a particular frequency). Superimposed on this background noise are certain peaks in the variance spectrum of

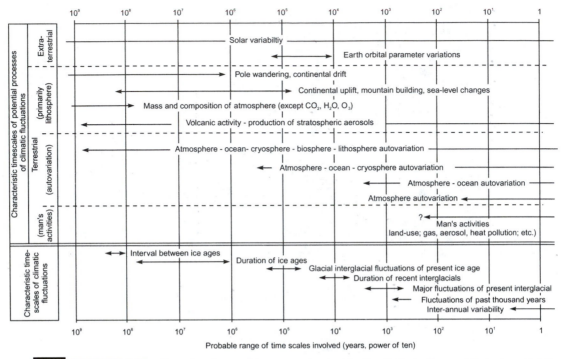

FIGURE 2.15 Examples of potential processes involved in climatic fluctuations and their characteristic timescales (Kutzbach, 1974).

climate that correspond to external forcing mechanisms operating over a restricted time domain (i.e., they are periodic or quasi-periodic phenomena). Such temporal variability may be associated with characteristic *spatial* variability (Mann *et al.,* 1996). For example, El Niño-Southern Oscillation (ENSO) events recur on timescales of 3–7 yr and have distinct spatial anomaly patterns (Diaz and Kiladis, 1992).

Deterministic forcing mechanisms are only known to operate at a few relatively narrow frequencies, and although very important to climatic variance at those frequencies their contribution to overall climatic variation is minor compared to the role of stochastic processes. This presents problems for both climatic predictability and the interpretation of past climatic changes (as seen in the paleoclimatic record) in terms of particular causative factors (Mitchell, 1976). Nevertheless, certain external forcing mechanisms have often been called upon to account for features of the paleoclimatic record. The most important of these for climatic fluctuations in the Quaternary Period are variations in the Earth's orbital parameters, which are the underlying cause of glacial-interglacial cycles over at least the last million years (Berger and Loutre, 1991). This is discussed further in the next Section and in Section 6.8.

2.6 VARIATIONS OF THE EARTH'S ORBITAL PARAMETERS

Although it has been known for over 2000 yr that the position and orientation of the Earth relative to the Sun has not been constant, it was not until the mid-nineteenth century that the significance of such variations for the Earth's climate was really appreciated. At that time, James Croll, a Scottish natural historian, developed a hypothesis in which the ultimate cause of glaciations in the past was considered to be changes in the Earth's orbital parameters (Croll, 1867a,b; 1875). The hypothesis was later elaborated by Milankovitch (1941) and more recently by A. Berger (1977a, 1978, 1979, 1988). An excellent account of the way in which this hypothesis developed into a crucial theory in paleoclimatology (the "astronomical theory") is given by Imbrie and Imbrie (1979).

The basic elements of the Earth's orbital motion around the Sun today are as follows: the Earth moves in a slightly elliptical path during its annual revolution around the Sun; because of the elliptical path, the Earth is closest to the Sun (perihelion) around January 3, and around July 5 it is farthest away from the Sun (aphelion). As a result, at perihelion the Earth receives ~3.5% more solar radiation than the annual mean (outside the atmosphere) and ~3.5% less at aphelion. The Earth is also tilted on its rotational axis 23.4° from a plane perpendicular to the plane of the ecliptic (the apparent surface over which it moves during its revolution around the Sun). None of these factors has remained constant through time due to gravitational effects of the Sun, Moon, and the other planets on the Earth. Variations have occurred in the degree of orbital eccentricity around the Sun, in the axial tilt (obliquity) of the Earth from the plane of the ecliptic, and in the timing of the perihelion with respect to seasons on the Earth (precession of the equinoxes) (Fig. 2.16).

Variations in orbital eccentricity are quasi-periodic with an average period length of ~95,800 yr over the past 5 million yr. The orbit has varied from almost circular (essentially no difference between perihelion and aphelion) to maximum eccentricity when solar radiation receipts (outside the atmosphere) varied by ~30% between aphelion and perihelion (e.g., at ~210,000 yr B.P.; Fig. 2.16). Eccentricity variations thus affect the relative intensities of the seasons, which implies an opposite effect in each hemisphere.

Changes in axial tilt are periodic with a mean period of 41,000 yr. The angle of inclination has varied from 21.8 to 24.4° with the most recent maximum occurring about 100,000 yr ago (see Fig. 2.16). The angle defines the latitudes of the polar circles (Arctic and Antarctic) and the tropics, which in turn delimit the area of daylong polar night in winter, and the maximum latitudes reached by the zenith sun in midsummer in each hemisphere. Changes in obliquity have relatively little effect on radiation receipts at low latitudes but the effect increases towards the poles. As obliquity increases, summer radiation receipts at high latitudes increase, but winter radiation totals decline. This is seen in the summer radiation variations over the last 250,000 yr for 65 and 80° N (see Fig. 2.16), which reflect mainly the periodic changes in axial tilt. As the tilt is the same in both hemispheres, changes in obliquity affect radiation receipts in the Southern and Northern Hemispheres equally.

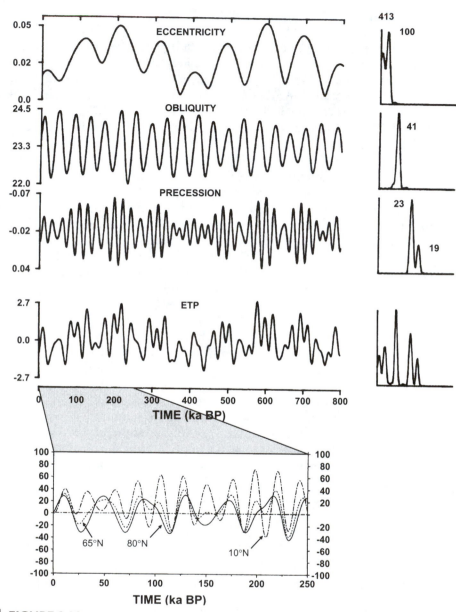

FIGURE 2.16 Variations of eccentricity, obliquity, precession, and the combination of all three factors (ETP) over the last 800,000 years with their principal periodic characteristics indicated by the power spectrum to the right of each time series (upper diagram). Below is the time series of July solar radiation at 10, 65, and 80°N (expressed as departures from A.D. 1950 values). Note that the radiation signal at high latitudes is dominated by the 41,000 year obliquity cycle whereas at lower latitudes the 23,000 precessional cycle is more significant (after Imbrie *et al.*, 1993b; lower diagram: data from Berger and Loutre, 1991).

Changes in the seasonal timing of perihelion and aphelion result from a slight wobble in the Earth's axis of rotation as it moves around the Sun (Fig. 2.17a). The effect of the wobble (which is independent of variations in axial tilt) is to change systematically the timing of the solstices and equinoxes relative to the extreme positions the Earth occupies on its elliptical path around the Sun (known as *precession of the equinoxes*) (Fig. 2.17b). Thus, 11,000 yr ago perihelion occurred when the Northern Hemisphere was tilted towards the Sun (mid-June) rather than in the Northern Hemisphere's midwinter, as is the case today. Precessional effects are opposite in the Northern and Southern Hemispheres and the change in precession occurs with a mean period of ~21,700 yr (see Fig. 2.16).

Clearly, the effects of precession of the equinoxes on radiation receipts will be modulated by the variations in eccentricity; when the orbit is near circular the seasonal timing of perihelion is inconsequential. However, at maximum eccentricity, when differences in solar radiation may amount to 30%, seasonal timing is crucial. The solar radiation receipts of low latitudes are affected mainly by variations in eccentricity and precession of the equinoxes, whereas higher latitudes are affected mainly by variations in obliquity. As the eccentricity and precessional effects in each hemisphere are opposite, but the obliquity effects are not, there is an asymmetry

 FIGURE 2.17a The Earth wobbles slightly on its axis (due to the gravitational pull of the Sun and Moon on the equatorial bulge of the Earth). In effect, the axis moves slowly around a circular path and completes one revolution every 23,000 yr. This results in precession of the equinoxes (Fig. 2.17b). This effect is independent of changes in the angle of tilt (obliquity) of the Earth, which changes with a period of ~41,000 years (from Imbrie and Imbrie, 1979).

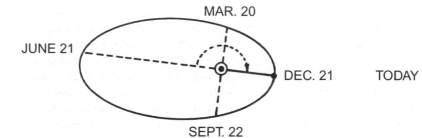

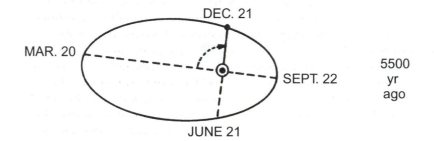

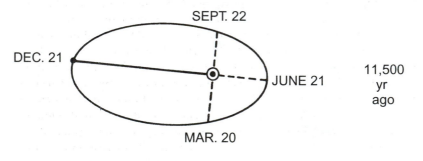

●   EARTH ON DEC. 21

◉   SUN

FIGURE 2.17b   As a result of a wobble in the Earth's axis (Fig. 2.17a) the position of the equinox (March 20 and September 22) and solstice (June 21 and December 21) change slowly around the Earth's elliptical orbit, with a period of ~23,000 yr. Thus 11,000 yr ago the Earth was at perihelion at the time of the summer solstice whereas today the summer solstice coincides with aphelion (from Imbrie and Imbrie, 1979).

between the two hemispheres, in terms of the combined orbital effects, which becomes minimal poleward of ~70°. It is also worth emphasizing that the orbital variations do not cause any significant overall (annual) change in solar radiation receipts; they simply result in a seasonal redistribution, such that a low summer radiation total is compensated for by a high winter total, and vice versa (A. Berger, 1980).

It is important to note that the periods mentioned for each orbital parameter (41,000, 95,800, and 21,700 yr for obliquity, eccentricity, and precession, respectively) are averages of the principal periodic terms in the equations used to calculate the long-term changes in orbital parameters. For the precessional parameter, for example, the most important terms in the series expansion of the equation correspond to periods of ~23,700 and ~22,400 yr; the next three terms are close to ~19,000 yr (A. Berger, 1977b). When the most important terms are averaged, the mean period is 21,700 yr, but some paleoclimatic records may be capable of resolving the principal ~19,000- and ~23,000-yr periods separately (Hays *et al.*, 1976). Similarly, the mean period of changes in eccentricity is 95,800 yr but it may be possible to detect separate periods of ~95,000 and ~123,000 yr in long high-resolution ocean core records corresponding to important terms (or "beats" produced by interactions of important terms) in the equation (Wigley, 1976). Eccentricity also has a longer-term periodicity of 412 ka, which has been identified in some marine sedimentary records (Imbrie *et al.*, 1993b). Furthermore, the relative importance of all these periods may have changed over time. For example, the 19 ka precessional and 100 ka eccentricity cycles were more significant prior to ~600 ka B.P. (Imbrie *et al.*, 1993b). This is one of the enigmas of the paleoclimatic spectrum; during the last one million years the 100 ka period in geological records increased in amplitude yet over the same interval of time the main period associated with eccentricity shifted to lower frequencies (~412 ka).

Orbital variations may also have significance for climatic variations on much shorter timescales. Loutre *et al.* (1992) calculated insolation changes over the last few thousand years, resulting from changes in precession, obliquity, and eccentricity. They found statistically significant periodicities in insolation (at 65° N in July) of 2.67, 3.98, 8.1, 18.6, 29.5, and 40.2 yr (Borisenkov *et al.*, 1983, 1985). At other seasons and locations, periodicities of 61, 245, and 830–900 yr are significant. These higher frequency variations are very small in amplitude compared to the orbital changes discussed earlier, but they may nevertheless be important for climatic variability on the decadal to millennial timescale. Interestingly, some of the periodicities in incoming insolation due to orbital effects are similar to those identified in sunspot data (which may relate to solar irradiance changes) so the cumulative effects may be significant for short-term climate variability. This matter has received relatively little attention so far.

Considered together, the superimposition of variations in eccentricity, obliquity, and precession produces a complex, ever-varying pattern of solar radiation receipts at the outer edge of the Earth's atmosphere[3]. To appreciate the magnitude of these variations and their spatial and temporal patterns, it is common to express the radiation receipts for a particular place and moment in time as a departure (or anomaly) from corresponding seasonal or monthly values in 1950. An example is shown in Fig. 2.18 for the month of July at all latitudes (90° N–90° S) from 0 to 200 ka B.P. (A. Berger, 1979). Of particular interest are the radiation anomalies at high northern latitudes (60–70° N), considered by Milankovitch (1941) to be critical for

---

[3] Values of midmonth insolation receipts for December/January and June/July, at 1000-yr intervals for the last 5 Ma, are given on a diskette accompanying the work of Berger and Loutre (1991).

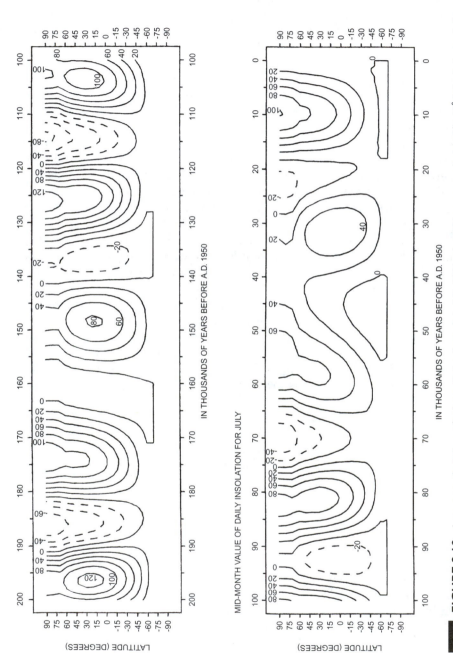

FIGURE 2.18 Long-term deviations of solar radiation from their A.D. 1950 values. Values were calculated at 10° latitude intervals from 100,000 yr B.P. to the present day (lower diagrams) and from 200,000 yr B.P. to 100,000 yr B.P. (upper diagram). Negative departures are shown by dotted lines. Values are given in calories per square centimeter per day (after Berger, 1979).

the growth of continental ice sheets. In this zone, periods of lower summer radiation receipts would have favored the persistence of winter snow into summer months, eventually leading to the persistence of snow cover throughout the year. Such conditions may have occurred at 185 ka, 115 ka, and 70 ka B.P. (see Fig. 2.18). At these times, there was the combination of conditions that Milankovitch suggested were most conducive to glaciation — minimum obliquity, relatively high eccentricity, and the Northern Hemisphere summer coinciding with aphelion (see Fig. 2.17b). At the same time, warmer winters (i.e., Northern Hemisphere winters coinciding with perihelion) would have favored increased evaporation from the subtropical oceans, thereby providing abundant moisture for precipitation (snowfall) at higher latitudes. Stronger Equator–Pole temperature gradients in summer and winter would have resulted in an intensified general circulation and more moisture transported to high latitudes to fuel the growing ice sheets. It is of great interest, therefore, that recent ocean core analyses point to these periods as being important times of ice growth on the continents (see Section 6.9).

Most of Milankovitch's attention focused on the radiation anomalies in summer and winter months, but it is noteworthy that transitional months appear to be most sensitive to changes in solar radiation receipts and to snow-cover expansion. In particular, autumn months appear to be especially critical for the build-up of snow in continental interiors (Kukla, 1975a). To examine the monthly pattern of solar radiation change through time, A. Berger (1979) computed month by month values of solar radiation departures from long-term means at 60° N, for the last 500,000 yr (Fig. 2.19). From these calculations it is clear that not only do the monthly departures vary greatly in amplitude but the seasonal timing of the anomalies may shift very rapidly from one part of the year to another. For example, a large positive anomaly of solar radiation in June and July at ~125 ka B.P. was replaced by a large negative anomaly in the same month by 120 ka B.P. Such features of the record have been termed insolation signatures (A. Berger, 1979) and are considered to be characteristic of a change from a relatively warm climate phase to a cooler one. During the last 500,000 yr such signatures are observed centered at 486, 465, 410, 335, 315, 290, 243, 220, 199, 127, 105, and 84 ka B.P., all periods which coincide remarkably well with geological evidence of deteriorating climatic conditions.

It is important to recognize that although the zone centered on 65° N may be of great importance in the actual mechanism of continental ice growth, a more fundamental control on glaciation is the atmospheric circulation, which is largely a function of the Equator–Pole temperature gradients at different times of the year (stronger radiation gradients produce stronger temperature gradients). When radiation gradients are strong, a more vigorous atmospheric circulation can be expected; subtropical high-pressure systems would tend to be displaced to lower latitudes (see Fig. 2.5) and a more intense circumpolar westerly flow would develop, leading to increased moisture flux to high latitudes. Weaker radiation gradients imply that the major axis of subtropical high-pressure cells would be displaced poleward, and a more sluggish westerly circulation would lead to a reduction in moisture flux to the continents at high latitudes. It is of interest, therefore, that the stronger summer and winter radiation gradients (resulting mainly

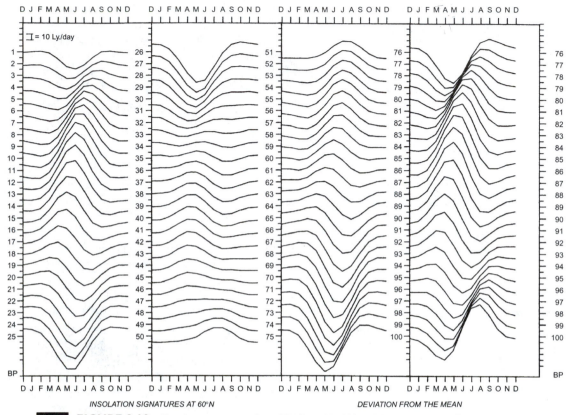

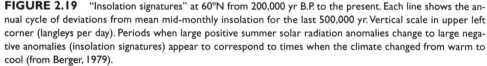

INSOLATION SIGNATURES AT 60°N   *DEVIATION FROM THE MEAN*

FIGURE 2.19 "Insolation signatures" at 60°N from 200,000 yr B.P. to the present. Each line shows the annual cycle of deviations from mean mid-monthly insolation for the last 500,000 yr. Vertical scale in upper left corner (langleys per day). Periods when large positive summer solar radiation anomalies change to large negative anomalies (insolation signatures) appear to correspond to times when the climate changed from warm to cool (from Berger, 1979).

from anomalously low radiation receipts at high latitudes) occurred during periods of major ice growth (e.g., 72,000 and 115,000 yr B.P.). By contrast, periods of deglaciation or interglacials correspond to weaker latitudinal radiation gradients, resulting mainly from higher radiation receipts, particularly at high latitudes (Fig. 2.20). Thus the resulting circulation intensities amplify the overall anomaly, whether positive or negative (Young and Bradley, 1984).

Finally, it must be recognized that the insolation changes calculated by Berger and others are for solar radiation *entering* the atmosphere (often stated as radiation at the top, or outside, the atmosphere). However, radiation passing through the atmosphere is reflected and absorbed differently from one region to another (depending to a large extent on the type and amount of cloud cover). Furthermore, surface albedo conditions also determine how much of the radiation reaching the surface will be absorbed (see Fig. 2.8). Such factors can minimize the significance of certain

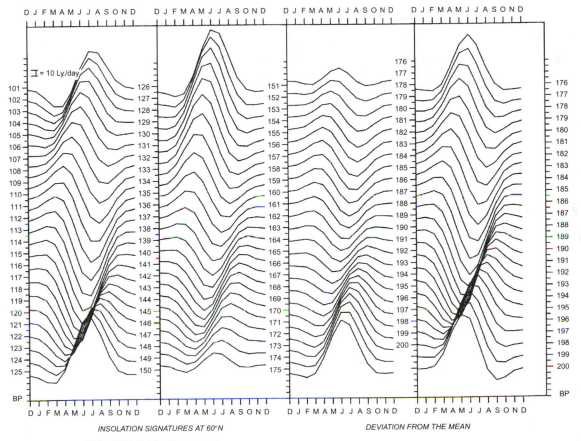

INSOLATION SIGNATURES AT 60°N

DEVIATION FROM THE MEAN

FIGURE 2.19 *(Continued)*

SEASONAL PATTERNS OF GRADIENT DEVIATIONS
(30°N - 90°N)

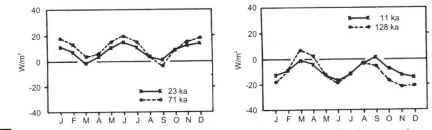

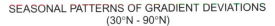

FIGURE 2.20 Variations in insolation gradients (monthly) expressed as departures from the last 150,000 yr averages for selected time periods. Periods of maximum ice growth (e.g., at 71,000 and 23,000 yr B.P.) correspond to periods of stronger than average insolation gradients in all months (left-hand diagram). Times of rapid ice decay (e.g., 128,000 and 11,000 yr B.P.) correspond to generally weaker than average gradients (right-hand diagram). Gradients calculated for Northern Hemisphere (30–90°N) (after Young and Bradley, 1984).

orbital frequencies. For example, Fig. 2.21 shows that the radiation gradient from 30–70° N in mid-July over the last 200 ka had a strong ~40 ka periodicity in extraterrestrial radiation. However, because of differential effects on radiation attenuation in the atmosphere, and latitudinal differences in surface albedo, the high latitude obliquity signal is reduced, leading to a dominant ~23 ka period in the latitudinal gradient of *absorbed* radiation (Tricot and Berger, 1988; Berger, 1988).

The astronomical theory of climatic change has tremendous implications for Quaternary paleoclimatology but there was little reliably dated field evidence to support or refute the idea until the mid-1970s. Since then many studies have demonstrated that variations in the Earth's orbital parameters are indeed fundamental factors in the growth and decay of continental ice sheets (e.g., Broecker *et al.*, 1968; Mesolella *et al.*, 1969; Hays *et al.*, 1976; Ruddiman and McIntyre, 1981a, 1984; Imbrie *et al.*, 1992, 1993a). This evidence is discussed in more detail in Section 6.12 but the major issues are summarized here in Fig. 2.22. Variations of incoming June solar radiation at 65° N are broken down into their component parts (precession, obliquity, and eccentricity) and compared to the same bandpass filtered components of the marine $\delta^{18}O$ record of the last 400 ka (representing changes in continental ice volume). Clearly, the frequency bands associated with precession and obliquity are similar (and coherent with) the $\delta^{18}O$ ice volume signal, but the 100 ka radiation signal is completely inadequate to explain the strong 100 ka cycle in ice volume. Sev-

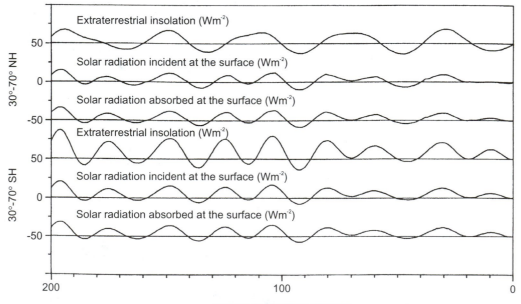

FIGURE 2.21　The gradient (from 30–70°N) of incoming solar radiation in mid-July over the last 200 ka (top) compared to the modeled gradient of radiation reaching the surface (middle) and of the gradient of radiation absorbed at the surface (bottom). Because of differential absorption and reflection with latitude, the dominant periodicity of radiation absorbed at the surface shifts from that of obliquity to that of precession, which is more characteristic of a lower latitude influence on the gradient (from Tricot and Berger, 1988).

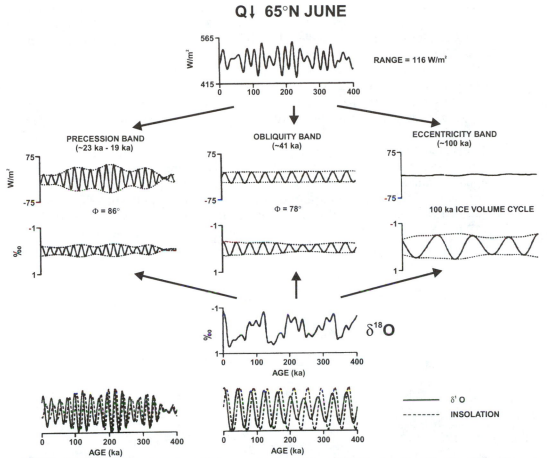

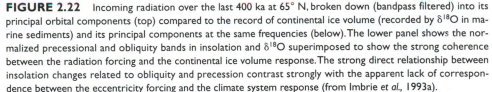

FIGURE 2.22 Incoming radiation over the last 400 ka at 65° N, broken down (bandpass filtered) into its principal orbital components (top) compared to the record of continental ice volume (recorded by $\delta^{18}O$ in marine sediments) and its principal components at the same frequencies (below). The lower panel shows the normalized precessional and obliquity bands in insolation and $\delta^{18}O$ superimposed to show the strong coherence between the radiation forcing and the continental ice volume response. The strong direct relationship between insolation changes related to obliquity and precession contrast strongly with the apparent lack of correspondence between the eccentricity forcing and the climate system response (from Imbrie *et al.,* 1993a).

eral possible reasons for this discrepancy have been proposed (Raymo, 1998). For example, there may be a nonlinear response within the climate system, perhaps involving some internal feedback mechanism, which sets up the observed 100 ka periodicity, or there may be oscillations within the climate system that in some way interact with precessional and obliquity changes to generate a cycle of 100 ka, in phase with eccentricity (Santer *et al.,* 1993). Others have suggested that the 100 ka period in paleoclimate data is unrelated to Milankovitch orbital forcing, and is due to the orbital plane of the earth passing through intergalactic dust clouds which reduce solar radiation on this timescale (Muller and MacDonald, 1997). Whichever mechanism is revealed as correct, it must also explain the shift towards much larger amplitude cycles in the climate system in the last ~800 ka (as recorded by ice volume changes, loess deposition, pollen records etc.) at a time when eccentricity forcing at the 100 ka period was declining in significance. Muller and MacDonald attribute this change to an increase in dust or meteoroids around that time, but so far there is only limited field evidence to support their idea.

In summary, the periodicities associated with orbital variations are prominent signals in many paleoclimate records and hence orbital forcing is undeniably an important factor in climatic fluctuations on the timescale of 10^4–10^6 years, and perhaps much longer (A. Berger et al., 1992). However, the precise mechanism of how such forcing is translated into a climate response remains unclear. The current emphasis is on computer modeling studies to bridge the gap between paleoclimatic theory and field data (e.g., Imbrie and Imbrie, 1980; Budd and Smith, 1981; Kutzbach and Otto-Bleisner, 1982; Kutzbach and Guetter, 1986; A. Berger, 1990; Gallée et al., 1991, 1992; Paillard, 1997). This topic is discussed further in Section 6.3.3.

3
■ DATING METHODS I

3.1 INTRODUCTION AND OVERVIEW

Accurate dating is of fundamental importance to paleoclimatic studies. Without reliable estimates on the age of events in the past it is impossible to investigate if they occurred synchronously or if certain events led or lagged others; neither is it possible to assess accurately the rate at which past environmental changes occurred. Strenuous efforts must therefore be made to date all proxy materials, to avoid sample contamination, and to ensure that the stratigraphic context of the sample is clearly understood. It is equally important that the assumptions and limitations of the dating procedure used are understood so that a realistic interpretation of the date obtained can be made. It is often just as important to know the margins of error associated with a date as to know the date itself. In this chapter, we discuss the main dating methods widely used for late Quaternary studies today. Further details can be found in Geyh and Schleicher (1990).

Dating methods fall into four basic categories (Fig. 3.1): (a) radioisotopic methods, which are based on the rate of atomic disintegration in a sample or its surrounding environment; (b) paleomagnetic (correlation) methods,[4] which rely on past

[4] One could argue that paleomagnetic changes do not constitute a method of dating but rather a method of stratigraphic correlation. Nevertheless, the development of a reliable timescale for paleomagnetic changes (Section 4.1.4) has meant that paleomagnetic changes are used, *de facto*, as dated reference horizons.

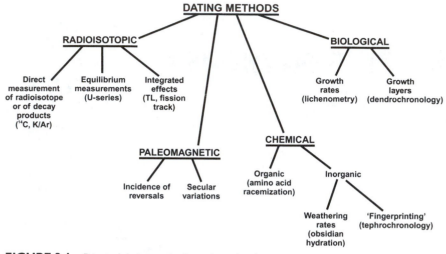

FIGURE 3.1   Principal dating methods used in paleoclimatic research.

reversals of the Earth's magnetic field and their effects on a sample; (c) organic and inorganic chemical methods, which are based on time-dependent chemical changes in the sample, or chemical characteristics of a sample; and (d) biological methods, which are based on the growth of an organism to date the substrate on which it is found.

Not all dating methods provide a reliable numerical age, but may give an indication of the *relative* age of different samples. In these cases, it may be possible to calibrate the "relative age" technique by numerical (e.g., radioisotopic) methods, as discussed for example in Section 4.2.1.3. Thus, there is a spectrum of approaches to dating: numerical age methods; calibrated age methods; relative age methods; and methods involving stratigraphic correlation (Colman *et al.,* 1987). In this and the following chapter, all of these approaches are discussed, beginning with numerical age methods.

3.2 RADIOISOTOPIC METHODS

Atoms are made up of neutrons, protons, and electrons. For any one element, the number of protons (the atomic number) is invariant, but the number of neutrons may vary, resulting in different isotopes of the same element. Carbon, for example, exists in the form of three isotopes; it always has six protons, but may have six, seven, or eight neutrons, giving atomic mass numbers (the total number of protons and neutrons) of 12, 13, and 14, designated ^{12}C, ^{13}C, and ^{14}C, respectively. Generally each element has one or more stable isotopes that accounts for the bulk of its occurrence on Earth. For example, in the case of carbon, ^{12}C and ^{13}C are the stable isotopes; ^{12}C is by far the more abundant form. It is estimated that the carbon exchange reservoir (atmosphere, biosphere, and the oceans) contains 42×10^{12} tons of ^{12}C, 47×10^{10} tons of ^{13}C, and only 62 tons of ^{14}C. Unstable atoms undergo spontaneous radioactive decay by the loss of nuclear particles (α or β particles) and, as a result, they may transmute into a new element.[5] For example, ^{14}C decays to ^{14}N, and ^{40}K decays to ^{40}Ar and ^{40}Ca. Furthermore, the decay rate is invariant so that a given quantity of the radioactive isotope will decay to its daughter product in a known interval of time; this is the basis of radioisotopic dating methods. Providing that the radioisotope "clock" is started close to the stratigraphically relevant date, measurement of the isotope concentration today will indicate the amount of time that has elapsed since the sample was emplaced. The amount of time it takes for a radioactive material to decay to half its original amount is termed its half-life. Table 3.1 lists the half-lives of some radioisotopes that have been used in the context of dating. In the case of radiocarbon (^{14}C) the half-life is 5730 ± 30 yr. Thus a plant that died 5730 yr ago has only half its original ^{14}C content remaining in it today.[6] After a further 5730 yr from today it will have only half as much again, that is, 25% of its original ^{14}C content, and so on (Fig. 3.2).

For a radioactive isotope to be directly useful for dating it must possess several attributes: (a) the isotope itself, or its daughter products, must occur in measurable quantities and be capable of being distinguished from other isotopes, or its rate of decay must be measurable; (b) its half-life must be of a length appropriate to the period being dated; (c) the initial concentration level of the isotope must be known; and (d) there must be some connection between the event being dated and the start

TABLE 3.1 Half-lives of radioisotopes used in dating — although the half-life of ^{14}C is calculated to be 5730 ± 40 yr, by convention the "Libby half-life" of 5568 ± 30 yr is used[6]

^{14}C	5.73×10^3 years
^{238}U	4.51×10^9 years
^{235}U	0.71×10^9 years
^{40}K	1.31×10^9 years

[5] An α particle is made up of two protons and two neutrons (i.e., a helium nucleus) and a β particle is an electron. Neutrons may decay to produce a β particle and a proton, thereby causing a transmutation of the element itself.

[6] When Libby (1955) expounded the principles of radiocarbon dating he calculated a half-life for ^{14}C of 5568 yr. This was the average of a number of estimates up to that time, and was adopted by all radiocarbon dating laboratories. By the early 1960s, further work had demonstrated that the original estimate was in error by 3%, and the half-life was closer to 5730 yr (Godwin, 1962). To avoid confusion, it was decided to continue using the "Libby half-life" (rounded to 5570 yr) and this practice has continued. For practical purposes it is not a significant problem as all dates are now reported in the journal *Radiocarbon* using the Libby half-life. However, when comparing "radiocarbon years" with calendar years (historical, archeological, and/or astronomical events) and dates obtained by other techniques, adjustments are necessary (see Section 3.2.1.5).

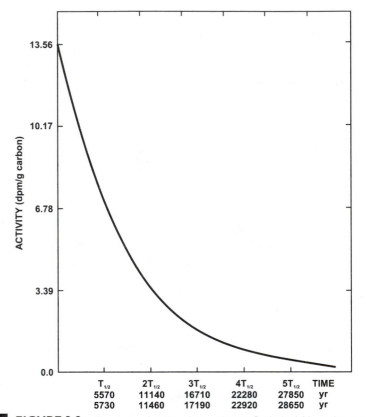

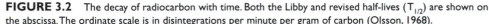
FIGURE 3.2   The decay of radiocarbon with time. Both the Libby and revised half-lives ($T_{1/2}$) are shown on the abscissa. The ordinate scale is in disintegrations per minute per gram of carbon (Olsson, 1968).

of the radioactive decay process (the "clock"). The relevance of these factors will be made clear in the ensuing sections.

In general terms, radioisotopic dating methods can be considered in three groups (see Fig. 3.1): those that measure (a) the quantity of a radioisotope as a fraction of a presumed initial level (e.g., ^{14}C dating) or the reciprocal build-up of a stable daughter product (e.g., potassium-argon, and argon-argon dating); (b) the degree to which members of a chain of radioactive decay are restored to equilibrium following some initial external perturbation (uranium-series dating); and (c) the integrated effect of some local radioactive process on the sample materials, compared to the value of the local (environmental) flux (fission-track and luminescence dating). Each of these methods will be considered separately.

3.2.1 Radiocarbon Dating

For studies of late Quaternary climatic fluctuations, ^{14}C or radiocarbon dating has proved to be by far the most useful. Because of the ubiquitous distribution of ^{14}C, the technique can be used throughout the world and has been used to date samples of peat, wood, bone, shell, paleosols, "old" sea water, marine and lacustrine sediments, and atmospheric CO_2 trapped in glacier ice. Furthermore, the useful time-frame for radiocarbon dating spans a period of major, global environmental change that would be virtually impossible to decipher in any detail without accurate dating control. Radiocarbon dating is also ideal for dating man's development from paleolithic time to the recent historical past and it has therefore proved invaluable in archeological studies. In addition, variations in the ^{14}C content of the atmosphere are of interest in themselves because of the implications these have for solar and/or geomagnetic variations through time and hence for climatic fluctuations.

3.2.1.1 Principles of ^{14}C Dating

Radiocarbon ($^{14}_{6}C$) is produced in the upper atmosphere by neutron bombardment of atmospheric nitrogen atoms:

$$^{14}_{7}N + ^{1}_{0}n \rightarrow ^{14}_{6}C + ^{1}_{1}H \qquad 3.1$$

The neutrons have a maximum concentration at around 15 km and are produced by cosmic radiation entering the upper atmosphere. Although cosmic rays are influenced by the Earth's magnetic field and tend to become concentrated near the geomagnetic poles (thus causing a similar distribution of neutrons and hence ^{14}C), rapid diffusion of ^{14}C atoms in the lower atmosphere obliterates any influence of this geographical variation in production. The ^{14}C atoms are rapidly oxidized to $^{14}CO_2$, which diffuses downwards and mixes with the rest of atmospheric carbon dioxide and hence enters into all pathways of the biosphere (Fig. 3.3). As Libby (1955) stated, "Since plants live off the carbon dioxide, all plants will be radioactive; since the animals on earth live off the plants, all animals will be radioactive. Thus . . . all living things will be rendered radioactive by the cosmic radiation."

During the course of geological time, an equilibrium has been achieved between the rate of new ^{14}C production in the upper atmosphere and the rate of decay of ^{14}C in the global carbon reservoir. This means that the 7.5 kg of new ^{14}C estimated to be produced each year in the upper atmosphere is approximately equal to the weight of ^{14}C lost throughout the world by the radioactive decay of ^{14}C to nitrogen, with the release of a β particle (an electron):

$$^{14}_{6}C \rightarrow ^{14}_{7}N + \beta + \text{neutrino} \qquad 3.2$$

The total weight of global ^{14}C thus remains constant.[7] This assumption of an essentially steady concentration of radiocarbon during the period useful for dating is fundamental to the method though, in detail, this assumption is invalid (see Section 3.2.1.5).

Plants and animals assimilate a certain amount of ^{14}C into their tissues through photosynthesis and respiration; the ^{14}C content of these tissues is in equilibrium with that of the atmosphere because there is a constant exchange of new ^{14}C as old

[7] Prior to atomic bomb explosions in the atmosphere the equilibrium quantity of ^{14}C was estimated to be ~62 metric tons. Since the 1950s, the amount of artificially produced ^{14}C has increased by perhaps 3–4%, though most of this has, as yet, remained in the atmosphere; consequently ^{14}C levels there have almost doubled (Aitken, 1974).

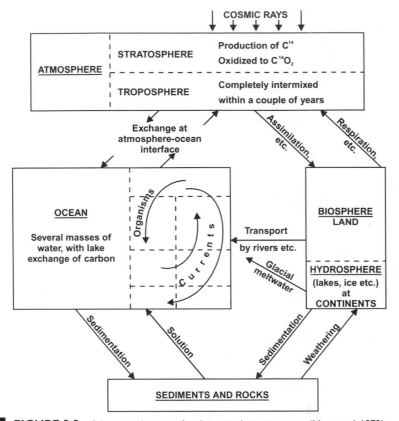

FIGURE 3.3   Schematic diagram of carbon circulation in nature (Mangerud, 1972).

cells die and are replaced. However, as soon as an organism dies this exchange and replacement of ^{14}C ceases. From that moment on the ^{14}C content of the organism declines as the ^{14}C decays to nitrogen, and the ^{14}C content is henceforth purely a function of time; the radioactive "clock" has been activated. Because the ^{14}C content declines at a negative exponential rate (see Fig. 3.2) by the time that ten half-lives have elapsed (57,300 yr) the sample contains less than 0.01% of the original ^{14}C content of the organism when it was alive. To put this in a more practical way, in a 1 g sample of carbon with a ^{14}C content equivalent to modern levels, decay of the radiocarbon atoms in the sample will produce about 15 β particles per minute, a rate that is relatively easy to count. By contrast, 57,300 yr after an organism has died, 1 g of its carbon will produce only about 21 β particles per day (Aitken, 1974). It is this ever-decreasing quantity of ^{14}C with increasing sample age that makes conventional radiocarbon dating so difficult; it simply becomes impossible to separate disintegrations of the sample from the extraneous background radiation (typically ~132 β particles per day in a modern counter) as the signal-to-noise ratio (S/N) becomes too small.

3.2.1.2 Measurement Procedures, Materials, and Problems

Until the early 1980s, nearly all radiocarbon-dating laboratories used so-called "conventional" methods — either proportional gas counters or liquid scintillation techniques. In the former method, carbon is converted into a gas (methane, carbon dioxide, or acetylene), that is then put into a "proportional counter" capable of detecting β particles (variations in output voltage pulses being proportional to the rate of β-particle emission). In liquid scintillation procedures, the carbon is converted into benzene or some other organic liquid and placed in an instrument that detects scintillations (flashes of light) produced by the interaction of β particles and a phosphor added to the organic liquid. In both methods, stringent measures are necessary to shield the sample counters from extraneous radioactivity in the instrument components, laboratory materials, and surrounding environment, including cosmic rays penetrating the Earth's atmosphere from outer space. Indeed, the difficulty of separating the sample β-particle signal from environmental "noise" was one of the major obstacles to the development of ^{14}C dating, particularly of older samples, which have very low levels of ^{14}C anyway (Libby, 1970). Lead shielding, electronic anticoincidence counters (to alert the counter to particles entering the counting chamber from outside) and construction of laboratories beneath the ground are common strategies to help keep background radiation levels as low as possible.

One of the problems of dating very old samples by conventional methods is the large sample size needed to obtain enough radiocarbon for its β activity to be counted. Technical difficulties place an upper limit on the volume of gas or liquid that can be accurately analyzed; hence, for very old samples, some means of concentrating the ^{14}C is needed to reduce the volume. One solution (no longer in common usage) is to concentrate a gas containing the ^{14}C (e.g., $^{14}CO_2$) by thermal diffusion, "enriching" the sample and reducing the required volume. Effectively, the gas containing the heavier isotope is encouraged to collect in the lower chamber of a thermal diffusion column; in this way, the radioactive component is concentrated, reducing the total volume of gas necessary for accurate counting. However, the procedure is very time consuming; a 6-fold enrichment may take up to 5 weeks! Nevertheless, this procedure has enabled samples as old as 75,000 yr (13 half-lives) to be dated (Stuiver *et al.*, 1978; Grootes *et al.*, 1980). The main limitation is that the initial sample must be large enough to yield 100 g of carbon for analysis, and there must be very low background ^{14}C activity to minimize any statistical uncertainty in the calculated age. Furthermore, if an infinitely old sample has even 0.1% contamination by modern carbon, it will yield a finite date of ~55,500 yr B.P., illustrating the inherent dangers of interpreting extremely old ^{14}C dates. In practice, ^{14}C dates of >45,000 yr B.P. should be viewed with caution.

Radiocarbon dating underwent a technological revolution in the late 1970s and early 1980s when a method for dating very small organic samples was developed, using an accelerator coupled to a mass spectrometer (AMS dating) (Muller, 1977; Nelson *et al.*, 1977; Litherland and Beukens, 1995). Instead of measuring the quantity of ^{14}C in a sample indirectly, by counting β-particle emissions, the concentrations of individual ions (^{12}C, ^{13}C, ^{14}C) are measured. Ions are accelerated in a tandem electrostatic accelerator to extremely high velocities; they then pass through a magnetic field that separates the different ions, enabling them to be distinguished (Stuiver, 1978a; Elmore and Phillips, 1987). Sample sizes used in this technique are much smaller than in conventional ^{14}C dating (only 1 mg of carbon is required) so that dates on small samples of foraminifera, pollen grains isolated from their surrounding matrix, or even individual seeds can be dated (Brown *et al.*, 1992; Regnall, 1992). There are now many accelerators designed specifically for ^{14}C dating applications and several labs can produce results within days of receiving a sample. This rapid turn-around time allows field workers to quickly reassess sampling strategies, thus making the most of time in the field. On the other hand, it may not be too long before a field-portable radiocarbon analyzer is available, at least for first-order estimates of sample age (Robertson and Grün, 1994).

3.2.1.3 Accuracy of Radiocarbon Dates

It is tempting to accept a radiocarbon date as the gospel truth, particularly if it confirms a preconceived notion of what the sample age should be! Radiocarbon dates are, however, statements of probability (as are all radiometric measurements). Radioactive disintegration varies randomly about a mean value; it is not possible to predict when a particular ^{14}C atom will decay, but for a sample containing 10^{10}–10^{12} atoms of ^{14}C a certain number of disintegrations will occur, on average, in a certain length of time. This statistical uncertainty in the sample radioactivity (together with similar uncertainty in the radioactive decay of calibration samples and "noise" due to background radiation) is inherent in all ^{14}C dates. A single "absolute" (i.e., numerical) age can therefore never be assigned to a sample. Rather, dates are reported as the midpoint of a Poisson probability distribution; together with its standard deviation, the date thus defines a known level of probability. A date of 5000 ± 100 yr B.P., for example, indicates a 68% probability that the true (radiocarbon) age is between 4900 and 5100 yr B.P., a 95% probability that it lies between 4800 and 5200 yr B.P., and a 99% probability of it being between 4700 and 5300 yr B.P. In conventional dating, the use of large samples, extended periods of counting, and the reduction of laboratory background noise will all improve the precision of the age determination. However, even rigorous ^{14}C analysis cannot account for all sources of error and these must be evaluated before putting a great deal of confidence in the date obtained. A sample age may be precisely determined (analytically) but it may not be an *accurate* reflection of the true age if the sample is contaminated, or if appropriate corrections are not made, as discussed in the following sections.

3.2.1.4 Sources of Error in ^{14}C Dating

(a) Problems of Sample Selection and Contamination

It is self-evident that a contaminated sample will give an erroneous date, but it is frequently very difficult to ascertain the extent to which a sample has been contaminated. Some forms of contamination are relatively straightforward: modern rootlets, for example, may penetrate deep into a peat section and without careful inspection of a sample and removal of such material, gross errors may occur. More abstruse problems arise when dating materials that contain carbonates (e.g., shell, coral, bone). These materials are particularly susceptible to contamination by modern carbon because they readily participate in chemical reactions with rainwater and/or groundwater. Most molluscs, for example, are primarily composed of calcium carbonate in the metastable crystal form, aragonite. This aragonite may dissolve and be redeposited in the stable crystal form of calcite. During the process of solution and recrystallization, exchange of modern carbon takes place and the sample is thereby contaminated (Grant-Taylor, 1972). This problem also exists in corals, which are all aragonite. Commonly, x-ray diffraction is used to identify different carbonate mineral species, and materials with a high degree of recrystallization are discarded. However, Chappell and Polach (1972) have noted that recrystallization can occur in two different modes: one open system and therefore susceptible to modern ^{14}C contamination; and one closed system, which is internal and involves no contamination. The former process tends to be concentrated around the sample margins, as one might expect, and so a common strategy is to dissolve away the surface 10–20% of the sample with hydrochloric acid and to date the remaining material. For shells thought to be very old, in which recrystallization may have affected a deep layer, the remaining inner fraction should ideally be dated in two fractions (an "outer inner" and an "inner inner" fraction) to test for consistency of results. However, even repeated leaching with hydrochloric acid may not produce a reliable result in cases where recrystallization has permeated the entire sample. It is worth noting in this connection that "infinite"-aged shells (those well beyond the range of ^{14}C dating techniques) contaminated by only 1% of modern carbon will have an apparent age of 37,000 yr (Olsson, 1974). Thus, even very small amounts of modern carbon can lead to gross errors, and many investigators consider that dates of >25,000 yr on shells should be thought of as essentially "infinite" in age. While such a conservative approach is often laudable, it does pose the danger that correct dates in the 20,000–30,000 year range may be overlooked. A possible remedy to this problem of dating old shells is to isolate a protein, conchiolin, present in very small quantities in shells (1–2%) and to date this, rather than the carbonate (Berger *et al.,* 1964). Carbon in conchiolin does not undergo exchange with carbon in the surrounding environment and hence is far less likely to be contaminated. Unfortunately, to isolate enough conchiolin, very large samples (>2 kg) are needed, and these are often not available.

Similar problems are encountered in dating bone; because of exchange reactions with modern carbon, dates on total inorganic carbon or on apatite carbonate are unreliable (Olsson *et al.,* 1974). As in the case of shells, a more reliable approach is to isolate and date carbon in the protein collagen. However, this slowly disappears under the influence of an enzyme, collagenase, so that in very old samples the amount of collagen available is extremely small and extra-large samples are needed to extract it. Furthermore, collagen is extremely difficult to extract without contamination and different extraction methods may give rise to different dates.

Another form of error concerns the "apparent age" or "hard-water effect" (Shotton, 1972). This problem arises when the materials to be dated, such as freshwater molluscs or aquatic plants, take up carbon from water containing bicarbonate derived from old, inert sources. This is a particularly difficult problem in areas where limestone and other calcareous rocks occur. In such regions, surface and ground water may have much lower $^{14}C/^{12}C$ ratios than that of the atmosphere due to solution of the essentially ^{14}C-free bedrock. Because plants and animals existing in these environments will assimilate carbon in equilibrium with their surrounding milieu rather than the atmosphere, they will appear older than they are in reality, sometimes by as much as several thousand years. This problem was well illustrated by Shotton (1972), who studied a late-glacial stratigraphic section in North Jutland, Denmark. Dates on contemporaneous twigs and a fine-grained vegetable residue, thought to be primarily algal, fell neatly into two groups, with the algal material consistently 1700 yr older than the terrestrial material. The difference was considered to be the result of a hard-water effect, the aquatic plants assimilating carbon in equilibrium with water containing bicarbonate from old, inert sources.

Other studies have demonstrated further complexities in that the degree of "old carbon" contamination may change over time. For example, Karrow and Anderson (1975) suggested that some lake sediments in southwestern New Brunswick, Canada, studied by Mott (1975) were contaminated with old carbon shortly after deglaciation. Initial sedimentation was mainly marl derived from carbonate-rich till and carbonate bedrock, but as the area became vegetated and soil development took place the sediments became more organic and less contaminated by "old carbon." Dates on the deepest lake sediments are thus anomalously old and would appear to give a date of deglaciation inconsistent with the regional stratigraphy. This points to the more general observation that the geochemical balance of lakes may have changed through time and that the modern water chemistry may not reflect former conditions. This is particularly likely in formerly glaciated areas where the local environment immediately following deglaciation would have been quite different from that of today. One should thus interpret basal dates on lake sediments or peat bogs with caution. Equally, dates on aquatic flora and fauna, in closed lake basins which have undergone great size changes, must also be viewed in the light of possible changes in the aqueous geochemistry of the site.

It is important to recognize that not all types of contamination are equally significant; contamination by modern carbon is far more important than that by old carbon because of its much higher activity (Olsson, 1974). Figures 3.4 and 3.5 show the errors associated with different percentage levels of contamination by modern and old material, respectively. It will be seen that a 5000-yr-old sample, 20% contaminated with 16,000-yr-old carbon, would give a date in error by only ~1300 years. By contrast, a 15,000-yr-old sample contaminated with only 3% of modern carbon would result in a dating error of about the same magnitude. Very careful sample selection is therefore needed; dating errors are most commonly the result of inadequate sampling.

(b) Variations in ^{14}C Content of the Oceanic Reservoir

In the preceding section, it was discussed that some freshwater aquatic plants or molluscs may be contaminated by water containing low levels of ^{14}C. In the case of marine organisms this problem is much more universal. First, when carbon dioxide is absorbed into the oceans a fractionation takes place that leads to an enrichment of 15‰ (equivalent to ~120 yr) in the ^{14}C activity of oceanic bicarbonate

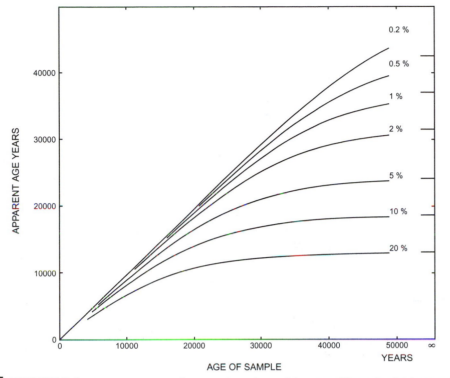

FIGURE 3.4 Apparent ages as a function of sample age if there are different levels of contamination by modern material in the sample. Thus a 20,000 yr-old sample contaminated with 10% of modern material would appear to be ~15,000 yr old (from Olsson and Eriksson, 1972).

relative to that of the atmosphere. However, ocean surface waters are not in isotopic equilibrium with the atmosphere because oceanic circulation brings ^{14}C-depleted water to the surface to mix with "modern" water. Consequently, the ^{14}C age of the surface water (the apparent age, or reservoir age) varies geographically (Fig. 3.6) (Bard, 1988). In the lower latitudes of all oceans, the mean reservoir age of surface waters is ~400 yr, which means that 400 yr must be added to a ^{14}C date on marine organic material from the mixed layer, in order to compare it with terrestrial material. At higher latitudes, this correction can be much larger due to up-welling of older water and the effect of sea ice, which limits the ocean-atmosphere exchange of CO_2. The modern North Atlantic is different from other high latitude regions because of advection of warmer waters (relatively enriched in ^{14}C) from lower latitudes, and strong convection (deepwater formation), which limits any up-welling of ^{14}C-depleted water.

The extent to which such ^{14}C gradients have been constant over time is of great significance for dating older events in the marine environment and comparing them with terrestrial records. If North Atlantic deepwater formation ceased during the Last Glacial (see Section 6.10) then it is likely that the surface water reservoir age would

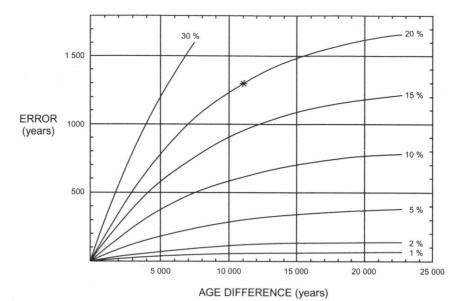

FIGURE 3.5   The error in a radiocarbon date if a certain fraction of the sample (indicated by each curve) is contaminated by *older* material (having a lower ^{14}C activity). Errors expressed as age differences (abscissa) between the sample and contaminant. For example, a 5000-yr old sample with 20% contamination by 16,000 yr-old material (i.e., material 11,000 yr older than the true age — see star on figure) will yield a date in error (too old) by 1300 yr (from Olsson, 1974).

be similar to other high latitude areas, as depicted by the dashed line in Fig. 3.6. Hence, late Glacial planktonic samples from this area might require an age adjustment of as much as 1000 yr, and if deepwater switched on and off rapidly, age corrections would be equally volatile, possibly even leading to age inversions in otherwise stratigraphically undisturbed sediments. The matter was addressed by Bard *et al.* (1994) and Austin *et al.* (1995), who compared marine and terrestrial ^{14}C-dated samples associated with the Vedde volcanic ash. This ash layer originated from an explosive Icelandic eruption and was widely distributed across the North Atlantic and adjacent land areas around 10,300 ^{14}C yr B.P., according to AMS dates on terrestrial samples from the time of ash deposition. However, all marine samples associated with the ash layer were dated ~11,000 ^{14}C yr B.P., indicating that the reservoir effect then was ~700 yr, vs ~400 yr today, probably as a result of reduced North Atlantic Deep Water (NADW) formation, and/or increased sea-ice cover at that time.

So far, we have focused primarily on surface water, but the same issues arise in deepwater changes. North Atlantic Deep Water and the Antarctic Bottom Water may remain out of contact with the atmosphere for centuries because of the overlying warmer water in mid and low latitudes. During this time, the ^{14}C content of the deep water decreases so that deepwater samples commonly give ^{14}C ages more than 1000 yr older than surface water (Fig. 3.7). Indeed, the gradual decline in ^{14}C activity of oceanic water has been used to assess the former "ventilation rate" of the ocean, by comparing the age of ^{14}C-dated planktonic and benthic foraminifera

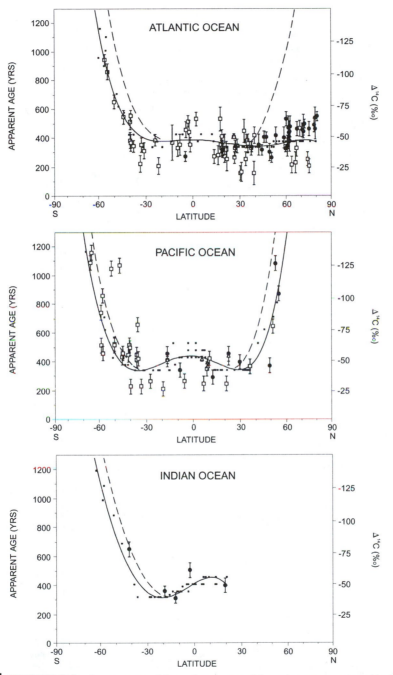

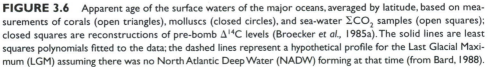

FIGURE 3.6 Apparent age of the surface waters of the major oceans, averaged by latitude, based on measurements of corals (open triangles), molluscs (closed circles), and sea-water ΣCO_2 samples (open squares); closed squares are reconstructions of pre-bomb $\Delta^{14}C$ levels (Broecker *et al.*, 1985a). The solid lines are least squares polynomials fitted to the data; the dashed lines represent a hypothetical profile for the Last Glacial Maximum (LGM) assuming there was no North Atlantic Deep Water (NADW) forming at that time (from Bard, 1988).

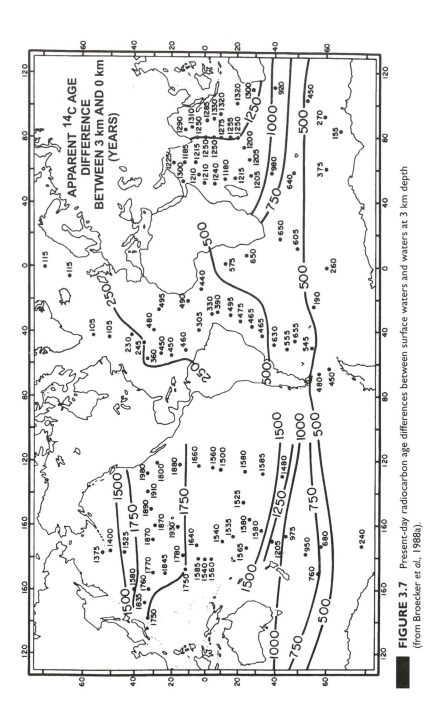

FIGURE 3.7 Present-day radiocarbon age differences between surface waters and waters at 3 km depth (from Broecker et al, 1988a).

(surface and deep-dwelling organisms, respectively) in marine sediments from the same site. Then, by comparing modern (core-top) samples with those from earlier periods, a record of changes in ventilation rate and ocean circulation can be established (see Section 6.9). In this way, Duplessy *et al.* (1989) found that the ventilation rate of the Pacific Ocean was less than today during the Last Glacial (resulting in a 1500–2500 yr age difference, vs ~1300 yr today) but during the deglaciation (15,000–10,000 ^{14}C yr B.P.) it was greater (200–1000 yr age difference) (Shackleton *et al.*, 1988). In the Atlantic Ocean, deepwater had a mean age of ~675 yr in the Last Glacial versus ~350 yr today (Peng and Broecker, 1995).

Because there are still uncertainties about the timing and extent of past changes in reservoir effects in both surface and deep waters, many investigators only make corrections for the observed modern oceanic reservoir effect, arguing that so far there is an insufficient basis of knowledge to do anything else. Nevertheless, this can lead to difficulties in trying to determine the sequence of terrestrial and marine events, particularly at times of rapid environmental change, such as occurred at the end of the Last Glacial and around the time of the Younger Dryas episode (Austin *et al.*, 1995). The problem is further compounded by the "plateau" found in ^{14}C ages around 10,000 ^{14}C yr B.P., which tends to make events that were in fact diachronous appear to be synchronous (see Section 3.2.1.5).

(c) Fractionation Effects

Basic to the principle of ^{14}C dating is the assumption that plants assimilate radiocarbon and other carbon isotopes in the same proportion as they exist in the atmosphere (i.e., the ^{14}C/^{12}C ratio of plant tissue is the same as that in the atmosphere). However, during photosynthesis, when CO_2 is converted to carbohydrates in plant cells, an isotopic fractionation occurs such that ^{12}C is more readily "fixed" than ^{14}C, resulting in a lower ^{14}C content in plants than that of the atmosphere (Olsson, 1974). This ^{14}C "depletion" may be as much as 5% below atmospheric levels but this is not consistent among all organisms. The magnitude of the fractionation effect varies from one plant species to another by a factor of two to three and depends on the particular biochemical pathways evolved by the plant for photosynthesis (Lerman, 1972). This is discussed in further detail in Appendix A. Fortunately, some assessment of the ^{14}C fractionation effect can be made relatively easily by measuring the ^{13}C content of a sample. The ^{14}C fractionation is very close to twice that of ^{13}C (Craig, 1953), a stable isotope which occurs in far greater quantities than ^{14}C and can hence be routinely measured by a mass spectrometer. The ^{13}C content is generally expressed as a departure from a Cretaceous limestone standard (Peedee belemnite, see Appendix A):

$$\delta^{14}C = 2\delta^{13}C = \frac{(^{13}C/^{12}C)_{sample} - (^{13}C/^{12}C)_{PDB}}{(^{13}C/^{12}C)_{PDB}} \times 10^{3}‰ \qquad 3.3$$

A change in δ^{14}C of only 1‰ (i.e., a change in δ^{13}C of 0.5‰) corresponds to an age difference of ~8 yr. Consequently, if two contemporaneous samples differed in δ^{13}C by 25‰, they would appear to have an age difference of ~400 years (Olsson and

Osadebe, 1974). To avoid such confusion, it has been recommended that the ^{13}C value of all samples be normalized to −25‰, the average value for wood. By adopting this reference value, comparability of dates is possible. This is particularly important in the case of marine shell samples, which characteristically have δ^{13}C values in the range +3 to −2‰. Standardization to δ^{13}C = −25‰ thus involves an age adjustment of as much as 450 yr (to be *added* to the uncorrected date).

This is further complicated because of the low ^{14}C content of the oceans, which gives modern seawater an "apparent age" of 400–2500 yr (see Section 3.2.1.4b) and results in a correction in the opposite direction from the correction for fractionation effects (a value to be *subtracted* from the uncorrected age). For the North Atlantic region, the apparent age of seawater is ~400 yr (Stuiver *et al.,* 1986) so the fractionation and oceanic effect more or less cancel out. In other areas, particularly at high latitudes, the oceanic adjustment is >450 yr, so the adjusted date will be lower (younger) than the original estimate, before correction for fractionation effects. Of course, for comparison of dates on similar materials within one area, these adjustments are irrelevant. However, if one wishes to compare, for example, dates on terrestrial peat with dates on marine shells, or to compare a shell date from high latitudes with one from another area, care must be taken to ascertain what corrections, if any, have been applied. Details of reservoir corrections for different locations can be found in Stuiver *et al.* (1986).

3.2.1.5 Long-term Changes in Atmospheric ^{14}C Content

Fundamental to the principles of radiocarbon dating is the assumption that atmospheric ^{14}C levels have remained constant during the period useful for ^{14}C dating. However, even in the early days of radiocarbon dating, comparisons between archeologically established Egyptian chronologies and ^{14}C dates suggested that the assumption of temporal constancy in ^{14}C levels might not be correct. It is now abundantly clear that ^{14}C levels have varied over time, though fortunately the magnitude of these variations can be assessed, at least since the Late Glacial. Variations in atmospheric ^{14}C concentration may result from a wide variety of factors, as indicated in Table 3.2, and it is worth noting that many of these factors may themselves be important influences on climate. It is a sobering thought that fluctuations in the concentration of radiocarbon may help to explain the very paleoclimatic events to which radiocarbon dating has been applied for so many years; there could be no better illustration of the essential unity of science (Damon, 1970).

Early work on carefully dated tree rings indicated that ^{14}C estimates showed systematic, time-dependent variations (de Vries, 1958). Both European and North American tree-ring samples spanning the last 400 yr showed departures from the average of up to 2%, with ^{14}C maxima around A.D. 1500 and A.D. 1700 (Fig. 3.8). These secular ^{14}C variations appear to be closely related to variations in solar activity, as discussed further in Section 3.2.1.6 (Suess, 1980). Also seen in Fig. 3.8 is the marked decline in ^{14}C activity during the last 100 yr. This resulted primarily from the combustion of fossil fuel during that period (Suess, 1965), causing a rapid increase in the abundance of "old" (essentially ^{14}C-free) carbon in the atmosphere (the so-called "Suess effect").

TABLE 3.2 **Possible Causes of Radiocarbon Fluctuations**

I. *Variations in the rate of radiocarbon production in the atmosphere*

    (1) Variations in the cosmic-ray flux throughout the solar system

        (a) Cosmic-ray bursts from supernovae and other stellar phenomena

        (b) Interstellar modulation of the cosmic-ray flux

    (2) Modulation of the cosmic-ray flux by solar activity

    (3) Modulation of the cosmic-ray flux by changes in the geomagnetic field

    (4) Production by antimatter meteorite collisions with the Earth

    (5) Production by nuclear weapons testing and nuclear technology

II. *Variations in the rate of exchange of radiocarbon between various geochemical reservoirs and changes in the relative carbon dioxide content of the reservoirs*

    (1) Control of CO_2 solubility and dissolution as well as residence times by temperature variations

    (2) Effect of sea-level variations on ocean circulation and capacity

    (3) Assimilation of CO_2 by the terrestrial biosphere in proportion to biomass and CO_2 concentration, and dependence of CO_2 on temperature, humidity, and human activity

    (4) Dependence of CO_2 assimilation by the marine biosphere upon ocean temperature and salinity, availability of nutrients, upwelling of CO_2-rich deep water, and turbidity of the mixed layer of the ocean

III. *Variations in the total amount of carbon dioxide in the atmosphere, biosphere, and hydrosphere*

    (1) Changes in the rate of introduction of CO_2 into the atmosphere by volcanism and other processes that result in CO_2 degassing of the lithosphere

    (2) The various sedimentary reservoirs serving as a sink of CO_2 and ^{14}C. Tendency for changes in the rate of sedimentation to cause changes in the total CO_2 content of the atmosphere

    (3) Combustion of fossil fuels by human industrial and domestic activity

From Damon *et al.* (1978).

These studies generated further interest in testing the assumptions of radiocarbon dating and led to hundreds of checks being made between ^{14}C dates and corresponding wood samples, each carefully dated according to dendrochronological principles (see Chapter 10). The longest tree-ring calibration set is the German oak and pine chronology of >11,000 yr, made up of more than 5000 different overlapping tree-ring sections from living trees, medieval housing timbers, and subfossil wood excavated from river gravels (Becker, 1993). Similar chronologies have been constructed from Irish oaks (to 5289 B.C.) and from Douglas fir and Bristlecone pine in the U.S. Pacific Northwest and California (to >6000 B.C.) Other "floating chronologies," derived from radiocarbon-dated subfossil wood, have now been accurately fixed in time by matching de Vries-type ^{14}C variations in the chronology with those observed in the well-dated continuous tree-ring records (Kuniholm *et al.*, 1996). By being able to precisely match such "wiggles," recorded in wood from

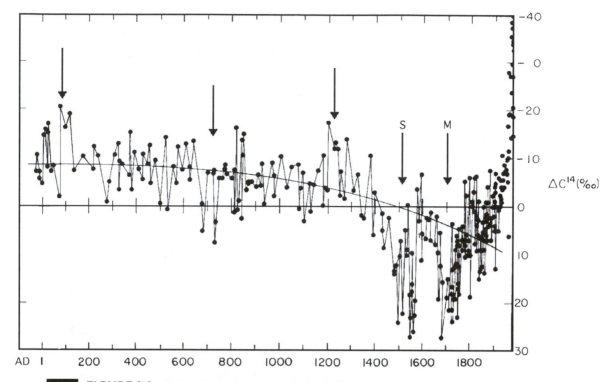

FIGURE 3.8 Radiocarbon variations over the last 2000 yr expressed as departures from the long-term average. Trend due to geomagnetic field variation shown by curved line. Positive ^{14}C anomalies around A.D. 1500 and A.D. 1700 correspond to periods of reduced solar activity (the Spörer and Maunder Minima, S and M, respectively). At these times the increased cosmic ray flux produces more ^{14}C in the upper atmosphere. Combustion of fossil fuel has contaminated the atmosphere with ^{14}C-free CO_2, hence the large negative departures shown since 1850 (from Eddy, 1977). Note that the ordinate is plotted with positive departures lowermost, corresponding to periods of *reduced* solar activity.

places thousands of kilometers apart, it is clear that the high-frequency variations have real geophysical significance and are not simply the result of noise in the radiocarbon chronology (de Jong *et al.*, 1980).

By radiocarbon dating wood of known age from different regions of the world, a very consistent picture of the relationship between ^{14}C age and calendar year age has been built up for the last ~11,400 yr (Stuiver and Pearson, 1993; Pearson and Stuiver, 1993; Pearson *et al.*, 1993; Kromer and Becker, 1993). This calibration is based on bi-decadal and decadal wood samples for most of the period, but for the last ~500 yr a year-by-year analysis of wood has provided a very detailed comparison of calendar year and ^{14}C ages (Stuiver, 1993). The ^{14}C age is very close to the dendrochronological age (± 100 yr) for the last ~2500 yr, but before that there was a systematic difference (^{14}C underestimating true age) increasing to ~1000 yr by ~10,000 calendar yr B.P. (Fig. 3.9).

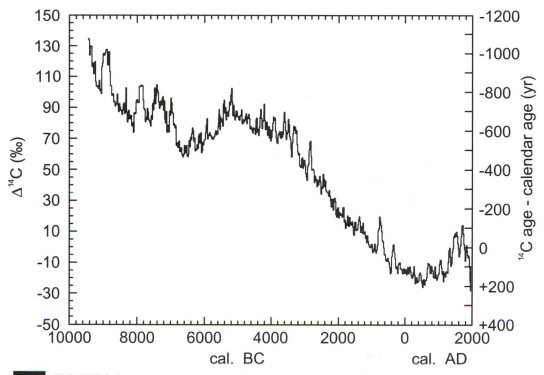

FIGURE 3.9 Bidecadal anomalies of $\Delta^{14}C$ (‰) (left axis) in relation to the calendar year age of the wood samples analyzed. An anomaly of +1‰ corresponds to the radiocarbon age underestimating the dendrochronological age by 8 yr (right axis) (from Stuiver and Reimer, 1993).

For the period before reliable dendrochronologically dated wood samples are available, other types of record must be used to calibrate the radiocarbon record. High precision uranium-series dates on Pacific and Atlantic corals, obtained by thermal ionization mass spectrometry (TIMS) are equivalent to calendar years and can be directly compared with ^{14}C dates on the same samples (Bard *et al.*, 1990, 1993; Edwards *et al.*, 1993). Such comparisons reveal that ^{14}C ages continue to be systematically younger than "true" ages at least as far back as 30,000 calendar yr B.P., with the deviation increasing to ~3500 years at that time (i.e., atmospheric $^{14}C/^{12}C$ was 400–500‰ above the modern reference level). Figure 3.10 shows the ^{14}C anomaly from present ($\Delta^{14}C$) for the combined coral and tree-ring calibration series. It is clear that the coral and tree-ring data fit quite well in the period of overlap, though the coral data are limited. For the interval 9000 to 15,000 calendar yr B.P., a Late Glacial/early Holocene varved (annually laminated) sediment record from the Cariaco Basin off the coast of Venezuela provides a far more detailed chronology, supported by dozens of AMS radiocarbon dates on forams extracted from the varves (Hughen *et al.*, 1996b, 1998). The floating varve chronology was first fixed in time by finding the best fit between "wiggles" in the radiocarbon

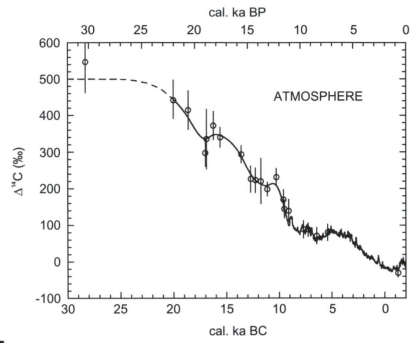

FIGURE 3.10   The atmospheric Δ^{14}C record of the last 30,000 yr derived from both dendrochronologically based wood samples and corals dated by ^{230}Th/^{234}U, with a 400-yr correction for the reservoir age of tropical surface waters. The coral data are shown as circles with 2σ error bars (from Stuiver and Braziunas, 1993).

versus calendar year age record of the German oak/pine chronology, and the AMS-dated varved sediments from the Cariaco Basin (Fig. 3.11). Once this section of the varved record was firmly anchored to the tree-ring series, the older varves could then be assigned calendar ages and compared to their radiocarbon age, derived from the AMS dates on forams (Fig. 3.12). In this way, the varved sediment record has been used to extend the radiocarbon calibration back to 15,000 (calendar) yr, and with farther varve studies it should be possible to push this back even farther in time. The results so far clearly support conclusions (based on the very sparse coral data) that radiocarbon and calendar ages diverged in pre-Holocene time. This evidence conflicts with several lake varve sediment studies suggesting that before ~12,000 calendar yr B.P., the ^{14}C ages and calendar ages may have converged (Wohlfarth *et al.*, 1995; Wohlfarth, 1996). However, *in toto* the lake varve studies provide an inconsistent picture, possibly due to redeposition of older material (which, when dated, makes the varve appear older than it is) or to errors in constructing the varve chronologies (Hajdas *et al.*, 1995a, 1995b; Zolitschka, 1996b). As independent studies of corals from Barbados, the South Pacific, and New Guinea are all strongly supportive of systematically higher levels of atmospheric ^{14}C in the period before 11,000 calendar yr B.P., this seems to fit well with the hypothesis that past geomagnetic field variations imposed a first-order control on ^{14}C

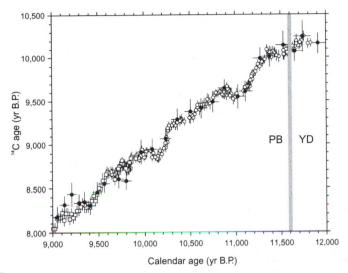

FIGURE 3.11   The very detailed radiocarbon chronology of tree rings from German oaks (squares) and pines (open circles) has been used to place the "floating" chronology of annually laminated (varved) sediments from the Cariaco Basin, north of Venezuela (solid circles) into a firm chronological framework (r = 0.99). By finding the optimum fit between both records, through matching the "wiggles" in the tree-ring record with those of the varve record, the varve sequence (which extends 3000 years further back in time) has been fixed in time. As the varves can be counted, giving calendar year ages, and they also have been AMS dated (using forams in the sediments), they can be used to further extend the calibration of the radiocarbon timescale. Timing of the Younger Dryas — Pre-boreal transition (recognized in the varves) is shown by the vertical shaded line (from Hughen *et al.*, 1998).

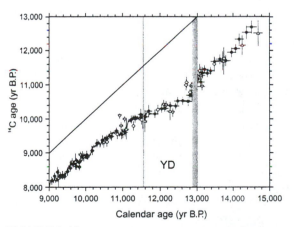

FIGURE 3.12   Calendar age vs radiocarbon age for Cariaco Basin varved sediments (solid circles) compared to uranium-series (TIMS) and radiocarbon ages of corals from the Atlantic and Pacific Oceans (open symbols) (coral data from Bard *et al.*, 1993, 1996; Edwards *et al.*, 1993). The thin diagonal line shows the expected relationship if no changes in atmospheric ^{14}C occurred. The vertical shaded lines show the time (and duration of each transition) at the start and end of the Younger Dryas episode, as recorded in the varves (from Hughen *et al.*, 1998).

levels (Edwards *et al.*, 1993). Indeed, models of past variations in the geomagnetic field reproduce remarkably well the observed ^{14}C variations manifested in the Cariaco varve sequence. This is discussed further in the next section.

A very significant feature of the radiocarbon-calendar year calibration is the presence of prolonged ^{14}C age plateaus centered at ~11,700, ~11,400 and ~9,600 radiocarbon yr B.P. (and to a lesser extent at ~8750 and ~8250 ^{14}C yr B.P.) (Becker *et al.*, 1991; Lotter, 1991; Kromer and Becker, 1993; Hughen *et al.*, 1998). There is also an interval from 10.6 to 10 ka ^{14}C B.P. when varve ages change by ~1600 yr (Fig. 3.12). These ^{14}C age plateaus document periods of time when atmospheric ^{14}C concentrations temporarily increased, so that organisms acquired higher levels of ^{14}C at those times. Consequently, they now appear to be the same age as organisms that are, in fact, several hundred years younger. In effect, this means that two samples with ^{14}C ages of 10,000 and 9600 yr B.P., for example, could in reality differ by as much as 860 years or as little as 90 years. The changes in atmospheric ^{14}C beginning at ~10.6 ka (radiocarbon) B.P. (+40–70‰ in <300 yr) correspond to the onset of the Younger Dryas cold episode (Goslar *et al.*, 1995). A sudden reduction in North Atlantic Deep Water (NADW) formation at that time (or more probably a shift from deep water transport to intermediate water) would have changed the equilibrium atmospheric ^{14}C concentration by reducing the ventilation rate of the deep ocean (Hughen *et al.*, 1996b, 1998).

3.2.1.6 Causes of Temporal Radiocarbon Variations

Table 3.2 lists some of the possible causes of ^{14}C fluctuations, and these are discussed in some detail by Damon *et al.* (1978). In general terms they can be considered in two groups: factors internal to the earth-ocean-atmosphere system (II and III in Table 3.2) and extraterrestrial factors (group I in Table 3.2). Although it is probable that all these different factors have played some part in influencing ^{14}C concentrations through time, it would appear that most of the variance in the record, as it is currently known, can be accounted for by changes in the intensity of the Earth's magnetic field (dipole moment) (Mazaud *et al.*, 1991; Tric *et al.*, 1992) and by changes in solar activity (Stuiver and Quay, 1980; Stuiver, 1994). The former factor is primarily related to low-frequency (long-term) ^{14}C fluctuations, and the latter factor to higher-frequency (de Vries-type) fluctuations in ^{14}C. Evidence for changes in magnetic field intensity has come mainly from archeological sites through studies of thermoremanent magnetism in the minerals of baked clay (Bucha, 1970; Aitken, 1974). Although there are uncertainties in this chronology, it appears that there is a strong inverse correlation between magnetic field variations and ^{14}C concentration, such that as the magnetic field strength decreases (thereby allowing more galactic cosmic rays to penetrate the upper atmosphere) ^{14}C concentration increases (Stuiver *et al.*, 1991; Sternberg, 1995).

On a shorter timescale, variations in solar activity also influence ^{14}C concentrations. This was well illustrated by Stuiver and Quay (1980), who used recorded sunspot data to demonstrate a convincing relationship between periods of low solar activity and high ^{14}C concentrations. Annual records of $\Delta^{14}C$ (and ^{10}Be, another cosmogenic isotope) also show very strong periodicities related to the sunspot cycle of solar activity (Stuiver, 1994; Beer *et al.*, 1994). Solar magnetic activity is reduced during periods of low sunspot number and this allows an increase in the intensity of galactic cosmic rays incident on the Earth's outer atmosphere, thereby increasing the neutron flux and ^{14}C (and ^{10}Be) production (Fig. 3.13). Thus, during the Maunder, Spörer, and Wolf periods of minimum solar activity (A.D. 1654–1714, 1416–1534, and ~1280–1350, respectively) ^{14}C concentrations were at their maximum levels for the past thousand years (Eddy, 1976; Stuiver and Quay, 1980). By subtracting the low frequency geomagnetic signal from the overall Δ^{14}C record, a residual component is revealed that primarily reflects solar (heliomagnetic) effects (Fig. 3.14). This shows a number of Maunder and Spörer-type Δ^{14}C maxima throughout the Holocene, each lasting 100–200 yr (Stuiver *et al.*, 1991).

Whatever the underlying reasons for the high frequency variations in Δ^{14}C, they are extremely important for the interpretation of ^{14}C dates in terms of calendar years. The ^{14}C variations or "wiggles" may not permit the assignment of a unique

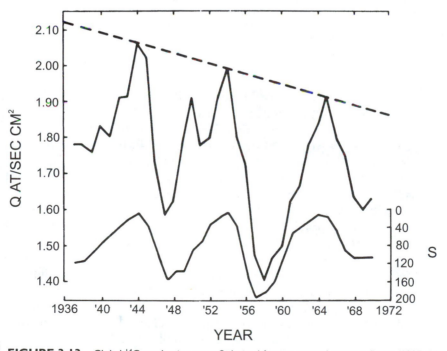

FIGURE 3.13 Global ^{14}C production rates Q derived from measured neutron fluxes (1937–70) in relation to sunspot numbers S (plotted inversely). During periods of higher solar activity cosmic-ray bombardment of the upper atmosphere is reduced, causing ^{14}C production to decrease. The broken line represents the long-term change in ^{14}C production during solar minima (Stuiver and Quay, 1980).

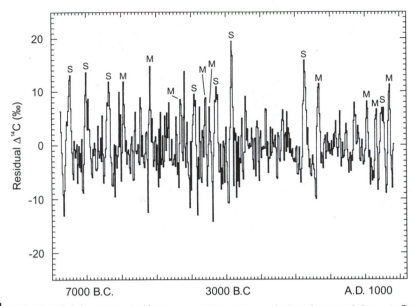

FIGURE 3.14   Residual $\Delta^{14}C$ from a ~400 yr spline applied to the record shown in Fig. 3.9. The spline (a type of low-frequency filter) is used to extract low-frequency changes thought to be the result of changes in the Earth's magnetic field. The residual represents changes due to solar activity. Prolonged episodes of reduced solar activity (and enhanced ^{14}C production) similar to the most recent Maunder and Spörer Minima are denoted by M or S (from Stuiver *et al.,* 1991).

calendar age to a sample. This is illustrated in Fig. 3.15; a sample radiocarbon dated at 220 ± 50 yr B.P. cannot be assigned a single age range, within the probability margin of one standard deviation. Because of ^{14}C fluctuations the actual calendar age of the sample could be from 150 to 210 yr B.P., from 280 to 320 yr B.P., or even from 410 to 420 yr B.P. (Porter, 1981a). In fact, only samples from a few decades around 300 years B.P. are likely to yield a unique radiocarbon date. In all other cases during the last 450 yr, multiple calendar dates, or a much broader spectrum of calendar ages, are derived from a single radiocarbon date (Stuiver, 1978b). This raises significant problems for studies attempting to resolve short-term environmental changes (such as glacier fluctuations; Porter, 1981a) and has important implications for the interpretation of radiocarbon dates at certain times in the Holocene (McCormac and Baillie, 1993). To help in taking such variations into account when calibrating ^{14}C ages in terms of calendar years, Stuiver and Reimer (1993) have prepared a computer program that provides all possible calendar year ages for a particular ^{14}C date and associated margins of error (1 or 2 σ). They stress that the margin of error associated with a dated sample should not only be the analytical error term, but should also include an "error multiplier" factor that varies from 1 to 2, reflecting the reproducibility of results in each laboratory (Scott *et al.,* 1990). This factor is readily accommodated in the program to compute the appropriate calendar year ages. Similar software has also been developed by Ramsey (1995).

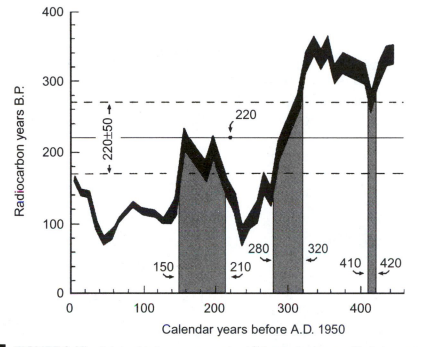

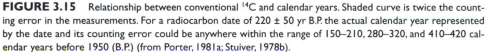

FIGURE 3.15 Relationship between conventional ^{14}C and calendar years. Shaded curve is twice the counting error in the measurements. For a radiocarbon date of 220 ± 50 yr B.P. the actual calendar year represented by the date and its counting error could be anywhere within the range of 150–210, 280–320, and 410–420 calendar years before 1950 (B.P.) (from Porter, 1981a; Stuiver, 1978b).

One consequence of the nonlinear calendar year $-^{14}$C age relationship is that a series of equally probable calendar dates may produce a histogram of strongly clustered ^{14}C dates (Fig. 3.16). Similarly, at certain times in the past, a histogram of ^{14}C dates showing two pronounced "events" may, in fact, correspond to a normal distribution around a single event (Bartlein *et al.,* 1995). This is not the case for the entire period calibrated so far, because there are times when there is less (local) variability around the ^{14}C-calendar year relationship, but certain key intervals (such as the Younger Dryas-Preboral transition) require very careful interpretation of radiocarbon dates to avoid misinterpretation of the true sequence of events. Similar caution is appropriate in dealing with estimates of rates of change during the late Glacial and early Holocene (Lotter *et al.,* 1992). Indeed, as the calibration of radiocarbon dates is extended back into the period before 11,400 yr ago, other periods of near-constant radiocarbon age may yet become apparent.

3.2.1.7 Radiocarbon Variations and Climate

A number of authors have observed that periods of low solar activity, such as the Maunder minimum, correspond to cooler periods in the past (Eddy, 1977; Lean *et al.,* 1995). As variations in radiocarbon production seem to be related to

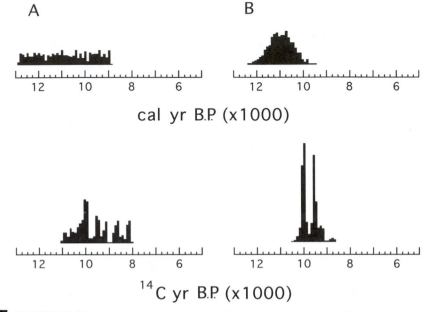

FIGURE 3.16 Simulations of the effect of differences between radiocarbon age and calendar year age. The top diagram shows a set of calendar years, and below are the equivalent ^{14}C ages, derived from the CALIB (ver. 3.03) program of Stuiver and Reimer (1993). In case A, the set of uniformly distributed calendar years leads to a set of clustered radiocarbon ages. In B, a cluster of dates around a mean of 11,000 calendar yr (and a standard deviation of 500 yr) is compared to the equivalent radiocarbon distribution. A bimodal distribution apparent in the ^{14}C data does not reflect reality (from Bartlein *et al.*, 1995).

solar activity, it has also been argued that ^{14}C variations are inversely related to worldwide temperature fluctuations (Wigley and Kelly, 1990). This implies that solar activity, radiocarbon variations, and surface temperature are all related, perhaps through fundamental variations in the solar constant (i.e., low solar activity = high ^{14}C production rate = low temperature). If so, then the ^{14}C record itself, as a proxy of solar activity, may provide important information on the causes of climatic change. However, this is a controversial topic; several authors have shown that the correlations between radiocarbon variations and paleotemperature records are very poor when the records are examined in detail (Williams *et al.*, 1981). This may be because atmospheric ^{14}C is only a small part of the global ^{14}C inventory and climate-related changes in ocean circulation and deep water formation may overwhelm the effect of solar activity changes. In this regard, ^{10}Be may be a better proxy of solar activity (Beer *et al.*, 1994, 1996). On the other hand, high resolution data from the Greenland ice sheet show a very strong 11-yr signal in δ^{18}O and spectral analysis reveals periodicities associated with those known from the spectra of radiocarbon variations (Stuiver *et al.*, 1995; Sonnett and Finney, 1990). Indeed the amplitude of the δ^{18}O signal is so large (~1.5‰) that it is very hard to imagine how such small irradiance changes (0.05%) could be amplified within the climate system to produce such a strong signal. There is some evidence from modeling experiments that larger (0.25% or more) reductions in irradiance that seem possible for extended periods of reduced solar activity, like the Maunder Minimum, can influence temperatures on a global scale (Rind and Overpeck, 1993; Lean, 1994). Still, it is not yet clear that radiocarbon variations or ^{10}Be can be used as an index of *irradiance* and at this stage, therefore, the evidence relating solar activity and radiocarbon variations to surface temperatures remains equivocal, an intriguing but so far unproven possibility.

3.2.2 Potassium-Argon Dating (^{40}K/^{40}Ar)

Compared to radiocarbon dating, potassium-argon dating is used far less in Quaternary paleoclimatic studies. However, potassium-argon and argon-argon dating have indirectly made major contributions to Quaternary studies. The techniques have proved to be invaluable in dating sea-floor basalts and enabling the geomagnetic polarity timescale to be accurately dated and correlated on a worldwide basis (Harland *et al.*, 1990; see also Section 4.1.4). Potassium-argon dating has also been used to date lava flows which, in some areas of the world, may be juxtaposed with glacial deposits. In this way, limiting dates on the age of the glacial event may be assigned (Löffler, 1976; Porter, 1979).

Potassium-argon dating is based on the decay of the radioisotope ^{40}K to a daughter isotope ^{40}Ar. Potassium is a common component of minerals and occurs in the form of three isotopes, ^{39}K and ^{41}K, both stable, and ^{40}K, which is unstable. The ^{40}K occurs in small amounts (0.012% of all potassium atoms) and decays to either ^{40}Ca or ^{40}Ar, with a half-life of 1.31×10^9 yr. Although the decay to ^{40}Ca is more common, the relative abundance of ^{40}Ca in rocks precludes the use of this isotope for dating purposes. Instead, the abundance of argon is measured and sample age is a function of the ^{40}K/^{40}Ar ratio. Argon is a gas that can be driven out of a sample by heating. Thus, the method is used for dating volcanic rocks that contain no argon after the molten lava has cooled, thereby setting the isotopic "clock" to zero. With the passage of time, ^{40}Ar is produced and retained within the mineral crystals, until driven off by heating in the laboratory during the dating process (Dalrymple and Lanphere, 1969). Unlike conventional ^{14}C dating, ^{40}K/^{40}Ar dating relies on measurements of the decay product ^{40}Ar; the parent isotope content (^{40}K) is measured in the sample.

As the abundance ratios of the isotopes of potassium are known, the ^{40}K content can be derived from a measurement of total potassium content, or by measurement of another isotope ^{39}K. Because of the relatively long half-life of ^{40}K, the production of argon is extremely slow. Hence, it is very difficult to apply the technique to samples younger than ~100,000 years and its primary use has been in dating volcanic rocks formed over the last 30 million years (though, theoretically, rocks as old as 10^9 years could be dated by this method). Dating is usually carried out on minerals such as sanidine, plagioclase, biotite, hornblende, and olivine in volcanic lavas and tuffs. It may also be useful in dating authigenic minerals (i.e., those formed at the time of deposition) such as glauconite, feldspar, and sylvite in sedimentary rocks (Dalrymple and Lanphere, 1969).

3.2.2.1 Problems of ^{40}K/^{40}Ar Dating

The fundamental assumptions in potassium–argon dating are that (a) no argon was left in the volcanic material after formation, and (b) the system has remained closed since the material was produced, so that no argon has either entered or left the sample since formation. The former assumption may be invalid in the case of some deep-sea basalts that retain previously formed argon during formation under high hydrostatic pressure. Similarly, certain rocks may have incorporated older "argon-rich" material during formation. Such factors result in the sample age being overestimated (Fitch, 1972). Similar errors result from modern argon being absorbed onto the surface and interior of the sample, thereby invalidating the second assumption. Fortunately, atmospheric argon contamination can be assessed by measurement of the different isotopes of argon present. Atmospheric argon occurs as three isotopes, ^{36}Ar, ^{38}Ar, and ^{40}Ar. As the ratio of ^{40}Ar/^{36}Ar in the atmosphere is known, the specific concentrations of ^{36}Ar and ^{40}Ar in a sample can be used as a measure of the degree of atmospheric contamination, and the apparent sample age appropriately adjusted (Miller, 1972).

A more common problem in ^{40}K/^{40}Ar dating is the (unknown) degree to which argon has been lost from the system since the time of the geological event to be dated. This may result from a number of factors, including diffusion, recrystallization, solution, and chemical reactions as the rock weathers (Fitch, 1972). Obviously, any argon loss will result in a minimum age estimate only. Fortunately, some assessment of these problems and their effect on dating may be possible.

3.2.2.2 ^{40}Ar/^{39}Ar Dating

One important disadvantage of the conventional ^{40}K/^{40}Ar dating technique is that potassium and argon measurements have to be made on different parts of the same sample; if the sample is not completely homogeneous, an erroneous age may be assigned. This problem can be circumvented by ^{40}Ar/^{39}Ar dating, in which measurements are made simultaneously, not only on the same sample, but on the same precise location within the crystal lattice where the ^{40}Ar is trapped. Instead of measuring ^{40}K directly, it is measured indirectly by irradiating the sample with neutrons in a nuclear reactor. This causes the stable isotope ^{39}K to transmute into ^{39}Ar; by collecting both the ^{40}Ar and ^{39}Ar, and knowing the ratio of ^{40}K to ^{39}K (which is a constant) the sample age can be calculated. Further details are given by Curtis (1975), McDougall and Harrison (1988), and McDougall (1995).

Actually, ^{40}Ar/^{39}Ar dating has no advantages over conventional ^{40}K/^{40}Ar dating for samples that have not been weathered, subjected to heating or metamorphism of any kind since formation, or are free of inherited or extraneous argon. In such cases, dates from ^{40}K/^{40}Ar methods would be identical to those from ^{40}Ar/^{39}Ar methods. In practice, however, there is no way of knowing the extent to which a sample has been modified or contaminated; hence the ^{40}Ar/^{39}Ar method has significant advantages over ^{40}K/^{40}Ar because it is often possible to identify the degree to which a sample has been altered or contaminated, and thus, to increase confidence in the date assigned. Furthermore, several dates can be obtained from one sample and the results treated statistically to yield a date of high precision (Curtis, 1975).

The advantages stem from the fact that the ^{40}K, which yields the ^{40}Ar by decay, occupies the same position in the crystal lattice of the mineral as the much more abundant ^{39}K that produces the ^{39}Ar on irradiation. Heating of the sample thus drives off the argon isotopes simultaneously. Any atmospheric argon contaminating the sample occurs close to the surface of the mineral grains, so it is liberated at low temperatures. Similarly, loss of radiogenic argon by weathering would be confined mainly to the outer surface of a mineral. In such cases the $^{40}Ar/^{39}Ar$ ratios on the initial gas samples would indicate an age that is too young (Fig. 3.17b). At higher temperatures, the deeper-seated argon from the unweathered, uncontaminated interiors of the crystals will be driven off and can be measured repeatedly as the temperature rises to fusion levels. If such gas increments indicate a stable and consistent age, considerable confidence can be placed in the result. By contrast, conventional $^{40}Ar/^{39}Ar$ dating on a sample such as that shown in Fig. 3.17b would yield a meaningless age, resulting from a mixture of the gases from different levels.

Plots of "apparent age" calculated from the ratios of ^{40}Ar to ^{39}Ar at different temperatures can indicate much information about the past history of the sample, including whether the sample has lost argon since formation or whether the rock was contaminated by excess radiogenic argon at the time of formation. Such interpretations are discussed further by Curtis (1975) and by Miller (1972). Thus $^{40}Ar/^{39}Ar$ dating possesses considerable advantages over conventional $^{40}K/^{40}Ar$ dating methods by providing more confidence in the resulting dates.

$^{40}Ar/^{39}Ar$ dating has been used to assess the age of the major geomagnetic polarity reversal in the Quaternary — the Brunhes-Matuyama (B/M) boundary. Early $^{40}K/^{40}Ar$ studies had placed the age at ~730 ka B.P. but this was questioned by Johnson (1982) and Shackleton et al. (1990), who found that a consequence of tuning the marine oxygen isotope record to maximize coherence with Milankovitch orbital frequencies was to push the B/M boundary back to ~780 ka B.P. They therefore argued rather boldly that the hitherto accepted age of 730 ka was probably incorrect. Several subsequent studies deriving $^{40}Ar/^{39}Ar$ dates from lava flows have supported this assertion (Spell and McDougall, 1992; Baksi et al., 1992; Izett and Obradovich, 1994) or at least demonstrated that the uncertainties in both approaches span the interval 730–780 ka B.P., making the two estimates statistically indistinguishable (Tauxe et al., 1992).

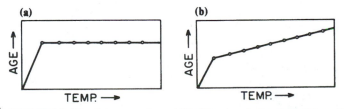

FIGURE 3.17 Schematic plots of $^{40}Ar/^{39}Ar$ data. Each point in **(a)** and **(b)** indicates the age obtained for that increment of argon released as the temperature is increased in steps from 0 °C to the fusion point (1000 °C). In **(a)** the data show uniform ages for all increments, the plateau indicating a precise age determination. In **(b)** the ages appear to be progressively older as the temperature rises, indicating loss of argon after original crystallization of the sample so that a precise age cannot be determined; even the oldest age obtained is probably too young (Curtis, 1975).

3.2.3 Uranium-series Dating

Uranium-series dating is a term that encompasses a range of dating methods, all based on various decay products of ^{238}U or ^{235}U. Figure 3.18 illustrates the principal decay series nuclides and their respective half-lives; some intermediate products with very short half-lives (in the order of seconds or minutes) have been omitted. The main isotopes of significance for dating are ^{238}U and ^{235}U, ^{230}Th (also known as ionium) and ^{231}Pa. The ultimate product of the uranium decay series is stable lead (^{206}Pb or ^{207}Pb).

In a system containing uranium, which is undisturbed for a long period of time ($\sim 10^6$ yr), a dynamic equilibrium will prevail in which each daughter product will be present in such an amount that it is decaying at the same rate as it is formed by its parent isotope (Broecker and Bender, 1972). The ratio of one isotope to another will be essentially constant. However, if the system is disturbed, this balance of production and loss will no longer prevail and the relative proportions of different isotopes will change. By measuring the degree to which a disturbed system of decay products has returned to a new equilibrium, an assessment of the amount of time elapsed since disturbance can be made (Ivanovich and Harmon 1982). Isotopic decay is expressed in terms of the activity ratios[8] of different isotopes, such as

NUCLIDE	HALF-LIFE	NUCLIDE	HALF-LIFE
uranium-238	4.51×10^9 years	uranium-235	7.13×10^8 years
uranium-234	2.5×10^5 years	protactinium-231	3.24×10^4 years
thorium-230 (ionium)	7.52×10^4 years	thorium-227	18.6 days
radium-226	1.62×10^3 years	radium-223	11.1 days
radon-222	3.83 days	lead-207	stable
lead-210	22 years		
polonium-210	138 days		
lead-206	stable		

FIGURE 3.18 Decay series of uranium-238 and uranium-235.

[8] Concentrations of radioactive isotopes are reported in units of decays per minute per gram of sample. In U-series dating, these rates are considered relative to each other and are referred to as activity ratios.

^{230}Th/^{238}U and ^{231}Pa/^{235}U; for the former, the useful dating range is from a few years to ~350,000 B.P., and for the latter 5000–150,000 B.P. (Fig. 3.19). Thermal ionization mass spectrometry (TIMS) revolutionized uranium series dating in the mid-1980s, making very precise analyses of small samples routine; subsequent refinements have made possible dates that are even more precise, enabling corals of last interglacial age to be dated to within ± 1 ka (2 σ error) (Edwards *et al.*, 1987b; 1993; Gallup *et al.*, 1994). Furthermore, in view of the inconstancy of atmospheric (and oceanic) radiocarbon content, ^{230}Th dating of corals can provide results that exceed the accuracy of ^{14}C dates, and have comparable accuracy to the counting of annual growth bands (Edwards *et al.*, 1987a).

In natural systems, disturbance of the decay series is common because of the different physical properties of the intermediate decay series products. Most important of these is the fact that ^{230}Th and ^{231}Pa are virtually insoluble in water. In natural waters these isotopes are precipitated from solution as the uranium decays, and collect in sedimentary deposits. As the isotope is buried beneath subsequent sedimentary accumulations, it decays at a known rate, "unsupported" by further decay of the parent isotopes (^{234}U and ^{235}U, respectively) from which it has been separated. This is known as the *daughter excess* or unsupported dating method (Blackwell and Schwarcz, 1995). In sediment that has been deposited at a uniform rate, the ^{230}Th and ^{231}Pa concentrations decrease exponentially with depth. Providing that this initial concentration of the isotopes is known, the extent to which they have decayed in sediment beneath the surface can be related to the amount of time elapsed since the sediment was first deposited (Fig. 3.20).

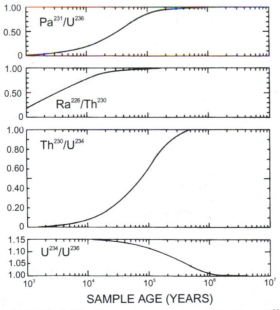

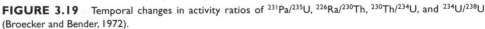

FIGURE 3.19 Temporal changes in activity ratios of ^{231}Pa/^{235}U, ^{226}Ra/^{230}Th, ^{230}Th/^{234}U, and ^{234}U/^{238}U (Broecker and Bender, 1972).

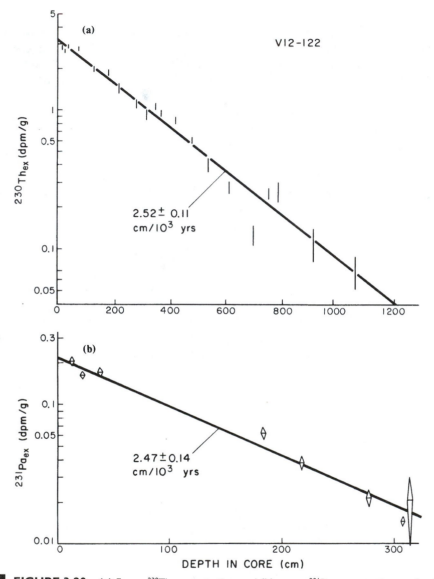

FIGURE 3.20 **(a)** Excess [230]Th concentrations and **(b)** excess [231]Pa concentrations vs depth in Caribbean core V12-122. As the original amounts of [230]Th and [231]Pa in freshly deposited sediment can be estimated, the extent to which they have been reduced with depth gives a measure of time since the sediment was deposited. Sedimentation rates are obtained from slopes of the best-fitting regression lines and a knowledge of the decay rate of each isotope (Ku, 1976).

This procedure is an example of dating based on the physical separation of the parent and daughter isotopes, with age calculated as a function of the decay rate of the unsupported daughter isotope (Ku, 1976). Another method relies on the growth of an isotope that is initially absent, towards equilibrium with its parent isotope;

this is known as *daughter deficiency* dating (Blackwell and Schwarcz, 1995) and is most commonly applied to carbonate materials (corals, molluscs, speleothems). It is based on the fact that uranium is co-precipitated with calcite or aragonite from natural waters that are essentially free of thorium and protactinium. Initial values of ^{230}Th and ^{231}Pa in the carbonates are thus negligible. Providing that the carbonate remains a closed system, the amounts of ^{230}Th and ^{231}Pa produced as the ^{234}U and ^{235}U decay will be a function of time, and of the initial uranium content of the sample. In the growth of corals, for example, uranium is co-precipitated from seawater to form part of the coral structure, but thorium concentrations are essentially zero. As the ^{234}U/^{238}U ratio in ocean water is constant at ~1.14 (and studies show it has been almost constant over long periods of time) the build-up of ^{230}Th in the coral as the uranium isotopes decay thus provides chronometric control on the time since the coral formed. Providing the coral has not undergone recrystallization (thereby incorporating anew more uranium) this approach can provide useful dating control from a few years to ~350,000 yr B.P. with extremely high precision (Edwards *et al.*, 1987b). New Guinea (Huon peninsula) coral samples demonstrate 2σ errors of only 30–80 yr on late Glacial/early Holocene materials, considerably smaller than the errors associated with AMS ^{14}C dates on the same samples (Edwards *et al.*, 1993). Indeed, U-series dating has been used to calibrate the radiocarbon timescale back to >25 ka B.P. (Bard *et al.*, 1993; see Section 3.2.1.5). The method has been widely used to date raised coral terraces and hence to provide a chronologically accurate assessment of glacio-eustatic changes of sea level, with broad implications for paleoclimatology (Bard *et al.*, 1990; Edwards *et al.*, 1993; Gallup *et al.*, 1994).

Attempts have also been made to date molluscs in the same way as coral but the results are generally inconsistent (Szabo, 1979a). The main problem is that molluscs appear to freely exchange uranium post-depositionally (i.e., they do not constitute a closed system) with the result that fossil molluscs commonly have higher uranium concentrations than their modern counterparts. Hence, the resulting thorium and protactinium concentrations are not simply a function of age (Kaufman *et al.*, 1971). ^{230}Th/^{234}U dates on bone have also been attempted (Szabo and Collins, 1975) but similar problems have been encountered. Repeated checks with different dating methods suggest that accurate dates have been obtained on only 50% of shell and bone samples to which ^{230}Th/^{234}U and ^{231}Pa/^{234}U dating methods have been applied (Ku, 1976). "Open system" models have been developed to compensate for post-depositional exchange problems (Szabo and Rosholt, 1969; Szabo, 1979b) but many assumptions are required that reduce confidence in the resultant dates. Indeed, Broecker and Bender (1972) categorically rejected dates obtained on any kind of molluscs and concluded that only corals can give reliable U-series dates. However, other work has shown that U-series dates on Arctic marine molluscs can provide valuable minimum age estimates when considered in relation to amino-acid data on the same samples (Szabo *et al.*, 1981).

Much more confidence can be placed in uranium-series dates obtained on carbonate samples from speleothems (stalactites and stalagmites). Such deposits are dense and not subject to post-depositional leaching. Because ^{230}Th is so insoluble, the water from which the speleothem carbonate is precipitated can be considered to be essentially thorium-free. Hence, providing that the initial uranium concentration is sufficient, measurement of the $^{230}Th/^{234}U$ ratio will indicate the build-up of ^{230}Th with the passage of time (Harmon *et al.*, 1975). The main problem is to determine reliably the initial $^{234}U/^{238}U$ ratio, and to ensure that detrital ^{230}Th has not contaminated the sample, thereby negating the assumption that the initial thorium content is zero (see Section 7.6.2).

On a much shorter timescale, unsupported ^{210}Pb may also be used as a chronological aid. The ^{210}Pb is derived from the decay of ^{222}Rn following the decay of ^{226}Ra from ^{230}Th (see Fig. 3.18). Both ^{226}Ra and ^{222}Rn escape from the Earth's surface and enter the atmosphere, where the ^{210}Pb is eventually produced. The ^{210}Pb is then washed out of the atmosphere by precipitation, or settles out as dry fallout, where it accumulates in sedimentary deposits and decays (with a half-life of 22 yr) to stable ^{206}Pb (Fig. 3.21). Assuming that the atmospheric flux of ^{210}Pb is constant, the decay rate of ^{210}Pb to ^{206}Pb with depth can be used to date sediment accumulation rates (Appleby and Oldfield, 1978, 1983). It is of value only in dating sediments over the last ~200 yr, but this may be of particular value in confirming that laminated sediments are true varves (i.e., annual) or in confirming that core-tops are undisturbed, enabling floral and faunal contents to be calibrated with instrumental climatic data to derive accurate transfer functions for paleoclimatic reconstructions. The ^{210}Pb has also proved useful in dating the upper sections of ice cores and hence allowing estimates of long-term accumulation rates to be made, though nowadays this is rarely carried out (Crozaz and Langway, 1966; Gäggeler *et al.*, 1983).

3.2.3.1 Problems of U-series Dating

The major problems in U-series dating have already been alluded to briefly. First, an assumption must be made as to the initial $^{230}Th/^{234}U$, $^{234}U/^{238}U$, and/or $^{231}Pa/^{235}U$ ratios in the sample. In the deep oceans this may not be a significant problem, as modern oceanic ratios are known to have been relatively constant over long periods of time, but in terrestrial environments such as closed inland lakes, this assumption is far less robust. The second problem concerns the extent to which the sample to be dated has remained a closed system through time. Recrystallization of aragonitic carbonate to calcite may provide some guidance, but as discussed in Section 3.2.1.4 this is not always reliable. At present only carbonate dates on coral seem to be consistently reliable. Reliability can be checked by obtaining activity ratios for different isotopes from the same sample. If the sample has remained "closed" the dates should all cross-check and be internally consistent (see Fig. 3.19).

3.2.4 Luminescence Dating: Principles and Applications

Luminescence is the light emitted from a mineral crystal (mainly quartz and feldspars) when subjected to heating or when exposed to light. Light emitted in response to heating is referred to as thermoluminescence or TL, light emitted in response to radiation in the visible or infrared parts of the spectrum is termed optically stimulated luminescence (OSL), or infrared stimulated luminescence (IRSL), respectively. In each case, the quantity of light emitted is related to the

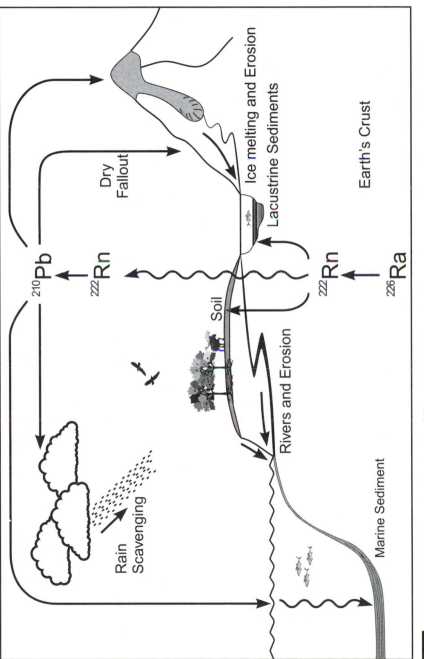

FIGURE 3.21 Schematic diagram of the ^{210}Pb global cycle, illustrating the source, transport, deposition, and post-depositional, redistribution processes (from Preiss et al., 1996).

amount of ionizing radiation that the sample has been exposed to over time, from surrounding sediments. The decay of radioisotopes in the surrounding matrix produces free electrons in the mineral grains, which become trapped at defects in the crystal lattice; the longer the mineral has been exposed to radiation, the higher will be its trapped electron population and the greater the resulting luminescence signal. Luminescence is thus a measure of the accumulated dose of ionizing radiation (expressed in units of Grays, Gy) which is a function of sample exposure age. Age is determined by exposing subsets of the sample to known doses of radiation and measuring the resulting luminescence signal. The amount of radiation needed to produce the same luminescence signal as that from the original sample is called the equivalent dose (ED) (sometimes known as the paleodose; Duller, 1996). If the amount of radiation a sample is exposed to each year (the dose rate) is known (by direct measurement of the radioactive properties of the surrounding matrix) the sample age can be calculated as:

$$\text{Age (a)} = \frac{\text{Equivalent dose (ED)}}{\text{Dose rate}} \qquad \frac{[\text{Gy}]}{[\text{Gy a}^{-1}]}$$

The key constraint is that the sample to be dated must either have no residual luminescence signal from the period before the event being dated, or any residual signal must be quantifiable. Because luminescence is released by either heating or optical bleaching of the sample, the event to be dated must have involved one of these processes that effectively set the "luminescence clock" to zero. Thus, TL dating has been widely used in archeology to date pottery or baked clay samples as well as baked flints from fire hearths attributable to early man (Wintle and Aitken, 1977). And TL has also been used to date sediment baked by contact with molten lava (Huxtable *et al.*, 1978) and inclusions within lava (Gillot *et al.*, 1979). By controlled heating experiments in the laboratory, measurements of TL can indicate the amount of time that has elapsed since the sample was last heated. The same principle applies to wind-blown and fluvially transported sediments. Exposure to sunlight can also dislodge electrons, resetting the luminescence clock. As shown in Fig. 3.22, the TL signal is reduced to an unbleachable residual signal, but the OSL signal is reduced to zero on exposure to sunlight. The TL or OSL signal at any time after deposition and burial will thus be a measure of the time that has elapsed since the grains were transported to their depositional site (Huntley *et al.*, 1985; Duller, 1996). Luminescence dating is thus extremely useful in studies of mainly inorganic sediments, such as loess and other aeolian deposits (Wintle, 1993; Zöller *et al.*, 1994) as well as water-lain sediments (Balescu and Lamothe, 1994). However, if zeroing of the sample luminescence is incomplete prior to burial, the age of the deposit will be overestimated. This is a common problem in fluvial or glacio-fluvial sediments where exposure to sunlight may be limited (in duration and/or wavelength) by turbid water (Forman *et al.*, 1994).

The useful timescale for TL dating depends on the radiation dose to which the sample has been exposed, and the capacity of the sample to continue to accumulate electrons, before becoming saturated. Samples with high quantities of potassium feldspars, for example, are potentially useful for dating older deposits, because such

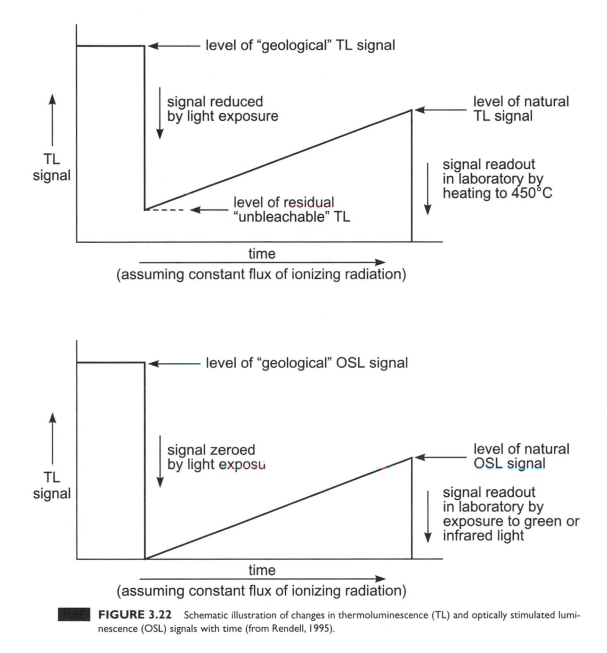

FIGURE 3.22 Schematic illustration of changes in thermoluminescence (TL) and optically stimulated luminescence (OSL) signals with time (from Rendell, 1995).

minerals are less easily saturated. Most analysts are reluctant to place much faith in TL age estimates of more than 200,000 yr, but ages of up to 800 ka B.P. have been reported from areas where dose rates are extremely low and saturation levels are potentially high; nevertheless, such dates are controversial (Berger *et al.,* 1992). Accuracy of luminescence dates may approach ±10% of the sample age, though comparison with $^{40}Ar/^{39}Ar$ dates suggests that TL dates on older samples are commonly 5–15% too young, perhaps due to saturation of the electron traps (see Section 3.2.4.2).

3.2.4.1 Thermoluminescence (TL) Dating

The thermoluminescence of a sample is a function of age. The older the sample, the greater will be the TL intensity. This is assessed by means of a glow curve, a plot of TL intensity vs temperature as the sample is heated (Fig. 3.23a). The TL emission at lower temperatures is not a reliable age indicator; such emissions correspond to shallow traps in the sample where electrons are not stable. The precise temperature necessary to dislodge the deeper "stable" electrons will depend on individual sample characteristics and is assessed by finding the point at which the ratio of natural TL

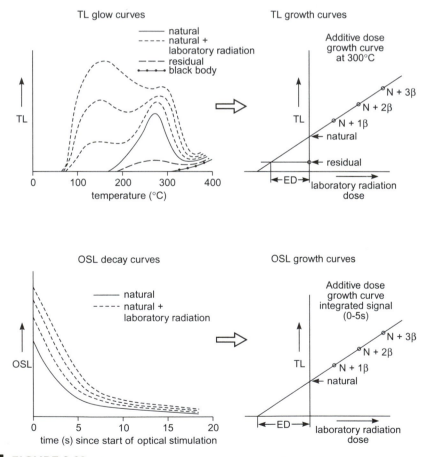

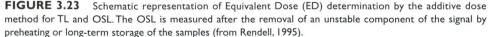

FIGURE 3.23 Schematic representation of Equivalent Dose (ED) determination by the additive dose method for TL and OSL. The OSL is measured after the removal of an unstable component of the signal by preheating or long-term storage of the samples (from Rendell, 1995).

to artificially induced TL becomes approximately constant. Generally this is in the range 300–450 °C, so TL intensity at these temperatures is used for age assessment.

To determine the age of a sample it is necessary first to know how much TL results from a given radiation dose, as not all materials produce the same amount of TL from a given radiation dose. In one approach (the additive dose method) subsets of the sample are exposed to a known quantity of radiation. Then TL is measured in each sample; the initial paleodose can then be determined by extrapolation to a laboratory-determined residual level (Fig. 3.23b). Another approach (the regenerative method) involves first bleaching subsamples of all their TL, then exposing them to different levels of radiation, followed by measurement of the TL emitted, corresponding to different levels of radiation exposure (Fig. 3.24). The TL in the original sample is then measured and compared with the artificially radiated samples to assess the corresponding paleodose (ED) as accurately as possible. Sample age can then be calculated simply by dividing the paleodose by the measured dose rate at the sample site. The dose rate is assessed by measuring the quantity of radioactive uranium, thorium, and potassium in the sample itself, and in the surrounding matrix. Alternatively, radiation-sensitive phosphors (such as calcium fluoride) may be buried at the sample site for a year or more to measure directly the environmental radiation dosage. A number of procedures have been devised to estimate the long-term dose rate and reduce the inherent uncertainty (Aitken, 1985).

3.2.4.2 Problems of Thermoluminescence Dating

In the age equation given in the preceding section, it is assumed that there is a linear relationship between radiation dose and the resulting TL. It is known, however, that this is not always the case at extremely low radiation dose levels, or at extremely high dose levels. The former problem (supralinearity) is most significant for relatively young samples (<5000 yr old) as the rate at which a sample acquires TL is reduced at relatively low radiation dose levels (or perhaps is nonexistent until a certain radiation threshold is exceeded). At the other end of the scale, very long exposures to radiation (high doses) may result in saturation of the available electron traps so that further exposure will not appreciably increase the sample TL. When this is likely to occur depends on sample age and mineral composition, but, in general, very old ages indicated by TL dating (greater than several hundred thousand years) are likely to be only minimum estimates.

A further difficulty in assessing the relationship between TL and radiation dose occurs when irradiated samples "lose" TL after very short periods of time, perhaps only a few weeks. This phenomenon is called anomalous fading (Wintle, 1973) and is common among certain minerals, particularly feldspars of volcanic origin. Unless corrected for, anomalous fading will result in underestimation of a sample age, but it can be identified relatively easily by storing irradiated samples in the dark and remeasuring TL periodically over a period of several months.

Perhaps the most significant problems in TL dating stem from variations in the environmental dose rate. Of particular importance is the mean water content of the sample and surrounding matrix during sample emplacement. Water greatly attenuates radiation, so a saturated sample will receive considerably less radiation in a given time period than a similar sample in a dry site; as a result the TL intensity will be much lower, giving an incorrect age indication. If the water content of the site can be assessed, this problem can be taken into account in the age calculation. Better still, if a radiation-sensitive phosphor can be placed in the environmental setting of the sample for a period of time, the effect of groundwater on the radiation dose may be assessed directly. However, there is always the problem of knowing how groundwater content has varied in the past and this uncertainty is a major barrier to more accurate TL dating. Groundwater may also leach away radioactive decay products, so long-term changes in groundwater content may further complicate the TL-dose relationship. Finally, it should be noted that one of the decay products of uranium is an inert gas (radon-222) which has a half-life of 3.8 days, long enough for it to escape from the sample site and effectively terminate the decay series (96% of the U-series γ-radiation energy is post-radon). Fortunately, laboratory studies have shown that relatively few soils exhibit significant radon loss either in the laboratory or *in situ*.

3.2.4.3 Optical and Infrared Stimulated Luminescence (OSL and IRSL) Dating

Although the TL method has been widely used for dating of sediments since its first application to deep sea sediments (Wintle and Huntley, 1979) it was always known that it would be more appropriate to use light as the stimulation mechanism rather than heat. This would allow measurement of the light-sensitive luminescence signal, as distinct from the light-insensitive signal that remains in the sample at deposition (Fig. 3.22). With TL dating, a relatively large light-insensitive TL signal found in modern samples prevents the dating of sediments less than 1000–2000 yr old. This problem was addressed by Huntley *et al.* (1985), who reported on the first samples dated by measurement of an optically stimulated luminescence (OSL) signal. The OSL signal was shown to be zero for modern sediments, thus opening up the possibility of dating sediments as young as a few decades in age. Subsequently, Godfrey-Smith *et al.* (1988) measured the OSL (stimulated with green light from a laser) from a number of quartz and feldspar samples extracted from sediments. They showed that not only was there a negligible residual OSL signal after a prolonged sunlight bleach, but the initial rate of luminescence signal loss was several orders of magnitude faster than that for the TL signal from the same samples.

The OSL signal is derived from electrons displaced from electron traps in the crystal by photons, such as those from a 514-nm argon laser (Fig. 3.25b). An electron thus released is able to recombine at a luminescence center, a process that results in the emission of a photon with a wavelength characteristic of the center. The ED is obtained by extrapolation as for TL (Fig. 3.24b). It is thus necessary to be able to observe this luminescence while totally rejecting the light from the stimulating light source. Fortunately, optical filters can be found which pass light from the violet and near ultraviolet emission, characteristic of quartz and feldspars, but reject the green laser light (Fig. 3.25b). Hütt *et al.* (1988) discovered that they were able to stimulate trapped electrons from feldspars (but not quartz) using near infrared radiation. This resulted in infrared stimulated luminescence (IRSL) signals

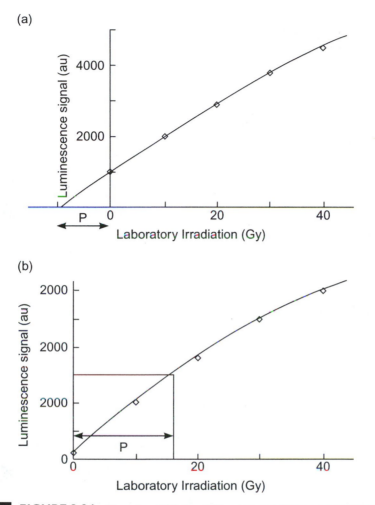

FIGURE 3.24 Examples of **(a)** the additive dose and **(b)** the regenerative methods of estimating paleo-dose in thermoluminescence dating. In the additive dose method, the samples are exposed to increasing levels of radiation, and the associated TL is measured. Extrapolation back to zero reveals the paleodose of the sample before it was exposed to further radiation exposure. In the regenerative method, subsamples are first bleached then exposed to radiation and the TL is measured for each exposure level. The radiation exposure corresponding to TL measured in the original sample is then readily obtained (Duller, 1996).

that could be observed in a wider wavelength range (blue and green) because an optical filter could be chosen to reject the stimulation wavelength region, around 850–880 nm (Fig. 3.25c).

The principal advantage of OSL and IRSL is that the laser-stimulated electron traps are very sensitive to light, so that a relatively brief exposure to sunlight (on the order of a few tens of seconds to a few minutes) is likely to have reduced the sample luminescence to near zero. Such brief exposures would certainly not reduce

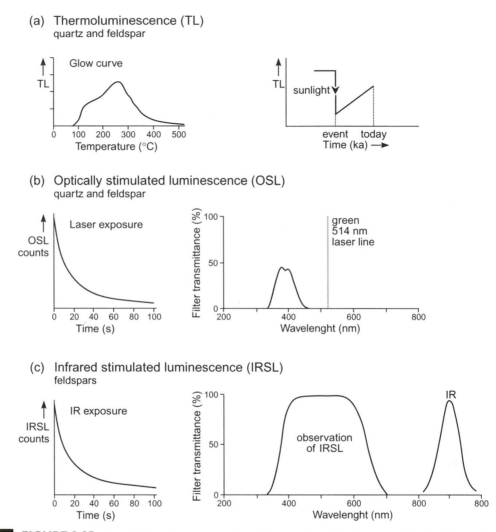

FIGURE 3.25 **(a–c)** Schematic representation of the principles of luminescence dating techniques (from Wintle *et al.*, 1993).

to zero the luminescence signal measured in TL studies; thus, OSL and IRSL open up the prospect of dating a wider variety of material, particularly quite young aeolian sands and silts (loess) as well as fluvial and lacustrine sediments where exposure to sunlight may have been brief (Wintle, 1993). For example, recent studies have dated aeolian sands deposited only a few decades to a few centuries ago, using IRSL techniques (Wintle *et al.*, 1994; Clarke *et al.*, 1996).

In TL studies, all of the luminescence in a sample is reduced to zero by heating, so repeat measurements are not possible. However, in OSL and IRSL dating, very brief exposures to the stimulation light source yield measurable signals that do not significantly deplete the potential luminescence in the sample, so multiple measurements can be made on the same sample (Duller, 1995). Indeed, multiple analyses on individual grains are now routinely possible (for grains with a high sensitivity), providing a set of results that helps to confirm whether the sample as a whole was adequately zeroed, thereby enhancing confidence that the event in question has been accurately dated (Lamothe *et al.*, 1994). However, it must be borne in mind that many of the problems facing TL dating described in the foregoing (saturation of traps, anomalous fading, and especially estimation of the dose rate) still apply to OSL and IRSL dating.

3.2.5 Fission-track Dating

As already discussed in Section 3.2.3, uranium isotopes decay slowly through a complex decay series, ultimately resulting in stable atoms of lead. In addition to this slow decay, via the emission of α and β particles, uranium atoms also undergo spontaneous fission, in which the nucleus splits into two fragments. The amount of energy released in this process is large, causing the two nuclear fragments to be ejected into the surrounding material. The resulting damage paths are called fission tracks, generally 10–20 μm in length. The number of fission tracks is simply a function of the uranium content of the sample and time (Naeser and Naeser, 1988). Rates of spontaneous fission are very low (for ^{238}U, 10^{-16} a^{-1}) but if there is enough uranium in a rock sample a statistically significant number of tracks may occur over periods useful for paleoclimatic research (Fleischer, 1975).

The value of fission-track counting as a dating technique stems from the fact that certain crystalline or glassy materials may lose their fission-track records when heated, through the process of annealing. Thus, igneous rocks and adjacent metamorphosed sediments contain fission tracks produced since the rock last cooled down. Similarly, archeological sites may yield rocks that were heated in a fire hearth, thereby annealing the samples and resetting the "geological" record of fission tracks to zero. In this respect, the environmental requirements of the sample are similar to those necessary for $^{40}K/^{40}Ar$ dating. Because different minerals anneal at different temperatures, careful selection is necessary; minerals with a low annealing temperature threshold, such as apatite, will be the most sensitive indicator of past thermal effects (Faul and Wagner, 1971).

Fission tracks can be counted under an optical microscope after polishing the sample and etching the surface with a suitable solvent; the damaged areas are preferentially attacked by solvents, revealing the fission tracks quite clearly (Fleischer and Hart, 1972). After these have been counted, the sample is heated to remove the "fossil" fission tracks and then irradiated by a slow neutron beam, which produces a new set of fission tracks as a result of the fission of ^{235}U. The number of induced fission tracks is proportional to the uranium content and this enables the ^{238}U content of the sample to be calculated. Sample age is then obtained from a knowledge of the spontaneous fission rate of ^{238}U. For a much more detailed discussion of the technique, and the problems of calibrating fission-track dates, see Hurford and Green (1982).

Fission-track dating may be undertaken on a wide variety of minerals in different rock types, though it has been most commonly carried out on apatite, micas, sphene and zircons in volcanic ashes, basalts, granites, tuffs, and carbonatites. It has also been widely used in dating amorphous (glassy) materials such as obsidian and is therefore useful in tephro-chronological studies (see Section 4.2.3; Westgate and Naeser, 1995). Its useful age range is large, from 10^3 to 10^8 years, but error margins are very difficult to assess and are rarely given. Microvariations in crystal uranium content may lead to large variations in the fission track count on different sections of the same sample (Fleming, 1976). This potential source of error may be reduced by repeated measurements, but for samples that are old and/or contain little uranium, the labor involved in counting precludes such checks being made. Tracks may also "fade" under the influence of mechanical deformation or in particular chemical environments and thus lead to underestimation of age (Fleischer, 1975). Fission-track dating is rarely used in paleoclimatology but its use in archeology and in tephrochronology is relatively common (Meyer *et al.,* 1991). In most cases where fission-track dating is used, ^{40}Ar/^{39}Ar dating would be preferable, but may not always be feasible.

4

▌DATING METHODS II

4.1 PALEOMAGNETISM

Variations in the Earth's magnetic field, as recorded by magnetic particles in rocks and sediments, may be used as a means of stratigraphic correlation. Major reversals of the Earth's magnetic field are now well known and have been independently dated in many localities throughout the world. Consequently, the record of these reversals in sediments can be used as time markers or chronostratigraphic horizons. In effect, the reversal is used to date the material by correlation with reversals dated independently elsewhere. However, as all episodes of "normal polarity" have the same magnetic signal, and all episodes of "reversed polarity" are similarly indistinguishable from just the polarity signal, it is necessary to know approximately the age of the material under study to avoid miscorrelations.

In addition to aperiodic global-scale geomagnetic reversals, smaller amplitude, quasi-periodic variations of the Earth's magnetic field have also occurred. These secular variations were regional in scale (over distances of 1000–3000 km) and can be used to correlate well-dated "master chronologies" with undated records exhibiting similar paleomagnetic variations.

4.1.1 The Earth's Magnetic Field

The magnetic field of the Earth is generated by electric currents within the Earth's molten core. The exact mechanism of its formation is not agreed upon, but for our purposes it is sufficient to consider the field as if it were produced by a bar magnet at the center of the Earth, inclined at ~11° to the axis of rotation (Fig. 4.1a). At the Earth's surface, we are familiar with this global field through magnetic compass variations. If a magnetized needle is allowed to swing freely, it will not only rotate laterally to point towards the magnetic pole, but also become inclined vertically, from the horizontal plane. The angle the needle makes with the horizontal is called the inclination (Fig. 4.1b). The inclination varies greatly, from near 0° at the Equator to 90° at the magnetic poles. If the needle is weighted, to maintain it in a horizontal plane, it will remain pointing towards magnetic north, and the angle it makes with true (geographical) north is called the declination (Fig. 4.1b).

The Earth's magnetic field is considered to be made up of two components — a primary and fairly stable component (the dipole field), which is represented by the bar magnet model, and a much smaller residual or secondary component (the non-dipole field), which is less stable and geographically more variable. Major changes in the Earth's magnetic field are the result of changes in the dipole field, but minor variations may be due to non-dipole factors (see Section 4.1.5).

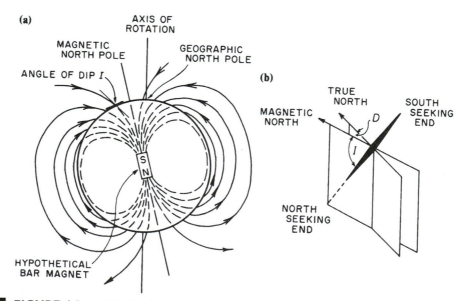

FIGURE 4.1 **(a)** The Earth's magnetic field. The main part of the Earth's magnetic field (the dipole field) can be thought of hypothetically as a bar magnet centered at the Earth's core. The lines of force represent, at any point, the direction in which a small magnetized needle tries to point. The concentration of these lines is a measure of the magnetic field strength. **(b)** Declination and inclination. Declination is a measure of the horizontal departure of the field from true north; inclination is a measure of dip from the horizontal. The resultant force is a vector representing declination, inclination, and field strength.

Because of the nature of the (dipole) field and the way in which it is generated, any change in its characteristics will affect all parts of the world. Records of significant magnetic field variations in a stratigraphic column (magnetostratigraphy) can thus be used directly to correlate sedimentary sequences in widely dispersed locations, regardless of whether they have common fossils or even similar facies. The broader significance of the magnetic characteristics of sediments, in a wide range of paleoenvironmental applications, is well documented by Thompson and Oldfield (1986).

4.1.2 Magnetization of Rocks and Sediments

So far we have referred to paleomagnetic variations and their usefulness without considering how such variations are recorded. It has been known for over 50 years that molten lava will acquire a magnetization parallel to the Earth's magnetic field at the time of its cooling. This is known as thermoremanent magnetization (TRM). The same phenomenon has been observed in baked clays from archeological sites; iron oxides in the clay, when heated above a certain temperature (the Curie point) realign their magnetic fields to those at the time the clay was baked. In this way, archeological sites of different ages have preserved a unique record of geomagnetic field variations over the last several thousand years (Aitken, 1974; Tarling, 1975).

Igneous rocks are not the only recorders of paleomagnetic field information; lake and ocean sediments may also register variations through the acquisition of detrital or depositional remanent magnetization (DRM). Magnetic particles become aligned in the direction of the ambient magnetic field as they settle through a water column. Providing that the sediment is not disturbed by currents, slumping, or bioturbation, the magnetic particles will provide a record of the magnetic field of the Earth at the time of deposition. Verosub (1977) considers that the acquisition of magnetization by sediments may occur after deposition due to the mobility of magnetic carriers within fluid-filled voids in the sediment. Once the water content of the sediment drops below a critical level (depending on the sediment characteristics) the magnetic particles can no longer rotate and magnetization becomes "locked in" to the sediment. This post-depositional DRM provides a more accurate record of the ambient magnetic field than simple depositional DRM, but in some circumstances it may lead to distinct regional differences because some sediments became realigned post-depositionally while others, perhaps more densely packed, did not (Coe and Liddicoat, 1994).

Unlike thermoremanent magnetization, detrital remanent magnetization is not an "instantaneous" event. Once the molten lava has cooled, perhaps in a matter of minutes, the ambient field becomes a permanent fixed record. In sediments, the record is subject to disturbance (e.g., by burrowing organisms) that may raise the water content of the sediment enough for magnetic particles to rotate again, changing the magnetization until the sediment is sufficiently dewatered to fix the record once more. Thus, the sedimentary record of the Earth's magnetic field, although continuous, should be considered as a smoothed or average record, unlikely to record short-term variations except in unusual circumstances where sedimentation rates are sufficiently high. Furthermore, it has been demonstrated by Verosub

(1975) that sediment disturbance may result in apparent reversals that would be hard, if not impossible, to detect in cores of non-laminated sediments (Fig. 4.2). This may have contributed to erroneous reports of short-term variations of the Earth's magnetic field (excursions; see Section 4.1.5).

Finally, it is now also recognized that iron minerals in some sediments undergo post-depositional chemical changes that result in a magnetization characteristic of the Earth's magnetic field long after initial deposition. This is known as chemical remanent magnetization (CRM); identification of the minerals typically affected in a sample can provide a warning that errors may be expected.

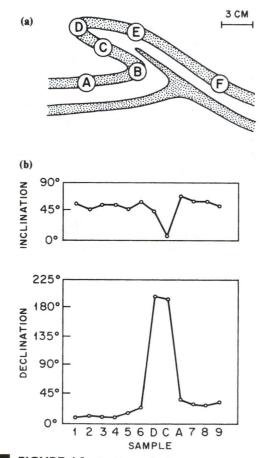

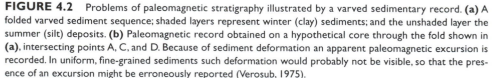

FIGURE 4.2   Problems of paleomagnetic stratigraphy illustrated by a varved sedimentary record. **(a)** A folded varved sediment sequence; shaded layers represent winter (clay) sediments; and the unshaded layer the summer (silt) deposits. **(b)** Paleomagnetic record obtained on a hypothetical core through the fold shown in **(a)**, intersecting points A, C, and D. Because of sediment deformation an apparent paleomagnetic excursion is recorded. In uniform, fine-grained sediments such deformation would probably not be visible, so that the presence of an excursion might be erroneously reported (Verosub, 1975).

4.1.3 The Paleomagnetic Timescale

Most of the early work on establishing a chronology or timescale of paleomagnetic events was carried out on lava flows. It was demonstrated that at times in the past the Earth's magnetic field has been the reverse of today's and that these periods of reversal (chrons) lasted hundreds of thousands of years. Potassium-argon dating methods enabled dates to be assigned to periods of "reversed" and "normal" fields so that eventually a complete chronology spanning several million years was constructed (Cox, 1969). Indeed the development of this chronology went hand-in-hand with the theory of plate tectonics because new lavas, produced at the centers of spreading (e.g., the Mid-Atlantic Ridge) were found to record identical paleomagnetic sequences on either side of the ridge (Opdyke and Channell, 1996). Careful study of lava flows and sea-floor paleomagnetic anomaly patterns has so far enabled a fairly accurate chronology of reversals to be constructed for the Cenozoic (Cande and Kent, 1992, 1995) and less certain chronologies have been constructed for even longer periods of time (Harland *et al.*, 1990). Major periods of normal or reversed polarity are termed polarity chrons or epochs, the most recent of which are named after early workers in the field. Thus we are currently in the Brunhes chron of "normal" polarity, which began ~780,000 yr B.P. Prior to that the Earth experienced a period of reversed polarity, the Matuyama chron, which began in late Pliocene times (Fig. 4.3).

In addition to major polarity epochs in which reversals persist for periods of ~10^6 yrs or more, the igneous record has also shown that reversals have occurred more frequently, but less persistently, for periods known as polarity events (or subchrons). These are intervals of a single geomagnetic polarity generally lasting 10^4–10^5 yrs within a polarity chron. During the last 2 million yrs, several such events are thought to have occurred, all within the Matuyama reversed polarity chron. Thus the Jaramillo (0.99–1.07 Ma B.P.) and the Olduvai (1.77–1.95 Ma B.P.) subchrons are periods of normal polarity, named after the locality of the lava samples studied. The dating of these relatively brief events is subject to change as new analyses are carried out (particularly by more accurate ^{40}Ar/^{39}Ar dating of samples). Figure 4.3 gives the current status of the polarity timescale.

All of the preceding discussion has referred to studies of polarity changes observed in lavas, but the widest application of paleomagnetism to paleoclimatic studies has been in the identification of reversals in sedimentary deposits, notably in marine sediments and in loess deposits (Hilgen, 1991; Rutter *et al.*, 1990). Studies of detrital remanent magnetization in ocean sediments may, in favorable circumstances, give a paleomagnetic record comparable even in detail with the terrestrial volcanic record (Opdyke, 1972). It is common in studies of undated ocean cores to plot the polarity sequence changes with depth and to assign an age of 0.78 Ma to the first major reversal (the Brunhes/Matuyama boundary). Younger ages are then derived by interpolation, assuming a zero age for the uppermost sediments and a constant sedimentation rate. This provides a first-order time-frame within which far more detailed radiometric or biostratigraphic checks and adjustments can be made (see Section 6.3.3). The Quaternary record of δ^{18}O in marine sediments, and its relationship to

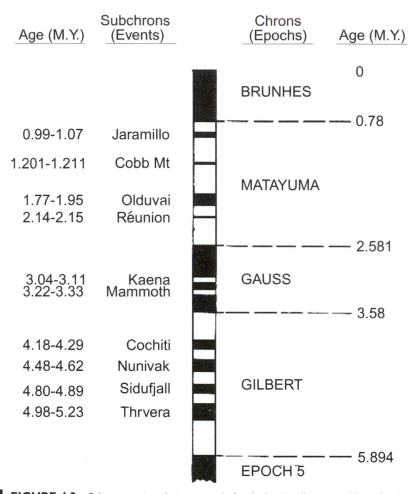

FIGURE 4.3 Paleomagnetic polarity timescale for the last 6 million years. Normal polarity periods in black. Dates are based on K/Ar dates on lava flows (Cande and Kent, 1995; Berggren *et al.*, 1995).

orbital forcing is now so well-established that it has been used to refine age estimates on the timing of paleomagnetic reversals (Shackleton *et al.*, 1990; Bassinot *et al.*, 1994; Tauxe *et al.*, 1996). Indeed, astrochronological estimates for the ages of magnetic reversals are now considered by some to provide the best framework for the late Quaternary geomagnetic timescale (Renne *et al.*, 1994; Cande and Kent, 1995; Berggren *et al.*, 1995).

4.1.4 Geomagnetic Excursions

In addition to polarity epochs and events (chrons and subchrons) there have been many reports of short-term geomagnetic fluctuations known as polarity excursions (or "cryptochrons"; Cande and Kent, 1992). Excursions are considered to last for a few thousand years (at most) and differ from events in that a fully reversed field is generally not observed, perhaps because the partial reversal is due to variations in the non-dipole component of the field. It is not yet clear whether excursions are "abortive reversals" or whether they represent normal geomagnetic behavior during a polarity epoch. In this regard it is interesting that Cox (1969) in a study of the spectrum of known reversal frequency formulated a statistical model of geomagnetic field behavior from which he inferred the existence of short events or excursions during the last 10 million yr, which were at that time undiscovered. Nevertheless, a great deal of controversy surrounds geomagnetic excursions and their usefulness in paleoclimatic studies is quite limited. Because of their short duration, they are only likely to be recorded in areas with high sedimentation rates and/or very frequent lava flows. Hence, it is often the case that an excursion apparently found in one location may not be registered nearby, where sedimentation rates at the critical time may not have been sufficient to record it. Such occurrences are common and lead to skepticism about the reality and significance of reported excursions (Verosub and Banerjee 1977; Lund and Banerjee, 1979). For an excursion to warrant recognition it should be based on observations of synchronous changes in both declination and inclination in several geographically separate cores, and the analysis should be restricted to fine-grained, homogeneous sediment. However, even if the individual excursions reported are correct, their absence from other sites in the same region mean that they are rarely of value in constructing regional chronologies or in magnetostratigraphic correlation.

In a recent review, Opdyke and Channell (1996) conclude that there were five times in the late Quaternary when there is credible evidence for excursions. These are named after the region where the record was first recognized: Mono Lake (27,000–28,000 yr B.P.); Laschamp (~42,000 yr B.P.); Blake (108,000–112,000 yr B.P.); Pringle Falls (218 ±10 ka B.P.); and Big Lost (~565 ka B.P.). Of these, the Blake excursion is the one most widely recorded, having been recognized in Chinese loess, as well as in marine sediments from the Caribbean, Atlantic, and Mediterranean. The Laschamp excursion has been recorded in Icelandic lava flows as well as in the Massif Central, France, where it was originally recognized (Condomes *et al.*, 1982; Levi *et al.*, 1990). The other excursions are all from the western United States where they have been seen in some (but not all!) lake sediments.

4.1.5 Secular Variations of the Earth's Magnetic Field

A number of studies of well-dated lake sediments from different parts of the world indicate that quasi-periodic changes in declination (and to a lesser extent inclination) have occurred during the Holocene (Mackereth, 1971; Verosub, 1988). These changes are of a smaller magnitude than those described as excursions and appear to be regional in extent (over distances of 1000–3000 km) presumably because they result from changes in the non-dipole component of the Earth's magnetic field.

If the changes observed can be accurately dated in one or more cores, it should be possible to construct a "master chronology" that would enable peaks and troughs in the declination record to be used as a chronostratigraphic template. Such a template would be valuable in estimating the age of highly inorganic sediments, which do not yield enough carbon for conventional ^{14}C dating, but which do contain a clear record of declination and inclination variations. However, it is first necessary to establish a reliable, well-dated magnetostratigraphic record for a "type-section" and to determine over what area this record can be considered to provide a master chronology. So far, this has been demonstrated for western Europe (Thompson, 1977) and the western United States (Verosub, 1988; Negrini and Davis, 1992). Secular paleomagnetic chronologies have also been proposed for other areas (Gillen and Evans, 1989; Ridge *et al.*, 1990). Undated sediments should have approximately the same sedimentation rate as that of the master chronology and not have experienced any significant disturbances, but even these problems may not be critical. Figure 4.4 shows the master declination and inclination chronology from Pleistocene Lake Russell (now Mono Lake, east-central California) for the period 12.5–30 ka B.P. Of particular note is the large excursion around 29–27 ka B.P. (named the Mono Lake geomagnetic excursion) which has been seen in other records from the region (Levi and Karlin, 1989). The chronology of the Lake Russell record was determined by 17 ^{14}C dates, but other

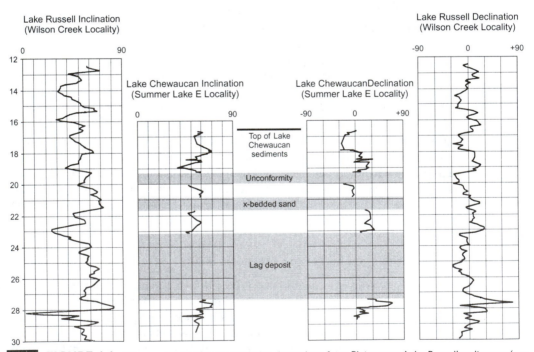

FIGURE 4.4 The master secular geomagnetic chronology from Pleistocene Lake Russell sediments (extreme left and right columns) and the correlated geomagnetic record from Lake Chewaucan, Oregon. Apparent episodes of nondeposition are shaded. The ^{14}C-dated master chronology provides a chronological template within which the undated sediments from Lake Chewaucan can be placed (Negrini and Davis, 1992).

lakes in the region have far less organic material; by using the Lake Russell secular geomagnetic record as a regional signal, other sedimentary records can be dated by correlating sequences of characteristic features. In this way, the secular paleomagnetic record from Pleistocene Lake Chewaucan (500 km north of Lake Russell) was mapped onto the master chronology, to fix the sequence in time (Fig. 4.4). In doing so, a number of periods of non-deposition were assumed to have occurred (with good supporting sedimentological evidence bracketing these apparent gaps). The chronological model for Lake Chewaucan, so defined, can be tested further because tephras embedded within the sediments have been found elsewhere in more organic-rich environments; preliminary results from ^{14}C dates associated with these tephras generally support the proposed geomagnetic timescale applied to the Lake Chewaucan sediments (Negrini and Davis, 1992).

Clearly, using secular variations from one record to date another can only be as accurate as the initial chronological control on the master record. Providing correlations between the master record and the undated target record are statistically significant, and changes in both inclination and declination support the match, it should be possible to build up regional templates that will be invaluable in many areas.

4.2 DATING METHODS INVOLVING CHEMICAL CHANGES

Two general categories of dating methods are based on chemical changes within the samples being studied since they were emplaced. The first involves amino-acid analysis of organic samples, generally used to assess the age of associated inorganic deposits. The method may also be used to estimate paleotemperatures from organic samples of known age. The second category encompasses a number of methods that assess the amount of weathering that an inorganic sample has experienced. They are primarily used to assess the relative age of freshly exposed rock surfaces in episodic deposits such as moraines or till sheets. There are a number of possible approaches (Colman and Dethier, 1986) but they must be calibrated by independent dating methods to convert the relative age to a numerical age estimate. Nevertheless, even without such calibration, weathering studies have proven to be useful in distinguishing and correlating glacial deposits in many alpine areas (e.g., Rodbell, 1993). One of the most widespread and well-tested methods is obsidian hydration dating (Section 4.2.2) — an example of the general group of methods that involve measurements of weathering rinds (Colman and Pierce, 1981; Chinn, 1981).

Another method involves the chemical "fingerprinting" of volcanic ashes that often blanket wide areas after a major eruption (Section 4.2.3). Chemical analyses of tephra deposits have proved to be successful in identifying unique geochemical signatures in ashes of different ages. Where the age of a tephra layer has been independently determined, the ash may be used as a chronostratigraphic marker to date the associated deposits and to correlate events over wide areas.

4.2.1 Amino-acid Dating

As all living organisms contain amino acids, a dating method based on amino acids offers a tremendous range of possible applications. Since the first use of amino acids in estimating the age of fossil mollusc shells (Hare and Mitterer, 1968), significant advances in the use of amino acids as a geochronological tool have been made. Relative ages are generally obtained, but where an independent age calibration is possible, and the thermal history of the sample can be estimated, more quantitative age estimates can be made. The method can be applied to material ranging in age from a few thousand to a few million years old and is therefore useful in dating organic material well beyond the range of radiocarbon dating.

Amino-acid analyses use very small samples (e.g., < 2 mg in the case of molluscs and foraminifera). However, for reliable results multiple analyses on a suite of samples is recommended (Miller and Brigham-Grette, 1989). The application of amino-acid dating is thus of particular significance in dating fragmentary hominid remains, where, in many cases, a conventional radiocarbon analysis would require destruction of the entire fossil to obtain a date (Bada, 1985). Analyses have also been carried out on samples of wood, speleothems, coral, foraminifera, and marine, freshwater, and terrestrial molluscs (Schroeder and Bada, 1976; Lauritzen *et al.*, 1994). Because amino-acid changes are both temperature and time-dependent, it is often difficult to be definitive in assigning an age to a sample. More often the method enables relative chronologies to be established and stratigraphic sequences to be checked (Miller *et al.*, 1979; Oches and McCoy, 1995a). It is perhaps in this application (aminostratigraphy) that amino-acid analysis offers the greatest potential (Miller and Hare, 1980).

4.2.1.1 Principles of Amino-acid Dating

Amino acids are so-called because they contain in their molecular structure an amino group (-NH2) and a carboxylic acid group (-COOH). These are attached to a central carbon atom, which is also linked to a hydrogen atom (-H) and a hydrocarbon group (-R) (Fig. 4.5a). If all atoms or groups of atoms attached to the central carbon atom are different, the molecule is said to be chiral or asymmetric. The significance of this is that chiral molecules can exist in two optically different forms (stereoisomers), each being the mirror image of the other (Fig. 4.5b). These optical isomers or enantiomers have the same physical properties and differ only in the way in which they rotate plane-polarized light. The relative configuration of enantiomers is designated, by convention, D or L (*dextro* [right] or *levo* [left]) and virtually all chiral amino acids in living organisms occur in the L configuration. Interconversion to the D configuration takes place by a process known as racemization. The extent of the racemization (expressed by the enantiomeric ratio, D:L) increases with time after the death of the organism. This can be measured by gas or liquid chromatographic methods.

Not all amino acids have only one chiral carbon atom. Some amino acids (e.g., isoleucine) contain two chiral carbon atoms, which means that they can exist as four stereoisomers — a set of mirror image isomers (enantiomers) and a set of non-mirror image isomers (diastereomers) (Fig. 4.5b). Interconversion of L-isoleucine

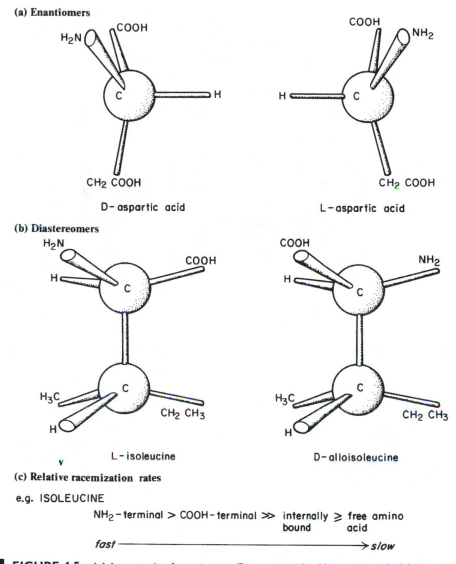

FIGURE 4.5 **(a)** An example of enantiomers (D-aspartic acid and L-aspartic acid). **(b)** An example of diastereomers (L-isoleucine and D-alloisoleucine). **(c)** Relative rates of racemization depending on whether the amino acid is internally bound, terminally bound, or free.

could thus theoretically produce all four stereoisomers. However, in diagenetic processes only one of the two chiral atoms undergoes interconversion, thereby producing only one other isomer (D-alloisoleucine, a diastereomer) by a process known as epimerization[9] (Schroeder and Bada, 1976; Rutter and Blackwell, 1995).

[9] For our purposes racemization and epimerization of amino acids can be considered as essentially equivalent processes.

The ratio of D-alloisoleucine to L-isoleucine (abbreviated as aIle/Ile, or D/L) increases on the death of an organism from near zero to ~1.3 at equilibrium. The time it takes to reach equilibrium varies with temperature (see in what follows) and may range from 150–300 ka in the Tropics, to ~2 Ma at mid-latitudes to >10 Ma in polar regions (Miller and Brigham-Grette, 1989). Diastereomers have physical properties sufficiently different enough that they can be separated by ion-exchange chromatography. Several different amino acids have been used to assess the age of a sample, particularly aspartic acid, leucine, and isoleucine. Epimerization of L-isoleucine to D-alloisoleucine is an order of magnitude slower than aspartic acid racemization so it is potentially of more value in dating older samples or those from warmer climates where epimerization and/or racemization rates are faster. Conversely, aspartic acid can be used in resolving differences between Arctic molluscs from the last glacial cycle, the temperature history of which is too low for them to have undergone significant isoleucine epimerization (Goodfriend *et al.*, 1996). Aspartic acid racemization has also provided excellent results in dating recent fossils, such as banded corals (Fig. 4.6) and land snails of Holocene age (Goodfriend, 1991, 1992; Goodfriend *et al.*, 1992).

Unlike radionuclide decay rates, racemization and epimerization rates are sensitive to a number of environmental factors, particularly temperature. In addition, racemization and epimerization rates vary depending on the type of matrix in which the amino acids are found (shell, wood, bone, etc.). In carbonate fossils, rates vary from one genus to another (Fig. 4.7) so it is important to compare amino-acid ratios derived from analyses on similar genera (Miller and Hare, 1975; King and Neville, 1977). Racemization rates also depend on how amino acids are bound to each other, or if they are free (unbound). When an amino acid is bound together with others, its position in the molecule may be internal or terminal (Fig. 4.5c). If it is terminally bound it may be attached to other amino acids by either a carbon or a nitrogen atom. Racemization rates are fastest when the amino acid is terminally bound and slowest when the individual amino acid is free, having been separated

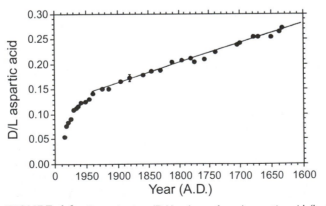

FIGURE 4.6 Racemization (D:L) values of total aspartic acid (hydrolyzed samples) in relation to the counted age of annual bands in the coral *Porites australiensis*, from the Great Barrier Reef, Australia. The linear regression for the period 1632–1985 is shown (Goodfriend *et al.*, 1992).

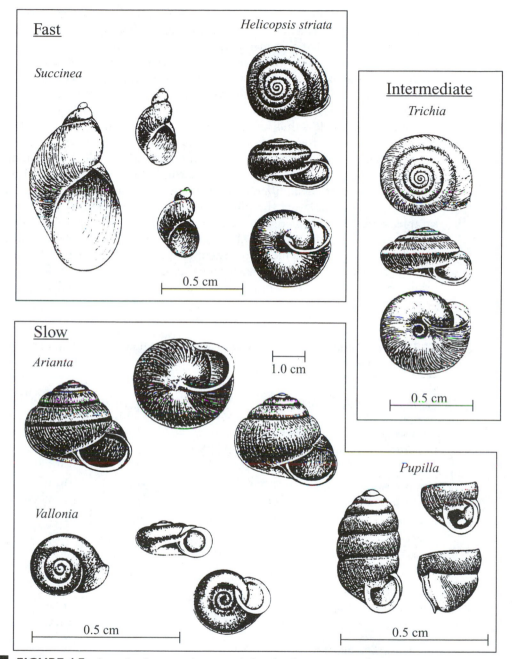

FIGURE 4.7 Some fossil gastropods commonly found in European loess deposits, grouped according to the relative racemization rates of isoleucine in the shell matrix. Note the variable scale (Oches and McCoy, 1995b).

from the rest of the molecule by hydrolysis. Racemization rates of internally bound amino acids are intermediate between rates of terminally bound and free amino acids (Fig. 4.5c). What this means is that, as the peptide is hydrolyzed, at some point each amino acid will become terminally bound before being eventually split off (free). In the terminally bound position, racemization rates are greatest, so the probability is relatively high that the free amino acid, when released, will already be in the D-form. Consequently the D:L ratios in the free fraction are higher than in the bound fraction. It is thus important to note whether analyses reported in the literature are based on free fraction or the total acid hydrolysate (free and bound), as the resulting ratios can vary by an order of magnitude (Table 4.1). Commonly, the D:L ratios from the free fraction and the total acid hydrolysate will be plotted against each other to help differentiate units of different age (Fig. 4.8). Recent studies have involved isolating the high molecular weight (HMW) polypeptides in a sample, which are considered to be less contaminated (by post-depositional bacterial degradation of the amino acids) (Kaufman and Sejrup, 1995). Analysis of the HMW fraction produces more consistent results (i.e., a lower standard deviation) and a much slower rate of racemization because of the presence of fewer terminally bound amino acids. This approach could extend the potential range of dating

TABLE 4.1   Temperature Sensitivity of Amino-acid Reactions in Dated Early Post-glacial Mollusc Samples

Location	^{14}C age (years)	MAT (° C)[a]	Species[b]	Allo: Iso[c] Total	Allo: Iso[c] Free
Washington	13,010	+10	*H.a.*	0.078	0.27
Denmark	13,000	+7.7	*H.a.*	0.053	0.21
Maine	12,230	+7	*H.a.*	0.050	0.21
New Brunswick	12,500	+5	*H.a.*	0.043	0.18
Southeastern Alaska	10,640		*H.a.*	0.040	0.15
Anchorage	14,160	+2.1	*M.t.*	0.034	0.16
Southern Greenland	13,380	-1	*H.a.*	0.027	<0.09
Southern Baffin Island	10,740	-7	*M.t.*	0.024	<0.1
Spitsbergen	11,000	-8	*M.t.*	0.022	<0.1
Northern Baffin Island	10,095	-12	*H.a.*	0.020	ND
Somerset Island	9000	-16	*H.a.*	0.018	ND
Modern	0		*H.a.*	0.018	ND

From Miller and Hare (1980).

[a] Mean annual temperature of the past one to five decades based on records of the nearest representative weather station.

[b] *H.a.* = *Hiatella arctica*; *M.t.* = *Mya truncata*. Hydrolysis rates in *Mya* are not directly comparable with those in *Hiatella*. For most localities, three or more separate values were analyzed; ratios given here are mean values.

[c] Ratio of D-alloisoleucine to L-isoleucine; ND = no detectable alloisoleucine.

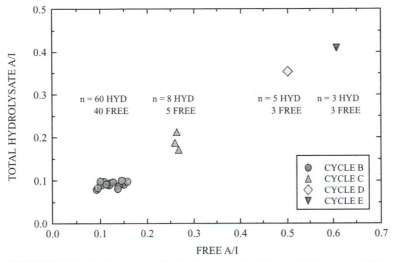

FIGURE 4.8 Total vs free alle/Ile for *Succinea* snails in loess units from central Europe. The sample means (shown by different symbols) fall into distinct clusters, which correspond to loess units of different age. Number of subsample preparations are indicated (Oches and McCoy, 1995b).

(which would be useful in some situations), but could be problematic in resolving age differences between samples from colder regions.

By far the most significant factor affecting the rate of racemization is temperature, specifically the effective diagenetic temperature (EDT), which is an integrated temperature of the sample since deposition. Racemization rates more or less double for a 4–5 °C increase in temperature, so the thermal history of a sample becomes of critical importance to the apparent age. An uncertainty of only ±2 °C is equivalent to an age uncertainty of ±50%, so this is clearly a major source of error in assessing the numerical age of a sample (McCoy, 1987a). Thermal histories are rarely known to within ±2 °C, even in isolated environments, such as in caves or in the deep oceans. However, the temperature dependence of racemization rates may be put to advantage if the sample age is known independently (e.g., by ^{14}C dating). In such cases, the relative amount of racemization can indicate the EDT of the sample since deposition (Bada *et al.*, 1973) or the extent of a step-change in temperature (Schroeder and Bada, 1973; see Section 4.2.1.4). It is important to note that because racemization rates increase exponentially with temperature, the amount of time a sample experiences high temperatures is much more significant than the time spent at low temperatures (Fig. 4.9). Thus EDT is not simply the long-term mean temperature of the site, and it will always be higher than the actual long-term mean. This means that at mid- to high-latitude sites, samples from the beginning or end of the last glacial period may be indistinguishable in terms of their D:L ratios, but they would be distinctly different from samples of the last interglacial, or preceding glacial period (which had passed through the last interglacial). This strong temperature influence on racemization also imposes strict sampling criteria, as samples

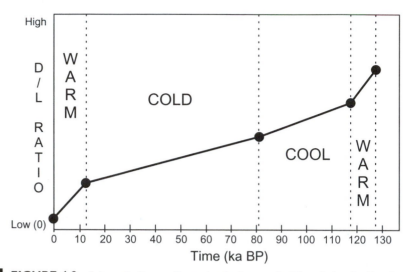

FIGURE 4.9   Schematic diagram illustrating the increase in D/L ratio in a fossil mollusc over the course of the last glacial-interglacial cycle. Most of the change in D:L values occurs during the warm interglacial episodes (Miller and Mangerud, 1985).

which have been close to the surface for a prolonged period (either the modern surface or a paleosurface) may give erroneous results because they were exposed to high temperatures, significantly raising their *effective* diagenetic temperature. Experiments suggest that samples should be deeply buried (>2m) to avoid such problems (Miller and Brigham-Grette, 1989).

There are basically three approaches used in amino-acid geochronological work. Two of these aim at producing an estimate of numerical age and the third uses enantiomeric ratios as simply a stratigraphic tool, to establish the relative age of two or more samples.

4.2.1.2 Numerical Age Estimates Based on Amino-acid Ratios

Numerical sample ages are estimated by either calibrated or uncalibrated methods (Williams and Smith, 1977). The *uncalibrated method* is based on high-temperature laboratory experiments, which attempt to simulate in a short period of time the much slower processes that occur in samples at the lower temperatures typical in nature. The general amino-acid racemization reaction is as follows:

$$\text{L-amino acid } k_1/k_2 \text{ D-amino acid}$$

where k_1 and k_2 are rate constants for the forward and reverse reactions. In the high-temperature studies, racemization rates are determined by sealing in a tube a sample of the same species as the fossil under investigation, and heating it for known lengths of time in a constant temperature bath. Providing the initial ratio of the sample is known, genus-specific rate constants for the amino acid in question can be determined in this way, at different (elevated) temperatures. These are then plotted on an Arrhenius plot in which the log of the rate constant forms the ordinate and the reciprocal of the absolute temperature the abscissa (Fig. 4.10). If the calculated rate constants fall on a straight line, extrapolation is made (beyond the experimental results) to obtain the rate constants applicable at lower temperatures (Miller and Hare, 1980). Providing that the EDT of the sample since its deposition is known (or can be closely estimated), racemization rate constants can be obtained for that temperature from the Arrhenius plot; sample age can then be calculated from the measured D:L ratio (for the appropriate equations, see Williams and Smith, 1977, p. 102). One might question whether such high-temperature, short-term laboratory kinetic studies accurately reflect the low-temperature, long-term diagenetic changes that occur in fossils. However, there is an increasing body of evidence that this is not a problem and that the high-temperature results can be extrapolated to

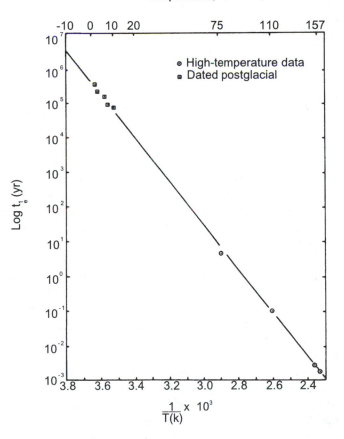

FIGURE 4.10 Arrhenius plot of isoleucine epimerization in *Hiatella arctica* derived from heating experiments at 75°, 110°, 152°, and 157°C, and ^{14}C-dated early postglacial samples (Miller and Hare, 1980).

real-life situations (see Fig. 4.10; Goodfriend and Meyer, 1991). The real difficulty concerns the problem of knowing accurately the thermal history of the sample, as slight errors in this parameter lead to large errors in numerical age estimates (Mc-Coy, 1987a). As a result, the ages derived by means of this uncalibrated method are considered to be the least reliable.

A more fruitful approach, though not entirely free of the problems already discussed here, is to derive rate constants empirically by the measurement of D:L ratios *in situ,* in fossil samples of known age (the *calibrated method*). Other samples at the same site can then be dated, if it is assumed that they have experienced essentially the same EDT as the fossil used for calibration (Bada and Schroeder, 1975). A Holocene fossil calibration sample is thus not suitable for assessing the age of older "glacial age" samples because their thermal histories will be quite different. Because of the reduction in racemization rate with sample age, and the sensitivity of the process to temperature, resolution of age becomes increasingly more uncertain in very old samples (Wehmiller, 1993). Miller and Brigham-Grette (1989) suggest that age estimates that are independently calibrated and are within the early "linear" stage of sample racemization (D:L < 0.3) should be reliable to ± 15–20%, but this is likely to increase to ± 30–40% for older samples. However, further cross-checks with independent age estimates can reduce such uncertainties.

Figure 4.11 illustrates the calibrated approach to estimating sample age. The three lines are calculated aIle/Ile ratios (based on a kinetic model, from heating experiments) for EDTs of 8, 11, and 14 °C (Wehmiller, 1993). The Peruvian samples are from a series of uplifted coastal terraces; sample IIa was independently dated at 100–130 ka B.P. (i.e., last interglacial, *sensu lato*) and thus provides an age calibration point. Assuming a long-term EDT of 14 °C for this area, the age of the older samples (IIb–V) can be estimated. A similar approach is also shown for samples of *Mercenaria* from the North Carolina and Virginia coastal plain, again using samples from the last interglacial as the single age calibration point. Greater confidence in the ages of older samples would be achieved by having more than one calibration point. Note that the estimation of EDT is critical for obtaining a "correct" age, and that its importance increases with sample age. For example, an aIle/Ile value of 0.6 could indicate a sample age of 0.4 Ma with an EDT of 11 °C, or 0.8 Ma with an EDT of 8 °C. Fortunately, since for most of the last million years the Earth was in a glacial mode, the EDT is primarily controlled by the temperature during relatively brief interglacial episodes; the long-term EDT is therefore assumed to be similar to that experienced by a sample since the time of the last interglacial. Such an assumption could be drastically in error if, for example, a sample that had been submerged below sea level for a significant amount of time was compared to a sample that had been subjected to air temperature changes. Table 4.2 illustrates this point; it shows aIle/Ile ratios on marine molluscs (*Hiatella arctica*) from Svalbard, Norway. Shells of the same ^{14}C age had significantly different values, depending on whether they had been continuously submerged for the entire Holocene, or exposed to much lower air temperatures (therefore reducing the racemization rate). Table 4.2 also shows that shells dated >61 ka B.P. from the same area gave aIle/Ile ratios identical

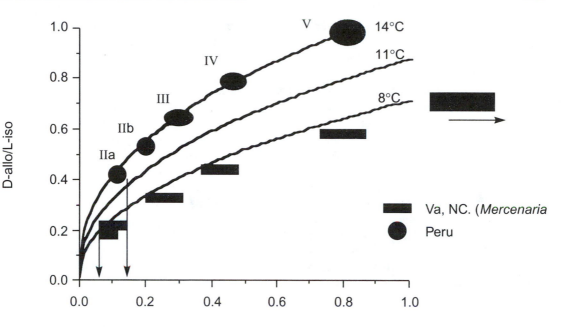

FIGURE 4.11 Isoleucine epimerization model curves for different EDTs, in relation to mean Alle/Ile ratios for samples from Peru (ovals) and from the U.S. Atlantic Coastal Plain (rectangles). The last interglacial samples used for calibration are shown by arrows (Wehmiller, 1993).

to the continuously submerged early Holocene samples, illustrating the potential for misinterpretation if the thermal history of a sample is not understood.

4.2.1.3 Relative Age Estimates Based on Amino-acid Ratios

In view of the numerous difficulties surrounding the assignment of numerical ages to fossil samples many investigators have found it prudent to use amino-acid ratios as relative age criteria only. By establishing a standard aminostratigraphic

TABLE 4.2 The Effect of Contrasting Thermal Histories on ^{14}C-dated *Hiatella arctica* from Western Spitsbergen, Svalbard, Norway

Thermal History	Current MAT (°C)	D:L ratio	[14]C age
Continuously submerged	+2.2	0.031	9900
Emerged shortly after deposition	-6.0	0.018	9940
Emerged shortly after deposition	-6.0	0.031	>61,000

From Miller and Brigham-Grette (1989).

framework for deposits in a region (where it is reasonable to assume a similar EDT history) other units can then be fitted into that relative age chronology (Wehmiller, 1993). For example, Oches and McCoy (1995a) showed that the conventional interpretation of loess stratigraphy in Hungary was incorrect, based on D:L ratios in fossil gastropod shells (snails) associated with each deposit. In some sections, units of quite different age had been assumed to be correlative, but they were clearly diachronous according to the aminostratigraphy. This approach has provided an independent means of testing the veracity of TL dates on loess across a wide swath of central Europe, from Germany to the Ukraine (Zöller *et al.*, 1994; Oches and McCoy, 1995b). Furthermore, by correlating the revised loess-paleosol sequences with marine isotope stages, as first suggested by Kukla (1977), it is possible to assign approximate ages to the D:L ratios in the snails of each loess unit (Fig. 4.12). A similar approach was taken by Miller and Mangerud (1985), who used aIle/Ile ratios in shallow water marine molluscs from European interglacial deposits to correlate deposits of similar age and thermal history, and to distinguish units of last interglacial age from older deposits. The results were helpful in resolving previous uncertainties in the age of many isolated and fragmentary stratigraphic sections. Bowen *et al.* (1989) also found that aminostratigraphic studies of non-marine molluscs were extremely useful in reassessing the chronology of Pleistocene depositional units in Great Britain; the revised stratigraphy could then be correlated with the SPECMAP standard marine chronostratigraphy. These are all relatively simple applications, using racemization of a single amino acid to solve a stratigraphic problem. More rigorous differentiation seems possible using several enantiomeric ratios and multivariate statistical techniques such as discriminant analysis.

At the present time, using amino-acid racemization and epimerization processes for establishing relative age seems to be the most practical application of the method. There may still be problems of contamination, leaching, and possibly thermal differences among sites, but these are relatively minor problems compared to those associated with numerical age determinations.

4.2.1.4 Paleotemperature Estimates from Amino-acid Racemization and Epimerization

Although amino-acid analyses are being increasingly used in stratigraphic studies, perhaps the most significant application of amino-acid ratios is in paleotemperature reconstruction. As already noted, a major barrier to accurate age estimates from amino-acid ratios is a knowledge of the integrated thermal history of the sample. However, the "age equation" (which includes the important thermal term) can be solved for temperature if the sample age is known. In the resulting "temperature equation," the time value is of relatively minor significance; as a result, for samples of known age, quite accurate estimates of the integrated thermal history of the depositional site can be achieved. Typically, for well-dated late Wisconsin or Holocene age samples, an uncertainty of ~3 °C (~1% of the absolute temperature of the site) can be expected in the paleotemperature estimates. However, if paleotemperatures are calculated from two samples of differing age, the temperature *difference* between the two periods can be estimated with considerably more accuracy (typically to within ± 1 °C). This is because many of the factors causing the initial uncertainty

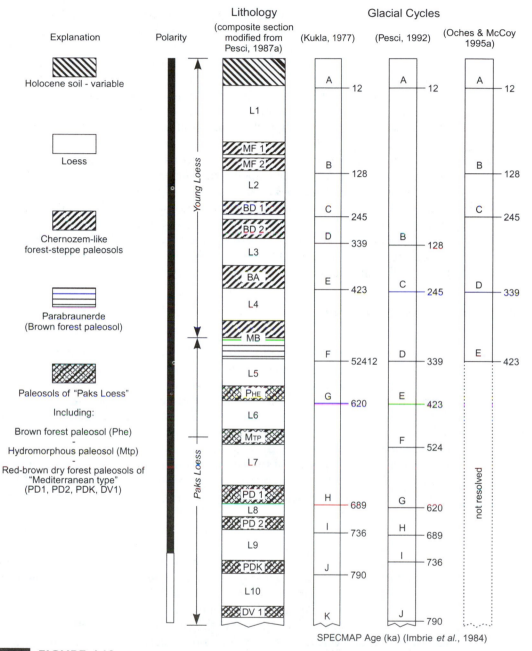

Lithology (composite section modified from Pesci, 1987a)

Glacial Cycles

FIGURE 4.12 The composite loess-paleosol stratigraphy characteristic of sites in Hungary is shown schematically in the center column (generally the sections are incomplete in any one area). Columns to the right show previous interpretations of the stratigraphy, subdivided into the main interglacial/glacial cycles (designated A to K), according to Kukla (1977) and Pecsi (1992). Their proposed correlations with the marine isotope stages are indicated by the age of glacial stage Terminations (based on SPECMAP dated records of Imbrie et al., 1984). Amino-acid relative age dating on snails in the loess suggests that the "correct" interpretation is that shown in the right-hand column (from Oches and McCoy, 1995a; Zöller et al., 1994).

in an individual paleotemperature estimate cancel out when temperature differences are computed (McCoy, 1987a). Using this approach McCoy (1987b) cast new light on the controversy of "cooler or wetter" conditions in the Great Basin during the late Wisconsin. Mean annual temperatures from 16,000 to 11,000 yr B.P. were estimated to have been 9 °C or more below the post-11,000 yr B.P. averages. If correct, no increase in regional precipitation would be required to explain the high levels of Lake Bonneville during the Late Glacial, with the maximum lake level phase occurring during a relatively cold, dry climate compared to the present (see Section 8.2.3). In another interesting application, Oches *et al.* (1996) used aIle/Ile ratios in gastropods from Mississippi valley Peoria loess (dating from the last glacial maximum) to estimate effective diagenetic temperature gradients from the Gulf of Mexico (30° N) to inland sites at 43° N. Today, the gradient is 0.9 °C/degree of latitude, whereas the amino-acid data point to gradients of only 0.3–0.6 °C/degree of latitude, and overall temperatures at least 7–13 °C lower than today. This suggests that SSTs in the Gulf of Mexico were significantly lower than today, otherwise the temperature depression would have been less in the lower Mississippi valley, and the overall temperature gradient would probably have been stronger, not weaker.

4.2.2 Obsidian Hydration Dating

Obsidian is one of the glassy products of volcanic activity, formed by the rapid cooling of silica-rich lava. Although its precise chemical composition varies from one extrusion to another, it always contains >70% silica by weight. Obsidian hydration dating is based on the fact that a fresh surface of obsidian will react with water from the air or surrounding soil, forming a hydration rind. The thickness of the hydration rind can be identified in thin sections cut normal to the surface; a distinct diffusion front can be recognized by an abrupt change in refractive index at the inner edge of the hydration rind. Hydration begins after any event that exposes a fresh surface (e.g., cracking of the lava flow on cooling, manufacture of an obsidian artifact, or glacial abrasion of an obsidian pebble); thus, providing one can identify the type of surface or crack in the rock, it is possible to date the event in question.

As one might expect, hydration rind thickness is a (non-linear) function of time; hydration rate is primarily a function of temperature, though chemical composition of the sample is also an important factor. For this reason, it is necessary to calibrate the samples within a limited geographical area against a sample of known age and similar chemical composition. These are difficult criteria to meet in a paleoclimatic context but are somewhat easier in archeological studies, where obsidian hydration dating has been most widely applied (Michels and Bebrich, 1971). Obsidian was widely traded in prehistoric time and often the precise source of the material can be identified and its diffusion throughout a geographical area can be traced. If samples can be found in a ^{14}C-dated stratigraphic sequence, hydration rinds can be calibrated, providing an empirically derived hydration scale for the site. This can then be used to clarify stratigraphy elsewhere, where radiocarbon-dated samples are unavailable. Alternatively, the hydration rate can be calibrated in the laboratory by heating experiments; if the effective hydration temperature of the sample can be estimated (i.e., its integrated temperature history), age can then be calculated (Lynch and Stevenson, 1992).

Obsidian hydration may also be used to date glacial events if obsidian has been fortuitously incorporated into the glacial deposits. Glacial abrasion of obsidian fragments creates radial pressure cracks normal to the surface and shear cracks sub-parallel to the surface. The formation of such "fresh" cracks allows new hydration surfaces to develop, and these effectively "date" the time of glacial activity. Hydration rinds resulting from glacial abrasion can then be compared with rinds that have developed on microfractures produced when the lava cooled initially. This event can be dated by potassium-argon isotopic methods (see Section 3.2.2), providing independent calibration for the primary hydration rind thicknesses. Pierce *et al.* (1976), for example, analyzed obsidian pebbles in two major moraine systems in the mountains of western Montana. Dates on two nearby lava flows indicated ages of 114,500 ± 7300 and 179,000 ± 3000 yr B.P. Hydration rinds on cracks produced during the initial cooling of these flows averaged 12 and 16 μm, respectively. These points enabled a graph of hydration thickness versus age to be plotted. It was then possible to estimate, by interpolation, the age of hydration rinds produced on glacially abraded cracks in the moraine samples. Two distinct clusters of hydration rind thicknesses enabled glacial events to be distinguished, at 35,000–20,000 and 155,000–130,000 yr B.P. Although the dates are by no means precise, they do at least indicate the important fact that the earlier glacial event predated the Sangamon interglacial (~125,000 yr B.P.), a point of some controversy in the glacial history of the western United States.

Obsidian hydration dating methods are limited by the problems of independent (radioisotopic) calibration, variations in sample composition, and temperature over time. Temperature effects are particularly difficult to evaluate. It is really necessary to produce a calibration curve for each area being studied, and this is not always possible. Nevertheless, where the right combination of conditions is found, obsidian hydration methods can provide a useful time-frame for events that might otherwise be impossible to date.

4.2.3 Tephrochronology

Tephra is a general term for airborne pyroclastic material ejected during the course of a volcanic eruption (Thorarinsson, 1981). Extremely explosive eruptions may produce a blanket of tephra covering vast areas, in a period which can be considered as instantaneous on a geological timescale. Tephra layers thus form regional isochronous stratigraphic markers. Tephras themselves may be dated directly, by potassium-argon or fission-track methods, or indirectly by closely bracketing radiocarbon dates on organic material above and below the tephra layer (Naeser *et al.*, 1981). In favorable circumstances, organic material incorporated within the tephra may provide quite precise time control on the eruption event (Lerbemko *et al.*, 1975; Blinman *et al.*, 1979). Providing that the dated tephra layer can be uniquely identified in different areas, it can be used as a chronostratigraphic marker horizon to provide limiting dates on the sediments with which it is associated. For example, a tephra layer of known age provides a *minimum* date on the material over which it lies and a *maximum* date on material superimposed on the tephra. If a deposit is sandwiched between two identifiable tephra layers of known age, they provide bracketing dates for the intervening deposit (Fig. 4.13). A prerequisite for such tephrochronological applications is that each tephra layer be precisely identified. This has been the subject of much study both in the field and in the laboratory. In the field, stratigraphic position, thickness, color, degree of weathering, and grain size are important distinguishing characteristics. In the laboratory, a combination of petrographic studies and chemical analyses are generally used to identify a unique tephra signature (Kittleman, 1979; Westgate and Gorton, 1981; Hunt and Hill, 1993). Multivariate analysis is commonly employed on the various parameters measured to provide optimum discrimination (or correlation) between the tephras being studied (Beget *et al.*, 1991; Shane and Froggatt, 1994).

In many volcanic regions of the world, tephrochronology is a very important tool in paleoclimatic studies. In northwestern North America, explosive eruptions have produced dozens of widely distributed tephra layers (Table 4.3). Some, such as the Pearlette "O" ash, covered almost the entire western United States and probably had a significant impact on hemispheric albedo (Bray, 1979). Others were more local in extent; around Mt. Rainier, for example, at least ten tephra layers have been identified spanning the interval from 8000 to 2000 yr B.P. (Mullineaux, 1974). Because of the eruption frequency and widespread distribution of tephra in this area, tephrochronological studies have proved to be invaluable in understanding its glacial history (Porter, 1979).

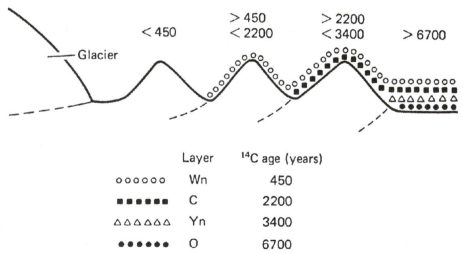

Layer	^{14}C age (years)
oooooo Wn	450
■■■■■■ C	2200
△△△△△△ Yn	3400
●●●●●● O	6700

 FIGURE 4.13 The use of tephra to date glacial deposits. If tephra age is known and tephra can be uniquely identified, ages can be used to "bracket" timing of glacial advance (Porter, 1981a).

TABLE 4.3 Some Important Tephra Layers in North America

Tephra layer	Source	Approximate age
Katmai	Mt Katmai, Alaska	A.D. 1912
Mt St Helens, Set T	Mt St Helens, Washington	A.D. 1800
Mt St Helens, Set W	Mt St Helens, Washington	450[a]
White River East	Mt Bona, South-eastern Alaska	1250[a]
White River North	Mt Bona, South-eastern Alaska	1890[a]
Bridge River	Plinth-Meager Mt, British Columbia	2600[a]
Mt St Helens, Set Y	Mt St Helens, Washington	3400[a]
Mazama	Crater Lake, Oregon	6600[a]
Glacier Peak B	Glacier Peak, Washington	11,200[a]
Glacier Peak G	Glacier Peak, Washington	12,750-12,000[a]
Old Crow	Alaska Peninsula	150,000
Pearlette O	Yellowstone National Park	600,000 ± 100,000
Bishop	Long Valley, California	700,000 ± 100,000
Pearlette S	Yellowstone National Park	1,200,000 ± 40,000
Pearlette B	Yellowstone National Park	2,000,000 ± 100,000

After Porter (1981b) and Westgate and Naeser (1995).

[a] Age given in radiocarbon years.

Tephrochronology has provided valuable time control in many paleoclimatic studies. For example, in the North Atlantic, the Icelandic Vedde ash has been identified in both lake and marine sediments (Mangerud *et al.*, 1984) and has recently been found in ice cores from Greenland (Grönvald *et al.*, 1995). This provides a very important chronostratigraphic marker (10,320 ± 50 ^{14}C yr B.P. or 11,980 ± 80 [ice core counted] calendar years B.P.) at a critical time for correlating the rapid environmental changes that were then taking place. Other important ash layers found in both marine sediments and ice cores are the Saksunarvatn ash (~10,300 calendar yr [ice cores] or ~9000 ^{14}C years B.P.) and the Z2 ash zone, which is dated in the GISP2 ice core at ~52,680 yr B.P. (Birks *et al.*, 1996; Zielinski *et al.*, 1997). Tephras have also been isolated from Holocene peat deposits (Pilcher *et al.*, 1995) opening up the prospect of more widespread applications of tephrochronology in paleoclimatic studies. It should also be noted that even when tephras are not present, geochemical signals in ice cores (principally excess sulphate washed out of the atmosphere following major explosive eruptions) are very important geochronological markers for dating ice at depth and hence for correlating different records (see Section 5.4.3.1).

Expanding our understanding of the frequency and extent of explosive eruptions in the past is extremely important (Beget *et al.*, 1996). There is abundant evidence to demonstrate that such eruptions lead to lower temperatures, at least for a limited period (Bradley, 1988; Palais and Sigurdsson, 1989). Whether periods with a high frequency of explosive eruptions in the past experienced a persistent temperature depression (possibly reinforced by additional positive feedbacks in the climate system, due to persistence of high albedo snow cover, or more extensive sea ice) remains controversial. However, there is persuasive circumstantial evidence that episodes of explosive volcanism have been associated with periods of glacier advance in the past, including those of the most recent neoglacial episodes, collectively known as the "Little Ice Age" (Bray, 1974; Porter, 1986; Grove, 1988). Numerous other examples of the importance of tephrochronological studies in paleoclimatic research are found in the two volumes edited by Sheets and Grayson (1979) and by Self and Sparks (1981).

4.3 BIOLOGICAL DATING METHODS

Biological dating methods generally use the size of an individual species of plant as an index of the age of the substrate on which it is growing. They may be used to provide minimum age estimates only, as there is inevitably a delay between the time a substrate is exposed and the time it is colonized by plants, particularly if the surface is unstable (e.g., in an ice-cored moraine). Fortunately this delay may be short and not significant, particularly if the objective is simply establishing a relative age.

4.3.1 Lichenometry

Lichens are made up of algal and fungal communities living together symbiotically. The algae provide carbohydrates via photosynthesis and the fungi provide a protective environment in which the algal cells can function. Morphologically, lichens range from those with small bush-like thalli (foliose lichens) to flat disc-like forms, which grow so close to a rock surface as to be inseparable from it. These crustose lichens commonly increase in size radially as they grow and this is the basis of lichenometry, the use of lichen size as an indicator of substrate age (Locke *et al.*, 1979). Lichenometry has been most widely used in dating glacial deposits in tundra environments where lichens often form the major vegetation cover and other types of dating methods are inapplicable (Beschel, 1961; Benedict, 1967). The technique may also be used to date lake-level (and perhaps even sea-level) changes, glacial outwash, and trim-lines, rockfalls, talus stabilization, and the former extent of permanent or very persistent snow cover.

4.3.1.1 Principles of Lichenometry

Lichenometry is based on the assumption that the largest lichen growing on a rock substrate is the oldest individual. If the growth rate of the particular species is known, the maximum lichen size will give a minimum age for the substrate, because all other thalli must be either late colonizers or slower growing individuals (i.e., those growing in less than optimum conditions). Lichen size dates the time at which the freshly deposited rocks become stable, because an unstable substrate will prevent uninterrupted lichen growth. Growth rates can be obtained by measuring maximum lichen sizes on substrates of known age, such as gravestones, historic or prehistoric rock buildings, or moraines of known age (perhaps dated independently by historical records or radiocarbon). It is also possible to measure growth directly by photographing or tracing lichens of varying sizes every few years on identifiable rock surfaces (Miller and Andrews, 1973; Ten Brink, 1973). Generally, the maximum diameter of the lichen thallus is measured on individuals that have shown fairly uniform radial growth.

Growth rates vary from one region to another so it is necessary to calibrate the technique for each study site, but the general form of the growth curve is now fairly well established. After initial colonization of the rock surface, growth is quite rapid (known as the great period); growth then slows to a more or less constant rate (Fig. 4.14; Beschel, 1950). Different lichens grow at different rates and indeed some species may approach senescence while other species are still in their great period of growth. The black foliose lichen *Alectoria minuscula,* for example, rarely exceeds 160 mm in diameter on rock surfaces in Baffin Island; lichens of this size represent a substrate age of ~500–600 yr B.P. By contrast, *Rhizocarpon geographicum* has only just entered its period of linear growth by this time (at ~30 mm diameter) and will continue to grow at a nearly constant rate for thousands of years after that. In fact, it has been estimated that a 280 mm thallus of *Rh. geographicum* on eastern Baffin Island dates its substrate at ~9500 ± 1500 yr B.P. (Miller and Andrews, 1973). Similarly, a 480 mm *Rh. alpicola* thallus in the Sarek mountains of Swedish Lapland is thought to have begun its growth following deglaciation of the region ~9000 yr B.P. (Denton and Karlen, 1973b). Different lichens may thus be selected to provide optimum dating resolution over different timescales. However, in view of its ubiquity, ease of recognition, and useful size variation over the last several thousand years, the lichen *Rh. geographicum* has been most commonly used in lichenometrical studies (Fig. 4.15; Locke *et al.*, 1979). Once a growth curve for the species in question has been established, measurements of maximum lichen sizes on moraines and other geomorphological features can be used to estimate substrate age (Fig. 4.16).

4.3.1.2 Problems of Lichenometry

There are three general areas of uncertainty in lichenometry, relating to biological, environmental, and sampling factors (Jochimsen, 1973).

(a) Biological Factors

Lichens are exceedingly difficult to identify to species level in the field and most users of lichenometry have no training in lichen taxonomy. Indeed, lichen taxonomy is itself a contentious subject, which compounds the users' difficulties. *Rhizocarpon geographicum* is exceedingly similar to *Rh. superficiale* and *Rh. alpicola* (King and Lehmann, 1973) and doubtless many investigations have been based on a mixture of observations (Denton and Karlen, 1973b). This presents no problem, of course, providing that the different species grow at similar rates, but generally such factors are not well known. What evidence there is suggests that growth rates may vary between species (Calkin and Ellis, 1980; Innes, 1982). Lichen dispersal and propagation rates

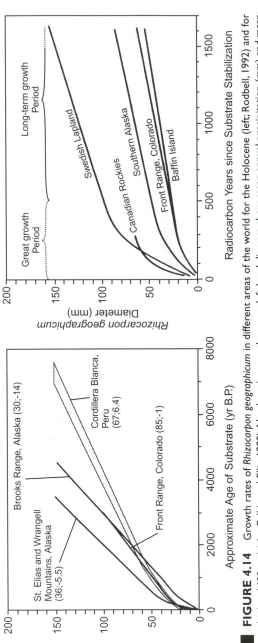

FIGURE 4.14 Growth rates of *Rhizocarpon geographicum* in different areas of the world for the Holocene (left; Rodbell, 1992) and for the last 1600 yr (right; Calkin and Ellis, 1980). Numbers in parentheses on left-hand diagram show mean annual precipitation (mm) and mean annual temperature (°C) in each region. Error bars on each curve are probably ±15–20%.

FIGURE 4.15 *Rhizocarpon geographicum,* a lichen that grows radially, enables substrates to be dated if the growth rate of the lichen is known. This specimen, growing on pyroxene-granulite gneiss in the glacier foreland of Høgvaglbreen (Jotunheimen, southern Norway) is growing at >1.1 mm a^{-1} (based on measurements from 1981–1996). The calipers are open at 10 cm (photograph kindly provided by J. Matthews, University of Wales, Swansea).

are also inadequately understood. Many lichens propagate their algal and fungal cells independently so that it may take some time for two individuals to find each other and form a new symbiotic union. In other cases, lichens are propagated when part of the parent breaks off the rock substrate and is blown or washed away to a new site. In either case, there may be a significant delay between the exposure of a fresh rock surface and colonization by lichens. Furthermore, even when lichen cells become established, decades may elapse before the thallus becomes visible to the naked eye. As time passes, rock surfaces may become virtually covered in lichens and inevitably this results in competition between individuals; indeed some lichens appear to secrete a chemical that inhibits growth in their immediate vicinity (Ten Brink, 1973). Such factors seem likely to reduce growth rates as rocks become heavily lichen-covered and this may give the erroneous impression of a relatively young age for the substrate.

Finally, as lichens become very old, growth rates may decline. Little information is available on senescence in lichens and unfortunately this corresponds to the part of the growth curve where there is the least dating control. Often growth rates beyond a certain age (i.e., the final dated control point) are assumed to continue at a constant rate, whereas in all probability the rate declines with increasing lichen age. This will lead to (possibly large) underestimates of substrate age; such errors can be avoided if extrapolation of growth rates is not attempted.

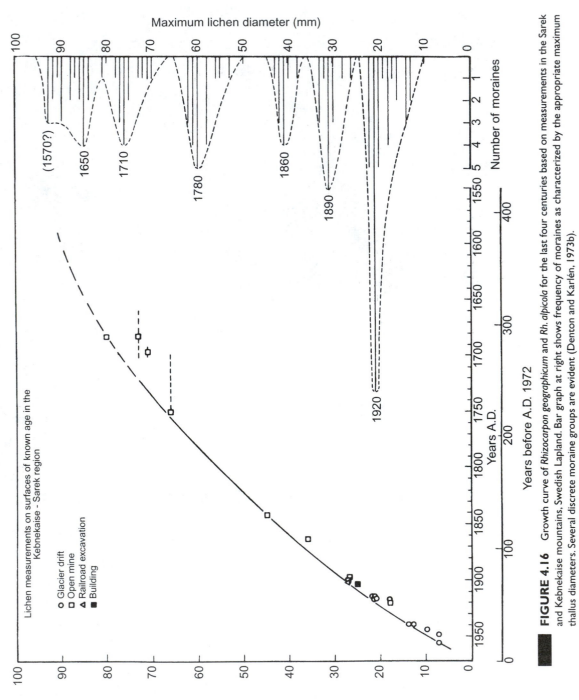

FIGURE 4.16 Growth curve of *Rhizocarpon geographicum* and *Rh. alpicola* for the last four centuries based on measurements in the Sarek and Kebnekaise mountains, Swedish Lapland. Bar graph at right shows frequency of moraines as characterized by the appropriate maximum thallus diameters. Several discrete moraine groups are evident (Denton and Karlén, 1973b).

(b) Environmental Factors

Lichen growth is dependent on substrate type (particularly surface texture) and chemical composition (Porter, 1981c). Rocks that weather easily, or are friable, may not remain stable long enough for a slow-growing lichen to reach maturity. Conversely, extremely smooth rock surfaces may preclude lichen colonization for centuries and possibly many never support lichens. Extremely calcareous rocks may also inhibit growth of certain lichens. Measurements should thus be restricted to lichens growing on similar lithologies whenever possible.

Climate is a major factor affecting lichen growth rates; comparison of growth rates from different areas suggests that slower growth rates are found in areas of low temperature, short growing seasons, and low precipitation (Fig. 4.16). However, both macro- and microclimatic factors are of significance. In particular, lichens require moisture for growth and the frequency of small precipitation amounts, even from fog and dew, may be of more significance than annual precipitation totals. Radiation receipts are also important because they largely determine rock temperatures. Generally it is impossible to equate such factors on those rocks used to calibrate the lichen growth curve with rocks that are eventually to be dated. Commonly, calibration will be carried out on buildings or gravestones in a valley bottom, whereas the features to be dated are hundreds of meters higher than the calibration site. Similar problems may be encountered along extensive fjord systems where conditions at the fjord mouth are less continental than at the fjord head. Lichen growth is far slower in the more continental locations, even over distances as short as 50 km, probably due to the lower frequency of coastal fogs and generally drier climates inland. Increasing elevation also appears to be significant in reducing growth rates, even though moisture availability might be expected to increase (Miller, 1973; Porter, 1981c). Presumably, this is offset by longer-lying snow and a reduced growing season due to lower temperatures (Flock, 1978). All these factors may complicate the construction of a simple growth curve for a limited geographical area. Further problems arise due to the possible influence of long-term climatic fluctuations. Apart from the general effect of lower temperatures in the past, it is quite probable that in the high elevation and/or high latitude sites where lichenometry is most widely used, periods of cooler climate resulted in the persistence of snow banks, which would have reduced lichen growth rates (Koerner, 1980; Benedict, 1993). Growth curves may thus not be linear, but rather made up of periods of reduced growth separated by periods of more rapid growth (Curry, 1969). Lack of resolution in calibration curves may obliterate such variations, but this could account for apparent "scatter" in some attempts at calibration. There is evidence that such factors have been of significance in some regions; on upland areas of Baffin Island, for example, persistence of snow cover during the Little Ice Age is thought to have resulted in "lichen-free zones," where lichen growth was either prevented altogether or severely reduced (Locke and Locke, 1977). These zones can be recognized today, even on satellite photographs, by the reduced lichen cover of the rocky substrate compared to lower elevations where snow cover was only seasonal (Andrews *et al.*, 1976). Similarly, attempts to date moraines that have periodically been covered by snow for long intervals would give erroneously young ages for the deposits (Karlén, 1979).

(c) Sampling Factors

It is of fundamental importance in lichenometric studies that the investigator locates the largest lichen on the substrate in question, but this is not always something one can be certain of doing (Locke *et al.*, 1979). Furthermore, very large lichens are often not circular and may sometimes be mistaken for two individuals that have grown together into one seemingly large and old thallus. It is also possible that a newly formed moraine may incorporate debris from rockfalls or from older glacial deposits; if such debris already supports lichens, and if they survive the disturbance, the deposit would appear to be older than it actually is (Jochimsen, 1973). A number of innovative methods to improve the reliability of lichenometry have been proposed (McCarroll, 1994) and these generally provide a firmer statistical basis for the sampling procedure.

Finally, in establishing a calibrated growth curve for lichens, reference points at the "older" end of the scale are often obtained from a radiocarbon date on organic material overridden by a moraine. This date is then equated with the maximum-sized lichen growing on the moraine today. Such an approach can lead to considerable uncertainty in the growth curve. First, dates on organic material in soils overridden by ice may be very difficult to interpret (Matthews, 1980). Secondly, there may be a gap of several hundred years between the time organic material is overridden by a glacial advance and the time the morainic debris becomes sufficiently stable for lichen growth to take place. This would lead to overestimation of lichen age in a calibration curve. Thirdly, ^{14}C dates must be converted to calendar years, which often results in very large margins of error, especially for the most recent Little Ice Age period, because of the nonlinearity between ^{14}C and calendar ages in this interval. This only amplifies the uncertainties associated with lichen growth curves such as those shown in Fig. 4.14. In reality, each curve should probably have an error bar of around 15–20%, perhaps even more for growth curves that are extrapolated beyond a dated control point (Bickerton and Mathews, 1992; Beget, 1994).

A consideration of all these factors indicates that caution is needed in using lichenometry as a dating method, even for establishing relative age. Nevertheless, if consideration is given to the possible pitfalls, it can provide useful age estimates. Most problems would result in only minimum-age estimates on substrate stability, but in some cases, overestimation of age could result. Although it is worth being aware of the potential difficulties of the method, it is unreasonable to expect that all of the problems, discussed already, would subvert the basic assumptions of the method all of the time, and often the potential errors can be eliminated in various ways. Lichenometry is thus likely to continue to play a role in dating rocky deposits in arctic and alpine areas, and hence make an important contribution to paleoclimatic studies in these regions.

4.3.2 Dendrochronology

Although not a widely used dating technique in paleoclimatology, the use of tree rings for dating environmental changes has proved useful in some cases (Luckman, 1994). The concepts and methods used in dendrochronological studies are discussed in more detail in Chapter 10 and need not be repeated here. Basically, dendrochronological studies are used in three ways: (a) to provide a minimum date for the substrate on which the tree is growing (e.g., an avalanche track or deglaciated surface); (b) to date an event that disrupted tree growth but did not terminate it; and (c) to date the time of tree growth, which was terminated by a glacier advance, or a climatic deterioration associated with a glacier advance (Luckman, 1988, 1995).

The first application is straightforward but to obtain a *close* minimum date assumes that the "new" surface is colonized very rapidly. This is highly probable in the case of avalanche zones (indeed, young saplings may survive the event) but in deglaciated areas surface instability due to subsurface ice melting and inadequate soil structure may delay colonization for several decades. Different approaches towards estimating the time delay before colonization of recently deglacierized terrain are described by Sigafoos and Hendricks (1961) and McCarthy and Luckman (1993). Unlike lichens, which may live for thousands of years, trees found on moraines are rarely more than a few hundred years old. Even when very old trees are located and dated, it cannot be assumed that they represent first generation growth. For example, Burbank (1981) found that a moraine on which the oldest tree was ~750 yr old (dated by dendrochronological methods) was in fact older than 2500 yr B.P. according to the local tephrochronology.

A more widespread application of dendrochronology involves the study of growth disturbance in trees. When trees are tilted during their development they respond by producing compression or reaction wood on the lower side of the tree in order to restore their natural stance. This causes rings to form eccentrically after the event that tilted the tree; the event can be accurately dated by identifying the year when growth changes from concentric to eccentric (Burrows and Burrows, 1976). Such techniques have been used in dating the former occurrence of avalanches (Potter, 1969; Carrara, 1979) and hurricanes (Pillow, 1931) and the timing of glacier recession (Lawrence, 1950). They have also been successfully applied in more strictly geomorphological applications, in studying stream erosion rates and soil movements on permafrost (Shroder, 1980).

Valuable insights into the history of glacier advances in the Canadian Rockies have been obtained by Luckman (1996) using tree snags (partially eroded or damaged trees) or tree stumps, from areas exposed by receding glaciers. By cross-dating these samples, evidence of formerly more extensive forest in areas only recently deglacierized has been documented, and the timing of climatic deterioration and glacier advance has been clearly revealed (Fig. 4.17).

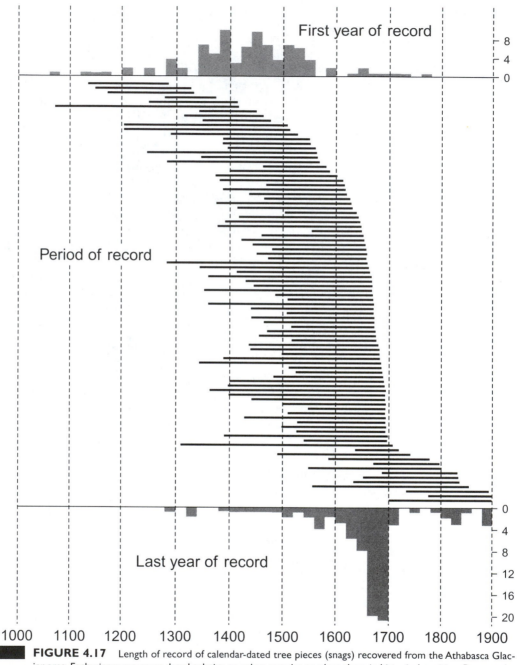

First year of record

Period of record

Last year of record

FIGURE 4.17   Length of record of calendar-dated tree pieces (snags) recovered from the Athabasca Glacier area. Each piece was cross-dated relative to other samples to place them in historical position. Because the
snags found are eroded and may not contain the full growth period, the length of record shown for each sample
may be only a minimum estimate of the life of the tree. Nevertheless, there is clearly a relationship between
termination of growth in many samples and an important Little Ice age glacier advance around A.D. 1714, as
shown by the lower histogram (Luckman, 1994).

5

ICE CORES

5.1 INTRODUCTION

The accumulation of past snowfall in the polar ice caps and ice sheets of the world provides an extraordinarily valuable record of paleoclimatic and paleo-environmental conditions. These conditions are studied by detailed physical and chemical analyses of ice and firn (snow that has survived the summer ablation season[10]) in cores recovered from very high elevations on the ice surface. In such locations (known as the dry snow zone; Benson, 1961) snow melt and sublimation are extremely low so that snow accumulation has been continuous, in some areas for as much as several hundred thousand years (Dansgaard *et al.*, 1973). The snowfall provides a unique record, not

[10] The metamorphism of snow crystals to firn, and eventually to ice, occurs as the weight of overlying material causes crystals to settle, deform, and recrystallize, leading to an overall increase in unit density. When firn is buried beneath subsequent snow accumulations, density increases as air spaces between the crystals are reduced by mechanical packing and plastic deformation until, at a unit density of about 830 kg m^{-3}, interconnected air passages between grains are sealed off into individual air bubbles (Herron and Langway, 1980). At this point, the resulting material is considered to be ice. The depth of this transition varies considerably from one ice body to another, depending on surface temperature and accumulation rate; for example, it does not occur until about 68m depth at Camp Century, Greenland, and ~100 m at Vostok, Antarctica; thus "ice cores" *sensu stricto* are actually firn cores near the surface (see Table 2.2 in Paterson, 1994). This distinction is not very important except in the reconstruction of past atmospheric composition (see Section 5.4.3) and the term ice core will henceforth be used to refer to both ice and firn core sections.

TABLE 5.1   Principal Sources of Paleoclimatic Information from Ice Cores

Parameter	Analysis
Paleotemperatures	
Summer	Melt layers
Annual? Days with snowfall?	δD, $\delta^{18}O$
Humidity	Deuterium excess (d)
Paleo-accumulation (net)	Seasonal signals, ^{10}Be
Volcanic activity	Conductivity, nss. SO_4
Tropospheric turbidity	ECM, microparticle content, trace elements
Wind speed	Particle size, concentration
Atmospheric composition: long-term and man-made changes	CO_2, CH_4, N_2O content, glaciochemistry
Atmospheric circulation	Glaciochemistry (major ions)
Solar activity	^{10}Be

only of precipitation amounts per se, but also of air temperature, atmospheric composition (including gaseous composition and soluble and insoluble particulates), the occurrence of explosive volcanic eruptions, and even of past variations in solar activity (Table 5.1). At present, several dozen cores spanning more than 1000 yr of record have been recovered from ice sheets, ice shelves, and glaciers in both hemispheres (Figs. 5.1 and 5.2). In a number of these cases, cores extend to bedrock and contain debris from the ice/sub-ice interface (Herron and Langway, 1979; Koerner and Fisher, 1979; Gow *et al.*, 1979, 1997). About 15 cores extend back into the last glaciation (Table 5.2) and several of these reached the penultimate glaciation.

Paleoclimatic information has been obtained from ice cores by four main approaches. These involve the analysis of (a) stable isotopes of water and of atmospheric O_2; (b) other gases from air bubbles in the ice; (c) dissolved and particulate matter in firn and ice; and (d) the physical characteristics of the firn and ice (Oeschger and Langway, 1989; Lorius, 1991; Delmas, 1992; Raynaud *et al.*, 1993). Each approach has also provided a means of estimating the age of ice at depth in ice cores (Section 5.3).

5.2 STABLE ISOTOPE ANALYSIS

The study of stable isotopes (primarily deuterium and ^{18}O) is a major focus of paleoclimatic research. Most work has been on stable isotope variations in ice and firn, and in the tests of marine fauna recovered from ocean cores. However, increasing attention is being placed on other natural isotope recorders, such as speleothems

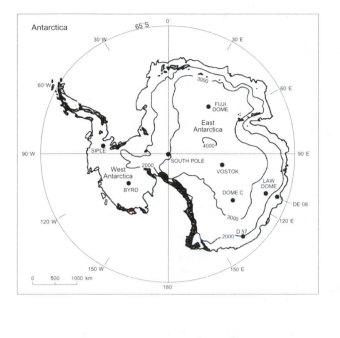

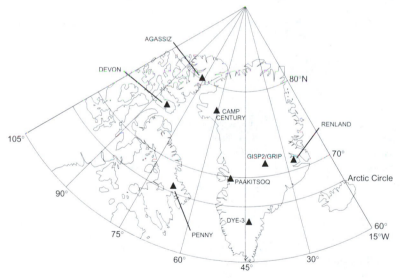

 **FIGURE 5.1** Location of principal ice-coring sites in the Canadian Arctic and Greenland, and in Antarctica.

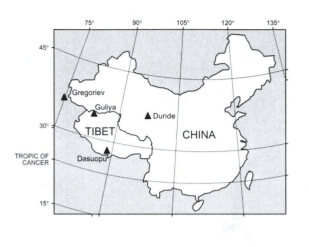

FIGURE 5.2 Location of principal ice-coring sites in low latitudes.

TABLE 5.2 Locations of Ice Cores with Records Going Back into the Last Glaciation[a]

Drill Site	Location[b]	Max. Depth (m)
Camp Century	N.W. Greenland	1387
GISP2 (Summit)	C. Greenland	3053
GRIP	C. Greenland	3029
Dye-3	S. Greenland	2037
Renland	E. Greenland	324
Agassiz	N. Ellesmere Island	338
Devon	Devon Island	299
Barnes	Baffin Island	
Penny	Baffin Island	334
Byrd	West Antarctica	2164
J9 (Ross ice shelf)	West Antarctica	
Dome C	East Antarctica	905
Vostok	East Antarctica	3350
Law Dome	East Antarctica	1203
Taylor Dome	East Antarctica	375
Dome Fuji	East Antarctica	2500
Dunde	Western China	140
Guliya	Western China	309
Huascarán	Peru	166
Sajama	Bolivia	133
Dasuopu	Western China	168

[a] Other short cores (<100 m) and surface samples from ice sheet margins in Greenland and Antarctica have also recovered ice from the last glacial period.

[b] See Figures 5.1 and 5.2 for locations.

(stalactites and stalagmites), tree rings, ostracods, and peat (Swart *et al.,* 1993). In this section a brief introduction to the theory behind stable isotope work is provided and applications to ice-core analysis are discussed. The importance of stable isotopes in other branches of paleoclimatic research are dealt with in Chapters 6, 7, and 10.

Water is the most abundant compound on Earth. The primary compound in all forms of life, it is perhaps the most important agent in weathering, erosion, and geological recycling of materials, and, of course, plays a crucial role in the global energy balance. The study of "fossil water," either directly in the form of firn and ice, or indirectly through materials deposited from solution in "fossil water" (e.g., speleothems) thus has important implications in many aspects of paleo-environmental reconstruction.

In common with most other naturally occurring elements, the constituents of water, oxygen, and hydrogen may exist in the form of different isotopes. Isotopes result from variations in mass of the atom in each element. Every atomic nucleus is made up of protons and neutrons. The number of protons in the nucleus of an element (the atomic number) is always the same, but the number of neutrons may vary, resulting in different isotopes of the same element. Thus, oxygen atoms (which always have 8 protons) may have 8, 9, or 10 neutrons, resulting in three isotopes with atomic mass numbers of 16, 17, and 18, respectively (^{16}O, ^{17}O, and ^{18}O). In nature these three stable isotopes occur in relative proportions of 99.76% (^{16}O), 0.04% (^{17}O), and 0.2% (^{18}O). Hydrogen has two stable isotopes, ^{1}H and ^{2}H (deuterium) with relative proportions of 99.984% and 0.016%, respectively. Consequently, water molecules may exist as any one of nine possible isotopic combinations with mass numbers ranging from 18 ($^{1}H_2{}^{16}O$) to 22 ($^{2}H_2{}^{18}O$). However, as water with more than one "heavy" isotope is very rare, generally only four major isotopic combinations are common, and only two are important in paleoclimatic research ($^{1}H^{2}H^{16}O$, generally written as HDO, and $^{1}H_2{}^{18}O$).

The basis for paleoclimatic interpretations of variations in the stable isotope content of water molecules is that the vapor pressure of $H_2{}^{16}O$ is higher than that of $HD^{16}O$ and $H_2{}^{18}O$ (10% higher than HDO, 1% higher than $H_2{}^{18}O$). Evaporation from a water body thus results in a vapor that is poorer in deuterium and ^{18}O than the initial water; conversely, the remaining water is (relatively speaking) *enriched* in deuterium and ^{18}O. At equilibrium, for example, atmospheric water vapor contains 10 ‰ (parts per thousand or per mil) less ^{18}O and 100 ‰ less deuterium than mean ocean water. When condensation occurs, the lower vapor pressure of HDO and $H_2{}^{18}O$ results in these two compounds passing from the vapor to the liquid state more readily than water made up of lighter isotopes. Hence, *compared to the vapor,* the condensation will be enriched in the heavy isotopes (Dansgaard, 1961). Further condensation of the vapor will continue this preferential removal of the heavier isotopes, leaving the vapor more and more depleted in HDO and $H_2{}^{18}O$ (Fig. 5.3). As a result, continued cooling will give rise to condensate with increasingly lower HDO and $H_2{}^{18}O$ concentrations than when the condensation process first began. The greater the fall in temperature, the more condensation will occur and the lower will be the heavy isotope concentration, relative to the original water

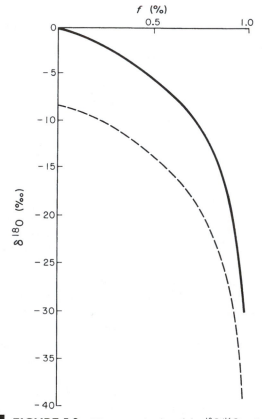

FIGURE 5.3 Diagrammatic plot of the $^{18}O/^{16}O$ ratio ($\delta^{18}O$) in a water (liquid-vapor) system in which the liquid is removed as it is formed by progressive condensation. Isotopic equilibrium between vapor and liquid is assumed (i.e., an equilibrium Rayleigh condensation process). ----, δ vapor; ———, δ liquid; f = percentage of the original water vapor condensed (Epstein and Sharp, *Journal of Geology,* **67**, © 1959 by the University of Chicago).

source (Fig. 5.4). Isotopic concentration in the condensate can thus be considered as a primary function of the temperature at which condensation occurs (subject to certain reservations to be noted in Section 5.2.2).

5.2.1 Stable Isotopes in Water: Measurement and Standardization

In the majority of paleoclimatic studies using stable isotopes, oxygen is generally the element of primary interest, though deuterium is important in ice core research. In oxygen isotope work, the water sample is isotopically exchanged with carbon dioxide of known isotopic composition:

$$^{1}H_2{}^{18}O + {}^{12}C^{16}O_2 \rightleftharpoons {}^{1}H_2{}^{16}O + {}^{12}C^{16}O^{18}O$$

The relative proportions of ^{16}O and ^{18}O in carbon dioxide from the sample are then compared with the isotopic composition of a water standard (Standard Mean

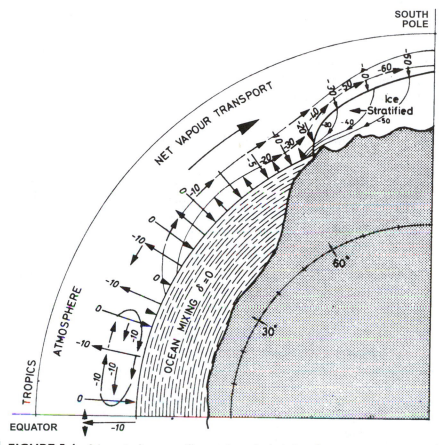

FIGURE 5.4 Schematic diagram to illustrate isotopic depletion of water vapor en route to the Antarctic ice sheet. As an air mass cools, precipitation produced is preferentially enriched in ^{18}O, leaving the remaining vapor relatively depleted. Consequently, with further condensation, the precipitation contains less and less ^{18}O (i.e., lower $\delta^{18}O$ values). This isobaric effect is accentuated by uplift (adiabatic) effects over the ice sheet itself, so that the lowest delta ^{18}O values are found in the ice-sheet interior (Dansgaard *et al.*, 1971; Robin, 1977).

Ocean Water or SMOW[11]) and the results expressed as a departure ($\delta^{18}O$) from this standard, thus

$$\delta^{18}O = \frac{(^{18}O/^{16}O)\text{sample} - (^{18}O/^{16}O)\text{SMOW}}{(^{18}O/^{16}O)\text{SMOW}} \times 10^3 \,‰$$

[11] In order that isotopic analyses in different laboratories be comparable, a universally accepted standard is used, known as SMOW (Standard Mean Ocean Water; Craig, 1961b). This is not an actual oceanic water sample, but is based on a US National Institute of Standards and Technology distilled water sample (NIST-1). However, the zero point on the SMOW scale has been adjusted so that it is more or less equivalent (−0.1‰) to the isotopic composition of real ocean water (measured in samples from depths of 200–500 m in the Atlantic, Pacific, and Indian Oceans; Epstein and Mayeda, 1953). Isotopic studies based on carbonate fossils use as a standard a Cretaceous belemnite from the Peedee Formation of North Carolina (PDB-1). Carbon dioxide released from PDB-1 is ≈ + 0.2‰ relative to CO_2 equilibrated with SMOW (Craig, 1961b). Recent updates to international reference standards are described by Coplen (1996).

All measurements are made using a mass spectrometer and reproducibility of results within ± 0.1‰ is generally possible.

A $\delta^{18}O$ value of –10 therefore indicates a sample with an $^{18}O/^{16}O$ ratio 1% or 10‰ less than SMOW. Under our present climate, the lowest $\delta^{18}O$ value recorded in natural waters is ~–58‰ (–454‰ in δD) in snow from the highest and most remote parts of Antarctica (Qin *et al.*, 1994).

5.2.2 Oxygen-18 Concentration in Atmospheric Precipitation[12]

In Section 5.2.1, and Fig. 5.3, the isotopic composition of water in equilibrium with water vapor was considered. In reality, we cannot consider the process to be always at equilibrium between vapor and condensate, nor can the process be considered to occur in isolation. Exchanges between atmospheric water vapor, water droplets in the air, and water at the surface (which may be isotopically "light") do occur continuously, so this complicates any simple temperature-isotope effect that we might expect to find (Koerner and Russell, 1979). There are also kinetic effects on fractionation that occur during evaporation and condensation, and the latter can be especially important at very low temperatures (see Section 5.2.5). Overall, the ^{18}O content of precipitation depends on

(a) the ^{18}O content of the water vapor at the start of condensation (this could be very low if evaporation occurred over an inland lake or ice body where ^{18}O concentrations are less than mean ocean water);

(b) the amount of moisture in the air compared to its initial moisture content;

(c) the degree to which water droplets undergo evaporation en route to the ground and whether any of this re-evaporated vapor re-enters the precipitating air mass (Ambach *et al*, 1968);

(d) the temperature at which the evaporation and condensation processes take place; and

(e) the extent to which clouds become supersaturated, with respect to ice, at very low temperatures.

In spite of these complications, empirical studies have demonstrated that geographical and temporal variations in isotopes do occur, reflecting temperature effects due to changing latitude, altitude, distance from moisture source, season, and long-term climatic fluctuations (Dansgaard *et al.*, 1973; Koerner and Russell, 1979; Petit *et al.*, 1991). As any interpretation of ice-core isotopic records is rooted in an evaluation of these factors, it is important to consider them in more detail.

5.2.3 Geographical Factors Affecting Stable Isotope Concentrations

For the last 30 years or so, precipitation samples from many locations throughout the world have been analyzed for their $\delta^{18}O$ content (Rozanski *et al.*, 1992, 1993). Figure 5.5 shows that $\delta^{18}O$ values of January and July precipitation generally reflect the distribution of temperature, decreasing at higher latitudes and at higher elevations

---

[12] In this section, $\delta^{18}O$ is generally referred to, but the same principles apply to variations in δD.

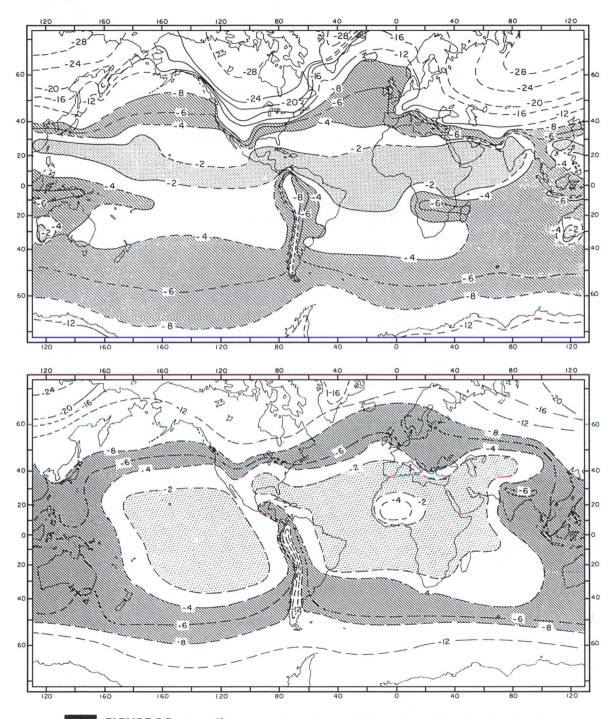

FIGURE 5.5 Mean δ¹⁸O in January (upper figure) and July precipitation (lower figure) collected at precipitation stations throughout the world, based on analyses by the International Atomic Energy Authority over the last few decades (Lawrence and White, 1991).

(e.g., in the high interior parts of Greenland and in the Andes). The influence of the Gulf Stream is also apparent on the January map (Lawrence and White, 1991). Changes in temperature from winter to summer are also reflected in $\delta^{18}O$, and this leads to an annual cycle in $\delta^{18}O$ of snowfall that can be used to count annual accumulation layers in ice cores (see Section 5.3.2).

Figure 5.5 indicates a strong latitudinal influence on $\delta^{18}O$. Lower $\delta^{18}O$ values are found at higher latitudes as a result of the loss of heavy isotopes in water condensed en route to those regions. This is sometimes referred to as an isobaric effect, implying a systematic change brought about by overall cooling at a particular level in the atmosphere, rather than cooling brought about by a change in elevation (adiabatic cooling). With increasing elevation, adiabatic cooling of the precipitating air mass leads to precipitation that is more and more depleted in ^{18}O due to preferential removal of the heavier isotope in the condensation process. For example, the Quelccaya Ice Cap in Peru receives moisture that has undergone adiabatic cooling as the air rises over the Andes from the Amazon Basin. This uplift reduces $\delta^{18}O$ in precipitation by ~11‰ (Grootes *et al.*, 1989). On large ice sheets, the adiabatic effect is superimposed on a "distance from moisture source" factor that results in lower $\delta^{18}O$ concentrations as the distance from oceanic moisture sources increases (Koerner, 1979). Hence, at high elevations in central Antarctica, thousands of kilometers from the southern oceans, atmospheric precipitation has the lowest heavy isotope concentrations of any natural water occurring today (Morgan, 1982; Qin *et al.*, 1994).

These different influences on $\delta^{18}O$ in precipitation lead to the geographical patterns seen in Fig. 5.5. There is a very strong correspondence between temperature and $\delta^{18}O$ in extra-tropical regions (Fig. 5.6). The overall $\delta^{18}O$-surface temperature

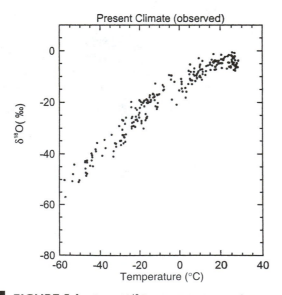

FIGURE 5.6   Annual $\delta^{18}O$ in precipitation in relation to mean annual temperature at the same site, based on data from the International Atomic Energy Authority (Jouzel *et al.*, 1994).

relationship for locations in the extra-tropics (regions with mean annual temperatures <15 °C) is

$$\delta^{18}O = 0.64T - 12.8$$

but this varies geographically (Jouzel *et al.*, 1987b; 1994). In the colder regions of Antarctica, the slope of this relationship is greater (~0.8, on average). At low latitudes, $\delta^{18}O$ is not related to temperature, but is more a function of precipitation amount (Dansgaard, 1964; Rozanski *et al.*, 1993). Significant departures from the regression may be expected in certain circumstances (Hage *et al.*, 1975): (a) if precipitation occurs in an area where very stable (i.e., inversion) conditions are common, surface temperatures will be lower than expected from the regression (and conversely, $\delta^{18}O$ values will appear anomalously high); and (b) if local precipitation is derived from water that was re-evaporated from a source with an already low $\delta^{18}O$ content (e.g. freshwater lake or snow cover) the mean annual $\delta^{18}O$ value may fall below the regression line, perhaps by as much as 10‰ for complete re-evaporation of precipitated water) (Koerner and Russell, 1979).

5.2.4 Calibrating $\delta^{18}O$ for Paleotemperature Reconstruction

A number of studies in Antarctica have shown that there is a strong relationship between $\delta^{18}O$ of snowfall and temperature on a daily (storm event) basis. However, strong surface-based temperature inversions essentially decouple the surface from the atmospheric circulation above the ice cap; temperatures are often much lower at the surface (Φ_g) than at the top of the inversion (Φ_i) and on average, $\Phi_i = 0.67\Phi_g - 1.2$ (Jouzel and Merlivat, 1984). Hence, there is a much stronger relationship between Φ_i (or cloud temperatures above the inversion) and $\delta^{18}O$, than between surface temperature and the isotopic content of snowfall. This was first demonstrated by Picciotto *et al.* (1960), who found a relationship between the mean temperature within precipitating clouds at King Baudouin Station, Antarctica (Fig. 5.7) and the isotopic composition of snowfall at the surface ($\delta^{18}O = 0.9T + 6.4$, where T is in the range +5 °C to –30 °C). Subsequently, Aldaz and Deutsch (1967) conducted a similar study at the South Pole, in which isotopes in snow samples collected during the course of a year were compared with temperatures at the surface and up to 500 mb. They found a relationship between $\delta^{18}O$ values and condensation level temperatures (t) such that $\delta^{18}O = 1.4t + 4.0$ (where t is in the range –25 to –50 °C). These results were confirmed by Jouzel *et al.* (1983), who found that correlations between mean annual δD and surface temperature at the South Pole were not as good as with temperatures just above the inversion layer.

These process-based studies are important in testing the statistically based relationships evident in long-term measurements from around the globe (see Fig. 5.6) with data from the polar ice sheets. However, with few exceptions (Steffensen, 1985) studies of $\delta^{18}O$ in polar regions rely on the *spatial* relationship, which has been observed between $\delta^{18}O$ in surface snowfall and mean annual temperature derived from 10-m depths (where the annual temperature cycle has been damped to near zero). Such data are derived from a geographically extensive network of ice

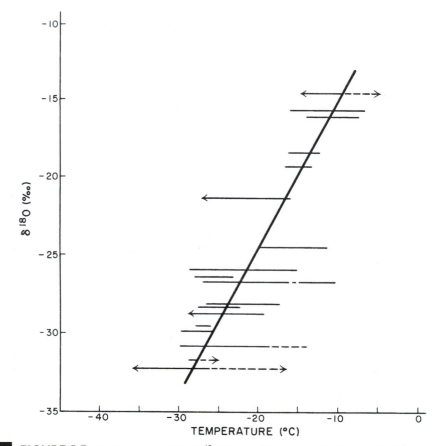

FIGURE 5.7   Isotopic composition ($\delta^{18}O$) in snowfall compared to the corresponding temperature in the precipitation cloud (Picciotto et al., 1960).

cap sites. In looking at mean annual or mean monthly temperatures and corresponding $\delta^{18}O$ values, the difficult problems associated with the development of precipitation and the processes occurring in clouds on a storm-to-storm basis are avoided. In effect, it is assumed that mean condensation temperature and mean annual temperature vary in parallel. Why this should be so is hard to understand; most snowfall on polar ice sheets results from a small number of synoptic events occurring on only a fraction of days per year (generally <50%) so *mean annual* temperature, which is greatly influenced by strong surface inversions in dry winter months, should have little in common with $\delta^{18}O$ values in the ice cores (Peel *et al.*, 1988). Nevertheless, empirical observations do show that mean annual $\delta^{18}O$ values and mean annual temperature are strongly correlated in the *spatial domain*, as shown in Fig. 5.6. Whether this relationship can be applied to the *temporal domain*, to convert variations in $\delta^{18}O$ or δD over time to changes in mean annual temperature, is an important issue (and one that also applies to many other paleoclimate proxies). Cuffey *et al.* (1994, 1995) used borehole temperatures at GISP2 to examine this question. The down-hole temperature profile represents the thermal history of the site that is, in a sense, buried with the accumulating snow. The down-core $\delta^{18}O$ record represents another measure of that thermal history. By developing a model that optimizes the fit between these two records, Cuffey *et al.* were able to resolve the long-term relationship (over the past 600 yr) between $\delta^{18}O$ and temperature ($\delta^{18}O = 0.53T - 18.2‰$). However, longer-term changes (over the last glacial-interglacial cycle) using a deeper borehole record produced a solution of $\delta^{18}O = 0.33T - 24.8‰$. These differences must be considered in terms of the temporal scale, or frequency domain of interest. The large range in estimates of the slope (*a*) of the $\delta^{18}O$-temperature relationship may be related to the various timescales being considered. On the timescale of individual storm events $a = \geq 1$; for monthly to annual values, $a = \sim 0.65$, and even lower values of *a* seem to be appropriate when considering changes over longer timescales. Thus, in comparing changes averaged over decades to millennia, $\delta^{18}O \approx 0.5T$ and over even longer periods of time (several to tens of millennia) $\delta^{18}O \approx 0.35T$ (Boyle, 1997). On this basis, Cuffey *et al.* (1995) interpreted the *overall* glacial to interglacial temperature change at GISP2 as +14–16 °C, considerably greater than previous estimates (which had relied on $\delta^{18}O \approx 0.65T$). For short-term (abrupt) changes in $\delta^{18}O$, higher values of *a* are probably more appropriate; thus Dansgaard *et al.* (1989) used $a = 0.65$ when interpreting abrupt changes in $\delta^{18}O$, giving mean annual temperature changes of ~ 7 °C within 50 yr. Such an interpretation implicitly assumes that the factors that influence modern patterns operated essentially unchanged (or at least with the same overall gradients) as they do today; changes in moisture source, or in the seasonal distribution of precipitation are not considered to be important. However, a change in the timing of precipitation could be particularly important, as mean temperatures change in the Spring and Fall by several degrees Celsius *per week*; a shift in precipitation events by only a few weeks could result in large changes in $\delta^{18}O$ without any real change in mean annual temperature (Steig *et al.*, 1994). Similarly, changing source regions for precipitation could also account for the abrupt isotopic shifts (Charles *et al.*, 1994; Kapsner *et al.*, 1995). Indeed, the rapid changes seen in Greenland ice cores during the last glaciation are associated with changes in other parameters (e.g., dust and Ca^{++} levels), which suggests that the precipitation source regions did vary.

Many other factors affecting $\delta^{18}O$ at a location today have not been constant over time. During glacial periods, ice thicknesses gradually increased on many ice sheets, resulting in lower $\delta^{18}O$ values at the surface because of the increase in elevation. However, this may not have been the case everywhere; there is evidence that both the GISP2 and Vostok sites were lower in the LGM (due to lower accumulation rates during colder times) so this could have led to less adiabatic cooling and higher $\delta^{18}O$ levels (Lorius *et al.*, 1984; Cuffey and Clow, 1997). More extensive sea ice during glacial periods would effectively have increased distance to moisture sources, leading to lower $\delta^{18}O$ values in isolated continental interiors (Kato, 1978; Bromwich and Weaver, 1983). Furthermore, during glacial periods the isotopic composition of ocean water itself changed (a $\delta^{18}O$ value of $\sim 1.1‰$ higher than today) due to the storage of water depleted in ^{18}O in large ice sheets (Labeyrie *et al.*, 1987; Shackleton, 1987). Finally, as temperatures fell to very low levels in some regions, $\delta^{18}O$ would decrease more rapidly for a given drop in temperature, because of the curvilinear nature of the $\delta^{18}O$-temperature relationship (Fig. 5.8). Any interpretation of isotopic values in ice cores must consider all these factors, which undoubtedly have had an effect on the isotopic composition of high-latitude precipitation over time.

One approach to understanding how these various factors have interacted to influence the isotopic content of precipitation in the past is to use general circulation models (GCMs) with isotopic tracers in the hydrological cycle (Joussaume *et al.*, 1984). Results from simulations with modern boundary conditions compare very well with observations (Jouzel *et al.*, 1987b, 1991). Running the models with glacial age boundary conditions suggests that there was little change in $\delta^{18}O$ equatorward of 40° N or 50° S at the last glacial maximum (LGM) but that there were large decreases in $\delta^{18}O$ at higher latitudes (Fig. 5.9) (Joussaume and Jouzel, 1993; Jouzel *et al.*, 1994; Charles *et al.*, 1994).

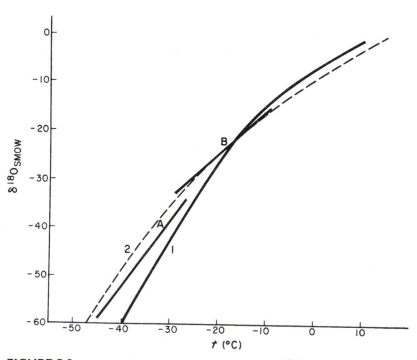

FIGURE 5.8 Relationship between oxygen isotope ratio ($\delta^{18}O$) and temperature of condensation level in samples of Antarctic precipitation. Curves A and B are based on empirical observations at King Baudoin base and Amundsen-Scott Station. Curves 1 and 2 are theoretical using different assumptions about the fractionation of ^{18}O at very low temperatures (Aldaz and Deutsch, 1967).

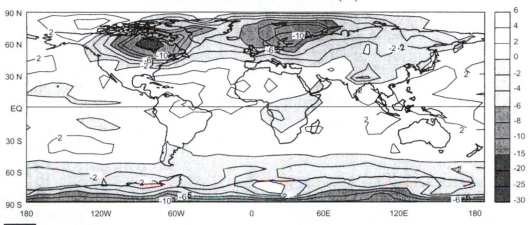

FIGURE 5.9 Differences between modern observed $\delta^{18}O$ in annual precipitation (using the same data as in Figs. 5.5 and 5.6) and simulated values for the last glacial maximum (LGM). Largest differences are associated with the polar ice sheets; in the intertropical zone, LGM values are actually higher than modern values, possibly reflecting a direct effect of higher $\delta^{18}O$ values in the ocean at that time. However, the tropical values are inconsistent with $\delta^{18}O$ in LGM groundwater in North Africa. Further data are needed to resolve this discrepancy (Jouzel *et al.*, 1994).

5.2.5 Deuterium Excess

On a global scale, fractionation of oxygen and hydrogen during evaporation and precipitation processes approximates a well-defined relationship, whereby $\delta D = 8\delta^{18}O + 10$. This defines what is termed the *meteoric water line* (Craig, 1961c), which in effect characterizes the "normal" equilibrium conditions that exist between $\delta^{18}O$ and δD. The offset value (10 in this case) is termed the deuterium excess (d, where $d = \delta D - 8\delta^{18}O$). However, deuterium excess varies under non-equilibrium conditions, providing information not available from $\delta^{18}O$ and δD alone.

Values of d vary because of kinetic effects occurring at both the evaporation and condensation stages of the water cycle. During evaporation, higher rates of diffusion to the water surface of molecules containing light isotopes leads to an increase in the light isotope content of water vapor, relative to the water source. This effect is in addition to the fractionation effect caused by the lower vapor pressure of water containing heavier isotopes (discussed in the preceding section). However, because of the different masses of HDO and $H_2^{18}O$, the kinetic effect is slightly greater for $H_2^{18}O$ than for HDO, resulting in changes of d when conditions deviate from equilibrium. Thus, when there is strong mixing of the surface waters (by higher wind speeds) or when relative humidities increase (reducing evaporation rates) or when water temperatures decrease (also reducing evaporation rates) the kinetic effect is reduced and values of d in precipitation will be lower. Thus, Jouzel *et al.* (1982) interpreted low values of deuterium excess (d = 4‰) in the pre-Holocene section of an ice core from Dome C, Antarctica, as indicative of higher relative humidities and/or higher wind speeds in the water vapor source regions than during the Holocene (when d = 8‰). Increased levels of Na^+ (from sea salt) and aeolian dust confirmed that wind speeds and humidities were probably higher in glacial time than in the Holocene.

A model of the isotopic fractionation of precipitation, based on the Rayleigh distillation process described earlier, works well at temperatures above ~ –10 °C, but under the colder conditions common in Greenland and Antarctica, the simple model cannot explain the observed δD – $\delta^{18}O$ relationship, or the isotope-temperature gradients. At such low temperatures, clouds become supersaturated with respect to ice and the transition from water vapor in clouds directly to ice crystals is the principal mode of precipitation formation. However, the lower molecular diffusivity of water molecules containing heavier isotopes (HDO, $H_2^{18}O$) leads to the preferential condensation of isotopically light water molecules on the ice crystals. This is analogous to the kinetic fractionation effect that occurs at the ocean surface during the process of evaporation. By taking this additional kinetic effect into account, a much better fit between modeled and observed deuterium excess values, and a realistic isotopic-temperature relationship is obtained for snow forming at very low temperatures (Jouzel and Merlivat, 1984). Unfortunately, snowfall deposited on polar ice sheets has generally undergone a complex history; the transition temperature at which clouds become primarily made up of ice crystals rather than water vapor is not well known, yet this significantly influences the value of d (Fisher 1991, 1992).

Models of isotopic fractionation taking into account kinetic effects during evaporation and snow formation can be used to constrain the uncertainties inherent in these processes, providing good simulations of observed variations (Jouzel and Merlivat, 1984; Petit *et al.*, 1991; Fisher, 1992). Petit *et al.* (1991), for example, show that the observed relationship between d and δD in East Antarctica could only be accounted for if initial sea surface temperature at the air mass source region was between 15 and 22 °C, corresponding to latitudes 30–40° S (Fig. 5.10). Changes in humidity in the source region are of secondary importance, but are of more significance at higher values of δD (warmer, lower elevation sites in Antarctica). Of course, it is unrealistic to expect that the moisture source for Antarctic snow is confined to only one latitudinal band, and this factor has also been modeled. Simulations in which moisture was added to the air mass as it passed from 30–40° S to Antarctica give results that are consistent with observations, suggesting that up to 30% of moisture originates from areas with SSTs <10 °C (Petit *et al.*, 1991). This model thus provides strong support that mid-latitudes are the primary moisture source for at least the interior of Antarctica, with smaller contributions from regions at higher latitudes.

On the other hand, experiments with the NASA/GISS general circulation model incorporating deuterium (which can be traced back to each ocean basin) suggest that the average source region SST for Antarctic mid-winter precipitation is considerably lower, in the range 9–14 °C (Koster *et al.*, 1992). Further experiments on total annual precipitation may help to resolve these differences. Whatever the solution

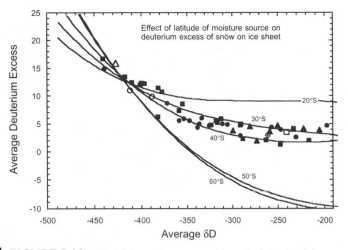

FIGURE 5.10 Model-derived estimates of how the latitude of the oceanic moisture source region influences the average δD and d values in Antarctic precipitation (solid lines) compared to the observed δD-d relationship in Antarctica (symbols — based on surface snow samples across a wide geographical region). Each model line uses a different supersaturation function (relating the supersaturation of water vapor [with respect to ice] to the snow surface temperature) optimized to produce the best fit with observed data. The latitudes 30–40° S correspond to SSTs of ~15–21° S (Petit *et al.,* 1991).

is, small seasonal variations in d (± 5‰) in snowfall at the South Pole seem to indicate that the process of delivering snow to this location involves fairly consistent pathways from the moisture source, year-round. Similar conclusions were reached by Johnsen *et al.* (1989) for the higher elevations of Greenland, though Fisher (1992) found that the isotopic content of snow at the summit region of the ice sheet (Crête) was not compatible with a single moisture source, and that a mixture of moisture from the east (2/3) and west (1/3) seemed probable. This conclusion is similar to that reached by Charles *et al.* (1994) using a GCM with isotopic tracers in the model's hydrological cycle. When glacial age boundary conditions were imposed on the model, source regions changed; the southern part of the ice sheet was dominated by North Atlantic moisture sources, and the northern part by North Pacific moisture. With the incorporation of isotopic tracers into other general circulation models, it should be possible to gain further insights into fractionation processes occurring today, providing greater confidence in the interpretation of past changes observed in ice core records (Jouzel *et al.,* 1993a).

Variations in deuterium excess in relation to $\delta^{18}O$ and δD also shed light on the abrupt changes observed in Greenland ice cores during the LGM. Unlike the situation in the Antarctic Dome C record, during the coldest episodes of the LGM, d was no lower than in the Holocene (d = 8‰), suggesting that conditions in the moisture source were no different than today, with most of the moisture originating in the subtropical Atlantic. However, when there were abrupt shifts to warmer conditions (higher values of $\delta^{18}O$) d became lower (by 4–5‰). Temperatures in the moisture source region must therefore have been *lower* in the milder episodes (and/or humidities and/or wind speeds were higher). Johnsen *et al.* (1989) explain this counter-intuitive conclusion by suggesting that the milder periods of the LGM were associated with an abrupt shift in oceanic conditions in which the sea ice boundary rapidly retreated northward, revealing colder waters that acted as a local moisture source near to the ice sheet. Subsequently, as water temperatures increased, deuterium excess values became higher.

Variations in d will be better understood when a transect or network of ice cores across the major ice sheets becomes available, because unique explanations are generally not possible with only one record. Nevertheless, it is clear that a consideration of deuterium excess together with δD and $\delta^{18}O$ will provide new insights into paleoclimatic conditions which cannot be obtained from dD or $\delta^{18}O$ alone.

5.3 DATING ICE CORES

One of the most important problems in any ice-core study is determining the age-depth relationship. Many different approaches have been used and it is now clear that very accurate timescales can generally be developed for *at least* the last 10,000–12,000 yr if accumulation rates are high enough. Prior to that, there is increasing uncertainty about the age of ice, but new approaches are constantly improving age estimates, allowing comparisons with other proxy records to be made with more confidence (see Section 5.4.5). Furthermore, many of the methods that have been investigated in order to improve the dating of ice cores have themselves produced important paleoclimatic information. Some of the principal methods used and their paleoclimatic implications are now reviewed.

5.3.1 Radioisotopic Methods

Several different radioactive isotopes have been analyzed in ice cores in an attempt to provide quantitative chronological methods for dating ice. These include: ^{10}Be, ^{14}C, ^{36}Cl, ^{39}Ar, ^{81}Kr, and ^{210}Pb (Stauffer, 1989). At present, however, apart from ^{210}Pb and ^{14}C analysis, radioisotopic dating of ice and firn is not a routine operation and other stratigraphic techniques are generally preferred.

The ^{210}Pb (half-life: 22.3 yr) is washed out from the atmosphere as a decay product of ^{222}Rn (see Fig. 3.21). It has been used successfully in studies of snow accumulation over the last 100–200 yr, providing an important perspective on the very short accumulation records otherwise available in remote parts of Antarctica and Greenland (Crozaz *et al.*, 1964; Dibb and Clausen, 1997). The AMS ^{14}C dates on CO_2 enclosed in air bubbles in ice can be obtained from ice samples as small as 10 kg (equivalent to a conventional ice core ~1.5 m in length) though precision is improved with larger samples (Andrée *et al.*, 1986). Unfortunately, the dates on CO_2 obtained may differ from the age of the enclosing ice by hundreds, or thousands, of years because of the time delay before gas bubbles become entirely sealed from the atmosphere (see Section 5.4.3). This problem limits the value of ^{14}C dates on ice core samples.

5.3.2 Seasonal Variations

Certain components of ice cores show quite distinct seasonal variations, which enable annual layers to be detected. These can then be counted to provide an extremely accurate timescale for as far back in time as these layers can be detected. Where uncertainties exist in one seasonal chronology, a comparison with other parameters enables accurate cross-checking to be accomplished, thereby reinforcing confidence in the timescale produced (Hammer *et al.*, 1978). For example, annual layer counts (back to 17,400 yr ago) have been carried out on the GISP2 ice core from Summit, Greenland, using a combination of, *inter alia*, visual stratigraphy, electrical conductivity measurements (ECM), laser light scattering (from dust) oxygen isotopes, and chemical variations in the ice (Meese *et al.*, 1995, 1997). When compared to the independently derived chronology from the nearby GRIP ice core, the two records match to within 200 yr back to 15,000 calendar yr B.P. (Taylor *et al.*, 1993a).[13] However, at greater depths the counts diverge significantly as the difficulty of unequivocally identifying annual layers increases. In this section, the different types of information used in layer counting are discussed.

Visual stratigraphy: visual stratigraphy provides a "first cut" at identifying annual increments in an ice core. Cores are examined on a light table to identify changes in crystal structure and the presence of dust layers. In the GISP2 ice cores, a distinctive coarse-grained depth hoar layer, characteristic of each summer, can be seen (Alley *et al.*, 1997a). In cores from the Quelccaya ice cap, Peru, a pronounced dust layer, which is diagnostic of conditions from May–August, permits the counting of annual layers (Thompson *et al.*, 1985).

$\delta^{18}O$: Because of the greater cooling that occurs in winter months, much lower $\delta^{18}O$ concentrations are found in winter snow than in summer snow. This results in a very strong seasonal signal that can be used as a chronological tool, providing accumulation rates are reasonably high (>25 cm water equivalent per year), wind scouring of snow is not severe, and no melting and refreezing of snow and firn has occurred. In effect, the annual layer thickness can be identified by counting each couplet of high and low $\delta^{18}O$ values from the top of the core downward (Fig. 5.11). Unfortunately, at increasing depths in polar ice sheets the amplitude of the seasonal signal is reduced until it is eventually obliterated. In the upper layers, where density is <0.55 g cm^{-3}, this results from isotopic exchange between water vapor and firn. In lower, denser layers, where air channels are closed off, obliteration results from diffusion of water molecules within the ice. This process is accelerated due to thinning by plastic deformation as the annual layers approach bedrock; thinning increases isotopic gradients in the ice, making molecular diffusion more effective in obliterating the seasonal variations (see Fig. 5.11).

In cores where seasonal isotopic differences are still preserved down to dense firn and ice layers, further smoothing due to molecular diffusion is so slow that the signal may then be preserved for thousands of years. This does not occur in most of

[13] For the GISP2 ice core, multi-parameter dating cross-checked with various independent reference horizons, suggests that the age of the ice is known to within <1% for the last 2000 yr, increasing to 2% by ~40,000 yr B.P., to 10% by 57,000 yr B.P. and up to 20% by ~110,000 yr B.P. (Meese *et al.*, 1997).

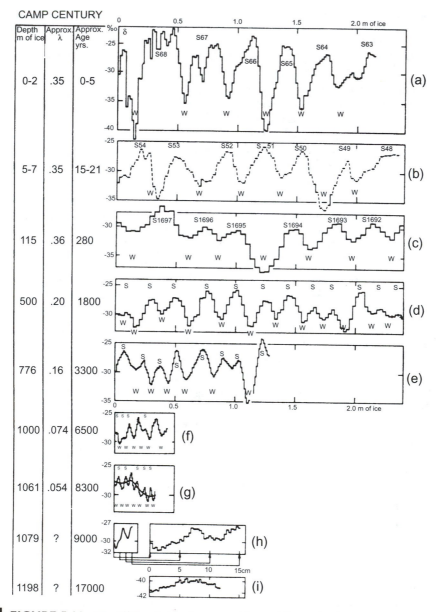

FIGURE 5.11   The $\delta^{18}O$ variations in snow and firn at different depths in ice core from Camp Century, Greenland. The S and W indicate interpretations of summer and winter layers, respectively. As the ice sinks towards the base of the ice sheet, the annual layer thickness (λ) is reduced due to plastic deformation. Within a few years, short-term $\delta^{18}O$ variations are obliterated by mass exchange in the porous firn. With increasing age, the amplitude of the seasonal delta cycle is reduced to 2‰. As annual layers become thinner, the seasonal $\delta^{18}O$ gradients increase and molecular diffusion in the ice smoothes out the intra-annual variations. Eventually, seasonal differences are obliterated entirely (Johnsen et al., 1972).

Antarctica, though, because of low accumulation rates (generally <25 cm water equivalent per year), which result in the seasonal signal being "lost" at relatively shallow depths. In many cases, removal of seasonal (or indeed annual) accumulation by wind scouring may occur, destroying any seasonal signal entirely. On temperate glaciers and ice caps, where snowmelt and percolation of meltwater takes place, it is also impossible to detect a reliable seasonal isotopic signal. In these conditions, seasonal differences in both δD and $\delta^{18}O$ are rapidly smoothed out (within a few meters of the surface) due to isotopic exchange as the ice recrystallizes (Arnason, 1969).

Microparticles and glaciochemistry: Detailed studies of microparticulate matter and ice chemistry (major ions and trace elements) in ice cores from Antarctica and Greenland reveal pronounced seasonal variations (Fig. 5.12). In Greenland, microparticles increase to a maximum in late winter-early spring, presumably as a result of a more vigorous atmospheric circulation at this time of year. Conversely, microparticle frequency minima are generally observed in autumn. There are similar seasonal variations in various cations and anions (e.g., sodium, calcium, nitrate, chloride) with spring concentrations of these ions commonly greater than at other times of the year (Hammer, 1989).

Compared to the diffusion rate of water molecules, which leads to obliteration of the seasonal $\delta^{18}O$ record at depth, diffusion of microparticles and metallic ions is essentially zero. Hence the counting of seasonal variations may allow dating of ice back to late Wisconsin time, or perhaps even earlier. This approach is particularly useful in areas where accumulation rates are so low that seasonal isotopic differences are rapidly lost at depth. For example, in parts of Antarctica, the concentration of sodium ions (Na^+) varies markedly, due to pronounced seasonal changes in the influx of marine aerosols (Herron and Langway, 1979; Warburton and Young, 1981). At Vostok, in eastern Antarctica, Na^+ concentrations reach a maximum in summer layers, due to sublimation of snow, leaving higher residual ionic concentrations (Wilson and Hendy, 1981). These variations are visible far below the level at which seasonal $\delta^{18}O$ variations become obliterated, and can even be detected at ~950 m depth in the Vostok ice core.

One difficulty in microparticle and trace element analysis is to ensure that the sample size selected is small enough to detect intra-annual changes. Near the surface, this is not a big problem, but in ice from very deep ice cores (where the actual thickness of an annual layer is not accurately known) intense lateral and vertical compressive strain may result in dust layers being merged together so that they cannot be adequately distinguished. This is particularly true if the strain rates of dirty ice and of clean ice are very different, as suggested by Koerner and Fisher (1979). Non-destructive laser light scattering can be used to produce a continuous record of dust variations in an ice core (Ram and Illing, 1995) but the problem of identifying each annual layer remains. This can lead to an underestimation of ice age at depth if the microparticle variations observed represent several years rather than seasonal variations. Fortunately, independent corroboration of age estimates can generally be achieved using multiple indicators, though difficulties increase greatly at depth.

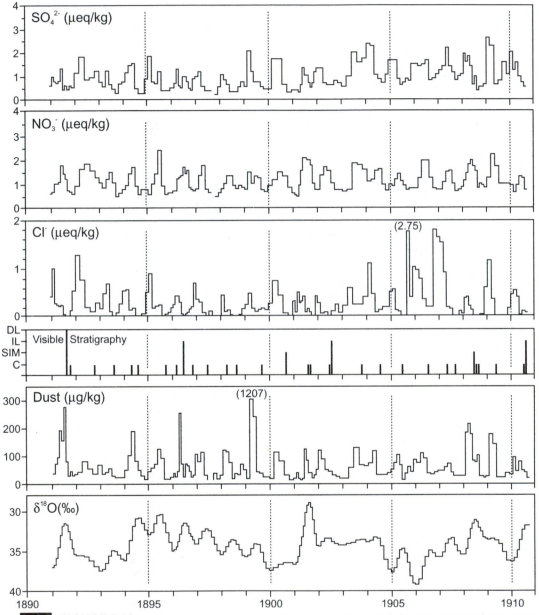

FIGURE 5.12   A section of an ice core from Site A, central Greenland, showing interannual variations in some parameters commonly used in dating. Seasonal peaks are seen in most records, though occasional uncertainties are apparent. Usually such uncertainties can be resolved by cross-referencing the records. In this way seasonal counting can be used to date the upper sections of ice cores (Steffensen, 1988).

Electrical conductivity measurements (ECM): ECM provides a continuous record of ice acidity by recording the ability of ice to conduct an electrical current. A current with a large potential difference is passed between two electrodes in contact with the surface of the ice core (1250 V was used on the GRIP and 2100 V on the GISP2 ice cores). When the ice contains strong acids from volcanic eruptions, ECM is high; layers containing alkaline continental dust, or ammonia (e.g., from biomass burning) have low ECM (Taylor *et al.*, 1993a, b). Changes in deposition of $CaCO_3$ dust are associated with large changes in ECM, reflecting changes in the source region and/or transport and deposition processes. Pronounced changes in ECM characterize the transitions from cold, glacial periods to warmer interstadials in the GISP2 ice core (see Figs. 5.26a and b). Figure 5.13 shows in detail the ECM record at the transitions marking the beginning and end of the Younger Dryas period (~12,850 and 11,670 calendar yr before present in this record).

5.3.3 Reference Horizons

Where characteristic layers of known age can be detected, these provide valuable chronostratigraphic markers against which other dating methods can be checked. On the short timescale, radioactive fallout from atmospheric nuclear bomb tests in the 1950s and 1960s can be detected in firn by measuring the tritium content (or gross β activity). As the timing of the first occurrence of these layers is fairly well

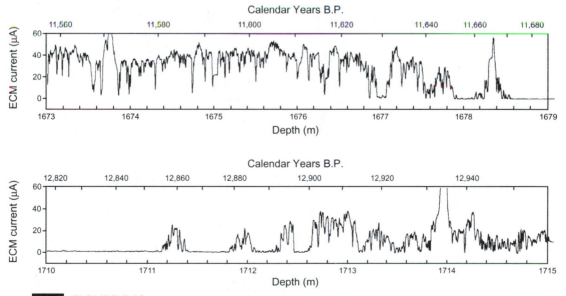

FIGURE 5.13 Electrical conductivity measurements (ECM) in the GISP2 ice core from Summit, Greenland, at the transitions to (below) and from (above) the Younger Dryas cold episode. Annual layer thickness varies from ~6 cm yr⁻¹ in warmer intervals (high ECM) to ~3 cm yr⁻¹ in colder intervals (low ECM). Counting of annual layers is not based on ECM alone, but involves multiple parameters (Taylor *et al.*, 1993b).

known (spring 1953 in Greenland and February 1955 over much of Antarctica, reaching maximum levels in 1963) they can be used as marker horizons for snow accumulation studies, facilitating regional surveys of net balance over the last few decades (Crozaz *et al.*, 1966; Picciotto *et al.*, 1971; Koerner and Taniguchi, 1976; Koide and Goldberg, 1985).

On a much longer timescale, other reference horizons have resulted from major explosive volcanic eruptions. Violent eruptions may inject large quantities of dust and gases (most importantly hydrogen sulfide and sulfur dioxide) into the stratosphere where they are rapidly dispersed around the hemisphere. The gases are oxidized photochemically and dissolve in water droplets to form sulfuric acid, which is eventually washed out in precipitation. Hence, after major explosive volcanic eruptions, the acidity of snowfall increases to levels significantly above background values (Hammer, 1977). By identifying highly acidic layers resulting from eruptions of known age, an excellent means of checking seasonally based chronologies is available (Fig. 5.14). For example, variations in electrical conductivity (a measure of acidity) along a 404 m core from Crête, Central Greenland, reveal a record that closely matches eruptions of known age (Hammer *et al.*, 1978, 1980). The core was originally dated by a combination of methods, primarily seasonal counting (Hammer *et al.*, 1978). This enabled the acidity record to be checked against historical evidence of major eruptions during the last 1000 yr (Lamb, 1970) confirming that the timescale developed was extremely accurate. Once major acidity peaks have been identified they can be used as critical reference levels over the entire ice sheet. For example, the highest acidity levels in the last 1000 yr in Greenland ice resulted from the eruption of Laki, Iceland, in 1783. At Crête, the only acidity peak of greater magnitude in the last 2000 yr resulted from another Icelandic eruption (Eldgja) at A.D. 934 ± 2, providing two very distinct reference layers (Hammer, 1980). Similarly, a major eruption of Huaynaputina (Peru) in February A.D. 1600 provides a diagnostic reference horizon in conductivity records from the Quelccaya and Huascarán (Peru) ice cores, as well as in Antarctica (Delmas *et al.*, 1992; Cole-Dai *et al.*, 1995) (see Fig. 5.38). Further discussion of the volcanic record in ice cores can be found in Section 5.4.4.

Volcanic dust (tephra) from large eruptions may also provide chronostratigraphic horizons if the chemical "fingerprint" of the layer can be correlated between the different sites. In the GISP2 (Greenland) ice core, for example, volcanic particles from an Icelandic eruption 52,680 ± 5000 yr ago can be matched with the Z2 tephra found in many marine sediment records from the North Atlantic (Ruddiman and Glover, 1972; Kvamme *et al.*, 1989; Zielinski *et al.*, 1997). Similarly, volcanic particles in the Younger Dryas section of the Dye-3 and GISP2 ice cores have the same geochemical signature as the Vedda ash (^{14}C-dated at 10,320 yr B.P.), which is widely distributed in northwest Europe and the North Atlantic (Mangerud *et al.*, 1984; Johnsen and Dansgaard, 1992; Birks *et al.*, 1996; Zielinski *et al.*, 1997).

Another important reference horizon is provided by "spikes" in the [10]Be record found in some polar ice cores. The [10]Be is a cosmogenic isotope, produced in the upper atmosphere, which eventually settles, or is washed out, to the earth's surface (see Section 5.4.1). Two large increases in [10]Be, far above background levels, are

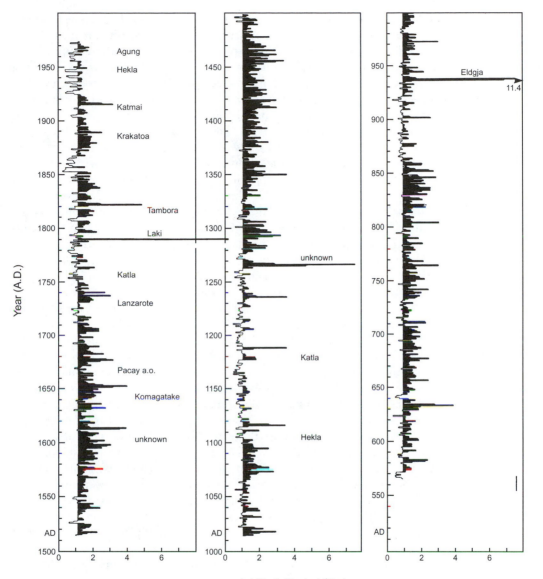

FIGURE 5.14 Mean acidity of annual layers from A.D. 553 to A.D. 1972 in the ice core from Crête, central Greenland. Acidities above the background (1.2 μequiv. H⁺ per kg of ice) are due to fallout of acids, mainly H_2SO_4, from volcanic eruptions north of 20°S. The ice core is dated with an uncertainty of ± 1 yr in the past 900 yr, increasing to ± 3 yr at A.D. 553, which makes possible the identification of several large eruptions known from historical sources (e.g., Laki, Iceland, 1783; Tambora, Indonesia, 1815; Hekla, Iceland, 1104). Also seen is the signal from the Icelandic volcano Eldgja, which was known to have erupted shortly after A.D. 930. Note the low level of volcanic activity recorded from A.D. 1100 to A.D. 1250 and from A.D. 1920 to A.D. 1960. Considerably higher levels of volcanic activity occurred from A.D. 550 to A.D. 850 and from A.D. 1250 to A.D. 1750 (Hammer *et al.,* 1980).

seen in the Vostok ice core around 35,000 and 60,000 yr B.P. (Raisbeck *et al.*, 1987). The reason for these peaks is not clear; they may have resulted from a change in primary cosmic ray flux, or a reduction in solar or geomagnetic modulation of cosmic rays penetrating the atmosphere (Baumgartner *et al.*, 1998), or even from a super nova. Whatever the cause, these anomalies can be seen in many ice cores and can be used as chronostratigraphic markers. For example, the 35 ka B.P. ^{10}Be spike is found in ice cores from Vostok, Dome C, and Byrd (Antarctica), enabling these records to be properly aligned (Figure 5.19). There is also a ^{10}Be peak in the Camp Century ice core from Greenland and a spike of ^{36}Cl (also a cosmogenic isotope) in the Guliya (western China) ice core at about the 35 ka level, confirming the age models applied to these records and allowing them to be aligned with those from Antarctica (Reeh, 1991; Beer *et al.*, 1992; Thompson *et al.*, 1997). However, the 60 ka ^{10}Be anomaly is less pronounced and has not been as useful a marker. Interestingly, a second ^{10}Be spike seen in the Camp Century ice core, if ascribed to the 60 ka event, would force a major revision in the chronology of this record, and of the Dye-3 record (in southern Greenland) with which it was correlated, making both series much shorter than envisioned by Dansgaard *et al.* (1982). Reeh (1991) argues that this is in fact the case, because the ice sheet was considerably smaller in the last interglacial, so that higher δ^{18}O values prior to 70 ka B.P. (in his revised chronology) were due to a smaller, lower ice sheet. This argument is supported by the studies of Koerner (1989) and Letréguilly *et al.* (1991). The counterargument is that the "60 ka" B.P. spike in Camp Century is not reliable (based on only a single high value) and that the chronology of Dansgaard *et al.* (1982) in fact fits much better with other proxy records and with reasonable flow model assumptions, which place the "60 ka" horizon closer to 95 ka B.P. (Beer *et al.*, 1992; Johnsen and Dansgaard, 1992; Reeh, 1991). This controversy nicely illustrates the difficulties of dating ice at depth, particularly in Greenland where there may have been dramatic changes in the ice sheet configuration over the last 150,000 yr. Clearly, having unequivocal stratigraphic markers would be extremely helpful in resolving such controversies. Other approaches, using gases in ice cores, are discussed further in Section 5.4.5.

As noted earlier, the best approach to identifying annual layers is a composite one, using δ^{18}O profiles, microparticles, variations in conductivity, and reference horizons. In this way, questionable sections of one record may be resolved by reference to the others. This multiparameter approach was adopted by Meese *et al.* (1995, 1997) in dating the upper section of the GISP2 ice core. Having established the annual chronology, it was then possible to calculate accumulation rate changes over time, given certain assumptions about vertical strain since deposition and density variations down core (Meese *et al.*, 1994). Figure 5.15 shows the accumulation record from Summit, Greenland, derived in this way for the last 11,500 yr. Accumulation was considerably lower in the last glacial period, increasing by >30% from ~12 ka to 9 ka B.P. Thereafter, the record (subjected here to 100 yr smoothing) reveals only minor changes (± 5% on this timescale). Interestingly, a drop in accumulation at ~8200 B.P. corresponds to very low δ^{18}O values in many ice cores, as well as a sharp reduction in CH_4 (see Fig. 5.33), indicating that a significant and abrupt, large-scale climatic change occurred at this time (Alley *et al.*, 1997c).

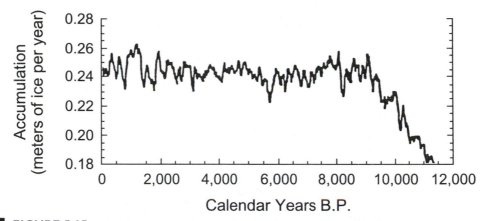

FIGURE 5.15   Accumulation at Summit, Greenland (100-yr running mean). Following an increase in accumulation from the LGM to the early Holocene, conditions became relatively stable, averaging 0.24 m ±5% on this timescale for the last ~9000 yr (see Fig. 5.33) (Meese *et al.,* 1994).

A more detailed examination of the last 200 yr reveals considerable variability, but no overall trend in accumulation. One problem with accumulation is that it is far less spatially coherent than temperature, making it difficult to correlate with other records. Hence it is perhaps not surprising that the GISP2 record shows few similarities with earlier studies of accumulation changes over the last 800–1500 yr at Dye-3, Milcent, or even Crête (Reeh *et al.,* 1978). Nevertheless, all studies seem to support the conclusion that there has not been any significant long-term change in accumulation over much of the Greenland Ice Sheet over (at least) the last 1400 yr. This is similar to the conclusion reached by Koerner (1977) for the Devon Island Ice Cap.

5.3.4 Theoretical Models

Dating ice at great depth poses severe problems which cannot be easily resolved by the methods previously described. At present, the method most widely used to date pre-Holocene ice is to calculate ice age at depth by means of a theoretical ice-flow model (Dansgaard and Johnsen, 1969; Reeh, 1989; Johnsen and Dansgaard, 1992). Such models describe mathematically the processes by which ice migrates through an ice sheet. Snow accumulating on an ice sheet is slowly transformed into ice during densification of the firn. As more snow accumulates the ice is subjected to vertical compressive strain in which each layer is forced to thin, and is advected laterally towards the margins of the ice sheet (Fig. 5.16). Hence, a core from any site, apart from the ice divide, will contain ice deposited up-slope, with the oldest and deepest ice originating at the summit. Because summit temperatures are cooler, if the core is not recovered from the highest part of the ice sheet, $\delta^{18}O$ values at depth must be corrected for this altitude effect, which will be present regardless of whether any long-term climatic fluctuation has occurred. Because of the nature of ice flow and

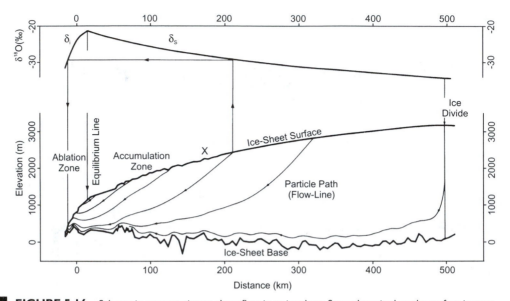

FIGURE 5.16 Schematic cross section to show flow in an ice sheet. Snow deposited on the surface is transformed to ice and follows the flowlines indicated. Ice thins by plastic deformation under compression by the overlying ice. Hence an ice core from site X will contain a record of ice originating upstream, requiring adjustment for the colder conditions in that region. An ice core from the summit will have fewer problems, providing the ice divide has not varied over time. Samples recovered at the surface in ice-sheet margins may represent the same paleo-environmental record as that from an ice core through the ice sheet (modified, Reeh, 1991).

deformation, most of the time period recorded in an ice core is found in the lowest 5-10% of the record. This means that even small differences in an age-depth model can result in large discrepancies in age estimates for the lowest part of deep ice cores (see e.g., revisions made in the Dye-3 chronology by Dansgaard *et al.*, 1982 and the discussion of age uncertainties in Reeh, 1991).

Simple models can provide a rough estimate of ice age at depth, but for more accurate age estimates, some knowledge of past changes in ice thickness and temperature, accumulation rates, flow patterns, and ice rheology (which changes with dust content) is required (Paterson, 1994). Many of these problems are minimized in the case of ice cores from ice divides (e.g., the GRIP core at Summit, Greenland) or in cores that penetrate very thick ice sheets to depths well above the bed (e.g., Vostok, Antarctica). Nevertheless, even in these cases, uncertainties related to past ice sheet dimensions and the stability of ice divides, changes in ice sheet thickness and especially changes in accumulation rate, can change age-depth relationships very significantly in the deepest sections of a core. On the other hand, if ice can be dated independently by some other means (see Sections 5.4.4 and 5.4.5), flow models can then be constrained and used to estimate changes in those parameters, such as accumulation rate, which would otherwise be problematical (Dahl-Jensen *et al.*, 1993). In this way, iterative changes in models, using best estimates of various parameters and how they might realistically have varied in the past, together with the ages of certain fixed points (such as tephras of known age, or the ~35,000 yr B.P. [10]Be anomaly) can be used to refine and improve an ice core chronology.

5.3.5 Stratigraphic Correlations

In addition to the methods described already, attempts have been made to correlate certain stratigraphic features in ice cores with other proxy paleoclimatic records that may have better chronological control. For example, a revised timescale for the Camp Century ice core was proposed by Dansgaard *et al.* (1982), who matched major (low-frequency) changes in the ice-core record with $\delta^{18}O$ changes in benthic foraminifera from the oceans. They assumed that a lowering of $\delta^{18}O$ values in the ice core indicates cooling and/or an increase in ice thickness, which corresponds to higher $\delta^{18}O$ values in the foraminifera due to reduced ocean volume as the ice sheets on land expanded (see Section 6.3.1). On this basis, they reinterpreted their original timescale (Dansgaard *et al.*, 1969), changing the estimated ~60 ka B.P. horizon to ~115 ka B.P. Interestingly, differences between their revised timescale and that originally predicted by a theoretical flow model imply that the accumulation rate was higher in the intervals 125,000–115,000, 80,000–60,000, and 40,000–30,000 yr B.P., all times when the oceanic $\delta^{18}O$ record indicates periods of major global ice volume increase.

One danger in the correlation approach is that certain "events" (e.g., the onset of colder conditions, broadly characterized as the Younger Dryas interval) may be used to align records from different regions. However, the "Younger Dryas transition" may not be synchronous and overly simplistic correlations may obscure potentially important leads and lags in the climate system, whereby changes in one area ultimately trigger a delayed response elsewhere. Thus, it has yet to be demonstrated that the "Younger Dryas" oscillation in Greenland ice cores is precisely synchronous with a similar oscillation seen in Antarctic ice (Jouzel *et al.*, 1987a). This has important implications for understanding the cause of this event and the mechanisms involved in its propagation.

Other approaches to stratigraphic correlations between ice cores in Antarctica and Greenland, and between ice cores and marine sediments, are discussed in Sections 5.4.4 and 5.4.5.

5.4 PALEOCLIMATIC RECONSTRUCTION FROM ICE CORES

Ice cores have revolutionized our understanding of Quaternary paleoclimatology by providing high resolution records of many different parameters, recorded simultaneously at each location. Here, we highlight the main results from the northern and southern hemispheres and show how these records are related to each other, and to changes in forcing.

5.4.1 Ice-core Records from Antarctica

A number of long ice-core records are available from Antarctica but the "crown jewel" of Antarctic ice cores is the record from Vostok on the East Antarctic ice plateau (78°28' S, 106°48' E, 3488 m above sea level). This is an important record, not only because it spans a very long interval of time (3350 m, ~426 ka) but because it has been recovered from an area where the ice is extremely thick (>3.5 km) and complications due to ice flow and disturbance at the bed are minimal. Furthermore, the relationship between isotopic fractionation and temperature is clear in this region, making a climatic interpretation of the isotopic record fairly straightforward. Thus, Vostok provides the longest well-resolved ice-core record on earth and a yardstick for comparison with other paleoclimatic records (Petit *et al.*, 1997).

Figure 5.17 shows the $\delta^{18}O$ record from Vostok over the uppermost 2083 m of the core (Lorius, *et al.*, 1985). Stages A to H designate the major features of the record, with stage A being the Holocene and Stage G being the last interglacial period. Four major cold periods (B, D, F, and H), each with $\delta^{18}O$ of around −62‰, are clearly recognizable. Establishing a precise chronology for this record is a fundamental problem (as with all ice cores) and requires several assumptions, including the original source of the snow (the local ice flow regime) and the accumulation rate. At Vostok sub-ice topographic effects are minor and the most significant factor is the change in accumulation rate over time. Today, accumulation is very low (~2.2 g cm^{-2} a^{-1}, compared to >50 g at Dye-3 in Greenland) but it was probably even lower during glacial times, as precipitation in Antarctica is closely related to temperature. This relationship has been used to assess precipitation changes by assuming that precipitation is a function of the ratio of the derivative of the saturation vapor pressure (s.v.p.) of water at time Z to the same parameter today. As the slope of the s.v.p. increases exponentially with temperature, so precipitation will

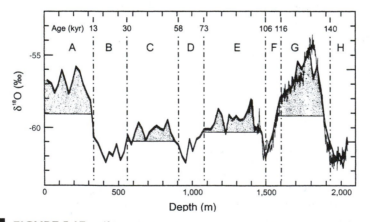

FIGURE 5.17 $\delta^{18}O$ vs depth in the Vostok ice core; climatic stages and their temporal limits are indicated. These are not directly comparable with the marine isotope stages (MIS) 1–6, though stages A–D approximate MIS 1–4, and H approximates MIS 6. Thick line is from continuous sampling, thin line from less detailed analysis (Lorius *et al.*, 1985). Stages A and G are interglacials.

change as a non-linear function of temperature. Temperature changes are in turn estimated from $\delta^{18}O$ (or δD) using empirical relationships observed in studies of contemporary snowfall (Jouzel *et al.*, 1983). Using this approach, precipitation rates at Vostok are estimated to have been 50–55% of modern values during glacial times, and, with this estimate, the chronology shown in Fig. 5.17 was established. It should be noted that even small differences in estimates of the "modern" accumulation rate at Vostok are amplified at depth; for example, a 10% difference at the surface produces an uncertainty of 10,000 yr in the chronology at 2000 m. Nevertheless, the proposed time-scale is supported by ^{10}Be data from the same core. The ^{10}Be is a cosmogenic isotope produced by cosmic-ray bombardment of the upper atmosphere. Assuming a constant production rate, any changes in ^{10}Be concentration in snowfall at Vostok would be due to changes in the accumulation rate (Yiou *et al.*, 1985). On this basis, precipitation at the last glacial maximum was ~50% of modern values (i.e., only ~1.1 g cm^{-2} a^{-1}). Both of these approaches to estimating paleo-precipitation rates yield surprisingly similar results (Fig. 5.18). Furthermore, using the s.v.p./precipitation rate relationship as the basis for calculating *independent* chronologies at Vostok and Dome C, a ^{10}Be "spike" is found to coincide (almost) in both cores at around 35,000 yr B.P. (a 3% correction is required in the Dome C s.v.p.-derived chronology), providing confidence that this approach to dating the record has validity (Raisbeck *et al.*, 1987). Additional support comes

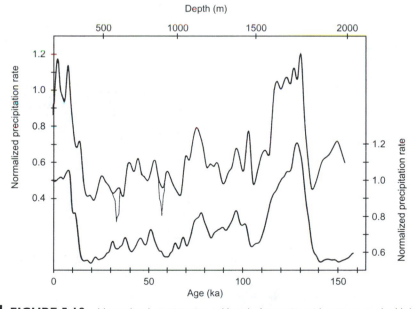

FIGURE 5.18 Normalized precipitation at Vostok, Antarctica with respect to the Holocene mean value (which is 1). Upper record is based on the assumption that ^{10}Be concentration in snowfall is a function only of changing accumulation rate (thick line not taking ^{10}Be peaks at ~35,000 and 60,000 B.P. into account). Lower record is based on the saturation vapor pressure (s.v.p)-temperature relationship, using δD to estimate changes in temperature, and assuming precipitation is directly related to s.v.p. (Jouzel *et al.*, 1989a).

from the Byrd ice core where accumulation changes can be estimated down-core by continuous acidity measurements (which enable seasonal cycles to be observed) (Jouzel *et al.*, 1989a). These show that accumulation during glacial time was 50% of Holocene levels (as at Dome C and Vostok) and with this taken into account, a [10]Be spike is also observed at ~35,000 yr B.P. in the Byrd record (Fig. 5.19) (Beer

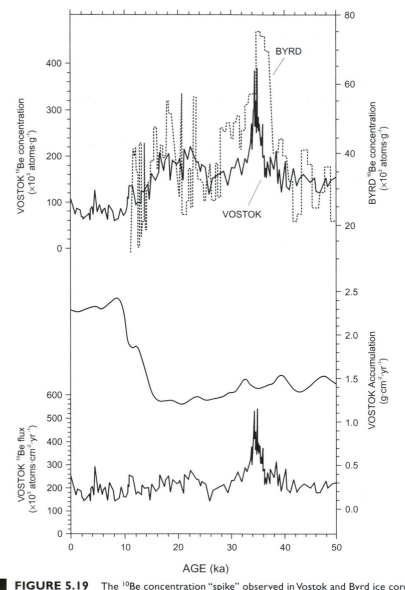

FIGURE 5.19   The [10]Be concentration "spike" observed in Vostok and Byrd ice cores at ~35,000 B.P., assuming precipitation was ~50% of Holocene levels at LGM (Raisbeck *et al.*, 1992, and Beer *et al.*, 1992).

et al., 1992). Hence through a series of procedures, each based on a somewhat arguable premise, there emerges strong support for the notion that precipitation amount (at least in East Antarctica) is directly related to temperature (as represented by isotopic changes) and that precipitation was much lower in glacial times than in warmer periods. This provides considerable confidence that the chronology shown in Figs. 5.17 and 5.18 is likely to be approximately correct; Lorius *et al.* (1988) suggest an uncertainty of 10,000–15,000 yr at around 160,000 yr in this chronology.

More recent studies of the Vostok record, including an extension of the record to >400,000 yr B.P., have focused on continuous measurement of δD (δ^2H) rather than $\delta^{18}O$ (Jouzel *et al.,* 1987b, 1989a, 1989b, 1993a). The δD changes by 6‰ per °C (at the ice surface) in East Antarctica, according to empirical observations and model-derived estimates (Jouzel *et al.,* 1983; Jouzel and Merlivat, 1984). This was discussed in more detail in Section 5.2.3. After correcting for higher δD levels in water vapor during glacial times (because of isotopic enrichment of the ocean by the heavier isotope of hydrogen) the δD change from the last glacial to the Holocene represents an increase in surface temperature of ~9°C at Vostok (Jouzel, *et al.,* 1987a). This compares with an independent estimate based on ice crystal growth rate changes with depth, of ~11 °C (Petit *et al.,* 1987) (though this approach is disputed by Alley *et al.,* 1988). The isotopic estimates assume no change in ice sheet thickness, though the lower accumulation rates of the last glacial period suggest that the ice sheet elevation may have been lower during stage B than Stage A (Jouzel *et al.,* 1989a). This would make the glacial-Holocene temperature estimate from $\delta^{18}O$ or δD a minimum estimate.

The δD values for the last interglacial (Stage G) indicate a period warmer than the Holocene by ~2 °C; interstadial Stages C and E were 4–6 °C warmer than the glacial maximum (Stage B). Of particular interest is the "two-step" change in temperature during the last deglaciation, when rapidly rising temperatures were interrupted by a cooling episode lasting ~1500 yr. It is estimated that *surface* temperatures fell by ~3–4 °C at Vostok during this "Antarctic Cold Reversal" (based on a peak-to-peak change in δD of 20‰, for ~25-yr means; Fig. 5.20) (Jouzel *et al.,* 1992; Mayewski *et al.,* 1996). Detailed analysis of the ice from Dome C indicate this cold episode lasted from ~13,500 to 11,700 (calendar) yr B.P. and seems to be related in some way to the Younger Dryas oscillation seen in many records from around the world (Wright, 1989; W. Berger, 1990; Peteet, 1992). Unlike the records from Greenland ice cores, this "reversal" is not associated with an increase in continental dust, or with a drop in CH_4 or CO_2 levels; CO_2 levels appear to have leveled off at this time, and CH_4 levels declined slightly later in the cold episode, possibly reflecting tropical aridification and/or cooling (or even freezing) of high latitude peatland (Jouzel *et al.,* 1992). However, levels of chloride at Taylor Dome during this cold episode indicate an increase in the flux of marine salts due to higher wind speeds at that time (Mayewski *et al.,* 1996). Temperatures at Vostok and other locations gradually declined during the Holocene, by ~1 °C at the surface (Ciais *et al.,* 1992) a pattern also seen in the Arctic (Koerner and Fisher, 1990; Bradley, 1990).

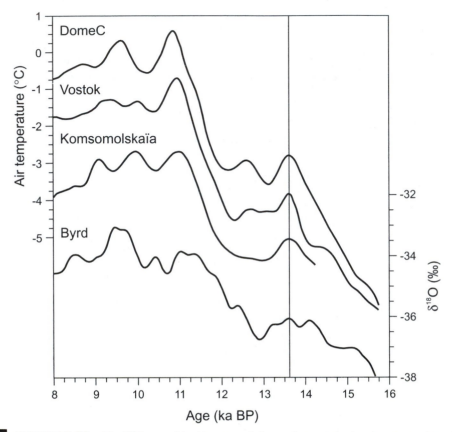

FIGURE 5.20   The $\delta^{18}O$ record (and interpreted changes in atmospheric temperature, above the surface inversion) at four Antarctic sites during the last deglaciation, showing a hiatus or reversal in the $\delta^{18}O$ increase, from ~13,500 to ~11,700 (calendar) yr B.P. Note that here the absolute chronology of each core is not well known; they are "matched" based on an optimum fit, relative to the Dome C record (Jouzel *et al.*, 1992). Recent work has fixed the Dome C chronology with respect to GISP2 (Mayewski *et al.*, 1996).

Further drilling at Vostok has yielded additional ice down to 3350 m. Comparison with the SPECMAP marine isotope record (see Section 6.3.3) strongly suggests that the ice core record extends to ~426,000 and thus spans the last four interglacial-glacial cycles (Fig. 5.21). Of particular note is the long cold episode from ~180–140 ka B.P. when δD values remained at levels comparable to the Last Glacial Maximum (Petit *et al.*, 1997).

5.4.2 Ice-core Records from Greenland

Four ice cores to bedrock have now been recovered from the Greenland ice sheet: from Camp Century, Dye-3, and two from Summit, the so-called GISP2 (Greenland Ice Sheet Project 2) and GRIP (Greenland Ice Core Project) cores. Three other long

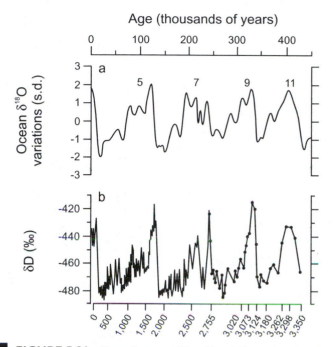

FIGURE 5.21   Deuterium record from Vostok, Antarctica, plotted with the SPECMAP $\delta^{18}O$ record of continental ice volume changes (above) (Petit *et al.*, 1997).

records have been obtained from the ice sheet margin (Reeh *et al.*, 1987, 1991, 1993; Johnsen and Dansgaard, 1992). The GISP2 and GRIP cores, in particular, have provided an enormous amount of information about the climatic history of Greenland and of processes that must have operated over a large area of the North Atlantic region, with effects of hemispheric or even global significance. Here, we discuss the long paleoclimatic series from these sites and examine their relationship to other records in the area.

Figure 5.22 shows the $\delta^{18}O$ record from the GRIP ice cores (Dansgaard *et al.*, 1993). The GISP2 and GRIP records are highly correlated down to ~2750 m (estimated in the GRIP core to be ~103,000 yr ago by means of a flow model) (Grootes *et al.*, 1993). A number of important characteristics of the $\delta^{18}O$ series can be clearly seen. First, the Holocene record was a period of relative stability with a mean $\delta^{18}O$ value of –34.9‰ at GRIP and –34.7‰ at GISP2. Fluctuations are small — on the order of ±1–2‰ — and show little correlation in detail between sites, probably due to local differences in accumulation and wind drifting of snow. A slight decline in $\delta^{18}O$ over the course of the Holocene is apparent in both records. Around 8250 yr ago (calendar years) a pronounced episode of low $\delta^{18}O$ values is observed (Fig. 5.22) and this has been seen at several other sites, including ice cores from the Canadian Arctic (Fisher *et al.*, 1995); it also corresponds to an abrupt drop in atmospheric methane levels (Blunier *et al.*, 1995; Alley *et al.*, 1997c).

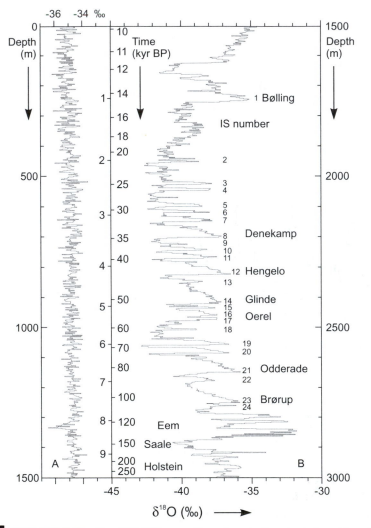

FIGURE 5.22   The GRIP δ¹⁸O record from Summit, Greenland, plotted linearly with respect to depth. Section A (left) is the Holocene section, showing only minor changes; section B (right) shows the preceding 250 kyr record at the same δ¹⁸O scale. Note the very large and rapid oscillations throughout the pre-Holocene record. Proposed interstadial isotope stages (IS) 1-24 are indicated, together with comparable European pollen stages. Dating was by annual layer counting to 14.5 kyr B.P. and beyond that by an ice-flow model (Dansgaard *et al.,* 1993).

All Greenland ice cores show that dramatically different climatic conditions prevailed in the late Pleistocene, compared to the last 10,000 yr (Dansgaard *et al.,* 1984; Johnsen *et al.,* 1992; Grootes *et al.,* 1993). In contrast to the relative stability of Holocene climate, the preceding ~100,000 yr were characterized by rapid changes between two (or more) modes. Dansgaard *et al.* (1993) recognize 24 interstadial episodes between 12,000 and 110,000 yr B.P. when isotopic values were as high as –37‰ at the GRIP site, separated by stadials, with values dropping precipitously to as low as −42 ‰ (see Fig. 5.22). These abrupt changes can be correlated between cores as far apart as Dye-3 (southern Greenland), Camp Century (northwest Greenland), and Renland (east-central Greenland) so, whatever their cause, the changes were geographically extensive (Fig. 5.23) (Johnsen and Dansgaard, 1992; Johnsen *et al.*, 1992). Indeed, they are well correlated to changes seen in North Atlantic marine sediments (Bond *et al.*, 1992, 1993) that represent large-scale shifts in water masses in that region. This is discussed in more detail in Section 6.10.

Figure 5.22 shows that the changes from low to high $\delta^{18}O$ were rapid, followed by a slower decline to low values once again (Dansgaard *et al.*, 1993). This "sawtooth" characteristic (Dansgaard *et al.*, 1984) is seen most clearly in the detailed studies that have been carried out on the most recent sequence of changes. The $\delta^{18}O$

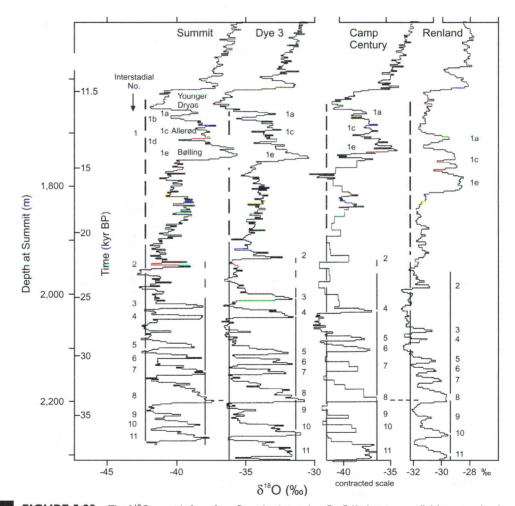

FIGURE 5.23   The $\delta^{18}O$ records from four Greenland sites (see Fig. 5.1) showing parallel, large-amplitude changes over very short periods of time. Changes are commonly "saw-tooth" in pattern (see Stage 8, for example) with an abrupt shift to higher $\delta^{18}O$ levels, followed by a slower return to lower values (Johnsen *et al.*, 1992).

rose very abruptly (over ~10 yr) from the Older Dryas (cold) phase to the Bølling/Allerød warm period, then fell slowly over the next 1700 yr to the very cold Younger Dryas episode (Dansgaard *et al.*, 1989) (Fig. 5.24). This lasted for 1250 ± 70 yr then ended very abruptly again (within a decade, around 11,640 ± 250 yr ago). The transition marked the beginning of Pre-Boreal conditions, which slowly led to higher Holocene $\delta^{18}O$ values. Dansgaard *et al.* (1989) argue that the Younger Dryas/Pre-Boreal shift in $\delta^{18}O$ of 5 ‰ at Dye-3 can be interpreted as an increase in mean annual surface temperature of 7 °C, more than half of the total Pleistocene/ Holocene change (estimated as at least ~12 °C in southern Greenland). Deuterium excess (d) also increased at the Younger Dryas/Pre-Boreal transition, suggesting the source area of moisture shifted rapidly northward at that time. The exposure of relatively cold seawater closer to the Summit site would provide both a moisture source (for the heavier accumulation of the Pre-Boreal period) and a lower evaporation temperature in the moisture source region, leading to lower d values in snowfall at Summit. Later, as the North Atlantic became warmer, d values slowly increased (Dansgaard *et al.*, 1989).

Pronounced isotopic changes were accompanied by equally dramatic changes in accumulation (Fig. 5.25). Accumulation more than doubled at the transition from the Older Dryas to the Bølling/Allerød, then declined to low values in the Younger Dryas, before doubling within only a few years at the start of the Pre-Boreal period (Alley *et al.*, 1993). This increase in accumulation must have been associated with an increase in temperature, because precipitation amount, $\delta^{18}O$, and

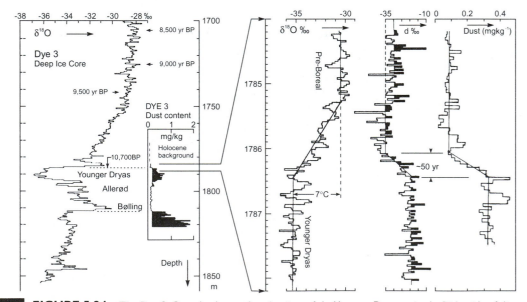

FIGURE 5.24 The Dye-3, Greenland record at the time of the Younger Dryas episode. Right side of diagram shows in detail the changes at the end of this episode, around 10,700 (calendar yr) B.P. $\delta^{18}O$ increased by ~6‰ within 50 yr, accompanied by an even more rapid decline in deuterium excess and in dust levels (Dansgaard *et al.*, 1989).

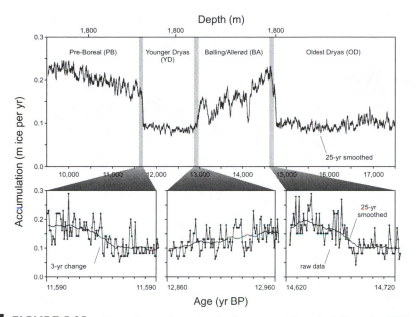

FIGURE 5.25 Accumulation changes at the Summit site, Greenland (from the GRIP ice core) between ~17,500 and 9500 calendar yr B.P. The very rapid changes at the transitions between climate stages are shown in detail in the bottom half of the figure. Colder stages were associated with much lower levels of accumulation (Alley *et al.,* 1993).

temperature are all positively correlated (Clausen *et al.,* 1988; Dahl-Jensen *et al.,* 1993). On this basis, Alley *et al.* (1993) also estimate that temperature changed by up to 7 °C from the Younger Dryas to the Pre-Boreal period.

Changes in atmospheric dust also occurred, with the colder periods being times when relatively alkaline (Ca^{++} rich) continental dust accumulated on the ice cap (Mayewski *et al.,* 1993, 1994). This is most clearly seen in the electrical conductivity (ECM) of the ice (Fig. 5.26a) where the cold dry periods are seen as having lower ECM values than the wetter interstadials (Taylor *et al.,* 1993). This is true for the Bølling/Allerød/Younger Dryas episodes as well as earlier stadial/interstadial events (Fig. 5.26b) and provides a vivid picture of how climatic conditions oscillated between different states before 27,000 yr ago, and again in late glacial time. The ECM in the Vostok ice core provides a similar indication of dust levels associated with a changing climate.

The deepest ~300 m of both GRIP and GISP2 ice cores poses a dilemma as there is very little correspondence between them (Taylor *et al.,* 1993a; Johnsen *et al.,* 1995). This is surprising in view of the excellent agreement between the two cores (and with other cores) above these levels. The differences appear to be related to disturbances in the ice (at one, or both, core sites) due to deformation at depth. The GRIP site is at the present day ice divide and, therefore, simple vertical strain is the primary factor in thinning each annual layer that accumulated, unless a shift in the position of the ice divide occurred. The GISP2 site is off to the side of the divide

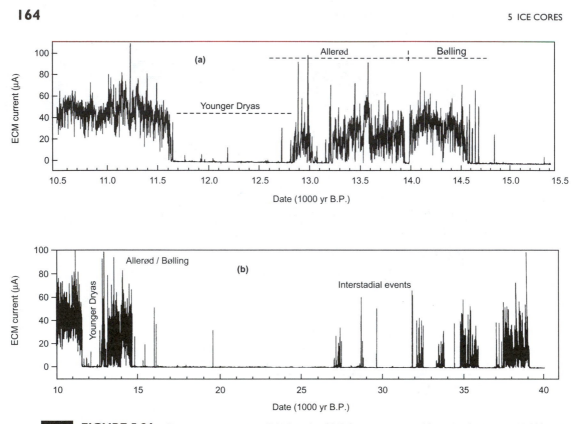

FIGURE 5.26 Electrical conductivity (ECM) in the GRIP Summit ice core (Greenland) between 15,500 and 10,500 and 10,000–40,000 calendar years B.P. Cold episodes such as the Younger Dryas are characterized by low ECM values, warmer episodes by higher values reflecting the relative change in atmospheric dust loading (Taylor *et al.*, 1993b).

and so the ice is more likely to have been subjected to shear (Alley *et al.*, 1995). At both locations it is possible that "boudinage effects" (Staffelbach *et al.*, 1988; Cunningham and Waddington, 1990) have caused some layers to be differentially thinned at great depths. Boudins are "pinch and swell" structures (Fig. 5.27) that can develop in materials where viscous layers are sandwiched between less viscous material. In such conditions, initially small surface irregularities in individual layers can become amplified as the layers thin. Because ice from the last interglacial period is relatively clean (like Holocene ice, it contains little wind-blown dust; Mayewski *et al.*, 1993) whereas glacial age ice is quite dust-laden (and therefore less viscous than the interglacial ice), conditions appear to be favorable for flow boudinage, leading to "swelling" of layers in some locations relative to others; indeed some parts of a particular layer may have been pinched out altogether. Obviously in the narrow diameter of an ice core only a tiny sample of any layer is obtained and it is impossible to tell if a "pinch" or "swell" is being sampled. However, theoretically such flow boudinage could lead to sharp discontinuities in the $\delta^{18}O$ record.

As the two Summit cores correlate well with each other down to ~2700 m, and with ice cores from Dye-3 and Camp Century, the abrupt changes in the last glacial

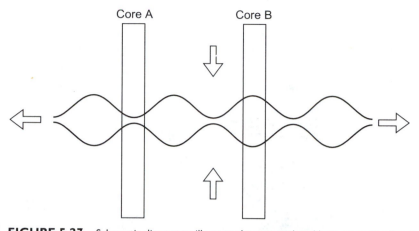

FIGURE 5.27 Schematic diagram to illustrate the potential problems caused by flow boudinage on the interpretation of two adjacent ice cores due to the "pinching out" of layers.

period certainly represent some sort of large-scale climatic process, but the differences between GRIP and GISP2 cores at greater depth point to possible disturbances in one or both of the cores. One way to resolve the problem is to look for evidence of inclined or disturbed layering in the core stratigraphy at depth (Meese *et al.*, 1997; Alley *et al.*, 1997b). In the GRIP core, such layering is seen at 2847 m but in the GISP2 site it starts around 2678 m (Grootes *et al.*, 1993). In the GRIP core there is no strong evidence of overturned folding of the ice, though there is a section (from 2900–2954 m) that is quite disturbed; above and below that, however, there appears to be a relatively undisturbed sequence (Johnsen *et al.*, 1995). This suggests that, overall, there may be a longer climatic record at the GRIP site, but the (undisturbed) record may be separated by sections that are uninterpretable. Chappellaz *et al.* (1997) examined this possibility by comparing CH_4 in the lower GRIP and GISP2 ice cores to the Vostok CH_4 record, which is undisturbed. This revealed that sections of both cores are very probably undisturbed and of interglacial age, whereas other sections are made up of either older or younger ice. Hence, the seemingly abrupt changes in $\delta^{18}O$ should not be interpreted as representing rapid changes in climate during the Eemian.

One final point worthy of note concerning ice core records from Greenland: remarkable records have been recovered by sampling surface ice along the ice sheet margin, from the equilibrium line to the ice edge (Reeh *et al.*, 1991). The $\delta^{18}O$ from these "horizontal ice cores" is highly correlated with that in deep ice cores (Fig. 5.28) because ice flow has advected ice originally deposited in the accumulation area to the ice sheet margin (see Fig. 5.16). The significance of this is that, potentially, large samples of quite old ice could be obtained by literally mining the ice margin, rather than coring the base of the ice sheet to obtain very small samples (Reeh *et al.*, 1987). Reeh *et al.* (1993) note that Summit cores yield less than 10 kg of ice per century for studies of ice older than 40,000 yr B.P., and less than 5 kg for ice of interglacial age. Larger samples from the ice margin might permit detailed investigations of, say, dust or pollen content in the distant past.

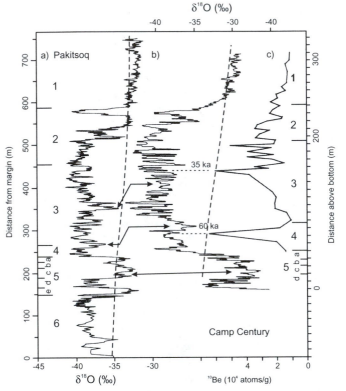

FIGURE 5.28   Comparison of δ^{18}O records from Pakitsoq (on the west Greenland ice sheet margin) and Camp Century. Arrows connect the points considered to be synchronous. The dashed lines show the expected change in δ^{18}O, even without a change in climate, due to ice flow to the sample site (from higher elevations, and, therefore, lower δ^{18}O) (modified from Reeh, 1991).

5.4.3 Past Atmospheric Composition from Polar Ice Cores

Ice cores are extremely important archives of past atmospheric composition. In particular, they contain records of how radiatively important trace gases — carbon dioxide, methane, and nitrous oxide — have varied both in the recent past, and over longer periods of time (Raynaud *et al.*, 1993). In addition, ice cores provide records of air mass characteristics (seen in total ion glaciochemistry) as well as the history of explosive volcanic eruptions and changes in atmospheric dust content that may have had significant effects on the global energy balance.

Instrumental measurements of radiatively important trace gases ("greenhouse gases") have a relatively brief history, generally providing a perspective on current gas concentrations of less than 40 yr. These measurements reveal dramatic increases in CH_4, CO_2, N_2O, and industrial chlorofluorocarbons. Over the same period, levels of heavy metals such as lead and vanadium, as well as anthropogenic sulfate and nitrate have also increased dramatically (Oeschger and Siegenthaler, 1988; Ehhalt, 1988; Stauffer and Neftel, 1988; Mayewski *et al.*, 1992). Ice cores enable the short instrumental records of these contaminants to be placed in a longer-term perspective, providing some measure of the background, preindustrial levels that prevailed before global-scale anthropogenic effects became important (Etheridge *et al.*, 1996). Fig. 5.29 shows the concentration of CO_2 and CH_4 in Antarctic ice cores over the last 150–250 yr; in 1995, CH_4 concentration reached 220% of its eighteenth century values whereas CO_2 was at 130% of its preindustrial level. The N_2O levels were 110% of what they were 250 yr ago. Collectively, these data unequivocally document the dramatic increases in greenhouse gases over the last 200–300 yr, to levels far higher than anything seen in records spanning the last 220,000 yr (although older records cannot provide the same time resolution of more recent ice cores). The extent to which these changes are responsible for recent changes in global temperature remains controversial (Lindzen, 1993; Karl 1993; Mann *et al.*, 1998) but there is little doubt that if current trends continue, significant changes in global climate will occur.

One of the more important results from the Vostok ice core is the evidence that atmospheric composition has not remained constant over glacial-interglacial cycles. In particular, the concentration of radiatively important trace gases — carbon dioxide, methane, and nitrous oxide — have all changed significantly. There is also evidence that aerosol concentrations changed dramatically and that changes in the global sulphur cycle may have had important consequences for global cloudiness and hence the earth's energy balance. These changes can also be used to link the chronologies of ice core records in both hemispheres because the mixing times of the important trace gases are short (1–2 yr), so that changes observed in one record should be essentially synchronous in both hemispheres. In this section, these issues are examined in more detail.

A fundamental problem in constructing a paleo-record of trace gas concentrations from ice cores is the fact that the air in ice bubbles is always younger than the age of the surrounding ice (Schwander and Stauffer, 1984). This is because as snow is buried by later snowfalls and slowly becomes transformed to firn and ice, the air between the snow crystals remains in contact with the atmosphere until the bubbles or pores of air

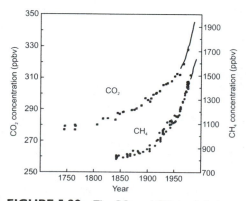

FIGURE 5.29 The CO_2 and CH_4 levels in Antarctic ice cores (Siple and DEO8, see Fig. 1). Solid lines are instrumentally recorded values (Raynaud *et al.*, 1993).

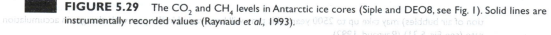

become sealed at the firn/ice transition (when density increases to 0.8–0.83 g cm⁻³). The sealed bubbles thus contain air that is representative of atmospheric conditions long after the time of deposition of the surrounding snow (Fig. 5.30). "Pore close-off" varies with accumulation rate, ranging from ~100 yr at high accumulation sites like Dye-3 in Greenland or Siple Station in Antarctica, to as much as 2600 yr at very low accumulation sites in central East Antarctica. Furthermore, this value will have changed over time because accumulation rates were much lower in glacial times and thus the density-depth profile (or time to "pore close-off") will have been considerably longer. At Vostok, the change in accumulation rate from interglacial to glacial time changed the air-ice age difference from ~2500 yr to ~6000 yr or more (Barnola *et al.*, 1991; Sowers *et al.*, 1992) (Fig. 5.31). Another consideration is that not all pores in a given stratum become sealed at the same time (perhaps closing over a period of ~50 yr at Dye-3, but ~500 years at Vostok, for example) so the air bubble gas record should be considered a "low pass" filtered record, with each sample of analyzed crushed ice, being representative of gas concentrations over several tens to several hundreds of years. The highest resolution records should therefore be found in high accumulation rate areas where snow is buried quickly and pore close-off is rapid. Such conditions tend to be found in warmer polar environments such as southern Greenland or in the more maritime sections of Antarctica. Unfortunately, this can create an additional

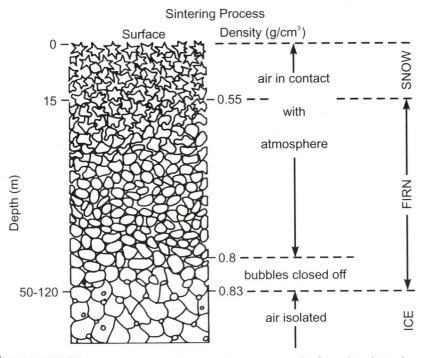

FIGURE 5.30 Schematic diagram illustrating how air is trapped in firn and ice during the process of sintering or recrystallization of snow crystals. Depending on the accumulation rate, "pore close off" (complete isolation of air bubbles) may take up to 2500 years; this time will have varied in the past with changes in accumulation rate (see Fig. 5.31) (Raynaud, 1992).

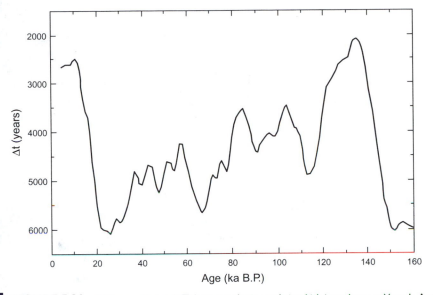

FIGURE 5.31   Difference in age (ΔT) between the ice and air which it encloses, at Vostok, Antarctica, based on a model of the air trapping process. Changes in accumulation rate over time result in variations in ΔT (Barnola *et al.,* 1991).

problem with some trace gas records if surface melting and refreezing has occurred. This can seriously affect CO_2 levels in the ice, as appears to have happened in ice cores from Dye-3, southern Greenland; earlier records from that site are now considered to be suspect (Jouzel *et al.,* 1992; Sowers and Bender, 1995).

Figure 5.32 shows the CO_2 and CH_4 records over the last 220,000 yr from Vostok, in comparison with the estimated temperature change in the atmosphere (above the surface inversion) derived from δD, taking into account the air-ice age difference and its changes with depth (Jouzel, *et al.,* 1993b). It is clear that there is a very high correlation between ΔT and ΔCH_4 ($r^2 \neq \sim 0.8$). During glacial times CO_2 levels were around 180–190 parts per million by volume (p.p.m.v.) compared to interglacial levels of 270–280 p.p.m.v. Similarly, CH_4 levels were around 350–400 parts per billion by volume (p.p.b.v.) in glacial times, versus $\sim$650 p.p.b.v. in interglacials. Of particular significance is the phase relationship between CO_2 levels and ΔT during the transition from glacial to interglacial climate and back again to glacial times. At the change from Stage H to Stage G (penultimate glacial stage to last interglacial), CO_2 was essentially in phase with ΔT, as far as can be determined, given the uncertainty in ice-air age difference as discussed already. Similar in-phase relationships are seen in the Byrd and Dome C records (Raynaud and Barnola, 1985; Neftel *et al.,* 1988). However, in the subsequent shift towards colder conditions (from $\sim$130 ka to 115 ka B.P.) CO_2 levels remained high while ΔT dropped by an estimated 7 °C (Barnola *et al.,* 1987). This change in temperature occurred before continental ice growth in the northern hemisphere started to influence oceanic $\delta^{18}O$ and sea level (Chappell and Shackleton, 1987). The most rapid decline in CO_2 occurred from $\sim$115–105 ka B.P., when levels fell from $\sim$265 to 230 p.p.m.v. This points to some mechanism first

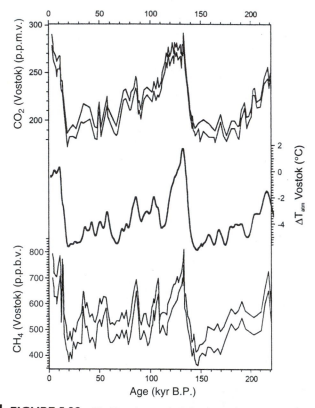

FIGURE 5.32   The Vostok record of changes in the concentration of carbon dioxide (top) and methane (bottom) with temperature *above the surface inversion* expressed as differences from present. Temperatures are estimated from changes in δD. The difference in age between the air and the enclosing ice has been taken into consideration, as have changes in this value, with variations in accumulation rate over time (Jouzel *et al.,* 1993b).

initiating a change in climate and subsequently leading to a situation in which atmospheric CO_2 levels were drawn down. The most probable mechanism for such a scenario is orbitally driven radiation changes, which brought about changes in the deep ocean circulation. This may have resulted in increased biological activity in areas of upwelling, which then brought about a reduction in atmospheric CO_2 levels. The radiative consequences of such a reduction would have reinforced any orbitally induced cooling, eventually leading to full glacial conditions. At a later stage, the rapid CO_2 increase at glacial-interglacial transitions may have been more related to changes in the surface ocean circulation (Barnola *et al.,* 1987).

The long-term record of CH_4 is broadly similar to that of CO_2 in the sense of large glacial-interglacial changes, but some important differences are nevertheless apparent (Fig. 5.32). Whereas CO_2 levels declined slowly from the last interglacial (Stage G) to Stage B, methane levels following Stage G remained generally low, punctuated by higher levels during interstadial times (Chappellaz *et al.,* 1990; Brook *et al.,* 1996). This difference in the CO_2 and CH_4 records reflects the fact that the primary driving forces for CO_2 and CH_4 are different. Atmospheric CO_2 levels are largely the result of oceanic changes, whereas there is relatively little CH_4 dissolved in the ocean and atmospheric levels are driven by changes in source areas on the continents. In particular, the extent of wetlands in the tropics (but also at high latitudes of the northern hemisphere) is of critical importance to CH_4 levels in the atmosphere. This points to the significance of monsoon circulations and their influence on the extent of low latitude wetland areas over glacial-interglacial cycles (Petit-Maire *et al.*, 1991). Considering that CH_4 is removed from the atmosphere largely by hydroxyl ion (OH^-) oxidation in the atmosphere, and that it is likely such a sink was more effective (with more abundant water vapor in the atmosphere) during warmer interglacial times, Raynaud *et al.* (1988) estimate that the global emissions of CH_4 increased by a factor of 2.3 from glacial to interglacial times. This compares with an observed increase (in the ice core record) of 1.8. Such an increase probably resulted from more extensive tropical wetlands and attendant anaerobic bacterial methanogenesis, and from higher rates of bacterial activity in high latitude peatlands during interglacials. Support for this hypothesis comes from detailed CH_4 measurements spanning the Holocene period in the GRIP (Summit) ice core from Greenland (Blunier *et al.*, 1995). Methane levels reached a minimum of ~590 p.p.b.v. at 5200 years B.P. compared to early and late Holocene levels around 730 p.p.b.v. (Fig. 5.33). This decline in CH_4 levels from the early to mid-Holocene corresponds to the well-recognized reduction in area of tropical wetlands over this interval with maximum aridity around 5000–6000 yr B.P. (Street-Perrott, 1993). The subsequent rise in CH_4 levels is thought to be due to growth of high latitude peatlands, since tropical regions in general remained relatively dry in the late Holocene.

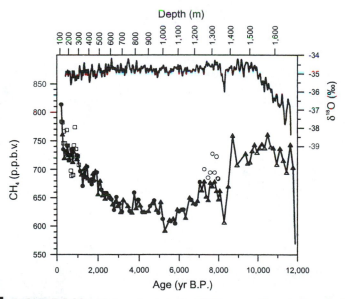

FIGURE 5.33 Methane levels in the GRIP ice core (Summit, Greenland) during the last 12,000 yr. A pronounced drop in $\delta^{18}O$ at ~8250 (calendar years) B.P. corresponds to a sudden reduction in CH_4. The decline in CH_4 to ~5200 B.P. is related to a reduction in tropical wetlands. The subsequent CH_4 increase is ascribed to high-latitude peatland expansion (Blunier *et al.*, 1995).

As noted earlier, the large difference in age between gas content and the age of the enclosing ice in areas of low accumulation like Vostok makes it difficult to compare directly the isotopic and gas records. This problem is minimized at Summit, Greenland where the CH_4 and $\delta^{18}O$ records over the period from 8000 to 40,000 yr B.P. provide clear evidence that large CH_4 and isotopic shifts have occurred essentially simultaneously (Chappellaz *et al.*, 1993; Brook *et al.*, 1996) (Fig. 5.34). Such rapid changes in the isotopic content of Greenland snow are considered to be linked to changes in North Atlantic themohaline circulation (deepwater production), which may thus provide the connection between tropical wetland extent and high latitude temperature. Alternatively, the higher CH_4 levels may reflect de-

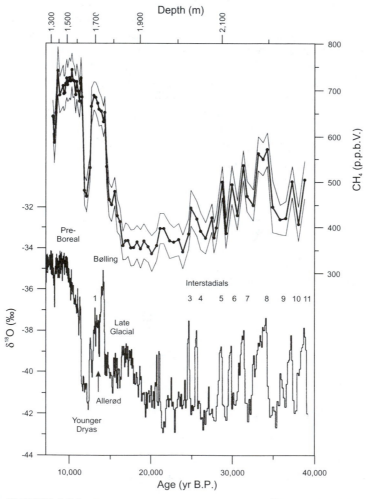

FIGURE 5.34 The record of CH_4 in air bubbles, and $\delta^{18}O$ of ice from Summit, Greenland (based on the GRIP ice core). The solid line in the upper figure is the mean concentration and the thin lines represent measurement uncertainty (2σ). Higher CH_4 levels are associated with higher $\delta^{18}O$ levels (Chappellaz *et al.* 1993).

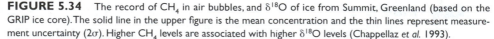

gassing of northern peatlands, in unglaciated areas (such as northern Eurasia and Alaska) during the warmer "interstadial" episodes seen in Fig. 5.34. Whatever the cause of these rapid oscillations, they will eventually provide a valuable chronological tool for matching the Antarctic and Greenland ice core records, at least in those areas of high accumulation where high-resolution trace gas data can be obtained.

Long-term changes in nitrous oxide (N_2O), another greenhouse gas, have also been determined from air bubbles in Antarctic ice — in this case from the Byrd ice core (Leuenberger and Siegenthaler, 1992). As with CO_2 and CH_4, N_2O levels were also much lower in glacial times, 30% lower than levels in the Holocene (~190 p.p.b.v. vs ~265 p.p.b.v.) (Fig. 5.35). Lower atmospheric N_2O concentration probably resulted from a reduction in the production rate in soils during glaciations, when the global terrestrial biomass was significantly reduced (estimated reductions

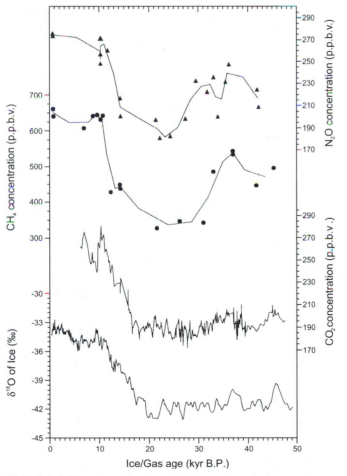

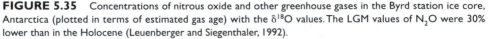

FIGURE 5.35 Concentrations of nitrous oxide and other greenhouse gases in the Byrd station ice core, Antarctica (plotted in terms of estimated gas age) with the $\delta^{18}O$ values. The LGM values of N_2O were 30% lower than in the Holocene (Leuenberger and Siegenthaler, 1992).

range from 22–57% of preindustrial conditions). Interestingly, the main sink for N_2O is in the stratosphere, where it reacts with ozone. Hence, stratospheric ozone levels may have been higher during the LGM, leading to lower ultraviolet (UV) radiation levels in the troposphere.

Spectral analysis of the CO_2 and CH_4 records in Vostok ice reveals periodicities that appear to be related to orbital forcing, though the signal is not simple. In the original 160 ka chronology of CO_2, for example, the strongest periodicity was in the precessional frequency band (~20 ka) but in the extended 220 ka record a strong obliquity (~41 ka) periodicity is apparent, as in the δD series. The CH_4 shows a strong precessional cycle over the last 160 ka but as can be seen in Figure 5.32 this is less significant over the last 220 ka. For reasons as yet unknown, CH_4 does not undergo strong variations in the penultimate glacial period and, indeed, there is no evidence of a lag between CO_2 and ΔT going from warm conditions around 220 ka to 160 ka B.P., like that observed in the subsequent glacial cycle.

If the δD or $\delta^{18}O$ record from Vostok is considered to be the "product" of various forcing factors, multivariate analysis can be used to try to identify the most important of these "independent" variables (Genthon et al., 1987). Such an analysis reveals that changes in greenhouse gases are of primary importance in explaining the variance of temperature changes in Antarctica over the last glacial-interglacial cycle, though it is certainly likely that orbitally induced changes in solar radiation distribution in some way modulate changes in greenhouse gases such as CO_2, CH_4, and N_2O. These considerations led Genthon et al. (1987) to conclude that "climatic changes [are] triggered by insolation changes, with the relatively weak orbital forcing being strongly amplified by possibly orbitally induced CO_2 changes." To this we could certainly add CH_4, N_2O and other atmospheric constituents (aerosols, DMS, etc.). This is of particular significance in understanding the synchronism of major glaciations in the northern and southern hemispheres, which are difficult to explain from orbital theory alone.

Although the radiative effects of greenhouse gases (and their associated feedbacks) are clearly important for the earth's energy balance, biogenic sulphur production (dimethyl sulphide, DMS, which oxidizes to sulphate in the atmosphere) also plays an important role (Charlson et al., 1987). With higher levels of DMS production, the number of cloud condensation nuclei increase, leading to more extensive stratiform clouds, a higher planetary albedo, and hence (perhaps) lower temperatures (Legrand et al., 1988). Measurements of methane sulphonic acid (an oxidation product of DMS) in Dome C and less specific measurements of non-sea salt sulphate (nss SO_4^{--}) in Vostok ice reveal important increases during cold periods (even after taking into account changes in accumulation rate and possible enhanced transportation of nss SO_4^{--} to Antarctica at those times) (Saigne and Legrand, 1987; Legrand et al., 1987, 1991). At Vostok, for example, in Stages B, C, and D, nss SO_4^{--} levels were 27–70% higher than in interglacial Stages A and H, suggesting greatly enhanced productivity of the oceanic biota that produce DMS during the period from ~70–18 ka B.P. (Legrand et al., 1991). This probably reflects higher biological activity in upwelling areas of the ocean during colder intervals (Sarnthein et al., 1987).

It is possible to estimate the direct radiative effects of the glacial-interglacial changes in greenhouse gases on global temperature change from radiative-convective models; for CO_2 this is around 0.5 °C; for CH_4, 0.08 °C; and for N_2O, 0.12 °C (Leuenberger and Siegenthaler, 1992). However, the more important issue concerns the associated feedbacks, involving clouds, snow cover, sea ice, etc., which may have resulted from (and amplified) these changes. The overall (equilibrium) temperature change due to doubling of CO_2, including radiative effects and feedbacks, is referred to as the climate sensitivity and the amplification effect as the net feedback factor (f). General circulation model experiments lead to estimates of f in the range of 1–4 so that the direct radiative effects of a doubling of CO_2 levels will be increased by a factor of 1 to 4. Using the observed changes in greenhouse gases, dust, non-sea salt sulphate and global ice volume, together with calculated orbitally induced changes in radiation, Lorius *et al.* (1990) attempted to account for the overall variance in Vostok temperature over the last glacial-interglacial cycle. Their analysis revealed that 50 ± 10% of the variance of the temperature record (~6 °C in the atmosphere above Vostok) is accounted for by greenhouse gas changes. If this figure can be applied to global glacial-interglacial temperature changes (estimated as ~4–5 °C; Rind and Peteet, 1985) then it suggests that changes in greenhouse gases were responsible for ~2 °C of the global temperature change over the last climatic cycle. Comparing this figure with the calculated direct radiative effect (0.7 °C) suggests a net feedback amplification value (f) of ~3. This is in line with model estimates, though at the high end of the general range. Possibly this reflects a higher climate sensitivity in glacial times, when "slow feedbacks" associated with ice extent on land, and semi-permanent sea ice and ice shelves, may have played a stronger role in amplifying the radiative forcing than they do at the present time.

One of the more remarkable features of ice cores from both hemispheres is the dramatic increase in aerosol concentration during the last glacial period. At Vostok, there were three main episodes of increased aluminum concentrations (an index of continental dust, which is primarily made up of aluminosilicates); these peaks are centered on 160 ka, 60 ka, and 20 ka B.P. (i.e., cold Stages H, D, and B) (De Angelis *et al.*, 1987; Petit *et al.*, 1990). Taking into account changes in accumulation rates over time, annual dust fluxes have been calculated (Fig. 5.36). These show that dust flux was 15 times higher in Stage B than during the Holocene. This is related to both an increase in mean wind speed (also leading to higher sea salt sodium levels in glacial times; De Angelis *et al.*, 1987) as well as drier conditions in many arid and semiarid areas of the world. Sources of dust in the southern hemisphere are the semiarid areas of Patagonia, and the extensive continental shelves that were exposed during glacial times. Isotopic studies on dust from the LGM in Dome C clearly point to Patagonia as the primary source region (Grousset *et al.*, 1992). Similarly, geochemical analysis of dust in Byrd, Vostock, and Dome C ice cores indicates a mixture of both marine carbonates and clays (mainly illite) from the Patagonia desert areas (Briat *et al.*, 1982; Delmas and Petit, 1994). Studies of the optical properties of LGM dust from Dome C suggest that there may have been important radiative effects on surface temperature due to the increased aerosol load (Royer *et*

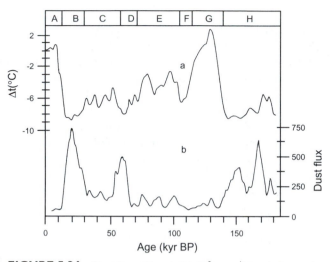

FIGURE 5.36   Dust flux at Vostok (× 10⁻⁹cm yr⁻¹) in relation to the estimated *surface* temperature change from the present, derived from changes in δD. Both records have been mathematically smoothed (see Fig. 5.32) (Petit *et al.*, 1990).

al., 1983). Although subject to considerable uncertainty, they estimate that the temperature effect resulted in a *warming* of ~2 °C over Antarctica, which would compensate somewhat for the lower levels of atmospheric greenhouse gases at that time (Overpeck *et al.*, 1996).

Glaciochemical analysis of ice cores from Greenland and Antarctica provides a comprehensive perspective on air mass characteristics that can be characterized in terms of the total chemistry in the ice. Thus, Mayewski *et al.* (1997) recognize that in the GISP2 ice core two major circulation regimes prevailed during the last glacial interglacial cycle. One regime is dominated by a polar/high-latitude air mass (with higher levels of continental dust and marine-derived ions) and a second mid-, low-latitude air mass (with high levels of biogenic nitrate and ammonium ions). Figure 5.37 shows the variations of these two regimes over the last 110,000 yr. As one might expect, this reveals that the abrupt changes in δ¹⁸O during the last glacial (Dansgaard-Oeschger oscillations) were associated with pronounced shifts in circulation regimes. During cold events (low δ¹⁸O) a polar/high latitude circulation prevailed whereas during the Holocene and mild interstadials, the mid-, low-latitude circulation pattern was more prevalent (Mayewski *et al.*, 1994). The changes in dominance of circulation regimes seen so prominently during the last glacial period are also identifiable (albeit more subtly) during the Holocene, enabling the principal circulation changes during recent millennia to be identified (O'Brien *et al.*, 1995). Although such changes are defined by the ice-core geochemistry in a remote part of Greenland, there is evidence that they may have significance far beyond this region because of teleconnections linking the atmospheric circulation over long distances (Stager and Mayewski, 1997).

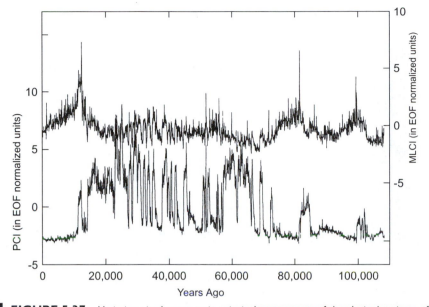

FIGURE 5.37 Variations in the two main principal components of the glaciochemistry of the GISP2 ice core over the last 110,000 yr (based on covariations in sodium, potassium, ammonium, calcium, magnesium, sulfate, nitrate, and chloride ions). The lower line is interpreted as changes in the relative importance of polar/high latitude circulation regimes and associated air masses, and the upper line as changes in mid-, low-latitude circulation regimes (Mayewski *et al.*, 1997).

5.4.4 Volcanic Eruptions Recorded in Ice Cores

Explosive volcanic eruptions may produce large quantities of sulfur and chlorine gases, which are converted to acids in the atmosphere and may be carried long distances from the eruption site (Devine *et al.*, 1984; Rampino and Self, 1984). When these acids are washed out at high latitudes, the resulting acidic snowfall (and dry deposition of acidic particles directly on the ice sheets) produces high levels of electrical conductivity, which appear as "spikes" above natural background levels in ice cores (Hammer, 1977; Hammer *et al.*, 1980). In most cases these acidity spikes result from excess sulfuric acid events (Figs. 5.14 and 5.38). Hammer's original studies showed a remarkable similarity between electrolytic conductivity in the Greenland Crête ice core and Lamb's Dust Veil Index (Lamb, 1970, 1983), indicating that the elevated acidity (above background levels) could be used as an index of volcanic explosivity. Large explosive eruptions are commonly associated with lower temperatures on a hemispheric (or sometimes even global) scale (Bradley, 1988) although circulation anomalies can lead to warmer conditions in some regions (Robock and Mao, 1995).

Volcanically induced conductivity variations have now been extensively investigated in other cores from Greenland, Antarctica and elsewhere, often with ionic analysis to determine the precise chemistry of the acidity spikes (Holdsworth and Peake, 1985; Mayewski *et al.*, 1986; Legrand and Delmas, 1987; Lyons *et al.*, 1990;

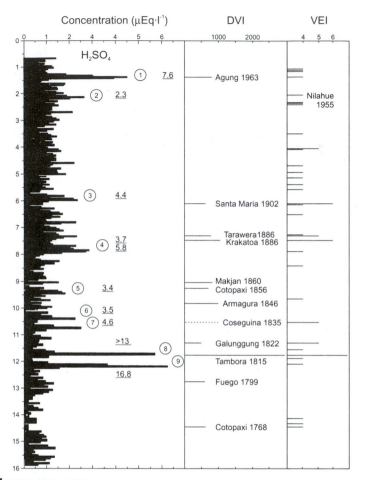

FIGURE 5.38 Sulfuric acid profile from Dome C, Antarctica over the last ~200–250 yr, compared to Lamb's Dust Veil Index (DVI: Lamb, 1970) and the volcanic explosivity index (VEI) of Newhall and Self (1982). The total sulfate deposition at Dome C (in kg km⁻²) for major volcanic eruptions is shown by the underlined values (Legrand and Delmas, 1987). Note the two major eruptions in the early nineteenth century, the earlier of which is not seen in the northern hemisphere ice cores (see Fig. 5.14).

Moore *et al.*, 1991; Delmas *et al.*, 1992; Zielinski *et al.*, 1994). Chemical analysis enables more precise "fingerprinting" of individual eruptions, some of which produce large amounts of HCl or HF, for example, rather than H_2SO_4 (Symonds *et al.*, 1988; Hammer *et al.*, 1997).

The longest record of SO_4^{2-} concentration in an ice core comes from the GISP2 site in Greenland and spans the last ~110,000 yr (Zielinski *et al.*, 1996). This provides a unique record of the volcanic aerosol, which is of most importance climatically. Temporal resolution of the analyses is ~2 yr for the last 11,700 yr, but then declines back in time (3–5 yr-samples back to ~14,800, 8–10 to ~18,200, 10–15 to ~50,000, and up to 50 yr per sample towards the oldest section of the core). This makes it difficult to compare these short-lived events directly because a single acidity spike readily seen in the Holocene would be effectively diluted over the longer periods represented by deeper samples. Nevertheless, there are many episodes when SO_4 concentrations exceeded those levels observed after recent major eruptions, suggesting that SO_4 levels may have been extraordinarily high at times in the past, and/or maintained for up to several decades at climatically significant levels. Thus the SO_4 record from GISP2 provides a very important perspective on volcanic events on the decade to millennial timescale (Zielinski *et al.*, 1996). Several periods stand out as having been particularly active, especially 8–15 ka and 22–35 ka B.P. (Fig. 5.39). It is interesting that these times correspond to the times of major ice growth and decay. This suggests that the greater frequency of eruptions may be a direct consequence of increased crustal stresses associated with glacial loading and unloading, and/or changes in water loading of ocean basins, especially in areas with a thin lithosphere (such as the volcanically active island arcs of the Pacific Rim (Zielinski *et al.*, 1996). Thus, there may be a direct feedback between continental ice growth and explosive volcanism, with orbitally driven ice volume changes driving shorter-term climatic changes associated with the eruptions.

The five largest sulfate anomalies at GISP2 over the last 2000 yr were in 1831 (Babuyan, Philippines), 1815 (Tambora, Indonesia), 1640 (Komataga-Take, Japan), 1600 (Huaynaputina, Peru), and 1259 (possibly El Chichón, Mexico or a near-equatorial source) (Zielinski *et al.*, 1994). Several major eruptions were recorded in cores from Antarctica as well as Greenland, indicating near-equatorial events from which dust and gases spread into both hemispheres (Langway *et al.*, 1988; Palais *et al.*, 1992; Delmas *et al.*, 1992). This points to a difficult problem in assessing the overall eruption size from measurements in an individual ice core. Ice cores from Greenland, for example, are likely to record Icelandic and Alaskan eruptions as larger than equivalent-sized eruptions at lower latitudes, simply because Greenland is located closer to the high latitude source regions (Hammer, 1984). However, even deposition from nearby eruptions will not be dispersed uniformly over the ice sheets, so estimates based on single (~10–15cm diameter!) ice-core samples may be misleading (Clausen and Hammer, 1988). Indeed, some major eruptions were not registered at all in some ice cores (Delmas *et al.*, 1985). Ideally, a suite of cores extending longitudinally along the major mountain ranges of the world is needed to get a more global picture of volcanic aerosol dispersal. However, in many high altitude, low latitude ice cores, deposition of alkaline aerosols neutralizes the volcanic acids and hence eliminates the eruption signal (also a problem during glacial times, when atmospheric dust levels were much higher than in the Holocene). Nevertheless, the collection of many short cores from a wide area of the larger ice sheets, and along polar/alpine transects (e.g., from the South Pole to Ecuador) will eventually enable a better assessment of the spatial pattern of acid deposition to be made, leading to a more reliable record of past explosive volcanism in both the northern and southern hemispheres (Clausen and Hammer, 1988; Mulvaney and Peel, 1987). Cores from Antarctica reveal major eruptions affecting the southern hemisphere that were not recorded in the northern hemisphere, including a second major event shortly before Tambora (possibly around 1809) (Delmas *et al.*, 1992; Cole-Dai *et al.*, 1995).

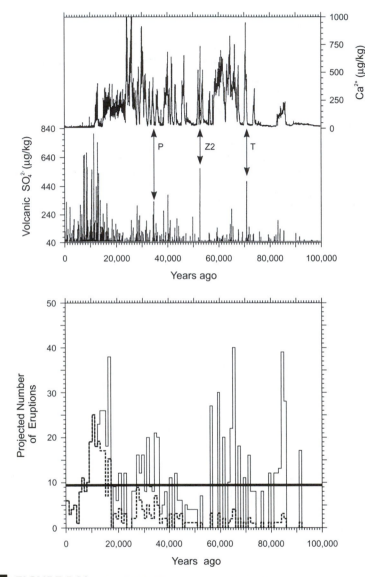

FIGURE 5.39 The GISP2 (Greenland) record of sulfate from volcanic eruptions with the terrestrial dust record, represented by Ca^{2+} (upper diagram). Time is in years before A.D. 2000. The eruptions of Toba, Indonesia (~71,000 yr B.P.), the Icelandic Z2 ash zone (~53,680 years B.P.), and the Phlegraean Fields eruption/Campanian ignimbrite (~34,500 yr B.P.) are indicated. The lower diagram shows the number of eruptions per millennium exceeding an SO$_4$ concentration of 74 p.p.b. (dashed line); this is approximately the magnitude of the largest historical eruption (Tambora) recorded at GISP2. Because of the decrease in temporal resolution with depth, the number of individual eruptions is underestimated back in time. The solid line is an estimate of the number of eruptions that may have occurred, if the sampling resolution was the same as in the Holocene throughout the record; thus the lines are coincident for the last ~12,000 yr (Zielinski *et al.*, 1997).

One approach that can be used to "scale" the ice-core sulfate (or acidity) record to compensate for long-distance dispersal of volcanic aerosols is to use the concentration of atomic bomb fallout on the Greenland Ice Sheet (from both high and low latitude sources) as a guide to how aerosols, dispersed via the stratosphere, are depleted en route from the source area (Clausen and Hammer, 1988). Using this idea, Zielinski (1995) estimated atmospheric optical depth changes resulting from major explosive eruptions of the last 2100 years. His analysis indicates that the overall impact of volcanic eruptions on atmospheric turbidity was significantly greater in the last 500 years than in the preceding 1600 years. In particular, multiple large eruptions in the interval A.D. 1588–1646 and 1784–1835 probably had a significant cumulative impact on atmospheric optical depth, leading to cooler conditions at those times (Bradley and Jones, 1995).

5.4.5 Correlation of Ice-core Records from Greenland and Antarctica

Because it is not yet possible to date deep ice cores directly by radiometric means (at least not beyond the radiocarbon timescale), establishing a precise chronology for long ice-core records relies on flow models, which are very sensitive to assumptions about past accumulation history and ice dynamics. Thus *absolute* chronologies are quite uncertain, making comparison of paleoclimatic records from one region to another quite difficult (Reeh, 1991). One means of correlating ice core records over long distances, at least in a relative sense, is to compare those constituents that vary on a global scale and are thus likely to have fluctuations that are essentially simultaneous. One such parameter has already been mentioned: ^{10}Be "spikes," resulting from some short-term increase in production rate (the cause of which is, as yet, unknown) can be seen in ice cores from Byrd, Vostok, and Dome C, Antarctica, and Camp Century, Greenland, providing chronostratigraphic horizons that can be used to correlate these records directly (Raisbeck *et al.*, 1987; Beer *et al.*, 1992). Another approach relies on changes in atmospheric gas composition. As the mixing time of the atmosphere is short (on the order of 1–2 yr), changes in gas content should be essentially synchronous from the Arctic to Antarctica, so the temporal record of gases should be in parallel. Bender *et al* (1994) have used this fact to good advantage in linking the chronologies of the Vostok and GISP2 ice cores. They use variations of $\delta^{18}O$ in gas bubbles in the ice as the common thread that ties the two records together. The $\delta^{18}O_{ATM}$ in the atmosphere today is +23.5‰ (relative to SMOW), a result of the balance between the fractionation that occurs during photosynthesis and that which occurs during respiration (the "Dole effect"). The $\delta^{18}O_{SW}$ in the ocean has changed over time due to removal of water relatively enriched in ^{16}O during times of continental ice build-up on the continents. Such changes are recorded in the $CaCO_3$ of benthic forams (see Chapter 6). However, if the isotopic content of the ocean changes, the atmospheric oxygen isotope content will undergo parallel changes, because all photosynthetically derived oxygen is affected, directly or indirectly, by the oceanic isotopic composition (i.e., directly, if the O_2 is produced by marine biota; indirectly, if O_2 is produced by terrestrial biota that are affected by isotopic changes in the hydrological cycle). By extracting O_2 from bubbles in the ice, a direct measure of these changes can be obtained (Bender *et al.*, 1985; Sowers *et al.*, 1991, 1993). After taking into account the air-ice age difference at each site, the very similar variations in $\delta^{18}O_{ATM}$ at Vostok and Summit, Greenland provide a compelling argument for aligning the two records relative to one another, though the absolute chronology remains imprecise (Fig. 5.40). The relatively low frequency changes in $\delta^{18}O_{ATM}$ (reflecting slow changes in oceanic isotope composition) contrast with the higher frequency changes in $\delta^{18}O_{ICE}$, which reflect changes in fractionation processes during the formation and delivery of precipitation to each site. During the last glacial period at Summit, very abrupt, large amplitude changes in $\delta^{18}O_{ICE}$ are characteristic of the record. Of the 22 "interstadials" recorded in the GISP2 ice core in the interval from 22–105 ka B.P., 8 are also identifiable in the Vostok δD record. It appears that only the longer episodes (those lasting >2ka in Greenland) are also seen in Antarctica, associated with increases in δD of >15‰. Because the isotopic changes in Greenland are more frequent and more rapid, Bender *et al.* (1994) argue that the warming events began in the northern hemisphere and eventually extended to Antarctica if they persisted for a long enough period of time. The most probable mechanism for such a linkage involves changes in heat flux brought about by oceanic circulation changes.

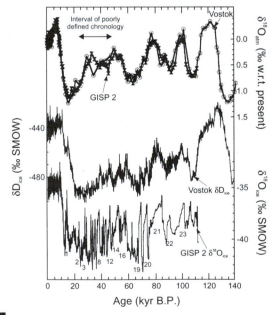

FIGURE 5.40 The $\delta^{18}O_{ATM}$ record in the GISP2 and Vostok ice cores (top) derived from gas bubbles in each core. The strong correlation has enabled the δD and $\delta^{18}O$ records in the ice (lower two series) to be aligned chronologically, though the *absolute* timescale remains uncertain. The two records show that interstadials in the Greenland record can also be identified in Antarctica if they lasted more than ~2 kyr (Bender *et al.*, 1994).

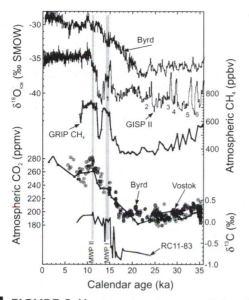

FIGURE 5.41 Isotopic and gas records from Byrd and Vostok (Antarctica) and Summit, Greenland (GISPII and GRIP). The records have been aligned by finding the optimum fit between $\delta^{18}O$ in gas bubbles in the ice (taking into account the time-varying age of gases trapped in the ice). With the chronologies phase-locked in this way, it is apparent that CO_2 increased just as $\delta^{18}O$ in ice at Byrd increased, whereas $\delta^{18}O$ in Greenland ice did not increase until ~3300 yr later. The methane record has more in common with the Greenland record, as it is largely controlled by northern hemisphere continental sources (Sowers and Bender, 1995).

An important by-product of being able to align the Antarctic and Greenland ice-core records together in time, using their respective $\delta^{18}O_{ATM}$ content, is the ability to then directly compare changes in other parameters such as CO_2 and CH_4 with temperature changes, as recorded by $\delta^{18}O_{ICE}$ (Fig. 5.41). By forcing the records of GISP2, Greenland and Byrd, Antarctica to match, Sowers and Bender (1995) showed that CO_2 levels began to increase by ~17 ka B.P., well before there were large increases in $\delta^{18}O_{ICE}$ in the Greenland record. This rise in CO_2 is more or less synchronous with an increase in $\delta^{18}O_{ICE}$ at Byrd, as well as with an increase in SSTs in the Southern Ocean and tropical Atlantic, and with eustatic sea-level rise. By contrast, Greenland $\delta^{18}O_{ICE}$ shows no comparable changes, with a rapid shift from late glacial conditions delayed until ~14.7 ka B.P. Hence, it appears that warming in Greenland and the North Atlantic was delayed until long after the southern hemisphere and the Tropics had begun to emerge from the last glacial period. A possible explanation is that the North Atlantic polar front was locked in position around ~45° N by the atmospheric circulation around the Laurentide Ice Sheet; only when the Laurentide was reduced in size to the point that it no longer had a major influence on circulation could the polar front and associated air masses shift northward (Keigwin *et al.*, 1991). This shift is documented by the pronounced increase in $\delta^{18}O_{ICE}$ at GISP2 around 14.7 ka B.P.

5.4.6 Correlation Between Ice Cores and Marine Sediments

Because changes in the isotopic content of the ocean due to continental ice sheet growth affect atmospheric $\delta^{18}O_{ATM}$, variations in ^{18}O in both marine sediments and ice cores represent a common denominator that can be used to link both types of record. There are several complicating factors that arise in such a comparison. Just as the time to pore close-off in ice cores acts as a time-varying lowpass filter on the ice bubble gas record, so bioturbation in marine sediments acts to smooth the changes that occurred in oceanic isotopic composition. Also, bottom water temperatures have changed over glacial-interglacial time, and such changes also influence the isotopic composition of benthic forams, requiring adjustments to be made to obtain an unbiased time series of $\delta^{18}O_{SW}$. The overall residence time of O_2, with respect to processes of photosynthesis and respiration, is around 2–3 ka, so that changes in $\delta^{18}O_{ATM}$ will lag those in the ocean by that amount of time. Finally, there are no compelling reasons to suppose that the "Dole effect" has remained constant over time; indeed it is likely that $\delta^{18}O_{ATM}$ has not simply followed $\delta^{18}O_{SW}$ variations in the same way at all times; changes in the relative primary productivity of the terrestrial biosphere versus the marine biosphere, changes in continental hydrology, etc. may have affected the $\delta^{18}O_{ATM}$–$\delta^{18}O_{SW}$ relationship over time. Nevertheless, in spite of these uncertainties, it is possible to make reasonable assumptions about each of these complications and then to correlate marine and ice core records over the last ~130,000 yr. Figure 5.42 shows such a comparison, using the SPECMAP

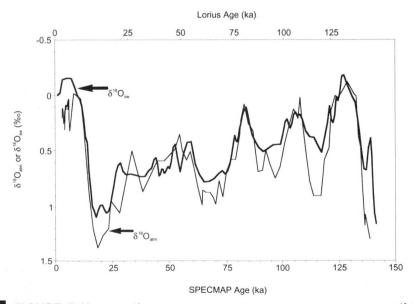

FIGURE 5.42 The $\delta^{18}O_{ATM}$ (from air bubbles in Vostok ice) compared to $\delta^{18}O_{SW}$ (from benthic foraminifera) plotted on the SPECMAP timescale (Martinson, *et al.*, 1987) (see Section 6.3.3). The $\delta^{18}O_{ATM}$ record has been forced to match the marine $\delta^{18}O$ record (Sowers *et al.*, 1993). The upper timescale is the ice core chronology estimated by Lorius *et al.* (1985).

timescale for the marine $\delta^{18}O_{SW}$ record, and optimizing the correlation with the ice core $\delta^{18}O_{ATM}$. Note that this comparison is between the age of the gas bubbles in the ice and the marine record; the age of the enclosing ice is older, the age difference being larger during low accumulation glacial periods than warmer interglacials. The overall correlation between these two records is excellent, though the fit is better at some times than at others, probably reflecting the fact that the Dole effect has *not* been constant over time. There is also very good agreement with the original Lorius *et al.* (1985) time-scale, which was based on fairly simple assumptions about past changes in accumulation rate at Vostok. However, the comparison suggests that the original ice-core chronology is "too old" at the penultimate glacial-interglacial transition (Termination 2) by 5–6 ka *relative to the estimated chronology of the marine record*. This argument has also been made by several other investigators who have compared marine records with the Vostok chronology (Pichon *et al.*, 1992; Shackleton *et al.*, 1992). This does not resolve the question of whether the absolute chronology is correct and it is conceivable that both records are still incorrect in absolute terms. Nevertheless, the alignment of the ice core and marine records is extremely valuable because it then enables other records to be compared, providing insight into how different parts of the climate system are related in time. Figure 5.43 shows such a comparison between the Vostok δD record and SSTs from a sub-Antarctic oceanic site (46° S). The δD at Vostok is a function of temperatures in the air above the surface inversion and can be expected to vary with air mass temperatures over the surrounding oceans. Clearly, when aligned on a common temporal

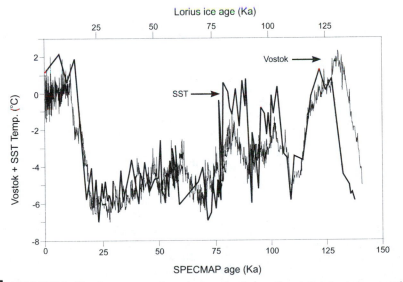

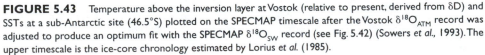

FIGURE 5.43 Temperature above the inversion layer at Vostok (relative to present, derived from δD) and SSTs at a sub-Antarctic site (46.5°S) plotted on the SPECMAP timescale after the Vostok $\delta^{18}O_{ATM}$ record was adjusted to produce an optimum fit with the SPECMAP $\delta^{18}O_{SW}$ record (see Fig. 5.42) (Sowers *et al.,* 1993). The upper timescale is the ice-core chronology estimated by Lorius *et al.* (1985).

basis, there is a strong correlation between these series, providing support for the many assumptions made in using the $\delta^{18}O_{ATM}-\delta^{18}O_{SW}$ relationship to relate the two records (Sowers *et al.*, 1993).

5.4.7 Ice-core Records from Low Latitudes

Ice caps are not confined to polar regions; they also occur at very high elevations in many mountainous regions, even near the Equator (Thompson *et al.*, 1985b). High elevation ice cores provide invaluable paleoenvironmental information to supplement and expand upon that obtained from polar regions. By 1998 six high altitude sites had yielded ice cores to bedrock — Quelccaya and Huascarán in Peru, Sajama in Bolivia, and Dunde, Guliya, and Dasuopu Ice Caps in western China (see Fig. 5.1). In the Dunde and Huascarán ice cores the glacial stage ice is thin and close to the base, making a detailed interpretation very difficult (Thompson *et al.*, 1988b, 1989, 1990, 1995). Nevertheless, even these short glacial sections can yield important information. For example, in ice cores from the col of Huascarán, Peru (6048 m) the lowest few meters contain ice from the last glacial maximum, with $\delta^{18}O$ ~8‰ lower than Holocene levels, and a much higher dust content (Thompson *et al.*, 1995). The lower $\delta^{18}O$ suggests that tropical temperatures were significantly reduced in the LGM (by ~8–12 °C), which supports arguments that changes in tropical SSTs were much lower than those indicated by the reconstructions of CLIMAP (1981), which have guided thinking on this matter for many years (see Section 6.6).

In the Guliya ice core, from the western Qinghai-Tibetan plateau, much of the ice appears to date from the last glaciation,[14] so details of conditions in this area at that time can be resolved in considerable detail (Thompson *et al.*, 1997). The long-term record reveals several stadial-interstadial oscillations since the last interglacial (Fig. 5.44) with $\delta^{18}O$ values in the "interstadials" reaching Holocene

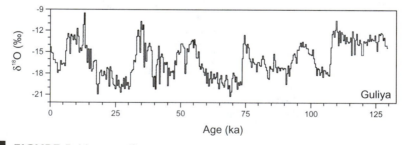

FIGURE 5.44 The $\delta^{18}O$ in an ice core from Guliya Ice Cap, western Tibet (~35 °N, 81 °E). The 308 m record was dated by correlating the CH_4 record from Vostok and GISP2 with the oscillations of $\delta^{18}O$. This assumes that the factors causing changes in $\delta^{18}O$ at Guliya would increase and decrease in phase with CH_4 (which is likely to be driven by low-latitude climatic changes) (Thompson *et al.*, 1997).

[14] ^{36}Cl in the core indicates that the very oldest section of the record (below 300 m) may be >500,000 yr old, making it the oldest ice ever recovered (Thompson *et al.*, 1997).

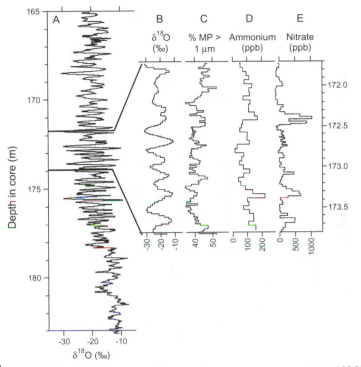

FIGURE 5.45   A detailed section of the Guliya ice core, centered around 25,000 yr B.P., showing large amplitude, rapid changes in $\delta^{18}O$, with a period averaging ~200 yr in length. For the central section, details of variations in $\delta^{18}O$, microparticles, NH_4, and NO_3 are shown; high values of $\delta^{18}O$ are associated with an increase in the percentage of dust particles >1 µm, and higher levels of ammonium and nitrate ions (Thompson et al., 1997).

levels, around −13‰. Abrupt, high amplitude changes in $\delta^{18}O$ occurred from ~15–33 ka B.P., with an average length of ~200 yr (Fig. 5.45). These oscillations are much shorter than the Dansgaard-Oeschger oscillations seen in GISP2 and are associated with higher levels of dust, NH_4, and nitrate during warm (higher $\delta^{18}O$) episodes (the opposite of what is seen in Greenland). This may indicate that warmer episodes occurred throughout the glacial stade, associated with less snow cover and more vegetation on the plateau. Whether the apparent periodicity is related to that seen in proxies of solar activity (~210 yr; Stuiver and Braziunas, 1992) remains to be seen.

Because of the high accumulation rates on mountain ice caps, high elevation ice cores can provide a high resolution record of the recent past, with considerable detail on how climate has varied over the last 1000–2000 yr, in particular (Thompson, 1991, 1992). The Quelccaya ice cores have been studied in most detail over this interval (Thompson et al., 1985, 1986; Thompson and Mosley-Thompson, 1987). Two cores extend back ~1500 yr (though only one can be reliably interpreted before ~A.D. 1200). These reveal a fairly consistent seasonal cycle of microparticles,

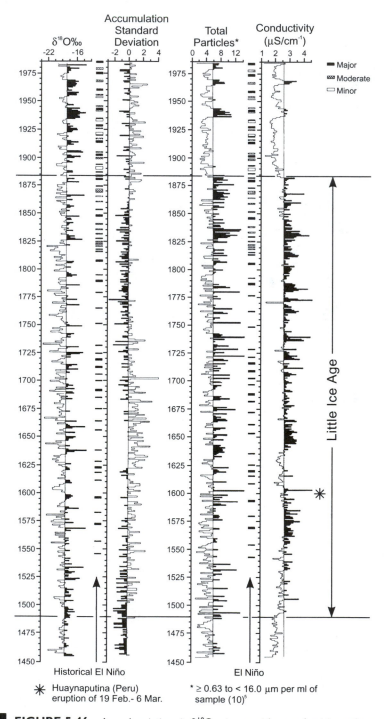

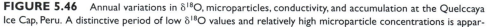

FIGURE 5.46   Annual variations in $\delta^{18}O$, microparticles, conductivity, and accumulation at the Quelccaya Ice Cap, Peru. A distinctive period of low $\delta^{18}O$ values and relatively high microparticle concentrations is appar-

conductivity and $\delta^{18}O$, which (collectively) have been used to identify and date each annual layer. Dust levels increase in the dry season (June–September) when $\delta^{18}O$ values and conductivity levels are highest, providing a strong annual signal. A prominent conductivity peak in A.D. 1600 (associated with a major eruption of the Peruvian volcano Huaynaputina in February–March, 1600) provides an excellent chronostratigraphic check on the annual layer counts.

Over the last 1000 yr $\delta^{18}O$ shows distinct variations in the Quelccaya core, with the lowest values from 1530–1900 (Fig. 5.46). This corresponds to the so-called "Little Ice Age" observed in many other parts of the world. The longer record from nearby Huascarán (9°S) provides a longer perspective on this episode; it had the lowest $\delta^{18}O$ values of the entire Holocene (Thompson *et al.*, 1995b). Accumulation was well above average for part of this time (1530–1700) but then fell to levels more typical of the preceding 500 yr (Fig. 5.47). Accumulation was also higher from A.D. ~600 to 1000. Archeological evidence shows that there was an expansion of highland cultural groups at that time. By contrast, during the subsequent dry episode in the mountains (A.D. ~1040–1490) highland groups declined while cultural groups in coastal Peru and Ecuador expanded (Thompson *et al.*, 1988a). This may reflect longer-term evidence for conditions that are common in El Niño years, when coastal areas are wet at the same time as the highlands of southern Peru are dry. Indeed, the Quelccaya record shows that El Niños are generally associated with low accumulation years, though there is no unique set of conditions observed in the ice core that permits unequivocal identification of an ENSO event (Thompson *et al.*, 1984a). Nevertheless, by incorporating ice-core data with other types of proxy record it may be possible to constrain long-term reconstructions of ENSO events (Baumgartner *et al.*, 1989).

High-altitude ice cores have experienced significant increases in temperature over the last few decades, resulting in glaciers and ice caps disappearing altogether in some places (Schubert, 1992). This is quite different from polar regions where temperatures have declined in many regions during the same period. At Quelccaya, temperatures in the last 20 yr have increased to the point that by the early 1990s melting had reached the Summit core site (5670 m), obscuring the detailed $\delta^{18}O$ profile that was clearly visible in cores recovered in 1976 and 1983 (Thompson *et al.*, 1993). In the entire 1500-yr record from Quelccaya, there is no comparable evidence for such melting at the Summit site. Similarly, at Huascarán, in northern Peru, $\delta^{18}O$ values increased markedly, from a "Little Ice Age" minimum in the seventeenth and eighteenth centuries, reaching a level for the last century that was higher than at any time in the last 3000 yr (Thompson *et al.*, 1995b). Ice cores from the Gregoriev Ice Cap (in the Pamirs) Guliya and Dunde, China also show evidence of recent warming (Lin *et al.*, 1995; Yao *et al.*, 1995);

ent, from ~A.D. 1490 to A.D. 1880. Accumulation in this period was initially high (to ~1700) then fell to low levels. A large peak in conductivity in A.D. 1600 was the result of the eruption of nearby Huaynaputina. The historical record of El Niños is also shown, based on Quinn *et al.* (1987). $\delta^{18}O$ is in ‰ relative to SMOW; accumulation is in units of the standard deviation over the last 500 yr (1σ = 34 cm of accumulation); total particulates are $\times 10^5$ ml^{-1}; conductivity is in microSiemens cm^{-1} (Thompson, 1992).

Net Accumulation (m ice eq.)

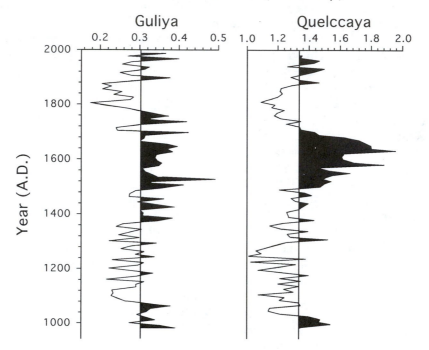

Decadal Averages

FIGURE 5.47 Decadal averages of net accumulation on the Quelccaya Ice Cap (Peru) and the Guliya Ice Cap (western Tibet) over the past 1000 yr. Broadly similar trends are seen in both areas, though the reasons are not clear. These two regions are linked by teleconnections on the timescale of ENSO events (3–7 yr) and there may be similar teleconnections at lower frequencies linking monsoon precipitation at Guliya with snowfall amounts in the mountains of Peru (Thompson et al., 1995a).

at Dunde, $\delta^{18}O$ values were higher in the last 50 years than in any other 50-yr period over the last 12,000 yr (though the record decreases in resolution with time). These records, plus evidence from other short ice cores from high altitudes (Hastenrath and Kruss, 1992) point to a dramatic climatic change in recent decades, prompting concern over the possible loss of these unique archives of paleoenvironmental history (Thompson et al., 1993). The cause of the recent warming remains controversial.

6 ■ MARINE SEDIMENTS AND CORALS

6.1 INTRODUCTION

Occupying more than 70% of the Earth's surface, the oceans are a very important source of paleoclimatic information. Between 6 and 11 billion metric tons of sediment accumulate in the ocean basins annually, and this provides an archive of climatic conditions near the ocean surface or on the adjacent continents. Sediments are composed of both biogenic and terrigenous materials (Fig. 6.1). The biogenic component includes the remains of planktic (near surface-dwelling) and benthic (bottom-dwelling) organisms, which provide a record of past climate and oceanic circulation (in terms of surface water temperature and salinity, dissolved oxygen in deep water, nutrient or trace element concentrations, etc.). By contrast, the nature and abundance of terrigenous material mainly provides a record of humidity-aridity variations on the continents, or the intensity and direction of winds blowing from land areas to the oceans, and other modes of sediment transport to, and within, the oceans (fluvial erosion, ice-rafting, turbidity currents, etc.).

The CLIMAP research group (Climate: Long-range Investigation, Mapping, and Prediction) contributed greatly to our understanding of past changes in ocean surface temperatures, generating many important new insights and hypotheses about orbital forcing and mechanisms of glaciation, such as the rates and timing of ice-sheet growth and decay (McIntyre *et al.*, 1975, 1976; CLIMAP Project Members,

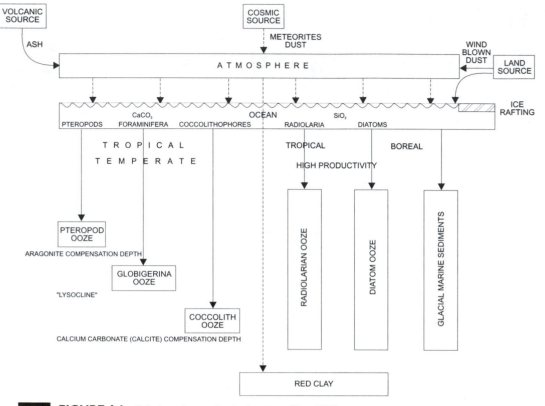

FIGURE 6.1   Pelagic sedimentation in the ocean (Hay, 1974).

1976, 1981, 1984; Hays *et al.*, 1976; Ruddiman and McIntyre, 1981a). Subsequently, the SPECMAP (Spectral Mapping Project) focused on determining the spectral characteristics of the ocean sediment-based paleoclimatic record, and establishing a basic timeframe for past climatic events (Imbrie *et al.*, 1984; Martinson *et al.*, 1987). The contribution of these projects to paleoclimatic research has been immense, but inevitably new research has raised questions about the validity of earlier results, especially the reconstructions of tropical and equatorial SSTs at the last glacial maximum (LGM). In the CLIMAP SST reconstructions, it appeared that SSTs at low latitudes changed very little at the time of the LGM, but current research indicates that, in some areas, temperatures may have been considerably lower (by 3–5 °C) at that time (Guilderson *et al.*, 1994; Stute *et al.*, 1995). This is a very controversial, but extremely important issue (discussed further in Sections 6.4 and 6.5) as it has significant implications for understanding not only past climatic changes, but also the potential magnitude of any future anthropogenic global warming. This is because low latitude SSTs play a key role in water vapor/cloud feedback within the climate system and are thus critical to assessing the overall climate sensitivity to particular forcing mechanisms. Current research also focuses on the three-dimensional structure of the world's oceans through time, in an effort to reconstruct not only surface water conditions in the past, but also the deepwater circulation that plays a critical role in energy transfer and the sequestration of carbon dioxide in the abyss.

6.2 PALEOCLIMATIC INFORMATION FROM BIOLOGICAL MATERIAL IN OCEAN CORES

Paleoclimatic inferences from biogenic material in ocean sediments derive from assemblages of dead organisms (thanatocoenoses), which make up the bulk of all but the deepest of deep-sea sediments (biogenic ooze). However, thanatocoenoses are generally not directly representative of the biocoenoses (assemblages of living organisms) in the overlying water column. For example, selective dissolution of thin-walled specimens at depth (see Section 6.6), differential removal of easily transported species by scouring bottom currents, and occasional contamination by exotic species transported over long distances by large-scale ocean currents all contribute elements of uncertainty. Because of these problems, sediments over much of the ocean floor are unsuitable for paleoclimatic reconstruction. This is illustrated in Fig. 6.2 for foraminiferal studies (Ruddiman, 1977a) though it should be noted that in some areas unsuitable for foraminiferal preservation the remains of other organisms, such as diatoms or radiolarians, may provide a useful record (Sancetta, 1979; Pichon *et al.*, 1992; Pisias *et al.*, 1997).

Biogenic oozes are made up primarily of the calcareous or siliceous skeletons (tests) of marine organisms. These may have been planktic (passively floating organisms living near the surface [0–200 m]) or benthic (bottom-dwelling). For paleoclimatic purposes, the most important calcareous materials are the tests of foraminifera (a form of zooplankton) and the much smaller tests, or test fragments, of coccolithophores (unicellular algae) known informally as coccoliths (Figs. 6.3 and 6.4). These are sometimes grouped with other minute forms of calcareous fossils and referred to as calcareous nannoplankton *(nanno* = dwarf) or simply nannoliths (Haq, 1978). Organic-walled dinoflagellate cysts are another important paleoceanographic indicator — of SSTs and sea-ice extent in high-latitude regions (de Vernal *et al.*, 1993, 1994, 1998). The most important siliceous materials are the remains of radiolarians (zooplankton), silicoflagellates, and diatoms (algae) (Haq and Boersma, 1978; Fig. 6.5). By studying the morphology of the tests, individuals can be identified to species level and their ocean-floor distribution can then be related to environmental conditions (generally temperature and salinity) in the overlying water column (Fig. 6.6). However, it should be noted that the species assemblage in the sediment is a composite of all the species living at different depths in the water column as well as species with only a seasonal distribution in that particular area. The depth habitats of many zooplankton species are still not well known, and it is believed that some species live at different water depths at different times in their life cycles. Phytoplankton, and zooplankton that possess symbiotic algae, are restricted to the euphotic zone, at least during the productive phase of their life cycles. Depth habitats are of particular significance to isotopic studies of tests

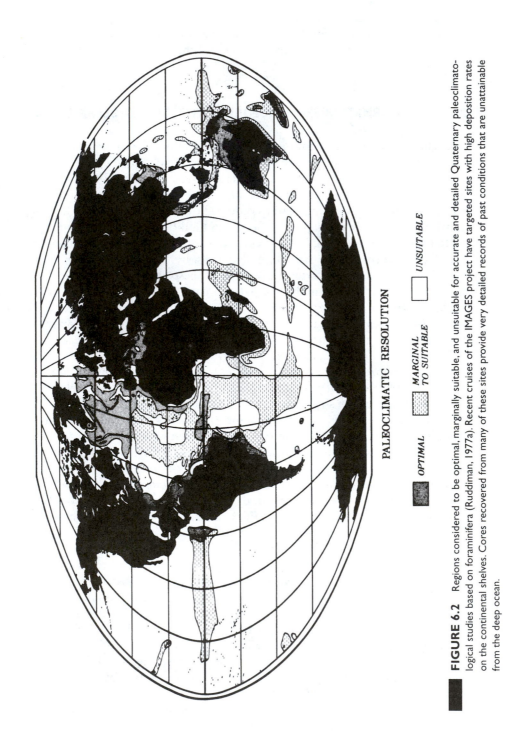

PALEOCLIMATIC RESOLUTION

▓ OPTIMAL   ▒ MARGINAL TO SUITABLE   ☐ UNSUITABLE

FIGURE 6.2   Regions considered to be optimal, marginally suitable, and unsuitable for accurate and detailed Quaternary paleoclimatological studies based on foraminifera (Ruddiman, 1977a). Recent cruises of the IMAGES project have targeted sites with high deposition rates on the continental shelves. Cores recovered from many of these sites provide very detailed records of past conditions that are unattainable from the deep ocean.

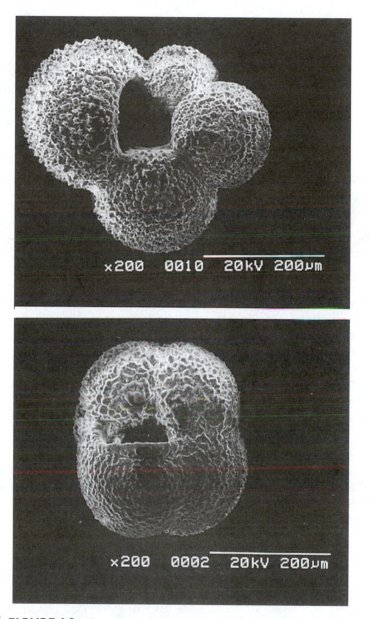

FIGURE 6.3   Two calcareous tests (200x magnification) commonly used in paleo-oceanographic studies. **Bottom:** the foraminifera *Neogloboquadrina pachyderma* (left coiling) (ventral view, 200x), from the North Atlantic (Irminger Basin). **Top:** *Globigerina bulloides* (ventral view, 200x) from the Labrador Sea (photographs kindly provided by Laurence Candon, GEOTOP, Université du Québec a Montréal).

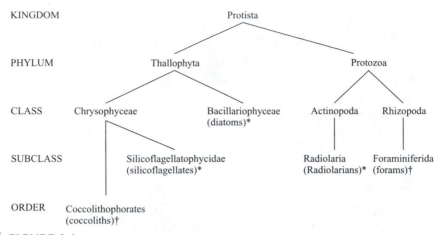

FIGURE 6.4 Taxonomic relationships of the main marine organisms used in paleoclimatic reconstructions. Asterisks indicate siliceous tests; the dagger indicates calcareous tests.

because oxygen isotope composition is a function of the water temperature (and to some extent salinity) at which the carbonate is secreted (see Section 6.3). If test walls are secreted at varying depths through the lifespan of an individual, simple correlations with surface water temperatures and salinities may not be meaningful (Duplessy *et al.,* 1981).

Paleoclimatic inferences from the remains of calcareous and siliceous organisms have resulted from basically three types of analysis:

(a) the oxygen isotopic composition of calcium carbonate in foram tests (Hecht, 1976; Mix, 1987);

(b) quantitative interpretations of species assemblages and their spatial variations through time (Imbrie and Kipp, 1971; Molfino *et al.,* 1982) and

(c) (of far less importance) morphological variations in a particular species resulting from environmental factors (ecophenotypic variations; Kennett, 1976).

Most work along these lines has concentrated on the Foraminiferida. In the following sections, therefore, the focus will be on foraminiferal studies. Paleoclimatic studies of coccoliths, radiolarians, diatoms, and silicoflagellates have mainly been in terms of changes in the relative abundance of assemblages (Pichon *et al.,* 1992) though isotopic studies of diatoms have also been carried out (Juillet-Leclerc and Labeyrie, 1987; Shemesh *et al.,* 1992). Oxygen isotope variations in coccoliths provide useful paleotemperature estimates and may provide even more reliable data than that based on forams alone (Margolis *et al.,* 1975; Dudley and Goodney, 1979; Anderson and Steinmetz, 1981). However, there are problems in isolating sufficiently pure samples of very small microfossils such as diatoms or coccoliths.

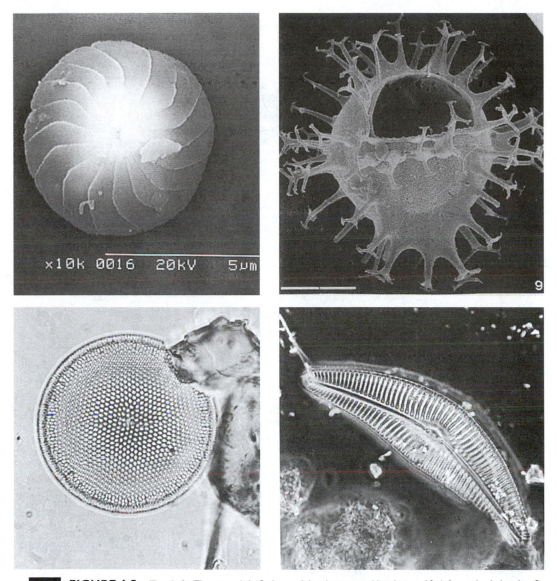

FIGURE 6.5   Top left: The coccolith *Cyclococcolithus leptoporus* (distal view, 10x) from the Labrador Sea; **top right:** dinoflagellate cyst *Spiniferites mirabilis* dorsal surface (scale bar = 20 μm) from the Gulf of Mexico (see de Vernal *et al.,* 1992). **Bottom left:** the centric diatom *Thalassiosira cf. nordenskioeldii* (100x) from the Gulf of St. Lawrence; **bottom right:** the pennate diatom *Cymbella proxima* (1300x) from the estuary of the Gulf of St. Lawrence (photographs kindly provided by Johanne Turgeon and Anne de Vernal, GEOTOP, Université du Québec a Montréal).

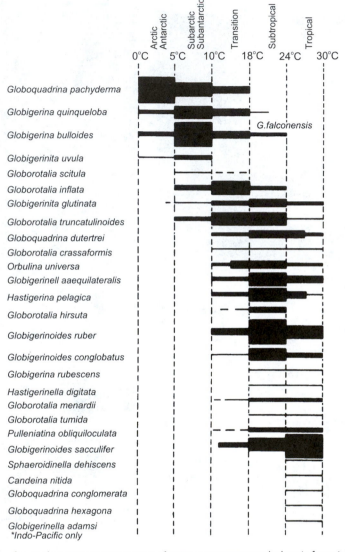

FIGURE 6.6   Sea-surface temperature ranges of some contemporary planktonic foraminifera, illustrating their temperature dependence. Width of lines indicates relative abundance (Boersma, 1978).

6.3 OXYGEN ISOTOPE STUDIES OF CALCAREOUS MARINE FAUNA

If calcium carbonate is crystallized slowly in water, ^{18}O is slightly concentrated in the calcium carbonate relative to that in the water. The process is temperature-dependent, with the concentrating effect diminishing as temperature increases. In a nutshell, this is the basis for a very important branch of paleoclimatic research — the analysis of oxygen isotopes in the calcareous tests of marine microfauna (principally foraminifera, but also coccoliths). The approach was first enunciated by Urey (1947, 1948) who noted, "If an animal deposits calcium carbonate in equilibrium with the water in which it lies, and the shell sinks to the bottom of the sea . . . it is only necessary to determine the ratio of the isotopes of oxygen in the shell today in order to know the temperature at which the animal lived" (Urey, 1948).

He then went on to calculate, from thermodynamic principles, the magnitude of this temperature-dependent isotopic fractionation. Although the principle of Urey's argument is correct, the numerous complications that arise in the real world have made direct paleotemperature estimates rather problematic. In fact, the oxygen isotope record in marine sediments varies *locally* (with temperature, and to a lesser extent salinity) and *globally* with variations in continental ice volume. This global signal is the single most important record of past climatic variations for the entire Cenozoic.

6.3.1 Oxygen Isotopic Composition of the Oceans

The oxygen isotopic composition of a sample is generally expressed as a departure of the $^{18}O/^{16}O$ ratio from an arbitrary standard[15]:

$$S = \frac{(^{18}O/^{16}O)\text{sample} - (^{18}O/^{16}O)\text{standard}}{(^{18}O/^{16}O)\text{standard}} \times 10^3 \qquad (6.1)$$

The resulting values are expressed in per mil (‰) units; negative values represent lower ratios in the sample (i.e., less $\delta^{18}O$ than ^{16}O and, therefore, isotopically lighter) and positive values represent higher ratios in the sample (more $\delta^{18}O$ than ^{16}O and, therefore, isotopically heavier).

Empirical studies relating the isotopic composition of calcium carbonate deposited by marine organisms to the temperature at the time of deposition have demonstrated a relationship that approximates the following[16]:

$$T = 16.9 - 4.2(\delta_c - \delta_w) + 0.13(\delta_c - \delta_w)^2, \qquad (6.2)$$

where T is water temperature in degrees Celsius, δ_c is the per mil difference between the sample carbonate and the SMOW standard, and δ_w is the per mil difference between the $\delta^{18}O$ of water in which the sample was precipitated and the SMOW standard (Epstein *et al.*, 1953; Craig, 1965).

For modern samples, δ_w can be measured directly in oceanic water samples; in fossil samples, however, the isotopic composition of the water is unknown and cannot be assumed to have been as it is at the site today. In particular, during glacial periods the removal of isotopically light water from the oceans to form continental ice sheets (Section 5.2) led to an increase in the $^{18}O/^{16}O$ ratio of the oceans as a whole by ~1.1 ± 0.25‰. Thus, the expected increase in δ_c of foraminiferal tests during glacial

---

[15] See footnote 11 on page 131, Chapter 5; Isotopic studies based on carbonate fossils use as a standard a Cretaceous belemnite from the PeeDee Formation of North Carolina (PDB-1) or a cross-referenced U.S. NIST sample. Carbon dioxide released from PDB-1 = +0.2‰ relative to CO_2 equilibrated with SMOW (Craig, 1961b).

[16] The precise form of the relationship depends on the particular technique used in analysis and on the temperature at which fractionation occurs (for further discussion, see Shackleton, 1974; Mix, 1987).

periods due to lower water temperatures is complicated by the increase in δ_w of the ocean water at these times. How much of the increase in δ_c is the result of variations in δ_w can be assessed by analyzing $\delta^{18}O$ in pore waters squeezed from sediments of the last glacial age (Schrag and DePaolo, 1993) or in the tests of benthic (bottom-dwelling) foraminifera. Bottom waters today (derived from cold, dense, polar water spreading through the deep ocean basins) are relatively close to the freezing point of seawater, so the bulk of the $\delta^{18}O$ increase in benthic forams in glacial times can not be due to significantly lower temperatures; rather, the evidence indicates that at least 70% of the increase results from the changing isotopic composition of the oceans (Duplessy, 1978; Mix, 1987). Hence, the isotopic changes recorded in benthic foraminiferal tests are primarily a record of changing terrestrial ice volumes, or a "paleo-glaciation" record (Shackleton, 1967; Dansgaard and Tauber, 1969). The benthic isotope record thus demonstrates that there have been more than 20 periods of major continental glaciation during the Quaternary, with the largest changes in ice volume (from glacials to interglacials) within the last 900,000 yr (Shackleton *et al.*, 1990). The record of changing $\delta^{18}O$ in relation to variations in continental ice volume (recorded in terms of eustatic sea level change) is discussed further in Section 6.3.4.

Changes in the isotopic composition of ocean water through time are not the only complications affecting a simple temperature interpretation of δ_c (Mix, 1987). Urey's initial hypothesis developed from a consideration of calcium carbonate precipitated inorganically, where the carbonate forms in isotopic equilibrium with the water. However, in the formation of carbonate tests by living organisms, metabolically produced carbon dioxide may be incorporated; in such cases, the carbonate would not be formed in isotopic equilibrium with the water and the resulting isotopic composition would differ from the thermodynamically predicted value, generally leading to $\delta^{18}O$ (and $\delta^{13}C$) lower than the expected equilibrium values (Duplessy *et al.*, 1970a; Vinot-Bertouille and Duplessy, 1973; Shackleton *et al.*, 1973). This was termed the vital effect by Urey (1947). The contribution of metabolic carbon dioxide to the test carbonate differs from one species of foraminifera to another (Grossman, 1987). Modern samples of *Globigerinoides ruber*, for example, give isotopic values 0.5‰ lighter than expected from thermodynamic principles alone (based on analysis of water from their modern habitat). This is equivalent to a temperature error of ~2.5 °C (Shackleton *et al.*, 1973). On the other hand, not all forams exhibit this unfortunate characteristic. For example, samples of *Pulleniatina obliquiloculata* and *Uvigerina* spp. (benthic forams) appear to be in isotopic equilibrium with surrounding water (Shackleton, 1974). In other species, where isotopic equilibrium is not achieved, there is evidence that the vital effect remains constant over time (Duplessy *et al.*, 1970a). It is thus possible to circumvent this particular problem by careful selection of the species being studied, or by assessing its specific vital effect, and adjusting the measured isotopic values accordingly.

Another complication in calculating water temperatures from the isotopic composition of carbonate tests is the problem of variations in depth habitat of planktic foraminifera. Even if the ice effect and vital effects are known, there is still some uncertainty as to whether foraminifera lived at the same depth from glacial to interglacial times. Water temperatures in the upper few hundred meters of the ocean change rapidly with depth, particularly outside the Tropics (Table 6.1), so small

TABLE 6.1 Mean Vertical Temperature Distribution (°C) and Temperature Gradients in the Three Oceans Between 40°N and 40°S

Depth(m)	Atlantic Ocean		Indian Ocean		Pacific Ocean		Mean	
	Temperature (°C)	Gradient (°C/100 m)	Temperature (°C)	Gradient (°C/100 m)	Temperature (°C)	Gradient (°C/100 m)	Temperature (°C)	Gradient (°C/100 m)
0	20.0		22.2		21.8		21.3	
		2.2		3.3		3.1		2.8
100	17.8		18.9		18.7		18.5	
		4.4		4.7[a]		4.4[a]		4.5[a]
200	13.4		14.3		14.3		14.0	
		1.8		1.6		2.6		2.0
400	9.9		11.0		9.0		10.0	
		1.5		1.2		1.2		1.3
600	7.0		8.7		6.4		7.4	
		0.7		0.9		0.65		0.75
800	5.6		6.9		5.1		5.9	
		0.35		0.7		0.4		0.5
1000	4.9		5.5		4.3		4.9	
		0.20		0.4		0.4		0.35
1200	4.5		4.7		3.5		4.2	
		0.15		0.3		0.2		0.22
1600	3.9		3.4		2.6		3.3	
		0.12		0.15		0.1		0.12
2000	3.4		2.8		2.15		2.8	
		0.08		0.09		0.05		0.07
3000	2.6		1.9		1.7		2.1	
		0.08		0.03		0.03		0.05
4000	1.8		1.6		1.45		1.6	

From Defant (1961).
[a] Maximum gradient.

variations in depth habitat can be equivalent to a change in temperature of several degrees Celsius (i.e., a change perhaps as large as the glacial to interglacial change at the surface of the ocean). It is thus critical to know what factors control depth habitat of foraminifera, and in particular the depths at which tests are secreted (Emiliani, 1971). Several studies have concluded that water density (a function of temperature and salinity; Fig. 6.7) is of prime importance to individual species, as the same species may be found in different areas living at different depths, but in water of the same temperature and salinity (Emiliani, 1954, 1969; Hecht and Savin, 1972). During glacial periods, when the oceans were more saline (due to the removal of water to the continental ice sheets), foraminifera may have migrated upwards in the water column, to a zone of warmer water, in order to maintain a constant density environment. Conversely, they may have migrated downwards (to cooler water) in interglacials (Fig. 6.7). Clearly, such vertical migrations would result in isotopic paleotemperature estimates of glacial to interglacial temperature differences, considerably *less* than the changes actually occurring in the water column (Savin and Stehli, 1974). Hence, if this model is correct, any residual paleotemperature signal obtained (after correcting for ice and vital effects) would have to be considered a minimum estimate only.[17] This problem may, however, be a relatively

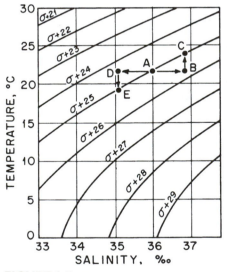

FIGURE 6.7 Temperature-salinity diagram: $\sigma = 10^3 (\rho_w - 1)$, where ρ_w = water density. Density of seawater is a function of both temperature and salinity. Lines of equal density are shown. During glacial periods, when the removal of water to the continental ice sheets would have made overall oceanic salinity higher, foraminifera may have migrated to warmer water (generally upwards in the water column) to maintain a constant-density environment (illustrated schematically as A - B - C). In an interglacial, the opposite situation may have prevailed and the response of foraminifera may have been to move downward in the water column to a cooler zone below (A - D - E).

[17] In addition, the effects of sediment mixing (bioturbation), which tend to smooth out extremes in the record, also contribute to making actual glacial-interglacial $\delta^{18}O$ differences appear smaller (Shackleton and Opdyke, 1976).

minor one compared to the important effect of variations in depth habitat of foraminifera during their life cycle. There is now convincing evidence that although the tests of living forams contain $CaCO_3$ that has been secreted in isotopic equilibrium with the upper mixed water layer, in certain species foram tests from the sea floor are significantly enriched with ^{18}O compared to their living counterparts (Duplessy *et al.*, 1981). This is apparently due to calcification of the tests at depths (>300 m) considerably below the upper mixed layer, during the process of gametogenesis (reproduction). Gametogenic calcification may account for ~20% of foram test weight in samples from the sea floor and, because calcium carbonate has been extracted from water that is much cooler than that nearer the surface, the overall $\delta^{18}O$ values indicate a mean temperature significantly lower than the near-surface temperature (Fig. 6.8). Obviously, the rate at which the organism descends through the water column and the relative extent of gametogenic calcification will greatly influence the final isotopic composition of the test calcite. Similarly, as certain species are distinctly seasonal, the water temperature at those times will be reflected in $\delta^{18}O$. In order to use $\delta^{18}O$ in foram tests as a temperature indicator, it is necessary to establish in some way exactly which temperatures (by depth and season) are being recorded. To determine empirically the optimum relationship, one approach

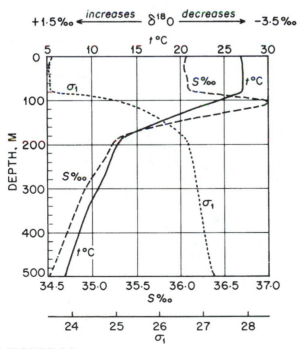

 FIGURE 6.8 A typical vertical temperature, salinity, and density profile from the tropical oceans. As the amount of calcium carbonate secreted at depths below the upper mixed layer (generally ~100 m) increases, so the ^{18}O content of the test carbonate increases. Isotopic temperature estimates from foram tests that have undergone gametogenic calcification at depth are considerably lower than those obtained from living foraminifera collected in the upper mixed layer (~0.2‰ per degree Celsius).

to this problem is to compare the $\delta^{18}O$ of forams in core-top assemblages or in sediment traps with oceanographic conditions in the overlying water column. For example, studies of $\delta^{18}O$ in forams collected in sediment traps at depth in the Sargasso Sea, compared to the temperature profile in the overlying water column have shed light on both the seasonal occurrence of forams and the relationship between depth habitat and $\delta^{18}O$ in the foram tests (Deuser and Ross, 1989). This revealed, *inter alia*, that *Globorotalia truncatulinoides* adds calcite to its test at a depth of ~800 m, resulting in a sample mean $\delta^{18}O$ that does not correspond to SSTs but to water temperatures at ~200 m. The sediment trap data also demonstrate that many species only occur at particular times of year. For example, *P. obliquiloculata* and *Neogloboquadrina dutertrei* live in the upper 100 m and give $\delta^{18}O$ values that are representative of the winter mixed layer, whereas *Globigerinoides ruber* (pink variety), which also live in the mixed layer (upper 25 m), occur only in summer months, resulting in different values of $\delta^{18}O$. This complication has some benefit; there is the potential of using seasonally occurring species of known depth habitat to reconstruct past changes in seasonality (and mixed layer depth) from foram assemblages (Reynolds-Sautter and Thunell, 1989).

One additional complexity in using $\delta^{18}O$ as a paleotemperature indicator is that $\delta^{18}O$ is also strongly related to salinity (Fig. 6.9). Hence any change in salinity due to large-scale dilution effects (because of ice sheet melting) or to local changes in the precipitation-evaporation (P-E) relationship will also be recorded in foraminifera so affected (Duplessy *et al.*, 1991). This can be put to good use in order to estimate salinity changes at the sea surface, providing all other effects can be determined independently. Thus, Duplessy *et al.* (1993) assessed local salinity changes off Portugal during the last deglaciation by first estimating SSTs (from micropaleontological transfer functions; see

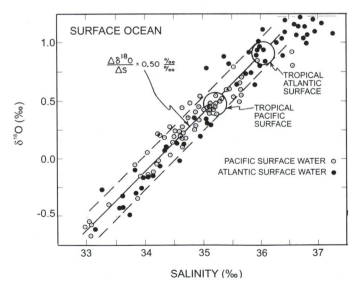

FIGURE 6.9   Relationship between $\delta^{18}O$ and salinity in oceanic surface waters, based on modern water samples (Broecker, 1989, using data of H. Craig).

Section 6.4) and taking into account the influence of ice sheet meltwater on surface salinity and oceanic $\delta^{18}O$. By comparing the $\delta^{18}O$ record "expected" from such variations with the observed record, they argue that the resulting series of differences must be due to local changes in salinity. In spite of the uncertainties in such an approach, the results strongly indicate that salinity was high during the time of maximum ice sheet melting (Meltwater Pulse 1, "mwp-IA," around 12 ka B.P.; Fairbanks, 1989) due to either local changes in P-E or to a shift in water masses (Fig. 6.10). The same conclusions were reached in studies of other cores from the North Atlantic and Norwegian Sea, providing no support for the notion of a low salinity lid on North Atlantic circulation during massive *low latitude* meltwater events, which might have been expected to inhibit the production of North Atlantic Deep Water (NADW). However, when meltwater was primarily discharged to the North Atlantic via the St. Lawrence River drainage, salinities were reduced, even though the overall freshwater discharge rates were lower. Apparently, additional feedbacks involving reductions in evaporation, and/or increased precipitation, or reduced poleward advection of salty subtropical water, led to lower surface salinities in the North Atlantic, and a corresponding

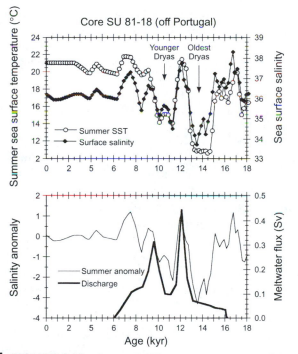

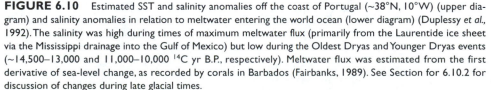

FIGURE 6.10 Estimated SST and salinity anomalies off the coast of Portugal (~38°N, 10°W) (upper diagram) and salinity anomalies in relation to meltwater entering the world ocean (lower diagram) (Duplessy *et al.*, 1992). The salinity was high during times of maximum meltwater flux (primarily from the Laurentide ice sheet via the Mississippi drainage into the Gulf of Mexico) but low during the Oldest Dryas and Younger Dryas events (~14,500–13,000 and 11,000–10,000 ^{14}C yr B.P., respectively). Meltwater flux was estimated from the first derivative of sea-level change, as recorded by corals in Barbados (Fairbanks, 1989). See Section for 6.10.2 for discussion of changes during late glacial times.

reduction in NADW production (Duplessy *et al.*, 1992). Changes in circulation and salinity are discussed further in Section 6.9.

A final problem affecting paleotemperature calculations from the isotopic composition of test carbonate concerns the effect of dissolution on species composition in the thanatocoenoses. This is a pervasive factor with implications not only for isotopic studies, but for all paleoclimatic studies based on floral and faunal assemblages. Because of its importance, it is discussed in more detail in Section 6.6.

6.3.2 Oxygen Isotope Stratigraphy

Oxygen isotope analyses have been carried out on cores from most of the important areas of calcareous sedimentation throughout the world and, in many cases, studies have been made of both planktic and benthic species (Shackleton, 1977). The overwhelming conclusion from such studies (after making due allowance for variations in sedimentation rates, vital effects, and other complicating factors mentioned in the previous section) is that similar isotopic ($\delta^{18}O$) variations are recorded in all areas (Mix, 1987). This is because the primary $\delta^{18}O$ signal being recorded is that of ice volume changes on the continents and concomitant changes in the isotopic composition of the oceans (see Section 6.3.1). Indeed, such changes also affect the $\delta^{18}O$ content of the atmosphere, as recorded in air bubbles from ice cores (see Section 5.4.5). Because the mixing time of the oceans is relatively short ($\sim 10^3$ yr) this global-scale phenomenon results in essentially synchronous isotopic variations in the sedimentary record (though bioturbation [mixing by burrowing organisms in the upper sediments] tends to smooth out fine details in the record). These synchronous variations enable correlations to be made between cores that may be thousands of kilometers apart (Pisias *et al.*, 1984; Prell *et al.*, 1986).

Because there is a consistent stable isotope signal in marine sediments from around the world, universally recognizable isotope stages can be defined (Emiliani, 1955, 1966; Pisias *et al.*, 1984). Warmer periods (interglacials and interstadials) are assigned odd numbers (the present interglacial being number 1) and colder (glacial) periods are assigned even numbers (Fig. 6.11). This provides a relative chronostratigraphic scheme, but absolute dating must then be based on a variety of techniques, such as radiocarbon, U-series dating, and paleomagnetism (see Chapters 3 and 4). A comparison of terrestrial chronostratigraphic markers of known age with equivalent horizons in the ocean sediments enables further checks to be made on the isotopic chronology. For example, the Barbados high sea-level stands (dated by uranium-series analyses of raised corals at 83,000, ~104,000, and ~125,000 yr B.P.; Gallup *et al.*, 1994; Stirling *et al.*, 1995) should correspond to isotopically light carbonate values in foraminiferal tests if interpolated timescales are correct. Such correspondence has been convincingly demonstrated by Shackleton and Opdyke (1973) and Shackleton and Matthews (1977), indicating that the chronology of the last interglacial-glacial cycle, at least, is fairly well established (Fig. 6.12). Five stages are recognized in the isotopic record of the last ~130,000 yr, with stage 5 being subdivided into several substages, bearing the letter designations 5a to 5e (centered on timelines 5.1 to 5.5 in the scheme proposed by Imbrie *et al.*, 1984). Stages 5a, 5c, and 5e were periods of reduced terrestrial ice volume and/or

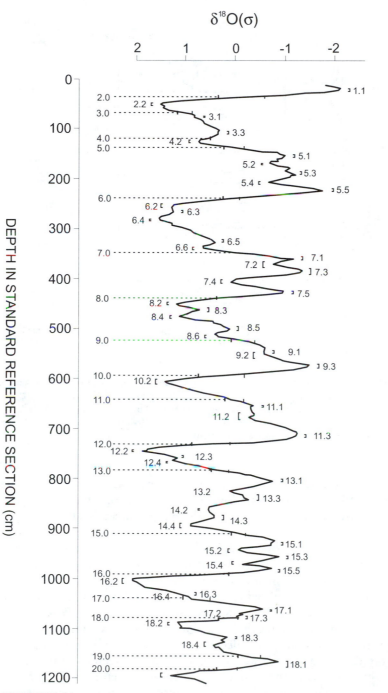

FIGURE 6.11 A composite oxygen isotope record for the Brunhes chron, derived by correlating the common features in 11 planktonic and 2 benthic formanifera records from different oceanic regions. Each record was normalized before they were combined, so the composite record is scaled in terms of standard deviation units. Isotope stage boundaries are shown (Prell et al., 1986).

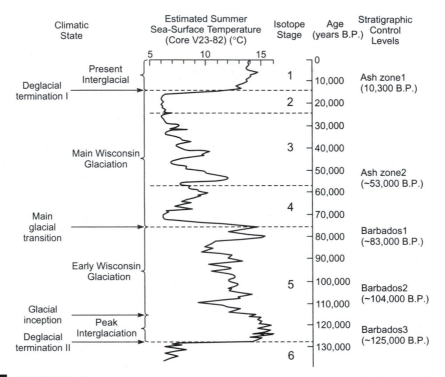

FIGURE 6.12   Summer sea-surface temperature reconstructions for the North Atlantic Ocean based on foraminiferal assemblage paleotemperature estimates, using core V23–82 from 53° N 22° W (Sancetta et al., 1973a,b). Chronological controls used in other cores are shown on the right (tephra layers and Barbados sea-level stands). Isotopic stages are after Shackleton and Opdyke (1973). Generalized climatic conditions and major changes are shown at left (Ruddiman, 1977b).

higher temperatures, with substage 5e being the peak of the last interglacial (Shackleton, 1969). Stages 5b and 5d were periods of cooler temperature and/or terrestrial ice growth, but on a smaller scale than occurred in stage 4. Interestingly, the change in benthic $\delta^{18}O$ commonly recorded between stages 5e and 5d is so large and so rapid that it is almost impossible to account for it only in terms of ice-sheet growth. Ice sheets take thousands of years to grow to such a size that they affect oceanic isotopic composition (Barry et al., 1975). It seems likely that at least part of this change reflects a rapid temperature decline (of ≥ 1.5 °C) in abyssal water temperature (Shackleton, 1969, 1987). Subsequent changes in $\delta^{18}O$ (in stages 5c to 1) were then primarily the result of changing ice volumes on the continents, according to this argument. However, paleo-sea-level data from New Guinea also points to a very rapid change in eustatic sea level (~60 m) between ~115,000 and ~105,000 yr B.P., which bears out the simple ice-growth interpretation of the $\delta^{18}O$ record (Fig. 6.13). If this change did indeed occur, it represents an extraordinary episode in late Quaternary history, with the water equivalent of one Greenland ice sheet being transferred from the oceans to the continents every 1000 yr during this interval (see also Section 6.34).

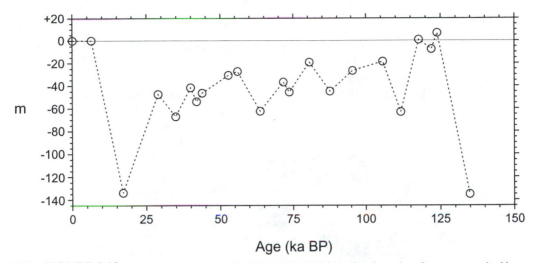

FIGURE 6.13   Sea levels of the last 150,000 yr estimated from dated coral reef terraces on the Huon peninsula of New Guinea (see also Fig. 6.17) (Chappell and Shackleton, 1986, with some additional data added based on high-precision radiometric dates). (Linsley, 1996).

Because the isotopic record provides an integrated summary of global ice volume changes, marine isotope stages are commonly used as standard reference units for both marine and terrestrial deposits (Shackleton and Opdyke, 1973). On the land, studies of continuous stratigraphic sequences spanning the last 150,000 yr or more are rare; more often, deposits studied are discontinuous both in time and space. In the oceans, the sedimentary record has been less disturbed so there are cogent reasons for using marine stratigraphic divisions to clarify and help understand the terrestrial record (Kukla, 1977, 1987b; Rutter *et al.*, 1991b). However, it should be emphasized that the isotopic signals in ocean cores contain both a temperature and an ice-volume component, which may not be synchronous. Furthermore, the ice-volume signal is an index of *global* ice volume and says nothing about ice extent in any one geographical area. This makes the application of a marine isotope stratigraphic frame of reference to many areas of questionable utility because local stratigraphy may bear little relationship to the marine isotopic record. For example, the Canadian High Arctic may not have had its most extensive ice cover during isotope stage 2 (the time of the global ice volume maximum) and may have experienced glacial advances and retreats out of phase with the larger continental ice sheets, variations of which dominate the oceanic isotope record (England, 1992). Furthermore, although there is a generally good correspondence between the marine isotope record of continental ice volume changes and glacial stratigraphy over North America, at least over the past ~1 M years (11 episodes of glaciation) nowhere in North America was the Laurentide Ice Sheet at its maximum extent at 18,000 yr B.P. (the marine isotope maximum); generally the ice sheet reached its greatest geographic extent 2000–3000 yr earlier (Fullerton and Richmond, 1986).

6.3.3 Orbital Tuning

An entirely different approach to dating marine sediments is to assume that orbital forcing has been the primary factor driving the growth and decay of ice sheets on land, and therefore of the $\delta^{18}O$ signal in benthic marine sediments. As the different periods associated with changes in eccentricity, precession, and obliquity are well-known, this provides an opportunity to "tune" the chronology of $\delta^{18}O$ in marine sediments to orbital frequencies (Imbrie, 1985). This strategy developed from the discovery of a strong orbital signal in several different marine sedimentary records, with changes in the earth's orbit acting as a "pacemaker" of ice ages (Hays *et al.*, 1976). In orbital tuning, the astronomically driven changes (the forcing) alters the climate system in some way, which is then registered in the proxy record (Fig. 6.14). Providing the phase lag between orbital forcing and climatic response is constant, tuning the proxy record to the original forcing will enable an absolute chronology to be established (Martinson *et al.*, 1987). In practice, tuning is carried out at only one or two frequencies and the resulting, tuned record is then analyzed to see if changes at other orbital frequency bands can be recognized. If there is coherency at other orbital frequencies, this provides strong support for the tuning strategy adopted. For example, Imbrie *et al.*, (1984) analyzed several records, in the 23,000 and 41,000 yr (precession and obliquity) frequency bands, initially using a rough chronology made up of points in each record, the ages of which were approximately known (e.g., the Brunhes-Matuyama boundary at 730,000 yr B.P., and the Stage 5–6 boundary at 127,000 yr B.P.). Assuming a time constant for large ice sheets of ~17,000 years (estimated from an ice-sheet growth model) an age model was constructed to maximize the coherency spectrum between the record of orbital forcing in these frequency bands and the sedimentary $\delta^{18}O$ record. The analysis proceeded iteratively, eventually dropping the initial assumption that certain points were of known age so that the final record was entirely based on orbital tuning. The ultimate test was

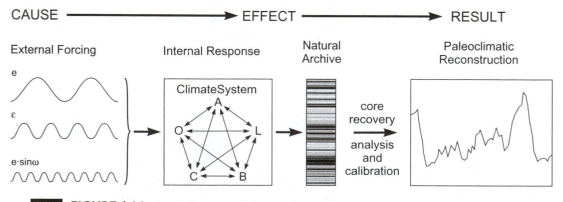

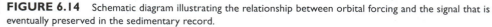

FIGURE 6.14 Schematic diagram illustrating the relationship between orbital forcing and the signal that is eventually preserved in the sedimentary record.

then to examine the coherency spectrum of the tuned record with the spectrum of orbital forcing at frequencies *other* than those used in the tuning procedure. In the tuned records obtained by Imbrie *et al.,* (1984), coherency was not only very high in the tuning bands but also in the ~100 ka eccentricity band, which was not used in tuning at all.

This tuning procedure was applied to a set of $\delta^{18}O$ records, which had previously been "stacked" (aligned with respect to important stratigraphic features, but without a timescale), to produce a standard reference chronostratigraphy for the late Quaternary (Fig. 6.15). The resulting "SPECMAP" (Spectral Mapping Project) record thus enables other $\delta^{18}O$ records to be adjusted to fit the reference chronology, even if no dates are available for the new records (Prell *et al.,* 1986; Martinson *et al.,* 1987). Table 6.2 gives the estimated ages of major stage boundaries over the last 250 ka based on the orbital (or astronomical) tuning approach. The isotopic record can be viewed (in terms of its major low-frequency components) as being

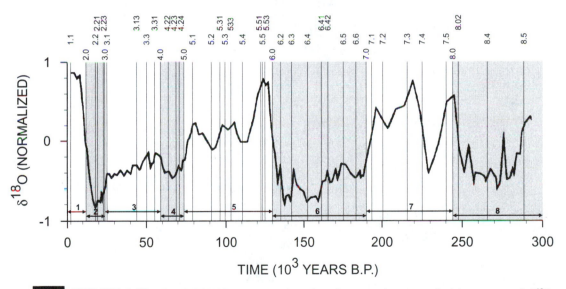

FIGURE 6.15 The SPECMAP composite chronology for a set of seven stacked (superimposed) $\delta^{18}O$ records from different ocean basins of the world. The records were stacked according to their common stratigraphic characteristics (like those used to derive Fig. 6.11) but without a reliable chronology. The chronology applied here is derived by orbital tuning that assumes that the primary forcing controlling the frequency characteristics of each sedimentary record is related to Milankovich orbital variations and that there has been a constant phase lag between the orbital forcing and system response over the last 300,000 yr. The tuning was first applied to subpolar Indian Ocean core RC11-120 and the chronology then transferred to the composite set of normalized records (i.e., each had a mean of 0 and a standard deviation of 1). Major stadial episodes (isotopic stages 2, 4, 6, and 8) are shaded. Vertical lines are times of transition between stages or substages as defined by Martinson *et al.* (1987); substages 5a to 5e are centered on lines 5.1 to 5.5, respectively. The age error estimates on the stacked stage boundaries average ~±5000 yr but this varies from one part of the record to another, as shown in Table 6.2 (Martinson *et al.,* 1987).

TABLE 6.2   Estimated Ages of Oxygen Isotope Stage Boundaries and Terminations

| Boundary[a] | Termination[b] | Estimated ages (× 10^3 years) | | | | |
		A[c]	B[d]	C[e]	D[f]	Error[g]
2.0	I	13	11	11	12.05	3.14
3.0		32	29	27	24.11	4.93
4.0		64	61	58	58.96	5.56
5.0		75	73	72	73.91	2.59
5.1					79.25	3.58
5.2					90.95	6.83
5.3					99.38	3.41
5.4					110.79	6.28
5.5					123.82	2.62
6.0	II	128 127	128		129.84	3.05
7.0		195 190	188		189.61	2.31
8.0	III	251 247	244		244.18	7.11
8-9		297 276	279			
9-10	IV	347 336	334			
10-11		367 352	347			
11-12	V	440 453	421			
12-13		472 480	475			
13-14		502 500	505			
14-15		542 551	517			
15-16	VI	592 619	579			
16-17		627 649	608			
17-18		647 662	671			
18-19		688 712	724			
Brunhes/Matuyama boundary		700 728				
Jaramillo (top)			908			
Jaramillo (bottom)			983			
Olduvai (top)			1640			
Olduvai (bottom)			1820			

[a] Isotope boundary designation 2.0 (Pisias *et al.*, 1984) is alternatively referred to as 1-2.

[b] Terminations from Broecker and van Donk (1970). They defined terminations on the basis of their interpretation of the saw-toothed character of the oxygen isotope record.

[c] Estimates from Shackleton and Opdyke (1973) by linear interpolation in core V28-238, using a mean sedimentation rate of 1.7 cm per thousand years.

[d] Estimates from Hays *et al.* (1976), Kominz *et al.* (1979), and Pisias and Moore (1981), based on the assumption that variations in the tilt of the Earth's axis (obliquity) have resulted in variations of global ice volume, and that the phase shift between the Earth's tilt and the 41,000-yr component of the isotopic record has remained fixed with time. See Section 6.8 for discussion.

[e] Estimations from Morley and Hays (1981) based on adjustments to maintain a constant-phase relationship between variations in oxygen isotope ratios and changes in obliquity and precession.

[f] Derived from orbital tuning by Martinson *et al.* (1987); paleomagnetic transition ages from Shackleton *et al.* (1990).

[g] Error estimates for the stacked record of Pisias *et al.* (1984) derived from the orbital tuning of Martinson *et al.* (1987) applied to those records; error averages ± 5000 yr over the last 300,000 yr, but is lower in some segments than in others.

made up of periods of gradually increasing $\delta^{18}O$ separated by shorter, relatively abrupt episodes when $\delta^{18}O$ values decrease. The curve is thus "saw-toothed" in character, the slow increase in $\delta^{18}O$ resulting from the gradual build-up of ice on the continents, followed by a period of rapid deglaciation when isotopically light water was returned to the oceans. Broecker and van Donk (1970) referred to the sharp decreases in $\delta^{18}O$ as Terminations, signifying the end of a glacial period, the most recent deglaciation being Termination I. Estimated ages of other Terminations are included in Table 6.2.

Orbital tuning has now been extended back over 2.5 Ma yr for selected cores (Fig. 6.16) (Shackleton *et al.*, 1990; Hilgen, 1991; Imbrie *et al.*, 1993b; Chen *et al.*, 1995). A number of important features are apparent in such records. First, $\delta^{18}O$ values have rarely been lower than Holocene levels, indicating that there have been very few occasions when there has been less continental ice than there is at present; these episodes were principally in isotope stages 5e, 9, 11, 31, and 37 (Raymo, 1992). From 3.1 to 2.6 Ma B.P. an increase in $\delta^{18}O$ indicates a progressive cooling, probably associated with the growth of the Antarctic ice sheet, and of continental ice sheets in the northern hemisphere. Around 2.7 Ma, a dramatic increase in the production of ice-rafted debris is seen in marine sediments from the North Atlantic. This increase in frequency corresponds with an increase in $\delta^{18}O$ values to levels typical of recent stadials. Only since isotope stage 22 (~800 ka ago) have there been episodes of continental glaciation comparable in magnitude to the most recent ice age (stage 2). Finally, the record provides a remarkable perspective on the increased importance of variance in the 100 ka (eccentricity) frequency

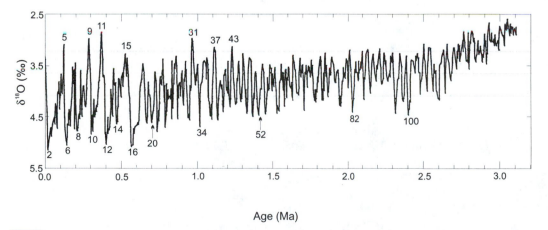

Age (Ma)

FIGURE 6.16 Benthic oxygen isotope record from equatorial Atlantic core ODP-607 for the past 3.2 Ma. Dashed horizontal lines indicate the $\delta^{18}O$ values for the Holocene (upper line), stage 5c (middle line), and the stage 2/1 boundary (lowest line), as recorded at this site. Note the increased importance of the ~100 ka cycle within the last 1 M yr (Raymo, 1992).

band over the last 1 M years, when continental ice sheets were much larger (as registered by higher levels of $\delta^{18}O$) compared to the preceding period. It seems unlikely that this is due simply to a change in the amplitude of eccentricity variations because over this interval there has actually been a shift *away* from variance in the 100 ka eccentricity frequency band, and an increase in variance at lower frequencies (~412 ka) (Imbrie *et al.*, 1993a). The 100 ka period in $\delta^{18}O$ is therefore likely to result from feedbacks internal to the climate system, which amplified the orbitally driven radiative forcing. From ice core records, greenhouse gases are clearly implicated in this amplification process (see Section 5.4.3). Ruddiman *et al.*, (1986) also point to the possibly critical effects of an increase in the elevation of major mountain ranges (Himalayas-Tibet and the north-south mountain ranges of western North America). As these ranges became higher, their effect on the general circulation would have been to establish a more meridional circulation regime, which may have favored faster ice sheet growth under certain orbitally driven radiation regimes.

An important by-product of orbitally tuning the $\delta^{18}O$ record is that it then provides age estimates on the timing of paleomagnetic polarity boundaries recorded within the sediments, which are independent of radiometric age determinations. Thus, if the chronology of the tuned ODP record by Shackleton *et al.* (1990) is correct, revisions in the paleomagnetic timescale are called for, such that the Brunhes-Matuyama boundary was ~780,000 B.P., the Jaramillo lasted from 1.07 M to 0.99 Ma B.P., the Olduvai from 1.95 to 1.77 Ma B.P. and the Matuyama-Gauss boundary was at ~2.6 Ma B.P. In fact, recent reassessments of radiometric data indicate that the orbitally derived dates are compatible with new high resolution dates on paleomagnetic reversals (Tauxe *et al.*, 1992; Chen *et al.*, 1995). The marine isotope chronology can also be used to date other stratigraphic events, such as the level at which particular species became extinct (its last appearance datum, or LAD) (Berggren *et al.*, 1980). These biostratigraphic events may then be used as chronostratigraphic markers in their own right, independent of both radioisotopic and stable isotope analyses on the sedimentary record in question. For example, extinction of the radiolarian *Stylatractus universus* has been found by stable isotope stratigraphy to have occurred throughout the Pacific and Atlantic Oceans at 425,000 ± 5000 yr B.P. (Hays and Shackleton, 1976, Morley and Shackleton, 1978). Similarly, the coccolith *Pseudoemiliania lacunosa* became globally extinct in the middle of isotope stage 12, at ~458,000 yr B.P. and the coccolith *Emiliania huxleyi* made its first appearance at ~268,000 yr B.P., late in isotope stage 8 (Thierstein *et al.*, 1977).

Stable isotope stratigraphy has also enabled volcanic ash horizons to be accurately pin-pointed in time (e.g., the ~53,000 year B.P. Z2 ash layer in the North Atlantic) so that they can also be used as independent chronostratigraphic markers (Kvamme *et al.*, 1989). This is particularly important where the ash can also be found in terrestrial deposits, such as ice cores, loess or lake sediments, enabling direct correlation of land and marine records to be carried out (Gronvald *et al.*, 1995).

6.3.4 Sea-level Changes and $\delta^{18}O$

It was noted earlier that the bulk of the $\delta^{18}O$ signal in benthic foraminifera is re-lated to changes in the isotopic composition of the oceans under the influence of changing continental ice volume. As ice sheets grew on the continents, the $\delta^{18}O$ of the ocean increased and global sea level fell. Hence, there should be some sort of re-lationship between $\delta^{18}O$ in forams, continental ice volume, and sea-level change. However, the connections between these three phenomena are complex; the mean isotopic composition of ice sheets no doubt changed over time, depending on the ice-sheet location (latitudinally) and its mean elevation. If ice sheets remained in a steady state for extended periods, the mean $\delta^{18}O$ of ice lost at the margin (repre-senting older ice formed at lower elevations) would probably have been higher than precipitation falling later on the high elevation accumulation zone of the ice sheet, leading to a systematic enrichment of the ocean in ^{18}O without any change in ice volume. There is therefore a non-linear relationship between ice-sheet volume and oceanic $\delta^{18}O$ composition (Mix and Ruddiman, 1984). Furthermore, the $\delta^{18}O$ of benthic forams is not only influenced by oceanic $\delta^{18}O$; if water temperature (or salinity) changed, this would affect $\delta^{18}O$ and such effects may have been more im-portant at certain times than at others. Finally, estimates of paleo-sea level are gen-erally based on coral reefs found on rising coastlines so that the record of former low sea-level stands are now found at locations well above present sea level. Under-standing how sea level at various times in the past relates to global ice volume changes requires not only a tectonic model for local uplift but also an understand-ing of global sea-level distribution, relative to the mean geoid. Temporally synchro-nous sea-level terraces may not be at equivalent heights above present sea level due to geophysical constraints on the distribution of water on the planet (Peltier, 1994).

Notwithstanding these difficulties, the relationship between $\delta^{18}O$ and sea level has been considered by Chappell and Shackleton (1986) and Shackleton (1987). By comparing paleo-sea-level estimates from U-series dated coral terraces on the Huon Peninsula of New Guinea (Fig. 6.17) with the record of $\delta^{18}O$ in benthic fora-minifera, it is apparent that the relationship between sea level change and $\delta^{18}O$ has not been constant over the last glacial-interglacial cycle. In fact, it seems likely that temperatures in the deep ocean must have been at least 1.5 °C cooler in glacial and interstadial times (~110–20 ka B.P.) than during the interglacials (isotope stages 1 and 5e). If it is further assumed that such a change has been typical of earlier glacial/interglacial cycles, it is then possible to estimate (from benthic $\delta^{18}O$) the rel-ative magnitudes of ice volumes in glacial and interglacial extremes back in time, for which no well-dated sea-level terraces exist. Such an analysis leads to the con-clusion that continental glaciation during marine isotope stage 6 was slightly greater than in stage 2, but that in stages 12 and 16 it was even larger (Shackleton, 1987). However, there is little evidence that previous interglacials 7, 13, 15, 17, or 19 were significantly warmer (i.e., had much less ice remaining on the continents) than the present one. Stages 1, 5e, 9, and 11 were all similar (isotopically) though independent evidence suggests that stages 5e and 11 (at least) were somewhat warmer with sea level ≥ 6 m above present levels.

FIGURE 6.17   Raised coral terraces, indicative of former sea levels, on the Huon Peninsula of Papua New Guinea. The coastline in this area has been continuously rising at the same time as sea level has been rising and falling due to ice volume changes on the land (eustatic sea-level changes). Upper Terrace VIIa (~140 ka B.P.) is slightly above Terrace VIIb (~125 ka B.P.). Terrace VI (~107 ka B.P.) is a small fringing reef above Terrace V (~85 ka B.P.). The dated sea-level record is shown in Fig. 6.13 (photograph courtesy of A.L. Bloom).

6.4 RELATIVE ABUNDANCE STUDIES

The possibility of reconstructing paleoclimates by using the relative abundance of a particular species, or species assemblage, in ocean sediment cores was first proposed by Schott (1935). Schott recognized that variations in the number of *Globorotalia menardii* (a foraminifera characteristic of subtropical and equatorial waters) are indicative of alternating cold and warm intervals in the past. However, 30 yr were to elapse before the availability of relatively long undisturbed cores and improved dating techniques enabled others to capitalize on Schott's work, and to develop his ideas further. For example, Ruddiman (1971) derived time series of the ratios of all warm-water species to cold-water species to obtain qualitative paleotemperature estimates that showed good correlations with oxygen isotope paleoglaciation curves. Although somewhat of an improvement over earlier studies based on individual species, Ruddiman recognized that the technique was still relatively simplistic in considering all species as equally "warm" or "cold" when gradations in their individual tolerances obviously exist.

In the early 1970s, major advances in paleoclimatic and paleo-oceanographic reconstructions were made by a number of workers. Multivariate statistical analyses of modern and fossil data were used to quantify former marine conditions ("marine climates") in an objective manner (Imbrie and Kipp, 1971; Hecht, 1973; Berger and Gardner, 1975; Williams and Johnson, 1975; Molfino *et al.*, 1982). The general approach in all of these studies is to calibrate the species composition of modern (core-top) samples in terms of modern environmental parameters (such as sea-surface temperatures in February and August). This is achieved by developing empirical equations that relate the two data sets together. These equations (transfer functions) are then applied to down-core faunal variations to reconstruct past environmental conditions (Fig. 6.18). Mathematically, the procedure can be simply expressed as follows:

$$T_m = XF_m \text{ and } T_p = XF_p \tag{6.3}$$

where T_m and T_p are modern and paleotemperature estimates, respectively, F_m and F_p are modern and fossil faunal assemblages, respectively, and X is a transfer coefficient (or set of coefficients).

A fundamental assumption in the use of transfer functions to reconstruct marine climates is that former biological and environmental conditions are within the "experience" of the modern (calibration) data set (as illustrated in Fig. 6.19). If this is not so, a no-analog condition exists and erroneous paleoclimatic estimates may result (Hutson, 1977). Another important assumption is that the relationship that currently exists between marine climate and marine fauna has not altered over time due, for example, to evolutionary changes of the species in question. However, perhaps the major uncertainty in these studies concerns the very nature of the calibration

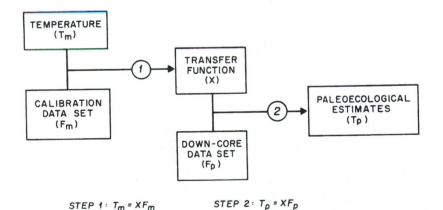

STEP 1: $T_m = XF_m$ STEP 2: $T_p = XF_p$

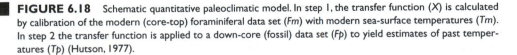

FIGURE 6.18 Schematic quantitative paleoclimatic model. In step 1, the transfer function (X) is calculated by calibration of the modern (core-top) foraminiferal data set (Fm) with modern sea-surface temperatures (Tm). In step 2 the transfer function is applied to a down-core (fossil) data set (Fp) to yield estimates of past temperatures (Tp) (Hutson, 1977).

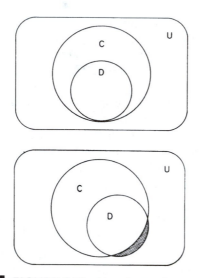

FIGURE 6.19 Venn diagram illustrating conditions that are ideal (top) and nonideal (bottom) for calibrating a transfer function. In an ideal situation, the calibration data set C encompasses the range of all biological and environmental conditions that exist in the down-core data set D. In a non-ideal situation the calibration data set C does not reflect all the biological and environmental conditions that are represented within the down-core data set D and a no-analog condition results (shaded area). U is the universe of all biological and environmental conditions both today and in the past (Hutson, 1977).

data set itself. The "modern" faunal assemblages are generally derived from core-top samples that may represent a depositional period of several thousand years, due to bioturbation and disturbance during core recovery (Emiliani and Ericson, 1991). Indeed, the chronological heterogeneity of core-top samples was considered by Imbrie and Kipp (1971) to be the largest single source of error in their paleoenvironmental reconstructions ("most [core-top samples] represent . . . the last 2000 to 4000 years and. . . . some may contain materials deposited in the age range 4000–8000 years B.P."; Imbrie and Kipp, 1971). Furthermore, it is not unusual for modern oceanographic parameters to be poorly known, commonly being based on interpolation between observations that are both short and geographically sparse (Levitus 1982). This is a particular problem in remote areas where sea-surface temperature and/or salinity gradients are strong, and may result in paleotemperature estimates for certain regions that are in error by several degrees. However, this is probably close to the magnitude of uncertainty associated with modern values, particularly in areas where significant changes of sea-surface temperature have occurred, even during the brief period of modern instrumental observations (Wahl and Bryson, 1975; Levitus, 1989). Under such circumstances the selection of a modern calibration value to equate with the core-top faunal assemblage is somewhat problematic, though by no means a problem unique to marine data (Bradley, 1991).

Of all the multivariate approaches to the quantification of former marine climates, the methodology of Imbrie and Kipp (1971) has been most widely applied.

In their original study, an attempt was made to reconstruct sea-surface temperature variations at a core site (V12-122) ~150 km south of Haiti. To achieve this, the species composition of core-top samples from 61 sites in the Atlantic Ocean (and part of the Indian Ocean) were used as the basic "modern fauna" data set. As a first step, Imbrie and Kipp reduced the number of independent variables in this data set by the use of principal components analysis. Principal components analysis is an objective way of combining the original variables into linear combinations (eigenvectors) that effectively describe the principal patterns of variation in a few primary orthogonal components, leaving the less coherent aspects ("noise") for the last few components (Sachs *et al.*, 1977). Thus, Imbrie and Kipp were able to condense much of the spatial variation of species abundance in 61 core-top samples from the Atlantic Ocean into five principal components or assemblages, which accounted for almost all of the variance in the original data set. By mapping the relative contribution of each component to the variance of each core-top sample, it was clear that four of these assemblages were related to temperature variations near the sea surface and could be simply described as subtropical, transitional, subpolar, and polar assemblages (Fig. 6.20). A fifth assemblage was more related to oceanic circulation around the subtropical high-pressure cells and was termed the gyre margin assemblage.

The next step was to utilize the relative weightings of each assemblage (factor scores) at each site to predict sea-surface temperatures. A stepwise multiple regression procedure was used, with temperature as the dependent variable and the factor scores as independent variables (predictors). In this way, an equation was derived that parsimoniously described sea-surface temperature in terms of the relative importance of the factor scores at each site. In the case of winter temperatures, for example, the following calibration equation was derived:

$$T_w = 23.6A + 10.4B + 2.7C + 3.7D + 2.0K \tag{6.4}$$

where A, B, C, and D refer to the four major assemblages (tropical, subtropical, subpolar, and polar) and K is a constant.[18] This equation explained 91% of variance in the modern winter sea-surface temperature observations (Fig. 6.21).

At this stage, the modern faunal data set had been calibrated in terms of sea-surface temperatures. It was then necessary to transform the fossil (down-core) faunal variations from core V12-122 into relative weightings of the major faunal assemblages already defined. Finally, these values were entered into the calibration equation to produce paleotemperature estimates. These are shown in Fig. 6.22 together with $\delta^{18}O$ measurements on the foraminifera *Globigerinoides ruber* from the same core (Imbrie *et al.*, 1973). A sequence of cooler episodes can be seen separated by warmer periods when temperatures approached modern (core-top) values and the cooler periods generally coincide with high $\delta^{18}O$ values (and vice versa). However, the changes in water temperatures can only account for a small fraction (~20%) of the isotopic change and, in fact, provide support for the view that global ice-volume changes are manifested mainly in the $\delta^{18}O$ record

---

[18] Gyre margin assemblage was not considered in this analysis.

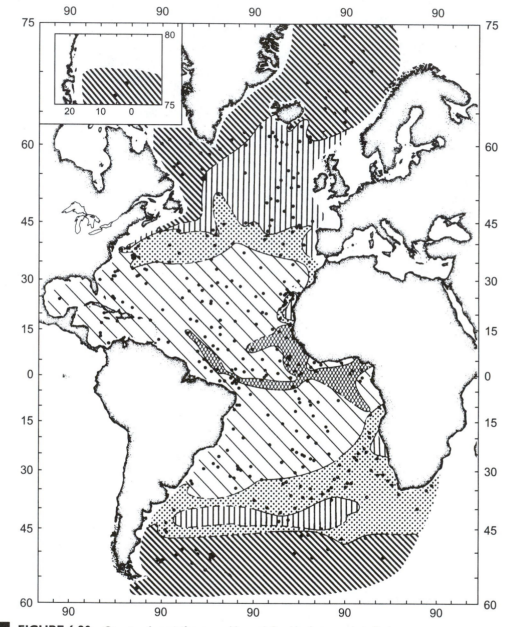

FIGURE 6.20 Core-top foraminifera assemblages, defined by factor analysis. Each zone represents an area dominated by a particular assemblage (polar, subpolar, transitional, subtropical, or gyre margin—off Equatorial West Africa) (Molfino et al., 1982).

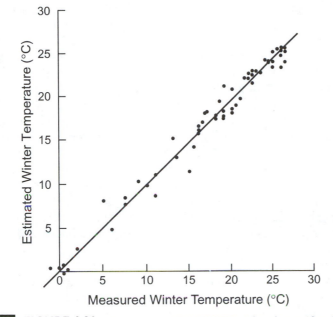

 FIGURE 6.21 Winter sea-surface temperature based on modern instrumental observations (including interpolated values) vs those estimated from faunal assemblages in 61 core-top samples using factor analysis and transfer function methods (Imbrie and Kipp, 1971).

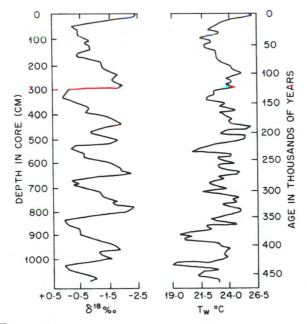

FIGURE 6.22 Winter sea-surface paleotemperature estimates (right) and $\delta^{18}O$ values (left) based on Caribbean core V12–122. The sea-surface temperature estimates are derived from transfer functions, in the manner shown schematically in Fig. 6.18 (Imbrie et al., 1973).

(see Section 6.3.1). From this reconstruction of Caribbean Sea paleotempera-tures, it appears that sea-surface temperatures in this region have been predomi-nantly cooler than today over the duration of the core record (~560,000 yr) with winter temperatures as much as ~7.5 °C lower than today around 430,000 yr B.P. (Imbrie *et al.*, 1973).

Since Imbrie and Kipp's pioneering work, there have been several attempts to refine and improve on their methodology (Ruddiman and Esmay, 1987; Dowsett and Poore, 1990). An alternative strategy was proposed by Prell (1985), who used a modern analog technique (MAT) to find the modern (core-top) assemblages that most closely resemble each fossil assemblage. Prell uses a statistical measure (a similarity coefficient) to quantify how closely each modern assemblage is to the fossil one. The modern sea-surface temperatures of the "top 10" modern assemblages are then used in a weighted average, to estimate the paleo-SST. This approach is de-veloped further for the Atlantic Ocean, by Pflaumann *et al.*, (1996), who demon-strate a very high degree of skill ($r^2 = 0.99$) in estimating modern SSTs for both summer and winter seasons over the entire range of Atlantic temperatures, from −1.4 °C (in winter at high latitudes) to +28.6 °C (in summer in equatorial regions). This augurs well for reliable paleo-SST estimates when this approach is applied to down-core foram assemblage records, providing differential dissolution of paleo-assemblages has not biased their representativeness.

One of the most rewarding and interesting applications of Imbrie and Kipp's methodology has been use of the technique to provide a synoptic view of paleo-oceanographic conditions in the past. By applying transfer functions to samples from a particular time horizon in many different cores, it is possible to reconstruct and map marine climates as they were at that time. This was one of the major ob-jectives of the CLIMAP project, which focused attention on marine conditions at 18,000 yr B.P. (CLIMAP Project Members 1976, 1981, 1984). The date of 18,000 (radiocarbon) yr B.P.[19] was selected as the time of the last maximum continental glaciation (LGM) defined by *maximum* $\delta^{18}O$ values during isotope stage 2 (Shack-leton and Opdyke, 1973). Using transfer functions derived for each of the major world oceans, February and August sea-surface temperatures have been recon-structed for this period (Table 6.3). Most studies relied mainly on foraminiferal as-semblage data, but in areas where siliceous fossils predominate (e.g., in the South Atlantic and Antarctic Oceans) the technique has also been applied to radiolarian assemblages (Lozano and Hays, 1976; Morley and Hays, 1979) and to diatoms (Pi-chon *et al.*, 1992; Koç Karpuz and Schrader, 1990). In the Pacific Ocean, where preservational characteristics of carbonate and siliceous fossils vary significantly from one area to another, it has been found advantageous to develop transfer func-tions based on four major microfossil groups (coccoliths, foraminifera, Radiolaria, and diatoms) to achieve optimum paleotemperature reconstructions (Geitzenauer *et al.*, 1976; Luz, 1977; Moore, 1978; Sancetta, 1979; Moore *et al.*, 1980). Although the paleotemperature maps for 18,000 yr B.P. are of interest alone, it is perhaps of most interest to use the reconstructions to produce maps of differences in tempera-

[19] This ^{14}C date is approximately equivalent to ~21,000 calendar yr B.P. (see Section 3.2.1.5).

TABLE 6.3 Sea-surface Paleotemperature Reconstructions for 18,000 Yr B.P.

Area	Principal faunal groups used	Major reference
North Atlantic	Foraminifera	Kipp (1976), McIntyre *et al.* (1976)
South Atlantic	Radiolaria	Morley and Hays (1979)
Norwegian and Greenland Seas	Foraminifera	T. Kellogg (1975, 1980)
	Diatoms	Koç Karpuz and Schrader (1992)
Caribbean and equatorial Atlantic	Foraminifera	Prell *et al.* (1976)
Western equatorial Atlantic	Foraminifera	Bé *et al.* (1976)
Eastern equatorial Atlantic	Foraminifera	Gardner and Hays (1976)
Indian Ocean	Foraminifera	Hutson (1978), Prell and Hutson (1979), Prell *et al.* (1980)
Antarctic Ocean	Radiolaria	Lozano and Hays (1976), Hays (1978)
	Diatoms	Pichon *et al.* (1992)
Pacific Ocean		
South	Foraminifera	Luz (1977)
North and South	Coccoliths	Geitzenauer *et al.* (1976)
North	Diatoms	Sancetta (1979, 1983)
North and South	Radiolaria	Moore (1978)
North and South	All four groups (synthesis)	Moore *et al.* (1980)
World Ocean (summary)	Foraminifera Radiolaria, Coccoliths	CLIMAP Project Members (1976)

ture between modern conditions and those at 18,000 yr B.P. Such maps are shown in Figs. 6.23–6.29 and are discussed briefly in the following sections. But first, a few caveats are neccessary.

One aspect of the CLIMAP SST reconstructions that has been controversial from the outset is the apparent lack of a significant temperature change in low latitude ocean surface temperatures at the last glacial maximum (LGM) (Prell, 1985). This conclusion does not fit well with other evidence from the tropics, such as the much lower snowlines recorded in mountainous areas (Selzer, 1990). If SSTs remained more or less constant, yet temperatures at an altitude of 4–6 km fell, an increase in lapse rate in the lower atmosphere is implied, but this is very difficult to envision (Webster and Streten, 1978). Furthermore, general circulation models can not reproduce an appropriate drop in snowline without sea-surface temperatures 5–6 °C lower than modern values (Rind and Peteet, 1985).

Recently, several independent lines of evidence have converged to challenge the veracity of low latitude CLIMAP SST estimates. Analysis of strontium/calcium ratios (see Section 6.8.5) in corals dating 10,200 B.P. from 16° S in the central Pacific indicate that temperatures were ~5 °C cooler at that time (Beck *et al.*, 1992). A similar result was obtained with corals of LGM age from Barbados (Guilderson *et al.*, 1994). Temperature changes of this magnitude are supported by studies of noble gases in groundwater from tropical Brazil. The concentration of noble gases (Ne, Ar, Kr, and Xe) in groundwater is largely a function of the temperature at the water table where dissolution occurs. By comparing noble gas concentrations in radiocarbon-dated groundwaters of Holocene and of glacial age, a temperature difference of 5.4 ± 0.6 °C was estimated. Additional evidence for lower tropical temperatures at the LGM comes from $\delta^{18}O$ measurements in an ice core from the high mountains of Peru (Huascarán: 6050 m elevation), which shows that snowfall was ~8‰ lower in $\delta^{18}O$ in glacial times compared to the Holocene (Thompson *et al.*, 1995b). This indicates much lower temperatures in the lower troposphere at that time. Finally, Miller *et al.*, (1997) estimate that temperatures in central Australia were at least 9 °C colder in glacial times compared to the Holocene, based on the extent of racemization of amino acids in ^{14}C-dated emu eggshells. As racemization is a function of age and temperature, by controlling for age, paleotemperatures can be calculated (see Section 4.2.1.4). These very diverse lines of evidence all point to significantly cooler temperatures in tropical and subtropical latitudes at the LGM.

Other lines of evidence are more in line with the original CLIMAP estimates (Table 6.4). For example, several recent studies use the temperature dependence of long chain alkenones (synthesized by planktic primnesiophyte algae — see Section 6.5) as a means of reconstructing SSTs (Brassell *et al.*, 1986). These studies generally find that LGM SSTs were only 1–2 °C lower in the western and eastern tropical Pacific (Ohkouchi *et al.*, 1994; Prahl *et al.*, 1988) and central equatorial Atlantic (Sikes and Keigwin, 1994) but up to ~2.5 °C lower in the central equatorial Indian Ocean (Rostek *et al.*, 1993; Bard *et al.*, 1997) and 2–3 °C lower in the eastern tropical Atlantic (with colder episodes as much as 4–5 °C lower for periods of a few hundred years at a time, corresponding to enhanced flow of the cool Canaries current during Heinrich events) (Zhao *et al.*, 1995). In addition, a reassessment of the LGM SSTs using a large database of modern core-top samples to identify optimum down-core analog assemblages (the "modern analog technique") led Prell (1985) to conclude that the original CLIMAP estimates were not biased by methodology and required no drastic revision. This was also the view of Thunell *et al.*, (1994) who used a high-quality core-top data set from the western Pacific; they found that LGM SSTs were generally within 1 °C of modern values between 20° N and 20° S, indicating that the western Pacific Warm Pool has existed since at least the LGM, and probably throughout the last glacial-interglacial cycle.

Such a view is strongly contested by Emiliani and Ericson (1991), who examine several lines of evidence which call into question the CLIMAP SST estimates at low latitudes. For example, they note that certain species of foram, with well-defined temperature tolerances, can be used as indicators of threshold temperatures in the past. Thus, the absence of *Pulleniantina obliquiloculata* and *Sphaeroidinella dehiscens* from equatorial Atlantic and Caribbean sediments of LGM age suggests that winter temperatures fell below 18.5 °C, a drop of at least 7–8 °C from modern val-

TABLE 6.4 Paleotemperature Estimates for the Last Glacial Maximum

Location	Lat	Long	ΔT (°C)	Reference
Based on planktonic forams[a]				
Western Pacific (many sites)	22° N-20° S	120° E-165° E	<2	Thunnell *et al.* (1994)
Based on alkenones[b]				
E. Equatorial Atlantic	0°	23° W	1.8	Sikes and Keigwin (1994)
N.E. Equatorial Pacific	1° N	139° W	1.3	Prahl *et al.* (1989)
W. Equatorial Pacific	3.5° N	142° E	<1.5	Ohkouchi *et al.* (1994)
Off N.W. Africa	19° N	20° W	3-4	Zhao *et al.* (1995)
Off N.W. Africa	19° N	20° W	~3	Chapman *et al.* (1996)
Off N.W. Africa	21° N	18.5° W	3-4	Eglington *et al.* (1992)
Indian Ocean (a transect)	20° N-20° S	~36° E	0.5-2.5	Bard *et al.* (1997)
N.E. Atlantic	48.3° N	25° W	3-4	Madureira *et al.* (1997)
N.E. Atlantic	56° N	12.5° W	5	Sikes and Keigwin (1996)
Central N. Atlantic	43.5° N	30.4° W	4-5	Villaneuva *et al.* (1998)
Based on Sr/Ca or U/Ca				
Huon Peninsula, New Guinea			>3	Aharon and Chappell (1986)
Huon Peninsula, New Guinea			>5-6	Min *et al.* (1995)
Vanuatu, southwestern Pacific			>4-5	Min *et al.* (1995)
Vanuatu, southwestern Pacific	15.5° S	167° E	>6.5	Beck *et al.* (1997)
Barbados			~5	Guilderson *et al.* (1994)
Barbados			4-5	Min *et al.* (1995)
Based on noble gases				
Brazil			5.4	Stute *et al.* (1995)
Based on amino acid racemization				
Central Australia			~9	Miller *et al.* (1997)

[a] Using Modern Analog Technique for foram assemblages.
[b] Up-to-date estimates can be found at http://NRG.NCLAC.UK:8080/CLIMATE/Art.htm

ues (see Fig. 6.6). They argue that the primary reason for the "incorrect" CLIMAP SST estimates is the probability (noted earlier) that many of the "modern" core-top samples were in fact not representative of truly modern sediments, but were contaminated by early Holocene faunas, which represent quite different oceanographic conditions in many tropical regions. If this is so, they argue, then not surprisingly any further analysis of this data set is only likely to confirm the original (erroneous) conclusions. However, this cannot be true of the high resolution data set used by Thunell *et al.,* (1994) in which core-tops were demonstrably modern.

Finally, a thorough review of planktic isotope data from the tropical oceans led Broecker (1986) to conclude that the LGM $\delta^{18}O$ values are consistent with the relatively small changes in SST revealed by the CLIMAP studies. This conclusion has been reinforced by Stott and Tang (1996), who examined $\delta^{18}O$ in individual planktic forams from the Holocene and the LGM in the tropical Atlantic. The majority of their samples showed that the tropical Atlantic SSTs were ~2 °C cooler at the LGM than in the Holocene. However, the presence of individuals with significantly higher $\delta^{18}O$ values (i.e. indicative of colder conditions) may reflect short-term episodes like those found by Zhao *et al.*, (1995) in their alkenone studies.

No doubt, this controversial subject will continue to be hotly debated as new evidence is brought to bear on the problem. But it is worth noting that the matter is of more than academic interest because the role of tropical SSTs is very important in understanding how global climates may change in the future. If large decreases in tropical SSTs did occur during glacial times, it implies that climate sensitivity to changes in greenhouse gases (and associated feedbacks) is at the high end of most estimates (~4.5 °C for a doubling of CO_2), otherwise the LGM forcing would not have produced such large changes (Crowley, 1994). Hence, resolving these differences in LGM SSTs is critical for a clearer understanding of the evolution of future climate. Meanwhile, when examining differences between modern SSTs and the CLIMAP reconstructions in Figs. 6.23–6.29 one should bear in mind that an ocean of controversy continues to surround LGM SST estimates at low latitudes.

One final point before the CLIMAP reconstructions are introduced; it is important to note that the maps are derived by interpolation between discrete sample points, which are often widely separated. This may lead to erroneous SST estimates being extrapolated over huge areas of the surrounding ocean (especially in the Pacific). This problem has been examined by Broccoli and Marciniak (1996), who find that the correspondence between GCM simulations of LGM paleotemperatures is significantly better when individual data points are compared, compared to using the interpolated paleo-SST maps. This should be borne in mind when examining Figs. 6.23–6.29; for example, in Fig. 6.23 the strongest gradients in temperature anomalies in the North Atlantic (located south of Newfoundland) are not well constrained by data (shown by the black dots) and result from the difference between an interpolated paleo-SST map and modern instrumentally measured conditions. This is not to say that they are wrong, but only that due caution is needed in interpretation of the maps.

6.4.1 North Atlantic Ocean

Situated between major ice sheets of the Northern Hemisphere (at 18,000 yr B.P.) the North Atlantic experienced the most significant changes in temperature of all oceanic areas (Figs. 6.23 and 6.24). At 18,000 yr B.P. August sea-surface temperatures were more than 10 °C cooler in a broad zone from 40–45° N in the west to 45–50° N in the east. This reflects a marked southward movement of polar and subpolar water at the time. In February, temperatures were less depressed off the North American coast ($\Delta T = 3$–5 °C) but in the eastern North Atlantic temperatures were 6–12 °C cooler than today in a triangular area stretching from Scandinavia to Portugal. In both seasons, strong upwelling off the coast of northwestern Africa (presumably related to stronger Trade winds, see Section 6.7) resulted in signifi-

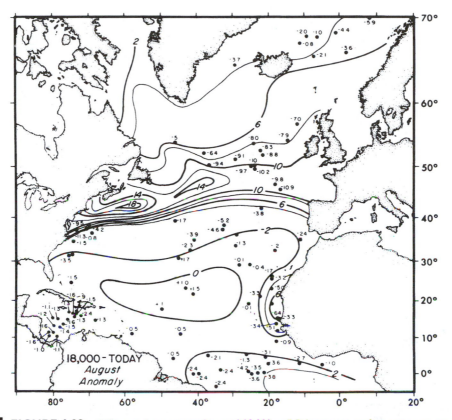

FIGURE 6.23 Difference between modern and 18,000 yr B.P. August sea-surface temperatures. Contour interval is 2 °C. Cores used in paleotemperature study shown as dots. Paleotemperatures were derived from faunal assemblage transfer functions. Figure derived by subtracting map of modern temperatures from map of paleotemperatures (both containing interpolated data). Hence the estimates of paleotemperature increase since 18,000 yr B.P. contain large values in some areas, even though core data may not have been available from those areas (e.g., the western North Atlantic between 42° and 50° N). See text for discussion of uncertainties in tropical SSTs (McIntyre *et al.*, 1976).

cantly cooler temperatures in that area (ΔT = 5–8 °C). Relatively minor temperature differences were apparent over most of the subtropical North Atlantic to the west, though February sea-surface temperatures in the Caribbean were 2–4 °C cooler than today at 18,000 yr B.P.

Considered together, the two maps suggest that the North Atlantic is made up of two zones, an area of dynamic change from ~40 to 50° N and a relatively stable zone to the south. Such a condition has apparently been characteristic of a much longer period than just the last 18,000 yr. McIntyre *et al.*, (1975) have reconstructed major water-mass boundaries along a meridional north-south transect through the Atlantic Ocean (at 20° W) for the past 130,000 yr (i.e., an interglacial-glacial cycle) and their study clearly indicates that the North Atlantic has been the most variable zone of

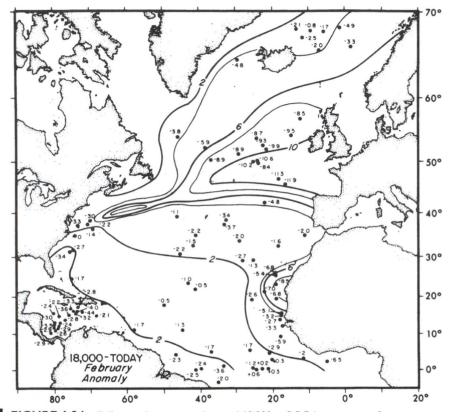

FIGURE 6.24   Difference between modern and 18,000 yr B.P. February sea-surface temperatures, derived as explained in Fig. 6.23. Contour interval is 2 °C (McIntyre *et al.,* 1976).

both hemispheres (Fig. 6.25). At 50° N, faunal assemblages have varied from being predominantly subtropical at 125,000 yr B.P. (the last interglacial) to polar from ~35,000 to 15,000 yr B.P. Furthermore, changes of a similar nature have been observed in cores reaching back over 600,000 yr, indicating that seven complete glacial-interglacial cycles have occurred during this period (Ruddiman and McIntyre, 1976).

6.4.2 Pacific Ocean

August sea-surface temperature differences were maximized in subarctic and equatorial regions (Fig. 6.26). In the area around Japan, temperatures were as much as 8 °C cooler at 18,000 yr B.P. due to a southward displacement of the warm Kuroshio current at that time and its replacement by subarctic water (Oyashio current). Less pronounced temperature depressions occurred in the Gulf of Alaska and southward along the California coast (ΔT = 2–4 °C), a fact which stands in marked contrast to conditions in the eastern Atlantic Basin. In equatorial regions, temperatures were cooler by 2–4 °C in a broad band, perhaps related to greater advection of cool waters into North and South Equatorial currents due to intensified tradewinds at the

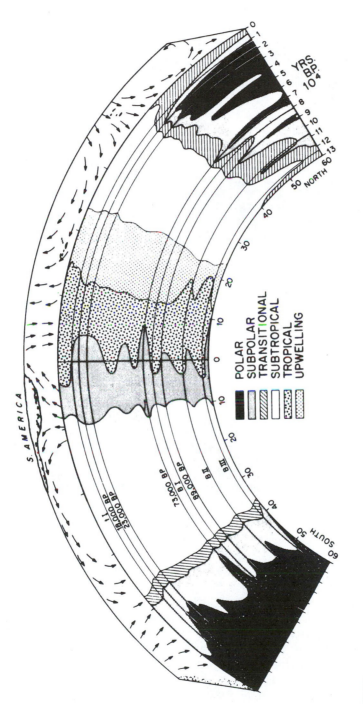

FIGURE 6.25 Variations of Atlantic surface water masses along the 20° W meridian, from 60° S (left) to 60° N (right) over the last 130,000 yr (vertical axis). Principal ocean currents along the 20° W meridian are shown schematically at top. Water mass variations shown cover the last interglacial-glacial cycle. Note the large latitudinal variations in water masses in the North Atlantic (McIntyre et al., 1975).

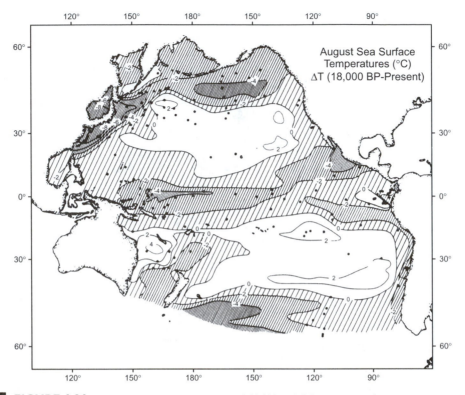

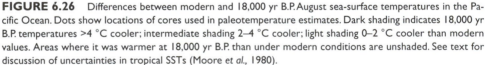

FIGURE 6.26 Differences between modern and 18,000 yr B.P. August sea-surface temperatures in the Pacific Ocean. Dots show locations of cores used in paleotemperature estimates. Dark shading indicates 18,000 yr B.P. temperatures >4 °C cooler; intermediate shading 2–4 °C cooler; light shading 0–2 °C cooler than modern values. Areas where it was warmer at 18,000 yr B.P. than under modern conditions are unshaded. See text for discussion of uncertainties in tropical SSTs (Moore *et al.,* 1980).

time. In the South Pacific, cooler waters adjacent to the coast of South America and west of New Zealand are noteworthy (ΔT = 2–4 °C). It is also of interest to note that, in addition to areas of major cooling, large parts of the Pacific Basin appear to have been warmer at 18,000 yr B.P. than at present. In particular, core regions of subtropical high-pressure centers are reconstructed as having been 1–2 °C warmer than modern values. Temperatures along the eastern coast of Australia were also warmer (by up to 4 °C) perhaps due to enhanced equatorial flow from the stronger Equatorial currents. However, this view is contested by Anderson *et al.,* (1989), who show that the CLIMAP paleo-SSTs in this region are too high. One potentially important problem in deriving paleo-SSTs from the tropical Pacific is the effect of differential dissolution, which tends to remove "warm" forams from the core-top assemblages. During the LGM, if dissolution in the Pacific was less, there may have been more "warm" forams preserved, leading to an erroneous view of overall temperature changes (Broccoli and Marciniak, 1996).

Similar patterns of difference are observed in many areas in the February sea-surface temperature reconstructions (Fig. 6.27) with maximum temperature changes

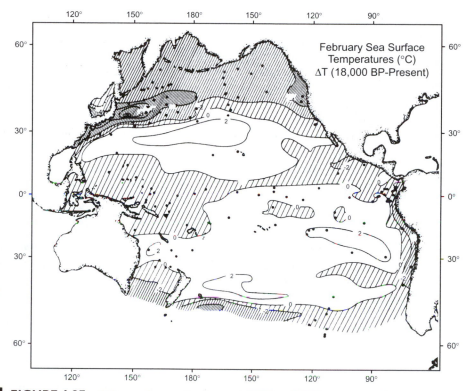

FIGURE 6.27 Difference between modern and 18,000 yr B.P. February sea-surface temperatures in the Pacific Ocean. Dots show location of cores used in paleotemperature estimates. Shading as in Fig. 6.26 (Moore et al., 1980).

in the area east and north-east of Japan (ΔT = 6–8 °C). Cooler temperatures at this season could have resulted in sea-ice formation over an extensive area (Sancetta, 1983). In the Southern Hemisphere, cooling was relatively minor at 18,000 yr B.P., except in the Peruvian current off the western coast of South America (ΔT = 2–4 °C) and in the extreme south due to an expanded subpolar water mass. Again, a noticeable feature is the extensive area of warmer sea-surface temperatures at 18,000 yr B.P. centered over the subtropical high-pressure cells. It is particularly interesting to note the large extent of this positive anomaly in the Southern Hemisphere, associated with the poleward movement of the subtropical high-pressure center at this time of year. A corresponding southward shift in the Northern Hemisphere positive anomaly field is also apparent in the February maps compared to those for August. Such a pattern suggests a more intense Hadley cell circulation at 18,000 years B.P., with well-developed subtropical high-pressure centers. In these areas, adiabatic warming and clear skies would favor warmer sea-surface temperatures and, on the subtropical margins, trade winds and gyre margin ocean currents would be strengthened. All these factors fit together quite coherently in relation to the reconstructed paleotemperatures, which demonstrates that the overall reconstructions are at least internally consistent.

6.4.3 Indian Ocean

August sea-surface temperature anomalies reveal relatively minor differences between 18,000 yr B.P. and today (Fig. 6.28). Apart from areas associated with the eastern and western boundary currents, off the western coast of Australia and off southeastern Africa (the Agulhas Current) most areas at 18,000 yr B.P. were within 1 or 2 °C of modern values. It is interesting that temperatures in the Arabian Sea were ~1 °C warmer at 18,000 yr B.P., suggesting a weaker Southwest Monsoon flow at that time, resulting in less upwelling of cool water (Prell *et al.*, 1980). However, this is not supported by alkenone evidence from off the Arabian Peninsula, which indicates glacial-interglacial temperature differences of >3 °C in spite of reduced upwelling at the LGM (which would tend to lessen the difference) (Emeis *et al.*, 1995).

February maps reveal larger temperature differences, particularly in the area centered on 40° S, where northward movement of the Antarctic Convergence zone and associated subpolar water caused temperatures to be lower by 4–6 °C at 18,000 yr B.P. (Fig. 6.29). Compared to the other ocean basins, however, temperature changes in the Indian Ocean were relatively small (overall cooling of only ~1.8 °C) and large areas off the coast of eastern Africa and in the Arabian Sea may have been slightly warmer at 18,000 yr B.P. than they have been in recent years.

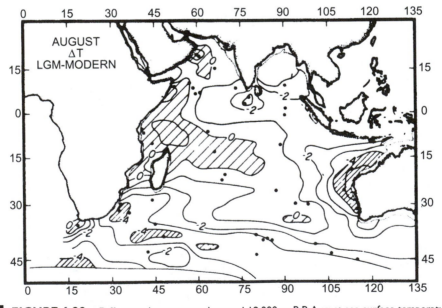

FIGURE 6.28 Difference between modern and 18,000 yr B.P. August sea-surface temperatures. LGM = last glacial maximum. Contour interval is 1 °C. Widely spaced diagonal lines indicate areas warmer at 18,000 yr BP than today. Closely spaced diagonal lines indicate areas at least 4 °C cooler than today at 18,000 yr B.P. See text for discussion of uncertainties in tropical SSTs (Prell *et al.*, 1980).

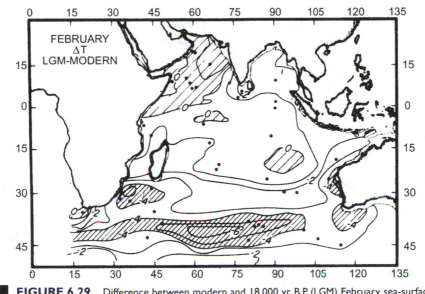

FIGURE 6.29 Difference between modern and 18,000 yr B.P. (LGM) February sea-surface temperatures. Contour interval is 1 °C. Shading as in Fig. 6.2 (Prell *et al.*, 1980).

6.5 PALEOTEMPERATURE RECORDS FROM ALKENONES

Certain marine phytoplankton of the class Prymnesiophyceae, most notably the coccolithophorid *Emilyania huxleyi*, respond to changes in water temperature by altering the molecular composition of their cell membranes. Specifically, as water temperature decreases, they increase the production of unsaturated alkenones (ketones). Cells contain a mixture of long-chain alkenones with 37, 38, or 39 carbon atoms (n-C_{37} to n-C_{39}), which are either di- or tri-unsaturated (designated for example, as $C_{37:2}$ and $C_{37:3}$, respectively). A temperature-dependent unsaturation index U_{37}^K is defined as:

$$U_{37}^K = \frac{[C_{37:2}] - [C_{37:4}]}{[C_{37:2} + C_{37:3} + C_{37:4}]} \qquad (6.5)$$

where $[C_{37:2}]$ represents the concentration of the di-unsaturated methyl ketone, alkadienone, containing 37 carbon atoms. The index varies from -1 (when all alkenones are $C_{37:4}$) to $+1$ (all $C_{37:2}$). However, as $C_{37:4}$ is absent in most sediments the index can be simplified to:

$$U_{37}^{K'} = \frac{[C_{37:2}]}{[C_{37:2} + C_{37:3}]} \qquad (6.6)$$

so that values are generally positive (~0.2 to 0.98) in Quaternary sediments (Brasell *et al.*, 1986). The importance of these organic biomarkers is that they do not appear to significantly degrade in marine sediments, nor are they influenced by changes in salinity or isotopic composition of the ocean. They thus provide a crucial complement to $\delta^{18}O$ and faunal composition studies of marine paleotemperatures, as discussed in what follows.

Studies of the algae *E. huxleyi* in controlled growth chambers, and of sediments accumulating beneath areas with different SSTs, show a strong signal relating water temperature to $U_{37}^{K'}$ (Prahl *et al.*, 1988; Sikes *et al.*, 1991; Rosell-Melé *et al.*, 1995; Müller *et al.*, 1998). From controlled experiments (Prahl *et al.*, 1988) the $U_{37}^{K'}$-temperature relationship is:

$$T\ (°C) = (U_{37}^{K'} - 0.039)/0.034 \qquad (6.7)$$

Others have found that this relationship varies somewhat, but generally the slope is the same over the range of 15–25 °C.

The potential of $U_{37}^{K'}$ as an SST paleothermometer is tremendous and is the basis of an emerging new field in paleoclimatology (molecular stratigraphy). A number of studies have now been carried out to reconstruct SSTs from the alkenone record in marine sediments. Almost uniformly, these studies show a smaller temperature difference between Holocene and LGM SSTs than estimates based on faunal composition (using modern analog or transfer functions). However, part of this difference is probably related to the fact that the two approaches are not dealing with precisely the same phenomena. Alkenone-based SSTs are derived from phytoplankton that live predominantly in the photic zone, with the bulk of the organisms inhabiting the upper 10 m of the water column. Furthermore, phytoplankton blooms commonly occur rapidly in the spring or early summer and the resulting organic sediment may thus represent a relatively short episode of only a few weeks (Sikes and Keigwin, 1994, 1996). By contrast, the foram-based paleotemperatures are based on a set of different organisms that may reach peak abundances at different times of the year, and live at different depth habitats. Consequently, such estimates are likely to represent a more time and depth-integrated measure of temperature change than that from alkenones. Furthermore, foram assemblage (and $\delta^{18}O$) data are subject to potential biases due to dissolution effects, whereas alkenone-based paleotemperatures are not (Sikes and Keigwin, 1994). However, in areas with a large annual range of SSTs, any shift in the seasonal timing of maximum phytoplankton productivity could result in a shift in alkenone-based paleotemperatures, even without any real change in oceanographic conditions (Chapman *et al.*, 1996).

By combining alkenone-based paleotemperature reconstructions with other approaches (such as faunal assemblage, or $\delta^{18}O$-based paleotemperature estimates) important new insights into paleo-oceanographic conditions can be obtained. For example, Zhao *et al.* (1995) studied sediments from off the northwest coast of Africa and found that paleotemperature minima at different times in the last 80 ka correspond closely to Heinrich events seen in North Atlantic sediments (see Section 6.10.1). These abrupt changes appear to represent times when cold meltwater, produced by ice-rafting events, was transferred southward by the Canary current, caus-

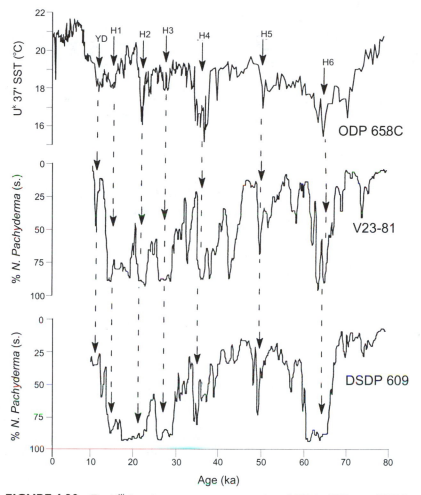

FIGURE 6.30 The $U^{K'}_{37}$-based temperature reconstruction of SSTs in ODP core 658C from off the north-west coast of Africa (upper panel) compared to percentages of the cold-water foram *N. pachyderma* (s.) in two cores from the North Atlantic. The strong relationship between Heinrich events and cold water episodes in the North Atlantic with episodes of low SSTs off the African coast suggests a linkage via the cold Canary current carrying cool, low salinity meltwater southward at these times (Zhao *et al.*, 1995).

ing temperatures to decline by 3–4° C in <100 yr (Fig. 6.30). Meltwater effects were also detected in northeastern Atlantic sediments by Sikes and Keigwin (1996) by comparing both alkenone and $\delta^{18}O$ records. This approach was used to good effect in "backing-out" past changes in salinity from an Indian Ocean sediment record by Rostek *et al.* (1993). They established SSTs using alkenones, then applied that to the $\delta^{18}O$ record to reconstruct paleosalinity. By subtracting the local temperature effect on $\delta^{18}O$, and knowing the effect of changing ice volume on $\delta^{18}O$, the residual change in $\delta^{18}O$ was interpreted as a record of changing salinity (Fig. 6.31). High salinity (by

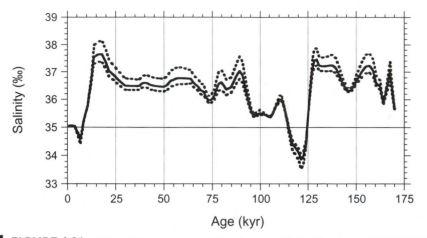

FIGURE 6.31   Paleosalinity reconstructed for the site of Indian Ocean core MD 9000963, (South south-west of India). This was derived from $\delta^{18}O$ by obtaining paleotemperatures from alkenones, then adjusting the $\delta^{18}O$ record for these changes, plus changes in global ice volume and salinity due to continental ice growth and decay. The residual $\delta^{18}O$ changes are interpreted as a record of paleosalinity. Dotted lines bracket the range of paleosalinity estimates (Rostek *et al.*, 1993).

+ 0.5–1‰) from 160–140 ka and 75–25 ka B.P. resulted from a weaker southwest monsoon (with less rainfall on the subcontinent, and hence less runoff to the Bay of Bengal) and/or a stronger northeastern (counter) monsoon airflow at those times.

Other studies have applied alkenone analysis to high-resolution paleotemperature reconstruction of both recent sediments (e.g., to examine ENSO events; Kennedy and Brassell, 1992) as well as to periods of rapid environmental change during glacial Terminations. "Younger Dryas-type" oscillations were found to have occurred during Terminations II and IV, suggesting that similar mechanisms involving rapid reorganization of North Atlantic deepwater formation had ocurred during earlier deglaciation events, as well as the most recent one (Eglinton *et al.*, 1992).

6.6 DISSOLUTION OF DEEP-SEA CARBONATES

Throughout the deep ocean basins of the world, a major factor affecting preservation of carbonate tests is the rate of dissolution at depth. The oceans are predominantly undersaturated with respect to calcium carbonate at all depths below the upper mixed layer (the zone above the thermocline; Olausson, 1965, 1967). After death of the organisms, deposition of the test on the ocean floor leads to dissolution in the undersaturated water (Adelsack and Berger, 1977). Pteropod tests (composed of calcium carbonate in the form of aragonite) are most susceptible to solution and are the first to disappear; pteropods are thus only found in relatively shallow waters where undersaturation is not pronounced (Berger, 1977). At greater depth, dissolution of tests made of calcite (e.g., foraminifera and coccoliths) becomes apparent. The level at which calcite dissolution is at a maximum is the lyso-cline (Berger, 1970, 1975) generally encountered at 2500–4000 m depth in the

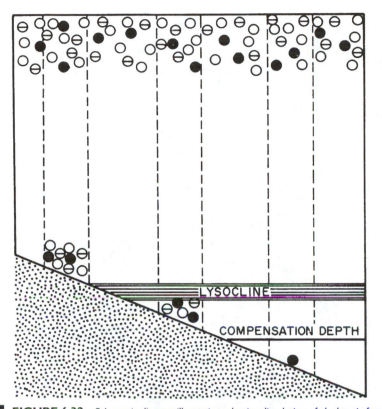

FIGURE 6.32 Schematic diagram illustrating selective dissolution of planktonic foraminiferal species at depth, due to undersaturation of the water with respect to calcium carbonate. Dark circles represent the resistant species *Globoratalia tumida*. Open circles represent *Globigerinoides ruber*, that is dissolved relatively easily. *Globigerina bulloides* (open circle with a line) is intermediate in resistance. Dissolution alters the species composition of the sediment so it may not be representative of species in the overlying water column. At depths below the compensation depth only the occasional *Globoratalia tumida* may survive. Changes in the depth of the lysocline and compensation depth through time may offset the sediment species composition, due to differential dissolution (Bé, 1977).

oceans (Fig. 6.32). In the Atlantic, there is evidence that this corresponds to the boundary between North Atlantic Deep Water and the deeper Antarctic Bottom Water (Berger, 1968). Below this level, calcite dissolution rates increase markedly, until at extreme depths the water is so corrosive to calcite that virtually no tests survive to be deposited. The depths at which the dissolution rate equals the rate of supply of carbonate tests from the overlying water column is the calcite compensation depth (CCD) (Berger, 1970). This can be envisioned as analogous to a snow-line on land; deep ocean basins below the compensation depth are devoid of carbonate sediments, and higher levels are increasingly blanketed by microfossil tests (Berger, 1971). Because the calcite compensation depth is a function of both the rate of supply of carbonate tests and the dissolution rate, its actual depth varies from one area to another (Fig. 6.33) though generally it is <4000 m (Berger and Winterer, 1974). As vast areas of the ocean floor are below 4000 m, particularly in

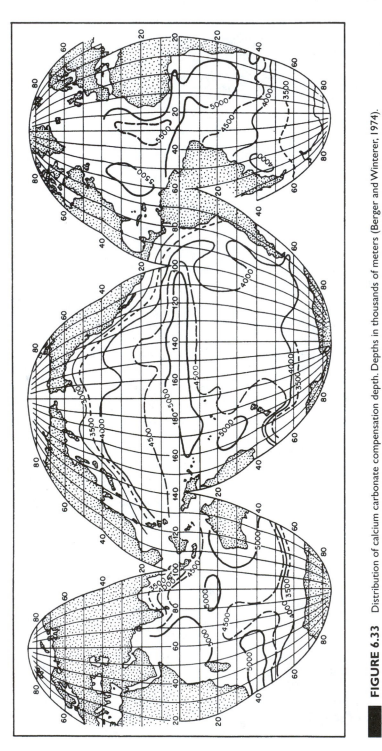

FIGURE 6.33   Distribution of calcium carbonate compensation depth. Depths in thousands of meters (Berger and Winterer, 1974).

the Pacific Basin, this phenomenon greatly restricts the area in which foraminiferal studies can be usefully carried out (see Fig. 6.2). Even in less deep areas of the ocean, sediments accumulating below the lysocline are subject to significant dissolution. Most importantly, dissolution does not affect all species uniformly; selective removal of the more fragile, thin-walled species may significantly alter the original assemblages (biocoenoses), leaving behind thanatocoenoses which are unrepresentative of productivity in the overlying water column. Assemblages may be enriched with resistant species, which tend to be deep-dwelling, secreting their relatively thick tests in water that is significantly cooler than that near the ocean surface (Ruddiman and Heezen, 1967; Berger, 1968). Similarly, in populations of a particular species, the thicker-walled, more robust individuals, which are preferentially preserved, tend to build their shells in deeper, colder water and are therefore isotopically heavier than their more fragile counterparts (Hecht and Savin, 1970, 1972; Berger, 1971).

Studies of the relative abundances of different foraminiferal species, in cores from various depths, have demonstrated these effects well (Fig. 6.34) and enabled species to be ranked according to their relative susceptibility to solution. Similar

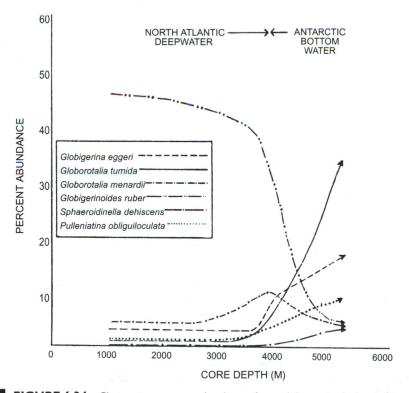

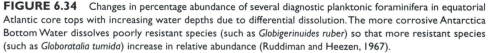

FIGURE 6.34 Changes in percentage abundance of several diagnostic planktonic foraminifera in equatorial Atlantic core tops with increasing water depths due to differential dissolution. The more corrosive Antarctica Bottom Water dissolves poorly resistant species (such as *Globigerinuides ruber*) so that more resistant species (such as *Globoratalia tumida*) increase in relative abundance (Ruddiman and Heezen, 1967).

studies of coccoliths indicate corresponding problems, with structurally solid, cold-water forms preferentially preserved in thanatocoenoses (Berger, 1973a).

Berger (1973b) suggests that the partially dissolved assemblages be designated taphocoenoses, to distinguish them clearly from assemblages more representative of the original biocoenoses. Clearly, paleoclimatic reconstructions based on tapho-coenoses require very careful interpretation. This is particularly so if the rate of dissolution has changed over time as suggested by a number of studies (Chen, 1968; Broecker, 1971; Berger, 1971, 1973b, 1977; Thompson and Saito, 1974; Ku and Oba, 1978). There is evidence that dissolution rates increased during inter-glacial times in the tropical Pacific and Indian Oceans, resulting in the removal of many less resistant species, and a relative concentration of individuals with a cold-water aspect (Wu and Berger, 1989).[20] Conversely, in glacial times, dissolution rates were reduced, giving rise to assemblages of both solution-susceptible and solution-resistant forms. In short, glacial-interglacial changes may be characterized by corresponding dissolution cycles in these areas (Berger, 1973b). Such effects would result in erroneous isotopic paleotemperature estimates as interglacial-age samples would have a higher abundance of cold-water individuals compared to glacial-age samples, thereby reducing the apparent glacial-interglacial temperature range (Berger, 1971; Emiliani, 1977; Berger and Killingley, 1977). Similarly, in foraminiferal assemblage studies (Section 6.4) dissolution cycles may result in taphocoenoses, which lead to quite erroneous paleotemperature estimates. Thus, Berger (1971) and Ruddiman (1977a) urge that all carbonate sediments be consid-ered residual, unless easily dissolved material is present.

One interesting, and potentially very important, aspect of dissolution rate changes over time is the presence of a "deglacial preservation spike" or strati-graphic zone representing a period when dissolution rates were markedly reduced (Broecker and Broecker, 1974; W. Berger, 1977). There is evidence that at around 14,000 yr B.P. (and during other terminations) there was a significant worldwide drop in the aragonite compensation depth and the lysocline, lasting for only a rela-tively short period (perhaps <1000 yr). This resulted in enhanced preservation of carbonate fossils at that time and hence a "spike" of well-preserved foraminifera and pteropods in the sedimentary record (Wu et al., 1990). Indeed, the preservation spike is particularly apparent because dissolution rates appear to have been even greater at ~12,000 yr B.P., directly following the time of the dissolution minimum (Berger and Killingley, 1977).

Another significant dissolution signal (the "Brunhes dissolution cycle") is seen in sediments from the equatorial Pacific and Indian Oceans (Wu and Berger, 1989). This is clearly shown in Fig. 6.35 in which the difference in oxygen isotope stra-tigraphies between records from the Ontong Java Plateau are plotted; one record (V28-238) is from 3120 m and the other (V28-239) is from 3490 m water depth. A generally positive difference in $\delta^{18}O$ (of ~0.3‰) between these records is expected because the core from the deeper site has been affected by dissolution (making it

[20] In the equatorial Atlantic and Gulf of Mexico, dissolution seems to have increased in glacial times (Gardner, 1975; Luz and Shackleton, 1975).

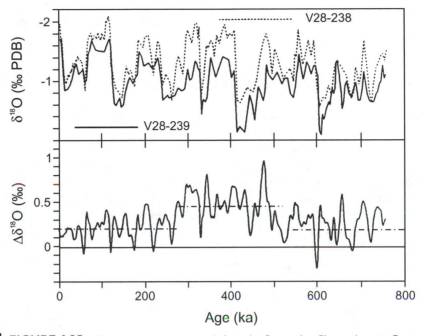

FIGURE 6.35   Two oxygen isotope records from the Ontong Java Plateau (eastern Equatorial Pacific) (upper diagram) and the difference between them (V28-239 minus V28-238) (lower diagram). The V28-239 record is from 3490 m and V28-238 from 3120 m water depth. Dissolution has affected the deeper record leading to the preservation of forams with a higher ^{18}O content. In addition, a pronounced episode of enhanced dissolution is seen in the lower diagram, from ~300–500 kyr B.P. (Wu and Berger, 1989).

isotopically heavier) and this effect is enhanced during the interglacials. However, from ~300–500 ka B.P. the dissolution effect is systematically greater, indicating some persistent influence on carbonate dissolution. This has also been seen in Indian Ocean sediments (Peterson and Prell, 1985) but the reason is not, as yet, fully understood.

The cause of these rapid shifts in compensation depth is not clear and, in fact, may result from a combination of many factors. Redeposition of carbonate from the continental shelves as sea level rose could have increased oceanic alkalinity and thereby reduced dissolution (W. Berger, 1977). However, as the sea level rose a low salinity upper water layer may have formed (from continental ice-sheet melting) creating a lid on the ocean and preventing vertical mixing (Worthington, 1968). Continued biological activity in the oceans might have led to the accumulation of CO_2 and hence increased dissolution (Berger *et al.*, 1977). Perhaps it is this signal that is observed following the dissolution minimum. Such a hypothesis has intriguing implications; if an extensive meltwater layer did exist during deglaciation (and if this resulted in a build-up of CO_2 in the subsurface waters), when ocean mixing was eventually restored an increase in atmospheric CO_2 concentrations would have ensued, resulting in an enhanced greenhouse effect. This may have been the sequence of feedbacks that contributed to the further decay of ice sheets and to early Holocene warmth.

6.7 PALEOCLIMATIC INFORMATION FROM INORGANIC MATERIAL IN OCEAN CORES

Weathering and erosion processes in different climatic zones may result in characteristic inorganic products. When these are carried to the oceans (by wind, rivers, or floating ice) and deposited in offshore sediments, they convey information about the climate of adjacent continental regions, or about the oceanic and/or atmospheric circulation, at the time of deposition (McManus, 1970; Kolla *et al.*, 1979). On continental margins, the bulk of sediment is deposited by rivers, but in remote areas of the ocean, far from land areas and the influence of floating ice, very fine wind-blown material washed out of the atmosphere may form a significant proportion of the total sediment accumulation (Windom, 1975). Modern observations show that total dust flux thousands of kilometers downwind of arid regions is mainly a function of conditions in the source region, whereas variations in grain size are more related to changes in (upper level) wind speed (Rea, 1994). Thus, by examining variations in the eolian fraction of marine sediment cores, an important index of continental aridity, modulated by changes in airflow patterns, can be obtained. For example, in a sediment core 2500 km east of the Chinese Loess Plateau, Hovan *et al.* (1989, 1991) found large variations in eolian accumulation rates that correspond to changes in the environment of the Loess Plateau. During interglacials, when loess accumulation slowed and soils formed on the plateau, eolian flux rates were low, but during glacial periods (defined by the benthic $\delta^{18}O$ record of the same core) eolian flux rates were many times higher (Fig. 6.36). By correlating the periods of high eolian flux in the marine sediments with episodes of loess accumulation and low magnetic susceptibility, improvements in the chronology of loess deposition could be made by taking advantage of the SPECMAP $\delta^{18}O$ chronological template.

Numerous studies of inorganic material in cores from off the coast of West Africa have enabled climatic fluctuations of the adjacent land mass to be deduced. In this area today, vast quantities of silt and clay-sized particles (>25 million tons per year) are transported from the Sahara desert westwards across the Atlantic by the northeast trade winds (Chester and Johnson, 1971). During late Quaternary glacial epochs, an even higher proportion of terrigenous material accumulated in the equatorial and tropical Atlantic off West Africa due to stronger tradewinds and a more extensive arid zone (Fig. 6.37) (Sarnthein *et al.*, 1981; Matthewson *et al.*, 1995). Further support for this scenario of drier conditions during glacial episodes is provided by studies of biogenic detritus in ocean cores. The concentration of freshwater diatoms (*Melosira*) and opal phytoliths (minute silica bodies derived from epidermal cells of land plants, particularly grasses) increased during glacial

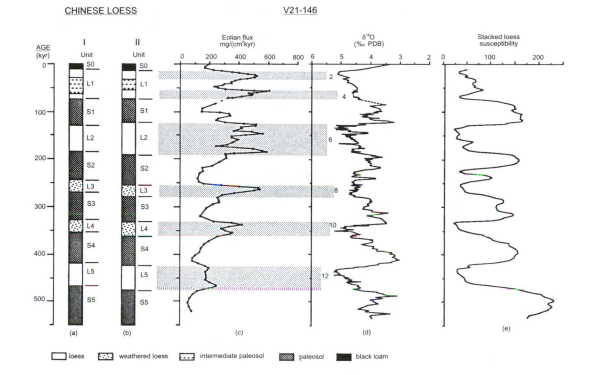

CHINESE LOESS   V21-146

FIGURE 6.36 Eolian flux recorded in North Pacific core V21-146 (from ~38° N, 163° E) and benthic isotope record in the same core (center graphs) compared to the magnetic susceptibility of loess in a stacked set of records from the Chinese Loess plateau (right graph). The chronology of loess and paleosol units described by Kukla (1987a) is shown on the extreme left (column I). Revised ages, based on correlations between the marine record and the loess susceptibility are shown in the second (column II) (Hovan et al., 1991).

periods in cores south of 20° N off the coast of West Africa. It is suggested that this results from deflation of lacustrine sediments by stronger tradewinds in relatively dry glacial times, following more humid interglacial periods when vegetation (grassland) was extensive and lakes were more common (Parmenter and Folger, 1974; Pokras and Mix, 1985).

Another region of major eolian dust flux to the oceans is off the coast of the Arabian Peninsula. Today, strong northwesterly winds carrying fine-grained sediments to the Indian Ocean sweep over low-level southwesterly (monsoon) winds from Somalia. In modern sediments, the 30% isoline of siliciclastic grains (>6 µm) approximates the location of the main convergence zone between the northwesterly and southwesterly airflows. Reconstructions of the position of this isoline for various time-slices in the past thus track variations in this convergence zone, showing that it was farther to the southeast during the LGM (i.e., stronger northwesterly

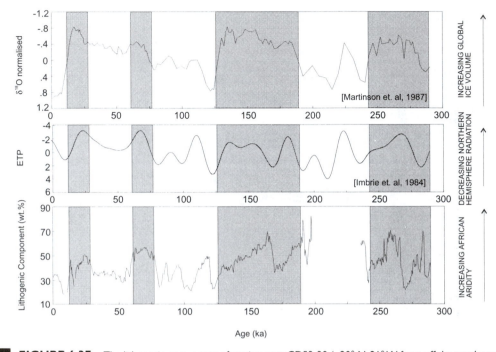

FIGURE 6.37 The lithogenic component of marine core CD53-30 (~20° N, 21° W, from off the northwest coast of Africa) (bottom panel) compared to the SPECMAP marine isotope record (upper panel) and the combined northern hemisphere record of precession, obliquity, and eccentricity forcing (middle panel, see Fig. 2.16). Glacial periods (shaded) correspond to higher levels of eolian dust flux from the Sahara to the adjacent ocean. The rapid increases in dust are generally associated with decreasing northern hemisphere radiation, driven by changes in precession, though the abrupt shifts in dust content suggest a nonlinear response to orbital forcing (Mathewson et al., 1995).

airflow) but much closer to the coast 6–9 ka B.P. (Fig. 6.38) (Sirocko and Sarnthein, 1989; Sirocko et al., 1991).

A final example comes from the southern hemisphere where eolian sediments in the Tasman Sea record variations in aridity in southeastern Australia (Hesse, 1994). During glacial periods, dust flux increased by 50–300% and the northern boundary of the main dust plume shifted equatorward by ~350 km. This cyclical pattern of increased dust flux during glacial periods and reduced dust flux in interglacials is superimposed on a longer-term increase in overall eolian sediment which began ~350–500 ka ago, reflecting the increasing aridification of southeastern Australia.

In each of these cases, there is strong evidence that eolian dust flux to the oceans was much greater during the last glaciation, as well as during earlier glacial events. Part of this increase was probably related to a stronger Pole-Equator temperature gradient and higher wind speeds (Wilson and Hendy, 1971) but there were also much larger arid areas in the intertropical zone during glacial times (Sarnthein, 1978). Both factors led to far higher levels of atmospheric turbidity during glacial periods, and this

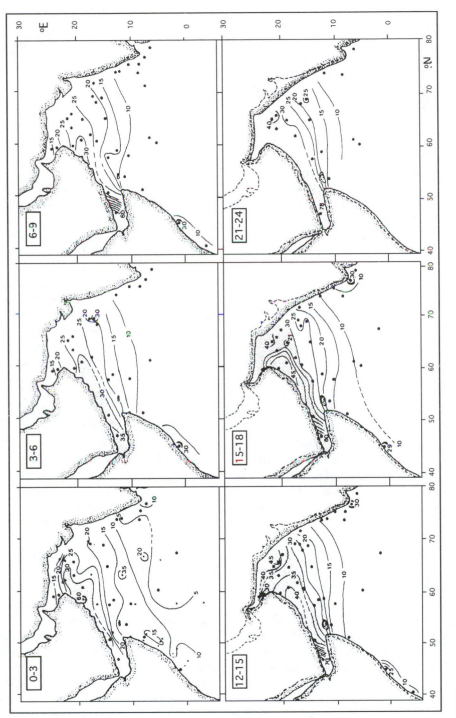

FIGURE 6.38 Percentages of siliciclastic grains >6μm in the siliciclastic fraction of cores off the Arabian Peninsula, at 3000-yr intervals from 21–24,000 yr B.P. (bottom right) to 0–3000 yr B.P. (top left). The 30% isoline corresponds to the main convergence zone between north-westerly airflow carrying dust from the Arabian Peninsula and southwesterly monsoon airflow from the horn of Africa (Somalia) (Sirocko et al., 1991).

is well recorded in remote polar (and high-elevation) ice cores as a pronounced increase in particulate matter (Petit *et al.,* 1990). There has also been speculation that the higher amounts of eolian material may have played a role in controlling carbon dioxide levels in the atmosphere during glacial periods. Because biological activity (particularly in the southern ocean) is limited by a lack of iron, it has been suggested that the additional iron deposited in the oceans during glacials may have increased oceanic photosynthetic activity to the point that carbon dioxide levels were reduced (Martin *et al.,* 1990). Recent experiments to "seed" large ocean areas with iron have indeed shown that productivity increases significantly when this limiting factor is removed (Price *et al.,* 1991; Coale *et al.,* 1996) but whether this effect can explain the lower carbon dioxide levels of glacial times remains controversial.

Coarse-grained sediments found in sediment cores in remote parts of the ocean provide evidence of former ice-rafting episodes (either from icebergs or formerly land-fast sea ice). One of the most compelling lines of evidence that continental glaciation began in late Pliocene time (~2.4 M yr ago) is ice-transported coarse sediment in cores from both the North Atlantic and the North Pacific (Shackleton *et al.,* 1984). In the late Quaternary, there were quasi-periodic episodes of major ice-rafting in the North Atlantic (above ambient background levels); these are now termed Heinrich events (Heinrich, 1988) (see Section 6.10.1). They appear to be correlated in some way with abrupt changes in $\delta^{18}O$ seen in ice cores from the Greenland ice sheet (Dansgaard-Oeschger events — see Section 5.42). At least some of these coarse layers contain an abundance of detrital carbonates, with a geographical distribution that indicates a source region in Foxe Basin (west of Baffin Island) or Hudson Bay. It appears that the material was transported to the North Atlantic through Hudson Strait, via icebergs that calved from a major ice stream of the Laurentide ice sheet (Andrews *et al.,* 1994; Dowdeswell *et al.,* 1995) (see Fig. 6.47).

Isotopic studies of individual mineral grains have been carried out in an attempt to pinpoint the provenance of material in different Heinrich events. Two approaches, one using lead isotopes, the other using neodynium and strontium isotope ratios, and strontium concentrations, to characterize the sediments and their potential source rocks have led to conflicting interpretations. Lead isotopic ratios point quite specifically to the Churchill Province of the Canadian Shield (northwest of Hudson Bay, Hudson Strait, and Baffin Island) as the source of debris in Heinrich event 2 (~21,000 yr B.P.) (Gwiazda *et al.,* 1996a). By contrast, Sr and Nd isotopes allow for the possibility of multiple sources of detrital material during this and other Heinrich events (including the Icelandic, Fennoscandian, and British Isles ice sheets) (Grousset *et al.,* 1993; Revel *et al.,* 1996). These different interpretations have important implications for identifying the forcing mechanisms that led to Heinrich events. If all Heinrich events involve only material from the Laurentide ice sheet (via Hudson Strait) this points to some sort of internal mechanism controlling ice discharge, such as the binge-purge model of MacAyeal (1993). However, if the Heinrich events result from ice being discharged from ice sheets all around the North Atlantic, this suggests a more pervasive climatic forcing mechanism that can influence both small (Icelandic) and large (Laurentide) ice sheets more or less simultaneously (Bond and Lotti, 1995). This is discussed further in Section 6.10.1.

6.8 CORAL RECORDS OF PAST CLIMATE

The term "coral" is generally applied to members of the order Scleractinia, which have hard calcareous skeletons supporting softer tissues (Wood, 1983; Veron, 1993). For paleoclimatic studies, the important subgroup is the reef-building, massive corals in which the coral polyp lives symbiotically with unicellular algae (zooxanthellae); these are known as hermatypic corals (as opposed to ahermatypic, which contain no algal symbionts and are not reef-builders). The algae produce carbohydrates by photosynthesis and thus are affected by water depth (most growing between 0–20 m) as well as water turbidity and cloudiness. Much of the organic carbon fixed by the algae diffuses from the algal cells, providing food for the coral polyps, which in turn provide a protective environment for the algae. Reef-building corals are limited mainly by temperature and most are found within the 20 °C mean sea-surface temperature (SST) isotherm (generally between 30° N and 30° S). When temperatures fall to 18 °C, the rate of calcification (skeletal growth) is significantly reduced and lower temperatures may lead to death of the colony.

Coral growth rates vary over the course of a year; when sectioned and x-rayed, an alteration of high- and low-density bands can be seen (Fig. 6.39). High density layers are produced during times of highest SSTs (Fairbanks and Dodge, 1979; Lough and Barnes, 1990) providing a chronological framework for subsequent analyses. Dating in this way is fairly accurate as shown by the close correspondence between large excursions in $\delta^{18}O$ and known El Niño-Southern Oscillation (ENSO) events (Cole *et al.*, 1992) and by very close similarity between annual counts and precise ^{230}Th dates on individual bands (Dunbar *et al.*, 1994). In some regions, exceptional runoff events from adjacent continents are recorded by fluorescent bands in corals (visible under UV light) and provide a further cross-check on coral chronologies (Isdale, 1984; Isdale *et al.*, 1998). These bands result from terrestrially derived fulvic acid being incorporated into the coral structure (Boto and Isdale, 1985). Such banding could also provide valuable estimates of the recurrence interval of extreme river discharge in certain regions.

Samples for analysis are generally drilled from the coral section at regularly spaced intervals along the coral growth axis. Assigning the precise seasonal time to each sample is problematic and is usually done by assuming a linear growth rate between the denser marker bands, the edge of which is assigned to the "onset" of high SSTs. However, if coral extension (growth) is nonlinear, very detailed sampling (6–10 samples per yr) is required or the samples may not cover the entire seasonal range, and provide only minimum estimates on total interannual variability. Furthermore, under extreme conditions (often associated with major El Niño-Southern Oscillation [ENSO] events) coral growth in some areas may even cease, so that the real extremes may go unrecorded in corals from those regions.

Coral studies have focused mainly on the environmental record in coral growth rates, isotopes, and trace elements. This has led to new information about paleo-SSTs, rainfall, river runoff, ocean circulation, and tropical wind systems. Many studies have been based on relatively short periods of time (the last few decades) to provide a better understanding or calibration of the parameter being analyzed, thereby increasing confidence in the paleoreconstructions. So far, only a few studies spanning more than

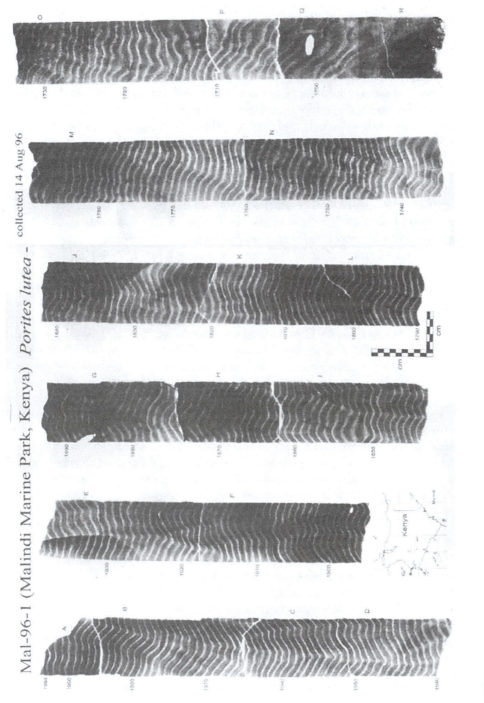

FIGURE 6.39 Positive x-ray photographs of slabs of coral (*Porites lutea*) from off the coast of Kenya showing annual banding. The slabs cover the period from 1994 (top left) back to <1700 (bottom right). (Photograph kindly provided by Rob Dunbar).

TABLE 6.5   Long Coral-based Records of Past Climate

Site	Latitude	Longitude	Record length	Parameter	Indicator of:	Reference
Bermuda	32° N	65° W	~1180–1986	Growth rate	SST/upwelling	Pätzold and Wefer, 1992
Cebu Island Philippines	10° N	124° E	~1860–1980	$\delta^{18}O$ $\delta^{13}C$	SST and rainfall/ cloudiness	Pätzold, 1986
Gulf of Chiriquí Panama	8° N	82° W	1707–1984	$\delta^{18}O$	Rainfall/ ITCZ position	Linsley et al., 1994
Tarawa Atoll Kiribati	1° N	172° E	1893–1989	$\delta^{18}O$	Rainfall	Cole et al., 1993
Isabela Island Galapagos Islands	0.4° S	91° W	1587–1953	$\delta^{18}O$	SST	Dunbar et al., 1994
Espiritu Santo Vanuatu	15° S	167° E	1806–1979	$\delta^{18}O$ $\delta^{13}C$	SST and rainfall/ cloudiness	Quinn et al., 1993
Great Barrier Reef Australia	22° S	153° E	1635–1957	$\Delta^{14}C$	Oceanic advection and/or upwelling	Druffel and Griffin, 1993
New Caledonia	22° S	166° E	1655–1990	$\delta^{18}O$	SST	Quinn et al., 1996

a century have been published (Table 6.5), but many more records are likely to be produced in the years ahead. Indeed, it is likely that the tropical oceans, having been almost totally unrepresented by high resolution paleoclimatic records in the past, may soon provide some of the best records, especially for the last few centuries (Dunbar and Cole, 1993). Furthermore, corals from raised marine terraces are found throughout the Tropics, some of which date back to the last interglacial, or even earlier. Providing diagenetic changes in the coral aragonite have not occurred (Bar-Matthews *et al.*, 1993) it may be possible to reconstruct SSTs (and annual variations in SSTs) for selected intervals over the last 130,000 yr or more (Beck *et al.*, 1992, 1997).

6.8.1 Paleoclimate from Coral Growth Rates

Coral growth rates are dependent on a variety of factors, including SSTs and nutrient availability (Lough *et al.*, 1996). The longest record of coral growth rate variations is that of Pätzold and Wefer (1992), who produced an 800-yr record from a massive coral head (*Montastrea cavernosa*) in Bermuda. In this region, growth rates are inversely related to SST, as cool upwelling water is nutrient-rich, which causes increased coral growth. The record shows that SSTs were generally above the long-term mean from ~1250–1470. Coolest conditions were experienced from ~1470–1710 and from ~1760 to the end of the nineteenth century, followed by twentieth century warming. This is broadly similar to estimates of northern hemisphere summer temperature change over this period (Bradley and Jones, 1993). By contrast, in the Galapagos Islands coral growth rates generally *increase* with SSTs; growth increased from ~1600 to the 1860s, then declined, reaching the lowest rates from ~1903–1940 (Dunbar *et al.*, 1994). However, this record bears little relation to the $\delta^{18}O$ record of SSTs in the same corals, suggesting that in this area other factors are probably involved in growth rate besides water temperature.

6.8.2 $\delta^{18}O$ in Corals

It has long been known that a temperature-dependent fractionation of oxygen isotopes occurs when biological carbonate is precipitated from solution (Epstein *et al.*, 1953). The $\delta^{18}O$ decreases by ~0.22‰ for each 1 °C increase in temperature. Seasonal variations in $\delta^{18}O$ along the growth axis of a coral and their relationship to seasonal SST variations were first reported by Fairbanks and Dodge (1979). Subsequently, Dunbar and Wellington (1981) also showed that, if finely sampled, corals can provide an intra-annual record of SST changes (Fig. 6.40c). Offsets from predicted equilibrium values may be caused by vital effects (see Section 6.3.1) but these are constant for a given genus (Weber and Woodhead, 1972).

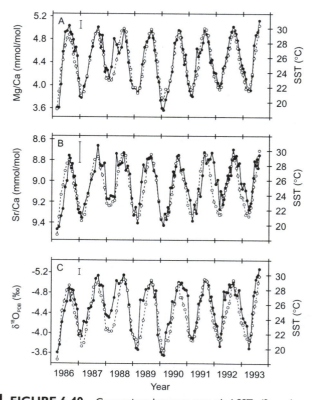

FIGURE 6.40   Comparison between recorded SSTs (3-week averages of daily temperatures) at a tidal station on the Ryukyu Islands, Japan (dashed lines) and geochemical variations in a nearby coral: **A**) Mg/Ca ratios (r = 0.92); **B**) Sr/Ca ratios (r = 0.85); **C**) $\delta^{18}O$ (r = 0.88). Analytical error bars are indicated; these amount to ±0.5, ±1.6, and ±0.4 °C, respectively (Mitsuguchi *et al.*, 1996).

In those areas of the Tropics where seasonal changes in the isotopic composition of seawater occur, a simple SST-δ^{18}O relationship is not found. In areas with seasonally heavy rainfall, which is depleted in δ^{18}O during convective activity, the ocean surface mixed layer becomes isotopically light during the wet season, producing a pronounced seasonal signal in coral δ^{18}O (Cole and Fairbanks, 1990; Linsley *et al.,* 1994). In some regions, this effect is brought about, or enhanced, by flooding of near-coastal waters by isotopically light river water discharged from the continents (McCulloch *et al.,* 1994). Conversely during prolonged hot, dry conditions, surface evaporation can increase sea-surface salinity (SSS) and lead to isotopic enrichment (more δ^{18}O) due to the preferential removal of ^{16}O. To avoid these complications, most studies either focus on areas with large annual changes in rainfall (Cole *et al.,* 1993; Linsley *et al., 1994*) or on areas with little change in SSS, but large SST changes (Dunbar *et al.,* 1994). For example, in parts of the western Pacific El Niños are associated with unusually heavy rainfall. At Tarawa Atoll (1° N, 172° E) negative δ^{18}O excursions of 0.6 ± 0.1‰ occur during ENSO events as a result of dilution of the mixed layer by isotopically depleted rainfall (Cole and Fairbanks, 1990). In this region, the anomalies provide a diagnostic signal of ENSO events over the past century. By identifying the appropriate ENSO signal in different parts of the Pacific Ocean, it should be possible to reconstruct the spatial and temporal characteristics of ENSO events (both "warm" and "cold") far back in time (Cole *et al.,* 1992).

In those areas that experience extreme SST anomalies during ENSO events, δ^{18}O in corals may provide a unique record of such occurrences (Carriquiry *et al.,* 1994). Using a δ^{18}O record in the coral *Pavona clavus* from the Galapagos Islands, Dunbar *et al.* (1994) reconstructed SST variations over the past 380 yr (Fig. 6.41). This record shows that, of the the 100 largest negative δ^{18}O anomalies over the last 350 yr (indicating extremely high SSTs in the region) 88 corresponded

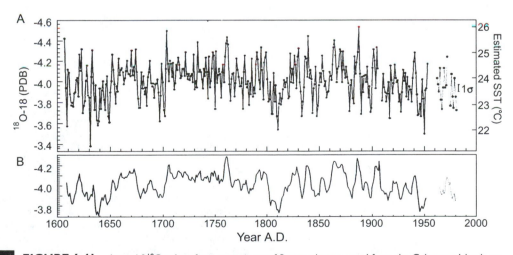

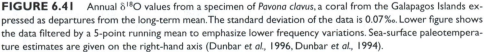

FIGURE 6.41 Annual δ^{18}O values from a specimen of *Pavona clavus,* a coral from the Galapagos Islands expressed as departures from the long-term mean. The standard deviation of the data is 0.07‰. Lower figure shows the data filtered by a 5-point running mean to emphasize lower frequency variations. Sea-surface paleotemperature estimates are given on the right-hand axis (Dunbar *et al.,* 1996, Dunbar *et al.,* 1994).

(±1 yr) to Quinn's (1992) chronology of El Niño events, derived from historical sources. They then examined the changing pattern of dominant perodicities in the record using evolutionary spectral analysis. This involves applying spectral analysis to the data sequentially, in overlapping intervals of time (in this case, 120 yr intervals, from 1610–1730 to 1862–1982) in order to map out the time/frequency response of SSTs in this area (Fig. 6.42). The analysis reveals several shifts in the dominant frequency modes; in the early 1700s, the quasi-periodic El Niño events shifted from the 4.6–7-yr band to 4.6 and 3 yr. A second shift occurred in the mid-1800s to a predominant period around 3.5 years. Similarly timed changes in the dominant lower frequency variance are also seen, especially in the mid-1800s, from ~33 to ~17 yr. It is interesting that a pronounced change (towards warmer and/or dryer conditions) also occurred in the mid 1800s (around 1866) in the South Pacific, as recorded in a coral record from Vanuatu (southwestern Pacific) (Quinn *et al.*, 1993). Whether such changes are coincidental or represent major reorganizations of the tropical ocean-atmosphere climate system (as Dunbar *et al.*, [1994] suggest) will become more apparent as new coral records are developed from throughout the tropical oceans.

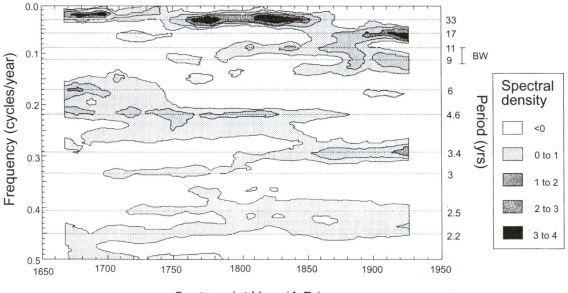

FIGURE 6.42 Evolutionary spectral diagram of the Galapagos coral record shown in Fig. 6.41. The diagram shows spectral density as a function of time (x-axis) and frequency (y-axis) based on analysis carried out in 120-yr segments, each offset by 10 yr. Shaded areas correspond to frequencies at which significant variance occurs, the darker the shading the greater the statistical significance. Lowest frequencies (secular trends) are shown in the upper part of the diagram, and increasingly higher frequencies (shorter periods — see right-hand axis) are shown in the lower part of the diagram. The record indicates a shift to higher frequencies occurred around A.D. 1750 and again around A.D. 1850 (Dunbar *et al.*, 1994).

6.8.3 $\delta^{13}C$ in Corals

The $\delta^{13}C$ in coralline carbonates is affected by a variety of factors, including the $\delta^{13}C$ of seawater (related, in part, to the relative contributions of surface and upwelled waters) and fractionation of carbon isotopes during algal photosynthesis. Algae preferentially take up ^{12}C from dissolved inorganic carbon (DIC) in ocean waters, so higher rates of photosynthesis lead to DIC becoming enriched in ^{13}C (less negative $\delta^{13}C$), which in turn affects the $\delta^{13}C$ of the skeletal carbonate being constructed (McConnaughey, 1989). Several studies have shown that $\delta^{13}C$ declines with water depth (Fairbanks and Dodge, 1979) and during cloudy months (i.e., as photosynthesis rates are reduced) (Shen *et al.*, 1992a; Quinn *et al.*, 1993), suggesting $\delta^{13}C$ in coral bands may provide a long-term index of cloudiness. However, complications related to coral geometry (varying growth rates and photosynthetic activity around a coral head) and possible nonlinear photosynthetic responses to changing light levels, have generally assigned $\delta^{13}C$ records a back seat to $d^{18}O$ in stable isotope paleoclimatic reconstructions from corals.

6.8.4 $\Delta^{14}C$ in Corals

Changes in oceanic mixed layer $\Delta^{14}C$ (that is ^{14}C anomalies from long-term trends) are related to either changes in atmospheric ^{14}C levels or to upwelling of ^{14}C-depleted waters from the deep ocean. The $\Delta^{14}C$ anomalies are also recorded in tree rings and, therefore, changes observed in corals, which are not seen in tree rings, are presumably related to changes in oceanic circulation, indicating either coral upwelling or advection of ^{14}C depleted (or enriched) waters from other regions. Thus, Druffel and Griffin (1993) related unusually large excursions of $\Delta^{14}C$ values in corals from the southwestern Great Barrier Reef between 1680 and 1730 to changes in the relative contributions of waters from the South Equatorial Current ($\Delta^{14}C\sim$ $-60‰$) and the East Australia current ($\Delta^{14}C\sim -38‰$).

6.8.5 *Trace Elements in Corals*

Because certain elements (Sr, Ba, Mn, Cd, Mg) are chemically similar to Ca, trace amounts of these elements may be found in coral skeletal carbonate. Many studies have shown that the relative concentration of such elements (expressed as the ratio of the trace element to calcium) often provides a paleoclimatic, or paleoceanographic signal (Shen and Sanford, 1990). For example, because Cd levels are generally much higher below the mixed layer, Cd/Ca ratios in Galapagos corals increase in association with seasonal upwelling (Shen *et al.*, 1987). The Ba/Ca ratios are inversely related to SSTs (Lea *et al.*, 1989) so low Cd/Ca and Ba/Ca ratios provide a useful index of El Niño events (in the Galapagos area) because such events are associated with very high SSTs and minimal upwelling. The Mn/Ca ratios also provide valuable information in some regions; for example, in the west-central Pacific, Mn is remobilized from lagoonal sediments during strong episodes of equatorial westerly winds, (associated with El Niños) and thus large Mn/Ca ratios in corals are indicative of such conditions (Shen *et al.*, 1992a, b). Elsewhere, Mn/Ca (and perhaps also Ba/Ca) ratios may provide information on runoff from continental regions because terrestrial material is rich in Mn and Ba compared to ambient levels in the oceanic mixed layer.

Other paleotemperature indicators are provided by Sr/Ca, U/Ca and Mg/Ca ratios in corals (see Fig. 6.40a, b) (McCulloch *et al.*, 1994; Min *et al.*, 1995; Mitsuguchi *et al.*, 1996). This opens up the prospect of using multiple parameters to reconstruct paleotemperatures in both recent and fossil corals with high accuracy. However, recent studies by de Villiers *et al.* (1995) indicate that estimates based on Sr/Ca may be in error by several degrees. Sr/Ca (and perhaps Mg/Ca) ratios are very dependent on coral growth rate, leading to lower paleotemperature estimates in coral sections with low growth rates compared to those derived from faster-growing sections of the same coral. If such problems can be resolved, perhaps by a combination of growth rate measurements and the analysis of Sr/Ca, as well as Mg/Ca and/or U/Ca ratios, it would be of great value in helping to resolve controversial tropical SST estimates from the last glacial period. So far, a number of coral studies point to much lower tropical SSTs in glacial time than other oceanic paleotemperature indicators (Table 6.4). For example, Beck *et al.* (1992) estimated paleotemperatures from Sr/Ca in corals from a paleoreef on Espiritu Santo, Vanuatu that were dated ~10,000 B.P. They calculated that SSTs were ~5.5 °C cooler than today, similar to the results of Min *et al.* (1995), who used U/Ca to estimate a temperature difference of 4–5 °C (LGM today) in the same area. Studies of Sr/Ca in corals from Barbados also indicate SSTs were ~5 °C lower at the LGM compared to today (Guilderson *et al.*, 1994). These results stand in stark contrast to both foram- and alkenone-based SST reconstructions (Table 6.4); determining which approach provides the correct answer (or if all are correct with respect to the *actual* temperature being recorded) is a key issue in paleoclimatology today.

6.9 THERMOHALINE CIRCULATION OF THE OCEANS

Circulation of water at the ocean surface is largely a response to the overlying atmospheric circulation which exerts drag at the surface. However, circulation of deeper waters in the oceans of the world is a consequence of density variations, which result from differences in temperature and salinity brought about by sensible and latent heat fluxes, precipitation, and runoff at the ocean surface; this is termed the thermohaline circulation. In areas where surface waters become relatively dense, due to cooling and/or evaporation (thereby increasing salinity) they will sink to the level at which they reach equilibrium (neutral buoyancy) with surrounding water masses.[21] Also, areas of sea ice formation, which result in salt expulsion from the ice, produce brine

[21] A water mass, like an air mass, is recognized by its distinctive physical properties (principally temperature and salinity), which enable it to be distinguished from adjacent water masses. As a water mass moves from its source region it will slowly mix with other waters and gradually lose its original identity.

which increases the water density, causing water to sink and form a cold dense water mass. Dense water masses flow away from their source regions as either Bottom Waters or Intermediate Waters, depending on their relative density. Much of the deepest sections of the world's oceans are filled by dense Antarctic Bottom Water (AABW), which originates from areas of sea-ice formation adjacent to the Antarctic continent. Antarctic Bottom Water (AABW) is characterized by a temperature of ~-0.4 °C and a salinity of ~34.7‰. In the Atlantic Ocean, much of the deepest sections (>2 km) are occupied by North Atlantic Deep Water (NADW), which forms mainly in the Norwegian Sea (~60° N, east of Iceland) and in the Greenland Sea (north and west of Iceland) (Kellogg, 1987; Hay, 1993). Deepwater does not form at present in the North Pacific, where waters are less saline than in the North Atlantic.

Overlying these dense water masses (at ~1 km depth) are Intermediate Waters, which generally have slightly lower salinities and/or higher temperatures. Much of the world ocean is occupied by Antarctic Intermediate Water (AAIW), which has a temperature of 2–4 °C and a salinity of ~34.2‰ and originates in the circum-Antarctic polar frontal zone. At high latitudes of the North Atlantic, intermediate water from the Labrador Sea (3–4 °C, 34.92‰) is found (sometimes referred to as Upper North Atlantic Deep Water, or Northwestern Atlantic Deep Water) and farther south saline water flowing from the Mediterranean Sea can also be traced at intermediate levels.

Changes in deepwater circulation are important for paleoclimatology because, as a result of deepwater formation and compensating fluxes of water in the upper mixed layer, large quantities of heat are carried around the globe. Of particular significance is the thermohaline circulation associated with the formation of North Atlantic Deep Water (Dickson and Brown, 1994). As noted earlier, NADW forms north of ~60° N as surface waters cool (by evaporation and sensible heat loss) and their salinity increases, thereby creating a dense water mass that moves southward at depth (Fig. 6.43) and carries saline water to the South Atlantic and other ocean basins (Fig. 6.44). The loss of water in this manner is compensated for by the poleward movement of warm, saline surface waters to the North Atlantic, in the Gulf Stream and associated North Atlantic Drift. These water masses are responsible for the relatively mild temperatures western Europe experiences, even in winter.

The movement of warm salty water to high latitudes of the North Atlantic, the formation of dense NADW, and the displacement of water as this water mass exits from the North Atlantic can be considered as a linked system or conveyor belt; disturbances to the system may cause it to change its speed (rate of water exchange) or even to cease operation altogether (Broecker, 1991). In fact, models indicate that the system is quite sensitive to disturbance, particularly by freshwater inflows into the North Atlantic (Manabe and Stauffer, 1988; Rahmstorf, 1994; Weaver and Hughes, 1994). Currently, the North Atlantic Basin loses slightly more freshwater via evaporation than it gains from either precipitation or river runoff (−1.21 m a^{-1}, v. +0.87 m and 0.21 m, respectively). It is this fact, together with a flux of saline Gulf Stream water and strong cooling (especially in winter) that leads to NADW formation. However, if the freshwater flux were to increase (as it may have when the major continental ice sheets melted) it would create an upper, low salinity water layer that

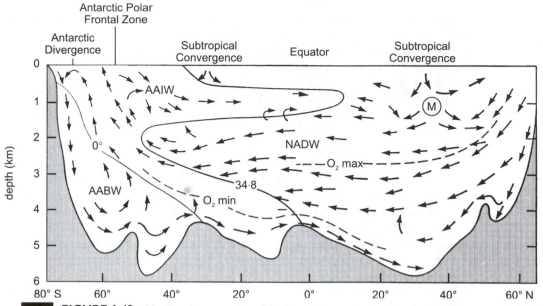

FIGURE 6.43 Meridional cross section of the Atlantic Ocean showing the principal water masses and their distribution today. NADW = North Atlantic Deep Water; AABW = Antarctic Bottom Water; AIW = Atlantic Intermediate Water; AAIW = Antarctic Intermediate Water; M = Mediterranean Intermediate Water. During the last glacial maximum, NADW was more limited in extent, whereas AABW penetrated farther north into the deep basins of the North Atlantic (see Fig. 6.45) (adapted from Brown *et. al.,* 1989).

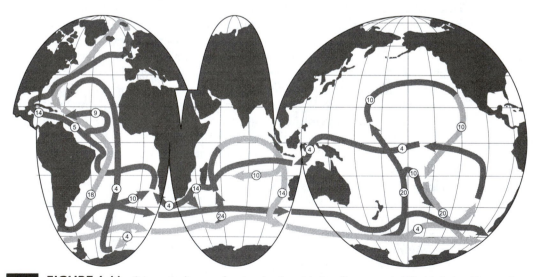

FIGURE 6.44 Schematic diagram showing the thermohaline ("conveyor belt") circulation. Near surface waters are shown by the darker shaded arrows, deepwater by the lighter colored arrows. Sinking of surface waters occurs in the North Atlantic, and to a much lesser extent in the North Pacific and Indian Ocean. Estimates of the volume of water transported in each section are shown in the circles (in Sverdrups; 1 sv = 10^6 m^3 s^{-1}). (Schmitz 1995).

would disrupt this process, and shut-down NADW formation. This in turn would "turn-off" the conveyor belt of NADW, eventually leading to a reduction of Gulf Stream water to replace that lost through sinking in the North Atlantic. This would mean less heat transported into the North Atlantic and generally colder climatic conditions leading to less freshwater runoff. Eventually the process of NADW formation would be restored, turning the conveyor back on and allowing the overall circulation to revert back to its former condition. Thus, the North Atlantic thermohaline circulation can be thought of as having two distinct modes ("conveyor on" or "conveyor off") controlled by the relative balance of freshwater flux to the surface waters of the North Atlantic Basin (Broecker *et al.*, 1985b; Broecker and Denton, 1989; Broecker *et al.*, 1990a, b; Broecker, 1994). When the conveyor is off, the rate of salt export from the North Atlantic (in NADW) is less than the rate of salt build-up resulting from evaporation and water vapor export to adjacent regions. Salinity gradually increases until some critical density threshold is reached, at which point the conveyor switches on, which brings more saline water to the North Atlantic via the Gulf Stream. Providing that meltwater flux to the North Atlantic is less than the salt build-up, NADW will continue to form. However if meltwater and/or salt export exceeds that threshold, NADW formation will be greatly reduced or eliminated altogether (Broecker *et al.*, 1990a). In this way the ocean-atmosphere-cryosphere systems are in dynamic equilibrium, in which disturbance of one part of any system may lead to a nonlinear response in another system (Broecker and Denton, 1989).

In a reassessment of this model, Boyle and Rosener (1990) question whether the coupled system really has only two modes or whether in fact there have been multiple stable circulation patterns as suggested by model simulations (Rahmstorf, 1994, 1995). They suggest that rather than being controlled by an "on-off switch" the system may be considered as responding more to a "valve" whereby there are many possible quasi-stable circulation states. Unfortunately, the resolution of deep-sea sediments is frequently insufficient (because of low sedimentation rates and bioturbation) to resolve which of these models is correct, though there is evidence that when NADW was not formed (e.g., in the Last Glacial Maximum) a distinct North Atlantic Intermediate Water was being produced (Boyle and Keigwin, 1987). This suggests an additional scenario, perhaps the result of some balance of factors neither at one extreme nor the other. Lehman and Keigwin (1992a, b) make the argument that the formation of deepwater in the Norwegian Sea ("Lower NADW") was disrupted quite often in Late Glacial times (see Section 6.10.2) but deepwater from the Labrador Sea ("Upper NADW") continued to form. Another scenario is proposed by Veum *et al.* (1992), who suggest that deepwater, formed by brine rejection during sea-ice formation, formed in the marginal ice zone of the Greenland-Iceland-Norwegian Seas throughout the Last Glacial Maximum, ventilating the deep basins of this region. By contrast, no deepwater formed to the south in the open North Atlantic Ocean, the deep basins of which were occupied by AABW until ~12,600 yr B.P. There may thus be many different states of deepwater circulation that have developed over time, with complete shutdown of both Lower and Upper NADW being the extreme end-member state of a whole range of possible conditions.

Before discussing further the evidence for such changes, it is first necessary to consider the means by which changes in deepwater circulation can be identified. Each water mass has certain geochemical characteristics that can be identified by analysis of benthic forams living in those waters. The geochemistry of benthic forams in marine sediments thus serves as a "tracer" of deepwater conditions at the time the forams were deposited. Of particular importance is the $^{13}C/^{12}C$ ratio ($\delta^{13}C$) in the carbonate tests, derived from dissolved CO_2 in the water column. The $\delta^{13}C$ value for the atmosphere is -7.2‰; because of fractionation effects, seawater in equilibrium with the atmosphere has a $\delta^{13}C$ value of ~+3.5‰ (at 2 °C) (Mook *et al.*, 1974). By contrast, organic matter has a $\delta^{13}C$ of –20 to –25‰. Oxidation of organic matter falling through the water column therefore causes the $\delta^{13}C$ of the water to decline. If surface waters are nutrient-rich and productive, the large input of organic material to the deep ocean will result in low $\delta^{13}C$ and reduced oxygen levels. The low $\delta^{13}C$ is balanced to some extent by dissolution of the carbonate tests of planktic forams as they fall through the water column, because these have a $\delta^{13}C$ value close to that of total dissolved CO_2 in the upper water column. Hence the overall $\delta^{13}C$ of a water mass reflects a balance between the amount of organic matter oxidized, and dissolution processes. Nevertheless, the global distribution of $\delta^{13}C$ strongly reflects the nutrient content and organic productivity of the water mass (Kroopnick, 1985). For example, Antarctic waters are nutrient-rich and productive, resulting in deepwater (AABW), which is depleted in ^{13}C; North Atlantic deepwater, on the other hand, has lower nutrient levels, is less productive, and has a higher $\delta^{13}C$ value (Duplessy and Shackleton, 1985). These characteristics, preserved in the tests of benthic forams, can be used to trace the presence and distribution of NADW and AABW over glacial-interglacial cycles (Curry *et al.*, 1988; Raymo *et al.*, 1990). Furthermore, in areas of deepwater *formation*, the vertical distribution of $\delta^{13}C$ is fairly homogeneous (due to convective mixing) so that the $\delta^{13}C$ signal in the calcareous tests of both benthic and planktic foraminifera are similar. With increasing distance from areas of deepwater formation, the difference between surface and deepwater $\delta^{13}C$ increases, and this is reflected in the tests of the deep-dwelling and surface forams (Duplessy *et al.*, 1988). Hence $\delta^{13}C$ can be used to identify changes in areas of deepwater formation, and to track their movement over time.

There are two slight complications to this neat approach. First, not all benthic foram species record the same values of $\delta^{13}C$, apparently due to a species-dependent habitat effect; this can be resolved by selecting only benthic forams (such as *Cibicidoides wuellerstorfi*) that do not show such an effect, or by using those whose effect is known (Zahn *et al.*, 1986). Second, the global mean $\delta^{13}C$ level decreased by 0.3–0.4‰ during glacial times due to a reduction in terrestrial biomass, and a remobilization of $\delta^{13}C$-depleted organic material on the exposed continental shelves (when sea level was up to 120 m lower) (Boyle and Keigwin, 1985; Duplessy *et al.*, 1988; Curry *et al.*, 1988; Keigwin *et al.*, 1994). This affects all records equally so it is easily accommodated; $\delta^{13}C$ thus provides a valuable proxy indicator of water masses in the past (Fig. 6.45).

Another useful tracer of deepwater is the Cd/Ca ratio in benthic foram tests (Boyle and Keigwin, 1982; Boyle, 1988). Cadmium is a proxy for oceanic nutrient

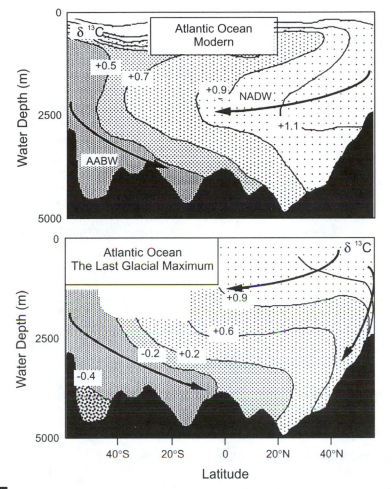

FIGURE 6.45 Cross section through the Atlantic Ocean showing the distribution of $\delta^{13}C$ today (GEOSECS data) and in benthic foraminifera from the last glacial maximum (LGM). The $\delta^{13}C$ is used to characterize particular water masses; thus the lowest values are indicative of deep water produced in the sub-Antarctic (AABW). At the LGM, NADW (which has higher $\delta^{13}C$ values) penetrated only to intermediate depths, whereas today it sinks to greater depths and occupies most of the deep Atlantic basins, as far south as the Equator (see Fig. 6.43). Changes in this and other tracers of deepwater circulation reflect important differences in the thermohaline circulation over time (Duplessy and Maier-Reimer, 1993).

levels; Antarctic Bottom Waters have relatively high Cd levels compared to NADW so Cd/Ca provides a useful index of these water masses (Boyle, 1992). Although Cd levels in the world ocean were higher during glacial time, sediments from the Bermuda Rise show that during the last glacial maximum (isotope stage 2) and during the Younger Dryas interval, Cd levels in the deep Atlantic Ocean increased even more, indicating that NADW flux was reduced and replaced by AABW at those times (Boyle and Keigwin, 1987). This supports the $\delta^{13}C$ data, which also points to a shift in deepwater circulation towards increased AABW flux (see Fig. 6.45) not only in the last glacial but also in isotope stage 6 (135 ka B.P.) and earlier glacial periods (Duplessy and Shackleton, 1985; Boyle and Keigwin, 1985; Oppo and Fairbanks, 1987; Curry *et al.,* 1988; Raymo *et al.,* 1990). The fairly rapid changes in deepwater recorded during the late glacial/Younger Dryas interval clearly link the well-documented changes in surface oceanic conditions and climate around the North Atlantic (mainly in western Europe) with deepwater variations.

Another important aspect of deepwater formation relates to the transport of oxygen into the deep ocean basins of the world. Water at the surface is generally well oxygenated but as deepwaters form and sink, oxygen levels decline as oxidation of organic matter falling through the water column proceeds. In effect, deepwater formation ventilates the deep ocean by carrying oxygenated water to great depths. Ventilation rates can be estimated by measuring the radiocarbon content of the water; once the water is isolated from the atmosphere, radiocarbon is no longer in equilibrium with the atmospheric reservoir and ^{14}C levels will decline. The radiocarbon age of deepwater therefore reflects the time since isolation from the surface; this obviously varies from one area to another (see Fig. 3.7) but typically deepwater in the Atlantic has a ^{14}C age of ~400 yr, in the Indian Ocean ~1200 yr, and in the Pacific Ocean, ~1600yr.[22] Broecker *et al.* (1988a) compared planktic and benthic forams from the last glacial maximum (LGM) in Atlantic and Pacific Ocean sediments. On average, the discrepancy between ages of forams in the upper and lower water columns increased during the LGM, from 400 to 600 yr in the Atlantic and from 1600 to 2100 yr in the Pacific, indicating that ventilation rates were significantly lower in glacial times compared to the present. A similar conclusion was reached by Bard *et al.* (1994), who found evidence for reduced North Atlantic ventilation during the cold Younger Dryas oscillation.

6.10 OCEAN CIRCULATION CHANGES AND CLIMATE OVER THE LAST GLACIAL-INTERGLACIAL CYCLE

Analysis of $\delta^{13}C$ and Cd/Ca in benthic foraminifera and $\delta^{18}O$ in planktic forams (reflecting temperature and/or salinity changes in surface waters) have enabled circulation changes to be reconstructed over the last glacial-interglacial cycle in some detail; longer-term changes in $\delta^{13}C$ are examined by Raymo *et al.* (1990). These studies indicate that significant changes in the thermohaline circulation of the oceans have occurred. Although production and circulation of NADW was similar to today in the last interglacial (5e), glacial periods were characterized by a reduction (or even cessation) of NADW production, or perhaps a change towards less dense Intermediate Water forming in the central North Atlantic and/or Labrador Sea (Duplessy *et al.,* 1988, 1991; Keigwin *et al.,* 1994; Oppo and Lehman, 1995). The extent to which deepwater formed in the Norwegian Sea in isotope stages 2 or 6 is unclear,

---

[22] Organisms living in upwelling areas where they develop organic tissue in equilibrium with "old water" will have an apparent radiocarbon age several hundred (and in some areas more than two thousand) years older than "modern."

with some studies suggesting none, others indicating intermittent production, or perhaps a shifting source area across the Greenland-Iceland-Norwegian Sea region, in close connection with the sea-ice margin (Veum *et al.*, 1992; Oppo and Lehman, 1995). Antarctic Bottom Water occupied most of the deep basins of the world oceans at those times (see Fig. 6.45) (Duplessy and Shackleton, 1985).

Deepwater changes are clearly related to changes in surface ocean conditions in the North Atlantic; when the surface waters were cold and less saline (as recorded by a high percentage of the cold-water foraminifera *Neogloboquadrina pachyderma* [sinistral]) deepwaters did not form, at least not in the areas where they form today. However, in the last interglacial when the marine polar front was far to the north, and North Atlantic surface waters were saltier and warmer, NADW filled the deep basins of the Atlantic Ocean (Broecker *et al.*, 1988b). What then could have brought the interglacial to a close? Of course, orbital variations were slowly bringing about a reduction in summer insolation and an increase in winter insolation (in the northern hemisphere) but other internal factors may have also played an important role. A higher interglacial sea level (+ 6 m) would have led to an increased flux of water through the Arctic Ocean from the North Pacific, bringing more low salinity water into the North Atlantic (Shaffer and Bendtsen, 1994). Higher temperatures may have increased evaporation and precipitation rates at high latitudes, adding additional freshwater to the North Atlantic and its surrounding drainage basins. If all of these factors were enough to lower surface water density, deepwater production (and its attendant changes in compensatory surface inflow) may have been reduced or even eliminated, setting the stage for renewed continental glaciation (Cortijo *et al.*, 1994). Once temperatures began to fall, evaporation rates would also have decreased, allowing salinity levels to remain low. The problem is that all of these factors are intimately related in a positive feedback loop that makes identifying "cause" and "effect" extremely difficult, especially when the resolution of sedimentary records is low and dating uncertainties are relatively high (W. Berger, 1990). This is particularly problematic during times of very rapid change, as occurred at the end of the last glacial period (see Section 6.10.2). Nevertheless, where high resolution records exist, there is evidence that changes in surface water temperatures and deepwater circulation often occurred simultaneously, and often very rapidly even within the coldest intervals (Boyle and Rosener, 1990; Lehman and Keigwin, 1992a; Oppo and Lehman, 1993; Keigwin *et al.*, 1994).

6.10.1 Heinrich Events

One of the most intriguing aspects of the Greenland ice-core records from the last glacial period is the evidence for very rapid changes in $\delta^{18}O$ during the last 75,000 yr (marine isotope stages 2 to 4). There is now convincing evidence that these rapid changes were also recorded in marine sediments from the North Atlantic (Rasmussen *et al.*, 1996). Heinrich (1988) first noted large changes in the percent of lithic materials in the >180 μm fraction of marine sediment cores from the northeastern North Atlantic during the last glacial period (Fig. 6.46). These percentage increases were to some extent related to an increase in the occurrence of

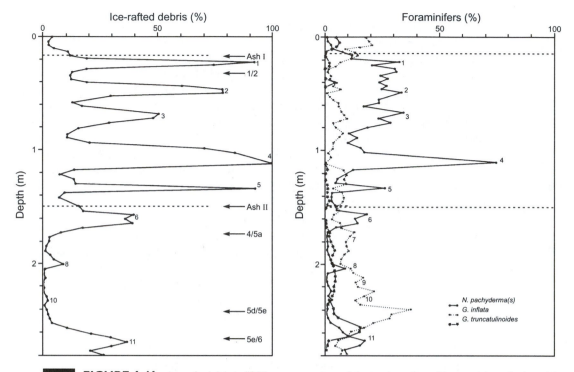

FIGURE 6.46   Ice-rafted debris (IRD) as a percentage of the total number of forams + ice-rafted particles in a sediment core from the North Atlantic (~47° N, 20° W). Isotopic stage boundaries and volcanic ash layers are shown (Ash I was deposited approximately 10,800 yr B.P. and Ash II ~54,000 B.P.). Times of maximum ice-rafted debris are now known as Heinrich events (see Table 6.5) (Heinrich, 1988).

N. pachyderma (sin.) which makes up the rest of this size fraction and, as noted earlier, is characteristic of cold polar waters. Further studies showed that these peaks of lithic material (now known as Heinrich events) could be traced over wide areas of the North Atlantic Ocean (Grousset *et al.*, 1993). Six events (H_1 to H_6) have been identified in numerous sediment cores; younger events were dated by bracketing AMS ^{14}C dates (Table 6.6). Each event appears to have been an episode of very rapidly accumulating ice-rafted detritus (IRD) at times associated with a drop in foram concentration, due to lower productivity and/or increased dissolution (Broecker *et al.*, 1992; Broecker, 1994). A further four events within marine isotope stage 5 have been recognized in two cores from the North Atlantic, as well as a peak of IRD in stage 6. Bond *et al.* (1993) also identify IRD in sediments of Younger Dryas age from the Labrador Sea, which they characterize as an additional Heinrich event (H_0 in Table 6.6) (Keigwin and Jones, 1995).

Heinrich originally described the IRD as mainly angular quartz grains, but cores from farther south and west have a relatively high percentage of distinctive limestone and dolomite in each detrital layer, suggesting a single source for these

TABLE 6.6   Age of Heinrich Events[a]

H_0	11,000
H_1	14,300
H_2	21,000
H_3	27,000
H_4	35,500
H_5	~52,000
H_6	~69,000
"H_7"	~71,000
"H_8"	~76,000
"H_9"	~85,000
"H_{10}"	~105,000
"H_{11}"	~133,000

[a] H_0 to H_3 bracketed by AMS ^{14}C dates on foraminifera. H_4 to H_6 based on extrapolation of sedimentation rates in upper sections of sediment cores and therefore subject to revision (Bond et al., 1992, 1993). Older events (designated here as "H_7" to "H_{11}") are based on two cores studied by McManus et al. (1994) with ages subject to uncertainties of probably ±5%.

materials. Also, Heinrich layers 1 and 2 increase in thickness westward, in a belt from 43–55° N, towards the Labrador Sea, suggesting that material originated from the Laurentide ice sheet and was dispersed across the Atlantic by icebergs (Fig. 6.47). Furthermore, Nd/Sr isotope ratios and K/Ar dates of ~900 Ma on detrital clays point to a source in the Pre-Cambrian shield rocks of northwest Greenland or northeastern Canada (Bond et al., 1992; Grousset et al., 1993; Andrews et al., 1994). Heinrich events 3 and 6 differ from the others in that the IRD distribution is largely confined to the western Atlantic, perhaps because these events occurred when the Laurentide ice sheet was smaller (at the start of Stages 2 and 4, respectively) so the delivery of icebergs and entrained debris would have been more limited (Gwiazda et al., 1996b).

It is significant that several of the Heinrich events occurred at the end of prolonged cooling episodes, as recorded by increased percentages of N. pachyderma leading up to the event (Fig. 6.48). Furthermore, these longer-term cooling cycles can be correlated with similar variations in $\delta^{18}O$ in the GRIP Summit ice core from Greenland, indicating direct links between the ocean and atmospheric systems, which each record primarily represents, and changes in ice sheet dynamics, recorded by IRD in the Heinrich layers. Following each Heinrich event, there is an abrupt shift to warmer conditions, which (from ice core evidence) apparently took

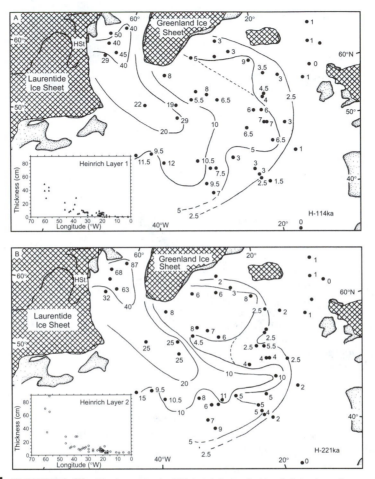

FIGURE 6.47 Thickness (cm) of Heinrich layers in North Atlantic sediments, based on whole-core magnetic susceptibility data. Cross-hatching denotes major ice sheets; HSt = Hudson Strait, thought to have been the major ice stream supplying material to the North Atlantic from the Laurentide ice sheet. Upper figure: Heinrich Event 1 (~14.3 ka ^{14}C yr B.P.; lower figure: Heinrich Event 2 (~21ka ^{14}C yr B.P.) (Dowdeswell, et al., 1995).

place over just a few decades. The cause of these large-scale changes in the ocean-atmosphere-ice systems is not known. Clearly, they took place on timescales far shorter than forcing related to orbital variations. Indeed, Bond and Lotti (1995) find evidence in high resolution marine sediments for even more episodes of ice-rafting between the major Heinrich events — 13 events from 38,000 to 10,000 B.P. — and these too seem to correspond to low δ^{18}O in the GRIP ice core (Fig. 6.49). As the detrital material clearly originates from ice-rafting, there must have been quasi-periodic increases in iceberg calving rates into the North Atlantic. High-resolution records show that not only detrital carbonate but also volcanic glass and hematite-

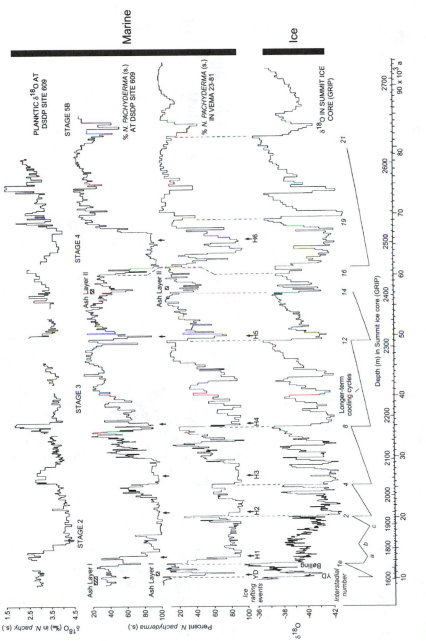
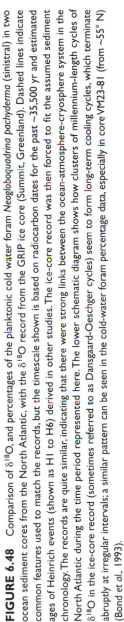

FIGURE 6.48 Comparison of $\delta^{18}O$, and percentages of the planktonic cold water foram *Neogloboquadrina pachyderma* (sinistral) in two ocean sediment cores from the North Atlantic, with the $\delta^{18}O$ record from the GRIP ice core (Summit, Greenland). Dashed lines indicate common features used to match the records, but the timescale shown is based on radiocarbon dates for the past ~35,500 yr and estimated ages of Heinrich events (shown as H1 to H6) derived in other studies. The ice-core record was then forced to fit the assumed sediment chronology. The records are quite similar, indicating that there were strong links between the ocean-atmosphere-cryosphere system in the North Atlantic during the time period represented here. The lower schematic diagram shows how clusters of millennium-length cycles of $\delta^{18}O$ in the ice-core record (sometimes referred to as Dansgaard-Oeschger cycles) seem to form long-term cooling cycles, which terminate abruptly at irregular intervals; a similar pattern can be seen in the cold-water foram percentage data, especially in core VM23-81 (from ~55° N) (Bond *et al.*, 1993).

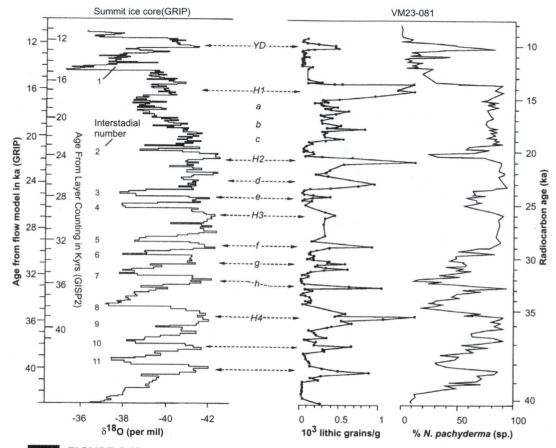

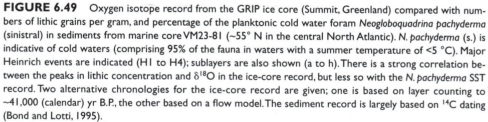

FIGURE 6.49 Oxygen isotope record from the GRIP ice core (Summit, Greenland) compared with numbers of lithic grains per gram, and percentage of the planktonic cold water foram *Neogloboquadrina pachyderma* (sinistral) in sediments from marine core VM23-81 (~55° N in the central North Atlantic). *N. pachyderma* (s.) is indicative of cold waters (comprising 95% of the fauna in waters with a summer temperature of <5 °C). Major Heinrich events are indicated (H1 to H4); sublayers are also shown (a to h). There is a strong correlation between the peaks in lithic concentration and δ18O in the ice-core record, but less so with the *N. pachyderma* SST record. Two alternative chronologies for the ice-core record are given; one is based on layer counting to ~41,000 (calendar) yr B.P., the other based on a flow model. The sediment record is largely based on 14C dating (Bond and Lotti, 1995).

coated grains are characteristic of many layers, suggesting more widespread sources of discharge (including Iceland) for the IRD than just the Laurentide ice sheet. Sediment cores from off Norway also show strong correlations between episodes of IRD and δ18O in Greenland ice cores, again with cold periods (low δ18O in the ice) associated with increased ice-rafting (Fronval *et al.,* 1995). Thus, there appear to have been numerous episodes when there was a massive draw-down of ice in one or more circum-Atlantic ice sheets, resulting in "armadas of icebergs" and entrained basal debris entering the cold waters of the North Atlantic (Broecker, 1994). Surprisingly, these episodes were not brought about by warmer conditions, but occurred when conditions were already cold (Madureira *et al.*, 1997), and much of the North Atlantic was covered by cold polar waters. One possible explanation, suggested by modeling experiments, is that ice sheets may grow to a point where they develop instabilities (basal ice melting), which cause rapid discharge (surges) into marine embayments via ice streams and ice shelves (MacAyeal, 1993; Alley and MacAyeal, 1994). Cold, slow-moving ice can freeze rock debris into its base, but during times of destabilization channelized ice streams develop where frictional heating allows ice to slide over a relatively warm and wet bed. Debris-laden icebergs are generated when the surging ice streams enter the marine environment, and this continues until the ice sheet becomes stable once again. This "binge-purge" model could account for the quasi-periodic character of the observed record, though just what climatic conditions are needed to bring about destabilization of the ice sheets is not clear. Apparently, there is a fairly delicate balance between conditions that: (a) maintain an ice sheet in a quasi-equilibrium state; (b) cause a periodic collapse, but then allow recovery; and (c) cause irreversible collapse (i.e., complete deglaciation). The interplay of ocean, atmosphere, and ice with the added complexities of eustatic sea-level change and glacio-isostatic adjustments as ice loads changed, allows for many possible scenarios of how the observed changes may have been brought about and provides a fruitful area for future research. One possible consequence of the increased calving rates was that the North Atlantic became flooded with a low salinity meltwater "lid"; this may have acted as a shallow mixed layer, with low thermal inertia, which could have warmed up fairly rapidly, leading to higher air temperatures and the observed $\delta^{18}O$ increase seen in GRIP ice cores (Fairbanks, 1989). This would have been a short-lived episode that came to an end as the meltwater layer became mixed with the deeper ocean. Alternatively, the iceberg flux associated with Heinrich events would presumably have been followed by a period with very little iceberg discharge and a rapid reduction in freshwater flux to the North Atlantic, allowing salinity to increase, and a return to the conveyor "on" mode, with increased advection of warmer water and air masses into the Greenland area (Paillard and Labeyrie, 1994).

One additional complexity to be considered in explaining the abrupt changes seen in marine sediments and ice cores during the last glacial period is that rapid $\delta^{18}O$ shifts in Greenland ice cores are associated with pronounced changes in atmospheric methane (cf. Figs. 5.34 and 6.49). This implicates (or necessitates an explanation involving) tropical and/or high latitude wetlands such that the warmer, high $\delta^{18}O$ episodes following Heinrich events (and other minor IRD events) are somehow associated with periods of CH_4 release. Possibly this in turn may have provided some positive feedback due to an enhanced greenhouse effect. Interestingly, there is some evidence that Heinrich events themselves may be a direct consequence of circulation changes in low latitudes brought about by orbital forcing (McIntyre and Molfino, 1996). Increases in the abundance of the coccolith *Florisphaera profunda* in the equatorial Atlantic coincide with Heinrich events in the North Atlantic, with a mean period of 8.4 ka. Variations in precession at this frequency cause changes in the strength of the zonal component of the tropical easterlies; when the easterlies diminish in strength (which favors an increase in the abundance of *F. profunda*) the "reservoir" of warm water within the Caribbean/Gulf of Mexico is no longer restricted and it rapidly "drains out" of these two basins, becoming entrained in the western boundary currents of the North Atlantic. In this way, warm salty waters would have periodically entered the subpolar Atlantic, causing rapid melting of ice sheets and the initiation of Heinrich events. This may help to explain the quasi-periodic nature of Heinrich events that has frequently been noted (Heinrich, 1988; Bond and Lotti, 1995).

6.10.2 Environmental Changes at the End of the Last Glaciation

It is now well-documented that numerous rapid changes in ocean circulation and atmospheric conditions took place throughout the last glacial period. Studies of well-dated high resolution sediments and uplifted coral reefs provide particularly valuable insights into such events at the very end of the last glaciation (Termination 1).

The $\delta^{18}O$ in benthic forams provides a broad-scale perspective on continental ice volume changes since the last glacial maximum (LGM) around 18,000 (^{14}C yrs) B.P. and many records demonstrate that deglaciation took place in two stages (Terminations Ia and Ib) (Duplessy *et al.*, 1986; Jensen and Veum, 1990). However, because of low sedimentation rates and bioturbation, the details of events during deglaciation are difficult to decipher accurately from the benthic record (Ruddiman, 1987). Fairbanks (1989) was able to address this situation by obtaining a detailed sea-level record from drowned coral reefs off the coast of Barbados (Fig. 6.50). His studies reveal that the maximum sea level lowering due to ice build-up on the continents was 121 ± 5 m at ~18,000 (^{14}C yr) B.P. As deglaciation set in, sea level slowly increased by ~20 m over the next 5000 yr, followed by a very rapid rise in sea level, centered on 12,000 (^{14}C yr) B.P. At that time (termed Meltwater Pulse IA, mwp-IA) the discharge of water from the continents to the world ocean reached ~ 14,000 km³ a⁻¹ and sea level rose a further ~24 m in <1000 yr (see Fig. 6.50). This was followed by a slower rate of sea-level rise, and then a second major meltwater pulse (mwp-IB) centered on 9500 (^{14}C yr) B.P., when sea level rose an additional 28 m in ~1500 yr. Thus, more than two thirds of the LGM continental ice had melted by the beginning of the Holocene,[23] and almost all of this was discharged to the world ocean (via the North Atlantic) in two short episodes, lasting a total of ~2500 yr. Fairbanks argues that the periods of most rapid meltwater flux resulted in a reduction of surface water salinity right across the North Atlantic, as seen in planktic forams from off the coast of Portugal where abrupt decreases in $\delta^{18}O$ (Terminations Ia and Ib) appear to be directly linked to the coral reef record of ice volume discharge and sea-level rise (Duplessy *et al.*, 1986; Bard *et al.*, 1989; Fairbanks, 1989).

One of the problems of understanding the exact sequence of events in Late Glacial time is the chronological uncertainty posed by ^{14}C production rate changes at the very time when rapid climatic changes were taking place. Indeed the "^{14}C

---

[23] If the Holocene is defined in calendar years, the $^{230}Th/^{234}U$ calibration of the ^{14}C timescale (Bard *et al.*, 1990), applied to Fairbanks (1989) sea-level record indicates that sea level had risen 85 m (of the 120 m LGM depression) by the start of the Holocene *sensu stricto* (see Fig. 6.50).

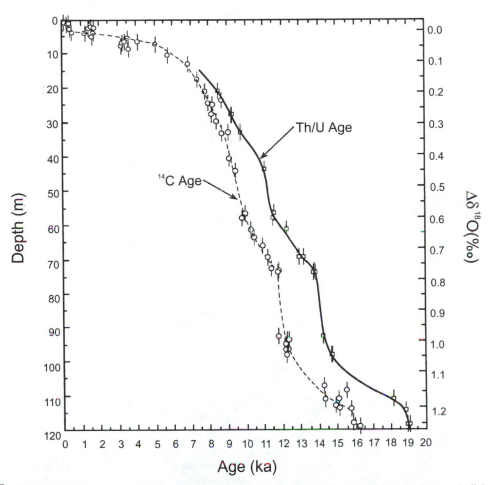

FIGURE 6.50 Paleo-sea level, as reconstructed from dated corals (*Acropora palmata*) submerged off the coast of Barbados. The left curve shows the ^{14}C-dated record and the right curve the [230]Th/[234]U-dated record (Bard *et al.,* 1990). The right-hand axis shows an estimate of the effect of sea-level change on the isotopic composition ($\delta^{18}O$) of the world ocean (Fairbanks, 1990).

plateau" is probably related to the very changes that are of interest, involving shifts in ocean circulation and de-gassing of "old" carbon from the deep ocean (Section 3.2.1.5). Bard *et al.* (1990, 1996) calibrated the ^{14}C record in this crucial interval by [230]Th/[234]U dates on corals, so that the Fairbanks (1989) meltwater record can be "corrected" to calendar yr (Fig. 6.51). This then places the peak of mwp-IA and mwp-IB at 14,000 and 11,300 calendar yr B.P., respectively (i.e., ~11,800 and ~10,000 ^{14}C yr B.P.).

A somewhat different scenario is presented by Keigwin *et al.* (1991) who link evidence for NADW reduction at 14,500, 13,500, 12,000, and 10,500 ^{14}C yr B.P., with times of increased meltwater flux to the North Atlantic. The 14,500 yr B.P.

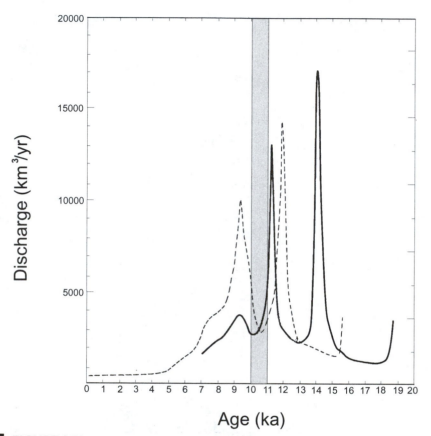

FIGURE 6.51 An estimate of the rate of glacial meltwater entering the world ocean, derived from the first derivative of the sea-level record shown in Fig. 6.50. Two chronologies are shown: one is in ^{14}C yr, derived from the ^{14}C-dated coral series (Fairbanks, 1990) and the other is in calendar years, based on a calibration of the ^{14}C record by [230]Th/[234]U-dates. Two major pulses of meltwater are seen (mwp-IA occurred first, mwp-IB occurred later) (Bard et al., 1990).

event was related to meltwater from the Fennoscandian and Barents Sea ice sheet, the 10,500 yr B.P. event to drainage of freshwater from the Baltic Ice Lake (Brunnberg, 1995). The 14,500 yr B.P. event is clearly documented by low $\delta^{18}O$ in planktic forams from the Norwegian and Greenland Sea (Lehman *et al.*, 1991) whereas the events at 13,500 and 12,000 yr appear to have been low salinity episodes in the subtropical Atlantic due to increased Mississippi River outflow at those times.[24] At all four times, Cd/Ca data from the Bermuda Rise show clear evi-

[24] Keigwin *et al.* (1991) argue that the Fairbanks sea-level record may be incomplete between 14,300 and 12,500 yr B.P. because sea level rose too rapidly for coral growth to keep pace, thereby "missing" the 13,500 yr B.P. discharge maximum. Broecker (1990) points out that Fairbanks's episodes of most rapid sea-level rise are actually gaps in the coral stratigraphy, perhaps also because of an overwhelming rate of sea-level rise at those times. Another possibility is that melting of marine-based ice sheets (e.g., the Barents and Kara Sea Ice Sheets) placed significant amounts of meltwater in the Atlantic *without* significantly changing sea level (Veum *et al.*, 1992).

dence of reduced NADW, suggesting that meltwater from both northern and southern sources entering the North Atlantic may have disturbed the thermohaline circulation system enough to reduce or shut down deepwater production at these times (Boyle and Keigwin, 1987).

This scenario is consistent with evidence from western Europe for a series of cold episodes punctuating the overall warming trend that brought about deglaciation. The coldest of these, (the Younger Dryas, YD), lasted from 11,000–10,000 (^{14}C yrs) B.P. (~13,000–11,700 calendar yrs B.P.). Evidence for this cold episode is recognized in continental deposits from many parts of the world but is most dramatically seen in pollen diagrams from western Europe where an abrupt reversal of post-glacial warming is well documented (Peteet, 1995). Broecker *et al.* (1988b, 1989) proposed that the Younger Dryas was the result of meltwater from the Laurentide Ice Sheet being diverted from the main Mississippi River drainage to the St. Lawrence River, thereby flooding the North Atlantic with low salinity meltwater that shut down the North Atlantic Deep Water production and consequently reduced the flow of warm subtropical waters in the Gulf Stream and North Atlantic Drift. This cessation of NADW flux in turn led to other parts of the world being affected, which would explain why many areas remote from the North Atlantic do appear to have experienced a correspondingly abrupt change in climate at that time (see the collection of papers in *Quaternary Science Reviews*, **12**, 5, 1993 and **14**, 9, 1995). The coral sea-level record poses a problem because it indicates that the YD occurred at a time when discharge to the world ocean was far less than in either the preceding or subsequent 1000 years (Bard *et al.*, 1996). Furthermore, de Vernal *et al.* (1996) showed that meltwater flux from the St. Lawrence River was *reduced* during the Younger Dryas episode. Broecker (1990) tried to reconcile these diverse factors by suggesting that the period of rapid sea-level rise prior to the YD (mwp-IA) "set-up" conditions in the North Atlantic for a shut down (or major reduction) of NADW by significantly lowering salinity in the mixed layer. This idea is supported by the modeling studies of Fanning and Weaver (1997). Subsequent changes in outflow from the Laurentide ice sheet, or possibly even changes in the precipitation/evaporation balance in the North Atlantic, then pushed the system "over the edge," leading to the cessation of NADW production (seen by Keigwin *et al.*, 1991) and very cold conditions around the Atlantic Basin. This in turn reduced continental ice sheet melting, which allowed oceanographic conditions to re-equilibrate, eventually restoring NADW production and terminating the YD episode. The Mwp-IB then followed shortly thereafter. This scenario is also supported by strong evidence that sea surface temperatures in the North Atlantic were indeed much lower in YD time; most of the North Atlantic was dominated by the cold water foram *G. pachyderma* (sin); (Ruddiman and McIntyre, 1981).

Another perspective is provided by very high-resolution sediment records from off the Norwegian coast (61–63° N) (Koç Karpuz and Jansen, 1992; Lehman and Keigwin, 1992). Diatom assemblages are a sensitive indicator of sea surface temperature (SSTs) in the area (Koç Karpuz and Schrader, 1990) and document a series of oscillations in water temperature over the last 15,000 yr. The first occurrence of diatoms was at 13,400 B.P., indicating ice-free conditions (at least seasonally). This corresponds to Termination 1A of Duplessy *et al.* (1986) and was a time of major changes in SST throughout the North Atlantic as seen in cores over a wide area, from 45 to 63° N. Five major climatic episodes can be recognized (Fig. 6.52): a Bolling-Allerød "Interstadial Complex" from 13,200–11,200 (^{14}C yr) B.P. when summer SSTs reached 4.5–7.5 °C. This interval was punctuated by a series of cold episodes each lasting 100–300 yr. Conditions changed abruptly at ~11,200 B.P.

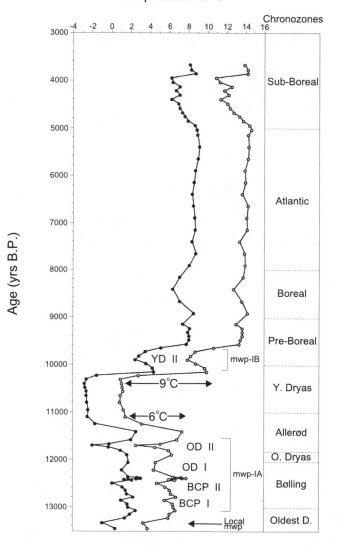

FIGURE 6.52 Reconstruction of SSTs in August (open circles) and February (dots) in the southeastern Norwegian Sea (~63° N) over the last 13,400 (^{14}C) yr, based on diatom assemblages calibrated with modern SSTs. Oscillations of temperature within the Bølling-Allerød chronozones (Older Dryas I and II, Bølling Cold Periods I and II) and meltwater peaks IA and IB are indicated (Koç Karpuz and Jansen, 1992).

(the start of the Younger Dryas period) with summer SSTs dropping to ~1 °C in <50 yr, and remaining low for the next 1000 yr. During this interval, glaciers advanced in many areas and quite dramatic changes in the distribution of flora and fauna took place throughout western Europe. An equally rapid warming ensued at ~10,200 B.P., followed by a short cool spell centered on 9800 B.P. at the beginning of the Pre-Boreal period. SSTs then rose to Holocene maxima (13–14.5 °C in summer) from 9500–5000 yr B.P. These changes are remarkably synchronous with the $\delta^{18}O$ record from Dye-3 in southern Greenland, which reflects the changes in oceanic heat flux and associated atmospheric temperature signal (Fig. 6.53). The $\delta^{18}O$ records from the same area indicate that meltwater from the Fennoscandian and Barents Sea ice sheets played a critical role in bringing about these oscillations. The strongest meltwater signal (resulting in low sea surface salinities, as seen in unusually low $\delta^{18}O$ values) began ~14,700 yr B.P., reaching a maximum at ~13,500 yr B.P. (Sarnthein *et al.,* 1992), at which point SSTs at 63° N rose to the level where diatoms could survive (Duplessy *et al.,* 1992). Thus, initial melting was probably related to the orbitally induced insolation increase, amplified by calving into a rising sea level, as the ice continued to melt back. Once warmer waters penetrated northward, the subsequent melting of the ice sheet was very much controlled by associated warm air advection. However, meltwater played a key role in controlling SSTs and led to the series of warm-cold oscillations in late Glacial and early Holocene time. Koç Karpuz and Jansen (1992) suggest that as meltwater from the Fennoscandian ice sheet flooded the Norwegian Sea, the low salinity surface waters would have frozen over in winter, minimizing mixing in the upper layers of the ocean and restricting the warmer Atlantic water to below the halocline. The ensuing cooler conditions would have reduced melting, eventually leading to erosion of the shallow low-salinity mixed layer and rapid warming as the warmer waters reappeared at the surface. This in turn led to renewed melting and the entire sequence began again, as recorded in the Bolling-Allerød Interstadial Complex of oscillating warm and cold episodes (see Fig. 6.52). The Younger Dryas cold event may have lasted longer because of major drainage of meltwater from the Laurentide ice sheet (mwp-IA), which in some way set up the conditions necessary to prolong disruption of circulation in the North Atlantic, as noted earlier. Similar conditions may have occurred during the final cool episode around 9800 yr B.P., which coincided with mwp-IB (Fig. 6.53). The $\delta^{13}C$ in benthic forams from the Bermuda rise record reductions in NADW flux from the Norwegian Sea during the Younger Dryas episode, though a less dense intermediate water ("upper NADW") from the Labrador Sea may have continued to form, as registered even as far away as the Southern Ocean at that time (Lehman and Keigwin, 1992a, b; Charles and Fairbanks, 1992).

An alternative scenario for the Younger Dryas involves Arctic Ocean sea ice. During the LGM, when sea level was 120 m lower and warm waters did not penetrate to the northern reaches of the North Atlantic Basin, the Arctic Ocean was extremely isolated. The exposed Bering Land bridge prevented North Pacific water from entering the Arctic Basin, and the only source of ice export from the Arctic Ocean was through a very restricted channel (a narrower Fram Strait, between northeast Greenland and Svalbard) as the Barents Sea region was occupied by a grounded ice

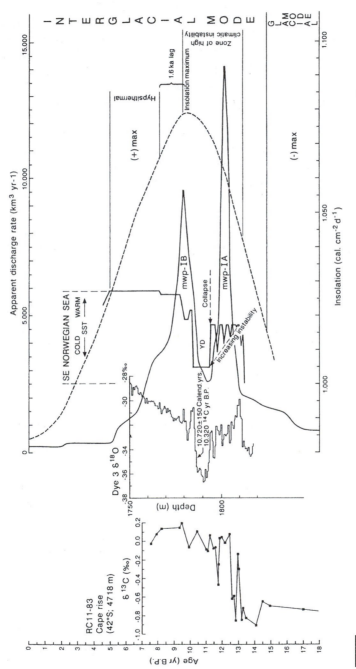

FIGURE 6.53 Schematic diagram showing the relationship between SSTs in the southeastern Norwegian Sea and δ^{18}O in the ice core from Dye-3, meltwater discharge from continental ice sheets (Fairbanks, 1989), insolation at 60° N and δ^{13}C in the Southern Ocean. A series of oscillations in temperature, related to meltwater events, preceded the cold conditions of the Younger Dryas. Deepwater production (registered as a rise in δ^{13}C) appears to have increased around the time of mwp-IA, and continued through the Younger Dryas chronozone (Koç Karpuz and Jansen, 1992).

sheet. Cold conditions with minimal melting and the continual accumulation of snow and superimposed ice during the LGM would have led to the build-up of extremely thick sea ice (tens of meters in thickness, like the ice shelves of northern Ellesmere Island today) which was essentially locked in place within this restricted basin. This composite of snow, superimposed ice, and sea ice would have remained more or less immobile until rising sea level, break-up of the Barents Sea ice sheet, and the return of warmer sub-surface waters to the Arctic Ocean induced some disturbance in (and heat flux to) the floating ice mass. The final trigger for a massive discharge of very thick, low salinity ice into the Atlantic Ocean may have come about when sea level rose to the point where Bering Strait was flooded (at −40 to −50 m) causing the Trans-Polar Current to become established. Recent ^{14}C dates on terrestrial peats from the Chukchi Shelf, north of Bering Strait (which were flooded by the transgressing sea) place the time of this crucial event at around 11,000 ^{14}C yr B.P. (Elias *et al.*, 1996b). This suggests a direct link to the onset of Younger Dryas conditions in western Europe as a result of the North Atlantic being flooded, not by icebergs from the Laurentide Ice Sheet, but by Arctic Ocean sea ice, which then reduced, or displaced, NADW production for the next 1000 yr.

Although the exact sequence of events that took place in late Glacial and early Holocene time is not yet entirely understood, it is clear that meltwater entering the North Atlantic is implicated as a critical factor influencing climatic fluctuations at that time (Manabe and Stouffer, 1995; Bard *et al.*, 1996). Very abrupt changes in thermohaline circulation took place, involving reductions in deepwater formation, or shifts in the pattern of deepwater production, with compensatory reductions in the flux of warm surface waters to the North Atlantic. Nevertheless, several enigmas remain. Perhaps because of ^{14}C dating problems, the timing of observed changes does not always fit together nicely, and sometimes the geochemical evidence is contradictory. Distinguishing deepwater produced in different areas is difficult, compounded by the fact that deepwater production regions probably shifted over time (Duplessy *et al.*, 1980, 1988). The timing and magnitude of meltwater discharges from the Arctic and Fennoscandia, and from both the northern and southern drainage systems of the Laurentide ice sheet remain problematic. Indeed, Clark *et al.* (1996) believe that meltwater from the Antarctic ice sheet must be implicated in the late glacial oceanographic changes of the North Atlantic; they argue that the estimated volumes of meltwater from the Laurentide ice sheet and the timing of discharge events (based on what is known from the geological record on land) do not fit with the various explanations so far proposed for changes in North Atlantic circulation. Further high resolution sediment studies will no doubt help to resolve many of these issues. Meanwhile, the important question to ask today is whether, in the absence of ice major sheets, changes in the North Atlantic salinity structure can be brought about by imbalances in the precipitation-evaporation-runoff relationship. This issue is especially relevant as greenhouse gases increase to unprecedented levels, with a possible consequence being higher precipitation and runoff into the North Atlantic Basin or a drastic reduction in Arctic Ocean ice cover. If the thermohaline circulation were to be disrupted by such changes, there may indeed be "unpleasant surprises in the greenhouse" (Broecker, 1987).

6.11 CHANGES IN ATMOSPHERIC CARBON DIOXIDE: THE ROLE OF THE OCEANS

It is clear from ice-core records that carbon dioxide levels in the atmosphere (pCO_2) were considerably lower (by 90–100 p.p.m.v.) at the Last Glacial Maximum than in the Holocene (see Section 5.4.3). Indeed, long-term carbon dioxide changes parallel $\delta^{18}O$ in ice and must have played a role in glacial-interglacial climatic changes, either directly or indirectly. How did such changes come about? Because the ocean carbon content is 50–60 times that of the atmosphere, even relatively small changes in the rates of ocean carbon dioxide uptake or loss (de-gassing) can have large effects on atmospheric CO_2 levels (Broecker, 1982). Changes in sea-surface temperature and salinity in glacial times (to a colder, more saline ocean) can account for ~10% of the observed pCO_2 changes (simply by the difference in CO_2 solubility under such conditions) but the bulk of the change must be related to an increase in biological productivity in the ocean during glacial periods. The CO_2 dissolved at the ocean surface is removed by biological activity in the photic zone, which forms the organic tissue and carbonate shells of marine organisms. As these die and fall through the water column, the carbon fixed near the surface is transferred to deeper layers of the ocean and may be deposited in the sediments. This process can be thought of as an "ocean carbon pump" whereby carbon is continually removed from the ocean surface (Volk and Hoffert, 1985); upwelling of carbon-rich deepwater returns CO_2 to the atmosphere. Thus, some areas of the ocean are carbon sinks and some are carbon sources for the atmosphere. Factors that alter the biological productivity of surface waters and the rate of upwelling, or which change the distribution of sources and sinks of carbon, can therefore have a significant effect on atmospheric CO_2 levels (Ennever and McElroy, 1985).

An important index of photosynthetic activity in the surface waters of the ocean is the relative proportion of ^{12}C to ^{13}C. During photosynthesis, ^{12}C is preferentially removed from the water to produce organic material with low $\delta^{13}C$. In many regions of the ocean, productivity is limited by a lack of nutrients (particularly phosphate and nitrate). Broecker (1982) suggested that during glacial periods, when sea level was lower, phosphates that had accumulated on the continental shelves would be eroded and dispersed into the ocean, increasing nutrient levels and hence productivity in the near-surface waters. This would lead to ^{12}C removal from near the surface and cause the $\delta^{13}C$ gradient between the upper and deep ocean to increase. At the same time, higher productivity in the photic zone would increase the rate of CO_2 removal from the ocean, causing an increase in the CO_2 flux to the atmosphere, thereby lowering pCO_2. Hence, there should be evidence of lower atmospheric CO_2 levels preserved in the skeletal remains of organisms that lived near the surface, compared to those that lived in the deep ocean. Shackleton *et al.* (1983) examined this question by measuring the difference between ^{13}C in the carbonate tests of planktonic and benthic forams in an equatorial Pacific core (Fig. 6.54a). The resulting record of the $\delta^{13}C$ gradient between near surface and deep waters ($\Delta\delta^{13}C$) provides a proxy of atmospheric CO_2 levels, with stronger gradients signifying increased productivity at the surface and hence lower pCO_2. This record is remarkably similar to that obtained from the Vostok ice core (Fig. 6.54b), suggesting

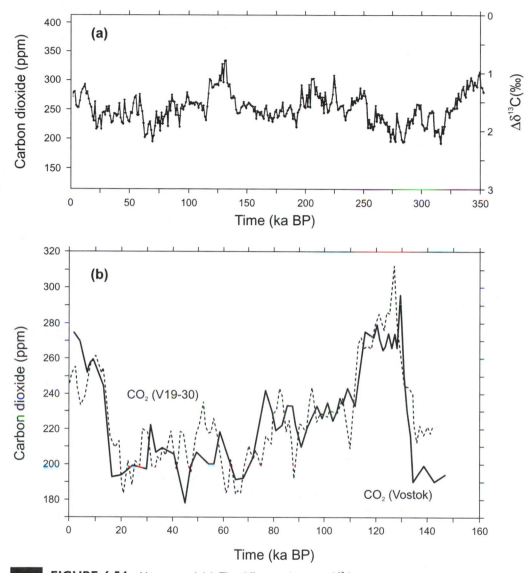

FIGURE 6.54 Upper panel (**a**): The difference between $\delta^{13}C$ in planktonic forams *Neogloboquadrina dutertrei*, and in benthic forams *Uvigerina senticosa*, from equatorial Pacific core V19-30 (Shackleton and Pisias, 1985). This difference ($\Delta\delta^{13}C$) shows the relative increase in biological productivity of surface waters in glacial periods compared to deep water that results in a higher $\delta^{13}C$ gradient at those times. Increased surface water productivity would have led to a reduction in atmospheric carbon dioxide, so the $\Delta\delta^{13}C$ can be interpreted as a paleo-CO_2 index (left scale). The lower panel (**b**) shows the same record plotted with the Vostok ice-core CO_2 record, although the exact temporal match may not be quite correct (Shackleton *et al.*, 1992).

that changes in the rate at which organic carbon is sequestered in the ocean (as distinct from changes in the carbonate content, which would not affect near surface $\delta^{13}C$) have been the dominant cause of atmospheric CO_2 changes on glacial to interglacial timescales. Furthermore, spectral analysis of this record in relation to benthic $\delta^{18}O$ (a proxy of continental ice volume) shows that CO_2 changes are strongly in phase with orbital forcing, but lead ice-volume changes at all orbital frequencies (Shackleton and Pisias, 1985). This points to CO_2 variations as playing a key role in the forcing of climatic changes involving ice sheet growth and decay, rather than being only a passive response to such changes. It also indicates that the phosphate mobilization idea of Broecker (1982) cannot be the main factor initiating productivity changes, as that mechanism is tied to sea-level changes (which lag by several thousand years both orbital forcing and CO_2 changes). It is of interest that much of the variance of the CO_2 record is in the frequency band related to obliquity, which has its main impact on radiation receipts at higher latitudes. This suggests that the subpolar ocean areas (perhaps via switches in thermohaline circulation) play a critical role in driving CO_2 changes and hence amplifying orbitally forced climatic change (Wenk and Siegenthaler, 1985).

One additional point to note is that the *overall* $\delta^{13}C$ content of the ocean was lower (by ~0.4‰) in glacial times because of the significant decrease in biomass on the continents. Approximately 500 Gton carbon (with a $\delta^{13}C$ of –25‰) was added to the ocean-atmosphere system at such times, lowering the ^{13}C content of the ocean accordingly (Siegenthaler, 1991). With less photosynthetic activity on land, one might expect higher pCO_2 levels during glaciations, but this was more than compensated for by enhanced CO_2 solubility (in cooler waters) and increased oceanic biological activity.

In a subject as complex as ocean geochemistry there is room for many alternative hypotheses, and in this field alternatives abound. Much attention has been paid to mechanisms that could have brought about changes in oceanic nutrient content and hence biological productivity. Martin (1990) for example points to the much higher levels of iron-rich atmospheric dust that was dispersed across the globe during the last glaciation. He suggests that this dust fertilized the sub-Antarctic Ocean, greatly increasing productivity and leading to a reduction in pCO_2. This idea has been controversial on biological (e.g., Dugdale and Wilkerson, 1990; Sunda *et al.,* 1991) as well as geological grounds, because there is conflicting evidence for increased productivity in this area during glacial periods (Mortlock *et al.,* 1991; Kumar *et al.,* 1995). Others have suggested that the general idea may have more merit in terms of changing productivity at lower latitudes, though distinguishing between increased productivity due to Fe fertilization, from a general increase in nutrient levels due to changing ocean circulation regimes, and increased surface wind stress and upwelling associated with glacial periods is very difficult (Sarnthein *et al.,* 1988; Berger and Wefer, 1991). Furthermore, there is evidence that in some upwelling areas, even when productivity increased, the net effect was insufficient to overcome the transfer of CO_2 from the ocean surface to the atmosphere. Hence, there may be some situations where higher productivity does not always equate with a reduction in atmospheric pCO_2

due to the even stronger influence of de-gassing from carbon-rich upwelling waters (Pedersen *et al.*, 1991).

One interesting argument in favor of increased high latitude biological activity was suggested by Kumar *et al.* (1995), who examined "excess" $^{231}Pa/^{230}Th$ and $^{10}Be/^{230}Th$ in sub-Antarctic sediments as a measure of past biological productivity. They point out that because most of the biomass initially deposited as sediment is not preserved, the measured sediment accumulation rate is not a reliable index of former productivity. The radionuclides protactinium-231 and thorium-230 (dispersed throughout the ocean from the decay of uranium in seawater) are removed from the water column by attachment to particles. However, ^{231}Pa is less easily removed than ^{230}Th, so as particle flux (productivity) increases the disparity in the scavenging rate of the two radionuclides increases (i.e, the $^{231}Pa/^{230}Th$ ratio increases). A similar index is provided by $^{10}Be/^{230}Th$; ^{10}Be is a cosmogenic isotope that is widely dispersed across ocean basins, but like ^{231}Pa, it has a longer residence time than ^{230}Th. Thus an increase in $^{10}Be/^{230}Th$ also provides a measure of particle flux. Both $^{231}Pa/^{230}Th$ and $^{10}Be/^{230}Th$ accumulate in the sediments even though the particles that transported them to the sea floor may no longer be present. An "excess" of these radionuclides (over that expected from a steady sedimentation rate) thus documents formerly high productivity in the overlying water column. Studies of these and other productivity indices reveal that sub-Antarctic waters were much more productive in glacial times, but this was not the case in the Southern Ocean, closer to the Antarctic continent (perhaps because of more extensive sea ice). The sediments provide evidence that perhaps 30–50% of the observed reduction in CO_2 can be explained by a more efficient biological pump in the cold waters surrounding Antarctica during glacial episodes.

Studies of nitrogen isotopes point to another mechanism with important implications for atmospheric CO_2 changes (Altabet *et al.*, 1995; Ganeshram *et al.*, 1995). As already noted, nitrate (NO_3^-) is a key nutrient limiting biological productivity in many parts of the ocean. Oceanic nitrate concentration is the result of the balance between upwelling of nitrate-rich waters and processes of denitrification carried out by bacteria in low oxygenated zones of the ocean. Denitrification produces gaseous nitrogen and nitrous oxide, which then escapes from the water column, thereby reducing the supply of fixed nitrogen available for plant growth, which in turn influences atmospheric CO_2 levels. During denitrification, fractionation results in the preferential loss of ^{14}N, enriching the waters in ^{15}N, which is then incorporated into whatever organic material is forming and settling to the sea floor. Consequently, sediments from times of reduced denitrification have a lower $\delta^{15}N$, as is evident during marine isotope stages 2, 4, and 6 in cores from off the northwestern coast of Mexico (Fig. 6.55) and from the Arabian Sea. Together with the eastern equatorial South Pacific, these areas are especially important for denitrification in the modern ocean, accounting for almost all of the water column denitrification going on today. The evidence for greatly reduced denitrification in glacial periods, plus the fact that lower sea level would have reduced the contribution of denitrification processes in continental shelf sediments, suggests that overall ocean nitrate levels would have been considerably higher at such times. This would have led to increased biological

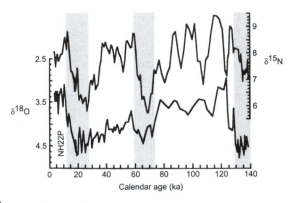

FIGURE 6.55   The record of $\delta^{18}O$ in benthic forams and $\delta^{15}N$ in bulk sediments from off the northwestern Mexican continental margin, spanning the last 140,000 yr. Principal glacial stages are shaded. The benthic record shows the familiar index of continental ice growth and decay; the $\delta^{15}N$ records denitrification processes occurring in the water column. Low values of δ^{15} indicate relatively low rates of denitrification, which imply higher levels of productivity and a reduction in atmospheric CO_2 levels (Ganeshram *et al.*, 1995).

activity in areas of the ocean that are today relatively unproductive (oligotrophic), causing a general decline in atmospheric pCO_2. Altabet *et al.* (1995) also point out that the reduced level of greenhouse gas N_2O, observed in glacial periods in the Vostok ice core (see Fig. 5.35) may be a direct consequence of reduced oceanic denitrification at those times.

From this brief summary of various lines of evidence, it is apparent that several factors were probably operating simultaneously to lower atmospheric carbon dioxide levels during glacial periods. Collectively, these factors increased ocean productivity to the point where there was a net flux of CO_2 from the atmosphere to the ocean, eventually leading to atmospheric pCO_2 levels ~100 p.p.m.v. lower than pre-industrial levels. The initial driving force for such changes seems to be orbital forcing, though no doubt additional feedbacks (involving SST changes, thermohaline circulation, surface wind stress, etc.) then amplified these changes. In view of the importance of fully understanding all the factors affecting atmospheric CO_2 levels (as well as other greenhouse gases such as N_2O) there will no doubt be a great deal more research on this topic in the future that may considerably revise our current understanding of this complex topic.

6.12 ORBITAL FORCING: EVIDENCE FROM THE MARINE RECORD

The availability of continuous paleoclimatic records from the ocean floor, spanning several hundred thousand years, has enabled hypotheses about the causes of climatic change to be tested, and has facilitated the development of new models. One of the most important hypotheses is that propounded by Milankovitch (1941), who argued that glaciations in the past were principally a function of variations in the

Earth's orbital parameters, and the resulting redistribution of solar radiation reaching the Earth (see Section 2.6 for a complete discussion). Emiliani (1955, 1966) was the first to note that $\delta^{18}O$ maxima in Caribbean and equatorial Atlantic cores closely matched summer insolation minima at 65° N, which was the latitude that Milankovitch had considered critical for the growth of continental ice sheets. Subsequently, Broecker and van Donk (1970) suggested revisions of Emiliani's timescale, but still concluded that insolation changes were a primary factor in continental glaciation. In addition, dates of coral terrace formation, indicative of a formerly higher sea level (lower global ice volume), were shown to be closely related to times of insolation maxima, again supporting the ideas of Milankovitch (Broecker *et al.*, 1968; Mesolella *et al.*, 1969; Veeh and Chappell, 1970).

The first rigorous attempt to assess the evidence for orbital changes in paleoclimatic data was made by Hays *et al.* (1976) using two ocean core records from the southern Indian Ocean (43° and 46° S). Three parameters were studied: $\delta^{18}O$ values in the foraminifera *Globigerina bulloides* (an index of global, but primarily northern hemisphere ice volume); summer sea-surface temperature (T_s) derived from radiolaria-based transfer functions (an index of sub-Antarctic temperatures); and abundance variations of the radiolaria *Cycladophora davisiana* (considered to be an index of Antarctic surface water structure). Using the ~450,000-yr record available, Hays *et al.* (1976) showed that much of the variance in these proxy records was concentrated at frequencies corresponding closely to those expected from an orbital forcing function. Specifically, spectral peaks were found at periods of ~100,000 yr, 40,000–43,000 yr, and 19,500–24,000 yr. Such periodicities closely match spectral peaks in orbital data (at ~100,000, ~41,000, and 19,000–23,000 yr, associated with variations in eccentricity, obliquity, and precession, respectively). Furthermore, not only are the proxy and orbital series closely matched in the frequency domain, but an examination of the time domain of each periodic component showed fairly consistent phase relationships (back to 300,000 yr) between orbital parameters and the "resultant" climatic signal. Such results are very improbable by chance and the work of Hays *et al.* thus provided the first really strong evidence that changes in the Earth's orbital geometry played an important role in causing glacial-interglacial variations over the past 300,000–400,000 yr.

Numerous other studies have subsequently confirmed these pioneering results and it is now quite clear that orbital forcing played a key role in pacing glaciations during the Quaternary period (Berger, 1990). However, the mechanisms involved in linking changes in insolation to changes in the climate system are not so clear. As noted in Section 2.6, the principal periodicity in the $\delta^{18}O$ marine sediment record lies in the 100 ka band, but this has relatively little power in the insolation record (see Fig. 2.22). Imbrie *et al.* (1992, 1993b) have examined this matter in great detail and propose a comprehensive model to account for this enigma. Their model builds on the earlier ideas of Weyl (1968), Broecker *et al.* (1985b), and Broecker and Denton (1989), different aspects of which were discussed in Sections 6.9 and 6.10. However, by focusing on how different parts of the climate system respond at the three distinct orbital frequencies (23 ka, 41 ka, and 100 ka), Imbrie *et al.* were able to demonstrate a recurrent geographical sequence of orbitally driven changes at all frequencies during the course of a glacial-interglacial cycle. By looking at the phase relationships between insolation, global ice volume, and other climate proxies, it was clear that certain parts of the climate system have a consistent early response to northern hemisphere high latitude insolation changes, whereas others respond later. By mapping the geographic distribution of these responses a mechanistic model of glacial-interglacial changes was constructed. This revealed that four key subsystems control the rate at which radiation changes are propagated through the climate system, each having a different level of inertia. Near-surface processes on the land and in the upper ocean, and in the deep waters of the southern ocean respond quickly (in < 1ka) whereas changes involving ice sheets, displaced wind systems, and deep ocean chemistry take longer (3–5 ka). Through the interaction of these different subsystems, the sequence of events leading to glaciation and deglaciation slowly unfold.

Like earlier researchers, Imbrie *et al.* (1992) determined that a critical factor driving glacial-interglacial changes is salinity-controlled convection in the Icelandic, Norwegian, and Greenland ("Nordic") Seas and in the Labrador Sea, which they refer to as the Nordic and Boreal heat pumps, respectively. During interglacials, both areas produced deepwater, which drove the thermohaline circulation of the Atlantic and carried heat to the Southern Ocean, thereby restricting sea ice around Antarctica. With both heat pumps operating, ventilation of the deep ocean was at its maximum. As summer insolation decreased at high northern latitudes the atmosphere and ocean surface cooled, reducing evaporation and increasing snowfall and sea-ice cover. Eventually, salinity in the Nordic Seas decreased, at first slowing and then entirely eliminating convective overturning via the Nordic heat pump, drastically curtailing warm water flux to the southern hemisphere. However, the Boreal heat pump continued to operate, producing intermediate water, but the net result was for a dramatic reduction in thermohaline circulation and, hence, in the return flow of warm water to the North Atlantic. As Antarctic sea ice expanded, Antarctic Bottom Water flux to the north increased and an equatorward shift in the southern Westerlies led to ocean circulation changes that sequestered carbon in the Southern Ocean. This led to a reduction in atmospheric CO_2 levels, which further reinforced the downward insolation trend. Further cooling in the northern hemisphere resulted in ice growth on land and a fall in sea level, allowing extensive marine-based ice sheets to develop in locations that were vulnerable to later sea-level increases. Ice sheet growth eventually disrupted the westerly wind system causing sea ice to form over large areas of the North Atlantic which increased the production of Intermediate Water and caused a slight increase in NADW, leading to a minor recession in Antarctic sea-ice extent. As the insolation cycle returned once more to higher northern hemisphere summer energy receipts, the northern oceans slowly warmed, and on the continents, snow melted and the ice margins retreated. This allowed the zone of maximum westerly winds to shift northward, and warmer waters and subtropical air masses could then be advected from the south, causing the ice sheets to rapidly melt, and sea level to rise very quickly with catastrophic collapse of marine-based ice sheets. At the same time, sea ice rapidly receded and this, together with the advected waters from the south, resulted in a sharp increase in the salinity of the

North Atlantic and a resumption of the Nordic heat pump. Large-scale melting events due to the rapid melting of continental and marine-based ice sheets may have resulted in short-term reversals of these trends (as discussed in Section 6.10.2) but eventually the primary sequence of events driven by orbital forcing prevailed, leading once more to a strong thermohaline circulation involving both the Boreal and Nordic heat pump systems.

Although this sequence of events appears to be causally linked to the 23 ka and 41 ka precession and obliquity cycles in a simple linear relationship, most of the variance in the ice volume record is actually in the 100 ka eccentricity cycle. As variations in radiation due to eccentricity changes are an order of magnitude smaller than those due to precession and obliquity changes, it is difficult to understand why changes at this frequency are so large. Several models have been proposed to explain this problem. One set of models generally views the climate system as having the ability to develop internal (free) oscillations or resonance in response to some external forcing, possibly unconnected to orbital forcing. Another set views the climate system as following the same sequence of responses noted for the 23 ka and 41 ka cycles, but in a non-linear manner; at some point a threshold is crossed that then gives rise to a much larger response (Imbrie *et al.*, 1993b). The critical factor driving this non-linear response appears to be the size of the continental ice sheets (primarily the Laurentide ice sheet). Large ice sheets significantly disrupt the westerlies, leading to strong meridional circulation, which then greatly amplifies the sequence of climate system changes associated with 23 and 41 ka cycles. Thus, when some combination of forcing at the 23 ka and 41 ka bands leads to the ice sheets exceeding some critical size, they then overtake the "normal" system response to orbital forcing. In effect the ice sheets themselves become the principal agents driving the climate system, producing a cycle of changes with a longer (100 ka) periodicity due to the large amount of inertia associated with ice-sheet growth and decay. This model provides an elegant explanation for the changes observed over the last few hundred thousand years, but why this pattern became more significant in the last million years remains to be fully explained.

7 NON-MARINE GEOLOGICAL EVIDENCE

7.1 INTRODUCTION

The range of non-marine geological studies providing information pertinent to paleo-climatology is vast; indeed, one could argue that virtually all continental sedimentary deposits convey a paleoclimatic signal to some degree. Aeolian, glacial, lacustrine, and fluvial deposits are, in large part, a function of climate, though it is often difficult to identify the particular combination of climatic conditions leading to the formation of the deposit. Similarly, erosional features such as lacustrine or marine shorelines, cirques, or other features of glacially eroded landscapes indicate in a general sense a particular type of climate, but quantitative paleoclimatic reconstructions based on this kind of information present real challenges (Flint, 1976). Commonly, climatic inferences drawn from such evidence are qualitative and even dating the features may be very difficult. Nevertheless, such evidence of past changes in climate is ubiquitous, and many innovative approaches to interpreting their paleoclimatic significance have been developed.

In this chapter no attempt will be made to review all of the possible paleoclimate-related geological and geomorphological phenomena. Instead, discussion will be limited to the smaller number of approaches that have either provided quantitative paleoclimatic data or helped in establishing the chronology of paleoclimatic events.

7.2 LOESS

Loess is a deposit of wind-blown silt that blankets large areas of the continents. It is characteristically creamy-brown in color and calcareous, consisting predominantly of quartz feldspars and micas (Pye, 1984, 1987). Geographically, loess is extensive in the North American Great Plains, south-central Europe, Ukraine, central Asia, China, and Argentina (Fig. 7.1). In North America, loess deposits are related to formerly extensive outwash deposits from the Laurentide ice sheet and to the floodplains of large, braided rivers. Similarly, in Europe loess deposits are common between the former Alpine and Scandinavian Ice Sheets though the thickest loess sections, in the Czech Republic, Slovakia, and Austria, may also have had a more local origin related to formerly extensive braided rivers at a time when vegetation cover was greatly reduced (Kukla, 1975b). In North America, loess deposits are up to 50 m thick in Alaska (Beget *et al.*, 1991) and sections along the Mississippi River valley may reach 20 m in thickness (Forman *et al.*, 1995; Oches *et al.*, 1996). Elsewhere, loess is related to desert conditions, especially the formerly extensive deserts of central Asia. In north central China (southeast of the Gobi Desert) loess deposits are extremely thick, up to 300 m in places, and completely cover the underlying topography, forming an extensive loess plateau (Liu *et al.*, 1985). It is there that the most comprehensive paleoclimatic studies of loess deposits have been carried out (Kukla, 1987a; Kukla and An, 1989). Loess has accumulated on the Loess Plateau for ~2.5 Ma and during episodes of warmer and wetter conditions weathering of the loess led to soil forma-

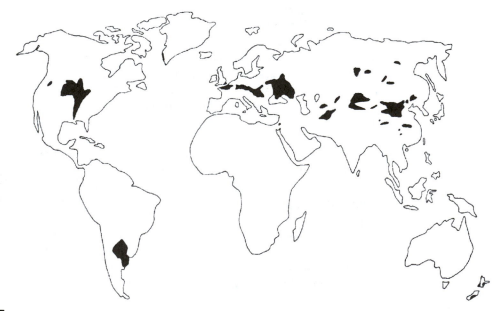

FIGURE 7.1 Location of principal loess deposits in the world (Pye, 1984).

tion. Today the alternating sequence of loess units and intervening paleosols (Fig. 7.2) form the most complete terrestrial records of Quaternary paleoclimatic conditions to be found on the continents (An *et al.*, 1990; Ding *et al.*, 1993).

7.2.1 Chronology of Loess-Paleosol Sequences

Early studies of loess-paleosol sections took a simple relative dating approach, counting the first well-developed soil below the surface (Holocene) soil as equivalent to the last interglacial, and older soils representing earlier interglacial episodes. Thus, in China the conventional terminology is based on counting back from the Holocene soil (S_0) to the last loess episode (L_1) then to the preceding interglacial soil (S_1) and underlying loess (L_2) and so on down the section. In this way, as many as 37 soils and associated loess episodes have been recognized at the Baoji type-section (on the southwestern margin of the Loess Plateau); of these, 32 (S_1–S_{32}) are at least as well developed as the Holocene soil, and document the alternation between glacial and interglacial conditions throughout the Quaternary (Rutter *et al.*, 1991a). In Europe, loess deposits are mostly restricted to river terraces and are, therefore, generally less continuous, leading to confusion over the chronology of loess-paleosol sequences and how they are related from one region to another. Recent aminostratigraphic studies (on snails embedded in the loess) have clarified the stratigraphic relationships between many different locations, improving the correlations between sections (Oches and McCoy, 1995b, 1995c). Thermoluminescence dating has proven useful in determining the chronology of loess deposition during the last glacial cycle but

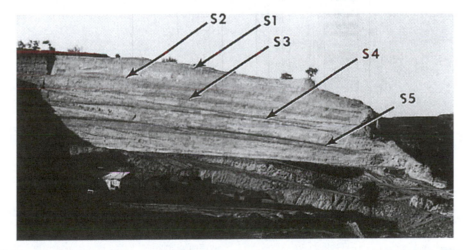

FIGURE 7.2 Part of the loess-paleosol sequence near Luochuan, north central China, showing an exposed section of loess with interbedded soil units. In this section, the upper loess unit (L_1) and Holocene soil (S_0) have been eroded away. Soils developed during more humid periods when monsoon rainfall was higher, but loess continued to accumulate in winter months albeit at a slower rate. During major loess deposition episodes, the summer monsoon rainfall rarely penetrated this far into China (photograph by R.S. Bradley).

dates on older deposits remain controversial (Wintle, 1990; Forman, 1991). For the longer Quaternary record, paleomagnetic studies have provided the fundamental chronology (Heller and Liu, 1984; Rolph *et al.*, 1989; Rutter *et al.*, 1990; Thistlewood and Sun, 1991) (Fig. 7.3). Once the basic reversal record has been determined some studies then simply interpolate between chron/subchron boundaries to obtain a chronology of the intervening loess and paleosol units (Rutter *et al.*, 1991a). However, this does not realistically take into account the changing loess accumulation rates from full glacial to interglacial conditions (when loess slowly accumulated as soils developed). A more realistic approach assumes a higher loess deposition rate in glacial times (Liu *et al.*, 1993 and Ding *et al.*, 1994 assumed loess accumulated at 1.5 to 2 times that during interglacial soil-forming intervals). This provides a first-

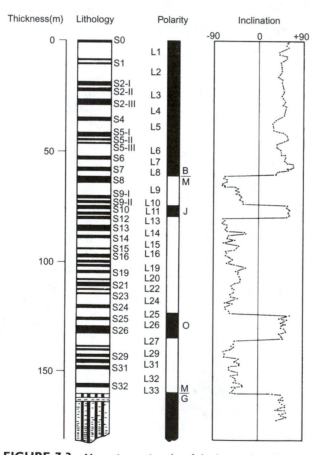

FIGURE 7.3 Magnetic stratigraphy of the loess-paleosol section at Baoji, Loess Plateau, north central China. Major paleosols (S) and loess units (L) are numbered from the top down (Holocene soil = S$_0$). Changes in polarity are clearly indicated by the inclination record on the right (B/M = Brunhes-Matuyama boundary; J = Jaramillo subchron; O = Olduvai subchron; M/G = Matuyama-Gauss boundary). Loess began to accumulate over a red clay deposit around the time of the Gauss-Matuyama transition (Liu *et al.*, 1993).

order approximation that reveals the strong relationship between the chronology of loess-paleosols in China and the marine isotope record (Fig. 7.4). Loess units correspond to even-numbered marine isotope stages (i.e., times of continental ice buildup) and the soil units correspond to interglacial (odd-numbered isotope stages). Weak soils appear to be related to interstadial events.

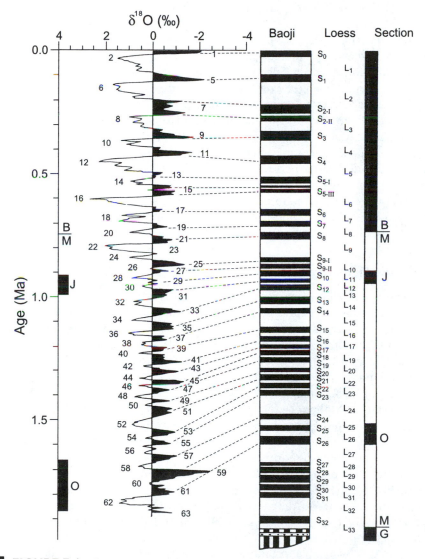

FIGURE 7.4 Composite marine oxygen isotope record (left) with stages indicated, and its proposed correlation with the loess-paleosol sequence at Baoji, Loess Plateau, north central China. Paleosol units (S_0–S_{32}) are indicated by dark bands, though not all soils are equally well developed. Magnetic stratigraphy of the records is indicated (Rutter et al., 1990).

This general correspondence suggests that a more accurate chronology of loess-paleosol sequences might be determined by tuning some diagnostic parameter to orbital forcing, just as the SPECMAP group did to resolve the chronology of the marine isotope record (Section 6.3.3). Using this approach, Ding *et al.* (1994) adjusted the record of grain-size variations at Baoji (a proxy of winter monsoon strength in north central China) to optimize power at frequencies corresponding to the principal orbital periodicities (Berger and Loutre, 1991). Following Imbrie *et al.* (1984), they used changes in orbital eccentricity, obliquity (lagged 8000 yr), and precession (lagged 5000 yr) as the combined target to which the grain size record was tuned (Fig. 7.5). Confidence in the resulting timescale is provided by a good match with predicted ages of K/Ar-dated paleomagnetic boundaries in the loess-paleosol sequence. Spectral analysis of the Baoji timescale, so derived, reveals some important changes over the course of the Quaternary. For the last 600 ka the grain-size record was dominated by the 100 ka eccentricity period, which accounts for 46% of the variance in this interval of time. However, from 0.8–1.6 Ma the record has a much stronger 41 ka obliquity cycle (30% of total variance) and from 1.6–2.5 Ma a more complex spectrum is apparent, with power concentrated mainly at 55 ka and 400 ka periodicities (35% of total variance). This shift towards a stronger obliquity signal around 1.6 Ma is not seen in the marine isotope record, suggesting some regional response to forcing, whereas the mid-Pleistocene change to a strong ~100 ka period is also clear in the marine isotope record and appears to be of global significance (Ding *et al.*, 1994).

7.2.2 Paleoclimatic Significance of Loess-paleosol Sequences

Several approaches to interpreting the alternation of loess deposits and paleosols have been made. At the simplest level, modern analogs of similar soils and loess accumulation areas provide an approximation to climatic conditions prevalent at different times in the past. Thus, Ding *et al.* (1992) suggest that the oldest thick loess units (L33 and L32) represent mean annual temperatures 12 °C lower than today and precipitation levels less than 25% of modern values, based on the modern climate of locations thought to be analogous to conditions prevailing when the loess accumulated. Maher *et al.* (1994) and Maher and Thompson (1995) focused on magnetic susceptibility as a proxy of rainfall. They argue that susceptibility is higher in soils due to the *in situ* formation of ultrafine (<0.02 μm) ferromagnetic (maghemite) grains by inorganic precipitation, aided by the presence of magneto-tactic (iron-reducing) bacteria (Maher and Thompson, 1992; Heller *et al.*, 1993; Verosub *et al.*, 1993; Liu *et al.*, 1994). Alternations of wetting and drying cycles favor this process so that susceptibility profiles can be considered a proxy for past rainfall variations. By establishing the relationship between modern rainfall amounts and susceptibility measurements on young soils throughout the Loess Plateau, they were able to calibrate the long record of susceptibility at Xifeng (spanning the past 1.1 M years) in terms of paleoprecipitation (Fig. 7.6). On this basis, rainfall has varied by a factor of ~2 over this interval (from ~400 to ~750 mm) and for 80% of the last 1.1 Ma, rainfall has been less than today at Xifeng. Liu *et al.* (1995) took a similar approach in reconstructing rainfall at Xifeng, but only for the

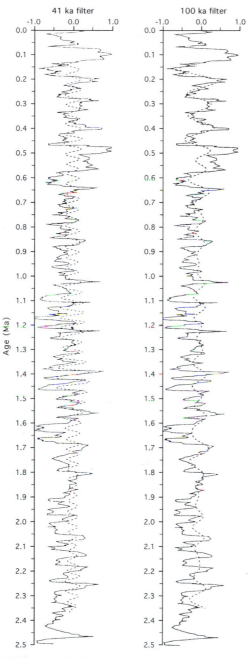

 FIGURE 7.5 Grain size variations at Baoji, Loess Plateau, north central China, compared to the 41,000- and 100,000-yr periodicities associated with variations in obliquity and eccentricity. Grain size variations are represented here by the ratio of the fraction of <2 μm grains to >10 μm; thus higher ratios represent a finer size distribution. Dashed lines show the filtered series bandpassed at 41 ka and 100 ka. A strong coherence between the loess record and the 100,000-yr period is noticeable over the last 0.6 Myr whereas prior to that the record has a stronger 41 kyr signal (back to ~1.6 Myr). The earliest 1 Myr of the record is poorly correlated with both frequencies (Ding *et al.*, 1994).

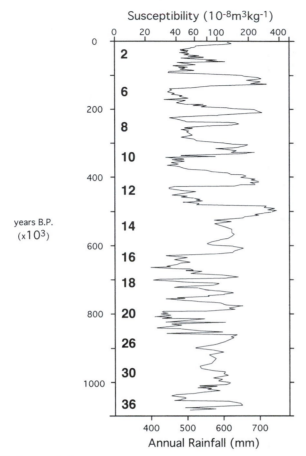

 FIGURE 7.6 The record of magnetic susceptibility at Xifeng, Loess Plateau, north central China, interpreted in terms of variations in annual rainfall. As rainfall increased the amount of magnetic material produced by weathering processes increased, leading to higher susceptibility levels (Maher and Thompson, 1995).

last glacial-interglacial cycle. They concluded that rainfall fell to much lower levels in glacial times (<200 mm a^{-1}) than estimated by Maher and Thompson (1995). Further studies with a much more extensive set of modern data may resolve these differences.

Increased wind speed, possibly coupled with an expanding desert margin during glacial periods, is reflected in variations of grain size in loess sections (An *et al.*, 1991). This is clearly seen in the general change from low median grain size during interglacials to higher values during glacial periods, but there are higher frequency variations superimposed on that pattern which may be of significance (Xiao *et al.*, 1995). Porter and An (1995) argue that many of these peaks correspond (within dating error limits) to Heinrich events (episodes of ice rafting) in the North Atlantic. They believe that the ice sheets of North America and Europe in some way are fundamentally linked to the winter monsoon regime, leading to stronger winds and large particles being deposited on the Loess Plateau during times when catastrophic draw-down of the ice sheets took place. It is difficult to envision the mechanism for such a linkage and it may be that the dating imprecision in both systems make the connection more apparent than real. The importance of precise dating in linking the loess record to other regions is well-illustrated by detailed studies of the Younger Dryas interval. In the North Atlantic area (at least) this is characterized by a return to near-glacial conditions. It might therefore be expected that this would correspond to an increase in loess deposition at that time in China. Indeed there was a late glacial oscillation on the Loess Plateau, but careful ^{14}C dating reveals that conditions actually became more humid during the Younger Dryas interval, though temperatures remained low (Zhou *et al.*, 1996). Evidently, the insolation maximum during that period led to higher levels of summer monsoon rainfall while cold northeasterly winter winds continued to deposit loess across the region (An *et al.*, 1993).

7.3 PERIGLACIAL FEATURES

The use of fossil periglacial phenomena as an index of former climatic conditions is limited by two basic problems. First, dating periglacial features directly is often difficult, if not impossible; generally they are dated by reference to the deposits within which they are found, thereby obtaining only a maximum age for the features. Secondly, although regions of modern periglacial activity can be circumscribed by particular isotherms, the occurrence of similar activity in the past can only indicate an upper limit to temperatures at the time, not a lower limit (R. Williams, 1975). Thus, in general terms, permafrost today only occurs in areas where the mean annual air temperature is <–2 °C and it is virtually ubiquitous north of the –6 to –8 °C isotherm in the Northern Hemisphere (Ives, 1974). Evidence of more extensive permafrost in the past, however, only demonstrates that temperatures were below these levels, and provides little information on how much lower. Mapping the distribution of relict periglacial features may indicate how far the southernmost boundary of the permafrost zone was displaced, but within this zone only the limiting maximum paleo-temperature estimates are possible. Nevertheless, periglacial features are of particular interest because they provide information about the periods of extreme temperature depression during past glacial episodes. They also provide information about areas close to the ice-sheet margins, for which there are few other sources of proxy paleoclimate data.

It has already been mentioned that permafrost only occurs in areas with mean annual temperatures below a certain level, but permafrost itself may leave no morphological evidence of its former existence. Paleoclimatic inferences can only be based on features which develop in regions of permafrost and disturb the sediments in a characteristic manner. In this way fossil or relict features can be identified and their distribution mapped (Fig. 7.7). The most useful and easily identified features include fossil ice wedges, pingos, sorted polygons, stone stripes, and

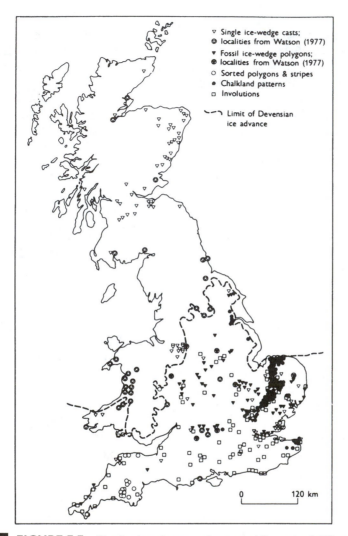

FIGURE 7.7 Distribution of patterned ground of Devensian (= Wisconsin) age in Great Britain (Washburn, 1979b).

periglacial involutions (Washburn, 1979a). The problem is to identify those climatic factors that are necessary for the formation of the features in question; commonly this can only be done in general terms (Table 7.1). Ice wedges, for example, result from thermal contraction at subfreezing temperatures. Winter temperatures of –15 to –20 °C (or less) are required before active frost cracking occurs, but the exact requirements depend on the material being considered. Cracking and ice-wedge formation will occur at higher temperatures in silts and fine-grained material than in gravels where mean annual temperatures of –12 °C may be necessary. Furthermore, the amount of snowfall is a significant factor because

TABLE 7.1   Climatic Threshold Values for the Distribution of Periglacial Geomorphic Features (after Karte and Liedtke 1981)

Periglacial geomorphic features†	Climatic threshold values‡	
	MAT(°C)§	MAP(mm)¤

1 Periglacial geomorphic features whose formation requires permafrost

1.1 Features connected with continuous permafrost Ice-wedge polygons

< -4 to $< -8°$ C

> 50-500 mm

Other climatic indication: rapid temperature drops in early winter

	MAT(°C)	MAP(mm)
Sand-wedge polygons	< -12 to $< -20°C$	< 100 mm
Closed system pingos	$< -5°C$	

1.2 Features connected with discontinuous permafrost
Open system pingos

$< -1°C$

1.3 Features which occur in connection with continuous, discontinuous, and sporadic permafrost

Depergelation forms ("thermokarst" forms, active layer failures, detachment failures, ground ice slumps, permafrost depressions, alas, "baydjarakhs," "dujodas," alas thermokarst valleys, beaded drainage, thaw lakes, oriented lakes, thermo-erosional niches, thermo-abrasional niches, degradation polygons, thermokarst mounds)

$< -1°C$

Other important indication: high ground-ice content

	MAT(°C)	MAP(mm)
Seasonal frost mounds (frost blisters, hydrolaccoliths, bugor)	< -1 to $< -3°C$	
Palsas	< 0 to $< -3°C$	
Rock glaciers	$< +2$ to $< 0°C$	< 1200 mm

Other climatic indications: continental climates with high incoming radiation, sublimation, evaporation and little snowfall

2 Features whose formation requires intense seasonally frozen ground but which also occur in connection with permafrost

2.1 Seasonal frost-crack polygons (ground wedges)

< 0 to $< -4°C$

Other climatic indication: mean temperature of coldest month

$< -8°C$

	MAT(°C)	MAP(mm)
2.2 Frost mounds (thufurs)		
Tundra hummocks (high latitude occurrences)	$< -10°C$	
Earth hummocks (high latitude occurrences)	$< -6°C$	
Earth hummocks (high altitude occurrences)	$< +3°C$	
2.3 Non-sorted circles (mud boils, mud circles)	$< -2°C$	> 400-800 mm
2.4 Sorted circles and stripes (>1 m)	$< -4°C$	

(Continues)

TABLE 7.1 (*Continued*)

2.5 Sorted circles and stripes (<1 m)	< +3°C
2.6 Gelisolifluction microforms (lobes, steps, ploughing blocks)	< −2°C
2.7 Nivation and cryoplanation features (nivation hollows, cryoplanation terraces, frost-riven cliffs)	< −1°C
3 Features which are linked to diurnally frozen ground and needle ice but which also occur in connection with seasonally frozen ground and permafrost	
3.1 Miniature polygons	
3.2 Miniature sorted forms and stripes	< +1°C
3.3 Microhummocks	

† For a description of these features, see Washburn (1979b).
‡ The thermal threshold values represent upper limits for development of features.
§ Mean annual air temperature.
¤ Mean annual precipitation.

snow will insulate the ground surface from the effects of severe cold. This has been demonstrated in many areas where today active ice-wedge formation does not normally occur; where snow is artificially removed (e.g., from roads or airport runways), frost cracks and ice wedges will develop. Paleoclimatic reconstructions based on such phenomena are thus subject to a certain amount of uncertainty, and similar problems have to be faced when dealing with other types of periglacial features. Nevertheless, it is possible to make conservative estimates of temperature change based on the former distribution of different types of periglacial features (Fig. 7.8). Accuracy is really limited by our understanding of the climatic controls on similar contemporary features. From this preliminary map, it would appear that mean annual temperatures in Europe were at least 14–17 °C below recent averages during the maximum phase of the last (Wisconsin/Würm/Weichsel) glaciation (Washburn, 1979b). Although many features used to compile the map are not well-dated, and often are simply considered to reflect conditions during the maximum stage of the last glaciation, it is worth considering the point made by Dylik (1975) that maximum temperature depression was generally not coincident with the maximum extent of ice (maximum "glaciation"). Dylik considers the extent of glacier ice to be more of an index of snowfall (i.e., of cold and humid conditions) than simply of low temperatures. Periglacial features may thus achieve their maximum development during periods of minimum temperature *prior to* the maximum extent of major ice sheets; this may explain the large discrepancy between botanically derived paleotemperatures for the last glacial maximum (sometimes indicating mean annual temperatures only 3–6 °C below those of today) compared with paleotemperature estimates derived from periglacial phenomena.

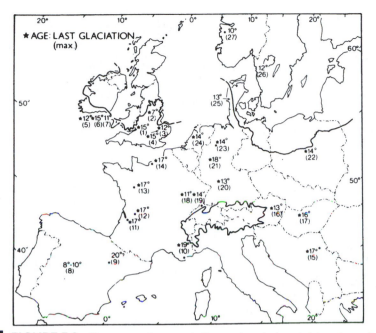

FIGURE 7.8   Temperature increases in Europe since the last glacial maximum, based on periglacial features. Increases are minimum estimates. Maximum ice limits are shown as heavy lines (Washburn, 1979a).

Where a variety of periglacial features of varying ages can be identified in a limited area, it may be possible to reconstruct paleotemperature through time. This has been attempted by Maarleveld (1976) using observations of relict periglacial features in the Netherlands (Fig. 7.9). Maarleveld associated each type of feature with particular temperature constraints; pingo remnants, for example, indicated maximum mean annual temperatures of –2 °C, whereas "extensive coarse snow meltwater deposits" were indicative of a range in mean annual temperature of –5 to –7 °C. Unfortunately, Maarleveld does not identify which sections of his graph are based on estimates of maximum temperature and which are based on a defined temperature range, so the graph may be more precise in some sections than in others. Nevertheless, as a first approximation to paleotemperature reconstruction through time the results compare well with other proxy data series (Fig. 7.9) and indicate the potential value of periglacial studies for paleoclimatic analysis.

7.4 SNOWLINES AND GLACIATION THRESHOLDS

In regions of permanent snow accumulation, it is possible to identify an altitudinal zone separating the lower region of seasonal snow accumulation from the upper region of permanent snow. The term zone is used because from year to year the actual boundary or snowline will vary in elevation, depending on the particular

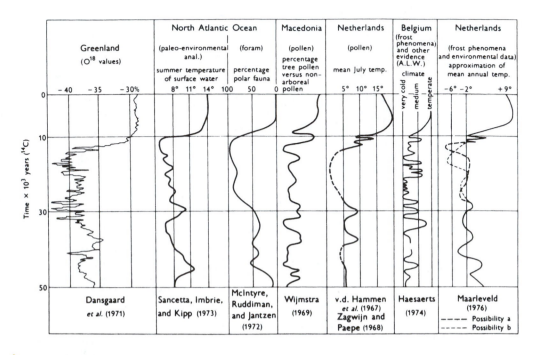

FIGURE 7.9 Paleotemperature reconstruction for Europe based on periglacial features (right column) compared to other long proxy data records (Maarleveld, 1976).

weather conditions during the accumulation and ablation seasons. On a glacier, this would be equivalent to the firn line or (on temperate glaciers where there is no superimposed ice zone) the equilibrium line altitude (ELA). If observations were made over a period of time, the average elevation of the snowline would be apparent, enabling the regional or climatic snowline to be identified (Østrem, 1974). Observations of modern glaciers indicate that the ELA approximates the height at which the accumulation area of the glacier occupies ~70% of its total area. To estimate paleo-snowlines, the common practice is to reconstruct the paleo-ELA by mapping the former glacier area (from moraine position) and then determining where the ELA would have been, based on an accumulation area ratio of ~0.7. This generally corresponds to the uppermost limit of lateral moraines. Mapping the difference between modern and paleo-ELAs constructed in this way can provide useful insights into past climatic conditions. For example, Péwé and Reger (1972) were able to demonstrate that the Arctic Ocean plays no significant role as a moisture source in the present-day glaciation of Alaska, because the snowline gradient along the north coast is not steep and the snowline does not fall to low elevations (Fig. 7.10). By contrast, snowline gradients in the south, adjacent to the Bering Sea, are very steep and increase in elevation rapidly away from the moisture source. A similar pattern of snowlines (though considerably lower) occurred in late Wisconsin times, indicating that the main moisture sources then were like those of today.

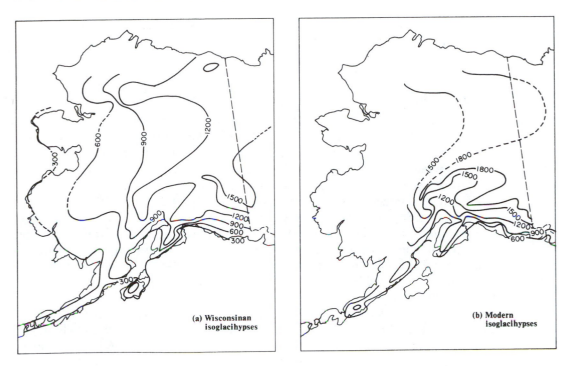

FIGURE 7.10 Modern and Wisconsin snowlines in Alaska (in meters). Snowlines today show the same pattern as glacial age snowlines, suggesting that no major change in moisture sources has occurred. The Gulf of Alaska, not the Beaufort Sea, was thus the most important source of moisture for glaciation (Péwé and Reger, 1972).

A similar index is provided by glaciation levels or glaciation thresholds that define the lowest limit at which glaciers or permanent ice fields can develop (Miller *et al.*, 1975; Porter, 1977). This is usually determined by identifying the highest unglacierized, and the lowest glacierized, mountain summits in a region and averaging the two elevations. Both snowlines and glaciation thresholds can be used in paleoclimatic reconstructions if similar features can be mapped for periods in the past (Osmaston, 1975), though obviously they only provide information about (extreme) glacial periods and can add nothing to our knowledge of warmer intervals. Although much effort has been expended in mapping both modern and former snowlines, with only a few exceptions, paleoclimatic reconstructions have been simplistic and the results often equivocal. This is due mainly to the following problems:

(a) Present-day snowlines have not been adequately studied in relation to present climate; climatic "controls" on modern snowline elevations are not well understood and cannot be assumed to be the same in all areas. Furthermore, atmospheric lapse rates have not been well-documented in mountain regions and are problematic in paleotemperature reconstructions.

(b) Paleosnowline reconstructions are often based on features of varying age, possibly accounting for the large variations in estimates of past snowline lowering (Reeves, 1965; Brakenridge, 1978).

7.4.1 The Climatic and Paleoclimatic Interpretation of Snowlines

It has commonly been assumed that snowlines are related to the height of the summer 0 °C isotherm, and, indeed, Leopold (1951) demonstrated a close correspondence between the two surfaces, in the western United States from 35° to 50° N. Paleosnowlines approximately 1000 m lower than today were thus interpreted as indicating lower summer temperatures. Using modern free air lapse rates of 0.6 °C per 100 m in summer months, Leopold concluded that July temperatures were 6 °C lower than today when snowlines were at the position of maximum depression. Other months were assumed to have cooled proportionately less, with midwinter (January) temperatures having remained the same as those today. As a result, he concluded that mean annual temperatures had been 4–5 °C lower. Using similar logic, but an assumed lapse rate of 0.75 °C per 100 m, Reeves (1965) concluded that the 1300 m lowering of the snowline in New Mexico was equivalent to a decrease in July temperature of 10 °C and in mean annual temperature of 5.1 °C. Considerable doubt was cast on these estimates by Brakenridge (1978), who found no strong relationship between present-day snowline in the American southwest and the July 0 °C isotherm, or indeed any July isotherm, as the latitudinal snowline gradient is considerably steeper. A better fit is obtained with the −6 °C mean annual isotherm; the "full glacial" snowline has a similar gradient, suggesting that there was a fall in mean annual temperature of 7 °C. In the central Andes (10°–30° S) snowlines are actually lower in areas with higher temperatures, because such areas have higher precipitation, which compensates for the warmer conditions; the highest snowlines are found in the cold, arid mountains of the western Altiplano (Sierra Occidental) (Fox, 1991).

These studies illustrate the basic problems of identifying how snowlines vary with temperature and, if they do, what lapse rate should be assumed in order to use former snowlines as an indicator of temperature change. It is probably not a reasonable assumption to use modern lapse rates in such calculations, as both temperature and moisture conditions in the past would have been different from today, perhaps resulting in lower lapse rates in many areas. However, this entire approach is extremely simplistic and neglects other important factors. Snowline is not only a function of lapse rate but also depends on the variation of accumulation with elevation (accumulation gradient), radiation balance, wind speed, humidity, and the variation of albedo with temperature (as temperature influences the frequency of snowfall vs rainfall events) (L. Williams, 1975; Seltzer, 1994).

The importance of precipitation in controlling snowline elevation and spatial variation has been noted for the Cordilleran mountains of both North and South America. In the Cascade Range of Washington, for example, 86% of the variance in glaciation thresholds is explained by accumulation season precipitation (though mean annual temperature is highly correlated with precipitation) (Porter, 1977). By assessing the amount of temperature change during the late Wisconsin (Fraser) glaciation from palynological evidence, Porter estimated that winter precipitation was 20–30% less than today, accompanied by ablation season temperatures 5.5 ± 1.5 °C lower than at present. Extending this approach to the Rocky Mountains, Porter *et al.* (1983) estimated that the lower late Pleistocene snowlines of that region required substantially lower temperatures than a simple lapse rate calculation would suggest (T = –10 to –15 °C vs –6 °C) because conditions were drier at the time. Unless both temperature and precipitation changes are taken into account, erroneous conclusions would be reached. However, such an approach is seen by Seltzer (1994) as only a partial solution because there were undoubtedly also changes in incoming radiation (due to orbital forcing) as well as regional effects due to changes in cloudiness and, hence, in the overall radiation balance. He proposes a model incorporating many of these factors, but estimating how the important variables may have changed remains a major challenge.

In the Andes, snowlines are highest (>6000 m) at the latitude of maximum aridity (Fig. 7.11). "Pleistocene" snowlines were lower by 650–1500 m, the depression being greatest in the hyperarid regions of southern Peru and northern Chile, and

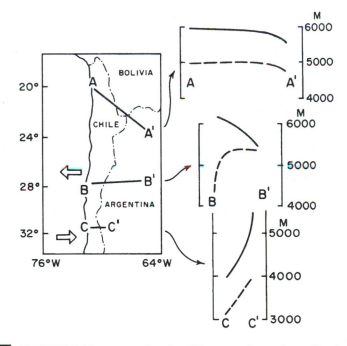

FIGURE 7.11 Modern (——) and Pleistocene (– – – –) snowline elevations along west-east transects in the South American Andes. Principal zonal wind components today are shown by arrows at diagram on left. At the northern and southern locations Pleistocene snowlines were uniformly lower throughout the transect. In the central location (28° S) a reversal of snowline gradient is apparent between Pleistocene and modern times. This resulted from a shift in the subtropical high-pressure cell and associated wind fields, changing from predominantly onshore flow in Pleistocene time to offshore today (Hastenrath, 1971).

least in the equatorial zone. This suggests a general temperature decrease over the entire region, but substantially higher precipitation amounts in the desert zone (Hastenrath, 1967). Further insight is provided by east-west transects across the mountain barrier, which indicate a reversal of modern snowline gradients between 28° and 32° S (Fig. 7.11). This is related to the prevailing circulation around subtropical high-pressure cells centered at ~30° S; prevailing winds are easterly to the north and westerly (onshore) to the south. During glacial periods the lower snowline gradients generally parallel modern snowlines, but at 28° S this is not the case. "Pleistocene" snowlines show a reversal of gradient, suggesting an equatorward shift of ~5° in the boundary between temperate latitude westerlies and tropical easterlies in the lower troposphere. Where westerly wind regimes became established, a marked eastward rise of Pleistocene snowline resulted (Fig. 7.11). Similar reversals of firn line gradient, from modern times to glacial periods, have also been noted in parts of East Africa (Hamilton and Perrott, 1979).

From this brief survey it is apparent that snowlines in different regions are controlled by different climatic parameters and that these must first be understood in order to use paleosnowlines in paleoclimatic reconstructions. However, some assessment of the relative importance of different climatic variables to snowline lowering can be made by the use of an energy balance model that takes into account many of the relevant variables and the interactions between them. For example, such a model has been applied to the question of what climatic conditions were necessary to bring about extensive glacierization of northern Canada (Williams, 1979). By calculating the regional snowline for varying climatic conditions it was possible to determine where perennial snow cover is most likely to have developed in the past (i.e., which areas are most susceptible to glacierization) and the extent of glacierization brought about by different changes in climatic conditions. Interestingly, Williams's model indicates that a substantial fall in summer temperatures (10–12 °C) is necessary to extensively glacierize Keewatin and Labrador, though smaller changes in temperature are sufficient to glacierize Baffin Island, to the north (Fig. 7.12). This confirms the view that Baffin Island is particularly sensitive to climatic fluctuations, and is likely to have been a primary site for ice-sheet initiation in the past (and probably in the future also) (Tarr, 1897; Bradley and Miller, 1972; Andrews *et al.*, 1972). It is also of interest that the model results indicate that increased snowfall has little impact on the areal extent of glacierization; temperature changes seem to be of most significance for regional snowline lowering and glacierization in this region (Williams, 1979).

7.4.2 The Age of Former Snowlines

In the preceding section, reference has frequently been made to "former" or "Pleistocene" snowlines without qualification. However, not all paleosnowlines are defined in the same way, further confusing the paleoclimatic picture. A common method is to estimate the average elevation of cirque floors occupied by glaciers during a particular glacial period (Péwé and Reger, 1972). Alternatively, paleosnowlines may be located at the median altitude between the terminal moraine of a given

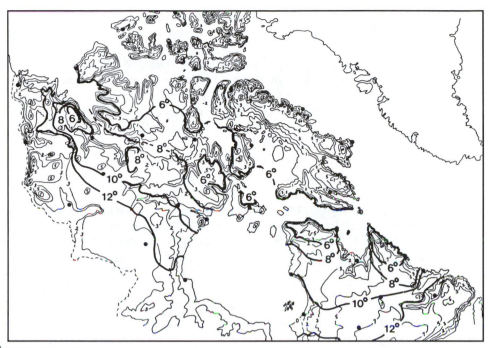

FIGURE 7.12 Boundaries of perennial snow cover predicted by an energy balance model for various amounts of uniform spring and summer temperature decrease, assuming "normal" (1931–1970) snow accumulation and Earth-orbital parameters of 116,000 yr B.P. for 31 March. For example, with a spring-summer lowering of 6–8 °C the perennial snow cover would have been confined mainly to the Arctic Islands and the northern tip of Labrador-Ungava; with a 10–12 °C temperature lowering extensive areas of Keewatin and Labrador would have been glacierized (L. D. Williams, 1979).

advance and the highest point on the cirque headwall (Richmond, 1965). In both approaches, orientation of the cirque is an important factor, not only in terms of radiation receipts, but also in terms of prevailing winds and enhanced precipitation catchment in the lee of mountain barriers. Markedly different paleosnowlines may result from studies based on different cirque populations of varying orientation. Furthermore, the use of cirque floor elevation without knowledge of the age of glacial deposits associated with the feature may lead to a mixed population of paleosnowline estimates. Thus, Hastenrath (1971) is only able to describe the paleosnowlines he mapped as "Pleistocene" and Péwé and Reger (1972) describe their paleosnowlines as "generalized Wisconsin age" features. Such factors may explain the large variations in snowline depression commonly reported; in New Mexico, for example, estimates of regional snowline lowering vary from 1000 to 1500 m (Brakenridge, 1978) and even larger variations are noted in other studies (Reeves, 1965). This imprecision in dating, and uncertainty over climatic controls on snowline elevation, severely limits the value of most snowline studies for paleoclimatic reconstruction. However, careful study of modern conditions and of well-dated glacial deposits can provide important insights into paleoclimatic conditions of mountain regions, as shown by Porter (1977). Indeed, there would appear to be considerable potential in a re-evaluation of this subject for many mountainous parts of the world.

An interesting example is provided by Dahl and Nesje (1996) in their study of Holocene paleoclimate in the mountains of southern Norway. They estimated changes in the size of Hardangerjøkulen from glaciofluvial and glaciolacustrine sediments, relying on the fact that as the outlet glacier advanced and retreated certain thresholds were passed, which were recorded as diagnostic signatures in downstream sediments. This enabled the ice cap area to be estimated, and from that the paleo-ELA could be calculated (Fig. 7.13). Changes in the upper limit of pine (*Pinus sylvestris* L.) provided an assessment of variations in summer temperature, and from these estimates of temperatures at the ELA could be obtained (using a lapse rate of 0.6 °C per 100 m). They then used a well-established relationship between winter accumulation (A) and temperature (t) at the equilibrium line of Norwegian glaciers ($A = 0.915 \ e^{0.339t}$) to reconstruct the change in winter accumulation at the ELA (assuming this relationship has remained constant over time) (Fig. 7.13). Their approach reveals that at certain times in the past (9500–8300, 7200–6200, and at ~5500 and ~4500 [calibrated ^{14}C] yr B.P.) the ice cap grew in size due to higher winter precipitation, in spite of warmer summer temperatures. The major "Little Ice Age" advance (culminating in the eighteenth century) was unusual in resulting from both higher winter precipitation and lower summer temperatures, which perhaps explains why the ELA at that time was lower than at any other time in the Holocene.

7.5 MOUNTAIN GLACIER FLUCTUATIONS[25]

Glacier fluctuations result from changes in the mass balance of glaciers; increases in net accumulation lead to glacier thickening, mass transfer, and advance of the glacier snout; increases in net ablation lead to glacier thinning and recession at the glacier front. A glacial advance therefore corresponds to a positive mass balance brought about by a climatic fluctuation, which favors accumulation over ablation (Fig. 7.14). However, there are many combinations of climatic conditions that might correspond to such a net change in mass balance (Oerlemans and Hoogendorn, 1989) so that evidence of a formerly more extensive ice position does not provide an unequivocal picture of climate at the time. However, if a check on one important variable (such as temperature) is available from an independent source of paleoclimatic data, it may be possible to calculate, or model, the overall change in climate which resulted in the advance or retreat of the glacier terminus (Allison and Kruss, 1977).

Changes in mass balance are not transformed immediately into changes in glacier front positions. There may be a period of down-wasting, during which the glacier

[25] The growth and decay of *continental* ice sheets, and associated changes in sea level and oceanic and atmospheric circulation are fundamental characteristics of the Quaternary Period (for a global overview, see *Quaternary Science Reviews*, 5, 1986 and 9, 2/3, 1990). Such studies provide important regional information on ice extent in response (implicitly) to large scale changes in climate.

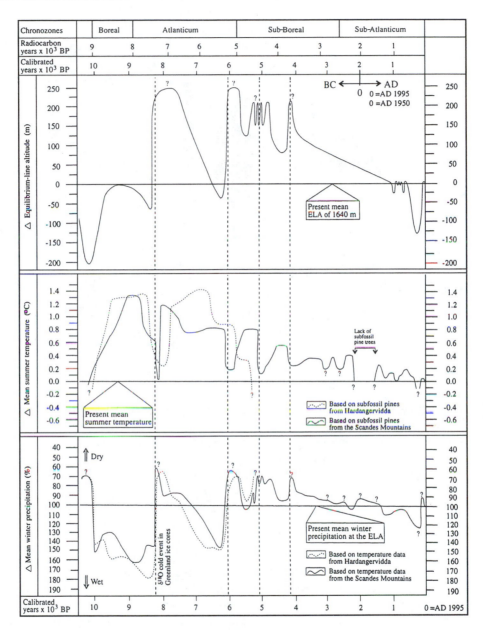

FIGURE 7.13 Variations in equilibrium line altitude (ELA) on Hardangerjøkulen, south central Norway (top panel), summer temperature based on subfossil wood (*Pinus sylvestris* L.) found above modern tree line (middle panel) and winter precipitation (accumulation) derived from modern relationships between accumulation and temperature at the ELA (lower panel: note scale is inverted). Values are expressed relative to modern conditions. Ages are given in calibrated and uncalibrated ^{14}C years at top. Summer temperatures were above present levels for most of the Holocene; glacier advances were associated with periods of higher winter accumulation. Only the "Little Ice Age" experienced both lower summer temperatures and higher winter precipitation, leading to the largest ice advance of the entire Holocene (Dahl and Nesje, 1996).

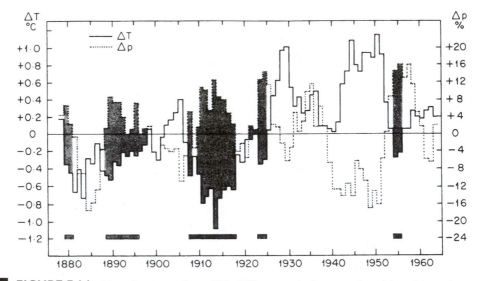

FIGURE 7.14   Mean departures (from 1851–1950 averages) of summer (June, July, and August) precipitation (%) and temperature (°C) at seven stations above 2000 m in the Swiss Alps. Shaded areas indicate times
when precipitation was above average *and* temperatures were below average. These correspond closely to the
times of principal glacier advances in the area (bars at bottom) (Hoinkes, 1968).

loses mass but does not recede. For example, the lower Khumbu Glacier, in the
Mount Everest massif, thinned by ~70 m between 1930 and 1956, but the glacier
snout remained stationary (Müller, 1958). Even when down-wasting is not a significant factor, glacier front positions will lag behind climatic fluctuations. Different
glaciers have different response times to mass balance variations (Oerlemans, 1989).
An increase in net mass results in a kinematic wave moving down the glacier at a rate
several times faster than the normal rate of ice flow. The time it takes for this wave
to reach the glacier snout is the response time of the glacier and depends on a number of factors including the glacier length, basal slope, ice thickness and temperature,
and overall geometry of the glacier itself (Nye, 1965; Paterson, 1994). The South
Cascade Glacier (Washington) for example, has a response time of only about 25–30
years, whereas large ice sheets have response times on the order of millennia.

There is some evidence that an excess of ablation over accumulation may
cause a more rapid response, with ice recession lagging very little behind the climatic fluctuation that caused the mass loss (Karlén, 1980). Glacier front variations
are thus a rather complex integration of both short- and long-term climatic fluctuations, so that one should not be surprised to see some larger glaciers advancing at
the same time that smaller glaciers, with shorter response times, are retreating. Indeed, many Arctic glaciers are still advancing today in response to cooler conditions of the last century, at the end of the Little Ice Age, whereas smaller mid-
latitude alpine glaciers over the same interval have advanced, receded (due to
the early twentieth century world-wide increase in temperatures), and subsequently
readvanced in response to cooler conditions over the last two or three decades.
Glaciers of different sizes and response times in different areas may therefore be

undergoing synchronous advances, but in response to different climatic events; the largest glacier systems respond to low-frequency climatic fluctuations whereas the smaller systems respond to higher-frequency fluctuations.

7.5.1 Evidence of Glacier Fluctuations

The complexity of climatic conditions and differing glacier response times makes glacier fluctuations, as viewed through the misty window of the paleoclimatic records, a rather complicated source of paleoclimatic data. This is compounded by the difficulties inherent in dating former glacier front positions in environments that generally have very little organic material for dating, and where weathering rates are extremely slow.

A record of changes in glacier front positions is generally derived from moraines produced during glacial advances; periods of glacier recession, and the magnitude of the recession, are much harder to identify in the field, though the record of down-valley glaciofluvial or glaciolacustrine sediments may assist in the interpretation of former glacier front positions (Dahl and Nesje, 1996). The problem with the record of moraines is that they are commonly incomplete, with recent advances (which were often the most extensive) obliterating evidence of earlier, less extensive advances. However, detailed stratigraphic studies of glacial deposits may reveal buried soils and weathered profiles indicative of former subaerial surfaces, subsequently buried by more recent morainic debris (Rothlisberger, 1976; and Schneebeli, 1976).

By far the greatest difficulty in the use of glacier front positions as a paleoclimatic index is the problem of dating the glacial deposits. Radiocarbon dates on organic material in soils that have developed on moraines may be obtained, but they provide only a minimum age on the glacial advance. There may be a delay of many hundreds of years between the time a moraine becomes relatively stable and the time it takes for a mature soil profile to develop. If a soil has been buried by debris from a subsequent glacial advance, a date on the soil would provide only a maximum age on the later glacial episode because the organic material in the soil may be hundreds of years older (at least) than the subsequent ice advance (Griffey and Matthews, 1978; Matthews, 1980). Past variations in radiocarbon production rates further impede accurate dating, especially in the last 500 yr or so, encompassing the critically important "Little Ice Age" glacier advances. For older glaciations in volcanic areas, tephrochronology (see Section 4.2.3) or the dating of interstratified lava flows may assist in the interpretation of glacial events (Loffler, 1976; Porter, 1979). Elsewhere, recent advances in cosmogenic isotope dating of exposed surfaces have opened up the possibility of dating the time that has elapsed since an area was ice-covered, or since which a moraine became stabilized (Phillips et al., 1996).

Lichenometry is commonly used to date moraines (see Section 4.3.1) but the technique is uncertain and probably no more reliable than ±20% before 1000 yr B.P.; in most cases, lack of a well-dated calibration curve, or the derivation of a curve based on observations in an environment totally unlike that of the glacial valley or cirque in question, makes the uncertainties even larger. In short, the chronology of glacier fluctuations is quite uncertain in most parts of the world.

7.5.2 The Record of Glacier Front Positions

Studies of glacier fluctuations have been conducted in virtually all mountainous parts of the world (Field, 1975). In spite of the difficulties of dating glacial deposits, mountain regions have provided some of the most comprehensive records of repeated glaciation throughout the Quaternary Period (Richmond, 1985). Glacial deposits from mountain glaciers indicate that there have been at least 11 episodes of major glaciation in the last 1 M years that were (broadly speaking) of global significance (Fig. 7.15) as evidenced by a corresponding enrichment of the oceans in ^{18}O. However, because of the forementioned dating problems, the most detailed work has focused on Postglacial (Holocene) glacier fluctuations. Early work in the Rocky Mountains led Matthes (1940, 1942) to suggest that many alpine glaciers disappeared during a mid-Holocene warm and dry period (termed the Altithermal by Antevs, 1948) only to be regenerated during subsequent cooler and/or wetter periods ("Neoglaciations"; Porter and Denton, 1967). Evidence that mountain glaciers and small ice caps may have disappeared entirely in the early to mid-Holocene is often circumstantial, but nevertheless quite compelling.

The most comprehensive studies have been carried out in Scandinavia, based on lake sediment studies and ^{14}C dated subfossil wood from above modern treeline that provide estimates of summer temperature for much of the Holocene (see Section 8.2.2). It seems likely that many alpine glaciers in the region could not have survived the warmest episodes from 8000 to 4000 yr B.P. (Karlén, 1981; Nesje *et al.*, 1991). Similar conclusions have been reached by other workers elsewhere (e.g., Brown [1990] for the equatorial glaciers of New Guinea). A post–mid-Holocene climatic deterioration is apparent in many areas, resulting in renewed glacierization and glacier advances between 5000 and 4000 yr B.P. (Fig. 7.16). Several episodes of glaciation occurred over the following few thousand years, culminating in the most recent Neoglacial episode (which occurred between the fourteenth and early nineteenth centuries) and which generally resulted in the most extensive Holocene glacial advances. This period is commonly referred to as the Little Ice Age (Bradley and Jones, 1993) and is particularly well-documented in western Europe (Grove, 1988). In the European Alps, glacier advances can be traced through historical records, paintings, and sketches, and this has greatly facilitated the interpretation of glacial deposits in the field. Detailed reconstructions of glacier front positions over the past 300–400 yr have been constructed for several Swiss and Austrian glaciers, most notably the Grindelwald Glacier, Switzerland (Zumbühl, 1980). Using a variety of such sources, periods of both glacier advance and recession can be determined (Fig. 7.17; Messerli *et al.*, 1978). However, such detailed records are rare and the advance/retreat record would have been impossible to construct with only geomorphological evidence to go on.

Because the Little Ice Age advances were most extensive (in many areas greater than at any time since the end of the last major glaciation) the record of

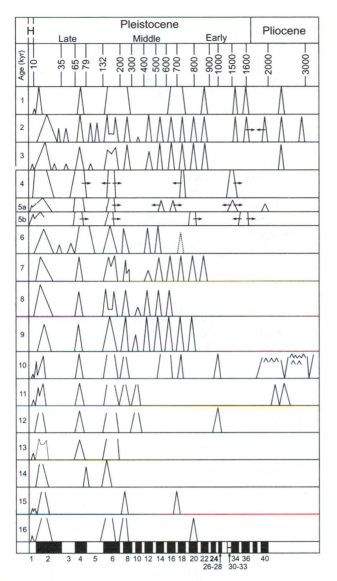

FIGURE 7.15 Summary diagram showing the principal episodes of glaciation in the northern and southern hemispheres over the past ~3 M years based on glacial geologic studies in the various regions indicated. Ice advances are indicated schematically by an upward pointing triangle, the relative dimensions of which signify the magnitude of each ice advance. Note that timescale (shown at top) is nonlinear; marine oxygen isotope stages are indicated at bottom with even-numbered stages (times of major ice accumulation on the continents) shaded (however, note that not all such stages are equal in magnitude — see section 6.3.2). Arrows indicate dating uncertainties; in spite of these, numerous episodes of globally significant ice advances are clearly identifiable. 1 = U.S. Cordilleran ice sheet; 2 = U.S. mountain glaciers; 3 = U.S. Laurentide ice sheet; 4 = Canadian Cordilleran ice sheet; 5 = Canadian Laurentide ice sheet (a = SW margin, b = NW margin); 6 = N.E. Russia; 7 = Poland/western (former) Soviet Union; 8 = N.W. Europe; 9 = European Alps; 10 = Southern Andes; 11 = New Zealand; 12 = Tasmania; 13 = Southern Ocean and sub-Antarctica; 14 = Antarctica (Ross Embayment); 15 = New Guinea; 16 = E. Africa (simplified from charts accompanying Bowen *et al.*, 1986; Clapperton, 1990).

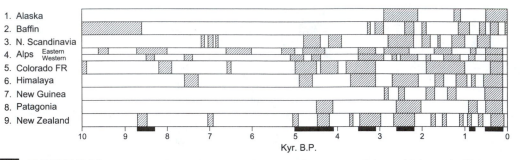

FIGURE 7.16 Summary of glacier expansion phases in different areas of the world during the Holocene. This compilation shows the complexity of the records and the difficulty of discerning worldwide synchronous episodes on this timescale (possible times of widespread advances are indicated by black bars at bottom of figure). This difficulty may be due to climatic fluctuations that are regional, not hemispheric or global in extent, or to poor dating, or to problems inherent in a discontinuous and incomplete data set. A general absence of glacier advances in the early to mid-Holocene is apparent, as is the onset of Neoglaciation after ~5000 yr B.P. (Grove, 1988).

earlier glacial events was often destroyed or buried. This has made worldwide correlations very difficult, compounded by the dating difficulties discussed earlier. Hence, whether there were globally synchronous glacial episodes is somewhat unclear. Certainly there were many periods of alpine glacier expansion throughout the Holocene (see Fig. 7.16) and these varied in magnitude from one area to another. However, on the basis of current evidence, there is little support for the notion that Holocene glacier fluctuations were synchronous throughout the world (Grove, 1979; Rothlisberger, 1986; Wigley and Kelly, 1990) and even less for a 2500-yr periodicity in mountain glacier fluctuations (as proposed by Denton and Karlén, 1973a). Until more detailed studies are carried out (along the lines of those by Patzelt [1974] and Rothlisberger [1976] in the European Alps) perhaps supplemented by the continuous time series offered by palynological and lake sediment studies (Karlén, 1981; Burga, 1988; Leeman and Niessen, 1994b), it seems probable that the record of glacier fluctuations in many areas will remain incomplete.

7.6 LAKE-LEVEL FLUCTUATIONS

Throughout the arid and semiarid part of the world, there is commonly no runoff (surface water discharge) to the oceans. Instead, surface drainage may be essentially nonexistent (in areic regions) or it may terminate in interior land-locked basins where water loss is almost entirely due to evaporation (endoreic regions). In these basins of inland drainage (Fig. 7.18), changes in the hydrological balance as a result of climatic fluctuations may have dramatic effects on water storage. During times of positive water budgets, lakes may develop and expand over large areas, only to recede and dry up during times of negative water balance. Studies of lake-level variations can thus provide important insights into paleoclimatic

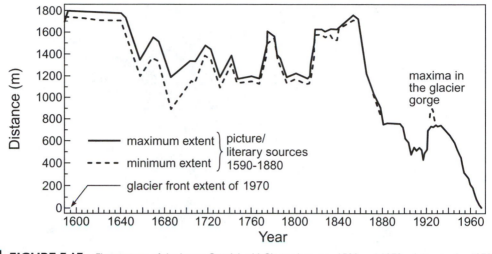

FIGURE 7.17 Fluctuations of the lower Grindelwald Glacier between 1590 and 1970 relative to the 1970 terminal position. The major recession since 1860 (with minor interruptions around 1880 and 1920) is clearly seen. Moraines are indicated by bold curves (Messerli et al., 1978).

conditions, particularly in arid and semiarid areas. In modern lake basins, periods of positive water balance are generally identified by abandoned wave-cut shorelines and beach deposits (Fig. 7.19) or perched deltas from tributary rivers and streams, and exposed lacustrine sediments at elevations above the present lake shoreline (Morrison, 1965; Butzer et al., 1972; Bowler, 1976). Periods of negative water balance (relative to today) are identifiable in lake sediment cores or by paleosols developed on exposed lake sediments (Street-Perrott and Harrison, 1985a). A study of the stratigraphy, geochemistry, and microfossil content of lake sediments from closed basins may be particularly valuable in deciphering lake history (Bradbury et al., 1981).

Most of the early studies of lake-level fluctuations focused on closed lakes where changes in the precipitation-evaporation balance led to volumetric changes and consequent adjustments in lake-level elevation. Recent work has attempted to expand beyond predominantly arid and semiarid regions, where such lake systems are commonly found, to higher latitudes where open (overflowing) lake systems are more common. In open system lakes, water balance changes are naturally compensated for by changes in outflow, but lake levels may also change somewhat, albeit far less than in closed lake systems. Where multiple sediment cores are available from a lake, it may be possible to detect in the sedimentary facies former shallow water conditions and to thereby date periods with lower lake levels (Harrison and Digerfeldt, 1993). Macrofossil and palynological evidence may also provide evidence of changes in lake level via shifts in the relative abundance of shallow and deepwater aquatic flora. Where such detailed studies have been carried out it may be possible to reconstruct water balance changes even in open system lakes (some of which may, of course, have become closed systems in drier periods). Although the vast majority of open system lake sediment studies do not

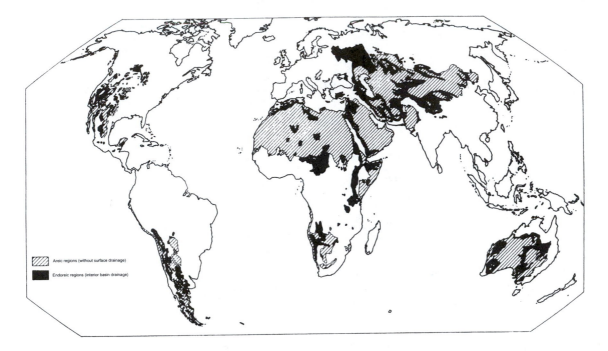

FIGURE 7.18 Areas of endoreic and areic drainage. Areic regions have no permanent surface drainage; endoreic areas are basins of inland drainage (Cooke and Warren, 1977; de Martonne and Aufrère, 1928).

have a suite of cores from deep to shallow water, detailed study of sedimentological, macrofossil, diatom, and palynological data has enabled an interpretation of lake level changes to be made at some locations (Yu and Harrison, 1995; Tarusov *et al.*, 1996). Lake-level changes derived in this way from open system lakes have then been used for large-scale hydrological reconstructions by Harrison (1989) and Harrison *et al.* (1996).

Lake-level fluctuations have been studied in dozens of closed basins throughout the world (for a list of principal works, see Street-Perrott *et al.*, 1983; lake-level data are also available from the World Data Center for Paleoclimatology: see Appendix B). The majority of these studies are stratigraphic and provide only qualitative estimates of climatic conditions. Periods of higher lake levels are commonly described as pluvials, but the question of whether such conditions result from increased precipitation or lower temperatures and more effective precipitation (via reduced evapotranspiration) is controversial (Brakenridge, 1978; Wells, 1979). In an attempt to resolve the controversy, a number of studies have attempted to use the geomorphological evidence, together with empirically derived equations relating climatic parameters today, to make quantitative estimates of paleoclimatic conditions associated with particular lake stages in the past. These can be considered in two general categories: hydrological balance models; and hydrological-energy balance models.

FIGURE 7.19   Late glacial lake-shore terraces, Lake Tauca, Salar Ujuni, southwestern Bolivia. The highest shorelines (~70 m above the present-day lake level, *see arrows*) were formed ~13,000 yr B.P. (photograph kindly provided by C. Ammann, University of Massachusetts).

7.6.1 Hydrological Balance Models

In a closed basin, variations in lake level are a function of water volume, which is in turn a reflection of the balance of water supply and water loss:

$$\frac{dV}{dt} = \frac{d(P + R + U)}{dt} - \frac{d(E + O)}{dt}$$

where V is water volume in the lake; P is precipitation over the lake; R is runoff from the tributary basin, into the lake; U is underground (subsurface) inflow to the lake; E is evaporation from the lake; and O is subsurface outflow from the lake. For any particular lake stage, if the hydrological balance is considered to be at equilibrium, such that

$$\frac{dV}{dt} = 0$$

then, $P + R + U = E + O$. Generally, subsurface inflow and outflow are considered to be negligible and are omitted from the equation, although in some cases they may be substantial; in the case of the Great Salt Lake, Utah, for example, subsurface inflow has been variously estimated at 3–15% of total lake input (Arnow, 1980). In most cases, however, the subsurface components are unknown even for modern lake levels and trying to estimate them for paleolakes would be extremely speculative. Assigning values of zero to these components, the hydrological balance equation for a particular basin is thus:

$$A_L P_L + A_T (P_T k) = A_L E_L$$

where A_L is the lake area; A_T is the area of the tributary basin from which water drains to the lake; P_T is mean precipitation per unit area over the tributary basin; k is a coefficient of runoff (hence $P_T k$ equals the runoff per unit area R_T from the tributary basin); P_L is mean precipitation per unit area, over the lake; and E_L is mean evaporation per unit area, from the lake. As only A_L and A_T are known for any given lake stage, the equation may be rearranged thus:

$$A_L = P_t k$$

$$A_T = A_L E_L$$

A solution of the equation therefore requires a knowledge of precipitation over the lake and adjacent catchment basin, runoff from the surrounding basin, and the amount of water that evaporates from the lake. All of these parameters are a function of many other variables that are also unknown, making a unique solution to the equation extremely difficult to say the least. To illustrate the uncertainties involved, Table 7.2 lists some of the principal factors affecting evaporation and runoff. Of major importance to both parameters is temperature, and this has enabled some limits to be placed on estimates of former runoff and evaporation values when reasonably good paleotemperature estimates are available. If paleotemperatures are known, empirically derived equations relating runoff, evaporation, and temperature (Figs. 7.20 and 7.21) can be used to solve the hydrological balance equation. However, these empirical relationships are often limited in the range of values considered, and constrained by inadequate or even nonexistent data. Consider, for example, the relationship between lake evaporation and temperature. Most empirical relationships are based on standard measurements of evaporation from metal pans 1.2 m in diameter, at different temperatures; lake evaporation is assumed to be less than pan evaporation by a factor of 0.7, based on empirical studies by Kohler *et al.* (1966). However, evaporation rates depend on a number of factors that have not been constant through time (Table 7.2), for example, lake volume and salinity variations. In large lakes, such as Lakes Superior and Ontario, there is a very poor correlation between monthly temperature and evaporation because much energy is used in raising the water temperature at depth (i.e., in heat storage).

Evaporation is at a maximum in fall and winter months when the lake surface eventually becomes warmer than the overlying air (Morton, 1967). Such an effect

TABLE 7.2 Factors Affecting Rates of Evaporation and Runoff

Evaporation	Runoff
Temperature (daily means and seasonal range)	Ground temperature
Cloudiness and solar radiation receipts	Vegetation cover and type
Wind speed	Soil type (infiltration capacity)
Humidity (vapor pressure gradient)	Precipitation frequency and seasonal distribution
Depth of water in lake and basin morphology (water volume)	Precipitation intensity (event magnitude and duration)
Duration of ice cover	Precipitation type (rain, snow, etc.)
Salinity of lake water	Slope gradients; stream size and number

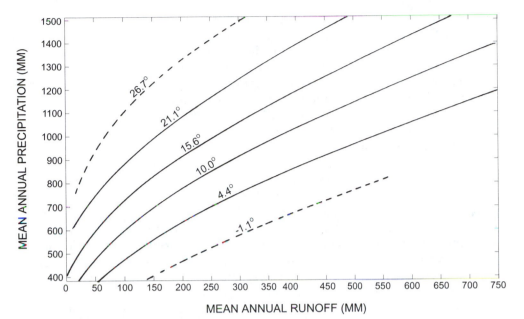

FIGURE 7.20 Relationship between mean annual runoff and mean annual precipitation for areas with different mean annual temperatures (in degrees Celsius). Temperatures are weighted by dividing the sum of the products of monthly precipitation and temperature by the annual precipitation. The quotient gives a mean annual temperature in which the temperature of each month is weighted in accordance with the precipitation during that month. A weighted mean annual temperature greater than the mean, which would normally be computed indicates that precipitation is concentrated in warm months (and vice versa) (Langbein *et al.*, 1949).

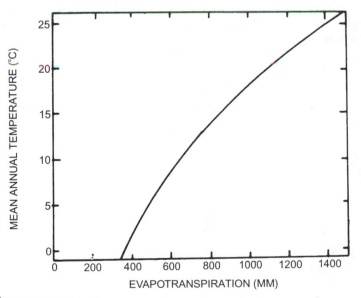

FIGURE 7.21 Relationship between mean annual temperature and evapotranspiration loss in humid areas, based on data from the eastern United States (Langbein *et al.*, 1949).

would be far less significant in tropical lakes, but could be important in midlatitude situations where relatively large lakes developed in the past (e.g., Lakes Bonneville and Lahontan). The formation and duration of ice cover would also significantly affect evaporation rates. At the other end of the spectrum, lakes that are drying up may have extremely high salt concentrations; as salinity increases, the evaporation rate will decline due to a lowering of vapor pressure. For example, in a lake with a salinity of 200‰, evaporation will be only 80% of that from a freshwater lake (Langbein, 1961). Such factors complicate any simple empirical relationship one might derive from instrumentally recorded data and point to the inherent difficulties involved in hydrological balance calculations for paleolakes.

Similar difficulties are encountered with precipitation-runoff relationships (see Table 7.2). Even if precise empirical relationships could be demonstrated, reliable paleotemperature estimates are required. Commonly, these too are fraught with uncertainty, and may, in fact, depend implicitly on assumptions about paleoprecipitation amounts. For example, paleotemperatures derived from studies of snowline depression depend on the assumption that precipitation amounts are similar to those of today. Any increase in precipitation would require a smaller fall in temperature to produce the same amount of snowline depression. Hence, the use of paleotemperature estimates based on snowline depression (Leopold 1951; Brakenridge, 1978) leads to suspiciously circular reasoning. Without accurate paleotemperatures, quite divergent conclusions may ensue. In Table 7.3, for example, three different studies of Paleolake Estancia, New Mexico are summarized. Although each used slightly different approaches and empirical relationships, the fundamental difference in their final paleoprecipitation estimates (ranging from 80–150% of today's values) lies in the different paleotemperatures assumed. The larger the change in temperature assumed, the smaller is the required increase in precipitation (Benson, 1981). Given a sufficiently large decrease in temperature, values of precipitation even smaller than today can be shown to balance the hydrological budget at times of relatively high lake levels (Galloway, 1970).

TABLE 7.3 Paleoprecipitation Estimates from Selected Western U.S. Hydrological Balance Studies

Study area	Author(s)	Paleotemperature change assumed (°C)		Paleoprecipitation/ modern precipitation
		July	Annual	
Lake Estancia, New Mexico	Leopold (1951)	−9	−4.5	1.5
Lake Estancia, New Mexico	Brakenridge (1978)	−8	> −7.5	1.0
Lake Estancia, New Mexico	Galloway (1970)	−10	−10.5	0.86
Spring Valley, Nevada	Snyder and Langbein (1962)	−7	−3.5	1.67
Various, in Nevada	Mifflin and Wheat (1979)		−2.8	1.68

It is unlikely that controversies over paleoprecipitation estimates will be resolved until (a) more detailed studies of modern relationships between evaporation and temperature, precipitation, and runoff are undertaken, to provide more reliable empirical equations, and (b) better (independent) paleotemperature estimates are available. Paleotemperatures calculated from the extent of amino-acid epimerization in the shells of freshwater gastropods of known age may help to resolve this issue (McCoy, 1987b; Oviatt *et al.*, 1994).

7.6.2 Hydrological-energy Balance Models

An alternative to the conventional hydrological balance models already described here has been proposed by Kutzbach (1980), who applied the method to Paleolake Chad in North Africa. Kutzbach utilized the climatonomic approach of Lettau (1969) by considering the hydrological balance of a lake basin in terms of energy fluxes at the surface. In simple terms, a positive hydrological balance results when there is insufficient energy available to evaporate precipitation falling on the basin. Instead of calculating paleoprecipitation amounts from estimates of runoff and evaporation (via paleotemperature estimates) a hydrological-energy balance model utilizes estimates of net radiation and sensible and latent heat fluxes over the lake and tributary basin. Modern values of these components are used, based on measurements from locations thought to characterize the paleo-environments of the basin being studied. Paleotemperature estimates are thus implicit in this "analog" approach; in the Paleolake Chad study, for example, the changes in vegetation that were assumed for 5000–10,000 yr B.P. correspond to an area-weighted fall in mean annual temperature of 1.5 °C, by analogy with areas of similar vegetation today. Precipitation was estimated to have been almost double modern values (~650 mm vs 350 mm today), a result that is similar to previous estimates of precipitation for the area at that time, based on a variety of paleoenvironmental data. Using a similar approach, but with somewhat different assumptions about the magnitude of the important parameters, Tetzlaff and Adams (1983) conclude that precipitation over the Lake Megachad basin was at least three times modern values in order to produce the observed increase in lake size.

Kutzbach's approach to the quantification of past climatic conditions from paleolake studies could be applied to many other lake systems. However, whether it represents a major improvement over conventional hydrological studies is debatable, as it involves at least as many assumptions, and may involve a good deal more (Benson, 1981). Nevertheless, this kind of modeling has the merit of being able to identify, via sensitivity tests, which climatic variables are likely to have been of most significance in bringing about the observed lake level changes.

7.6.3 Regional Patterns of Lake-Level Fluctuations

A number of relatively well-dated stratigraphic and geomorphological studies of lake-level fluctuations have enabled maps of relative lake levels to be constructed for selected time intervals for many parts of the world (Street-Perrott and Harrison,

1985a, 1985b; Harrison, 1993). Although individual errors in dating lake levels no doubt occur, mapping relative lake levels for discrete time periods has the advantage that regionally coherent patterns may be discerned, even if isolated "anomalies" occur. Thus, relative lake levels in much of the arid and semiarid world can be mapped for selected time intervals over the last 25,000 yr. Street-Perrott and Grove (1976, 1979) established a basic methodology that has been widely followed ever since. They identified the total range of lake-level fluctuation at each site, from the level of complete desiccation to the known maximum level (or overflow) and defined three categories: low levels were when lakes were at no more than 15% of their maximum range; intermediate levels were when lake levels fluctuated between 15 and 70% of their range; high levels were when lake levels exceeded 70% of the total altitudinal range. Maps were then prepared to show the spatial distribution of low, intermediate, or high lake levels, so defined, at each time interval (Figs. 7.22 and 7.23). These maps demonstrate remarkable coherence in the spatial and temporal patterns of lake-level fluctuations. During the Last Glacial Maximum (18,000–17,000 yr B.P.) most of the evidence from the intertropical zone (largely dominated by data from Africa) indicates that the area was relatively dry (see Fig. 7.22); only in extratropical regions (the North American Great Basin, in particular) is there an abundance of evidence for extensive lake stages at that time, related to

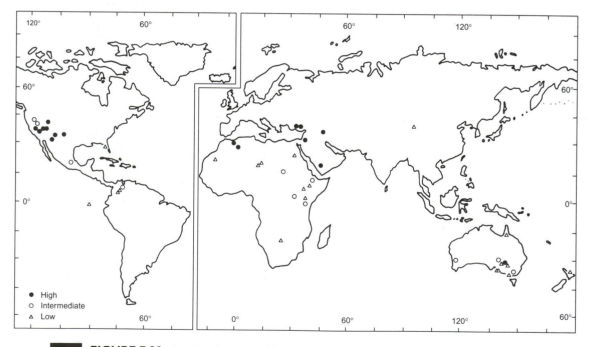

FIGURE 7.22   Lake-level status at ~18,000 yr B.P. Lakes were low over most of Africa, but high in the western United States. See text for definition of high, intermediate, and low lake status (Street-Perrott and Harrison, 1985a).

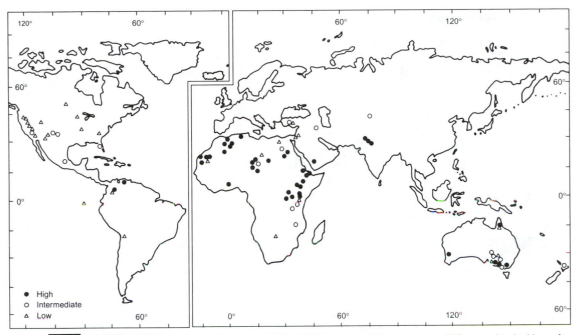

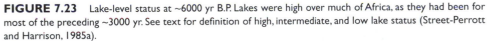

FIGURE 7.23 Lake-level status at ~6000 yr B.P. Lakes were high over much of Africa, as they had been for most of the preceding ~3000 yr. See text for definition of high, intermediate, and low lake status (Street-Perrott and Harrison, 1985a).

the displacement of storm tracks around the southern margins of the Laurentide ice sheet (Webb *et al.*, 1993a). Thus, the so-called "pluvial" climate of the western United States at the glacial maximum is not a viable model for tropical and equatorial regions (Butzer *et al.*, 1972; Nicholson and Flohn, 1980; see Section 9.7.4). During this arid phase, dune systems on the equatorward margins of the Sahara and Kalahari deserts greatly expanded (Sarnthein, 1978); in the southwestern Sahara, for example, dunes even blocked the much-reduced flow of the Senegal River in Mali (Michel, 1973). Low lake levels persisted in these regions until ~12,000 yr B.P., when many lakes began to fill, commonly spilling over from their enclosed catchment basins and initiating new drainage systems. Although maximum lake development throughout the intertropical zone peaked in the early Holocene (Rognon and Williams, 1977; Harrison, 1993; Figs. 7.23 and 7.24), it is of interest that lake levels were generally low in midlatitudes at this time, suggesting a poleward displacement of the Westerlies. In sub-Saharan Africa lake expansion was particularly spectacular, with Lake Chad expanding in area to a size comparable with the Caspian Sea today (Grove and Warren, 1968; Rognon, 1976; Street and Grove, 1976). The high lake phase continued until 4500–4000 yr B.P., interrupted in most of Africa, at least, by an episode of increased aridity and somewhat lower lake levels (though still relatively high) around 7000 yr B.P. (Nicholson and Flohn, 1980). Profound ecological changes accompanied these periods of more effective precipitation,

Global Intertropical (IT)

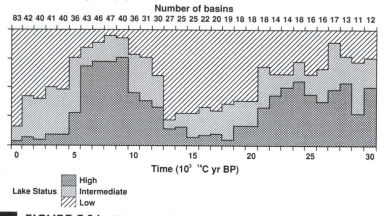

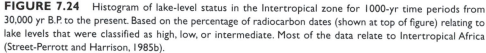

FIGURE 7.24 Histogram of lake-level status in the Intertropical zone for 1000-yr time periods from 30,000 yr B.P. to the present. Based on the percentage of radiocarbon dates (shown at top of figure) relating to lake levels that were classified as high, low, or intermediate. Most of the data relate to Intertropical Africa (Street-Perrott and Harrison, 1985b).

enabling human occupancy and cultural activities to take place in Saharan Africa at a scale almost inconceivable today (see Section 9.7.4). Since ~4500 yr B.P. there has been a relatively steady decrease in lake size, leading to a situation over the last 1000 yr in which lakes over most of the Tropics and lower midlatitudes are virtually all at low stages (Fig. 7.25). In many areas, it appears that lake levels have rarely been lower during the past 20,000 yr.

In spite of the excellent spatial coherence displayed in the maps shown, caution must be used in interpreting lake-level data in this way. Once a lake has desiccated, there is no way of knowing just how much drier the conditions may have been during the period of desiccation. Dry periods are thus likely to be underestimated. Comparison of lake basins of vastly different sizes can lead to erroneous conclusions. Small volume lakes respond much more rapidly to hydrological variations than large deepwater lakes and are likely to record higher frequency climatic variations than large volume lakes. A good analogy to this would be the problems encountered in trying to correlate fluctuations of a small alpine glacier with those of a large ice sheet; clearly the response times of the two systems are different by perhaps an order of magnitude. The problem is not quite as profound in lake systems, however, as mass turnover rates in lakes are rarely longer than a few decades even for very large lakes (Langbein, 1961). Thus, providing a coarse enough time interval is used, and major low-frequency components of hydrological changes are being considered, it should be possible to make broad regional comparisons. When regional patterns show little spatial coherence for a particular interval, it may be that the climate was fluctuating fairly rapidly over a wide range so that no major low-frequency signal dominates the record. Finally, it should be noted that the 15 and 70% category boundaries (for low and high stages) used by Street and Grove (1976, 1979) may represent quite different surface area states, depending on the morphol-

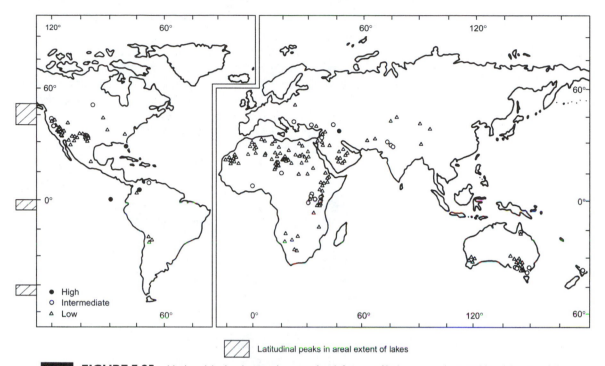

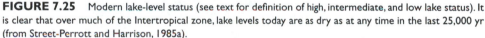

Latitudinal peaks in areal extent of lakes

FIGURE 7.25 Modern lake-level status (see text for definition of high, intermediate, and low lake status). It is clear that over much of the Intertropical zone, lake levels today are as dry as at any time in the last 25,000 yr (from Street-Perrott and Harrison, 1985a).

ogy of the basin. Consider, for example a lake that overflows from a deep narrow basin to a broad shallow plain at some level; the increase in lake depth represents a vastly greater change in surface area (and hence in evaporation from the surface) than a fall in lake level of comparable magnitude. When evaporation from the expanded lake area balances inflow, a new equilibrium is reached. Thus lake surface area is the critical variable controlling lake depth, and this in turn is a function of the basin morphology. Unfortunately, there are not yet enough reliable data on lake area changes to make a global-scale study feasible.

In spite of these caveats, lake-level data from Africa, together with palynological, geomorphological, and archeological data, have enabled a fairly detailed picture of paleoclimatic fluctuations over the last 20,000 yr to be obtained. This led Nicholson and Flohn (1980) to speculate on what the major circulation features over the continent were like at different periods in the past. Figures 7.26 and 7.27 show the principal differences in circulation that they envision during the main arid phase (20,000–12,000 yr B.P.) and the subsequent period of high lake levels (10,000–8000 yr B.P.). Major changes in position of the subtropical high pressure centers are evident in their reconstructions; at 20,000–12,000 yr B.P. a much stronger Hadley cell circulation would have resulted in increased subsidence in the subtropical high pressure cells and intensified upwelling of cooler equatorial water, thereby reducing oceanic evaporation rates in those regions. The seasonal migration of the intertropical convergence zone (ITCZ) would have been greatly reduced, preventing moisture-bearing winds from the Gulf of Guinea reaching southern Saharan regions. Displaced westerly flow (due to a strong baroclinic zone along the ice-sheet margin over northern Europe) would have brought relatively frequent depressions and, hence, relatively moist conditions, to North Africa. By contrast, from 10,000 to 8000 yr B.P. subtropical high-pressure zones may have been displaced poleward as the ice sheet over Scandinavia diminished in size and Equator-Pole temperature gradients (in both hemispheres) were reduced. An increase in interhemispheric temperature differences could have resulted in a northward dis-

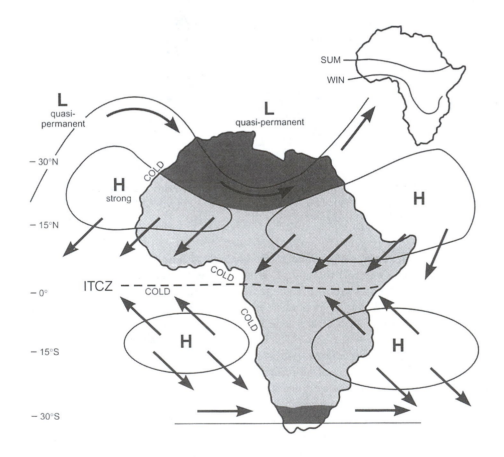

18,000 B.P.

FIGURE 7.26 Conceptual model of atmospheric circulation at ~18,000 yr B.P. (and prevailing circulation pattern from 20,000–12,000 yr B.P.) based on geological and palynological data. Dark shading = areas more humid than today; light shading = areas drier than today. Inset (top right) shows present position of intertropical convergence zone (ITCZ) in summer and winter months (Nicholson and Flohn, 1980).

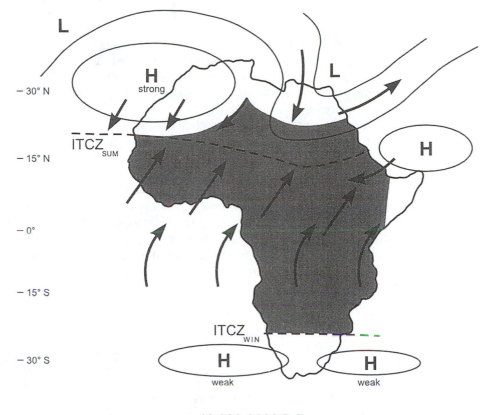

10,000-8000 B.P.

FIGURE 7.27 Conceptual model of atmospheric circulation at 10,000–8000 yr B.P. Dark shading = areas more humid than today; light shading = areas drier than today. ITCZ$_{win}$ refers only to the position over southern Africa (Nicholson and Flohn, 1980).

placement of the ITCZ and increased moisture flux to the continent (facilitated by a warmer equatorial ocean). Evaporation rates from the ocean may have increased by as much as 50% in areas where upwelling of cool water was no longer occurring. Finally, the interaction of upper level troughs with low-level tropical disturbances may have led to increased cyclogenesis and a significant contribution to Sahara rainfall totals from the resultant Sudano-Saharan depressions (Flohn, 1975; Nicholson and Flohn, 1980).

General circulation model experiments by Kutzbach and Otto-Bliesner (1982), Kutzbach and Street-Perrott (1985), and Kutzbach and Guetter (1986) largely support the conceptual models of Nicholson and Flohn. However, their studies indicate that the most important driver of the observed hydrological changes is orbital forcing. At 9000 yr B.P. July radiation from 0° to 30° N was 7% higher (at the top of the atmosphere) compared to today, resulting in an increase in net radiation at the continental surface of ~11%. This led to an increase in the pressure gradient between the oceans and land areas, causing enhanced monsoonal airflow and an increase in precipitation of >4 mm/day in July (compared to today) over much of the Saharan region (Kutzbach, 1983; Street-Perrott and Perrott, 1993). The effectiveness of the moisture increase was further enhanced by reduced winter radiation receipts, and hence less evaporation. Positive feedbacks due to the change in vegetation from desert to grasslands (and associated changes in soils) may also have led to increased precipitation at the boundaries of the arid zone (Kutzbach *et al.*, 1996). These large-scale changes in forcing seem to explain the first-order changes in lake levels and environment recorded by the sedimentary and archeological evidence from Africa. However, there is also much evidence for very rapid changes in lake levels that occurred at a higher frequency than can be accounted for simply by orbital forcing (Street-Perrott and Roberts, 1983; Gasse and Van Campo, 1994). Such changes may be related to sea-surface temperature changes in the Atlantic (Street-Perrott and Perrott, 1990), but further studies of such linkages are needed to fully understand the mechanisms involved.

7.7 LAKE SEDIMENTS

Lakes accumulate sediments from their surrounding environment and so sediment cores recovered from lakes can provide a record of environmental change. Accumulation rates in lakes are often high, so lake sediments offer the potential for high-resolution records of past climate, providing they can be adequately dated. Lake sediments are made up of two basic components: allochthonous material, originating from outside the lake basin; and autochthonous material, produced within the lake itself. Allochthonous material is transported to lakes by rivers and streams, overland flow, aeolian activity, and (in some cases) subsurface drainage. It is made up of varying amounts of fluvial or aeolian clastic sediments, dissolved salts, terrestrial macrofossils, and pollen. Autochthonous material is either biogenic in origin or it may result from inorganic precipitation within the water column (often as a consequence of seasonally varying biological productivity that can significantly alter the water chemistry). Both allochthonous and autochthonous material can be useful in paleoclimatic reconstruction.

Pollen is the most widely studied component of lake sediments and is discussed at length in Chapter 9. Plant macrofossils can be especially helpful in corroborating vegetation reconstructions based on pollen (Hannon and Gaillard, 1997; Jackson *et al.*, 1997). Insect parts, both terrestrial and aquatic, are often found in lake sediments and these may provide additional quantitative paleoclimatic data (see Section 8.3). The character of the inorganic (clastic) sediments transported to lakes can also provide useful paleoenvironmental information, based on sediment geochemistry, grain size variations, magnetic properties etc., though interpretation of such data can be complicated due to post-depositional diagenetic changes in the sediments. In some cases, the thickness of annually laminated sediments (varves) may be of climatic significance, reflecting climatic controls on sediment flux into a lake; this is discussed in what follows.

Biological productivity in lakes is, in part, climatically dependent and so the remains of organisms that lived in the water column can be of paleoclimatic significance. For example, different species of ostracods (small crustaceans, typically ~1 mm in size) are characteristic of specific salinity conditions and so ostracods in lake sediments can be useful indicators of paleosalinity, which may reflect the overall water balance of a lake. Furthermore, geochemical changes in the carbonate shells of ostracods reflect variations in lake water chemistry (De Decker and Forester, 1988; Chivas *et al.*, 1993). Thus, a change from freshwater to brackish (athalassic) conditions, resulting from a shift in the precipitation-evaporation balance, would be reflected in the chemistry of the ostracod shells and/or in the species composition (Holmes, 1996; Xia *et al.*, 1997).

Changes in diatoms (unicellular algae of the class *Bacillariophyceae*) may also reflect water chemistry (Moser *et al.*, 1996). Certain diatoms favor particular salinity ranges so a down-core shift in species composition could reflect a change in water-balance, particularly in arid and semiarid environments (Fritz *et al.*, 1991, 1993; Gasse *et al.*, 1997). In arctic and subarctic freshwater lakes of northwest Canada, modern diatom assemblages have been related to lake water temperature in summer (or to temperature-dependent variables), enabling paleotemperatures to be estimated from diatoms in lake sediments (Pienitz *et al.*, 1995). Oxygen isotopes in the silica of diatoms (biogenic opal) has also been used to derive paleotemperature estimates (Shemesh and Peteet, 1998).

Inorganic precipitation within carbonate-rich lakes can provide an isotopic record that may reflect the varying composition of meteoric waters entering the drainage basin over time (Eicher and Siegenthaler, 1976; McKenzie and Hollander, 1993). For example, in Lake Gerzensee, Switzerland, changes in $\delta^{18}O$ of lake carbonates appear to vary in parallel with isotopic changes seen in the GRIP ice core during late glacial time, suggesting that there were widespread changes in the isotopic composition of precipitation at that time (Eicher, 1980; Oeschger *et al.*, 1984).

In many parts of the world, the annual climatic cycle is the strongest part of the overall spectrum of climate variability. This is commonly reflected in the seasonal deposition of sediments in lakes. However, this cyclicity is only rarely identifiable in lacustrine sediments because a variety of processes within lakes act to mix or disturb the seasonally varying flux of material to the lake floor. In particular, benthic organisms mix the sediments and this may prevent the identification of annually deposited sequences. Where the input of sediment is large enough to overwhelm any benthic disturbance and/or where anoxic conditions exist at depth in a lake (thereby eliminating benthic organisms), the seasonal cycle of sediment deposition may be preserved; the resulting annual layers are called varves[26] (O'Sullivan, 1983; Saarnisto, 1986). Varves

---

[26] In the marine environment, varved sediments are much rarer, but outstanding examples have been found in areas of high biological productivity (upwelling regions) and where deep anoxic basins allow sediment preservation (e.g., the Cariaco Basin off the coast of Venezuela [Hughen *et al.*, 1996b] and the Santa Barbara Basin off southern California [Sancetta, 1995; Pike and Kemp, 1996]).

are most common in cold-temperate environments, particularly in deep anoxic lakes that are frozen over for part of the year and where sediment input is strongly seasonal (Zolitschka, 1996a). In some lakes, varved sediments have accumulated throughout the Holocene and provide a chronological yardstick that can be used in calibrating the radiocarbon timescale (Anderson *et al.*, 1993; Zolitschka, 1991) (see Section 3.2.1.5). Varve thickness variations can provide useful paleoclimatic information, if carefully calibrated against instrumentally recorded climatic data (Zolitschka, 1996b; Overpeck, 1996). For example, detailed studies of Lake C2 in the Canadian High Arctic showed that summer temperature above the local surface inversion controlled sediment flux to the lake; this was confirmed by using long-term climatic data to "predict" how varve thickness would have changed over the previous 40 yr, producing a good fit with the recorded varve thickness variations (Hardy, 1996; Hardy *et al.*, 1996). Studies of varved sediments in the Swiss Alps have also demonstrated the importance of summer temperatures in controlling sedimentation in that region, enabling a late Holocene paleotemperature record to be reconstructed (Leeman and Niessen, 1994a, 1994b). However, in glacierized basins, changes in the extent of glaciers can alter sediment flux downstream, making the interpretation of long-term varve thickness variations more complex (Leonard, 1997).

7.8 SPELEOTHEMS

Speleothems are mineral formations occurring in limestone caves, most commonly as stalagmites and stalactites, or slab-like deposits known as flowstones. They are composed primarily of calcium carbonate, precipitated from ground water that has percolated through the adjacent carbonate host rock. Certain trace elements may also be present (often giving the deposit a characteristic color); one of these, uranium, can be used to determine the age of a speleothem, as discussed in what follows. Deposition of a speleothem may result from evaporation of water or by degassing of carbon dioxide from water droplets. Evaporation is normally only an important process near cave entrances; most speleothems therefore result from the degassing process. Water that has percolated through soil and been in contact with decaying organic matter usually accrues a partial pressure of carbon dioxide exceeding that of the cave atmosphere. Thus, when water enters the cave, degassing of carbon dioxide occurs, causing the water to become supersaturated with calcite, which is thus precipitated (Atkinson *et al.*, 1978).

　　The deposition of speleothems is dependent on a number of factors — geological, hydrological, chemical, and climatic. A change in any one of these factors could cause water percolation to cease, terminating speleothem growth at a particular drip site. However, cessation of speleothem growth over a large geographical area is more likely to be due to a climatic factor than anything else, so dating periods of speleothem growth can provide useful paleoclimatic information (Harmon *et al.*, 1977). Uninterrupted speleothem growth is recognizable in a polished section as a series of very fine growth layers; major hiatuses in deposition are usually marked by erosional surfaces, desiccation and chalkification, dirt bands, and sometimes color changes. In speleothems deposited close to sea level, a rise in relative sea level may result in an overgrowth of marine aragonite on the deposit.

7.8.1 Paleoclimatic Information from Periods of Speleothem Growth

Speleothems grow in sheltered environments that often have escaped the radical surface alterations resulting from glaciation. For example, speleothems dated at up to 350,000 yr B.P. are found beneath the present-day Columbia Icefield of Alberta, whereas the geomorphology of the surface has been repeatedly altered by glacial events. Speleothems may thus provide a long, often continuous record of past environmental conditions. Furthermore, the extensive distribution of karst landscapes (Fig. 7.28) means that studies can be undertaken on a worldwide basis.

Paleoclimatic studies have focused on the timing of speleothem growth periods, their isotopic composition (of both the minerals and fluid inclusions) and their relationship to sea-level fluctuations (Gascoyne, 1992). These are discussed separately in the following sections. Pollen extracted from speleothems may also provide a record of regional vegetation change related to past climatic conditions (Burney *et al.*, 1994).

7.8.2 Dating of Speleothems and the Significance of Depositional Intervals

Speleothems are most commonly dated by uranium-series disequilibrium methods (generally $^{230}Th/^{234}U$) described in Chapter 3. Isotopes of uranium leached from the carbonate bedrock are co-precipitated as uranyl carbonate with the calcite of the speleothems. Normally, the precipitating solution contains no ^{230}Th because thorium ions are either adsorbed onto clay minerals or remain in place as insoluble hydrolysates (Harmon *et al.*, 1975). Thus, providing the speleothem contains no clay or other insoluble detritus, which are carriers of detrital thorium, the activity

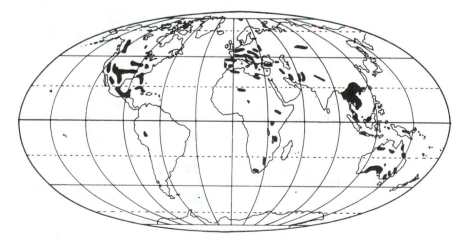

FIGURE 7.28 Distribution of karst in the world, showing the potential sources of paleoclimatic information from speleothems.

ratio of ^{234}U to its decay product ^{230}Th will give the sample age (Harmon et al., 1975). The method is useful over the time range 350,000–10,000 yr B.P. A number of precautions are taken to ensure a reliable age estimate, most notably that any samples containing more than 1% of acid-insoluble detritus are rejected. Also, any indication that recrystallization has occurred (suggesting that the sample may not have remained a closed system) would cause it to be rejected. Recent developments in U-series dating by thermal ionization mass spectrometry (TIMS) have meant that smaller samples (3–5 g of calcite) can be used, yielding more precise dates (a 1 σ precision of <1%) and extending the potentially useful dating horizon to >400,000 yr (Edwards et al., 1987a; Li et al., 1989). The TIMS dating also opens up the prospect of high-resolution studies of samples spanning episodes of rapid environmental change (Baker et al., 1995; Goede et al., 1996). However, the temporal limit of such studies is limited by the residence time of water in the aquifers, which link surface climatic conditions to subsurface speleothem growth (Schwarcz, 1996). This effectively acts as a lowpass filter on the environmental record in speleothems, but if flow rates are high enough annual or even subannual variations may be resolved (Baker et al., 1993; Shopov et al., 1994).

Dating the onset and termination of speleothem growth, using samples from a wide area enables regional chronologies to be built up and may indicate large-scale (climatically related) controls on periods of speleothem growth (Hennig et al., 1983). This is most successful when dating samples from subalpine or subarctic sites that are currently marginal for speleothem growth. During glacial periods, colder conditions, less snowmelt, and more extensive permafrost would result in a marked reduction, even a cessation, of groundwater percolation. Furthermore, decreased biotic activity would lead to a decrease in the partial pressure of carbon dioxide in the soil atmosphere, and therefore less carbonate in solution; consequently speleothem growth might cease (Harmon et al., 1977; Atkinson et al., 1978). Uranium-series dates on alpine speleothems from western Canada, collected at sites that currently have a surface mean annual temperature close to 0 °C and are hence marginal for speleothem growth today reveal four periods of speleothem deposition. These are assumed to represent interglacials, when conditions for speleothem growth were most favorable. Periods of relatively warm climate occurred from ~320,000 to 285,000 yr B.P., from ~235,000 to 185,000 yr B.P., from 150,000 to 90,000 yr B.P., and from ~15,000 yr B.P. to the present (Harmon et al., 1977). In addition, a brief interval, possibly an interstadial, was identified at ~60,000 yr B.P. The earliest period may, in fact, have begun >350,000 yr B.P., representing a long warm interval of massive spleothem deposition (Harmon, 1976) possibly correlative with the prolonged interglacial episode from 460–560,000 yr B.P. represented in China by a very thick paleosol S_5 (An et al., 1987).

These dates compare favorably with periods of speleothem growth observed in caves in the North of England from which a large number of U-series dates have been obtained (Fig. 7.29) (Gascoyne et al., 1983). Speleothems formed extensively during the periods 130–90 ka B.P. and from 15 ka B.P. to the present (Atkinson et al., 1978, 1986a; Gordon et al., 1989). Deposition appears to have ceased entirely from 165–140 ka B.P. and from 30–15 ka B.P., when the area was too cold for water movement in the caves (Gascoyne, 1992). Limited speleothem deposition occurred during an interstadial episode, from ~80–30 ka B.P. It is interesting that these periods of speleothem growth correspond reasonably well with warm intervals noted in the $\delta^{18}O$ record of foraminifera in ocean sediments and also with dates on corals that grew during interglacial periods of high relative sea-level stands (at 10,000 yr B.P. to the present and between 145,000 and 85,000 yr B.P.). The period from ~300–165 ka B.P. seems to have been warm enough for cave waters to circulate, although the larger error bars on many older dates make it difficult to resolve any short cold episodes. More dates on speleothems from selected localities are needed to clarify the record further.

7.8.3 Isotopic Variations in Speleothems

In addition to using periods of speleothem growth as a rather crude index of paleoclimatic conditions, attempts have also been made to use oxygen isotope variations along the speleothem growth axis as an indicator of paleotemperatures. When air and water movement in a cave is relatively slow, a thermal equilibrium is established between the bedrock temperature and that of the air in the cave, approximating the mean annual surface temperature. During deposition of calcite from seepage (drip) water, as CO_2 is lost, fractionation of oxygen isotopes occurs that is dependent on the temperature of deposition. Thus, in theory oxygen isotopic variations in the speleothem calcite ($\delta^{18}O_c$) should provide a proxy of surface temperature through time. Unfortunately, the situation is not quite so simple. First, isotopic paleotemperatures are recorded only if the calcite (or aragonite) is deposited in isotopic equilibrium with the drip-water solution. This can be assessed by determining if $\delta^{18}O_c$ is constant along a growth layer; if values vary for the same depositional interval, it indicates that deposition was affected by evaporation, not just the slow degassing of CO_2; this would alter the simple temperature-dependent fractionation relationship.

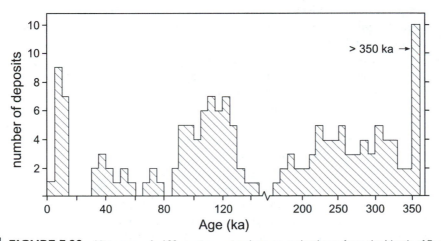

FIGURE 7.29 Histogram of ~180 uranium series dates on speleothems from the North of England (Yorkshire Dales). Speleothem formation is associated with warm interglacial or interstadial conditions when groundwater movement was not impeded by temperatures below freezing. No samples dated between ~140–165 kyr B.P. or from 15–30 kyr B.P., indicating cold, glacial conditions at those times (Gascoyne, 1992).

Another test of isotopic equilibrium involves comparing variations in carbon and oxygen isotopes along individual growth layers (Hendy and Wilson, 1968; Hendy, 1970). If a nonequilibrium situation existed, the isotopic composition would be controlled by kinetic factors and the same fluctuations would be found for both carbon and oxygen isotopes. If no correlation between these two isotopes is found, it can be assumed that the carbonate speleothem was deposited in equilibrium. Some indication of the likelihood of equilibrium conditions being present in the past can be obtained by analyzing the $^{18}O/^{16}O$ ratios in present-day ground water and in calcite deposited from it, which should indicate deposition in isotopic equilibrium.

Interpretation of $^{18}O/^{16}O$ variations in speleothems is not easy because a number of climatic factors other than cave temperature can influence observed $\delta^{18}O_c$ values. First, with a *decrease* in cave temperature, the fractionation factor between calcite and water increases, causing an *increase* in the $\delta^{18}O_c$ values. However, as air temperature at the surface decreases, so the $^{18}O/^{16}O$ of precipitation, and thus the $\delta^{18}O$ value of drip water tends to decrease. Finally, during glacial periods, the growth of ^{18}O-depleted continental ice sheets results in an increase of $\delta^{18}O$ values of oceanic water and hence also of precipitation (see Section 5.2.3). Thus, for a given climatic shift several opposing factors come into play, and it is difficult to assess *a priori* in which direction the $\delta^{18}O_c$ will change (Thompson *et al.*, 1976; Harmon *et al.*; 1978a); indeed the calcite $\delta^{18}O$-temperature relationship may not even be constant through time. This has led to diametrically opposite interpretations of $\delta^{18}O_c$ variations in speleothems. Duplessy *et al.* (1970b, 1971) for example, assumed that measured $\delta^{18}O$ variations were the result of variations in the ^{18}O content of precipitation; hence lower $\delta^{18}O$ values were interpreted as indicating colder conditions. This was disputed by Emiliani (1972), who observed that the speleothem $\delta^{18}O_c$ record, as interpreted by Duplessy *et al.* was the inverse of paleotemperatures derived from oceanic foraminifera. He therefore concluded that the speleothem $\delta^{18}O_c$ variations were not controlled by variations in the $\delta^{18}O$ of precipitation, but due to the dominant effect of temperature-dependent fractionation. This has subsequently been confirmed in other (but not all) localities by the analysis of the isotopic composition of drip water trapped as tiny liquid inclusions as the speleothem grew (Schwarcz *et al.*, 1976). These inclusions vary in abundance and, when present in large amounts (>1% by weight), give speleothems a milky appearance; by isolating and analyzing the liquid at successive levels along the growth axis of a speleothem it is possible to assess, directly, whether isotopic variations of precipitation have occurred (Thompson *et al.*, 1976; Harmon *et al.*, 1979).

Because it is possible that the inclusion water may have continued to exchange oxygen isotopes with the surrounding calcite following its entrapment, it is of no value to measure oxygen isotopes directly. A measure of the oxygen isotope fractionation between calcite and inclusion water would probably give a temperature close to present-day ambient temperature levels. Instead, the deuterium-hydrogen (D/H) ratio is measured because there is no hydrogen in the calcite with which hydrogen in the water might have exchanged. It is assumed that the relationship between δD and $\delta^{18}O$ in drip water approximates that noted in meteoric water by Dansgaard (1964):

$$\delta D = 8\, \delta^{18}O + 10$$

In this somewhat circuitous manner it is possible to estimate the former $\delta^{18}O$ values of meteoric water over very long periods of time. By thus controlling for changes in the isotopic composition of precipitation, it is then possible to use the $\delta^{18}O_c$ values of the surrounding calcite to estimate paleotemperatures. Thompson *et al.* (1976) have carried out such studies on speleothem calcite and inclusion waters from caves in West Virginia. Their results show that the oxygen isotopic composition of inclusion water at this site has changed very little over time, supporting the view of Emiliani (1972) that changes in $\delta^{18}O$ were largely controlled by variations in calcite-water fractionation factors (i.e., temperature changes at the site). Thus, at least in this case, the $\delta^{18}O_c$ values of speleothem calcite *increase* with falling temperatures. Using this interpretation, the $\delta^{18}O_c$ data indicate that West Virginia experienced three major warm episodes in the last 200,000 yr — at <10,000 yr B.P., at 110,000–100,000 yr B.P., and at 175,000 ± 10,000 yr B.P. Cold intervals appear to have occurred prior to 200,000 yr B.P. and at ~180,000, 165,000–110,000, and 95,000–15,000 yr B.P., the last-mentioned period perhaps interrupted by a warmer interval at ~50,000 yr B.P. These records, although incomplete, do show some similarities with other isotopic records from sites in Alberta, Canada; Iowa, Kentucky, and Bermuda (Harmon *et al.*, 1978b) and are in reasonable agreement with marine isotopic records.

A very detailed $\delta^{18}O$ analysis of a stalagmite from northern Norway (at a sampling interval of 20–30 yr for the last 5000 yr) has been interpreted in terms of paleotemperature by Lauritzen (1996) (Fig. 7.30). He "calibrated" the $\delta^{18}O$ variations by reference to both modern temperatures (compared to $\delta^{18}O$ in calcite forming today) and to conditions in the mid-18th century when other data indicate temperatures were 1.5 °C lower. At that time (one of the coldest in the entire Holocene), speleothem growth ceased. These points provided a crude yardstick on which to hang the overall Holocene temperature changes, assuming the $\delta^{18}O$ record primarily reflects temperature-related shifts in the isotopic composition of meteoric waters (precipitation). Some reassurance that this may well be the case is provided by a very similar (though less detailed) Holocene paleotemperature record for west-central Norway, based on entirely separate and independent data (Nesje and Kvamme, 1991). However, further studies are required before this speleothem paleotemperature reconstruction can be accepted with confidence.

Elsewhere, $\delta^{18}O_c$ data indicate that temperature-related effects on the isotopic composition of precipitation are more important than the temperature-dependent fractionation effects in calcite deposition (Dorale *et al.*, 1992). In such cases, $\delta^{18}O_c$ and temperature are *positively* correlated (that is, colder temperatures are indicated by isotopically lighter calcite). Thus, Goede (1994) suggested that declining values of $\delta^{18}O_c$ in a Tasmanian speleothem were indicative of a cooling trend from 98–60,000 yr B.P., followed by an increase in temperatures to ~55,000 yr B.P. (Fig. 7.31). Other proxy data suggest that the overall change in temperature over this interval was ~6 °C, which would mean that $\delta^{18}O_c$ in this cave varied by + 0.26‰ /°C.

The longest isotopic speleothem record currently available is from Devil's Hole, Nevada, where a calcite vein precipitated from groundwater supersaturated in calcite spans almost 0.5 Ma (60 ka to ~560 ka B.P.)(Ludwig *et al.*, 1992). In this system, $\delta^{18}O$ in calcite is considered to reflect changes in the isotopic composition of precipitation feeding the groundwater, so that higher values reflect warmer conditions

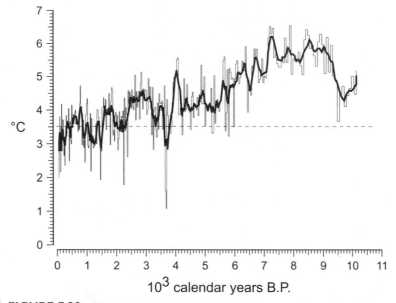

FIGURE 7.30 Paleotemperature reconstruction from oxygen isotopes in calcite sampled along the growth axis of a stalagmite from a cave at Mo i Rana, in northern Norway. Samples were taken every 1 mm at the top, corresponding to a resolution of ~25–30 yr. Linear interpolation between 12 TIMS U-series dates along the stalagmite provide a timescale in calendar years, with an average error of 10–50 yr. The record is very similar to that reconstructed independently from pollen and glaciological data (Lauritzen, 1996; Lauritzen and Lundberg, 1998).

(Winograd *et al.*, 1988, 1992; Johnson and Wright, 1989). The record shows very similar variations to the SPECMAP marine isotope record (Fig. 7.32) as well as to δD in the Vostok ice core suggesting that the signal recorded is of more than regional significance. However, the Devil's Hole record has generated considerable controversy because the timing of the transition from the penultimate glaciation to the last interglacial occurred earlier at Devil's Hole than in the marine isotope record; further the interglacial maximum was ~140 ka B.P. at Devil's Hole, compared to the SPECMAP isotopic minima at ~128 ka B.P. It is worth noting that the Devil's Hole calcite is extremely well dated (by U-series, including many high-precision dates derived by TIMS) whereas the SPECMAP record was tuned to orbital frequencies, assuming orbital forcing was the dominant control on continental ice volume changes.[27] Furthermore, the Devil's Hole calcite was only deposited after precipitation had been transferred through the groundwater system, which must have taken at least several thousand years (and by some estimates >10 ka), making the discrepancy between the two records even larger. Winograd *et al.* (1992) argue that the warming that led to

[27] However, the last interglacial isotopic minimum in marine sediments corresponds to the high sea-level stand, recorded by corals (uplifted by tectonic activity) in several locations around the world and consistently dated at ~125,000 ± 2500 yr B.P. by U-series TIMS (Gallup *et al.*, 1994; Stirling *et al.*, 1995). Direct U-series dating of marine sediments also confirms this time (123,500 ± 4500 yr B.P.) as the last interglacial peak, when continental ice volume was at a minimum (Slowey *et al.*, 1996).

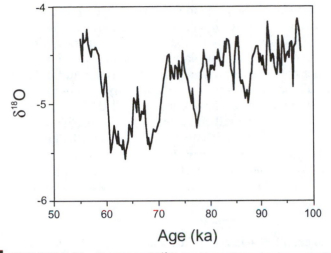

FIGURE 7.31 Variations of $\delta^{18}O_c$ in a Tasmanian speleothem spanning the interval from 98,000–55,000 yr B.P., based on uranium-series dating methods. In this record temperature and $\delta^{18}O_c$ are positively correlated so temperatures generally declined from 98,000 to ~60,000 years B.P.; independent evidence suggests that this temperature change was ~6 °C (Goede, 1994).

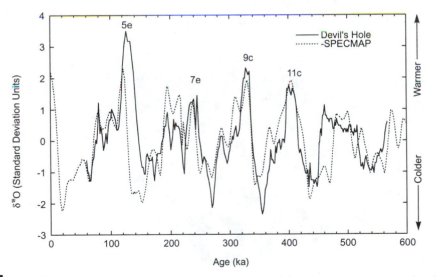

FIGURE 7.32 The $\delta^{18}O_c$ record in a calcite vein from Devil's Hole, Nevada compared to the SPECMAP marine isotope record. Selected marine isotope substages are numbered. Values are expressed as departures from the overall mean of each series, in standard deviation units (computed over the full record). The value of zero thus represents the mean for each record. Note that the sign of the SPECMAP time series has been reversed so that interglaciations appear as peaks (Winograd et al., 1997).

the last interglacial maximum (stage 5e in the marine isotope record) actually began around 150 ka B.P. and eustatic sea level had reached modern levels by ~135 ka B.P. The critical implication of this argument is that sea level started to rise when orbitally driven northern hemisphere insolation anomalies were low (and were actually declining) so that the conventional view of solar insolation forcing deglaciations is thus not tenable. This assault on the Milankovitch hypothesis launched numerous counterarguments, none of which have entirely settled the matter (Johnson and Wright, 1989 and reply by Winograd and Coplen, 1993; Shackleton, 1993 and reply by Ludwig *et al.*, 1992; Edwards and Gallup, 1993 and reply by Ludwig *et al.*, 1992; Imbrie *et al.*, 1993c and reply by Winograd and Landwehr, 1993). Additional precisely dated records are needed to duplicate and establish the global significance of the Devil's Hole record, and to resolve the important questions this controversy has raised (Hamelin *et al.*, 1991).

7.8.4 Speleothems as Indicators of Sea-Level Variations

Speleothem growth in carbonate island locations close to sea level can provide an extremely valuable indicator of former sea-level position, and hence ice volume changes on land. As speleothems must form above sea level in air-filled caves, the occurrence of speleothems in locations presently below sea level provides an upper limit to sea level at the time of formation (Fig. 7.33). Similarly, those speleothems

FIGURE 7.33   Cross-section through a stalagmite (center) and calcareous serpulid overgrowth from the Mediterranean Sea, off Toscana (Central Italy). The stalagmite formed during glacial time (22,670 ± 460 yrs cal BP) when sea-level was up to 120m lower, but was subsequently submerged due to eustatic sea-level rise. Dating the inner parts of the overgrowths on several submarine stalagmites, sampled at different depths, has enabled a well-constrained sea-level history to be established (Alessio *et al.*, 1998). (photograph kindly provided by Fabrizio Antonioli, ENEA-Environmental Dept., Roma Italy).

exposed today that have overgrowths of marine aragonite indicate unequivocally that sea level was formerly higher. Uranium-series dating of the speleothems and coral deposits enables a picture of relative sea-level variations through time to be built up. Thus, Harmon *et al.* (1978b) were able to conclude from studies of Bermuda speleothems that interglacial conditions (high sea-level stands) occurred at around 120,000 and 97,000 yr B.P.; between these events, a lower sea-level stand (–8 m) occurred at ~114,000 yr B.P. Li *et al.* (1989) and Lundberg and Ford (1994) extended this approach by applying high-precision (TIMS) dating to a flow-stone from the Bahamas (at –15 m water depth), which revealed growth hiatuses attributable to pre-Holocene high sea-level episodes at >280 ka, ~230 ka, ~215 ka, ~125 ka, and ~100 ka B.P. Not all hiatuses are attributable to sea-level rise; a reduction in precipitation could limit groundwater flow and lead to cessation of calcite deposition (Richards *et al.*, 1994). Nevertheless, evidence of continuous speleothem growth below current sea level in areas of platform stability is a definitive indication that sea level could not have been above the threshold level during that time. Thus, Richards *et al.* (1994) were able to eliminate the possibility that sea level rose above –18 m at any time between 93 ka and 15 ka B.P. Furthermore, their data constrains sea level at marine isotope stage 5a as having been between –15 and –18 m, based on the onset of speleothem growth at these levels at 93 ka and 80 ka B.P., respectively. Further (submarine) sampling holds the potential of revealing in considerable detail the precise magnitude of former eustatic sea-level changes, which in turn has important implications for interpretation of ice sheet growth and decay rates.

8

NON-MARINE BIOLOGICAL EVIDENCE

8.1 INTRODUCTION

The study of non-marine biological material as a proxy of climate spans a wide range of subdisciplines, of which two are so large that separate chapters must be devoted to them (see Chapter 9, Pollen Analysis; Chapter 10, Dendroclimatology). Here, those topics that have less spatial and/or temporal coverage or that deal with relatively new areas of research will be discussed.

8.2 FORMER VEGETATION DISTRIBUTION FROM PLANT MACROFOSSILS

It is not uncommon for plant macrofossils to be found far beyond the range of the particular species today. Where climatic controls on present-day plant distributions are known, their former distribution may be interpreted paleoclimatically, from dated macrofossils. Fluctuations of three major biogeographical boundaries have been studied in considerable detail using macrofossils: the arctic treeline; the alpine treeline; and the lower or "dryness" treeline of semiarid and arid regions. In each case, the precise definition of treeline poses considerable problems as there is rarely a clearly demarcated boundary. Commonly, there is a gradual transition from mature dense forest through more open, discontinuous woodland to isolated trees or groups of trees, which may include dwarf or krummholz (deformed) forms, particularly in the alpine case (LaMarche and Mooney, 1972). Topoclimatic factors are particularly important in determining the precise limit of trees. It is not necessary to dwell on this at length here, but sufficient to note that the location of the modern treeline is itself often problematic, and may make the interpretation of macrofossils somewhat difficult. For further discussion of the problem, see Larsen (1974) and Wardie (1974).

8.2.1 Arctic Treeline Fluctuations

Macrofossil evidence of formerly more extensive boreal forests has been found throughout the Northern Hemisphere, in tundra regions of Alaska, northern Canada, and the former USSR (Miroshnikov, 1958; Tikhomirov, 1961; McCulloch and Hopkins, 1966; Ritchie, 1987). In addition, paleopodsols (relict forest soils) and charcoal layers (relating to forest fire episodes) have been found in many tundra areas of Keewatin, north-central Canada (Bryson *et al.*, 1965; Sorenson and Knox, 1974). In most areas the macrofossil evidence is episodic in nature, made up of radiocarbon dates on isolated tree stumps located north of the modern treeline. In Keewatin a number of dates (also on organic material in paleopodsols, and on charcoal layers) have enabled a time series of the forest/tundra boundary during the latter part of the Holocene to be constructed (Fig. 8.1; Sorenson, 1977). According to these data, the northern treeline was 250 km or more north of the modern treeline between 6000 and 3500 yr B.P. (Moser and MacDonald, 1990; Gajewski and Garralla, 1992). Less extensive northward migrations occurred around 2700–2200 yr B.P. and 1600–1000 yr B.P. By contrast, the presence of arctic brown paleosols (relict tundra soils) buried beneath more recent podsols south of the modern treeline, suggests that the treeline was at least 80 km farther south around 2900, 1800, and 800 yr B.P. (see Fig. 8.1).

What paleoclimatic significance can be ascribed to such fluctuations? Several authors have noted the correspondence of northern treelines with isotherms of summer or July mean temperatures (Larsen, 1974), so a northward migration of the ecotone may indicate warmer summer conditions. A reconstruction of treeline migration in terms of July temperatures was made by Nichols (1967). Nichols assumed that, when the forest limit moved northward 250 km, July temperatures at the modern treeline were similar to locations 250 km south of the treeline today. In this way, July paleotemperatures were reconstructed for Keewatin, using paleosol, macrofossil, and palynological evidence. Modern treeline is also closely related spatially to the mean or modal position of the arctic front in summer over North America and the median front position over northern Eurasia (Fig. 8.2; Bryson, 1966; Krebs and Barry, 1970). Whether this is a causative factor in the location of the northern forest border or whether the vegetation boundary itself largely determines the climatic differences noted across the vegetation boundary is difficult to assess, although general circulation model experiments show that the treeline has strong feedback effects on climate (Foley *et al.*, 1994). Mid-Holocene warming of the 60–90° N zone due to orbital forcing alone was about 2°C, but experiments with a

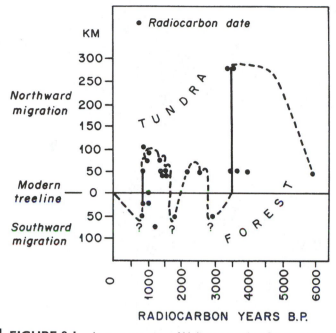

FIGURE 8.1 A reconstruction of Holocene treeline fluctuations in southwestern Keewatin, Northwest Territories, Canada. Treeline position is based on radiocarbon-dated tree macrofossils *in situ* north of the present treeline, and on dates on buried forest and tundra soils north and south of the modern treeline (Sorenson and Knox, 1974).

more extensive boreal forest zone show an additional warming of 1 °C (in summer) and 4 °C in spring. This was the result of higher net radiation due to lower albedo under forest cover (compared to tundra), which was especially critical in the snow-covered spring season.

If air mass boundaries are a determinant of the forest border, then mapping paleoforest limits may provide an important insight into the dynamic climatology of the past (Ritchie and Hare, 1971). Unfortunately, a number of factors make such interpretations difficult. Northward treeline migration during a climatic amelioration is more rapid than southward treeline migration in response to a climatic deterioration. Once established, trees may survive periods of adverse climate, and the treeline will only slowly "recede" as the trees that die are not replaced (see alpine treelines; LaMarche and Mooney, 1967). This process may differ from one location to another. For example, in the forest-tundra zone of northeastern Quebec (east of Hudson Bay), charcoal from formerly extensive coniferous forests indicates that the treeline (i.e., the limit of continuous forest) does not appear to have migrated north-south en masse. Rather, the evidence suggests that the modern tundra-forest ecotone in this area is the product of a formerly more extensive (mid-Holocene) forest that experienced lower temperatures and periodic destruction by fire in the late Holocene, leading to the predominantly tree-less landscape, with the isolated stands

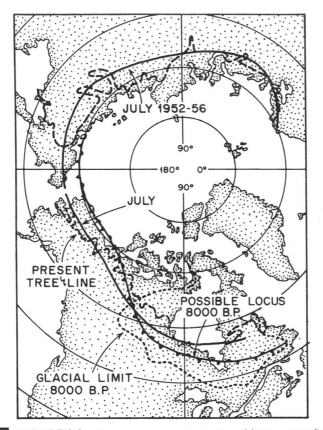

FIGURE 8.2   Modern treeline in relation to modal, mean, or median position of Arctic Front in recent years (for definitions of positions, see Krebs and Barry, 1970; Bryson, 1966). Proposed location of front at 8000 yr B.P. is shown, based on macrofossil and palynological evidence. The 8000 yr B.P. position implies a higher amplitude upper level westerly flow pattern at that time (Ritchie and Hare, 1971).

of trees seen today (Payette and Gagnon, 1985). Trees were unable to reproduce in the cooler conditions after 3000 yr B.P. (and especially from ~650–450 yr B.P.) so the forest-tundra boundary seen today reflects the combined influence of climate and fire history (Payette and Morneau, 1993). Furthermore, detailed studies show the importance of changing growth form in response to climatic change, particularly changes in snow depth. Black spruce (*Picea mariana*) at the northern treeline in Quebec can grow in both upright and prostrate (krummholz) forms, as is typical of many species at their arctic and alpine range limits. In the warmer interval from ~A.D. 1435–1570, spruce was growing mainly as erect trees, but after 1570 cold conditions led to stem-dieback so only the snow-protected krummholz forms survived (Payette *et al.*, 1989). Milder temperatures (and possibly heavier snowfall) in the eighteenth century allowed more shoots to form, but very severe winters from ~1801–1880 again killed many exposed stems (Lavoie and Payette, 1992). Thus, trees at the northern forest boundary may adapt to cooler climatic conditions by adopting a more prostrate growth form, awaiting more favorable conditions to assume erect growth. In such areas, the idea of a north-south migration of treeline is clearly too simplistic. No doubt further studies of ^{14}C-dated macrofossils, paleosols, growth form analysis, and tree ring variations will produce a more coherent picture of paleoclimate at this important ecotone.

8.2.2 Alpine Treeline Fluctuations

In a survey of upper treelines in different climatic zones, C. Troll remarked: "It is absolutely clear that upper timberlines in different parts of the world cannot be climatically equivalent, not even in a relatively small mountain system such as the Alps or Tatra mountains" (Troll, 1973, p. A6). In the paleoclimatic interpretation of macrofossils from above the modern treeline, we must therefore recognize that treeline variations in one area may result from different climatic factors than in another area, and that controls on tree growth may be complex.

In the arid subtropics, treeline is influenced by both temperature and the availability of moisture. In the humid tropics, where seasonal temperature differences are extremely small, the transition to a treeless zone is generally quite abrupt, apparently related to a critical temperature threshold. In mid to high latitudes (particularly in the Northern Hemisphere) treelines are more diffuse, often reflecting strong topoclimatic controls. Climatic studies generally point to summer or July temperature as the major controlling factor in these latitudes (Wardle, 1974), although mean temperatures are only a convenient proxy for the actual controls, which probably involve the frequency and timing of extreme events (Tranquillini, 1993; Holtmeier, 1994). Nevertheless, evidence of higher treeline in the past is generally interpreted as indicating warmer summer temperatures, a lapse rate of 0.6–0.7 °C per 100 m commonly being used to assess the magnitude of temperature change. For example, Dahl and Nesje (1996) estimated that summer temperatures in northern Sweden have varied by ~1.5 °C over the Holocene, based on treeline variations of <220 m (Fig. 8.3). However, Karlén (1976) pointed out numerous problems in interpreting such data, among them the following:

(a) The incomplete nature of the macrofossil record; the highest trees may not have been found, or indeed may not have been preserved. Furthermore, in some areas, mountain summits may not extend far enough beyond the modern treeline to give a maximum estimate on former treeline extent (LaMarche, 1973). Thus, paleotreeline evidence should be considered as providing a minimum paleotemperature estimate only.

(b) The present altitude of the treeline is often not precisely determined and may not be in equilibrium with modern climate (Ives, 1978); a recent history of fire, overgrazing, avalanches, gales, or insect infestation may have resulted in a treeline well below the potential maximum for modern climatic conditions (Griggs, 1938).

(c) Trees take many years to become established during periods of favorable climate and may not have reached their highest position until after the temperature maximum; again this factor would tend to make any paleotemperature estimates based on treeline fluctuations minimum only.

(d) Treeline in some areas may have been affected by regional isostatic uplift following deglaciation of the region; such an effect must be taken into account when assessing the treeline record, particularly from the early Holocene.

It is also worth noting that the time of treeline establishment may be of more significance than the time of tree death (which may not have been climatically related). Denton and Karlén (1977), for example, have dated wood from trees growing above the modern treeline that were killed by explosive volcanic eruptions around 1800 and 1150 yr B.P. Ideally then, one should attempt to date the innermost wood fraction either by radiocarbon analysis, or dendrochronology, or both (LaMarche and Mooney, 1967, 1972). Often, the inner wood has weathered away, or decayed, making this impossible and providing only a minimum estimate on the timing of alpine treeline advances. Karlén (1976) considers that higher treelines in the past probably indicate mean temperatures above contemporary values for periods of 50–100 yr, in order for young seedlings to become established and for the treeline as a whole to "advance." Once established, trees can survive periods of adverse climate that are perhaps as long as favorable climate periods; thus treeline variations are a low frequency record of past climate and are biased towards recording warmer intervals of >50 years in duration.

Bearing all these factors in mind, it is perhaps not surprising to find that the alpine treeline record, considered on a worldwide basis, is extremely complex. Nevertheless, there is evidence that treelines were higher during the early to mid-Holocene in many areas, perhaps reflecting a globally extensive warmer interval. The most comprehensive studies have been carried out in Scandinavia where hundreds of birch, alder, and pine macrofossils from above the modern treeline have been ^{14}C-dated. These show unequivocally that the altitudinal limit of trees was well above modern treeline elevation from soon after deglaciation of the mountains (~9000 yr B.P.) to the mid-Holocene. In northern Fennoscandia the maximum extent of pine was ~5 ka B.P., becoming lower especially after ~3.5 ka B.P. (Eronen and Huttunen, 1987; Karlén, 1993). In central Sweden (Fig. 8.3), pine grew at least 220 m above its mid-twentieth century limit from 9–7 ka B.P., but was replaced by birch as the highest elevation species around 6 ka. At that time pine reached its maximum Holocene abundance, and birch forest extended 200 m above modern limits (Kullman, 1989, 1993). This suggests July temperatures were ~1–2 °C above mid-twentieth century levels (using a lapse rate of ~0.65 °C/100 m). Cooling set in after ~3500 B.P., at which time many glaciers that had completely disappeared in the early Holocene reformed (Kvamme, 1993; Karlén, 1993; Matthews, 1993). This marked the first of several neoglaciation episodes, culminating in the most recent cold episode from the sixteenth–nineteenth century A.D. (the "Little Ice Age").

Detailed studies of dead larch trees above the modern treeline in the northern Urals showed that prior to the Little Ice Age trees grew 60–80 m above the modern limit (from the ninth to the thirteenth century A.D.) but nothing was established after that date until the mid-twentieth century (Shiyatov, 1993). Trees from this medieval warm period did survive into the subsequent colder period, but many died in the late thirteenth and early fourteenth century, in the early 1500s, and in the early and late nineteenth century (Kullman, 1987). This pattern echoes the views of many

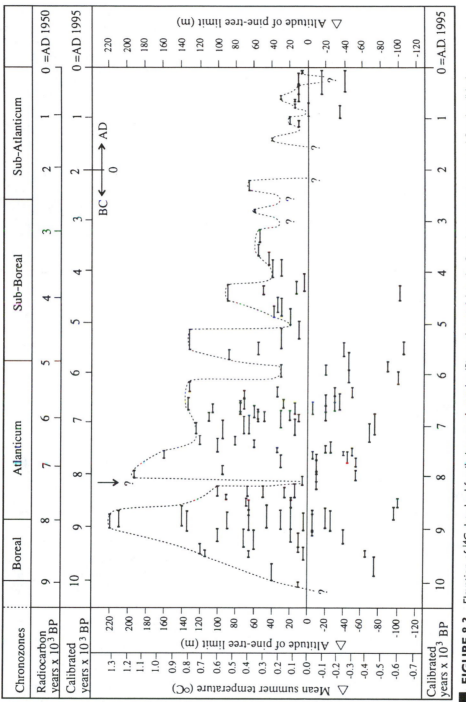

FIGURE 8.3 Elevation of ^{14}C dated subfossil pine wood samples (*Pinus sylvestris* L.) in the Scandes mountains, central Sweden (black bars) relative to modern pine limit in the region (after adjustment for isostatic rebound during the Holocene). Upper limit of pine growth is indicated by the dashed line, but this view may change somewhat as new samples are recovered. Changes in temperature are estimated by assuming a lapse rate of 0.6 °C 100 m⁻¹. Wood samples were ^{14}C dated close to the core to obtain an age near the time of germination (Dahl and Nesje, 1996, based on samples collected by L. Kullman and G. and J. Lundqvist).

paleoclimatologists that there was an earlier onset to the Little Ice Age (in the fourteenth century) and that the subsequent 500 yr experienced a succession of both mild and sharply colder episodes, culminating in the coldest period in the early nineteenth century (Jones and Bradley, 1992; Bradley and Jones, 1992, 1993; Mann et al., 1998).

Alpine treeline studies elsewhere mirror to a large extent the picture emerging from Scandinavia (see e.g., the parallel changes found in the Carpathian mountains by Rybníčková and Rybníček, 1993; and in the Swiss Alps by Tinner et al., 1996). Similarly, in western North America, several studies indicate treeline was above modern limits in the early Holocene. In the San Juan Mountains, Carrara et al. (1991) dated >50 macrofossils of conifer wood from above treeline and found evidence for trees up to 140 m above modern limits from 9600–5400 yr B.P. indicating July temperatures were up to 0.9 °C warmer than present. From 5400–3500 the treeline was near modern limits, then a climatic deterioration led to a decline in treeline after 3500 B.P. This is similar to evidence from the Canadian Rockies (Luckman and Kearney, 1986; Clague and Mathewes, 1989) and from other parts of the western U.S. (Rochefort et al., 1994). In some areas the absence of radiocarbon-dated wood at certain intervals suggests brief colder episodes within the early Holocene but more extensive sampling may revise that picture (Kullman, 1988). Macrofossil evidence is a rather blunt instrument and apparent "data gaps" are difficult to interpret. Karlén (1993) supports his interpretation of cooler episodes with lake sediment evidence, and this points to the need for integration of many different proxies to fully comprehend the complexity of Holocene climates in mountain areas.

8.2.3 Lower Treeline Fluctuations

Throughout the arid and semiarid regions of the southwestern United States, there is an altitudinal zonation of vegetation with xerophytic desert scrub (commonly sagebrush, evergreen creosote bush, and evergreen blackbrush) at lower elevations, grading into mesophytic woodland (juniper, piñon pine, and live oak) at successively higher elevations (Fig. 8.4). The precise elevation of the woodland/desert scrub boundary varies more or less with latitude, being lowest in the Chihuahuan Desert of Mexico and highest in the interior Great Basin of Nevada. This elevational gradient, decreasing to the south, is related to distance from the source of summer moisture (the Gulf of Mexico and the tropical Pacific); the interior Great Basin is both farthest from these source regions of tropical maritime air and isolated from temperate Pacific moisture sources by the mountain ranges to the west (Wells, 1979). The lower treeline elevational gradient thus strongly reflects the importance of moisture for tree growth and for this reason it is sometimes referred to as the "dryness treeline."

Fluctuations of the lower treeline have been greatly facilitated by the analysis of fossil middens of the packrat (genus Neotoma) from caves throughout the southwestern United States (Wells and Jorgensen, 1964; Wells and Berger, 1967). Packrats forage incessantly within a very limited range (~1 ha) of their dens, which are constructed of plant material from the surrounding site. Because of their propensity for

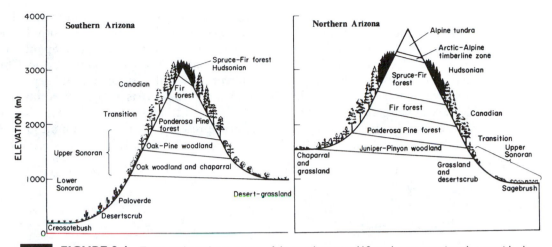

FIGURE 8.4 Transect through mountains of the southwestern U.S. to show vegetation change with elevation. During glacial times, vegetation boundaries were generally lower due to increased effective moisture (resulting from lower temperatures and/or higher precipitation).

collecting items at random, and not simply food stocks, the dens or middens effectively provide a remarkably complete inventory of the local flora (Wells, 1976; Spaulding *et al.*, 1990; Vaughan, 1990). Middens are cemented together into hard, fibrous masses by a dark brown, varnish-like coating of dried *Neotoma* urine (known as amberat). The amberat cements the deposit to rocky crevices in caves and prevents its destruction by fungi and bacteria. Because the cave sites are so dry, *Neotoma* middens may remain preserved for tens of thousands of years; in fact over a thousand macrofossils from *Neotoma* middens in the western U.S. have so far been dated, ranging in age from the late Holocene to >40,000 yr B.P. (Webb and Betancourt, 1990). Studies of rat middens in other arid and semiarid regions of the world are just beginning (see Part IV of Betancourt *et al.*, 1990; Pearson and Dodson, 1993).

Middens are constructed over relatively short intervals, until the rock crevice is filled, so continuous stratigraphic records are not available; rather, they represent samples of vegetation near the site, from discrete time intervals in the past. Macrofossils recovered include branches, twigs, leaves, bark, seeds, fruits, grasses, invertebrates such as snails and beetles, and even the bones of vertebrate animals. Such prolific inventories of macrofossils have enabled quite detailed pictures of the local vegetation around the midden site to be reconstructed, although the material collected may not represent a random sample of plants in the area. Because packrats may collect certain types of material preferentially, middens do not necessarily give a complete picture of the relative abundance of plants in the collection area. Furthermore, different species of packrat have different collection preferences, so sequential occupancy of the same site by different species might give the erroneous impression of a change in local vegetation (Dial and Czaplewski, 1990). In spite of these potential problems, regional comparisons of midden composition for different time intervals have enabled the broad-scale patterns of vegetation change since the last glaciation to be established. Most significantly, the results demonstrate a dramatic increase in the area of piñon-juniper woodland throughout the South during the late Wisconsin period of maximum glaciation (Van Devender and Spaulding, 1979; Van Devender, 1990a). In the Great Basin, Mojave, Sonoran, and Chihuahuan deserts of today (Fig. 8.5), such woodlands are restricted to higher elevations, often on isolated peaks surrounded by vast areas of desert scrubland. However, over periods ranging from >40,000 to ~12,000 yr B.P., *Neotoma* middens from currently hyperarid sites document the presence of juniper or piñon-juniper woodlands over a vastly enlarged area. In the Great Basin, forest vegetation (now

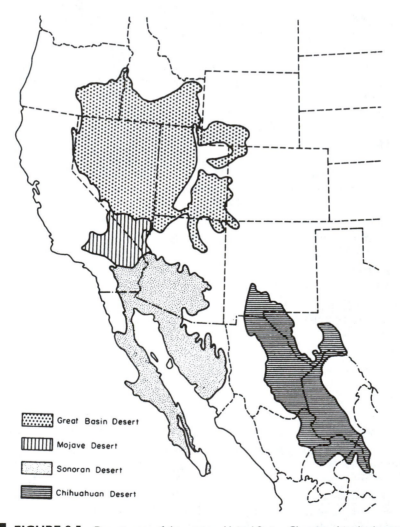

Great Basin Desert

Mojave Desert

Sonoran Desert

Chihuahuan Desert

FIGURE 8.5   Desert areas of the western United States. Changing altitudinal vegetation zonation within desert areas is recorded in fossil packrat middens.

restricted to isolated mountain peaks) extended well below its present range; for example, subalpine conifers grew as much as 1000 m below modern limits (Thompson, 1990). In the Mojave Desert, piñon-juniper woodland occupied terrain down to ~900 m, in areas that today support only thermophilous desert scrub. Similarly, in the Sonoran and Chihuahuan Desert, piñon-juniper-oak woodland extended down to 500–600 m, into areas which are extensive deserts today (Van Devender, 1990a, 1990b). Interestingly, in these southern desert regions the vegetation communities that predominated have no modern analog in the region today that has both forest vegetation and frost-sensitive desert succulents. This suggests a lower frequency of winter cold air outbreaks leading to killing frosts; possibly the forest cover also helped by reducing night-time radiational cooling.

The patterns of glacial period vegetation change in the region all indicate lower summer temperatures, by 6–10 °C, reduced evaporation, and enhanced winter rainfall (by up to 50%) (Table 8.1). This is related to the displacement of winter storms southward into the region, and a reduction in summer monsoon airflow from the Gulf of Mexico. Such conditions appear to have prevailed, with little change, from at least 22,000 yr B.P. to 12,000 yr B.P. Rapid changes in climate and vegetation took place in the following few thousand years, so that by 8000–9000 yr B.P., vegetation patterns had begun to look more like those seen today; the winter rainfall regime had ended, temperatures had risen to within a few degrees of modern levels, and rainfall was largely delivered by summer monsoons (Fig. 8.6). Maximum aridity appears to have occurred in the mid-Holocene in some areas (e.g., the Grand Canyon and Mojave Deserts; Cole, 1990; Spaulding, 1990, 1991) but elsewhere there is evidence for a stronger mid-Holocene monsoon rainfall regime, with drier conditions and the most extensive desert vegetation developing later in the Holocene (Van Devender, 1990a; Van Devender *et al.*, 1994). In the Mojave Desert, evidence for increased effective moisture in the late Holocene (3.8–1.5 ka B.P.) suggests a "neopluvial" of cooler and/or wetter conditions, perhaps correlative with neoglacial episodes in the mountains to the north and east (Spaulding, 1990).

TABLE 8.1 **Estimates of Climatic Conditions in the Western and Southwestern United States During the Last Glacial Maximum, Relative to Today, Based on Fossils Found in Packrat Middens**

Location	ΔT (°C)	Rainfall	Notes	Source
Colorado Plateau	>6.3 (Su)	Higher (Wi.)	Drier summers	Betancourt, 1990
Grand Canyon	~6.7 (Ann)	+24–41%	Wi. rainfall max	Cole, 1990
Great Basin	10 (Ann)	Higher		Thompson, 1990
Sonoran Desert	8 (July)	+50%	Wi. rainfall max	Van Devender, 1990b
Chihuahuan Desert	Strong Su. cooling	Higher	Winter rainfall regime; few freezes	Van Devender, 1990a
Mojave Desert	6 (Ann)	+<40%	Wi. rainfall max	Spaulding, 1990

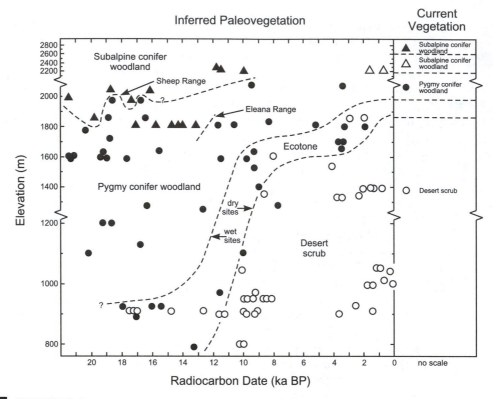

FIGURE 8.6   Changes in major vegetation zones in rocky areas of central-south Nevada over the past 22,000 yr, as recorded by macrofossils in the middens of packrats. Dashed lines indicate the approximate elevation of ecotones. The pygmy conifer (piñon-juniper) woodland/desert scrub transition varied from wet to dry sites, as indicated by the broad ecotone (Spaulding, 1990).

8.3 INSECTS

Insects are the most abundant class of animals on Earth and representatives of the group can be found in virtually every type of environment, from polar desert to tropical rainforest. Naturally this ubiquitous distribution is only possible because of the great diversity of insect types, each of which has adapted to particular environmental conditions. Of overriding significance to the distribution of an individual species is the climate, and in particular the temperature conditions, of an area. Species that are restricted to specific climatic zones are said to be stenothermic, whereas species with less rigorous climatic requirements are eurythermic; clearly the former group are of most value in paleoclimatic reconstructions and it is these on which paleoclimatic inferences are based. It would be unwise, however, to place too much faith in the presence of any particular individual insect as a climatic indicator, as insects are often extremely mobile and inevitably individuals will be blown far from their optimum habitat. More reliable interpretations can be placed on assemblages of insects that are commonly found in associations characteristic of a particular climatic regime. Such assemblages are observed today and it is reasonable to assume that similar fossil assemblages represent similar climatic conditions in the past. In this respect, the approach resembles that of palynology, but in insect studies abundance is not of major significance. Abundance is considered to be more indicative of local conditions rather than the macroclimate, which is of primary interest. It is the characteristic fossil assemblage that provides the climatic information (Coope, 1967). An excellent, comprehensive guide to the study of Quaternary insects and their use in paleo-environmental reconstruction is provided by Elias (1994).

Most paleoclimatic work utilizing insects has involved the study of fossil beetles (Coleoptera; Coope, 1977 a,b), but other insects such as flies (Diptera), caddis flies (Trichoptera), and wasps and ants (Hymenoptera) have also provided additional information (Morgan and Morgan, 1979). Insect fossils are commonly found in sedimentary deposits such as lake sediments or peat, where their chitinous exoskeletons may be extremely well preserved. This is of great value because taxonomic differentiation of the class is primarily based on exoskeleton morphology. Fossils can, therefore, often be identified down to species level by an examination of microscale features in the exoskeleton. One result of this work has been the demonstration of morphological constancy for many species throughout the Quaternary. This is considered to be evidence that they have also exhibited physiological constancy; in other words, they have not altered their ecological requirements, at least over the last 2 million years or so. Although no direct evidence of this can be obtained, the fact that fossil assemblages are often so similar to modern assemblages, in what are assumed to be similar environmental conditions, suggests that radical changes in physiological development have not occurred. This is a fundamental assumption in using fossil insects as paleoclimatic indices, as any change in their climatic tolerances would, of course, invalidate any conclusions that might be drawn from their presence. However, this problem is no different from that facing palynologists or marine microfaunal analysts, and indeed entomologists have considerably more evidence for genotypic stability in their fossils than can be provided in many other branches of biology.

From a paleoclimatic viewpoint, one of the most important attributes of insects is their ability to occupy new territory fairly rapidly following a climatic amelioration. They thus provide a more sensitive index of climate variation than plants, which have much slower migration rates. Indeed, Coleoptera may occupy and abandon a new territory in response to a marked but brief warm interval, whereas there may be no evidence for such an event in the pollen record because of the lag in vegetation response time (Coope and Brophy, 1972; Morgan, 1973). In short, "this combination of sensitivity and rapidity of response to climatic changes, coupled with their demonstrated evolutionary stability, makes the Coleoptera one of the most climatically significant components of the whole terrestrial biota" (Coope, 1977a). Evidence for the extreme mobility of insect populations is found by comparing the modern distribution of a species with its fossil occurrence. For example, *Tachinus caelatus* has been found in glacial-age sediments in Great Britain, but today it appears to be restricted to the mountains of Mongolia where an extreme continental climate prevails (Coope, 1994). Numerous other species from glacial deposits in the United Kingdom are today found only in tundra regions of Siberia.

Thus the insects have responded to climatic change by mass migration, effectively maintaining a more or less constant environment for themselves by shifting geographically as the global climate changes of the Quaternary ebbed and flowed around them.

8.3.1 Paleoclimatic Reconstructions Based on Fossil Coleoptera

A great deal of paleoclimatic work using insects has been carried out in Europe, especially in Great Britain, where the temperate assemblages of Coleoptera today were replaced in the past by an alternation of boreal or polar assemblages during glacial and stadial events, and by more southern or subtropical assemblages during interglacials and interstadials (Coope, 1975a, 1977b). A large number of sites have been studied, ranging in age from interglacial to postglacial (Flandrian). Figure 8.7 illustrates the estimated average July temperature record of the last 120,000 yr, since the last (Ipswichian [= Sangamon = Eemian]) interglacial. July temperatures are assumed to be a major control on insect distribution as the northern limits of most thermophilous (warmth-loving) species more closely parallel July or summer season isotherms than isotherms of winter months (Morgan, 1973). Nevertheless, some estimate of winter temperatures can be made by considering the occurrence of species that are today characteristic of continental Eurasia. A species may be an arctic stenotherm (having a northern distribution) but it may live in continental areas where July temperatures are relatively high and winter temperatures extremely low. Bearing such factors in mind, and considering modern climate in areas where the (fossil) species are found today, it is possible to assess the annual temperature range at intervals in the past (see Fig. 8.7; Coope, 1977b).

Over the last 125,000 yr, there appear to have been three distinct periods when temperatures in central England were at least as warm, or warmer, than they are at present: the Ipswichian interglacial; the Upton Warren interstadial; and the Lake Windermere interstadial. The last interglacial was, by definition, the warmest of these episodes, with Coleopteran assemblages characteristic of southern Europe today, present in lowland England; July temperatures are estimated to have been ~3 °C higher than today (Coope, 1974). Between 50,000 and 25,000 yr B.P. the climate of Great Britain appears to have fluctuated rapidly between temperate and cold continental conditions. This inference is based on the occurrence of climatically contrasting Coleopteran assemblages that follow one another quite abruptly in stratigraphic sequences. The period has been termed the Upton Warren interstadial complex, and includes one brief period (~43,000 yr B.P.) when temperatures seem to have been warmer (by 1–2 °C) than the present day (see Fig. 8.7). The duration of this interval is uncertain (indeed uncertain radiocarbon dates, close to the limit of the method, may account for the apparently rapid temperature fluctuations of this period) but it may have lasted for 1000–2000 yr, followed by a gradual fall in temperature. This cooling was accompanied by more continental conditions, as evidenced by a beetle assemblage typical of parts of Eurasia today; average February temperatures of –20 °C and July temperatures of only +10 °C seem probable. In spite of periods of relative warmth during the "interstadial complex," central England was devoid of trees all the time and there is little palynological evidence for

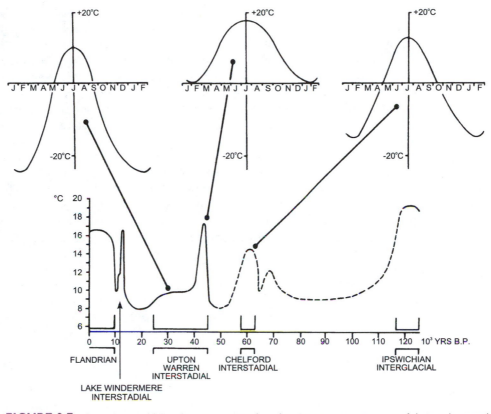

FIGURE 8.7 Reconstructed July paleotemperatures based on insect remains in areas of the southern and central British Isles since the last (Ipswichian) interglacial. Annual temperature ranges are also shown (Coope, 1977b). The period before 50,000 yr B.P. (dashed line) is very uncertain and there may have been a more gradual, monotonic decline in temperature from 120,000 yr B.P. to 60,000 yr B.P.

any climatic amelioration (Coope, 1975b). Evidently the Coleoptera were sufficiently mobile that they could rapidly move northward as the climate improved, whereas certain plants could not migrate northward fast enough to become established in Great Britain before the climate once again deteriorated. A similar situation occurred at the end of the Devensian (Weichselian) cold phase when temperatures again rose abruptly, but for only a relatively short period (the Lake Windermere Interstadial; Coope and Pennington, 1977). At this time, an abrupt change from arctic to thermophilous beetle assemblages took place, with maximum warmth occurring around 12,500–12,000 yr B.P. Shortly thereafter (when from pollen data it is apparent that birch began to colonize the north of England), the interstadial peak of warmth had passed and a significantly cooler episode was already beginning. The newly established birch forest declined and the thermophilous beetle assemblage was replaced by a northern assemblage typical of tundra regions today. By 9500 yr B.P. this sequence had been entirely reversed and thermophilous species again rapidly replaced the arctic stenotherms that had been abundant only

500 yr earlier (Osborne, 1974, 1980). Again, the more mobile insects were in advance of the vegetation and provide a more accurate assessment of paleoclimatic conditions than could be obtained simply from palynological data.

It is perhaps appropriate to note that coleopteran and palynological data do not always appear to be out of phase; such situations are probably the exception rather than the rule. The Chelford interstadial (radiocarbon dated at ~60,000 yr B.P.), for example, must have lasted long enough for trees to migrate northward into Great Britain following the cold, early Devensian period. Coleopteran assemblages are in complete accord with palynological evidence for a cool but quite continental climate at this time (see Fig. 8.7); conditions in central England were similar to those in southern Finland today (Simpson and West, 1958; Coope, 1959, 1977b).

A more rigorous, quantitative approach to paleoclimatic reconstruction has been applied to coleopteran fauna from a set of ^{14}C-dated sites in Great Britain, spanning the last 22,000 yr. The "Mutual Climatic Range" method is based on climatic conditions found across the modern range of particular species, which defines their tolerance limits (Grichuk, 1969). The climate of locations where several fossil species once coexisted is defined by the overlapping range of climatic conditions that is compatible with their occurrence together in one place (Fig. 8.8). For beetles, this "mutual climatic range" is defined in terms of the mean temperature of the warmest and coldest months (Atkinson *et al.*, 1986b, 1987). Figure 8.9 shows the temperatures reconstructed in this way for Great Britain (50–55° N) for the last

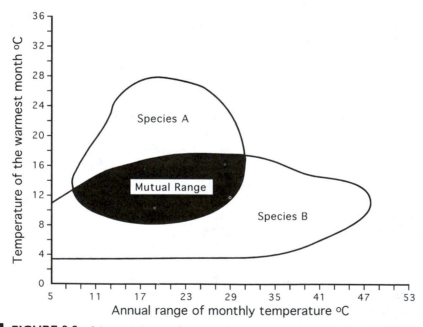

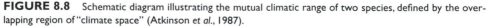

FIGURE 8.8   Schematic diagram illustrating the mutual climatic range of two species, defined by the overlapping region of "climate space" (Atkinson *et al.*, 1987).

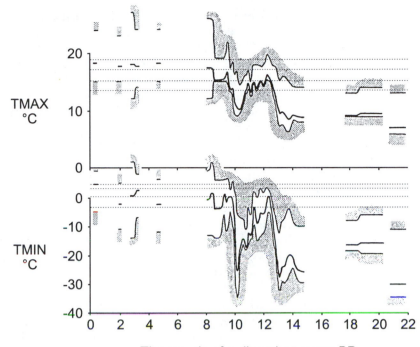

Thousands of radiocarbon years BP

FIGURE 8.9 Mean temperature of the warmest and coldest months of the year (TMAX and TMIN) for intervals over the last 22,000 yr, according to beetle remains in sediments from various sites across the British Isles, calibrated using the mutual climatic range method. The dark center lines give the best estimate of temperature, and the upper and lower lines, the extreme ranges of estimates (based on the average of samples of similar ages). The inner horizontal lines give the range of decadal mean temperatures, in the warmest and coldest months, recorded in central England over the interval A.D. 1659–1980; the outer horizontal lines give the range of warmest and coldest individual years over the same interval (Atkinson et al., 1987).

22,000 yr, compared to the observed range of temperatures in central England from A.D. 1659–1980. Of particular note is the extremely low winter temperatures during the last Glacial Maximum (22–18 ka B.P.) and also from ~14.5 to 13 ka B.P., when the mean temperature of the coldest month was –16 °C and <–20 °C, respectively. Such low temperatures must certainly have been associated with extensive sea ice in the Atlantic, west of Great Britain at those times, or the oceanic influence would have produced a much more moderate climatic regime. From 13.3–12.5 ka B.P., rapid warming took place (by +25 °C in winter and +7–8 °C in summer), indicating northward migration of the sea–ice front at that time. For a short interval around 12,000 yr B.P., temperatures were similar to those of today, but in the ensuing Younger Dryas cold event the region was plunged back into glacial-like conditions, followed by abrupt warming to a more equable maritime climate by the early Holocene. One note of caution is required regarding the apparent abrupt changes: radiocarbon "plateau x" (see Section 3.2.1.5) make the definition of a precise chronology difficult at certain times (especially around 10,000 ^{14}C yr B.P.). This may tend to exaggerate the rapidity of environmental changes at these times.

The longest insect-based paleotemperature reconstructions using the mutual climatic range approach are those of Ponel (1995), who examined insects in cores from the Grande Pile (France). This record extends back to before the last interglacial and provides insect-based paleotemperature estimates from ~135 ka to ~25 ka B.P. (see Figs. 9.16 and 9.20). The analysis indicates that in the coldest part of the last glaciation, mean temperatures in the warmest month were ~10 °C lower than in the Eemian optimum, and in the coldest month of the year they may have been >20 °C colder in glacial times than in the last interglacial.

The mutual climatic range approach has not yet been widely employed beyond Europe, though applications to North American data have begun (Elias, 1996; Elias *et al.*, 1996a). A number of sites that have been studied in North America are summarized in Elias (1994). However, it has not yet been possible to reconstruct long-term temperature variations in any detail, as Coope has done for Great Britain. This is due primarily to two factors: (a) The insect fauna of North America is significantly larger than that of Europe and systematic relationships between the modern faunal elements are not as well known. (b) The distribution and ecology of modern insects are also not well known in North America and many areas remain entomologically unexplored (Ashworth, 1980; Morgan and Morgan, 1981). Consequently, it is more difficult in North America to identify paleoenvironmental conditions precisely by fossil insect faunal assemblages. As more studies of both modern and fossil assemblages are carried out, this situation should improve significantly. Similar database problems exist elsewhere, but important results are nevertheless possible in some areas. For example, studies of late glacial sites in Chile suggest that there was no Younger Dryas episode in that area, shedding new light on an on-going controversy (Hoganson and Ashworth, 1992).

8.3.2 Paleoclimatic Reconstruction Based on Aquatic Insects

In studies of lake sediments, certain aquatic insects have proven useful in paleoclimatic reconstruction. Midge flies (Order: Diptera; Family: Chironomidae) can be identified by the characteristic chitinous head capsules that are often preserved in sediments (Hofmann, 1986; Walker, 1987). Walker *et al.* (1991a) showed that assemblages of chironomids in a suite of lakes from Labrador are related to the surface water temperature of the lakes in summer. Although the relationship may involve other factors (Hann *et al.*, 1992) down-core analysis of chironomid remains provides an estimate of former lake surface temperatures (Walker *et al.*, 1991b). Applying this relationship to chironomid remains in lake sediments from Maine, Cwynar and Levesque (1995) found strong evidence of a pronounced climatic reversal, which they correlated with the Younger Dryas episode (Fig. 8.10). Changes in chironomid assemblages indicate an abrupt drop in temperature of ~10 °C at around 11,000 yr B.P., with temperatures then remaining low for 1–1.5 ka, followed by a similarly abrupt warming phase. This change has not been recognized in pollen data from the region (though there was an increase in *Alnus* and *Picea* at that time) so in this case the chironomid record appears to be a more sensitive climatic indicator and adds

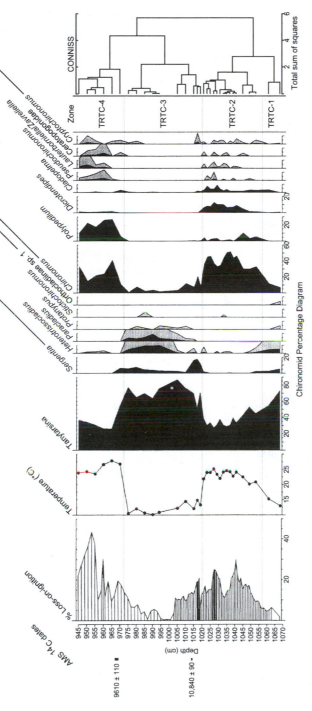

FIGURE 8.10 Chironomid percentage diagram for the late Pleistocene/early Holocene section of a sediment core from Trout Pond, Maine. Radiocarbon dates are indicated on the left. The down-core changes in chironomid taxa have been calibrated in terms of summer lake surface temperature, providing the paleotemperature estimates in the second column. The pronounced drop in temperature from ~11 to 9.8 kyr B.P. is correlative with the Younger Dryas episode, widely seen in European sediments (Cwynar and Levesque, 1995).

new insight into conditions prevailing in the region at the late Pleistocene/Holocene transition. Brooks *et al.* (1997) also found that the chironomid record from southeastern Scotland provided considerably more insight into rapidly changing late glacial conditions than could be obtained from pollen data alone.

In certain situations, chironomid remains may be indicative of salinity conditions in a lake, so that some insight into the precipitation-evaporation balance (P-E) over time may be possible, provided an appropriate set of lakes is selected. Studies of lakes with a wide range of salinity in a small area of British Columbia demonstrate a strong statistical relationship between water salinity and chironomid assemblages (Walker *et al.*, 1995). Hence, in areas where pronounced changes in salinity are likely to have occurred, chironomids may be useful paleo-ecological indicators. However, in most situations they will be of secondary importance to diatoms, which are more diagnostic of salinity conditions (Fritz *et al.*, 1991).

Apart from chironomids, the use of aquatic insects in paleo-environmental reconstruction has been somewhat exploratory. Williams and Eyles (1995) suggested that the occurrence of different caddisflies (Trichoptera) in deposits of the last interglacial and early Wisconsin age from near Toronto, Canada indicate a change in temperature of 4–5 °C from 80 to 55 ka B.P. This estimate is based on modern climatic conditions in the present day distribution of the different species identified. Such studies provide complementary information to support more comprehensive paleo-environmental reconstructions (N. E. Williams *et al.*, 1981).

9

POLLEN ANALYSIS

9.1 INTRODUCTION

Every year millions of tons of organic material are dispersed into the atmosphere by flowering plants and cryptogams (plants without true flowers or seeds) in an effort to reproduce. The higher plants (angiosperms and gymnosperms) produce pollen grains containing the male genetic material; sexual reproductive success is assured only if this material reaches a female receptacle of the same plant species. The lower plants or cryptogams produce spores containing the necessary genetic material for the growth of an independent generation of plants. Pollen grains and spores are the basis of an important aspect of paleoclimatic reconstruction — pollen analysis, or palynology, the study of pollen and spores.[28] Where pollen has been preserved over time, in lakes, bogs, estuaries etc., it provides a record of past vegetation changes that may be due to changes of climate. Pollen analysis is one of the most important branches of Quaternary paleoclimatology, providing information from the continents to complement that derived from marine sediments and ice cores.[29] Pollen

[28] As pollen grains have been studied far more than spores for reconstructing past climates (i.e., the emphasis has been on higher plants; see Table 9.1), the following sections will focus on pollen grains. However, from a methodological viewpoint, most of the problems of studying pollen grains apply equally to the study of spores.

[29] In some cases, pollen extracted from marine sediments has enabled direct correlations to be made between terrestrial and marine records, providing a check on the terrestrial chronologies (Hooghiemstra *et al.*, 1992, 1993). Pollen has also aided in the chronological interpretation of tropical ice cores (Liu *et al.*, 1998).

records from some sites span the entire Quaternary; more commonly they span the Holocene and/or late glacial period. Because pollen records from lakes and bogs are more or less ubiquitous, they can provide an important link between the higher resolution tree-ring records of the late Holocene and other longer but lower resolution records from the land and oceans (see Table 1.1).

An important consideration in pollen analysis is the concept of scale (Webb, 1991; Bradshaw, 1994). Vegetation changes occur on a variety of scales (both temporal and spatial). Not all such changes are necessarily due to a change in climate; fire, insect infestation, plant successional changes, and interference by man, as well as changes in factors leading to the accumulation and preservation of the fossil material itself often make interpretation of the pollen record complex. However, by selecting the appropriate spatial and temporal scale for analysis, pollen studies can isolate the important climatic signal from the nonclimatic noise (Fig. 9.1). It is here that the most important advances in Quaternary palynology have been made over the last 20 yr. This chapter focuses on some of the methods used to achieve this goal of quantitative paleoclimatic reconstruction from pollen, preserved mainly in lakes and bogs. Studies that pertain to the reconstruction of past vegetation composition or plant succession *sensu stricto* are not dealt with in any detail (Huntley and Webb, 1988; Jackson, 1994). Chapter 9 concludes with some examples of palynological studies from different regions that have shed light on important paleoclimatic problems. The emphasis in this final section is on long records, which provide information about past climatic changes on the continents at mid and low latitudes, for comparison with long ice-core and marine sedimentary records.

9.2 THE BASIS OF POLLEN ANALYSIS

Paleoclimatic reconstruction by pollen analysis is possible thanks to four basic attributes of pollen grains: (1) they possess morphological characteristics that are specific to a particular genus or species of plant; (2) they are produced in vast quantities by wind-pollinated plants, and are distributed widely from their sources; (3) they are extremely resistant to decay in certain sedimentary environments; and (4) they reflect the natural vegetation at the time of pollen deposition, which (if viewed at the right scale) can yield information about past climatic conditions.

9.2.1 Pollen Grain Characteristics

Pollen grains range in size from 10–150 μm and are protected by a chemically resistant outer layer, the exine. Because pollen grains of many plant families are different morphologically, they can be recognized by their distinct shape, size, sculpturing, and number of apertures (Fig. 9.2). In some cases, identification to species level may be possible (Faegri and Iversen, 1975; Moore and Webb, 1978). The exine is made of sporopollenin, a complex polymer resistant to all but the most extreme oxidizing or reducing agents. Thus, the organic or inorganic matrix in which the pollen grains are trapped can be removed by chemical means without de-

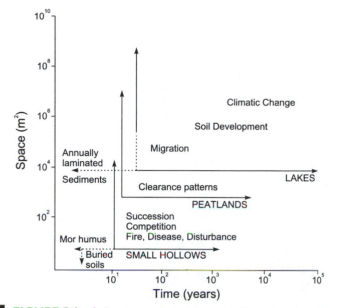

FIGURE 9.1 Scale is important in defining the information that can be obtained from pollen analysis. Information about climate is obtained at the larger temporal and spatial scales, mainly from lakes and peatlands. At smaller spatial and temporal scales, nonclimatic effects dominate the pollen signal (Bradshaw, 1994).

stroying the pollen itself. There is some evidence, however, that in certain sedimentary environments not all pollen grains will be equally well preserved (Cushing, 1967). For example, pollen grains are more subject to corrosion in moss peat than in silt deposits and this may be due to the activities of phycomycetes, bacteria, and other micro-organisms. Furthermore, the pollen of some species (e.g., *Populus*) may begin to disintegrate even before reaching a deposition site (Davis, 1973).

9.2.2 Pollen Productivity and Dispersal: the Pollen Rain

All plants that participate in sexual reproduction produce pollen grains, dispersing them by various mechanisms in an endeavor to reach and fertilize the female reproductive organs of other plants. The amount of pollen produced is generally inversely proportional to the probability of success in fertilization; thus plants using insects or animals as a dispersal agent (entomophilous or zoophilous species) produce orders of magnitude less pollen than those dispersing pollen by wind (anemophilous species). By the same token, plants that are self-fertilizing (autogamous or cleistogamous species) produce only minute quantities of pollen compared to anemophilous species. Because of these factors, even though the vast majority of flowering plants are insect-pollinated, the accumulation of pollen grains at any given site will usually be dominated by pollen of anemophilous species. A single oak tree may produce and disperse by wind more than 10^8 pollen grains per yr, and hence the pollen from an entire forest (the pollen rain) assumes astronomical proportions. Pollen accumulation in a

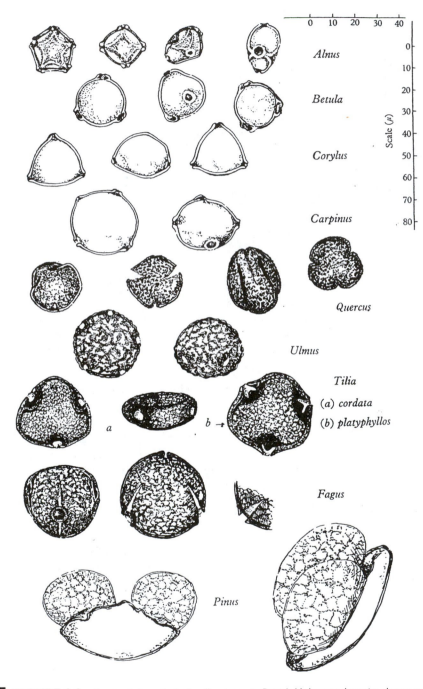

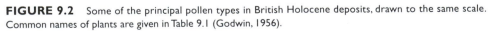

FIGURE 9.2 Some of the principal pollen types in British Holocene deposits, drawn to the same scale. Common names of plants are given in Table 9.1 (Godwin, 1956).

TABLE 9.1 Some Important Plant Taxa in North American and European Quaternary Palynology (see Fig. 9.1).

Genus	Family	Common name
Abies		Fir
Acer		Maple
Alnus		Alder
Ambrosia		Ragweed
Artemisia		Wormwood/Sage
Betula		Birch
Carpinus[a]		Ironwood
Carya		Hickory
	Chenopodiaceae	Goosefoot
Corylus		Hazel
	Cyperaceae	Sedges
Ephedra		Horsetail
Eucalyptus[b]		Eucalyptus
Fagus		Beech
Fraxinus		Ash
	Gramineae	Grasses
Juglans		Walnut
Juniperus		Juniper
Larix		Larch
Liquidambar		Sweet gum
Lycopodium[b]		Clubmoss
Nyssa		Tupelo
Ostrya[a]		Hornbeam
Picea		Spruce
Pinus		Pine
Populus		Poplar
Pseudotsuga		Douglas fir
Quercus		Oak
Salix		Willow
Taxodium		Bald cypress
Taxus		Yew
Tilia		Basswood/lime
Tsuga		Hemlock
Ulmus		Elm

[a] *Ostrya* and *Carpinus* pollen are indistinguishable and are generally considered together.

[b] Exotic pollen added to samples for pollen influx calculations (see Section 9.2.4).

northern hardwood forest may reach 80 kg ha^{-1} a^{-1} (Faegri and Iversen, 1975). Pollen production by entomophilous species is generally several orders of magnitude lower and autogamous species produce even less. In some cases, an entomophilous species, such as *Tilia*, may produce fairly large amounts of pollen but the relatively efficient dispersal mechanism (via insects) means that pollen grains are rarely found in large numbers, even in forests where *Tilia* is abundant (Janssen, 1966).

9.2.3 Sources of Fossil Pollen

As pollen is an aeolian sediment, pollen falling on sites where organic or inorganic sediments are accumulating will become part of the stratigraphic record (Traverse, 1994). Pollen has thus been recovered from peat, lake sediments, alluvial deposits, estuarine and marine sediments, and glacial ice. Pollen has also been recovered from archaeological sites (Dimbleby, 1985), rat middens (King and Van Devender, 1977), and coprolites (fossilized fecal matter of animals; Martin *et al.*, 1961). In Quaternary palynology, the principal sources of paleoclimatic information are peat from bogs and marshes, and sediments from relatively shallow lakes (Jacobson and Bradshaw, 1981). In many lakes, sedimentation rates are often quite high, providing an opportunity to recover samples with a high temporal resolution, generally an order of magnitude greater than in marine sediments. Terrestrial pollen studies can thus provide a temporal perspective on climatic changes rarely possible in a marine setting.

The vast majority of pollen grains dispersed by wind are not carried more than 0.5 km beyond their source. Dispersal by wind is a function of grain size, the larger and heavier grains falling to the ground sooner than the smaller, lighter grains (Dyakowska, 1936). Pollen grains of beech (*Fagus*) and larch (*Larix*) for example, are relatively heavy and settle out close to their source. Consequently, the occurrence of fossil beech or larch grains in a deposit would indicate the former growth of a species in the immediate vicinity of the site. Field measurements of pollen dispersal from artificial sources and from isolated stands of vegetation indicate that pollen produced by an individual plant is not identifiable above background levels (the regional pollen rain) beyond a few hundred meters. This is also indicated by theoretical dispersal models (Tauber, 1965). Many investigators thus favor the analysis of sediments from fairly large lakes (>1 km^2) because they act as catchment basins for the regional pollen rain and are not unduly influenced by vegetation in the immediate vicinity of the sampling site (Prentice, 1985).

Much work has been conducted on the problems of pollen transport and sedimentation in lake basins (Pennington, 1973; Holmes, 1994). Just as in the atmosphere, differential settling of pollen grains occurs in water also, with the result that the original ratios in which pollen enters the lake from the air may be distorted, with the lighter pollen grains preferentially deposited in the littoral zone. Pollen is also concentrated in lake basins by inflowing streams, especially during periods of heavy runoff. Furthermore, resuspension and redeposition of pollen grains during periods of turbulent mixing, particularly in shallow water, also occur, thereby smoothing out yearly variations in pollen and sediment inputs to the lake. Further smoothing may result from the activities of burrowing worms and other mud-dwellers

(Davis, 1974). Thus, very high resolution studies (annual to decadal) are not practicable except perhaps in annually laminated (varved) sediments (Swain, 1978). In any case, this temporal scale is not appropriate for climatic reconstruction from pollen, as the climatic signal (recorded through changes in vegetation) will not be strong at this scale. Other nonclimatic factors would likely overwhelm any climatic signal. As the sampling timescale increases, the climatic signal begins to predominate over nonclimatic noise (Bradshaw, 1994). By careful interpretation of the record at the appropriate temporal scale and aggregation of data at the regional scale, the fundamental climatic signal can be distilled from the array of other factors affecting the accumulation of pollen in lakes.

9.2.4 Preparation of the Samples

In order to isolate pollen grains and spores from the matrix of organic or inorganic sediment, rigorous chemical treatment by hydrochloric, sulfuric, and hydrofluoric acid is generally required, as well as acetolysis by a mixture of acetic anhydride and sulfuric acid (for details, see Moore and Webb, 1978; or Faegri *et al.*, 1989). Removal of the matrix enables the remaining pollen grains and spores to be seen clearly when stained and mounted on slides for microscopic analysis. Generally, the original core is sampled at intervals of a few centimeters (depending on the sedimentation rate) and slides are prepared of pollen and spores at each level. These are then examined and the number of different grains in each sample are noted. Although the total number of grains counted at each level would depend on the purpose of the study and the source of material being studied (Moore and Webb, 1978), at least 200 grains are usually counted.

For pollen flux density calculations (see Section 9.4), the number of pollen grains counted on each slide must be related to the total pollen content of the sample from the level being considered. The most widely used method is to add a known quantity of exotic pollen or spores (e.g., *Eucalyptus* or *Lycopodium*) to the sample initially and then to count the number of these grains that occur on the final slide preparation. The ratio of exotic pollen counted to the number of exotic pollen added originally can be used to estimate the total pollen content of the original sample (Stockmarr, 1971; Bonny, 1972).

9.2.5 Pollen Rain as a Representation of Vegetation Composition and Climate

Differences in pollen productivity and dispersal rates pose a significant problem for the reconstruction of vegetation composition because the relative abundance of pollen grains in a deposit cannot be directly interpreted in terms of species abundance in the area. It is necessary to know the relationship between plant frequency in an area and the total pollen rain from that plant species in order to use pollen data to calculate the actual composition of the surrounding vegetation. For example, a vegetation community composed of 10% pine, 35% maple, and 65% beech may be represented in a deposit by approximately equal amounts of pine, maple, and beech pollen, because of the differences in their pollen productivity and dispersal rates. However, for paleoclimatic reconstruction, these matters are of less significance. What the paleoclimatologist needs to know is whether there are *patterns* in the pollen data that can be calibrated in terms of climate. Again, the question comes down to the appropriate scale of analysis. At the smallest spatial scale, and the shortest temporal scale, the signal represented by pollen will be dominated on the one hand by short-term (synoptic scale) factors affecting dispersal of pollen, and on the other by local factors affecting plant growth. Stepping back from this complexity reveals the desired climate information.

Palynologists adopt the uniformitarian principle: *the present is the key to the past*. By using spatial relationships in modern pollen distribution and their relationship to modern climate as a guide to interpreting pollen patterns recorded in the past, paleoclimatic reconstructions can be made. Modern vegetation communities can be considered as analogs of former vegetation cover; if pollen assemblages in the modern pollen rain resemble fossil pollen assemblages, the former vegetation and its associated climate is assumed to be similar to that in the analog region today. If similar pollen assemblages cannot be found today, presumably no modern analog for the former vegetation cover and climate exists (Ritchie, 1976). The main problem with this approach is the difficulty of sampling the vast array of possible vegetation assemblages in the modern landscape in order to find a good analog, and the fact that much "natural vegetation" in both the Old and New Worlds has been destroyed or greatly modified. Nevertheless, even in such modified environments, on the regional scale (~10^3 km) modern pollen still contains a climatic signal that allows differentiation of *regional-scale* climatic conditions.

Pollen rain in mountainous regions poses particular problems of interpretation, because vegetation communities may be confined to narrow climatic zones along mountainsides (Maher, 1963). In parts of South America, for example, vegetation may grade all the way from equatorial rainforest below 500 m to tundra-like páramo (or the drier puña) above 3500 m, over a horizontal distance of less than 100 km. Nevertheless, modern pollen rain studies demonstrate that good discrimination is possible between different vegetation zones, even where complex spatial interfingering of different vegetation communities occurs (Salgado-Labouriau, 1979; Gaudreau *et al.*, 1989; Lynch, 1996). Pollen from lower vegetation zones is commonly carried upwards to higher elevations by daytime upslope winds. However, with increasing elevation, gradient winds predominate, so that there may be an upper limit to upslope pollen transport (Markgraf, 1980). Pollen from high elevations is not dispersed very far beyond the high-altitude vegetation zone (Hamilton and Perrott, 1980). Characteristic pollen rain assemblages corresponding to different elevations can thus be identified and used to reconstruct altitudinal changes in vegetation through time (Salgado-Labouriau *et al.*, 1978). Such changes may be converted to paleoclimatic estimates if modern climatic controls on the altitudinal limits of different taxa are known, although consideration must be given to constraints imposed on vegetation by changing CO_2 levels during glacial periods (Jolly and Haxeltine, 1997). At those times, vegetation may have changed more in response to CO_2 levels than to a simple drop in temperature.

9.2.6 Maps of Modern Pollen Data

Extensive studies of modern pollen rain and modern vegetation have been made for North America by Davis and Webb (1975), Webb and McAndrews (1976), Webb *et al.*, (1978), and Delcourt *et al.*, (1984), and for Europe by Huntley and Birks (1983). In these studies, the amount of modern pollen of a particular genus was expressed as a percentage of the total pollen accumulation at each site; isopolls (lines of equal percentage pollen representation) were then mapped for each major genus and superimposed on maps of the regional vegetation (Fig. 9.3). Often, the zero isopoll corresponds well with the range limit of the taxa (though this varies by genus) and isopoll maxima coincide with the zone of maximum frequency of the taxa in the vegetation. However, the exact isopoll corresponding to the limits of a particular taxa needs to be determined by detailed regional studies as it may vary from one area to another. For example, the limit of *Picea* in the boreal forest of Northeastern North America corresponds to the 20% isopoll because spruce pollen is carried far north of the forest limit by frequent southwesterly winds. In northern Alaska the limit of *Picea* corresponds to the 10% isopoll; there the Brooks Range creates a strong topographic barrier to spruce growth farther north, and prevailing westerly winds reduce the spread of pollen northward. Spruce distribution in both areas is related to similar climatic controls, but the same isopolls do not provide that information (Anderson *et al.*, 1991).

One of the most convincing demonstrations of the correspondence between modern vegetation and modern pollen is that of Webb (1974), who compared pollen content in the upper 2 cm of 64 lake sediment cores from throughout lower Michigan with detailed forest inventory records. Maps showing the percentage distribution of individual tree genera were compared with corresponding isopoll maps (Fig. 9.4). Clearly, the spatial distribution of pollen of each genus closely resembles the percentage cover of that genus in the state (Prentice, 1978; Solomon and Webb, 1985). Furthermore, the forest composition (i.e., the covariation of individual genera) when summarized by principal components is also revealed by principal component analysis of the pollen rain data (Fig. 9.5). In this analysis the first principal component reflects the dominant climatic control on vegetation (and hence pollen) and the second principal component reflects regional variations in soil type.

These studies clearly indicate that in spite of the problems involved in pollen dispersal, preservation, and accumulation (Sections 9.2.2 and 9.2.3), the broad geographical patterns of vegetation are closely mirrored by the composition and amount of the pollen rain. Thus, at this scale of analysis, fossil pollen has much to offer for reconstructing paleoclimatic conditions.

9.3 HOW RAPIDLY DOES VEGETATION RESPOND TO CHANGES IN CLIMATE?

An important question that arises in studies of high resolution (e.g., varved) sediments is: What is the lag response of vegetation to climate change? Can the pollen record provide information on short period, large amplitude changes in climate, or to put the question more generally: What are the frequency response characteristics

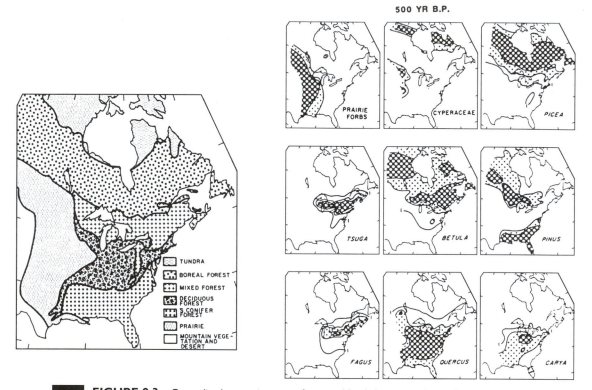

FIGURE 9.3 Generalized vegetation map of eastern North America and isopoll maps of selected taxa and groups of taxa at 500 yr B.P. Contours shown are 5 and 10% for forbs, 1, 5, and 10% for Cyperaceae, *Fagus*, and *Tsuga*, 1, 5, and 20% for *Picea* and *Quercus*, 1, 10, and 20% for *Betula*, 20 and 40% for *Pinus*, and 1, 3, and 6% for *Carya*. Forb pollen is the sum of *Ambrosia*, *Artemisia* and other Compositae, Chenopodiaceae, and Amaranthaceae pollen (Webb, 1988).

of the pollen record? Davis and Botkin (1985) attempted to answer this question by the use of a forest growth model to simulate changes in forest composition (basal areas of particular tree species) after steplike changes in temperature. They compared large amplitude, short duration events with smaller amplitude, longer-term changes and concluded that the forest response to climatic cooling lagged 100–150 yr behind, due to "community inertia." There is a natural delay in colonization by new species, due mainly to the shading effects of mature canopy trees. Consequently, they anticipate that changes in forest composition in response to the warming from ~1850–1990 will continue for at least another century (Overpeck *et al.*, 1990). They also found that there were similar vegetation responses to large, short events as there were to longer, smaller amplitude changes (Fig. 9.6). The smallest change necessary for a detectable forest response was a ~50-yr change in mean annual temperature of 2 °C, or a 1 °C change sustained over ~200 yr. In short, these model experiments suggest that one should not expect the pollen record of (mid-latitude forest) vegetation to resolve the influence of climatic changes "more closely than within a century or two"; rather it provides a "running mean of climatic variation" (Davis and Botkin, 1985).

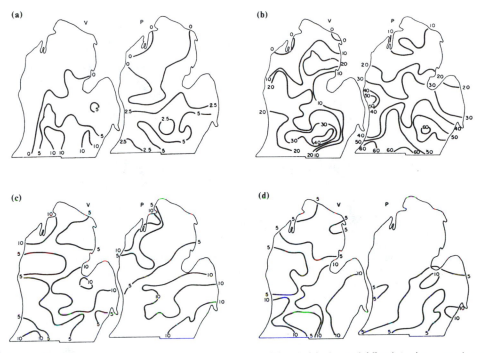

FIGURE 9.4 Maps of the percentages of (**a**) hickory, (**b**) oak, (**c**) elm, and (**d**) ash in the vegetation of Michigan (V) compared to the percentages of pollen (P) from the same trees in the modern pollen rain (based on arboreal pollen sum). Modern pollen data based on analysis of uppermost lake sediments (Webb, 1974).

The question of vegetation response to climatic change also involves the notion of ecosystems: Have they always been the same as we see them in the landscape today? Webb (1988), Huntley and Webb (1989), and Huntley (1990a) provide persuasive arguments that modern vegetation should not be viewed as fixed units with a constant composition, which moved regimentally across a region in response to changes in climate. Rather, individual taxa respond differently, leading to a constant change in vegetation composition, at times producing vegetation formations with no modern analogs (e.g., in Late Glacial time in much of western Europe; Huntley, 1990b). This is due partly to the individualistic responses of taxa to climatic changes and their different abilities to migrate, but also due to the fact that in many areas late Pleistocene and early Holocene climates were quite unlike anything seen today (COHMAP members, 1988). According to Webb (1988) "ecosystems and plant assemblages are to the biosphere what clouds, fronts and storms are to the atmosphere . . . features that come and go . . . [with] . . . internal dynamics . . . [but] . . . not of sufficient strength to overcome major changes from the outside." This has important implications for anticipating future changes in ecosystems that may accompany global warming. New communities may well be unlike those of today, or those in the past (Overpeck *et al.*, 1990; Davis, 1991).

With such a view of vegetation change, we inevitably turn to the controversial issue of whether vegetation can ever be considered "in equilibrium" with climate.

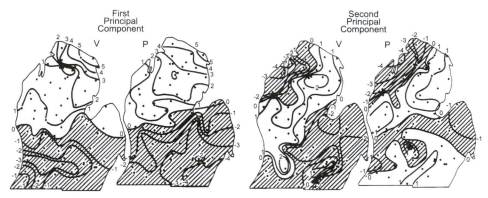

FIGURE 9.5   Map of the first two principal components (PC) of vegetation percentages (**V**) and pollen type percentages (**P**) in Michigan. The principal components of vegetation reflect major vegetation formations in the state. PC1 has a distribution reflecting the change from deciduous forest in the south to mixed coniferous-hardwood forest in the north. It accounts for 25% of variance in the original data set. PC2 depicts primary divisions within the two major vegetation formations, differentiating the northern hardwoods from the pine-birch-aspen forests in the north and the beech-maple and elm-ash-cottonwood forests from the oak-hickory forests in the south. It accounts for a further 35% of variance in the original data set. The first two principal components of pollen mirror the principal components of vegetation, indicating that the spatial pollen data may be used as a reliable indicator of vegetation distribution. Lakes shown by dots (Webb, 1974).

This seems to be very much a question of one's definition of equilibrium, and the scale at which one examines the question. Webb (1986, 1987, 1988) argues that vegetation is in *dynamic equilibrium* with climate; on the timescale of 10^3–10^5 yr, climate has changed continuously, and vegetation (viewed on subcontinental scales, at ~10^3 yr intervals) has kept pace with these changes. Changing the temporal and spatial focus to shorter intervals and smaller areas would no doubt reveal "disequilibria" related to migrational, successional, or edaphic influences (Prentice, 1986; Davis *et al.*, 1986). However, this issue is further complicated by differences related to particular types of vegetation or environment (e.g., in a forest environment, if a species can tolerate shade by canopy trees, or if the landscape being invaded is open). In particular, the extent of disturbance in an ecosystem can have an important effect on the ability of a species to occupy new environments even under the pressure of a change in climate (Davis, 1991). Furthermore, in some environments such as desert and semiarid regions, where vegetation has adapted to a high variability of rainfall, a sustained change in precipitation may generate an almost immediate response in vegetation cover[30] (Ritchie, 1986). Notwithstanding these specific problems, the evidence is that in midlatitudes, at least, a synoptic view of vegetation at ~2000–3000 yr intervals reveals changes, which are consistent with the changes that occurred in climate, as we currently understand them (Webb *et al.*, 1987).

[30] This has implications for vegetation-induced changes in trace gas concentrations (CH_4, CO_2) in relation to climatic forcings, as a rapid response in vegetation over large areas (such as the tropical desert margins) could have more or less immediate feedback effects on the climate system (Petit-Maire *et al.*, 1991).

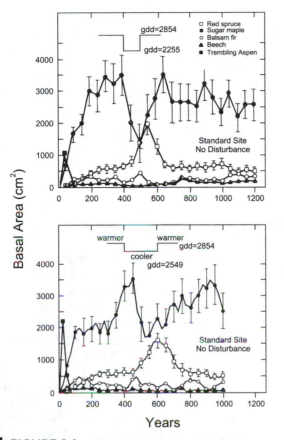

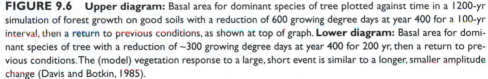

FIGURE 9.6 **Upper diagram:** Basal area for dominant species of tree plotted against time in a 1200-yr simulation of forest growth on good soils with a reduction of 600 growing degree days at year 400 for a 100-yr interval, then a return to previous conditions, as shown at top of graph. **Lower diagram:** Basal area for dominant species of tree with a reduction of ~300 growing degree days at year 400 for 200 yr, then a return to previous conditions. The (model) vegetation response to a large, short event is similar to a longer, smaller amplitude change (Davis and Botkin, 1985).

As a test of the hypothesis that vegetation is in dynamic equilibrium with climate, Bartlein *et al.* (1986) and Webb *et al.* (1987) used equations that relate modern pollen rain to contemporary climate (see Section 9.6) and applied them to climatic conditions in the past (derived from a general circulation model) to predict what the pollen rain should have been if vegetation was in equilibrium with climate. This approach assumes (a) the equations adequately characterize the pollen-climate relationship; and (b) the models accurately reconstruct past climate. The results show fairly good correspondence between the simulated and observed pollen rain, demonstrating that vegetational lags are not significant on the space and timescales being considered. In another approach, Prentice *et al.* (1991) used "response surfaces" (see Section 9.6) to predict climate (from 18 ka to 3 ka B.P.) directly from the fossil pollen data. The results were compared with (independently derived) model-simulated climate, and good agreement was found. As there were no major anomalies between the pollen-derived climate and that derived from the model, they concluded that "continental scale vegetation patterns have responded to continuous climatic changes during the past 18 ka, with lags no greater than ~1500 yr" (Prentice *et al.*, 1991). In fact, pollen can clearly register changes over much shorter intervals. For example, many sites show a pronounced change in the pollen spectra during the Younger Dryas interval, which seems to have lasted <1000 years. It seems safe to conclude that in some locations pollen may be a sensitive indicator of abrupt, short-lived climatic changes (lasting perhaps a few hundred years). Elsewhere, the site may be poorly located or the sedimentation rate may be too low to reveal such transient changes in climate.

9.4 POLLEN ANALYSIS OF A SITE: THE POLLEN DIAGRAM

Pollen data from a stratigraphic sequence are generally presented in the form of a pollen diagram composed of "pollen spectra" from each level sampled (Fig. 9.7). A pollen spectrum consists of the number of different pollen grains at a particular level expressed as a percentage of the total pollen count (the pollen sum). Actually, the pollen sum is not always made up of all pollen types counted. For paleoclimatic purposes, the objective of the analysis is to depict climatically significant regional vegetation change, so both arboreal and nonarboreal (shrub and herb) species are included in the pollen sum. Species that commonly grow in wet (lowland) environments around the sample site are usually excluded, though difficulties arise when a particular genus has different species (not easily distinguished by their pollen) that grow in both wet lowland and drier upland environments (e.g., *Picea mariana*, black spruce, and *Picea glauca*, white spruce, respectively; Wright and Patten, 1963).

In the pollen diagram, changes in the percentage of one species are assumed to reflect similar changes in the vegetation composition (due consideration being given to the factors of over- and under-representation already discussed here). The problem with this is that apparent changes in one species may occur in the percentage data as a result of changing receipts of pollen from other species because the total must always equal 100% (what Prentice and Webb [1986], call the "Fagerlind effect"). The following example from Faegri and Iversen (1975, p. 160) succinctly summarizes the difficulty:

> If we visualize a forest consisting of equal parts of oak and pine, and we use the pollen production figures quoted, we find that the corresponding spectrum will contain 15 percent oak, 85 percent pine. If beech is substituted for pine (apart from the botanical improbability of that succession) the same quantity of oak will give 60 percent of the pollen as against 40 percent beech. If the beech is then replaced by a tree, e.g. *Acer* spp. or *Populus balsamifera*, which is scarcely, or not at all, registered in the spectra, we shall find almost 100 percent oak pollen, although the quantity of oak has not changed at all. It is necessary to take into account not only the curve under discussion, but the others as well.

To circumvent such problems, palynologists may calculate pollen flux density (sometimes, incorrectly, called absolute pollen influx values), which is the number

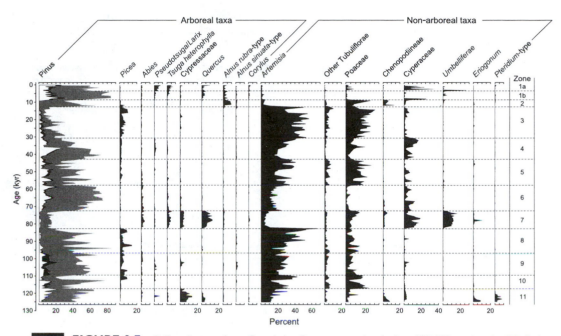

FIGURE 9.7 Pollen diagram from Carp Lake, Oregon, spanning the last 125,000 yr, showing 11 distinct zones, which were derived objectively from the spectrum of pollen represented. Three types of pine are recorded in the first column: *P. contorta* or *P. ponderosa* (white); *P. monticola* or *P. albicaulis* (black) and indeterminate (shaded) (Whitlock and Bartlein, 1997).

of grains accumulating on a unit of the sediment surface per unit time (Fig. 9.8). However, to do this the sediment accumulation rates must be known. Generally samples are ^{14}C dated at close intervals to establish a mean sedimentation rate. In North America, the rise of *Ambrosia* (ragweed) pollen at the time of colonial settlement (when forests were being cleared and herbaceous plants were increasing rapidly in numbers) is clearly seen in lake sediments; because the dates of settlement are known, sedimentation rates since then are readily calculated (Bassett and Terasmae, 1962; McAndrews, 1966; Davis *et al.*, 1973). Clearly the latter method is useful only for obtaining modern pollen flux statistics whereas ^{14}C dating enables influx to be calculated over earlier periods.

Pollen flux values can often clarify a stratigraphic record but they also have some important disadvantages. Sediment focusing in lakes may result in unrealistically high values of total pollen flux, leading to erroneous conclusions. Most importantly, flux calculations require that sediments be closely and accurately dated so that reliable sedimentation rates can be obtained. Relatively few records are sufficiently well-dated, and for large regional climate reconstructions, in which dozens if not hundreds of sites are used, percentage pollen values are always employed. There is strong evidence that in spite of their inherent limitations, such data provide reliable, reproducible, and verifiable paleoclimate reconstructions (T. Webb *et al.*,

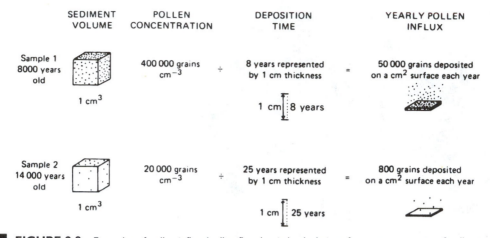

FIGURE 9.8   Examples of pollen influx (pollen flux density) calculations from measurements of pollen concentration and the rate of accumulation of the sediment matrix (Davis, 1963).

1993b; R. S. Webb *et al.*, 1993). Consequently, pollen flux density studies have taken a back seat to pollen percentages in most modern paleoclimatic studies.

9.4.1 Zonation of the Pollen Diagram

Pollen diagrams contain a large amount of information on the covariance of different pollen types through time. In order to facilitate comparison between different sites, the stratigraphic record in pollen diagrams is commonly subdivided into pollen zones; these are biostratigraphic units defined on the basis of the characteristic fossil pollen assemblage (see Fig. 9.7). Generally pollen zones contain a homogeneous assemblage of pollen and spores, but some investigators may recognize a zone that is characterized by abrupt changes. Needless to say, the definition of what constitutes a zone is a rather subjective decision and may not be agreed upon by different scientists (Tzedakis, 1994). Human nature also compels us to look for correlations with zonations previously "identified," thereby reinforcing systems which may not justify such blind faith! To avoid these problems, a number of more objective, computer-based methods have been suggested (Birks and Gordon, 1985). These are capable of identifying both major and minor zone boundaries (i.e., of defining zones and subzones) and permit objective comparisons to be made between sites. Objective computer-based zonation can also be applied to other variables in a sedimentary sequence (e.g., macrofossils, diatoms, sediment characteristics) to shed further light on the major climate-related features of the stratigraphic record (Birks, 1978; Birks and Birks, 1980). If similar local pollen assemblage zones can be identified over a large geographical area, it may be possible to define regional pollen assemblage zones, reflecting regional vegetation changes of broad paleoclimatic significance (Gordon and Birks, 1974; Birks and Berglund, 1979). The process of identifying such large-scale changes is effectively that of applying a lowpass filter to the spatial pattern of temporal change (T. Webb, *personal communication*). Only the major patterns of regional significance survive such scrutiny.

9.5 MAPPING VEGETATION CHANGE: ISOPOLLS AND ISOCHRONES

Studies of modern pollen rain and the distribution of different taxa in the landscape today (see Section 9.2.3) indicate that there is a fairly good spatial correspondence between them. Maps of modern pollen data can reproduce broad-scale patterns of individual taxa over large areas (Davis and Webb, 1975; Webb and McAndrews, 1976; Delcourt *et al.*, 1984; Huntley and Birks, 1983). Studies such as these paved the way for synoptic mapping of vegetation distribution at discrete periods in the past. This approach was first suggested by Szafer (1935), who defined the term isopoll as lines of equal percentage representation of a particular pollen type in the pollen sum. Using isopolls, Szafer constructed maps showing the distribution of beech and spruce across East Germany and Poland at five intervals from late-glacial to late Holocene time. However, Szafer's time framework was speculative because, at that time, there was no means of accurately dating organic material. It was only with the extensive use of ^{14}C dating that identical stratigraphic horizons could be identified (by interpolation between dates) at sites over large geographical areas, thus facilitating the preparation of time-sequential maps, such as those compiled by Huntley and Birks (1983), Delcourt *et al.* (1984), and Jacobsen *et al.* (1987). As an example, Fig. 9.9 shows isopolls of spruce (*Picea*), pine (*Pinus*), and oak (*Quercus*) over eastern North America at intervals from 18 ka B.P. to the present (T. Webb *et al.*, 1993b). These maps indicate the vegetation response to warming, and to the retreat of the Laurentide ice sheet, with rapid migration of individual taxa in the 12–9 ka B.P. period. By 9 ka B.P., the principal features of the modern pollen distribution (0 ka) can be seen for the first time, and by 6 ka B.P. most taxa had achieved their northernmost postglacial limit. Note, however, that spruce pollen percentages show a small southward shift after 6 ka B.P., possibly reflecting a change to higher levels of available soil moisture in the northeastern U.S. (which favors spruce over pine) after the early Holocene (R. S. Webb *et al.*, 1993).

Isopoll maps of individual taxa implicitly acknowledge that vegetation formations as we now know them have been impermanent features of the landscape (as discussed in Section 9.3). This is clearly illustrated by Huntley (1990b), who showed maps of "vegetation units," identified by associations of different pollen taxa at times in the past. Many of the vegetation units typical of western Europe today did not exist in Europe before the early Holocene and, indeed, even as recently as 1 ka B.P. there were areas of vegetation with no analog in the modern vegetation formations of Europe (Huntley, 1990a, 1990b).

Isopoll maps can be reinterpreted to show the migration of a particular genus or ecotone through time by isochrones (equal time lines). Figure 9.10a, for example, shows the location of the 15% spruce isopoll at intervals from 11,500 to 8000 yr B.P., a line thought to approximate the southern boundary of the late-glacial boreal forest in this area. Similarly, Fig. 9.10b shows the position of the conifer-

Spruce

Northern Pines

Oak

Southern Pines

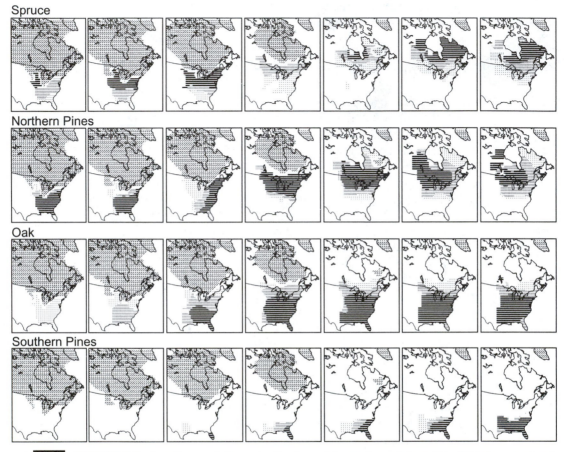

FIGURE 9.9   Isopoll maps of observed pollen data at 3 ka year intervals, from 18 ka B.P. (left-most column) to the present (at right). Three levels of shading indicate pollen percentages >1% (lightest), >5%, and >20% (darkest) (Webb *et al.*, 1987).

hardwood/deciduous forest ecotone derived by analogy with modern pollen, which indicates that this boundary coincides with the 20–30% isopolls for oak and pine. Both maps indicate a rapid northward migration of forests in late-glacial times. A final map, Fig. 9.10c, illustrates the position of the "prairie border," the grassland/ forest ecotone of midwestern North America based on 30% isopolls for herb pollen. Following rapid eastward shift in the early Holocene, this boundary regressed westwards after ~7000 yr B.P., indicating that the period of minimum precipitation and maximum warmth in the area had already passed.

Isopoll and isochrone maps depicting former vegetation patterns provide a qualitative perspective on past climate because vegetation formations on a broad scale are clearly determined by climate (Bryson and Wendland, 1967; Prentice *et al.*, 1992). However, more quantitative reconstructions of past climate can be obtained by mathematically relating modern climatic conditions to modern pollen rain, and using these relationships to convert the fossil pollen record into specific paleoclimate estimates.

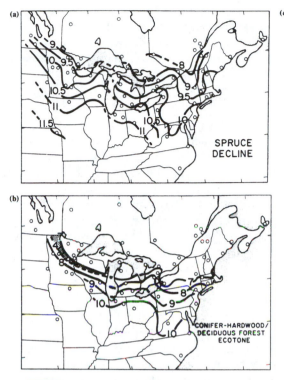
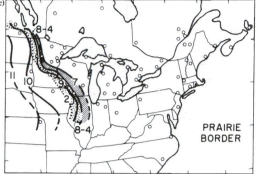

FIGURE 9.10   Isochrones (in thousands of years) on the migration of various vegetation taxa or ecotones, based on the position of diagnostic isopolls considered to be characteristic of the vegetation boundary. For example, in the modern boreal forest isopolls of spruce exceed 15%. In **(a)** isochrones indicate the position of the 15% spruce isopoll at different times and are considered to reflect the southern margin of the boreal forest as it migrated northward. In **(b)** the conifer-hardwood/deciduous forest ecotone is identified by the 20% isopoll for pine and the 30% isopoll for oak, reflecting a change from dominance of oak to dominance of pine (in a northward direction). In **(c)** the "prairie border" is delimited by the 30% isopoll for herbaceous pollen. Shading indicates the area across which the Prairie border first expanded (to 7000 yr BP) then retreated (westward) as conditions became more moist in the region in the mid- to late Holocene (Bernabo and Webb, 1977).

9.6 QUANTITATIVE PALEOCLIMATIC RECONSTRUCTIONS BASED ON POLLEN ANALYSIS

Paleoclimatic reconstruction from fossil pollen spectra is based on the notion that, as vegetation distribution is largely determined by climate, it should be possible to use that distribution (as represented in the fossil pollen spectra) to reconstruct past climate. Large databases of surface pollen samples (generally from the surface sediments of lakes) are now available for many parts of the world (see Appendix B) and these have enabled pollen assemblages to be calibrated directly in terms of climate. Although in many regions natural vegetation has been dramatically reduced, the broad-scale relationships between pollen rain and climate are sufficiently robust that they can be used to make reliable paleoclimatic reconstructions. This has been demonstrated several times (Bartlein *et al.*, 1984; Huntley, 1990b; Huntley and Prentice,

1993), but Guiot (1990) believes this factor (the reduction in natural vegetation) creates a great deal of noise in paleoclimate estimates (see discussion that follows).

The simplest approach to determining a relationship between modern pollen rain assemblages and contemporary climate is through multiple linear regression, such that:

$$C_m = T_m P_m$$

where C_m is the modern climatic data, P_m is the modern pollen rain, and T_m is a functional coefficient or set of coefficients ("transfer functions") derived from the relationship between modern climate and pollen data. Former climatic conditions (C_f) are then derived by using the fossil pollen assemblage (P_f) and the modern transfer function (T_m). In studies of pollen-climate relationships in eastern North America the equations were derived for different regions, defined in such a way that a "clear and monotonic" relationship could be recognized between climate and pollen percentages (generally using a pollen sum of the major forest taxa, plus *Cyperaceae* and prairie forbs [herbs]) (Bartlein and Webb, 1985). Those species associated with human settlement (e.g., *Ambrosia*) were eliminated to minimize anthropogenic influences on the pollen sum.

Scatter plots of pollen percentages and climate variables (e.g., % oak and July mean temperature across a region) commonly show nonlinear relationships between the pollen percentage and the climate variable. Such non-linearities can often be resolved by transforming the pollen data by some power function (Fig. 9.11). The transformed data are then used in developing an equation in which climate is the dependent variable and pollen percentages are the independent (predictor) variables. Thus, for the New England region of the United States, the following equation was constructed ($R^2 = 0.77$) (Bartlein and Webb, 1985):

$$\text{July } T_{\text{mean}} \text{ (°C)} = 17.76 + 1.73(\text{Quercus})^{0.25} + 0.09(\text{Juniperus}) + 0.51(\text{Tsuga})^{0.25}$$
$$- 0.41(\text{Pinus})^{0.5} - 0.12(\text{Acer}) - 0.04 \text{ (Fagus)}$$

Using this approach, Bartlein and Webb (1985) estimated that at 6 ka B.P. July temperatures over the north-central and eastern U.S. and southern Canada were 1–2 °C warmer than modern temperatures. Using a very similar approach, Huntley and Prentice (1988) estimated that July temperatures in central and southern Europe were as much as 4 °C warmer at 6 ka B.P. as compared to modern temperatures. The assumptions underlying this method (both ecological and statistical) are discussed at length in Howe and Webb (1983). Most important is the key uniformitarian assumption that any changes seen in pollen percentages of the past can be interpreted in terms of modern climate-pollen relationships. Indeed, this assumption underlies almost all paleoclimatic research involving calibration of paleorecords by means of modern data. One must also accept that paleoclimatic reconstructions based on transfer functions are only reliable if the modern calibration data set is extensive enough to be representative of all (or nearly all) conditions occurring in the past, otherwise unwarranted extrapolations may be made. Hence, transfer functions will not help to explain in climatic terms fossil pollen assemblages in the past that bear no relation to modern experience (the no-analog situations).

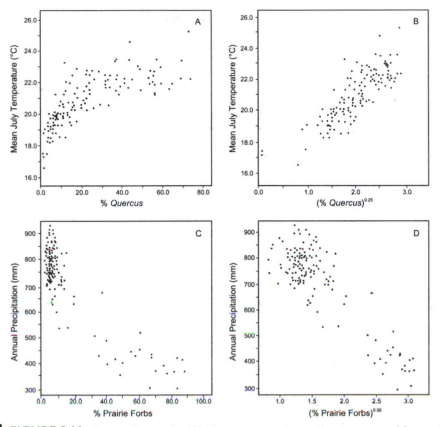

FIGURE 9.11 Scatter diagrams for **(A)** July mean temperature vs the percentages of *Quercus* (oak) pollen; **(B)** July mean temperature vs the percentages of *Quercus* pollen raised to the 0.25 power; **(C)** annual precipitation vs the percent of prairie-forb pollen (excluding *Ambrosia*) and **(D)** annual precipitation vs the percent of prairie-forb pollen raised to the 0.5 power (Bartlein *et al.*, 1984).

The no-analog problem was encountered by Overpeck *et al.* (1985), who quantified the relationship between modern pollen rain and fossil pollen spectra by calculating "dissimilarity coefficients" for each level in a pollen diagram (Prell, 1985; Bartlein and Whitlock, 1993). By matching the locations that were most similar (or least dissimilar) to a given pollen spectra in the past, they were able to characterize Holocene vegetation change in the eastern U.S. in terms of modern analogs, thereby making deductions about former climatic conditions. This approach worked well for most Holocene samples, but for the period from ~11–9 ka in the Upper Midwest, no close analogs could be found in the modern pollen rain. This was probably because climate at that time was rapidly changing and individual species responded to those changes at different rates, creating transient ecosystems not observed today.

Another mutivariate approach to paleoclimatic reconstruction involves the computation of "response surfaces." Pollen distribution is considered as occupying "climate space" defined by a three-dimensional (3D) array of climate variables.

Figure 9.12a–c illustrates the concept, beginning with simple bivariate graphs, relating spruce (*Picea*) pollen percentages to mean January and July temperatures, and to annual precipitation. The graphs are based on over 1000 modern pollen samples from across northern North America and Greenland and climatic data from the nearest weather stations (Anderson *et al.*, 1991). Clearly, it would be difficult to use any one of these graphs alone in interpreting pollen data. By plotting spruce pollen percentages in relation to both January and July mean temperatures (Fig. 9.12d) a more coherent picture emerges with the highest pollen percentages associated with mean July temperatures of 12–15 °C and mean January temperatures of –17 °C to –20 °C. Similarly, maximum spruce pollen percentages are found where annual precipitation is around 900 mm and July temperatures are 11–13 °C. Spruce pollen percentages thus vary within the 3D "climate space" defined by the two monthly temperatures and annual precipitation. This can be envisaged by considering how pollen percentages vary

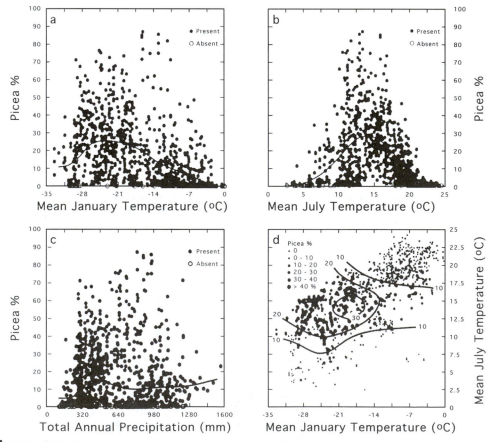

FIGURE 9.12 Scatter diagrams and response surfaces of *Picea* (spruce) pollen percentages and **(a)** mean January temperature, **(b)** mean July temperature, **(c)** total annual precipitation, and **(d)** mean July temperature and mean January temperature (Anderson *et al.*, 1991).

in relation to both January and July temperatures, as annual precipitation changes along a third axis — represented in Fig. 9.13 as a series of slices at selected precipitation intervals. Spruce pollen is highest in areas with July temperatures of 10–13 °C, January temperatures of –12 to –18 °C and annual precipitation of 880–1600 mm. These figures show schematically that the pollen percentage data define a body of variable density (i.e., the % data) suspended in climate space (Prentice *et al.*, 1991).

Response surfaces are mathematical expressions of these relationships, obtained by locally weighted regression techniques. Each pollen taxa is described in this way so that with a set of several equations it is possible to define the combination of climatic conditions to which a given pollen spectra (made up of many different pollen types) corresponds. This is made clear by considering what climatic conditions might be represented (in a fossil pollen spectra) by, say, 20% spruce. Clearly, there are many possible combinations of climate variables where 20% spruce pollen would be

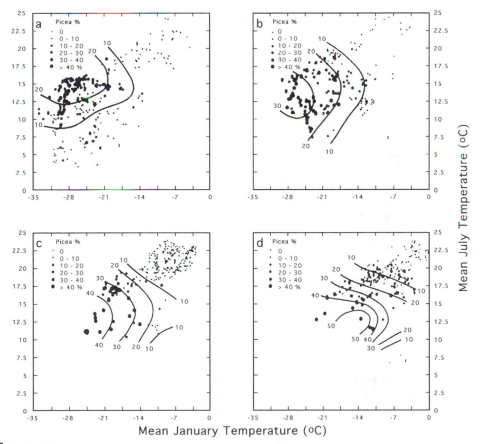

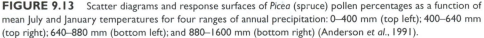

FIGURE 9.13 Scatter diagrams and response surfaces of *Picea* (spruce) pollen percentages as a function of mean July and January temperatures for four ranges of annual precipitation: 0–400 mm (top left); 400–640 mm (top right); 640–880 mm (bottom left); and 880–1600 mm (bottom right) (Anderson *et al.*, 1991).

expected (Fig. 9.13d). But those options would be more limited if the spectra had 20% spruce and 10% pine (which has its own constraints in climate space), and as more pollen types are added, each one uniquely delimited in climate space, the number of possible options becomes more and more limited. Eventually, the appropriate climatic conditions corresponding to the *combination* of pollen types is isolated, or at least the range of options is minimized, and the centroid of the final climate space, so defined, could be taken as the best estimate of the climate.

Prentice *et al.* (1991) demonstrate that, in eastern North America, at least 6 major pollen taxa (spruce, birch, northern pines, southern pines, oak, and prairie forbs) are needed to avoid non-unique or indeterminate solutions. These six types are sufficient to define past climatic conditions at fossil pollen sites across eastern North America (Fig. 9.14) although it must be recognized that not all values on the map are as reliable as others; this depends on the goodness of fit of the response surface for the modern climate-pollen relationship and the "size" of the area in climate space to which the fossil pollen spectra correspond. The maps would be improved if some indication of relative error or reliability was represented (in time and space). Using more taxa should decrease the uncertainty in the paleoclimatic estimates (T. Webb *et al.*, 1993b, who used 14 taxa). In the example given, pollen rain is assumed to reflect January and July mean temperatures and annual precipitation. However, other parameters such as soil moisture or an index of continentality might be more discriminating variables. For example, R. Webb *et al.* (1993) used a soil moisture index to define response surfaces, then reconstructed soil moisture changes over the northeastern U.S. since 12 ka B.P. Changes in precipitation were obtained from another set of response functions. This analysis revealed that although precipitation was lowest at 12 ka B.P., effective soil moisture was lowest at 9 ka B.P., a time when pine became most abundant in this region. The reconstruction is supported by lake-level data that point to the early Holocene as a time of significantly drier conditions.

Guiot (1987) argues that in many areas (such as in Europe) the variable topography and a long history of human impact make it very difficult to capture, in the contemporary pollen rain, all the information necessary to interpret fossil pollen spectra. These problems introduce considerable noise into paleoclimate estimates. Furthermore, the response of vegetation to a given change in climate will not always be represented in the pollen spectra in the same way; it will depend on the preceding vegetation state. There will be a certain autocorrelation in the pollen series that is not accommodated by the transfer or response function approaches (though Webb *et al.*, 1987 and Prentice *et al.*, 1991, implicitly take this into account by reconstructing climate only at 3 ka intervals).

Guiot proposes a method to reduce the variability (noise) in a paleoclimate reconstruction based on transfer or response functions (which he terms "analog climates"). First, the changes in fossil pollen spectra from a limited region are analyzed to extract those variations that are common to the different locations. This is achieved by a method akin to principal components analysis. The main component might account for, say, 80% of the common variance in two or more records and it is assumed that this common signal (termed the "paleobioclimate") represents the overriding influence of climatic variations on vegetation changes at all the

Mean July Temperature (°C)

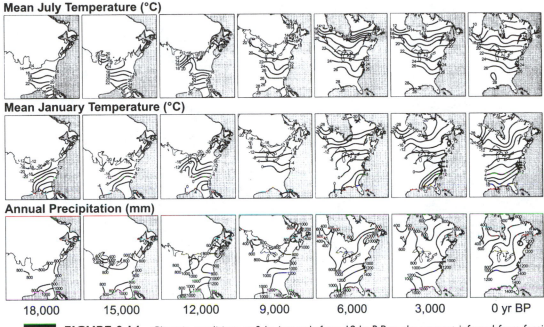

Mean January Temperature (°C)

Annual Precipitation (mm)

| 18,000 | 15,000 | 12,000 | 9,000 | 6,000 | 3,000 | 0 yr BP |

FIGURE 9.14 Climatic conditions at 3 ka intervals from 18 ka B.P. to the present inferred from fossil pollen and the modern abundances of six pollen types (spruce, birch, northern pines, southern pines, oak, and prairie forbs) using response surfaces. The blank area is the Laurentide ice sheet (Prentice *et al.*, 1991).

sites. This series is then used in identifying the most appropriate modern analog climate, based on a measure of similarity between fossil and modern pollen spectra (Guiot *et al.*, 1989). By reducing what is assumed to be non-climatic noise in the original series, the final paleoclimate estimates have a larger climate signal-to-noise ratio than would otherwise be obtained. Using this approach, Guiot *et al.* (1989) were able to estimate mean annual temperature and precipitation variations over the last 140,000 yr at two sites in France (Fig. 9.15). These show temperatures similar to or slightly higher than Holocene levels at the height of the last interglacial in Europe (the Eemian); two subsequent interstadials (St. Germain I and II) were almost as warm. Extremely cold and dry conditions were first experienced around ~65 ka B.P. Three or four other cold, dry intervals occurred in the main (Würm/Weichselian) glacial period, followed by a change to much warmer and wetter conditions in the Holocene. These reconstructions, and their limitations, are further discussed in Section 9.7.1.

One way of examining the coherency of paleoclimatic estimates quantitatively was suggested by Webb *et al.* (1987). They used the NCAR community climate model (CCM1) simulations of past climate (at 3 ka intervals, from 18 ka to 3 ka B.P.) to "predict" the pollen rain expected at those times in the past, using simulated climatic conditions applied to the response surface equations derived from modern calibration studies. These predictions of the expected pollen rain can then be

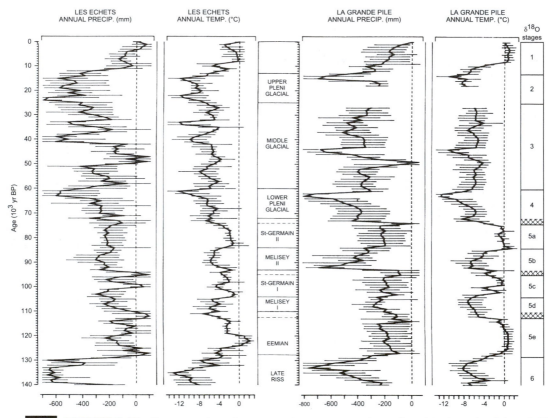

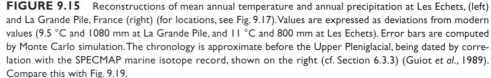

FIGURE 9.15　Reconstructions of mean annual temperature and annual precipitation at Les Echets, (left) and La Grande Pile, France (right) (for locations, see Fig. 9.17). Values are expressed as deviations from modern values (9.5 °C and 1080 mm at La Grande Pile, and 11 °C and 800 mm at Les Echets). Error bars are computed by Monte Carlo simulation. The chronology is approximate before the Upper Pleniglacial, being dated by correlation with the SPECMAP marine isotope record, shown on the right (cf. Section 6.3.3) (Guiot et al., 1989). Compare this with Fig. 9.19.

compared with observed data. This is both a test of the model's ability to simulate the paleoclimate correctly, and of the accuracy with which the response surfaces can reproduce climatic conditions. The results (for eastern North America) showed good broad-scale agreement between observed and simulated pollen distributions, though late glacial conditions were less reliably simulated, presumably reflecting the difficult problems of finding appropriate analogs in the modern data network. More recently, T. Webb *et al.* have directly compared January and July paleotemperatures, and annual precipitation, derived from response surfaces ("inferred") with paleoclimatic reconstructions "simulated" by CCM1 (Webb *et al.*, 1993b, 1997) (Fig. 9.16). The results show good agreement for the Holocene time slices between these quite independent approaches (Fig. 9.16e); differences in July paleotemperature recon-

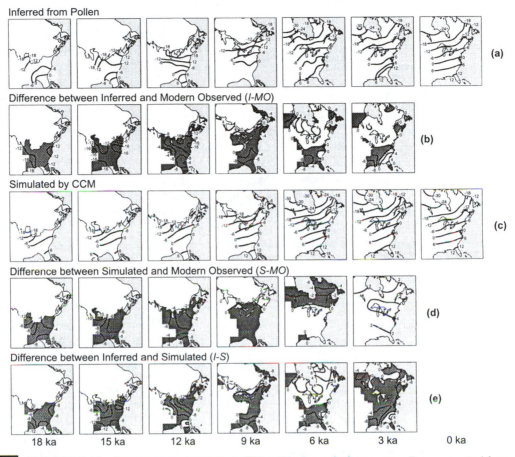

Inferred from Pollen *(a)*

Difference between Inferred and Modern Observed (*I-MO*) *(b)*

Simulated by CCM *(c)*

Difference between Simulated and Modern Observed (*S-MO*) *(d)*

Difference between Inferred and Simulated (*I-S*) *(e)*

18 ka   15 ka   12 ka   9 ka   6 ka   3 ka   0 ka

FIGURE 9.16 January mean temperatures (°C) at 3 ka intervals: **(upper panel)** reconstructed from pollen by the use of response surfaces and **(second panel)** these values expressed as departures from modern conditions; **(third panel)** temperatures simulated by the NCAR general circulation model (CCM1) and **(fourth panel)** those values expressed as differences from modern conditions. The lowermost panel shows differences between the pollen-based reconstructions and those of the model (first and third panels) (Webb et al., 1993b).

structions are generally <2 °C. However, for the late glacial period, discrepancies are much larger, generally reflecting the considerably lower paleotemperatures estimated from pollen data (Fig. 9.16a, b) than from the general circulation model (Fig. 9.16c, d). Similar discrepancies are seen in the January paleotemperature estimates. Further refinements in both model simulations and pollen calibrations are in progress, as is the development of independent late glacial paleotemperature indicators to resolve these differences and provide greater confidence in the reconstructions (see Webb and Kutzbach, 1998 and accompanying papers).

9.7 PALEOCLIMATIC RECONSTRUCTION FROM LONG QUATERNARY POLLEN RECORDS

There are now numerous palynological records that span the late Quaternary and, in some cases, extend back continuously into the Pliocene. Dating is, of course, problematic beyond the range of radiocarbon dating (~40,000 yr in most cases) and often the records are simply assumed to extend back to the last interglacial, and beyond, based on the character of the pollen record itself. In some cases, correlation with marine oxygen isotope stratigraphy, either directly (in marine sedimentary records) or indirectly (by comparisons between terrestrial and nearby marine records) can prove helpful in constructing a chronology. Here, a selected number of long palynological records from different areas of the world are presented to provide an overview of how such records, even if not well-dated or calibrated quantitatively, can reveal important paleoclimatic changes of large-scale significance. They provide important insights into changes in terrestrial climates on timescales comparable to the important ice core and marine sedimentary records, thereby completing the global picture of climatic variations over the Quaternary period.

9.7.1 Europe

Several long records extending >100 ka have been recorded from lakes and bogs in Europe (Fig. 9.17). Of these, La Grande Pile in the French Vosges mountains has been studied in the most detail, with over 20 cores recovered for pollen, plant macrofossils, and sedimentological and faunal (insect) analysis (Woillard, 1978; Woillard and Mook, 1982; Beaulieu and Reille, 1992; Seret *et al.*, 1992; Pons *et al.*, 1992; Guiot *et al.*, 1992, 1993; Ponel, 1995). The pollen record from this location is well-documented, enabling the overall sequence of vegetation changes over the last interglacial-glacial cycle to be established (Fig. 9.18). The penultimate glaciation (Riss) was characterized by an open grassland with few trees. The transition to peak interglacial conditions in the Eemian is marked by a clear sequence of taxa,[31] first *Juniperus*, then *Pinus* and *Betula*, *Ulmus* and *Quercus* with *Corylus,* and, finally, *Taxus*, indicative of the interglacial climatic optimum. Subsequent climatic deterioration is marked by a rise in *Abies* and *Carpinus*, then *Picea*, *Pinus*, *Betula*, and *Juniperus* once again. This short cool episode is the first of two such periods (termed Melisey I and II) separated by more temperate conditions (St Germain I and II). This is clearly seen in the relative proportions of arboreal to nonarboreal pollen (AP/NAP) with the colder periods characterized by sharp increases in the NAP fraction of the pollen sum. The onset of full glacial conditions (the Pleniglacial) began ~70 ka B.P. when *Artemisia* and other cool steppe taxa increased in abundance. The following 50 ka was dominated by NAP, although occasional fluctuations in the abundance of arboreal taxa suggest that conditions were not completely static. Coldest and driest conditions occurred in the late glacial (Tardiglacial) as shown by lowest NAP values and highest levels of *Artemisia* (Fig. 9.18).

These interpretations have been quantified by Guiot *et al.* (1989, 1992) as discussed earlier (Fig. 9.15) based on calibration with modern pollen assemblages. However, the reconstructions are problematic because there are no good modern

[31] For common names, see Table 9.1.

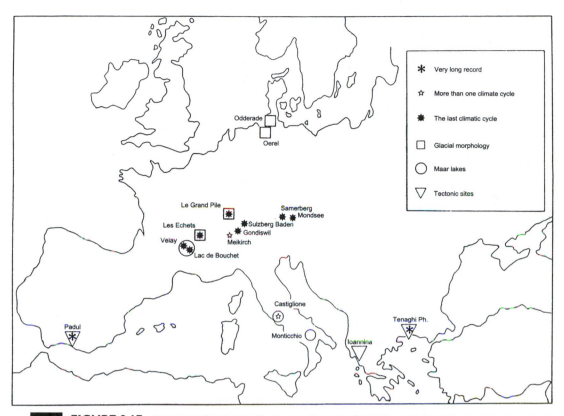

FIGURE 9.17   Locations of the longest European pollen records, which span at least the last glacial-interglacial cycle (Guiot *et al.*, 1993).

analogs (at least not in Europe) for pollen assemblages recorded in the coldest parts of the last glaciation. Indeed, the full glacial vegetation assemblage may have more in common with the cold, arid steppes of interior Asia and Tibet today than anywhere in Europe. Furthermore, periods of rapid change may not be adequately recorded in the pollen record because the overall inertia in vegetation assemblages makes pollen a poor indicator of short, abrupt climatic changes. These concerns have led to the incorporation of other climatic indicators, such as sedimentary characteristics (Seret *et al.*, 1992) and insects (Ponel, 1995; Guiot *et al.*, 1993) together with pollen in recent reconstructions. For example, the loess content of the sediment is highest during the coldest intervals of the Pleniglacial and organic matter content is lowest. Taking such factors into account results in significant differences of interpretation (up to 6 °C) with more variability of temperature during the full glacial period. Further studies have combined pollen with insect fauna to refine the reconstruction of mean annual temperature (Fig. 9.19). Insects respond quickly to climatic fluctuations (see Section 8.4) and so are especially useful in a period when climate is unstable (Ponel, 1995). Combining both

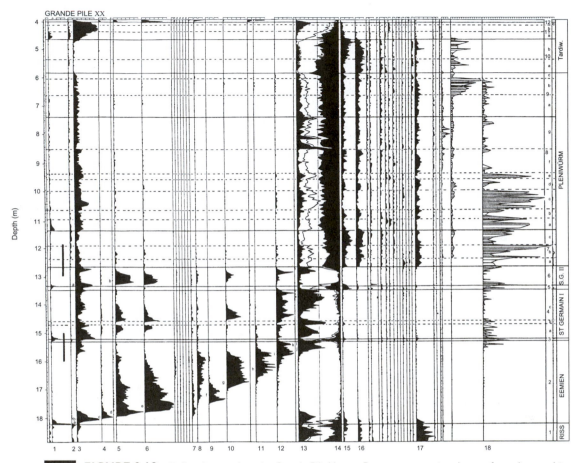

FIGURE 9.18   Pollen diagram from La Grande Pile, Vosges, France representing changes from the penultimate glaciation (Riss) to the late glacial ("TardiWürm" – see pollen zones indicated on the right). Numbers across bottom correspond to principal pollen types; the sequential change in vegetation during a complete glacial-interglacial cycle is clearly seen. 1 = *Juniperus*; 2 = *Salix*; 3 = *Betula*; 4 = *Ulmus*; 5 = Deciduous *Quercus*; 6 = *Corylus*; 7 = *Fraxinus*; 8 = *Alnus*; 9 = *Taxus*; 10 = *Carpinus*; 11 = *Abies*; 12 = *Picea*; 13 = *Pinus*; 14 = Poaceae; 15 = *Artemisia*; 16 = Heliophytes (various); 17 = Cyperaceae; 18 = *Isoetes*. The thin line between 13 and 14 represents the overall arboreal/nonarboreal pollen ratio, with arboreal pollen increasing to the right (de Beaulieu and Reille, 1992).

insect and pollen data reveals many rapid changes in climate during the Pleniglacial, some of which correspond to Dansgaard-Oeschger oscillations seen in the Greenland ice cores, and possibly also to North Atlantic Heinrich events. This multivariate approach to paleoclimate reconstruction has much to offer as it is clear that no one variable can provide an accurate view of past climate in Europe during full glacial conditions. By pooling the information each proxy provides, a more reliable reconstruction can be obtained.

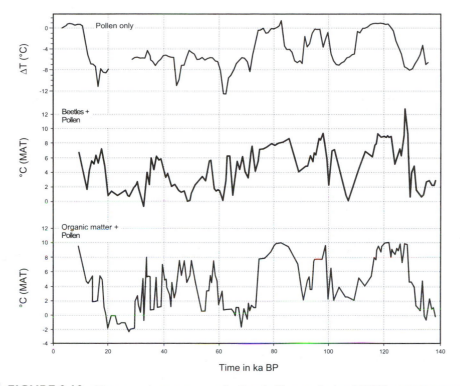

FIGURE 9.19   Mean annual temperature at La Grande Pile over the last 140,000 yr, reconstructed from pollen alone (top), and pollen constrained by the additional consideration of organic matter variations (lower diagram) or insects (Coleoptera) (middle) (cf. Fig. 9.15). Modern pollen analogs are not good descriptors of climatic conditions during the main glacial phase (~70–20 ka B.P.) or when climate changed abruptly. By considering insect fauna and sedimentological changes, additional constraints on paleotemperature estimates are introduced (Guiot et al., 1993).

9.7.2 Sabana de Bogotá, Colombia

The longest continuous sedimentary record from South America comes from a mountain-rimmed basin in the Colombian Andes (Hooghiemstra, 1984). Known as the Sabana de Bogotá, it is a former lake basin, currently at ~2550 m above sea level. The record (from near the village of Funza) extends back to late Miocene time and indicates that the region experienced ~2 km of uplift between ~5 and 3 Ma B.P. Vegetation in the area is zoned altitudinally, ranging from tropical forest below ~1000 m, to Andean forest above ~2300 m, to páramo above ~3500 m. At the highest elevations, perennial snow is found. At times in the past, vegetation has migrated along the mountain slopes but maintained its characteristic floristic zonation as the climate has changed (Fig. 9.20) (Van der Hammen, 1974; Hooghiemstra and Ran, 1994). By grouping the pollen types characteristic of each major zone, it has been possible to reconstruct the altitudinal limits of these zones over time (Fig. 9.21). The chronological framework for the record has been provided by correlation with a marine

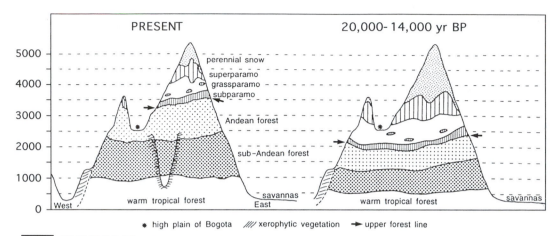

FIGURE 9.20 Altitudinal distribution of vegetation zones in the eastern Cordillera of Colombia at the present time and during the last glacial maximum (Van der Hammen, 1974, modified by Andriessen *et al.*, 1993).

isotope record from ODP site 677 (Hooghiemstra *et al.*, 1993) and by fission track dates on zircons from tephras in the sediments (Andriessen *et al.*, 1993). It has been estimated that the altitudinal extent of the Andean forest can be correlated with threshold values of the overall arboreal pollen (AP) sum recorded at Funza; thus when total AP exceeded 75% (during the last ~263 ka) the forest zone extended across the basin to above 3000 m in elevation. When the AP sum fell below 40% (in

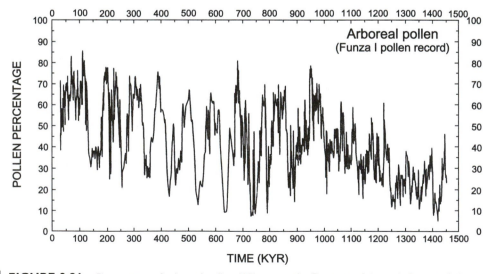

FIGURE 9.21 Fluctuations of arboreal pollen (AP) sum at the Funza site, Sabana de Bogotá, Colombia, over the past 1.5 Ma. The former lake of Bogota desiccated at ~27 ka B.P., terminating this sequence. The development of a pronounced ~100 ka climatic cycle within the last 1 Ma is clearly seen (Hooghiemstra and Ran, 1994).

the same time interval) the forest zone was limited to elevations of less than 2000 m, below the intermontane basin. When viewed over very long periods of time (Fig. 9.21) other factors complicate this simple idea (e.g., the establishment of *Quercus* [which produces a lot of pollen] in the forest at around ~263 ka B.P.) and it is then necessary to adjust the AP "threshold values" accordingly. Taking such factors into account, if the upper forest limits are controlled by temperature, AP variations range from ~1800 m in glacial times to ~3500 m in interglacials and suggest overall temperature changes on the order of ~10 °C at Funza (i.e., Δ1700 m at a moist adiabatic lapse rate of 0.6 °C 100 m^{-1}). Whether such changes in vegetation can be interpreted simply in terms of temperature is debatable; studies of vegetation change in other montane environments suggest that the lower concentrations of carbon dioxide in glacial times may have had a particularly strong impact on vegetation at high elevations. Ice-core evidence indicates CO_2 levels were 80–100 ppm lower at the last glacial maximum compared to pre-industrial levels, so that at elevations above 3500 m, the partial pressure of CO_2 may have been <130 ppm, preventing C_3 plants such as trees from surviving (Street-Perrott, 1994; Street-Perrott *et al.*, 1997, 1998). Lower carbon dioxide levels would cause an increase in stomatal gas exchange, thereby raising transpiration rates and drought stress on plants (Jolly and Haxeltine, 1997). Those C_4 plants such as grasses are generally more efficient in utilizing CO_2 and water, so evidence for lower treelines and more extensive savannas or grassy tundra may have less to do with temperature change than with CO_2 levels. It is also of interest that the overall late Cenozoic record indicates that a shift in the frequency of climatic change occurred around 1 Ma B.P., from relatively low-amplitude high-frequency variations to higher amplitude fluctuations with a period close to 100 ka (see Fig. 9.21) (Hooghiemstra *et al.*, 1993). However, as the record was dated by tuning it to the marine isotope record, this may not be an entirely independent result.

9.7.3 Amazonia

Although it is commonly assumed that the large numbers of species of flora and fauna in the equatorial lowlands of South America are the result of long periods of stable climate, a number of biogeographical studies have cast doubt on that assumption. In a comprehensive study of different species of birds in and around the Amazon Basin, Haffer (1969, 1974) identified a number of regions he believed had been refugia for groups of birds during drier periods in the past when the extensive tropical forests of today were reduced to discrete forest enclaves separated by savanna vegetation. Forest-dwelling species, which were isolated in this way, differentiated (developed new species) independently from members of the same species, which had been separated into *other* forest enclaves. Haffer argued that when wetter conditions returned and the forests reoccupied the savanna region, forest-dwelling species also expanded their ranges, coming into contact with other population groups in the intervening areas. In these areas of "secondary contact," hybridization of species took place so that the discrete morphological characteristics of species that had evolved within the forest refugia were no longer obvious.

According to proponents of the refuge hypothesis, the results of these changes can be seen today in contemporary biogeographical distribution patterns. Within the extensive tropical forests, zones of relatively high species diversity (i.e., zones containing extreme concentrations of different plant and animal species) can be identified. These are sometimes referred to as centers of endemism (Brown and Ab'Saber, 1979). Within these zones, individual species may exhibit very uniform morphological characteristics (Vanzolini and Williams, 1970). Such regions are considered by many to be the former forest refuges that served as survival centers for forest-dwellers during drier intervals. Between these centers of endemism, contact areas or "suture zones" are found, characterized by far fewer species than in the refuges and by more diverse morphological characteristics in the population of a particular species.

The refuge hypothesis is controversial. On the one hand there is considerable biogeographical evidence that there are indeed certain regions where species diversity is extraordinarily high. Such regions are generally identified by first mapping the ranges of individual species, then superimposing the distributions, and selecting those areas that exhibit very high levels of species diversity (Haffer, 1982). In this way, studies of rainforest trees, butterflies, and lizards have been undertaken, and all reveal geographically similar core areas to those suggested by Haffer (1974) on the basis of his detailed studies of tropical birds (Vanzolini and Williams, 1970; Vanzolini, 1973; Brown et al., 1974; Prance, 1974, 1982; Brown, 1982). It is interesting that there is also linguistic and ethnographic evidence that points to the existence of similarly distributed forest refuges in prehistoric Amazonia (Migliazza, 1982; Meggers, 1982). The general coincidence of all these regions is quite impressive, considering the range of evidence involved. However, it could be argued that the distribution patterns observed do not reflect former refugia at all, but merely reflect modern ecological units that have evolved together in response to contemporary edaphic and climatic conditions, the uniqueness of which may or may not be immediately obvious (Endler, 1982; Colinvaux, 1996). Similarly, zones of "secondary contact" may simply reflect significant environmental gradients (Benson, 1982).

Clearly, these arguments can only be satisfactorily resolved by well-dated stratigraphic evidence demonstrating that certain areas contained savanna *at the same time* as other areas (i.e., the postulated refugia) were under forest cover (Livingstone, 1982). So far, there are simply not enough records to resolve this matter unequivocally, but many lines of evidence strongly suggest that the lowland Amazon Basin remained extensively forested throughout the last 40,000 yr (at least). Of particular relevance is a long pollen record from the lowlands of northwestern Brazil, in the heart of the dense tropical rain forest ecosystem (*selva*). This record is continuous for >40,000 yr and shows no evidence for any savanna phase in this area; arboreal pollen remained at 70–90% of the pollen sum throughout the period (Colinvaux et al., 1996a). Interestingly, the pollen record shows an increase in montane species, such as *Podocarpus*, in the last glacial period (marine isotope stage 2), so at that time a unique forest assemblage made up of both lowland and montane elements was present. There is no analog for such an assemblage today; it seems to represent a migration of montane plants into a lowland environment that was cooler than today by 5–6 °C (based on a descent of *Podocarpus* by ~800–1000 m and assuming an adiabatic lapse rate of 0.6 °C 100 m^{-1}). Taken on its own, this one site does not provide a totally convincing argument for dismissing the notion of greater aridity and more widespread savanna in glacial times. However, several other studies support the idea of cooler, not drier, conditions, at least in the core of the evergreen tropical forest zone. In eastern Ecuador, a section dating from the last glaciation also shows an unusual mixture of montane and lowland rainforest pollen types, as well as associated macrofossils. In this area, trees such as *Alnus* and *Podocarpus* descended ~1500 m from their modern range limits (Liu and Colinvaux, 1985; Bush *et al.*, 1990). Such evidence has also been found in Panama, where oak trees (*Quercus*) grew 1000 m lower than today in glacial times (Bush and Colinvaux, 1990). In southeastern Brazil, palynological data also suggest cooler, wetter conditions during the Last Glacial Maximum, with temperatures lower by 6–9 °C. Collectively, such evidence provides a compelling argument for an extensive lowland rain forest throughout the last glacial period and into the Holocene, albeit a forest with a quite different composition than that seen in Amazonia today (Colinvaux *et al.*, 1996b). That is not to say that some peripheral areas that currently experience seasonal moisture deficits were not drier (Markgraf, 1989; Van der Hammen and Absy, 1994) but there is currently little stratigraphic evidence to support the idea that *most* of the Amazon Basin was occupied by savanna in glacial times (Clapperton, 1993a, b). Indeed, recent studies of pollen in marine sediments from off the mouth of the Amazon reveal little change in arboreal pollen percentages throughout the last 100,000 yr; if savanna had been extensive, there ought to be a clear signal in the pollen carried down the Amazon and deposited offshore, but there is not (Haberle, 1997). It therefore seems likely that the centers of endemism noted in so many biogeographical studies reflect a complex of conditions (climatic, topographic, geological, geomorphological, etc.), which have distinguished these regions over long periods of time, even as climate fluctuated from cooler glacial to warmer interglacial conditions.

9.7.4 Equatorial Africa

As in South America, biogeographers have long held the view that the extensive tropical forests of the Congo Basin and adjacent coastal regions in the Gulf of Guinea were formerly more limited in extent, confined to areas where climatic conditions have remained favorable over long periods of time. These refugia are recognized today by the higher number of endemic taxa and rich species diversity compared to other areas (Hamilton 1976; Sosef, 1991; Maley, 1996). Unlike South America, direct evidence *in support* of the biogeographical arguments is provided by palynological data from lake sediments. In Ghana, a 27,000-yr long record from crater Lake Bosumtwi shows clearly that the semi-deciduous equatorial forests that have occupied this area for most of the Holocene were not present in glacial times (Fig. 9.22). From 19–15 ka B.P., arboreal pollen fell to as low as 5% of the pollen sum (compared to >75% today) and herbaceous plants (grasses and sedges) occupied the area. This is confirmed by $\delta^{13}C$ in sediment cores; values were –10 to –20‰ during the time when levels of Gramineae pollen were high (consistent with the dominance of C$_4$ plants) compared to Holocene values of –28‰, typical of forest

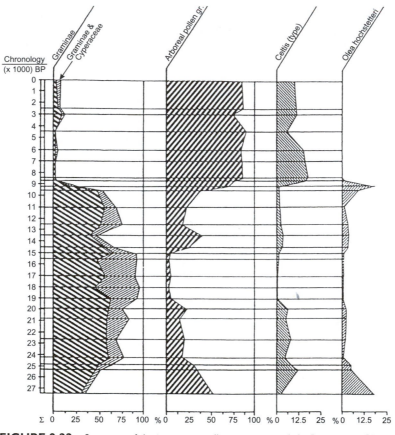

FIGURE 9.22 Summary of the important pollen variations in Lake Bosumptwi, Ghana, over the last 27,000 yr. Prior to ~10 ka B.P., grasses (Gramineae) characteristic of open savanna environments, dominated this location and arboreal pollen was low, especially from ~19–15.5 ka B.P. The presence of mountain olive (*Olea hochstetteri*) in this lowland environment indicates cooler conditions prevailed during glacial times (Maley, 1996).

(C_3) plants (Talbot and Johannessen, 1992; Giresse *et al.*, 1994). Certain montane plants (such as mountain olive, *Olea hochstetteri*) also migrated into the area; today they are found only far to the west, above ~1200 m, suggesting temperatures were cooler by 3–4 °C during glacial times. Similar evidence has been reported from sites in West Cameroon, Plateau Batéké in Congo and farther east in Burundi, where data of the modern pollen rain were used to quantify paleotemperature changes (Fig. 9.23) (Maley, 1991; Bonnefille *et al.*, 1992). By piecing together all the palynological and biogeographical evidence, Maley (1996) constructed a map showing the limits of lowland rain forest refugia in equatorial Africa during the last cold, dry period in the region, which corresponds to the Last Glacial Maximum of higher latitudes (Fig. 9.24). This reveals how dramatically different the region was at that time, with savanna and grasslands covering vast areas that are forested today. Conditions changed abruptly at ~9500 yr B.P. with the rapid expansion of

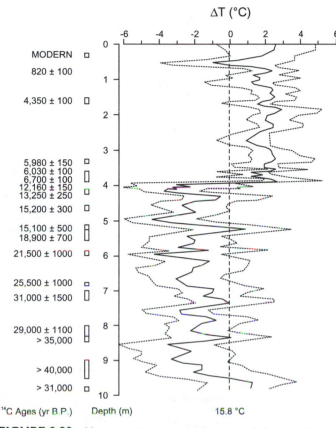

FIGURE 9.23 Mean annual temperature departures (relative to the present-day temperature of 15.8 °C) derived from pollen in a peat bog at Kashiru, Burundi, East Africa. The site is at 2240 m in the humid montane forest. Paleotemperature estimates were obtained using a network of modern pollen rain samples over a wide area of East Africa, calibrated against modern climatic data from the region. Dashed lines indicate confidence interval. The ^{14}C dates are shown on the left (note that the data are plotted linearly with respect to depth, not age). Paleotemperature estimates from the surface samples are affected by human-induced changes in vegetation and hence the modern pollen spectrum provides a temperature estimate warmer than the instrumentally recorded value (Bonnefille et al., 1992).

the rainforest to cover an area even larger than today within 2 ka (Maley, 1991). This change is thought to be related to rapid warming of SSTs in the Gulf of Guinea (as a result of reduced upwelling), leading to a longer wet season and higher total rainfall amounts (Maley, 1989a), but dramatic changes also occurred in East Africa around this time (see Fig. 9.23). Jolly and Haxeltine (1997) argue that significant changes in tropical vegetation would have occurred (even at low elevations) regardless of changes in temperature due to the lower CO_2 levels during glacial times, which favored grasses and sedges (C_4 plants) over trees. If this proves to be correct, at least some part of the observed changes may be related to physiological rather than climatic factors (Street-Perrott, 1994).

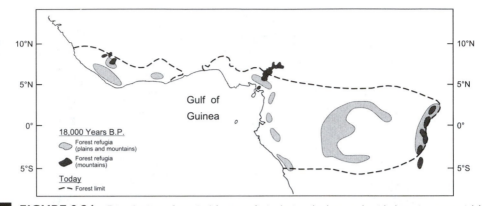

FIGURE 9.24 Distribution of tropical forest refugia during the last cool, arid phase in equatorial Africa (~20–16 ka B.P.) and the present day forest limits. Much of the area now occupied by tropical forest was covered by savanna during the Last Glacial Maximum (LGM) (Maley, 1996).

At the same time as lowland forest vegetation was expanding in the early Holocene, climatic conditions in sub-Saharan Africa became less arid, enabling the semiarid savanna vegetation belt to extend farther northward (see Section 7.6.3). As a result, the range of large herbivores (such as giraffe, elephant, hippopotamus, and gazelle) was also more extensive and today the bleached bones of these animals provide a mute reminder of the remarkably different climatic conditions that existed there in the early Holocene. Accompanying the animal migration into sub-Saharan Africa were aboriginal hunters who recorded their way of life on magnificent rock paintings and carvings (Lhoté, 1959; Monod, 1963; Lajoux, 1963). Today these are found hundreds of kilometers from the nearest permanent settlements.

Of particular significance during the period of more extensive savanna and tropical forests were the much more extensive riverine and lacustrine environments, which effectively provided water connections across the entire sub-Saharan region, from the Nile to Senegal (Beadle, 1974). Fauna of the Lake Chad Basin, for example, provide unequivocal evidence for recent connections, not only with the Niger and Congo basins, but also with the Nile drainage system over 1000 km to the east. Even today, relict populations of animals and plants are found isolated in topographically favorable environments, far removed from their nearest adjacent populations. For example, the Eurasian green frog (*Rana ridibunda*) has been found in the streams of the Ahaggar mountains at least 1000 km from adjacent population groups. Furthermore, and perhaps most remarkably, a Nile crocodile (*Crocodilus niloticus*) was found in a pool in the Tassili-N-Ajjer Mountains, separated by vast stretches of desert from major population centers to the east (Seurat, 1934; Beadle, 1974). Such disjunct species bring climatic changes to life.

Subtropical Africa is probably more arid today than it has been during most of the Holocene and was only more arid during the late Wisconsin glacial maximum (see Figs. 7.23–7.25). However, the record is still far from complete in either time or space, so brief periods of relatively moist or even drier conditions (on the order of 10^2

years) may have occurred. Indeed, lake level evidence does show that abrupt changes in climate (probably related to upwelling) did occur in the late Holocene (e.g., ~3500–4000 yr B.P., as recorded by lake-level changes in Lake Bosumptwi) but these oscillations did not persist and no significant change in forest cover resulted (Talbot and Delibrias, 1977; Maley, 1991). Nevertheless, such episodes may have had significant consequences for human populations in the region (Maley, 1989b, 1997).

9.7.5 Florida

Because much of North America was ice-covered during the last glacial period, there are few sites where long records (extending back tens of thousands of years) have been recovered. In Florida, lower sea level and generally drier conditions led to lower water tables during the Wisconsin glaciation, so only the deepest lakes seem to have sedimentary records from that period (Watts and Hansen, 1994). Lake Tulane in south-central Florida is one example; an 18.5 m core appears to span the last ~50 ka (though dating of the record before ~35 ka is somewhat problematic). The pollen record from this site (Fig. 9.25) shows pronounced oscillations in the percentage of *Pinus* (pine) pollen, which varies in opposition to the percentages of

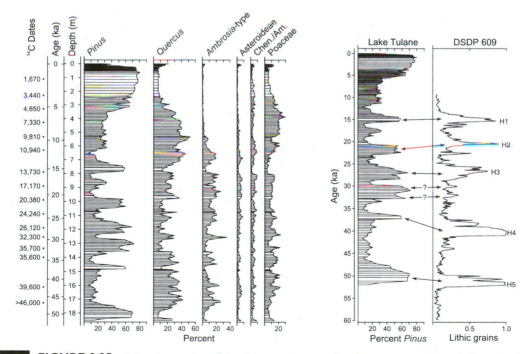

FIGURE 9.25   Pollen diagram from Tulane Lake, south-central Florida, showing oscillations in *Pinus* (pine) pollen alternating with intervals when *Quercus* (oak) and *Ambrosia* (ragweed) pollen percentages were high. Also shown is the relationship between pine pollen and Heinrich events, represented here by the fraction of lithic grains in samples from North Atlantic marine sediment core DSDP 609. Dating of the lake sediments prior to ~35 ka B.P. is uncertain. There appears to be a correspondence between the timing of Heinrich events and pine pollen increases, though the mechanism linking these two records is not clear (Grimm *et al.*, 1993).

Quercus (oak) and *Ambrosia*-type (ragweed and marsh-elder). This is interpreted as a shift from pine to oak-savanna or open grassland-type vegetation, with abrupt transitions between the two (Grimm *et al.*, 1993). The changes represent a shift in moisture availability, with the pine phase indicating wetter intervals separating drier episodes. Of particular interest is the similarity in timing of the pine expansion with Heinrich events in the North Atlantic (see Section 6.10.1), indicating the possibility that the terrestrial and marine events are causally linked. However, the mechanism of such a linkage is not clear. Heinrich events represent large-scale discharge of icebergs into the North Atlantic following prolonged periods of SST cooling. Cooler and/or wetter conditions in Florida (represented by increases in pine pollen) are related to cool waters in the Gulf of Mexico, which shifts the P-E balance towards positive values. One scenario is that the large-scale reorganization of North Atlantic circulation associated with Heinrich events led to cooler conditions in the Gulf of Mexico (via a reduction in the inflow of warm Caribbean/Gulf Stream waters). Alternatively, the drainage of glacial meltwater through the Mississippi River to the Gulf may have cooled surface waters in the Florida region. Neither scenario is clearly supported by the existing literature, so the exact mechanism linking the North Atlantic and Florida records remains enigmatic. Indeed, Watts and Hansen (1994) suggest that the higher pine percentages may be related to *warmer* SSTs in the Gulf, leading to stronger convective activity and higher rainfall amounts. This is one area where modeling could shed some light on the various alternatives. Further studies of other long continental records from along the Atlantic and Caribbean coasts are also needed to help resolve the matter. Because lake sediments have the potential for much higher resolution records than marine sediments, there are good prospects that detailed palynological studies can resolve in considerable detail the evolution of climatic changes in North America prior to, during and after glacial stage Heinrich events.

IO

DENDROCLIMATOLOGY

10.1 INTRODUCTION

Variations in tree-ring widths from one year to the next have long been recognized as an important source of chronological and climatic information. In Europe, studies of tree rings as a potential source of paleoclimatic information go back to the early eighteenth century when several authors commented on the narrowness of tree rings (some with frost damage) dating from the severe winter of 1708–1709. In North America, Twining (1833) first drew attention to the great potential of tree rings as a paleoclimatic index (for historical reviews, see Studhalter, 1955; Robinson et al., 1990; Schweingruber, 1996, p. 537). However, in the English-speaking world, the "father of tree-ring studies" is generally considered to be A.E. Douglass, an astronomer who was interested in the relationship between sunspot activity and rainfall. To test the idea of a sunspot-climate link, Douglass needed long climatic records and he recognized that ring-width variations in trees of the arid southwestern United States might provide a long, proxy record of rainfall variation (Douglass, 1914, 1919). His efforts to build long-term records of tree growth were facilitated by the availability of wood from archeological sites, as well as from modern trees (Robinson, 1976). Douglass' early work was crucial for the development of dendrochronology (the use of tree rings for dating) and for dendroclimatology (the use of tree rings as a proxy indicator of climate).

10.2 FUNDAMENTALS OF DENDROCLIMATOLOGY

A cross section of most temperate forest trees will show an alternation of lighter and darker bands, each of which is usually continuous around the tree circumference. These are seasonal growth increments produced by meristematic tissues in the tree's cambium. When viewed in detail (Fig. 10.1) it is clear that they are made up of sequences of large, thin-walled cells (earlywood) and more densely packed, thick-walled cells (latewood). Collectively, each couplet of earlywood and latewood comprises an annual growth increment, more commonly called a tree ring. The mean width of a ring in any one tree is a function of many variables, including the tree species, tree age, availability of stored food within the tree and of important nutrients in the soil, and a whole complex of climatic factors (sunshine, precipitation,

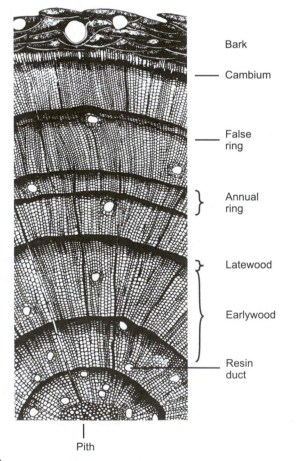

Bark

Cambium

False ring

Annual ring

Latewood

Earlywood

Resin duct

Pith

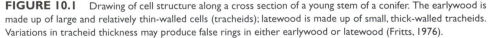

FIGURE 10.1 Drawing of cell structure along a cross section of a young stem of a conifer. The earlywood is made up of large and relatively thin-walled cells (tracheids); latewood is made up of small, thick-walled tracheids. Variations in tracheid thickness may produce false rings in either earlywood or latewood (Fritts, 1976).

temperature, wind speed, humidity, and their distribution throughout the year). The problem facing dendroclimatologists is to extract whatever climatic signal is available in the tree ring data and to distinguish this signal from the background noise. Furthermore, the dendroclimatologist must know precisely the age of each tree ring if the climatic signal is to be chronologically useful. From the point of view of paleoclimatology, it is perhaps useful to consider the tree as a filter or transducer which, through various physiological processes, converts a given climatic input signal into a certain ring width output that is stored and can be studied in detail, even thousands of years later (Fritts, 1976; Schweingruber, 1988, 1996).

Climatic information has most often been gleaned from interannual variations in ring width, but there has also been a great deal of work carried out on the use of density variations, both inter- and intra-annually (densitometric dendroclimatology). Wood density is an integrated measure of several properties, including cell wall thickness, lumen diameter, size and density of vessels or ducts, proportion of fibers, etc. (Polge, 1970). Tree rings are made up of both earlywood and latewood, which vary markedly in average density and these density variations can be used, like ring-width measurements, to identify annual growth increments and to cross-date samples (Parker, 1971). It has also been shown empirically that density variations contain a strong climatic signal and can be used to estimate long-term climatic variations over wide areas (Schweingruber *et al.*, 1979, 1993). Density variations are measured on x-ray negatives of prepared core sections (Fig. 10.2) and the optical density of the negatives is inversely proportional to wood density (Schweingruber *et al.*, 1978).

Density variations are particularly valuable in dendroclimatology because they have a relatively simple growth function (often close to linear with age). Hence standardization of density data may allow more low-frequency climatic information to be retained than is the case with standardized ring-width data (see Section 10.2.3). Generally, two values are measured in each growth ring: minimum density and maximum density (representing locations within the earlywood and latewood layers, respectively), although maximum density values seem to be a better climatic indicator than minimum density values. For example, Schweingruber *et al.* (1993) showed that maximum density values were strongly correlated with April-August mean temperature in trees across the entire boreal forest, from Alaska to Labrador, whereas minimum and mean density values and ring widths had a much less consistent relationship with summer temperature at the sites sampled (D'Arrigo *et al.*, 1992). Maximum latewood density values are calibrated in the same way as with the ring-width data using the statistical procedures described in Section 10.2.4. However, optimum climatic reconstructions may be achieved by using both ring widths and densitometric data to maximize the climatic signal in each sample (Briffa *et al.*, 1995).

Isotopic variations in wood have been studied as a possible proxy of temperature variations through time, but the complexities of fractionation both within the hydrological system, and in the trees themselves, make simple interpretations very difficult (see Section 10.4). Ring-width and densitometric and isotopic approaches to paleoclimatic reconstruction are complementary and, in some situations, could be used independently to check paleoclimatic reconstructions based on only one of

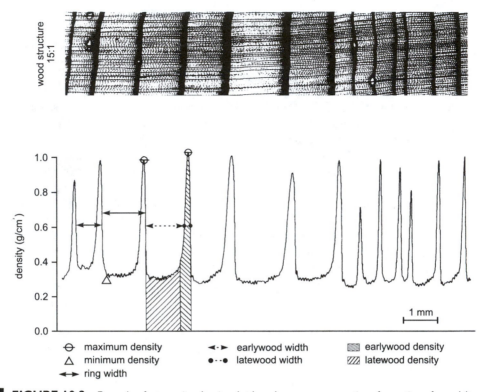

FIGURE 10.2   Example of a tree-ring density plot based on an x-ray negative of a section of wood (top of figure). Minimum and maximum densities in each annual ring are clearly seen, enabling the annual ring width to be measured as well as the width of both the earlywood and latewood (courtesy of F. Schweingruber).

the methods, or collectively to provide more accurate reconstructions (Briffa *et al.*, 1992a).

10.2.1 Sample Selection

In conventional dendroclimatological studies, where ring-width variations are the source of climatic information, trees are sampled in sites where they are under stress; commonly, this involves selection of trees that are growing close to their extreme ecological range. In such situations, climatic variations will greatly influence annual growth increments and the trees are said to be sensitive. In more beneficent situations, perhaps nearer the middle of a species range, or in a site where the tree has access to abundant groundwater, tree growth may not be noticeably influenced by climate, and this will be reflected in the low interannual variability of ring widths (Fig. 10.3). Such tree rings are said to be complacent. There is thus a spectrum of possible sampling situations, ranging from those where trees are extremely sensitive to climate to those where trees are virtually unaffected by interannual climatic variations. Clearly, for

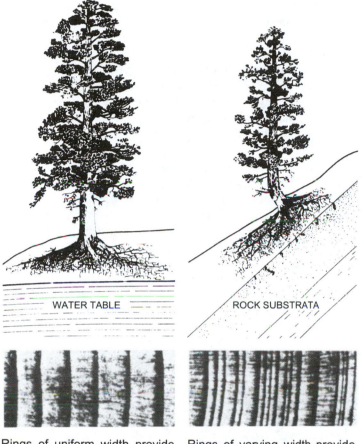

WATER TABLE

ROCK SUBSTRATA

Rings of uniform width provide little or no record of variations in climate.

Rings of varying width provide a record of variations in climate.

FIGURE 10.3 Trees growing on sites where climate seldom limits growth processes produce rings that are uniformly wide **(left)**. Such rings provide little or no record of variations in climate and are termed *complacent.* **(right)**: Trees growing on sites where climatic factors are frequently limiting produce rings that vary in width from year to year depending on how severely limiting climate has been to growth. These are termed *sensitive* (Fritts, 1971).

useful dendroclimatic reconstructions, samples close to the sensitive end of the spectrum are favored as these would contain the strongest climatic signal. Often, therefore, tree-ring studies at the range limit of trees are favored (e.g., alpine or arctic treeline sites). However, climatic information may also be obtained from trees that are not under such obvious climatic stress, providing the climatic signal common to all the samples can be successfully isolated (LaMarche, 1982). For example, ring widths of bald cypress trees from swamps in the southeastern United States have been used to reconstruct the drought and precipitation history of the area over the last 1000 years or more (Stahle *et al.*, 1988; Stahle and Cleaveland, 1992).

Paleoclimatic reconstructions have also been achieved using teak from equatorial forests in Indonesia (D'Arrigo *et al.*, 1994) as well as from mesic forest trees in Tasmania (Cook *et al.*, 1992a). For isotope dendroclimatic studies (Section 10.4), the sensitivity requirement does not seem to be as critical and it may, in fact, be preferable to use complacent tree rings for analysis (Gray and Thompson, 1978). Sensitivity is also less significant in densitometric studies and good relationships between maximum latewood density and temperature have been found in "normal" trees, which are growing on both moist and well-drained sites (Schweingruber *et al.*, 1991, 1993).

In marginal environments, two types of climatic stress are commonly recognized, moisture stress and temperature stress. Trees growing in semiarid areas are frequently limited by the availability of water, and ring-width variations primarily reflect this variable. Trees growing near to the latitudinal or altitudinal treeline are mainly under growth limitations imposed by temperature and hence ring-width variations in such trees contain a strong temperature signal. However, other climatic factors may be indirectly involved. Biological processes within the tree are extremely complex (Fig. 10.4) and similar growth increments may result from quite different combinations of climatic conditions. Furthermore, climatic conditions prior to the growth period may "precondition" physiological processes within the tree and hence strongly influence subsequent growth (Fig. 10.5). For the same reason, tree growth and food production in one year may influence growth in the following year, and lead to a strong serial correlation or autocorrelation in the tree-ring record. Tree growth in marginal environments is thus commonly corre-

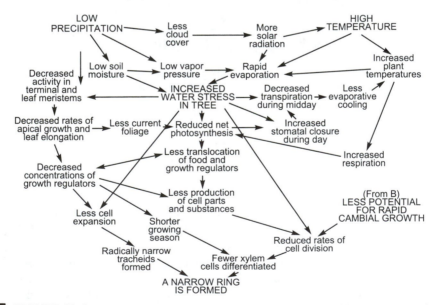

FIGURE 10.4 A schematic diagram showing how low precipitation and high temperature during the growing season may lead to the formation of a narrow tree ring in arid-site trees. Arrows indicate the net effects and include various processes and their interactions. It is implied that the effects of high precipitation and low temperature are the opposite and may lead to an increase in ring widths (Fritts, 1971).

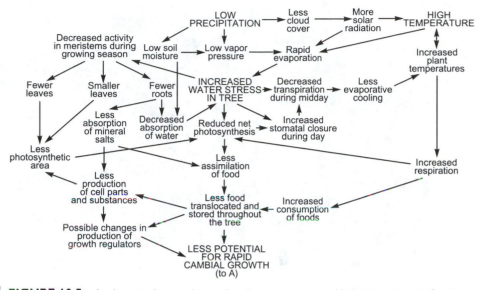

FIGURE 10.5 A schematic diagram showing how low precipitation and high temperature *before* the growing season may lead to a narrow tree ring in arid-site trees (Fritts, 1971).

lated with a number of different climatic factors in both the growth season (year t_0) and in the preceding months, as well as with the record of prior growth itself (generally in the preceding growth years, t_{-1} and t_{-2}). Indeed in some dendroclimatic reconstructions, tree growth in subsequent years (t_{+1}, t_{+2}, etc.) may also be included as they also contain climatic information about year t_0.

Trees are sampled radially using an increment borer, which removes a core of wood (generally 4–5 mm in diameter) and leaves the tree unharmed. It is important to realize that dendroclimatic studies are unreliable unless an adequate number of samples are recovered; two or three cores should be taken from each tree and at least 20 trees should be sampled at an individual site, though this is not always possible. Eventually, as will be discussed, all of the cores are used to compile a master chronology of ring-width variation for the site and it is this that is used to derive climatic information.

10.2.2 Cross-Dating

For tree-ring data to be used for paleoclimatic studies, it is essential that the age of each ring be precisely known. This is necessary in constructing the master chronology from a site where ring widths from modern trees of similar age are being compared, and equally necessary when matching up sequences of overlapping records from modern and archeological specimens to extend the chronology back in time (Stokes and Smiley, 1968). Great care is needed because occasionally trees will produce false rings or intra-annual growth bands, which may be confused with the actual earlywood/latewood transition (Fig. 10.6). Furthermore, in extreme years some

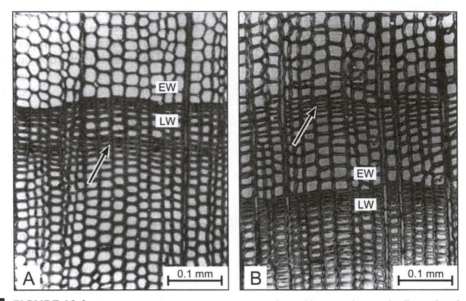

FIGURE 10.6   Annual growth increments or rings are formed because the wood cells produced early in the growing season (earlywood, EW) are large, thin-walled, and less dense, while the cells formed at the end of the season (latewood, LW) are smaller, thick-walled, and more dense. An abrupt change in cell size between the last-formed cells of one ring (LW) and the first-formed cells of the next (EW) marks the boundary between annual rings. Sometimes growing conditions temporarily become severe before the end of the growing season and may lead to the production of thick-walled cells within an annual growth layer (arrows). This may make it difficult to distinguish where the actual growth increment ends, which could lead to errors in dating. Usually these intra-annual bands or false rings can be identified, but where they cannot the problem must be resolved by cross-dating (Fritts, 1976).

trees may not produce an annual growth layer at all, or it may be discontinuous around the tree or so thin as to be indistinguishable from adjacent latewood (i.e., a partial or missing ring) (Fig. 10.7). Clearly, such circumstances would create havoc with climatic data correlation and reconstruction, so careful cross-dating of tree ring series is necessary. This involves comparing ring width sequences from each core so that characteristic patterns of ring-width variation (ring-width "signatures") are correctly matched (Fig. 10.8). If a false ring is present, or if a ring is missing, it will thus be immediately apparent (Holmes, 1983). The same procedure can be used with archeological material; the earliest records from living trees are matched or cross-dated with archeological material of the same age, which may, in turn, be matched with older material. This procedure is repeated many times to establish a thoroughly reliable chronology. In the southwestern United States, the ubiquity of beams or logs of wood used in Indian pueblos has enabled chronologies of up to 2000 years to be constructed in this way. In fact, accomplished dendrochronologists can quickly pinpoint the age of a dwelling by comparing the tree-ring sequence in supporting timbers with master chronologies for the area (Robinson, 1976). Similarly, archeologically important chronologies have been established in western Eu-

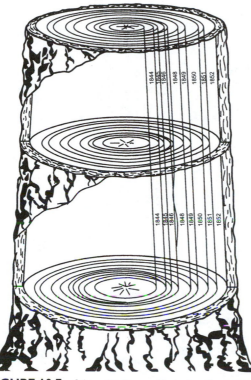

FIGURE 10.7   Schematic diagram illustrating the potential difficulty presented by the formation of a partial ring (in 1847). In the lowest two sections the ring might not be sampled by an increment borer, which removes only a narrow wood sample. In the upper section the ring is thin, but present all around the tree circumference. Such missing or partially absent rings are identified by careful cross-dating of multiple samples (Glock, 1937).

rope. Hoffsummer (1996) for example, has used beams of wood from buildings in southeastern Belgium to establish an oak chronology extending back to A.D. 672, and in several regions of France chronologies of over 1000 yr in length have been assembled from construction timbers (Lambert *et al.*, 1996). Dendrochronological studies have also been used in studies of important works of art, for example in dating wooden panels used for paintings, furniture, and even the coverboards of early books (Eckstein *et al.*, 1986; Lavier and Lambert, 1996). Finally, tree stumps recovered from alluvial sediments and bogs have been cross-dated to form composite chronologies extending back through the entire Holocene. Long tree-ring series such as these are so accurate that they are used to calibrate the radiocarbon timescale (see Section 3.2.1.5). Tree rings are unique among paleoclimatic proxies in that, through cross-dating of multiple cores, the absolute age of a sample can be established. This attribute distinguishes tree rings from other high resolution proxies (ice cores, varved sediments, corals, banded speleothems, etc.) because in such proxies, comparable replication of records and cross-dating of many samples across multiple sites is rarely possible. Consequently, with the exception of specific marker

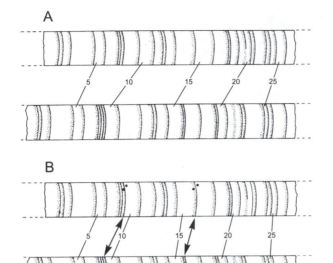

FIGURE 10.8   Cross dating of tree rings. Comparison of tree-ring widths makes it possible to identify false rings or where rings are locally absent. For example, in **(A)** strict counting shows a clear lack of synchrony in the patterns. In the lower specimen of **(A)**, rings 9 and 16 can be seen as very narrow, and they do not appear at all in the upper specimen. Also, rings 21 (lower) and 20 (upper) show intra-annual growth bands. In **(B)**, the positions of inferred absence are designated by dots (upper section), the intra-annual band in ring 20 is recognized, and the patterns in all ring widths are synchronously matched (Fritts, 1976).

horizons (e.g., those associated with a volcanic event of known age), dating of such proxies is always more uncertain than in dendroclimatic studies (Stahle, 1996).

Once the samples have been cross-dated and a reliable chronology has been established there are three important steps to produce a dendroclimatic reconstruction:

1. standardization of the tree ring parameters to produce a site chronology;
2. calibration of the site chronology with instrumentally recorded climatic data, and production of a climatic reconstruction based on the calibration equations; and
3. verification of the reconstruction with data from an independent period not used in the initial calibration.

In the next three sections, each of these steps is discussed in some detail.

10.2.3 Standardization of Ring-Width Data

Once the chronology for each core has been established, individual ring widths are measured and plotted to establish the general form of the data (Fig. 10.9). It is common for time series of ring widths to contain a low frequency component resulting entirely from the tree growth itself, with wider rings generally produced during the early life of the tree. In order that ring-width variations from different cores can be

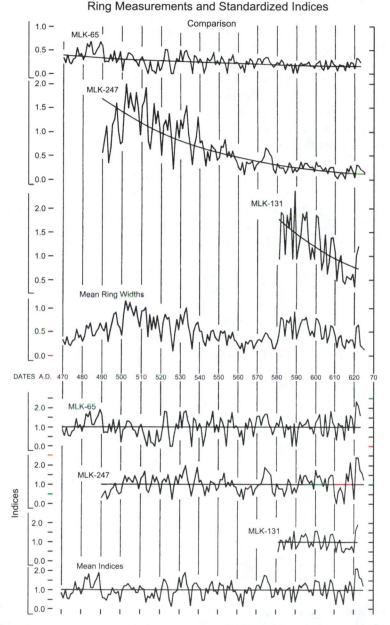

FIGURE 10.9 Standardization of ring-width measurements is necessary to remove the decrease in size associated with increasing age of the tree. If the ring widths for the three specimens shown in the upper figure are simply averaged by year, without removing the effect of the tree's age, the mean ring-width chronology shown below them exhibits intervals of high and low growth, associated with the varying age of the samples. This age variability is generally removed by fitting a curve to each ring-width series, and dividing each ring width by the corresponding value of the curve. The resulting values, shown in the lower half of the figure, are referred to as indices, and may be averaged among specimens differing in age to produce a mean chronology for a site (lowermost record) (Fritts, 1971).

compared, it is first necessary to remove the growth function peculiar to that particular tree. Only then can a master chronology be constructed from multiple cores. Growth functions are removed by fitting a curve to the data and dividing each measured ring-width value by the "expected" value on the growth curve (Fig. 10.9). Commonly, a negative exponential function, or a lowpass digital filter is applied to the data. Cook *et al.* (1990) recommend that a cubic-smoothing spline be used, in which the 50% frequency response equals ~75% of the record length (n). This means that low frequency variations in the data (with a period $>0.75\ n$) are largely removed from the standardized data, so the analyst then has an explicit understanding of the frequency domain that the resulting series represents (Cook and Peters, 1981).

The standardization procedure leads to a new time series of ring-width indices, with a mean of one and a variance that is fairly constant through time (Fritts, 1971). Ring-width indices from individual cores are then averaged, year by year, to produce a master chronology for the sample site, independent of growth function and differing sample age (Fig. 10.9, lowest graph). Averaging the standardized indices also increases the (climatic) signal-to-noise ratio (S/N). This is because climatically related variance, common to all records, is not lost by averaging, whereas non-climatic "noise," which varies from tree to tree, will be partially cancelled in the averaging process. It is thus important that a large enough number of cores be obtained initially to help enhance the climatic signal common to all the samples (Cook *et al.*, 1990).

Standardization is an essential prerequisite to the use of ring-width data in dendroclimatic reconstruction but it poses significant methodological problems. Consider, for example, the ring-width chronologies shown in Fig. 10.10. Drought-sensitive conifers from the southwestern United States characteristically show ring-width variations like those in Fig. 10.10a. For most of the chronology a negative exponential function, of the form $y = ae^{-bt} + k$, fits the data well. However, this is not the case for the early section of the record, which must be either discarded or fitted by a different mathematical function. Obviously, the precise functions selected will have an important influence on the resulting values of the ring-width indices. In the case of trees growing in a closed canopy forest, the growth curve is often quite variable and unlike the negative exponential values characteristic of arid-site conifers. Periods of growth enhancement or suppression related to non-climatic factors such as competition, management, insect infestation, etc., are often apparent in the records. In such cases (Fig. 10.10b) some other function might be fitted to the data and individual ring widths would then be divided by the local value of this curve to produce a series of ring-width indices. Care must be exercised not to select a function (such as a complex polynomial) that describes the raw data too precisely, or *all* of the (low frequency) climatic information may be removed; most analysts select the simplest function possible to avoid this problem but inevitably the procedure selected is somewhat arbitrary. Further problems arise when complex growth functions are observed, such as those in Fig. 10.10c. In this case it would be difficult to decide on the use of a polynomial function (dashed line) or a negative exponential function (solid line) and in either case the first few observations should perhaps be discarded. Such difficulties are less significant in densitometric or isotope dendroclimatic studies because there is generally less of a growth trend in density

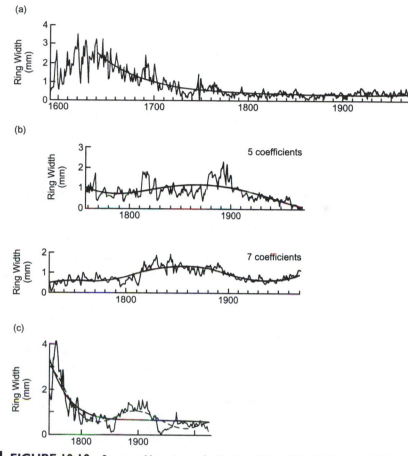

FIGURE 10.10 Some problems in standardization of ring widths. In **(a)** most of the tree-ring series can be fitted by the exponential function shown. However, the early part of the record must be discarded. In **(b)** the two ring-width series required higher-order polynomials to fit the lower frequency variations of each record (the greater the number of coefficients for each equation, the greater the degree of complexity in the shape of the curve). In **(c)** the series could be standardized using either a polynomial (dashed) or exponential function (solid line). Depending on the function selected and its complexity, low-frequency climatic information may be eliminated. The final ring-width indices depend very much on the standardization procedure employed (examples selected from Fritts, 1976).

and isotope data; hence these approaches may yield more low-frequency climatic information than is possible in the measurement of ring widths alone (Schweingruber and Briffa, 1996).

It is clear that standardization procedures are not easy to apply and may actually remove important low-frequency climatic information. It is not possible, *a priori*, to decide if part of the long-term change in ring width is due to a coincident climatic trend. The problem is exacerbated if one is attempting to construct a long-term dendrochronological record, when only tree fragments or historical timbers spanning limited time intervals are available, and the corresponding growth function may not be apparent.

The consequences of different approaches to standardization are well illustrated by the studies of long tree-ring series (Scots pine, *Pinus sylvestris*) from northern Fennoscandia by Briffa *et al.* (1990, 1992a). In order to produce a long dendroclimatic reconstruction extending over 1500 yr, Briffa *et al.* (1990) constructed a composite chronology made up of many overlapping cores which varied in their individual length, from less than 100 to more than 200 yr (Figs. 10.11 and 10.13). In the shorter segments, the growth function is significant over the entire segment length, but in longer segments the growth factor becomes less significant (see Fig. 10.9, upper panel). In Briffa *et al.* (1990) each segment was standardized individually (the procedure used in almost all dendroclimatic studies), in this case by the use of a cubic spline function that retains variance at periods less than ~2/3 of the record length. Thus, in a 100 yr segment, variance at periods >66 yr would be removed, whereas in a 300 yr segment, variance at periods up to 200 yr would be retained. All standardized cores were then averaged together, producing the record shown in Fig. 10.12c. This shows considerable interannual to decadal scale variability, but little long-term low frequency variability. In fact, as the mean segment length varies over time (Fig. 10.11) so too will the low frequency variance represented in the composite series.

In Briffa *et al.* (1992a) the standardization procedure was revised by first aligning all core segments by their relative age, then averaging them (i.e., all values of the first year in each segment (t_1) were averaged, then all values of t_2 etc. . . . . to t_n). This assumes that in each segment used, t_1 was at, or very close to, the center of the tree and that there is a tree-growth function (dependent only on biological age) that is

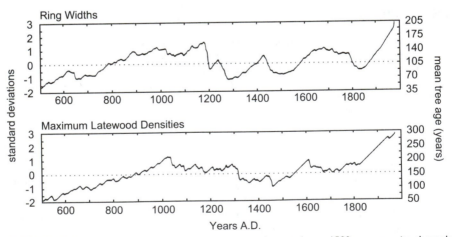

FIGURE 10.11   Average age of the core segments used to produce a 1500-yr composite chronology from trees in northern Fennoscandia. Samples representing the recent period are generally longest because samples are usually selected from the oldest living trees, whereas older samples (dead tree stumps) may be from trees of any age (Briffa *et al.*, 1992a).

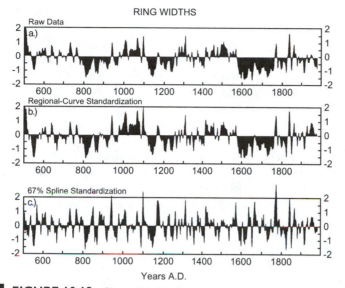

FIGURE 10.12   Ring-width data from trees in northern Fennoscandia showing **(a)** the mean indices without any standardization, **(b)** indices derived from standardization using the regional curve standardization, or **(c)** indices derived from standardization using a cubic-smoothing spline function (see text for discussion) (Briffa et al., 1992a).

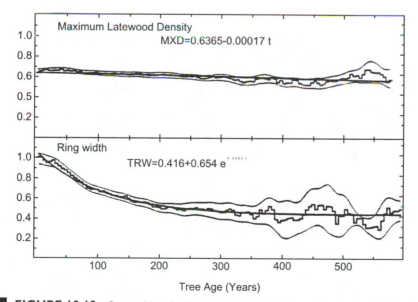

FIGURE 10.13   Regional standardization curve (RCS) of tree-ring samples from northern Fennoscandia based on a least squares fit to the mean values of all series, after they were aligned according to their biological age (Briffa et al., 1992a). Ring-width data commonly has a more pronounced growth function than maximum latewood density data.

common to all samples. The resulting "regional curve" provided a target for deriving a *mean* growth function, which could be applied to all of the individual core segments regardless of length (Fig. 10.13). Averaging together the core segments, standardized in this way by the regional curve, produced the record shown in Fig. 10.12b. This has far more low frequency information than the record produced from individually standardized cores (Fig. 10.12c) and retains many of the characteristics seen in the original data (Fig. 10.12a). From this series, low growth from the late 1500s to the early 1800s is clearly seen, corresponding to other European records that record a "Little Ice Age" during this interval. Also seen is a period of enhanced growth from A.D. ~950–1100, during a period that Lamb (1965) characterized as the "Medieval Warm Epoch." It is apparent from a comparison of Figs. 10.12b and 10.12c that any conclusions drawn about which were the warmest or coldest years and decades of the past can be greatly altered by the standardization procedure employed. All of the high frequency variance of Fig. 10.12c is still represented in the record produced by regional curve standardization but potentially important climatic information at lower frequencies is also retained. The problem of extracting low frequency climatic information from long composite records made up of many individual short segments is addressed explicitly by Briffa *et al.* (1996) and Cook *et al.* (1995) who refer to this as the "segment length curse"! Although it is of particular concern in dendroclimatology, it is in fact an important problem in all long-term paleoclimatic reconstructions that utilize limited duration records to build up a longer composite series (e.g., historical data).

10.2.4 Calibration of Tree-Ring Data

Once a master chronology of standardized ring-width indices has been obtained, the next step is to develop a model relating variations in these indices to variations in climatic data. This process is known as calibration, whereby a statistical procedure is used to find the optimum solution for converting growth measurements into climatic estimates. If an equation can be developed that accurately describes instrumentally observed climatic variability in terms of tree growth over the same interval, then paleoclimatic reconstructions can be made using only the tree-ring data. In this section, a brief summary of the methods used in tree-ring calibration is given. For a more exhaustive treatment of the statistics involved, with examples of how they have been used, the reader is referred to Hughes *et al.* (1982) and Fritts *et al.* (1990).

The first step in calibration is selection of the climatic parameters that primarily control tree growth. This procedure, known as response function analysis, involves regression of tree-ring data (the predictand) against monthly climatic data (the predictors, usually temperature and precipitation) to identify which months, or combination of months, are most highly correlated with tree growth. Usually months during and prior to the growing season are selected but the relationship between tree growth in year t_0, t_{-1} may also be examined as tree growth in year t_0 is influenced by conditions in the preceding year. If a sufficiently long set of climatic data is available, the analysis may be repeated for two separate intervals to determine if the relationships are similar in both periods (Fig. 10.14). This then leads to selection of the

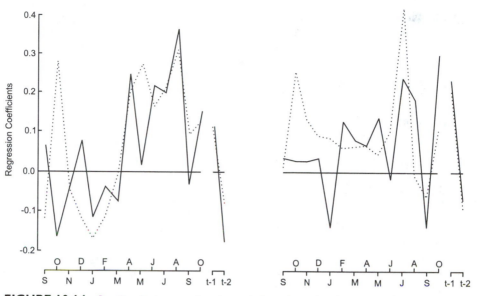

FIGURE 10.14 Results of a response function analysis to determine the pattern of growth response in Scots pine (from northern Fennoscandia) to temperature in the months September (in year t_{-1}) to October in year t_0. Values plotted are coefficients from multiple regression analysis of tree-ring maximum latewood density (left) and ring widths (right) in relation to instrumentally recorded temperatures in the region. Two periods were used for the analysis to examine the stability of the relationships: 1876–1925 (dotted lines) and 1926–1975 (solid line). In addition to the monthly climate variables, the ring width values from the two preceding years (t_{-1} and t_{-2}) are also included to assess the importance of biological preconditioning of growth in one summer by conditions in preceding years. The analysis shows that maximum latewood density is increased by warm conditions from April–August of year t_0, and a similar, but less strong signal is found in tree-ring widths. Growth in the previous year is also important (Briffa *et al.*, 1990).

month or months on which the tree-ring records are dependent, and which the tree rings can therefore be expected to usefully reconstruct. For example, by this approach Jacoby and D'Arrigo (1989) found that the ring widths of white spruce (*Picea glauca*) at the northern treeline in North America are strongly related to mean annual temperature, whereas Briffa *et al.* (1988, 1990) showed that summer temperature (April/May to August/September) is the major control on ring widths and maximum latewood density in Scots pine (*Pinus sylvestris*) in northern Fennoscandia.

Once the climatic parameters that influence tree rings have been identified, tree-ring data can be used as predictors of these conditions. Various levels of complexity may be involved in the reconstructions (Table 10.1). The basic level uses simple linear regression in which variations in growth indices at a single site are related to a single climatic parameter, such as mean summer temperature or total summer precipitation. An example of this approach is the work of Jacoby and Ulan (1982), where the date of the first complete freezing of the Churchill River estuary on Hudson Bay was reconstructed from a single chronology located near Churchill, Manitoba. Similarly, Cleaveland and Duvick (1992) reconstructed July hydrological drought indices for Iowa from a single regional chronology which was an average

TABLE 10.1   Different Levels of Complexity in the Methods Used to Determine Relationship Between Tree Ring Parameter and Climate

| Level | Number of variables of | | Main statistical procedures used |
	Tree growth	Climate	
I	1	1	Simple linear regression analysis
IIa	n	1	Multiple linear regression (MLR)
IIb	nP	1	Principal components analysis (PCA)
IIIa	nP	nP	Orthogonal spatial regression (PCA and MLR)
IIIb	nP	nP	Canonical regression analysis (with PCA)

1 = a temporal array of data.

n = a spatial and temporal array of data.

nP = number of variables after discarding unwanted ones from PCA.

From Bradley and Jones (1995).

of 17 site chronologies of the same species. More commonly, multivariate regression is used to define the relationship between the selected climate variables and a set of tree-ring chronologies within a geographical area where there is a common climate signal. The tree ring data may include both ring widths and density values. Equations that relate tree rings (the predictors) to climate (the predictand) are termed transfer functions with a basic form (assuming linear relationships) of:

$$y_t = a_1 x_{1t} + a_2 x_{2t} + a_3 x_{3t} \ldots + a_m x_{mt} + b + e_t$$

where y_t is the climate parameter of interest (for year t); $x_{1t}, \ldots, x_{mt}$ are tree-ring variables (e.g., from different sites) in year t; $a_1, \ldots, a_m$ are weights or regression coefficients assigned to each tree-ring variable, b is a constant, and e_t is the error or residual. In effect, the equation is simply an expansion of the linear equation, $y_t = ax_t + b + e_t$, to incorporate a larger number of terms, each additional variable accounting for more of the variance in the climate data (Ferguson, 1977). Theoretically, it would be possible to construct an equation to predict the value of y_t precisely. However, adding too many coefficients simply widens the confidence limits about the reconstruction estimates so that eventually the uncertainty become so large that the reconstruction is virtually worthless. What is needed is an equation that uses the minimum number of tree ring variables to account for the maximum amount of variance in the climate record. Commonly, the procedure of stepwise multiple regression is used to achieve this aim (Fritts, 1962, 1965). From a matrix of potentially influential predictor variables, the one that accounts for most of the climate variance is selected; next, the predictor that accounts for the largest proportion of the remaining climatic variance is identified and added to the equation, and so on in a stepwise manner. Tests of statistical significance, as each variable is selected, enable the procedure to be terminated when a further increase in the number of variables in the equation results in an insignificant increase in variance explanation. In this way, only the most important variables are selected, objectively, from the large array of potential predictors. An example of this approach is the reconstruction of drought in Southern California by Meko *et al.* (1980).

A major problem with stepwise regression is that intercorrelations between the tree-ring predictors can lead to instability in the prediction equation. In statistical terms this is referred to as multicollinearity. To overcome this, a common procedure is to transform the predictor variables into their principal modes (or empirical orthogonal functions, EOFs) and use them as predictors in the regression procedure. Principal components analysis involves mathematical transformations of the original data set to produce a set of orthogonal (i.e., uncorrelated) eigenvectors that describe the main modes of variance in the multiple parameters making up the data set (Grimmer, 1963; Stidd, 1967; Daultrey, 1976). Each eigenvector is a variable that expresses part of the overall variance in the data set. Although there are as many eigenvectors as original variables, most of the original variance will be accounted for by only a few of the eigenvectors. The first eigenvector represents the primary mode of distribution of the data set and accounts for the largest percentage of its variance (Mitchell *et al.*, 1966). Subsequent eigenvectors account for lesser and lesser amounts of the remaining variance (Fig. 10.15). Usually only the first few eigenvectors are considered, as they will have captured most of the total variance. The value or amplitude of each eigenvector varies from year to year, being highest in the year when that particular combination of conditions represented by the eigenvector is most apparent. Conversely, it will be lowest in the year when the inverse of this combination is most apparent in the data. Eigenvector amplitudes can then be used as orthogonal predictor variables in the regression procedure, generally accounting for a higher proportion of the dependent data variance with fewer variables than would be possible using the "raw" data themselves. A time series of eigenvector values (amplitudes) is referred to as a principal component (PC), the dominant eigenvector being PC1, the next most common pattern PC2, etc.

Apart from reducing the number of potential predictors, principal components analysis also simplifies multiple regression considerably. It is not necessary to use the stepwise procedure because the new potential predictors are all orthogonal. This approach was used in the reconstruction of July drought in New York's Hudson Valley by Cook and Jacoby (1979). They selected series of ring-width indices from six different sites, and calculated eigenvectors of their principal characteristics. These were then used as predictors in a multiple regression analysis with Palmer Drought Severity Indices (Palmer, 1965)[32] as the dependent variable. The resulting equation, based on climatic data for the period 1931–1970, was then used to reconstruct Palmer indices back to 1694 when the tree-ring records began (Fig. 10.16). This reconstruction showed that the drought of the early 1960s, which affected the entire northeastern United States, was the most severe the area has experienced in the last three centuries.

---

[32] Palmer indices are measures of the relative intensity of precipitation abundance or deficit and take into account soil-moisture storage and evapotranspiration as well as prior precipitation history. Thus they provide, in one variable, an integrated measure of many complex climatic factors. They are scaled from +4 or more (extreme wetness) to −4 or less (extreme drought) and are widely used by agronomists in the United States as a guide to climatic conditions relevant to crop production.

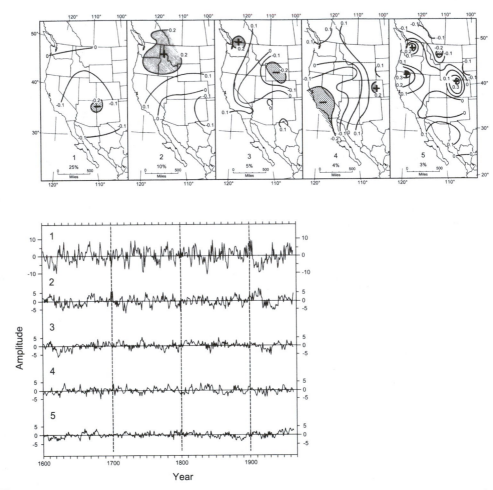

FIGURE 10.15 The first five eigenvectors of tree-ring widths based on a network of 65 chronologies distributed across the western United States, northern Mexico, and southwestern Canada. These represent the major patterns of growth anomalies in the region. Eigenvector 1 accounts for 25% of the overall variance; each subsequent eigenvector accounts for progressively less. The lower diagram shows the relative amplitudes of these five eigenvectors since A.D. 1600 (i.e., the principal components, [PCs] 1–5). These and other PCs were used in canonical regressions with gridded temperature, precipitation, and pressure data, first to calibrate the tree-ring data and then to reconstruct maps of each climatic parameter back to A.D. 1600 (Fritts, 1991).

Simple univariate transfer functions express the relationship between one climatic variable and multiple tree-ring variables. A more complex step is to relate the variance in multiple growth records to that in a multiple array of climatic variables (e.g., summer temperature over a large geographical region) (Table 10.1). To do so, each matrix of data (representing variations in both time and space) is converted into its principal modes or eigenvectors; these are then related using canonical regression or orthogonal spatial regression techniques (Clark, 1975; Cook *et al.*, 1994). These

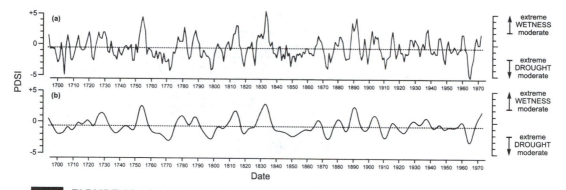

FIGURE 10.16 July Palmer drought indices for the Hudson Valley, New York, from 1694 to 1972 reconstructed from tree rings. **(a)** Unsmoothed estimates; **(b)** a lowpass filtered version of the unsmoothed series that emphasizes periods of ≥10 yr (Cook and Jacoby, 1979).

techniques involve identifying the variance that is common to individual eigenvectors of the two different data sets and defining the relationship between them. The techniques are important in that they allow spatial arrays (maps) of tree-ring indices to be used to reconstruct maps of climatic variation through time (Fritts *et al.*, 1971; Fritts, 1991; Fritts and Shao, 1992; Briffa *et al.*, 1992b).

The most comprehensive work of this sort is that of Fritts (1991). Ring-width indices from 65 sites across western North America (i.e., 65 variables) were transformed into eigenvectors, where each one represented a spatial pattern of growth covariance among the sites (Fig. 10.15). The first ten eigenvectors accounted for 58% of the joint space-time variance of growth anomaly over the site network. Eigenvectors were also derived for seasonal pressure data at grid points over an area extending from the eastern Pacific (100° E) to the eastern U.S. (80° W) and from 20 to 70° N; the first three eigenvectors of pressure accounted for ~56% of variance in the data. Using amplitudes of all these eigenvectors for the years common to both data sets (1901–1962), canonical weights were computed for the growth eigenvectors to give maximum correlations with pressure anomalies. Amplitudes of these weighted eigenvectors were then used as predictors of normalized pressure departures at each point in the pressure grid network, by applying the canonical weights to the standardized ring-width data (the level IIIb approach in Table 10.1). This resulted in estimates of pressure anomaly values at each point in the grid network, for each season, for each year of the ring-width network. Maps of mean pressure anomaly could thus be produced for any interval by simply averaging the individual anomaly values (Fritts and Shao, 1992). The same procedure was used for a network of gridded temperature and precipitation data across the United States. Figure 10.17 gives the mean pressure, temperature and precipitation anomalies for 1602–1900 for each gridded region compared to twentieth century means. This ~300 yr interval spans much of what is commonly referred to as the "Little Ice Age" and it is interesting that the reconstruction suggests there was a stronger ridge over western North America, with higher pressure over Alaska during this time. This was associated with

Reconstructions
1602-1900
compared to
instrumental record
1901-1970

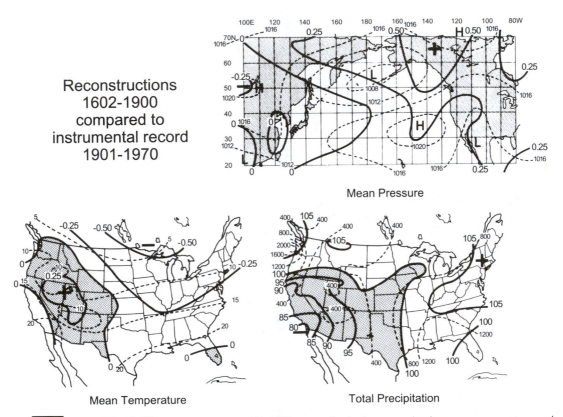

FIGURE 10.17 Anomalies (from 1901–1970 mean values) of mean sea-level pressure, temperature, and precipitation (in % of mean) for the period 1602–1900. Values were reconstructed from a 65-chronology network of tree-ring width data (see Fig. 10.15). Shaded areas on the temperature and precipitation maps are warm or dry anomalies, respectively (Fritts, 1991).

higher temperatures and lower precipitation over much of the western and southwestern United States, but cooler and wetter conditions in the east and northeast. Precipitation was also higher in the Pacific Northwest, presumably reflecting more frequent depressions affecting this area. In effect, the reconstruction points to an amplification of the Rossby wave pattern over North America, with an increase in cold airflow from central Canada into the central and eastern U.S. (Fritts, 1991; Fritts and Shao, 1992). Reconciling these reconstructions with evidence for extensive glacier advances in the Rockies and other mountain ranges of western North America during this period is clearly very difficult (Luckman, 1996).

Related procedures have been used by other workers. Briffa *et al.* (1988) for example, used orthogonal spatial regression to reconstruct April–September temperatures over Europe west of 30° E using densitometric information from conifers over Europe. In their procedure both the spatial array of temperature and the spatial array of densitometric data were first reduced to their principal components. Only significant components in each set were retained. Each retained PC of climate was then regressed in turn against the set of retained densitometric PCs. This procedure can be thought of as repeating the level IIb approach (Table 10.1) m times, where m is the number of retained climate PCs. Having found all of the significant regression coefficients, the set of equations relating the climate PCs to the tree-growth PCs were then transformed back to original variable space, resulting in an equation for each temperature location in terms of all the densitometric chronologies. A similar approach was used by Schweingruber *et al.* (1991) and Briffa and Schweingruber (1992) to derive temperature reconstructions for Europe, and by Briffa *et al.* (1992b), who derived temperature anomaly maps for the western United States from a network of tree-ring density data.

Fig. 10.18 shows some examples of these reconstructions in comparison with instrumental data for the same years, providing a qualitative impression of how good the pre-instrumental reconstructions might be. In fact, verification statistics over an independent period are generally good, providing a more quantitative assessment that earlier reconstructions are likely to be reliable. Of particular interest is the reconstruction of temperatures in the early nineteenth century, around the time of the eruption of Tambora (Fig. 10.19). Tambora (8° S, 118° E) exploded in April 1815; it is considered to have been the largest eruption in the last thousand years if not the entire Holocene (Rampino and Self, 1982; Stothers, 1984). Contemporary accounts

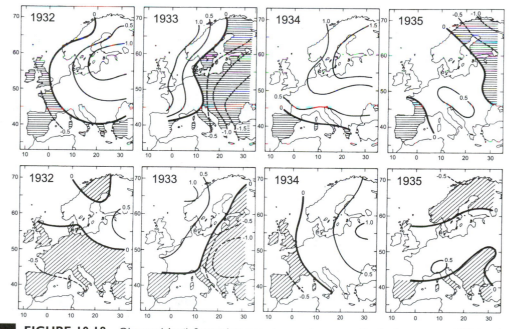

FIGURE 10.18 Observed April–September mean temperature anomalies in the summers of the early 1930s (expressed as departures from the 1951–1970 mean) compared with (bottom panel) the corresponding reconstruction based on a tree-ring density network made up of 37 chronologies distributed across the region. (Schweingruber *et al.*, 1991).

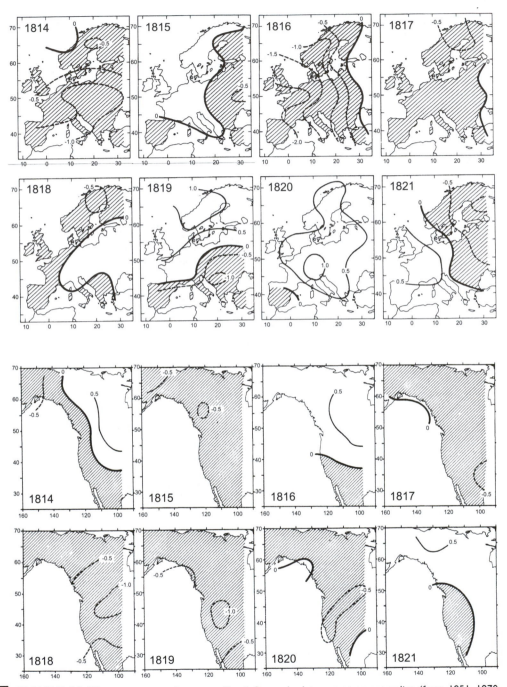

FIGURE 10.19 Reconstructed summer (April–September) temperature anomalies (from 1951–1970 means) for western Europe and western North America for the early nineteenth century based on a 37 site tree-ring density network in Europe and a 53 site tree-ring density network in North America. Conditions were cold in both regions in the years following the eruption of Tambora, though 1816 itself was mostly warm in western North America (Schweingruber *et al.*, 1991).

document the severe climatic conditions that were experienced in western Europe and in the eastern United States in the following year, which became known as "the year without a summer" (Harington, 1992). The dendroclimatic reconstruction for summer temperatures (April–September) in western Europe indeed show extremely cold conditions in 1816 with cold summers also in 1817 and 1818 over most of western Europe (and northwestern Russia; Shiyatov, 1996) continuing into 1819 in southern Europe. Over most of the western United States and Alaska, 1816 was not cold, but the following four summers were uniformly cooler than the 1951–1970 reference period. The coldest summer in the last few centuries in the western United States (1601) was also associated with an eruption, probably Huaynaputina in Peru (Briffa *et al.*, 1992b, 1994). Temperatures across the region averaged 2.2 °C below the 1881–1982 mean as a result of that event.

Before concluding this section on calibration, it is worth noting that tree-ring indices need not be calibrated only with climatic data. The ring-width variations contain a climatic signal and this may also be true of other natural phenomena that are in some way dependent on climate. It is thus possible to calibrate such data directly with tree rings and to use the long tree-ring records to reconstruct the other climate-related series. In this way, dendroclimatic analysis has been used to reconstruct runoff records (Stockton, 1975; Stockton and Boggess, 1980) and lake-level variations (Stockton and Fritts, 1973). Some of these applications are discussed in more detail in Section 10.3.3.

10.2.5 Verification of Climatic Reconstructions

An essential step in dendroclimatic analysis (indeed in all paleoclimatic studies) is to test or verify the paleoclimatic reconstruction in some way. The purpose of verification is to test if the transfer function model (derived from data in the calibration period) is stable over time, usually by comparing part of the reconstruction with independent data from a different period. Inevitably, when the prediction estimates are tested against an independent data set the amount of explained variance will almost always be less than in the calibration period. To quantify how good the reconstruction is, in comparison with independent data, various statistical tests are generally performed (Gordon, 1982; Fritts *et al.*, 1990; Fritts, 1991, Appendix 1). These statistics then provide some level of confidence in the rest of the reconstruction; the performance of the transfer function over the verification period is the best guide to the likely quality of the reconstruction for periods when there are simply no instrumental data.

Two approaches to verification are generally adopted. First, when calibrating the tree-ring data, very long instrumental records for the area are sought. Only part of these records are then used in the calibration, leaving the remaining early instrumental data as an independent check on the dendroclimatic reconstruction. If the reconstruction is in the form of a map, several records from different areas may be used to verify the reconstruction, perhaps indicating geographical regions where the reconstructions appear to be most accurate (Briffa *et al.*, 1992b). This approach is difficult in some areas where tree-ring studies have been carried out (e.g., the western United States and northern treelines) because these are areas with very few early instrumental records (Bradley, 1976). Dendroclimatic studies in western Europe (Serre-Bachet *et al.*, 1992; Briffa and Schweingruber, 1992) can be more exhaustively tested because of the much longer instrumental records in that area. Indeed, it is sometimes possible to conduct two calibrations, with both tree-ring and climatic data from different time periods and to compare the resulting dendroclimatic reconstructions for earlier periods derived independently from the two data sets (Briffa *et al.*, 1988). This provides a vivid illustration of the stability of the derived paleoclimatic reconstructions (Fig. 10.20).

A second approach is to use other proxy data as a means of verification. This may involve comparisons with historical records or with other climate-dependent phenomena such as glacier advances (LaMarche and Fritts, 1971b) or pollen variations in varved lake sediments (Fritts *et al.*, 1979) etc. It may even be possible to use an independent tree-ring data set to compare observed growth anomalies with those expected from paleoclimatic reconstructions. Blasing and Fritts (1975), for exam-

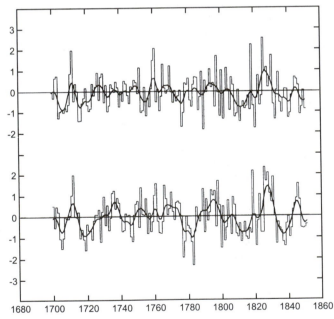

FIGURE 10.20 Two reconstructions of northern Fennoscandinavian temperatures for July–August using the same transfer function model, but with calibration based on 1852–1925 in the upper diagram, and 1891–1964 in the lower diagram. Ordinate axis is in standard deviation units with one unit approximately equal to 1 °C. The two calibrations accounted for 69% and 56% of the variance of instrumental temperature over the same (calibration) interval, respectively. Both gave statistically significant statistics when verified against data from the other "independent" interval. Thus, both reconstructions could be viewed as statistically reliable. Although the two series are strongly correlated ($r = 0.87$) there are important differences between them, which warns against overinterpreting the reconstructions for individual years. For example, the lower diagram shows an extreme in 1783 that does not appear in the upper diagram, and even the lower frequency variations show important differences requiring careful examination of the regression weights used for each reconstruction (Briffa *et al.*, 1988).

ple, used a network of trees from an area between northern Mexico and southern British Columbia to reconstruct maps of sea-level pressure anomalies over the eastern Pacific and western North America. A separate temperature-sensitive data set from Alaska and the Northwest Territories of Canada was then used to test the reconstructions. Periods of anomalously low growth in the northern trees were associated with increased northerly airflow as predicted by the pressure reconstructions.

In all verification tests, one is inevitably faced with two questions:

(a) If the verification is poor, does the fault lie with the dendroclimatic reconstruction (and hence the model from which it was derived) or with the proxy or instrumental data used as a test (which may itself be of poor quality and subject to different interpretations)? In such cases, re-evaluation of the tree-ring data, the model, and the test data must be made before a definitive conclusion can be reached.

(b) Is the dendroclimatic reconstruction for the period when no independent checks are possible as reliable as for the period when verification checks can be made? This might seem an insoluble problem but it is particularly important when one considers the standardizing procedure employed in the derivation of tree-growth indices (Section 10.2.3). Errors are most likely to occur in the earliest part of the record, whereas tests using instrumental data are generally made near the end of the tree-growth record (where replication is generally highest and the slope of the standardization function is generally lowest) and least likely to involve the incorporation of large error. The optimum solution is for both instrumental and proxy data checks to be made on reconstructions at intervals throughout the record, thereby increasing confidence in the overall paleoclimatic estimates.

Dendroclimatologists have set the pace for other paleoclimatologists by developing methods for rigorously testing their reconstructions of climate. Many other fields would benefit by adopting similar procedures.

10.3 DENDROCLIMATIC RECONSTRUCTIONS

The following sections provide selected examples of how tree rings have been used to reconstruct climatic parameters in different regions of the world. This is not by any means an exhaustive review and for further information the reader should examine the sections on dendroclimatology in Bradley and Jones (1995), Jones *et al.* (1996), and Dean *et al.* (1996).

10.3.1 Temperature Reconstruction from Trees at the Northern Treeline

Many studies have shown that tree growth at the northern treeline is limited by temperature; consequently, both tree-ring widths and density provide a record of past variations in temperature. A 1500-yr long summer temperature reconstruction using Scots pine (*Pinus sylvestris*) from northern Scandinavia has already been discussed (see Fig. 10.12). Even longer records may be possible because in this region logs of Holocene age have been dredged from lakes and bogs at or beyond the present treeline. By cross-dating these samples, it should eventually be possible to construct a well-replicated chronology extending back over most of the Holocene (Zetterberg *et al.*, 1995, 1996). Similarly, in the northern Ural mountains of Russia, the dendroclimatic record of living trees has been extended back over 1000 yr by overlapping cores from dead larch trees (*Larix sibirica*) found close to or above present treeline (Graybill and Shiyatov, 1992; Briffa *et al.*, 1995) (see Section 8.2.1). Ring-width and density studies of these samples show warm conditions in the fourteenth and fifteenth centuries, declining to uniformly low summer temperatures in the sixteenth and seventeenth centuries. There is little similarity between Fennoscandian and northern Urals data prior to ~1600, after which time both records show a gradual increase in temperatures through the 20th century. In fact, in the northern Urals, the 20th century (1901–1990) is the warmest period since (at least) 914 AD, although this does not appear to be the case in northern Scandinavia. Care is needed in the interpretation of these apparent long-term changes because of the number of cores and the individual core segment lengths contributing to the overall record vary over time (Briffa *et al.*, 1996). In the northern Urals data set, the replication (sample depth) and mean core length are at a minimum from ~1400–1650 and ~1500–1700, respectively. For similar reasons, reliability of the mean chronology is poor before ~1100. The overall summer temperature reconstruction is therefore very sensitive to the standardization procedure used to correct for growth effects in each core segment; quite different temperature reconstructions can be produced depending on whether cores are individually standardized (by a cubic spline function) or if a regional curve standardization method is employed (Fig. 10.21). This again points to the dangers implicit in long-term paleotemperature reconstructions built up from multiple short cores when sample replication and mean record length are limited.

Shorter records, derived only from living trees have been recovered all along the Eurasian treeline by Schweingruber and colleagues (Schweingruber and Briffa,

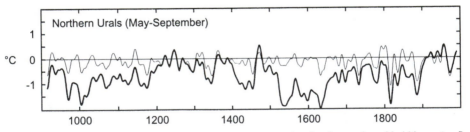

FIGURE 10.21   May–August mean temperature reconstruction for the northern Ural Mountains, Russia, based on tree-ring widths standardized by a cubic-smoothing spline (thin line) and maximum latewood density data standardized by deriving a regional curve for all samples (see Fig. 10.13) (thick line). Data are shown in °C anomalies from the 1951–1970 mean and are smoothed with a 25-yr lowpass filter. The large differences between the two reconstructions indicate that low frequency information is lost in the reconstruction in which only ring widths are used (Briffa *et al.*, 1996). Note that both approaches give comparable reconstructions in the calibration and verification periods, thus providing no *a priori* warning that there might be a problem with the spline-standardized data set.

1996; Vaganov *et al.*, 1996). Both ring-width and density studies of these samples enable a picture of regional variations to be built-up, from Karelia in the west to Chukotka in the east. This shows that the temperature signal is not uniform across the region; for example, from ~80–150° E the early 19th century was uniformly cold, but this cool period was far less pronounced or as persistent farther west, from ~50–80° E. Such variations point to possible shifts in the upper level Rossby wave circulation, which may have been amplified or displaced during this interval. Building up networks of tree-ring data in this way will eventually lead to continental-scale maps of Eurasian temperature anomalies through time (see Fig. 10.19; Schweingruber *et al.*, 1991).

Ring widths in spruce (mainly *Picea glauca*) from the northern treeline of North America contain a strong record of mean annual temperature (Jacoby and D'Arrigo, 1989; D'Arrigo and Jacoby, 1992). Indices averaged over many sites along the treeline, from Alaska to Labrador, reveal a strong correlation with both North American and northern hemisphere temperatures over the last century (Jacoby and D'Arrigo, 1993). Based on these calibrations, long-term variations in temperature have been reconstructed (Fig. 10.22). This indicates low temperatures throughout the seventeenth century with a particularly cold episode in the early 1700s. The eighteenth century was relatively warm, followed by very cold conditions (the coldest of the last 400 years) in the early to mid-1800s. Temperatures then increased to the 1950s, but have since declined. This record is broadly similar to that derived from trees in the northern Ural mountains of Russia, but has less in common with the Fennoscandian summer temperature reconstruction discussed earlier.

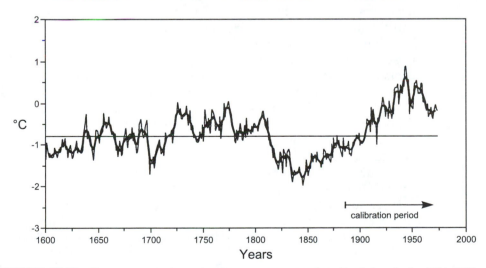

Reconstructed Annual Temperatures for Northern North America

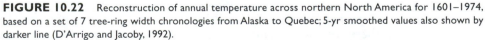

FIGURE 10.22 Reconstruction of annual temperature across northern North America for 1601–1974, based on a set of 7 tree-ring width chronologies from Alaska to Quebec; 5-yr smoothed values also shown by darker line (D'Arrigo and Jacoby, 1992).

Using a different network of trees sampled for maximum latewood density, Schweingruber *et al.* (1993) and Briffa *et al.* (1994) identify distinct regional signals along the North American treeline; by grouping records with common signals they reconstructed summer (April–September) temperatures back to A.D. 1670 for the Alaska/Yukon, Mackenzie Valley, and Quebec/Labrador regions. These series do not capture low frequency variability very well, but do show significant anomalies associated with major volcanic eruptions known from historical records (Jones *et al.*, 1995). However, not all regions are affected in the same way; for example, 1783 (the year that Laki, a major Icelandic volcano erupted) was a very cold summer in Alaska (also noted by Jacoby and D'Arrigo, 1995) but it was relatively warm in the Mackenzie valley. By contrast, 1816 (following the eruption of the Indonesian volcano Tambora in 1815) was exceptionally cold in Quebec and Labrador, but was warm in Alaska/Yukon. Such variations indicate that volcanic aerosols do not always produce uniform climatic responses across the globe and may, in fact, produce strong meridional temperature gradients (Groisman, 1992).

10.3.2 Drought Reconstruction from Mid-latitude Trees

Drought frequency is of critical importance to agriculture in the central and eastern United States. Several dendroclimatic studies have attempted to place the limited observational record in a longer-term perspective by reconstructing precipitation or Palmer Drought Severity Indices (PDSI) for the last few centuries. Cook *et al.* (1992b) used a network of over 150 tree-ring chronologies from across the eastern United States to reconstruct summer PDSI for each of 24 separate regions. On average, 46% of the summer PDSI variance was explained by the tree-ring data. The entire set of reconstructed PDSI for all 24 regions was then subjected to principal components analysis to identify the principal patterns of drought distribution across the entire eastern United States. Six main patterns were identified, with each one corresponding to a different subregion affected by drought (Fig. 10.23). By examining the time series of each principal component, the drought history of each region could be assessed (Fig. 10.24). This revealed that while drought is common in some areas, it is quite rare for drought to extend over many different regions simultaneously. One exception was the period 1814–1822, when severe drought devastated the entire northeastern quadrant of the United States, from New England to Georgia and from Missouri to Wisconsin. This may have been related to circulation anomalies resulting from the major eruption of Tambora in 1815 and/or to four moderate-to-strong El Niño events in quick succession (1814, 1817, 1819, and 1821) (Quinn and Neal, 1992).

Further analysis of drought history in Texas (see Factor 4 in Fig. 10.24) using ring-width chronologies from post oak trees (*Quercus stellata*) has enabled the PDSI to be reconstructed separately for northern and southern Texas (Stahle and Cleaveland, 1988). From these series, the recurrence interval of different levels of drought severity can be calculated (Fig. 10.25). This shows that while the probability of moderate drought (a PDSI of −2 to −3) is high every decade in both north and south Texas, there is a >50% chance of severe drought (PDSI<-4) once a decade in south Texas; in north Texas, however, such droughts are less likely (once in 15 yr).

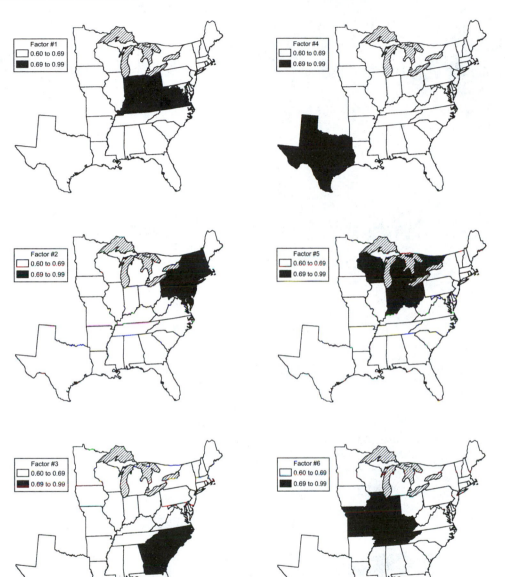

FIGURE 10.23 Major drought patterns across the eastern United States, based on principal components analysis. Shaded areas are where 49% or more of the variance of each drought factor (1 to 6) is explained. The temporal history of pattern 2, for example, will primarily reflect drought in New England, whereas pattern 4 will reflect conditions in Texas (Cook *et al.*, 1992b).

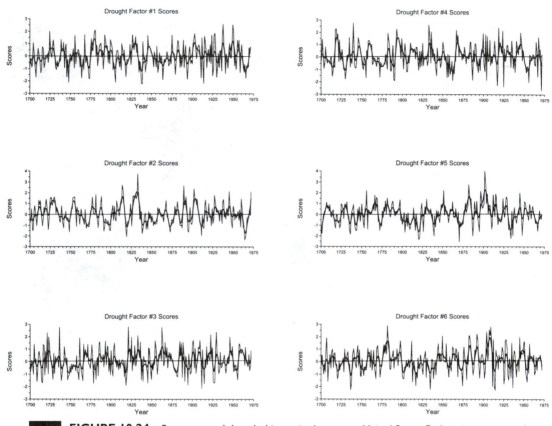

FIGURE 10.24   Reconstructed drought history in the eastern United States. Each series corresponds to a drought pattern shown in Fig. 10.23. The drought in New England in the 1960s (drought factor 2) was unprecedented in the last 270 yr (Cook *et al.*, 1992b).

In the southeastern United States, bald cypress trees (*Taxodium distichum*) growing in swamps have provided a surprisingly good record of precipitation and drought history extending back over 1000 years in some places (Stahle *et al.*, 1985; 1988; Stahle and Cleaveland, 1992). Apparently, tree growth is affected by the water level and water quality in the swamps (oxygenation levels, pH etc.) so tree-ring records show excellent correlation with spring and early summer rainfall in the region. In fact, individual tree-ring chronologies in Georgia and the Carolinas can explain almost as much of the variance in state-wide rainfall averages as a similar network of individual rainfall records (Stahle and Cleaveland, 1992). Long-term reconstructions (>1000 yr) show that non-periodic, multidecadal fluctuations from predominantly wet to dry conditions have characterized the southeastern United States throughout the last millennium (Fig. 10.26). However, since 1650, rainfall in North Carolina appears to have systematically increased, with the period 1956–1984 being one of the wettest periods in the last 370 yr. This trend was

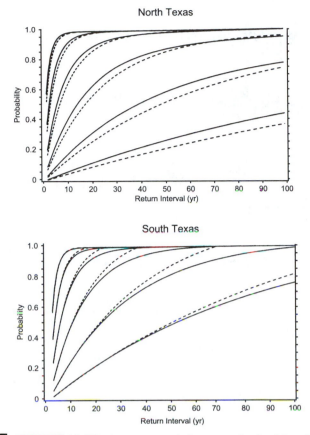

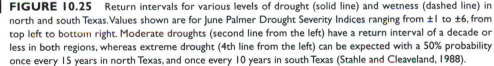

FIGURE 10.25 Return intervals for various levels of drought (solid line) and wetness (dashed line) in north and south Texas. Values shown are for June Palmer Drought Severity Indices ranging from ±1 to ±6, from top left to bottom right. Moderate droughts (second line from the left) have a return interval of a decade or less in both regions, whereas extreme drought (4th line from the left) can be expected with a 50% probability once every 15 years in north Texas, and once every 10 years in south Texas (Stahle and Cleaveland, 1988).

abruptly ended by two remarkable, consecutive drought years (1986 and 1987); only five other comparable two-year droughts have been registered by the bald cypress trees since A.D. 372 (Stahle *et al.*, 1988).

Many meteorological studies have noted the relationship between El Niño/ Southern Oscillation (ENSO) events in the Pacific (or their cold event equivalents, La Niñas) and anomalous rainfall patterns in different parts of the world. These teleconnections result from large-scale displacements of major pressure systems and consequent disruptions of precipitation-bearing storm systems (Ropelewski and Halpert, 1987, 1989; Diaz and Kiladis, 1992). Several attempts have been made to identify an ENSO signal in tree-ring data in order to reconstruct a long-term ENSO index (Lough and Fritts, 1985; Lough, 1992; Meko, 1992; Cleaveland *et al.*, 1992; D'Arrigo *et al.*, 1994). The strongest regional signal in North America is in the

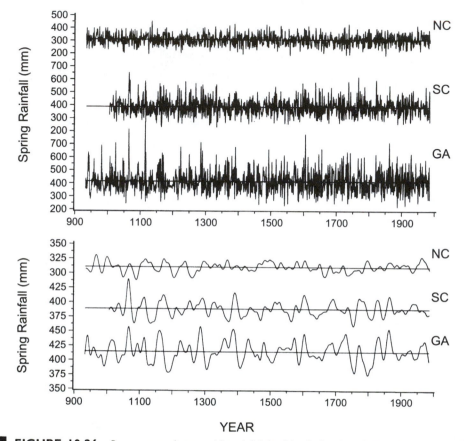

FIGURE 10.26   Reconstructed state-wide rainfall for North Carolina (NC) in April–June and South Carolina and Georgia (SC and GA) in March–June. The upper diagram shows the annual values reconstructed; the lower diagram shows the same data smoothed to emphasize low frequency variations more clearly (with periods >30 yr) (Stahle and Cleaveland, 1992).

southwestern U.S. and northern Mexico, where warm events tend to be associated with higher winter and spring rainfall, which leads to increased tree growth (Fig. 10.27). This led Stahle and Cleaveland (1993) to focus on trees from that area to reconstruct a long-term South Oscillation Index (SOI) back to 1699. Although their results showed considerable skill in identifying many major ENSO events in the past (in comparison with those known from historical records) they estimate that only half of the total number of extremes were clearly defined over the last 300 yr. Similar problems were encountered by Lough and Fritts (1985) using a network of arid-site trees from throughout the western United States. The principal SOI extremes they identify for the last few centuries are not the same as those selected by Stahle and Cleaveland's analysis. This points to the problem of characterizing ENSOs from teleconnection patterns, which are spatially quite variable in relation to both positive and negative extremes of the Southern Oscillation. Consequently, although

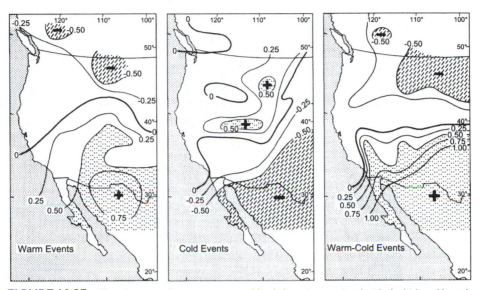

FIGURE 10.27 Tree-ring anomalies across western North America associated with the high and low phases of the Southern Oscillation. The pattern on the left is associated with El Niños; teleconnections lead to heavier winter and spring rainfall in northern Mexico and the U.S. Southwest. By contrast, cold events in the Pacific (La Niñas) are associated with drier conditions and reduced tree growth in the same region (Lough, 1992).

there may be an *overall* signal related to ENSO in one region, precipitation patterns vary enough from event to event that precise, yearly ENSO reconstructions are very difficult (Lough, 1992; D'Arrigo *et al.*, 1994).

One additional factor related to drought is the frequency of fire in some areas (Swetnam, 1993). Fire scars damage the cambium of trees and are clearly visible in tree sections. By building up a chronology of fire history from non-contiguous regions, the frequency of large-scale fires affecting wide areas (related to regional drought episodes) can be identified. In California, fires affecting widely separated groves of giant Sequoia are associated with significant negative winter/spring precipitation anomalies. Over the last 1500 yr, fire frequency was low from A.D. 500–800, reached a maximum ~A.D. 1000–1300, then generally declined. Interestingly, the relationship with temperature in the region is not significant on an annual basis, but over the long-term fire frequency and temperature are positively correlated. Swetnam (1993) attributes this to long-term temperature fluctuations controlling vegetation changes on decadal to century timescales, whereas fire activity from year to year is more related to fuel moisture levels, which are highly correlated with recent precipitation amounts. In the southwestern United States, dry springs and extensive fires are associated with "cold events" in the Pacific (La Niñas) as a result of related circulation anomalies that block moisture-bearing winds from entering the region (Swetnam and Betancourt, 1990). Thus, tree-ring widths and fire occurrence are manifestations of large-scale teleconnections linking sea-surface temperatures in the tropical Pacific to rainfall deficits in the arid Southwest.

10.3.3 Paleohydrology from Tree Rings

Tree rings can be used to reconstruct climate-related phenomena that in some way integrate the effects of the climate fluctuations affecting tree growth. In particular, much work has been devoted to paleohydrological reconstructions involving streamflow. Stockton (1975) was interested in reconstructing long-term variations in runoff from the Colorado River Basin, where runoff records date back only to 1896. As runoff, like tree growth, is a function of precipitation, temperature, and evapotranspiration, both during the summer and in the preceding months, it was thought that direct calibration of tree-ring widths in terms of runoff might be possible. Using 17 tree-ring chronologies from throughout the watershed, eigenvectors of ring-width variation were computed. Stepwise multiple regression analysis was then used to relate runoff over the period 1896–1960 to eigenvector amplitudes over the same interval. Optimum prediction was obtained using eigenvectors of ring width in the growth year (t_0) and also in years t_{-2}, t_{-1}, and t_{+1}, each of which contained climatic information related to runoff in year t_0. In this way an equation accounting for 82% of variance in the dependent data set was obtained; the reconstructed and measured runoff values are thus very similar for the calibration period (Fig. 10.28). The equation was then used to reconstruct runoff back to 1564, using the eigenvector amplitudes of ring widths over this period (Fig. 10.29). The reconstruction indicates that the long-term average runoff for 1564–1961 was ~13 million acre-feet (~16 $\times$ 10^9 m^3) over 2 million acre-feet (~2.47 $\times$ 10^9 m^3) less than during the period of instrumental measurements. Furthermore, it would appear that droughts were more common in this earlier period than during the last century, and the relatively long period of above average runoff from 1905 to 1930 has only one comparable period (1601–1621) in the last 400 yr. Stockton argues that these estimates, based on a longer time period than the instrumental observations, should be seriously considered in river management plans, particularly in regulating flow through Lake Powell, a large reservoir constructed on the Colorado River. In this

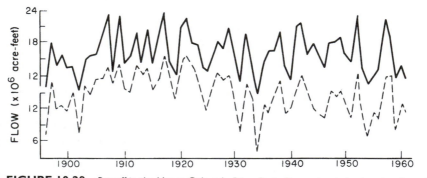

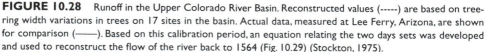

FIGURE 10.28   Runoff in the Upper Colorado River Basin. Reconstructed values (-----) are based on tree-ring width variations in trees on 17 sites in the basin. Actual data, measured at Lee Ferry, Arizona, are shown for comparison (———). Based on this calibration period, an equation relating the two days sets was developed and used to reconstruct the flow of the river back to 1564 (Fig. 10.29) (Stockton, 1975).

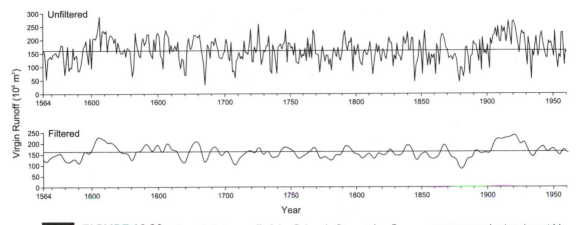

FIGURE 10.29 Annual virgin runoff of the Colorado River at Lee Ferry, as reconstructed using ring-width index variation, calibrated as shown in Fig. 10.28. Growth for each year, and the three following years, was used to estimate water flow statistically. Smooth curve (below) represents essentially a 10-yr running mean. Runoff in this period ~1905–1925 was exceptional when viewed in the context of the last 400 yr (Stockton, 1975).

case, dendrohydrological analysis provided a valuable long-term perspective on the relatively short instrumental record. Similar work has been accomplished by Stockton and Fritts (1973), who used tree-ring eigenvectors calibrated against lake-level data to reconstruct former levels of Lake Athabasca, Alberta, back to 1810 (Figure 10.30). Their reconstruction indicated that although the long-term average lake level is similar to that recorded over the last 40 yr, the long-term variability of lake levels is far greater than could be expected from the short instrumental record. To preserve this pattern of periodic flooding, essential to the ecology of the region, the area is now artificially flooded at intervals that the dendroclimatic analysis suggests have been typical of the last 160 yr.

A few studies have attempted to reconstruct streamflow in humid environments, but they have been less successful than dendroclimatic reconstructions in more arid areas (Jones *et al.*, 1984). In humid regions, periods of low flow are generally more reliably reconstructed than high flows, as trees are more likely to register prolonged drought than heavy precipitation events that may lead to high streamflow and flooding. Cook and Jacoby (1983) reconstructed summer discharge of the Potomac River near Washington, DC, back to 1730 using a set of 5 tree-ring width chronologies; their results indicate that there was a shift in the character of runoff around 1820. Before this time, short-term oscillations about the median flow were common, generally lasting only a few years. After 1820 more persistent, larger amplitude anomalies became common. For example, runoff was persistently below the long-term median from 1850–1873. If such an event were to be repeated in the future with all of the modern demands on water from the Potomac River, the consequences would be extremely severe. Such a perspective on natural variability of the hydrological system is thus invaluable for water supply management and planning.

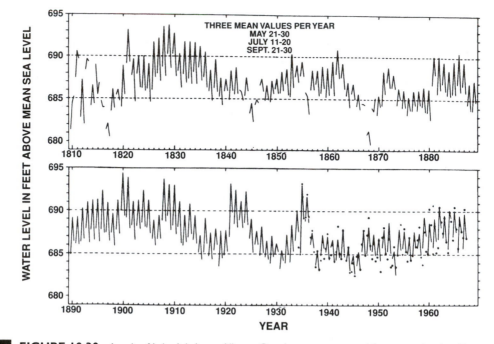

FIGURE 10.30 Levels of Lake Athabasca, Alberta, Canada, as reconstructed from tree-ring data. Tree rings indicate that prior to 1935 there was greater variability in lake levels during May and July, but there was less variability in lake levels for September than during the recent calibration period. Dots indicate actual lake levels used for calibration. Lines connect the three estimates from tree rings, representing mean lake level for May 21–30, July 11–20, and September 21–30. Points are not connected over the winter season, as calibrations of levels for the frozen lake could not be made (Stockton and Fritts, 1973).

10.4 ISOTOPE DENDROCLIMATOLOGY

Many studies have demonstrated empirically that variations in the isotopic content of tree rings ($\delta^{13}C$, $\delta^{18}O$, and δ^2H) are in some way related to climate.[33] Early studies used whole wood samples but some studies now suggest that latewood may need to be isolated from earlywood to get a clear signal of climatic conditions in the growth year (Switsur *et al.*, 1995; Robertson *et al.*, 1995). In mid- and high latitudes, there is a positive relationship (*sensu lato*) between the oxygen and hydrogen isotopic composition of rainfall and temperature (Rozanski *et al.*, 1993) so it is reasonable to expect that the isotopic content of wood in trees might preserve a record of past temperature variations in such regions (Epstein *et al.*, 1976). The problem is

[33] To avoid the difficulties of chemical heterogeneity in wood samples, a single component, α-cellulose (polymerized glucose), is extracted for isotopic analysis; α-cellulose contains both carbon-bound and oxygen-bound (hydroxyl) hydrogen atoms, but the latter exchange readily within the plant. It is therefore necessary to remove all hydroxyl hydrogen atoms (by producing nitrated cellulose) to avoid problems of isotopic exchange after the initial period of biosynthesis. For a discussion of isotopes, deuterium/hydrogen (D/H) and $^{18}O/^{16}O$ ratios, see Sections 5.2.1 and 5.2.2.

that additional isotopic fractionation (of hydrogen, oxygen, and carbon atoms) occurs within trees during the synthesis of woody material and these biological fractionations are themselves dependent on many factors, including temperature, relative humidity, and wind speed (evapotranspiration effects) (Burk and Stuiver, 1981; Edwards, 1993). Nevertheless, in spite of such complications, a number of studies have found very strong positive correlations between $\delta^{13}C$, $\delta^{18}O$, and δ^2H and temperature, and negative correlations with relative humidity. The exact temperature relationship varies, but in several studies of $\delta^{13}C$ and $\delta^{18}O$, it is commonly in the range of 0.3–0.4‰ per °C (Burk and Stuiver, 1981; Yapp and Epstein, 1982; Stuiver and Braziunas, 1987; Lipp *et al.*, 1991; Switsur *et al.*, 1995). Most studies have focused on only the last few decades, but a few have attempted longer paleoclimatic reconstructions. For example, based on the strong correlation observed in recent data between August temperature and $\delta^{13}C$ in fir (*Abies alba*) from the Black Forest of southern Germany, Lipp *et al.* (1991) reconstructed temperature back to A.D. 1000 (Fig. 10.31). This record suggests that there was a steady decline in temperature from the early fourteenth century to ~1850, with an earlier warm episode centered on A.D. 1130. Other long-term temperature reconstructions include a 2000 yr $\delta^{13}C$-based record from the western United States (Stuiver and Braziunas, 1987) and a 1500 yr δ^2H-based record from California (Epstein and Yapp, 1976). There is little similarity between these two records, possibly reflecting real temperature differences, but perhaps also highlighting the many complications that may confound any simple interpretation. Certainly, the processes involved are complex. For example, Lawrence and White (1984) found that δD in trees from the northeastern United States does not contain a strong temperature signal, as might

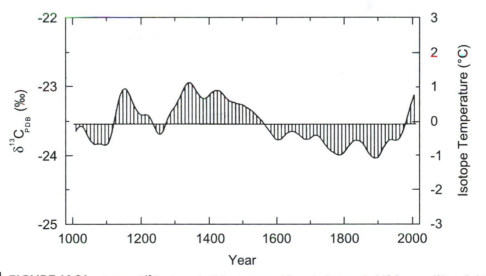

FIGURE 10.31 Average $\delta^{13}C$ values of cellulose extracted from the latewood of 19 fir trees (*Abies alba*) in the Black Forest, from A.D. 1004–1980. Values shown are smoothed by a 100-yr Gaussian lowpass filter; values from 1850–1980 are corrected for the effects of contamination by fossil fuel CO_2. Based on calibration over recent decades, the scale of August mean temperature is shown on the right, relative to the long-term mean (Trimborn *et al.*, 1995).

be expected, but is inversely correlated with summer rainfall amount. By contrast, Ramesh *et al.* (1989) found δD in teak from western India was positively correlated with rainfall. In each case it is possible to construct an explanation for the observed relationship, but one feels that this approach is a little too *ad hoc*. There is a real need to build on the empirical approach by constructing comprehensive models to help in understanding all the processes involved (Edwards and Fritz, 1986; Clague *et al.*, 1992; Edwards, 1993). Isotopic dendroclimatology has much potential, but significant efforts are still needed to develop reliable paleoclimatic records to complement ring-width and densitometric studies.

10.4.1 Isotopic Studies of Subfossil Wood

Yapp and Epstein (1977) showed that δD in wood was strongly correlated with δD in associated environmental waters, which were consistently 20–22‰ *higher* than the nitrated cellulose (Epstein *et al.*, 1976; Epstein and Yapp, 1977). As δD of annual precipitation is correlated (geographically) with mean annual temperature (Yapp and Epstein, 1982) long-term records of δD from trees may be a useful proxy for temperature variations. This idea is supported by mapping δD values from modern plants and comparing them with measured δD values of meteoric waters (Fig. 10.32). It is clear

FIGURE 10.32 Isolines of δD based on modern meteoric waters (i.e. in precipitation) compared to δD values inferred from cellulose C-H hydrogen (cellulose nitrate) in modern plants (underlined values). With the exception of only two Sierra Nevada samples, the inferred values differ by an average of ~ 4‰ from the values measured in precipitation samples (Yapp and Epstein, 1977).

that the δD in plants provides a good proxy measure of spatial variations in δD of precipitation. Assuming that this relationship has not changed over time, it is possible to reconstruct former δD values of meteoric water by the analysis of radiocarbon-dated subfossil wood samples (Yapp and Epstein, 1977). Figure 10.33 shows such a reconstruction, for "glacial age" wood (dated at 22,000–14,000 yr B.P.). Surprisingly, the ancient δD values at all sites are consistently *higher* than modern values (an average of +19‰). The higher δD of meteoric waters implies that temperatures over the ice-free area of North America were warmer in late Wisconsin times than today. However, there are many other factors that could account for the high δD values observed. In particular, δD values in the wood may reflect rainfall in warm growing seasons and so would be isotopically heavier than the annual values mapped in Fig. 10.32. This would be especially true if more of the extracted cellulose came from late-wood rather than earlywood.

Other factors which could help to explain this surprising set of data include: (a) a reduction in the temperature gradient between the ocean surface and the adjacent precipitation site on land; (b) a change in δD of the ocean waters as a result of ice growth on land (probably corresponding to an increase in oceanic δD of 4–9‰); (c) a change in the ratio of summer to winter precipitation; and (d) a positive shift in the average δD value of oceanic water vapor, which at present is not generally evaporating in isotopic equilibrium with the oceans.

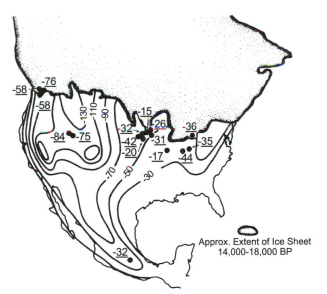

FIGURE 10.33 Distribution of 16 δD values of meteoric waters from 15 different sites during the late Wisconsin glacial maximum, as inferred from δD values of tree cellulose C-H hydrogen (underlined). The approximate position of the southern margin of the ice sheet is shown at its maximum extent (hatched line). The "glacial age" meteoric waters of the 15 sites have, on average, δD values which are 19‰ more positive than the corresponding modern meteoric waters at those sites as deduced from the data shown in Fig. 10.32. The "glacial age" distribution pattern of δD values is similar to the modern pattern, but is systematically shifted by the positive bias of the ancient waters. The North American coastline shown is that of today and does not take into account the lower sea level at the time of glacial maximum (Yapp and Epstein, 1977).

Whatever the reason, the isotopic composition of glacial age precipitation has important implications for the isotopic composition of the Laurentide Ice Sheet, which is generally assumed to have been composed of ice that was very depleted in ^{18}O and deuterium. Using δD values from trees growing along the shores of Glacial Lakes, Agassiz and Whittlesey, Yapp and Epstein (1977) calculated that the $\delta^{18}O$ value of waters draining from the former Laurentide ice sheet probably averaged around −12 to −15‰, far higher than would be expected by analogy with glacial age ice in cores from major ice sheets. Measured δD values from glacial age ice in cores from Greenland and Antarctica (Chapter 5) average 80‰ *below* those of modern precipitation in the same area. These results are thus somewhat enigmatic and it would be extremely valuable to extend this work to other formerly glaciated areas (particularly Scandinavia) to study "glacial age" δD values in more detail because they have an important bearing on the interpretation of other paleoclimatic records, particularly ice and ocean cores. On the other hand, there may have been other factors operating to alter the δD/temperature relationship observed today, or to bias the wood isotopic signal in some way, illustrating once again the difficulty of interpreting isotopic signal in tree rings.

11
██ DOCUMENTARY DATA

11.1 INTRODUCTION

Some of the most diverse and invaluable sources of proxy data are historical documentary records. These data are particularly important as they deal with short-term (high-frequency) climatic fluctuations during the most recent past. In terms of the climatic future, it is this timescale and frequency domain that is often of most interest to planners and decision makers. A great deal can be learned about the probability of extreme events by reference to historical records and this provides a more realistic perspective on the likelihood of similar events recurring in the future (Bryson, 1974; Ingram *et al.*, 1978).

The use of historical documents to reconstruct past climates is intimately linked with the debate over the extent to which climate and climatic fluctuations have played a role in human history. Three volumes devoted to climate and history address this isssue (Wigley *et al.*, 1981, Rotberg and Rabb, 1981; Delano Smith and Parry, 1981). Much of the discussion in these volumes revolves around the effects (if any) that climate and climatic variations have had on various aspects of human activity, particularly economic activity and its social and political consequences (Ingram *et al.*, 1981a; Salvesen, 1992). For the purposes of this chapter it is not necessary to review the arguments *pro* or *con*, except to note the danger of using historical data as a proxy for climate when no cause-and-effect relationship has been clearly demonstrated. Usually this problem is tackled empirically by establishing a relationship between the historical time series and a brief record of overlapping climatic data. However, without a model demonstrating the logic of such a relationship, erroneous conclusions may result (de Vries, 1981). Models should be comprehensive, recognizing not just the role of climate but also the potential significance of other non-climatic factors (Parry, 1981). In this way, not only will there be a stronger basis for historical paleoclimatic reconstruction but also a better understanding of the role of climate and climatic fluctuations in history.

Paleoclimatic data from historical sources rely, of course, on written observations. This means that certain parts of the world have been endowed with a much richer heritage of historical paleoclimatic information than other regions (Table 11.1). The longest records come from Egypt, where stone inscriptions relating to the Nile flood levels are available from mid-Holocene times (~5000 yr B.P.) indicating higher rainfall amounts from East African summer monsoons at that time (Bell, 1970; Henfling and Pflaumbaum, 1991). Arabic chronicles also provide intermittent observations (mainly on rainfall) in other parts of the Middle East (Iraq, Syria, Palestine) going back over 1000 yr (Grotzfeld, 1991). In China, the earliest inscriptions on oracle bones date back to the time of the Shang dynasty (~3700–3100 yr B.P.) when conditions appear to have been slightly warmer than in modern times (Wittfogel, 1940; Chu, 1973). Such records are, of course, few and far between and for the vast majority of the continental land areas, as well as the oceans, historical observations are generally only available for a few hundred years at the most.

Historical data can be grouped into four major categories. First, there are observations of weather phenomena *per se*, for example, the frequency and timing of frosts or the occurrence of snowfall recorded by early diarists. Second, there are records of weather-dependent natural phenomena (sometimes termed parameteorological phenomena) such as droughts, floods, lake or river freeze-up and break-up, etc. Third, there are phenological records, which deal with the timing of recurrent weather-dependent biological phenomena, such as the dates of flowering of shrubs and trees,

TABLE 11.1   Earliest Historical Records of Climate

Area	Earliest written evidence (approximate dates)
Egypt	3000 B.C.
China	1750 B.C.
Southern Europe	500 B.C.
Northern Europe	0
Japan	A.D. 500
Iceland	A.D. 1000
North America	A.D. 1500
South America	A.D. 1550
Australia	A.D. 1800

Modified from Ingram *et al.* (1978).

or the arrival of migrant birds in the spring. We will include in this group of records observations of the former spatial extent of particular climate-dependent species. A fourth category involves records of forcing factors that may have had an influence on climatic conditions in the past. Within each of these categories there is a wide range of potential sources and an equally wide range of possible climate-related phenomena. These are discussed in more detail here.

11.2 HISTORICAL RECORDS AND THEIR INTERPRETATION

Potential sources of historical paleoclimatic information include:

(a) ancient inscriptions;
(b) annals, chronicles; etc.,
(c) governmental records;
(d) private estate records;
(e) maritime and commercial records;
(f) personal papers, such as diaries or correspondence; and
(g) scientific or protoscientific writings, such as (non-instrumental) weather journals (Ingram *et al.*, 1978).

In all these sources, the historical climatologist is faced with the difficulty of ascertaining exactly what qualitative descriptions from the past are equivalent to in terms of modern-day observations. What do the terms "drought," "frost," or "frozen over" really mean? How can qualifying terms (e.g., "extreme" frost) be interpreted? For example, Baker (1932) notes that one seventeenth century diarist recorded three droughts of "unprecedented severity" in the space of only five years! An approach to solving this problem has been to use content analysis (Baron, 1982) to assess in quantitative terms, and as rigorously as possible, climatic information in the historical source. Historical sources are examined for the frequency with which key descriptive words were used (e.g., "snow," "frost," "blizzard," etc.) and the use the writer may have made of modifying language (e.g., "severe frost," "devastating frost," "mild frost," etc.). In this way an assessment can be made of the range of descriptive terms that were used, so they can be ranked in order of increasing severity as perceived by the original writer. The ranked terms may then be given numerical values so that statistical analyses can be performed on the data. This may involve simple frequency counts of one variable (e.g., snow) or more complex calculations using combinations of variables. The original qualitative information may thus be transformed into more useful quantitative data on the climate of different periods in the past. Non-climatic information is often required to interpret the climatic aspects of the source: Where did the event take place (was the event only locally important; was the diarist itinerant or sedentary?) and precisely when did it occur and for how long? This last question may involve difficulties connected with changing calendar conventions as well as trying to define what is meant by terms such as "summer" or "winter" and what time span might be represented by a phrase such as "the coldest winter in living memory." Not all of these problems may be soluble, but content analysis can help to isolate the most pertinent and unequivocal aspects of the historical source (Moody and Catchpole, 1975).

Historical sources rarely give a complete picture of former climatic conditions. More commonly, they are discontinuous observations, very much biased towards the recording of extreme events, and even these may pass unrecorded if they fail to impress the observer. Furthermore, long-term trends may go unnoticed, as they are beyond the temporal perspective of one individual. In a sense, the human observer acts as a high-pass filter, recording short-term fluctuations about an ever-changing norm (Ingram *et al.*, 1981b). Nevertheless, long-term records of natural phenomena (such as the freezing of a lake) can convey relevant low frequency information even if the time series is made up of many short-term observers.

Historical records that appear to be rich in parameteorological information were probably not produced with the historical climatologist in mind, and care must be taken to assess the purpose of the writer in making the record. For example, Chinese dynastic histories contain abundant references to floods and droughts, but the purpose of the notation was not to record the vagaries of climate, but to make a record of human suffering and damage caused by changes in rainfall. Such events were considered to convey a supernatural warning to the Emperor of the day (Yao, 1944). Because of this, there is a strong seasonal bias towards reporting flood and drought events in the vitally important growing season, as well as a spatial bias in the data towards settled areas with large populations. For example, in the Ming dynasty (1368–1643) the number of recorded droughts and floods in the metropolitan province exceeds that of any other province (Chu, 1926; Yao, 1943). Also, because floods and droughts resulted in tax exemptions in the region afflicted, it is not unlikely that local officials may have been tempted to exaggerate the severity of extreme events. This, of course, is true of many areas throughout history.

References to unusual temperatures may be strongly skewed, depending on the perspective of the author. This bias poses a difficult challenge to reconstructing reliable paleotemperature estimates. For example, in the more than 20,000 "climatic calamities" noted in Chinese documents, only 100 concerned warm anomalies (Wang, 1991b). Cold episodes were mentioned in 35% of years, but warm episodes only figured in 5% of the years. This bias reflects the significance of cold conditions to agricultural activities in China. In cooler climates, it is warm conditions that are of particular significance; Icelandic archives commonly record mild spells and these characterized the warm decades of the 1570s, 1640s–1650s, and 1700s (Fig. 11.1) (Ogilvie, 1992).

Finally, it is worth noting that not all historical sources are equally reliable. It may be difficult in some cases to determine if the author is writing about events of which he has first-hand experience, or if events have been distorted by rumor or the passage of time. Ideally, sources should be original documents rather than compilations; many erroneous conclusions about past climate have resulted from climatologists relying on poorly compiled secondary sources that have proved to be quite erroneous when traced back to the original data (Bell and Ogilvie, 1978; Wigley, 1978; Ingram *et al.*, 1981b). Nevertheless, some major compilations such as the voluminous Chinese encyclopedia, the *T'u-shu Chi-cheng*, have proved to be invaluable sources of paleoclimatic information (Chu, 1926, 1973; Yao, 1942, 1943).

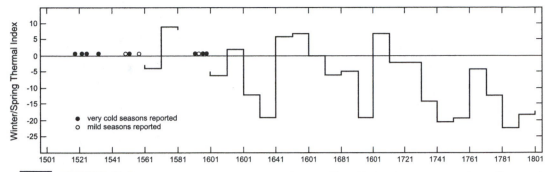

FIGURE 11.1 Winter-Spring thermal index for Iceland 1501–1801 based on content analysis of historical documents (annals, correspondence, and diaries). Isolated reports of very cold or very mild seasons are also shown (Ogilvie, 1992).

As with all proxy data, historical observations need to be calibrated in some way, in order to make comparisons with recent data possible. This is commonly done by utilizing early instrumental data that may overlap with the proxy record, to develop an equation relating the two data sets. Thus, Bergthorsson (1969) regressed observations of sea-ice frequency off the coast of Iceland with mean annual temperatures during the nineteenth century, and then used the resulting equation to reconstruct long-term temperature fluctuations over the preceding 300 yr from sea-ice observations. Similarly, observations of Dutch canal freezing frequencies were calibrated with instrumentally recorded winter temperature data by de Vries (1977) (Fig. 11.2). The calibration equations were then used to reconstruct paleotemperatures prior to the period of instrumental records in each area.

In many locations, precipitation occurrence (daily frequency) was recorded by diarists together with remarks on rainfall intensity, rather than rainfall totals. Several studies have shown that statistically significant relationships exist between rainfall frequency (and/or rainfall duration) and daily rainfall amounts, enabling estimates of daily rainfall to be made, and then for monthly and seasonal totals to be accumulated. In the lower Yangtze River valley of eastern China, for example, local officials maintained detailed weather observations during the Qing Dynasty (1636–1910). This set of records (*Qing Yu Lu*, the "Clear and Rain Records") provide daily records on sky conditions, wind directions, precipitation type (rain or snow, light, heavy, torrential, etc.) and the duration of precipitation events (Wang and Zhang, 1988). By relating precipitation totals to precipitation frequency and intensity during recent decades (using instrumental data), Wang and Zhang (1992) were able to establish regression equations that could then be applied to the historical data to reconstruct past precipitation amounts. Others have noted that rainfall frequency (number of rainy days and/or the length of rainy episodes) is often inversely related to temperature because cloud cover tends to reduce direct radiation, and rainfall itself will lead to evaporative cooling. W.C. Wang *et al.* (1992) applied this line of reasoning to the *Qing Yu Lu* in order to extend the instrumental record of summer temperature in Beijing from 1855 back to 1724. Mikami (1992b) also noted the strong relationship between the number

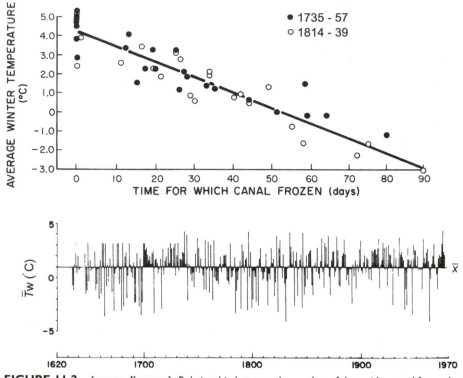

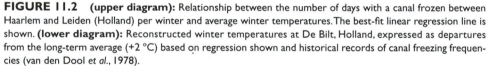

FIGURE 11.2 **(upper diagram):** Relationship between the number of days with a canal frozen between Haarlem and Leiden (Holland) per winter and average winter temperatures. The best-fit linear regression line is shown. **(lower diagram):** Reconstructed winter temperatures at De Bilt, Holland, expressed as departures from the long-term average (+2 °C) based on regression shown and historical records of canal freezing frequencies (van den Dool *et al.*, 1978).

of rain days in parts of Japan and August temperature (based on instrumental data). He then used this to reconstruct August temperature from 1770–1840 using the historical records of rainfall frequency (Fig. 11.3).

Some observations may not need direct calibration if recent comparable contemporary observations are available. This applies to such things as rain/snow frequency, dates of first and last snowfall, river freeze/thaw dates, etc., providing urban heat island effects, or technological changes (such as river canalization) have not resulted in a non-homogeneous record.

11.2.1 Historical Weather Observations

The most widely recorded meteorological phenomenon in historical documents is snow, in terms of the date of the first and last snowfalls of each year, or the number of days with snow on the ground or the frequency of days with snowfall. Some of the earliest snowfall observations are from Hangchow, China, for the period A.D.

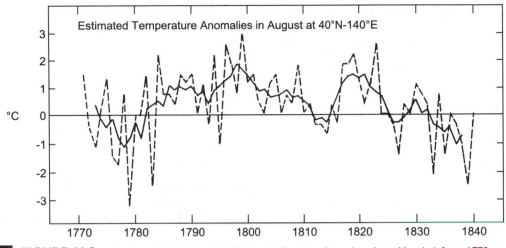

FIGURE 11.3   August temperature anomalies in northeastern Japan (northern Honshu) from 1770 to 1840, derived from historical records of rainfall frequency. The reference period (zero line) is the mean for 1951–1970 (Mikami, 1992b).

1131–1210, when discontinuous records were made of the dates of the last snowfall in spring. Comparison with Hangchow records for the period 1905–1914 indicates that snowfalls commonly occurred 3–4 weeks later in the spring months during the twelfth century than was typical of the early twentieth century (Chu, 1926). This suggests a more prolonged, and probably more severe, winter during the Southern Song Dynasty (A.D. 1127–1279).

More modern and more complete records of snowfall (dates of first and last occurrences) were compiled for the London area by Manley (1969). These extend back to 1811 and indicate a reduction in the snow season of approximately 6 weeks between 1811–1840 and 1931–1960, though most of this change has occurred during the twentieth century, perhaps due to a marked increase in the urban heat island effect. A similar problem may have affected recent observations of first snow cover in the Tokyo area. During the period from 1632–1633 to 1869–1870 the average date of the first snow cover was January 6, whereas from 1876–1877 to 1954–1955 the mean was January 15 (Arakawa, 1956a). For both periods, the standard deviations were similar (approximately 20 days). How much of the change is due to a warmer Tokyo metropolitan area is difficult to assess, and points to the particular value of data from more rural locations. Pfister, for example, has compared the number of days with snow cover (extensive snow on the ground) at various places with similar elevations on the upper and lower plateaus of Switzerland. During the eighteenth century, snow cover was often far more persistent than in even the harshest winter of the twentieth century, 1962–1963 (Table 11.2) and this provides strong support for the record of "days with snow lying" in the Zurich-Winterthur area (Fig. 11.4). These data point to the decade 1691–1700 as having had the highest number of days with snow on the ground (> 65 annually) compared to recent

TABLE 11.2   Number of Days With Snow Cover per Year (March to May Values in Parentheses) for Selected Localities in Switzerland

Winter of	Lower plateau	Upper plateau
1769–1770		126 (45)
1784–1785	134 (51)	154 (60)
1788–1789	112 (35)	
1962–1963[a]	59 (0)	86 (12)

From Pfister (1978a).

[a] Harshest winter of the twentieth century. Severe winters were also noted in 1684–1685, 1715–1716, 1730–1731, and 1969–1970.

averages of only half that number (Pfister, 1985). It is also of significance that this decade was the coldest in central England, according to long-term instrumental records of temperature (Manley, 1974). Thus, urban growth effects, while undoubtedly exerting a progressive influence on temperature, cannot account for all the marked changes observed.

An interesting method of assessing past winter temperatures has been demonstrated by Flohn (1949), who observed that the ratios of snow days to rain days in winter months correlate well with winter temperatures during the instrumental period. Using sixteenth century observations of Tycho Brahe in Hven, Denmark (1582–1597) and of Wolfgang Haller in Zurich (1546–1576), Flohn was able to show that winters after 1564 were increasingly severe, leading to a marked increase in glacial advances in the Alps in the early seventeenth century. At Hven, winter

FIGURE 11.4   Fluctuations in the number of days with snow lying in Zürich (note break in abscissa scale). Prior to 1800, figures have been estimated based on daily non-instrumental observations. Observations for 1721–1738 made in Winterthur, 20 km northeast of Zürich. Value plotted as 10-yr running means with value at year n corresponding to the decade n to (n + 9) (Pfister, 1978b).

temperatures from 1582 to 1597 averaged 1.5 °C below those around the early twentieth century. It is also interesting that sixteenth century weather singularities (synoptic events that recur at the same time each year) continue to be observed in the twentieth century, indicating an underlying cyclicity in the general circulation that has persisted in spite of fundamental changes in the climate of the region.

In the Kanazawa area of Japan (west central Honshu) historical records of snowfall were maintained by the ruling Maeda family from 1583 until ~1870. Based on detailed analysis of these and subsequent records, Yamamoto (1971) constructed an index of snowfall variation that corresponds reasonably well with instrumentally recorded winter temperatures in recent years. Although no precise calibration has been attempted, the index gives an overall impression of snowier winters, particularly in the first half of the nineteenth century, and this conclusion seems to be supported by other snowfall indices derived from Japanese historical sources.

Another observation frequently noted in historical records is the incidence of frost during the growing season, an occurrence of particular significance to agriculturalists. As with snowfall, long records of frost occurrence are available from China, in particular from the farming regions of the mid and lower sections of the Yellow River (Huang-he). Higher frequencies of frost were recorded during the periods 1551–1600, 1621–1700, 1731–1780, and 1811–1910. However, the range is small (from 1 to 6 events per decade) and the significance of such changes in the frequency of extremes for the overall growing season temperature is not clear. Other historical records indicate that the *frost-free* period in Inner Mongolia and northeastern China from 1440 to 1900 was ~2 months shorter on average than during the twentieth century. Similarly, in southern China the frost-free season in recent decades has been 5–6 weeks longer than the long-term mean from 1440 to 1900 (Zhang and Gong, 1979).

European historical archives have provided a variety of non-instrumental records of value to the climatologist. Many of these sources were evaluated by Lamb to produce an index of winter severity and summer wetness spanning the last 1000 years (Lamb, 1961, 1963, 1977). Lamb concluded that the period A.D. 1080–1200 was characterized by dry summers throughout Europe, the like of which has not been seen since. He therefore designated this interval the "Medieval Warm Epoch" (Lamb, 1965, 1988). Recently, this has generated considerable interest as a possible analog for future, greenhouse-gas induced climates. However, subsequent studies have provided mixed support for this concept (Hughes and Diaz, 1994); the available evidence is limited (geographically) and equivocal. A number of records do indeed show evidence for warmer conditions at some time during this interval, especially in the eleventh and twelfth centuries in parts of Europe, as Lamb pointed out. However other records show no such evidence, or indicate that warmer conditions prevailed, but at different times. Indeed, not all seasons may have been warm; for example winters at this time were relatively harsh in western Europe, at least until A.D. ~1170 (Alexandre, 1977). This rather incoherent picture may be due to an inadequate number of records, and a clearer picture may emerge as more and better calibrated proxy records are produced. However, it is not yet possible at this point to say whether the notion of a Medieval Warm Epoch should be considered as a worldwide episode or of no more than regional significance.

One of the most interesting and unusual studies of historical material is that of Neuberger (1970), who examined the changing climate of the "Little Ice Age" (sixteenth–nineteenth centuries) in Western Europe through artists' perceptions of their climatic environment, as depicted in contemporary paintings. Over 12,000 paintings from the period 1400–1967 were examined and wherever possible the intensity of the blue sky, visibility depicted, percentage cloudiness, and cloud type were categorized for each painting. More than half of the paintings contained some sort of meteorological information and the basic characteristics were averaged for different periods within the last 570 yr, as shown in Fig. 11.5. During the period 1400–1549, paintings have a high percentage of blue sky, good visibility, and little cloud cover. Paintings completed during the next 300 yr were generally darker, with less blue sky, showed lower visibilities, and had a much higher percentage of cloud cover and a greater frequency of low and convective type clouds. Over the last 100 yr there has been some reduction in cloudiness depicted and a drop in low and convective cloud frequency, though visibilities remain low, perhaps reflecting increasing atmospheric turbidity due to industrial and agricultural activity. It is remarkable that, in spite of many changes in style, the artists have captured in their paintings a significant record of climatic variation through time. Perhaps this perceptual record, more than any cold statistic, indicates the degree to which life during the "Little Ice Age" was affected by a deteriorating climate.

11.2.2 Historical Records of Weather-Dependent Natural Phenomena

Of all the weather-related natural catastrophes, floods and droughts appear to have had the most widespread and persistent impact on human communities, as records of these events are found in historical documents from all over the world (Pfister and Hächler, 1991; Pavese *et al.*, 1992; Barriendos, 1997). The longest and most detailed records come from China, where regional gazetteers have been kept in many provinces and districts since the fourteenth century and many records are available for the last 2000 yr (Chu, 1973). Each of these gazetteers recorded local facts of historical or geographical interest as well as climatological events of significance to agriculture and the local economy (for example, droughts, floods, severe cold snaps, heavy snowfalls, unseasonable frosts, etc.). Not surprisingly, the gazetteers have been the focus of much interest, though there are often many difficulties in interpreting the data (see Section 11.2). In particular, technological improvements such as the building of irrigation channels or drainage ditches may drastically reduce the frequency of climatological disasters. For example, the Chinese province of Sichuan is unusual in that recorded occurrences of floods are rare, yet droughts are common. Usually, one finds over long periods of time a similar number of extremely wet and extremely dry events. The reason for this anomaly appears to be the particularly efficient flood-control measures introduced by the administrator Li Ping 2100 yr ago, which were able to reduce flood hazards but did little to alleviate the perils of droughts (Yao, 1943). In a similar way, the rise in flood and drought frequency during the Yuan Dynasty (A.D. 1234–1367) may be partly due to the destruction of irrigation and drainage systems by the Mongol invaders of the time (Chu, 1926).

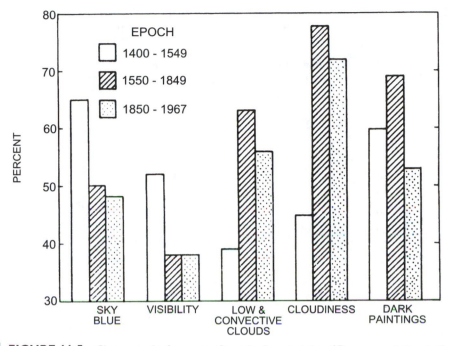

FIGURE 11.5 Changes in the frequency of certain characteristics of European paintings in the periods 1400–1549, 1550–1849, and 1850–1967 (Neuberger, 1970).

Early work attempted to reduce the bias due to technological changes and increasing observations in recent history by assuming that such changes affected drought and flood observations equally. An index of "raininess" or precipitation anomaly was obtained by expressing the number of droughts (D) and floods (F) as a ratio (D/F), a common approach to reducing data errors in historical climatology. Thus, Yao (1942) was able to characterize each century over the last 2200 yr as "wet" or "dry" compared to the mean ratio for different regions of China (Table 11.3).

A detailed spatial-temporal data base of floods and droughts in China over the last 500 yr was compiled by the State Meteorological Administration (1981); each regional record, year by year, was classified into one of five anomaly classes (wet to dry). Wang and Zhao (1981) subjected these matrices to principal components analysis. One hundred and eighteen stations from almost the whole of eastern China were used in the analysis of data from 1470 to 1977. The resulting eigenvectors indicated broad-scale patterns of "precipitation anomalies" and these were then compared with eigenvectors derived from instrumentally recorded summer precipitation data for the period 1951–1974 (Fig. 11.6). Summer was chosen because it is the time corresponding to most recorded droughts and floods (Yao, 1942). Similarities between the principal modes of precipitation anomaly in both the instrumental and the historical periods indicate strongly that the historical records can indeed provide a valuable proxy of the patterns of summer precipitation variation over long periods of time. One advantage of such long time series is that periodic or quasi-periodic

■ **TABLE 11.3   Ratio of Droughts to Floods in Northern, Central, and Southern China**

| Century | Northern China | | | | Central China | | | | Southern China | | | |
	D	F	R	Remarks	D	F	R	Remarks	D	F	R	Remarks
B.C.												
2	-	11	0.0	Wet?	-	2	0.0	Wet?	-	2	0.0	Wet?
1	-	12	0.0	Wet?	-	1	0.0	Wet?	-	-	-	-
A.D.												
1	5	6	0.83	Dry	-	-	-	-	-	1	0.0	Wet?
2	12	17	0.71	Dry	1	3	0.33	Wet	-	1	0.0	Wet?
3	9	27	0.33	Wet	-	27	0.0	Wet?	1	4	0.25	Wet
4	4	6	0.67	-	4	17	0.24	Wet	-	6	0.0	Wet?
5	7	20	0.35	Wet	8	39	0.21	Wet	-	-	-	-
6	13	27	0.48	Wet	4	14	0.29	Wet	-	-	-	-
7	26	57	0.46	Wet	11	26	0.42	Wet	1	3	0.33	Wet
8	15	67	0.22	Wet	5	21	0.24	Wet	2	6	0.33	Wet
9	22	56	0.39	Wet	27	60	0.45	Wet	9	12	0.75	Dry
10	74	98	0.76	Dry	20	46	0.44	Wet	8	16	0.50	Wet
11	56	118	0.47	Wet	23	50	0.46	Wet	2	15	0.13	Wet
12	37	34	1.09	Dry	103	131	0.79	Dry	36	48	0.75	Dry
13	60	83	0.72	Dry	55	112	0.49	Wet	25	33	0.76	Dry
14	110	199	0.55	Wet	65	108	0.60	Dry	32	50	0.64	-
15	40	75	0.53	Wet	42	115	0.37	Wet	33	46	0.72	Dry
16	52	49	1.06	Dry	60	104	0.58	-	66	104	0.63	-
17	77	107	0.72	Dry	96	145	0.66	Dry	55	81	0.68	Dry
18	116	131	0.89	Dry	107	164	0.65	Dry	42	72	0.58	-
19	100	89	1.12	Dry	85	114	0.75	Dry	6	16	0.37	Wet
Totals	835	1289	0.65		716	1299	0.55		318	516	0.62	

D = number of droughts.

F = number of floods.

$R = D/F$.

Yao (1942).

variations in climate may be observed, which could not be resolved in the shorter instrumental records. For example, Wang *et al.* (1981) used the long-term precipitation anomaly data for different latitude zones in China to construct a time-space diagram that points to a recurrent pattern of precipitation anomaly beginning in northern China and migrating southward, with a period of ~80 yr. More detailed studies of records from different areas have used spectral analysis to isolate statistically significant periodicities in the data. Thus, in the precipitation anomaly series for the Shanghai region a periodicity of 36.7 yr is apparent (Wang and Zhao, 1979). This periodicity is also seen in other records from southwestern China and the east-

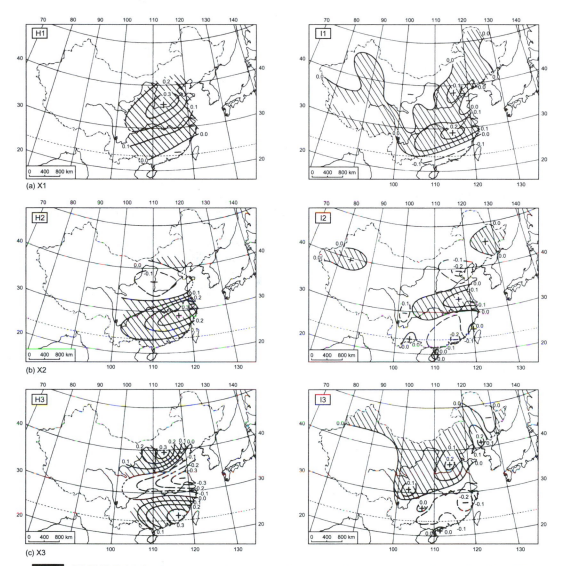

FIGURE 11.6 Eigenvectors of drought and flood data using historical records for A.D. 1470–1977 (left) and instrumentally recorded precipitation data from 1951–1974 (right). The first three eigenvectors of historical data (H1 to H3) account for 15, 11, and 7% of variance in the data respectively. The first three eigenvectors of instrumental data (I1 to I3) account for 18, 13, and 11% of variance in the data, respectively. Eigenvectors H1 and I1 are similar; so are H2 and I3 and H3 and I2 (inverse patterns). This indicates that historical data can be reliable indicators of climatic anomaly patterns (Wang and Zhao, 1981).

ern part of the Yangtze River Basin. Elsewhere, other periodic variations have occurred, most notably a quasi-biennial oscillation (2–2.5 yr) in the Yangtze River area and north of the Yellow River (Wang and Zhao, 1981). Both the short-term and long-term periodicities appear to be related to large, synoptic-scale pressure anomalies over eastern Asia and adjacent equatorial regions.

Apart from floods and droughts, the parameteorological phenomenon that seems to have attracted most attention in historical records is the freezing of lakes and rivers. The longest continuous series is that of Lake Suwa (near Kyoto) in Japan. Data on the time of freezing of this small (~15 km²) lake are available almost annually from 1444 to the present, although the dates are not very reliable from 1680 to 1740 (Arakawa, 1954, 1957). Gray (1974) calibrated this record with instrumental data from Tokyo for the period 1876–1953 and found the best correlation with mean December to February temperature data. Using the regression equation relating Lake Suwa freezing dates to temperatures since 1876, Gray was able to reconstruct Tokyo midwinter temperatures back to 1450. The coldest periods appear to have been ~1450–1500 and ~1600–1700, when winter temperatures were about 0.5 °C below the mean of the last 100 yr.

Long historical records of river and lake freezings are also available from China. Using local gazetteers and diaries, Zhang and Gong (1979) compiled records of the frequency of freezings of lakes in the mid and lower reaches of the Yangtze River, freezings of rivers and wells in the lower Yellow River Basin, the occurrence of sea ice in the Gulf of Chihli and Jiangsu Province (31–41° N), and snowfall in tropical areas of southern China. Using all this information they calculated the number of exceptionally cold winters per decade from 1500 to 1978 (Fig. 11.7). The highest frequency of cold winters occurred during the periods 1500–1550, 1601–1720, and 1830–1900, with the decade 1711–1720 being the most severely cold period in the last 480 yr. By mapping the areas most affected by severe winters it was also possible to recognize two main patterns of the anomaly — periods when cold winters predominated east of ~115° E and periods when areas to the west were colder. From modern meteorological studies it appears that such large-scale anomaly patterns result from changes in the position of the upper level trough over East Asia. When the trough is in a more westerly position and fairly deep, cold air sweeps down more frequently over the area west of 115° E. Westerly flow prevails when the trough is weaker and less extensive, leading to milder conditions in the west, with cold air outbreaks more common to the east. Generally speaking, the colder periods shown in Fig. 11.7 had more frequent cold outbreaks west of 115° E, indicating that such periods were characterized by a strongly developed upper air trough over eastern Asia. Conversely, the warmer

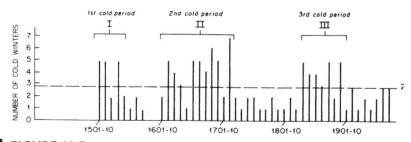

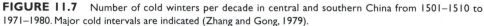

FIGURE 11.7 Number of cold winters per decade in central and southern China from 1501–1510 to 1971–1980. Major cold intervals are indicated (Zhang and Gong, 1979).

periods were times of stronger westerly flow in winter and weaker upper level trough development (Zhang and Gong, 1979).

Historically, one of the most important effects of severe winter temperatures was on water transportation systems, and many records exist regarding the disruptions caused by canals and rivers freezing over for prolonged periods. In the Netherlands, for example, canals were built in the early seventeenth century to connect major cities, and records of transportation on the canals, including times of freezing over, have been kept since 1633 (de Vries, 1977). Using instrumental winter temperature data from De Bilt (Labrijn, 1945) the number of days on which the Haarlem-Leiden canal was frozen each winter was calibrated (see Fig. 11.2a), enabling De Bilt winter temperatures to be reconstructed back to 1657. Further temperature estimates, back to 1634, were possible by calibration of the canal freezing data with barge trip frequency between Haarlem and Amsterdam (1634–1682) a service that was commonly suspended due to ice cover on the canal (van den Dool *et al.*, 1978). In this way, a complete winter temperature reconstruction for De Bilt has been obtained back to 1634 (Fig. 11.2b). Note however that this record conveys no information about how warm winters may have been in years when the record shows only that the canal was never frozen over (Fig. 11.2a).

In more northern latitudes, rivers freeze over every year, and in historical time the dates of freeze-up and break-up were both economically and psychologically important. Consequently, diaries and journals from these regions commonly contain frequent reference to the state of icing on nearby rivers and estuaries. Ironically, remote regions of northern Canada are relatively well-endowed with historical records, thanks to the efforts of Hudson Bay Company managers at various company posts around Hudson Bay and points to the west (Ball, 1992). A valuable analysis of such data from western Hudson Bay has been made by Catchpole *et al.* (1976). Using content analysis they analyzed journals kept by Hudson Bay Company trading post managers from the early eighteenth to the late nineteenth century. Although reference to the state of ice on nearby rivers and estuaries was often imprecise, content analysis enabled quite reliable estimates to be made of the dates of freeze-up and break-up (Fig. 11.8) and these provide a unique index of overall "winter duration" in this remote region (Moody and Catchpole, 1975). The prolonged period of both early freeze-up and late break-up in the early part of the nineteenth century is particularly noteworthy. Comparisons with modern data are difficult because the sites are no longer inhabited, but where comparisons can be made it appears that the "freeze season" (the time between freeze-up and break-up) averaged 2–3 weeks longer during the eighteenth and nineteenth centuries than in recent years (Table 11.4). Recently, the mid-eighteenth century record of first freeze-up dates has been used to calibrate white spruce tree-ring records from the area, enabling a 300-yr record of first freeze-up dates to be reconstructed (Jacoby and Ulan, 1982). Although only a limited amount of modern data was available for verification, the results were reasonably good, suggesting that some confidence can be placed in the long-term reconstruction. This is an interesting example of how one proxy data set may be used to calibrate (or verify) another.

TABLE 11.4 Comparison of Historical (H) and Modern (M) Dates of Freeze-up and Break-up (in Days after December 31)

Site	First partial freezing (H) or first permanent ice (M)			First complete freezing (H) or complete freezing (M)			First breaking (H) or first deterioration of ice (M)		
	Earliest	Mean	Latest	Earliest	Mean	Latest	Earliest	Mean	Latest
Churchill River at Churchill (M)	273	291	318	288	319	336	141	160	169
Fort Prince of Wales	273	292	319	295	321	345	150	168	187
Hudson Bay at Churchill (M)	292	305	319	313		340	124	159	180
Moose Factory (H)	281	304	335	290	319	341	105	126	145
Moose River at Moosonee(M)	304	316	331	217	330	347	103	116	126

Catchpole *et al.* (1976)

11.2.3 Phenological and Biological Records

In this section consideration will be given to purely phenological data, that is, data on the timing of recurrent biological phenomena (such as the blossoming and leafing of plants, crop maturation, animal migrations, etc.), as well as historical observations on the former distribution of particular climate-sensitive plant species. The value of phenological records as a proxy of climate is illustrated in Fig. 11.9. From 1923 to 1953, the flowering dates of 51 different species of plants in a Bluffton, Indiana, garden were noted. For each species, the average date of flowering was computed, and individual years expressed as a departure from the 30-yr mean (Lindsey and Newman, 1956). Yearly departure values for all species were then averaged to give an overall departure index for the 51 species; in Fig. 11.9, this is plotted against the mean temperature of the period March 1 to May 16 (i.e., the start of the growth period). Clearly, the phenological data are an excellent index of spring temperatures, cool periods corresponding closely to late flowering dates and vice versa. This example illustrates well the potential paleoclimatic value of phenological observations; if they can be calibrated, they may provide an excellent proxy record of past climatic variation.

One of the longest and best known phenological records comes, like so many other long historical records, from the Far East. At Kyoto (the capital of Japan until 1869) the Governor or Emperor used to hold a party under the flowering cherry blossoms of his estate, when they were in full bloom (Arakawa, 1956b, 1957). The blooming dates can be considered as an index of spring warmth (February and March) as shown by Sekiguti (1969) using modern phenological records and instrumental data. Higher spring temperatures result in earlier blooming dates; according to Kawamura (1992) an increase in March temperature of 1 °C changes the mean date of cherry blossom flowering by 2–3 days. The Kyoto record is extremely sparse, but nevertheless is of interest as it spans such a long period of time (Table 11.5). It appears from this record that the eleventh to fourteenth centuries were relatively

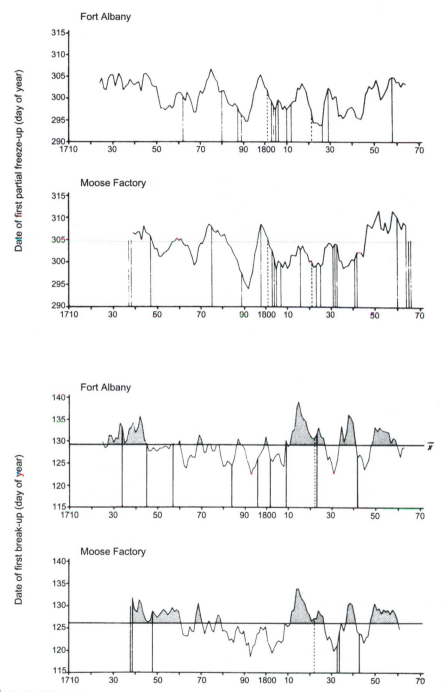

FIGURE 11.8 Seven-year running means of dates of first partial freeze-up (above) and first break-up (below) of ice in estuaries at the locations indicated (all on west coast of Hudson Bay, Canada). Data obtained by content analysis of historical sources. See also Table 11.4. Comparable dates for modern conditions shown as horizontal lines. Dates are given in days after December 31 (Catchpole et al., 1976).

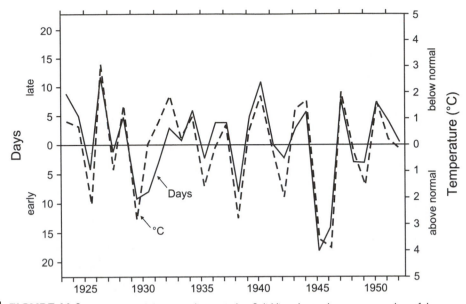

FIGURE 11.9 Phenological data as a climatic index. Solid line shows the average number of days per year in which the flowering dates of 51 species were earlier (below zero, left ordinate) or later (above zero, left ordinate) than the average. The dashed line shows the average departure of mean daily temperature from March 1 to May 16 for each year (right ordinate scale). Observations from Bluffton, Indiana (Lindsey and Newman, 1956).

cool, though this impression could be due entirely to the inadequate statistics (only 30 reliable dates in 400 yr!). Were it not for similar phenological observations from China, tending to support this idea, particularly of a cooler twelfth century in the Far East, one could place little faith in the Kyoto data alone (Chu, 1973).

The most important phenological records from Europe concern the date of the grape harvest. Grape harvests are, of course, not only determined by climatic factors but also by economic considerations. For example, an increasing demand for brandy may cause the vineyard owner to delay the harvest in order to obtain a liquor richer in sugar and more desirable from the point of view of spirits production. However, such factors are unlikely to be of general significance, and,

TABLE 11.5 Average Cherry Blossom Blooming Dates at Kyoto, Japan, by Century. Mean date, April 14.6; n = 171

Century	9th	10th	11th	12th	13th	14th	15th	16th	17th	18th	19th	20th[a]
Day in April	11	12	18	17	15	17	13	17	12	-	12	14
No. observations	7	14	5	4	8	13	30	39	10	-	5	36

From Arakawa (1956b).

[a] Chu (1973) notes 20th century data (1917–1953) for blossoms "in full bloom."

providing variation in the dates of grape harvests shows regional similarities, it is reasonable to infer that climate is a controlling factor. Thus, Le Roy Ladurie and Baulant (1981) have produced a regionally homogeneous index of grape harvest dates for central and northern France, based on over 100 local harvest-date series, extending back to 1484. For the period of overlap with instrumental records from Paris (1797–1879), the index had a correlation coefficient with mean April to September temperatures of +0.86, indicating that it provides a good proxy of the overall warmth of the growth season (Fig. 11.10; Garnier, 1955; Le Roy Ladurie, 1971). Indeed, Bray (1982) has shown that the reconstructed summer paleotemperatures also show a strong correlation with the record of alpine glacier advances in western Europe. Periods characterized by temperatures consistently below the median value are generally followed by glacier advances.

The deterioration of climate during the Little Ice Age also had important geographical consequences for plants and animals, particularly in marginal environments. In the Lammermuir hills of southeastern Scotland, for example, oats were cultivated up to elevations of over 450 m during the warm interval from A.D. 1150 to A.D. 1250. However, by A.D. 1300, the *uppermost* limit had fallen to 400 m and by A.D. 1600 to only ~265 m, more than doubling the area of uncultivable land (Parry, 1975, 1981). These changes probably resulted from reduced summer warmth, wetter conditions, and earlier snowfalls in winter months, factors which all combined to increase the probability of crop failure from only 1 yr in 20 in the Middle Ages to 1 yr in 2 or 3 during the Little Ice Age. The abandonment of upland field cultivation was also accompanied by the abandonment of upland settlements and resulted in a considerable redistribution of population in the area.

Former plant and animal distributions can provide useful indices of climatic fluctuation, though it is not always possible to quantify the significance of the change in species range. Harper (1961), for example, has documented significant changes in the distribution of flora and fauna in subarctic Canada during the twentieth century as a result of the widespread increase in temperature during this period. Similar observations have been made in Finland by Kalela (1952) and the change is not confined to terrestrial species; northward migration of fish in the North Atlantic as a result of increasing water temperatures in the North Atlantic has also been noted (Halme, 1952). Trading post records of animal catches around the coasts of Greenland point to the close dependence of animal populations on climatic fluctuations and associated changes in the distribution of sea ice (Vibe, 1967). These records rarely extend back beyond the mid-nineteenth century, however, and merely point to the biological significance of climatic fluctuations, which instrumental records have documented in considerable detail. Nevertheless, there is great potential in using historical records of former plant and animal distributions, the timing of migrations, etc., to document climatic variations during periods for which no instrumental records exist. The possible value of such work is well illustrated by the wide-ranging surveys of former plant and animal populations in China reported by Chu (1973).

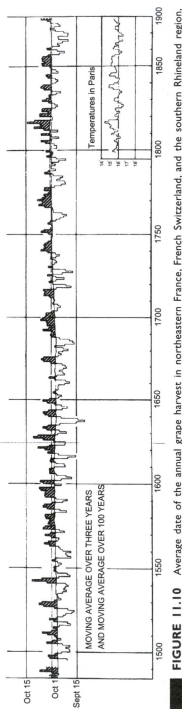

FIGURE 11.10 Average date of the annual grape harvest in northeastern France, French Switzerland, and the southern Rhineland region, 1484–1880. At lower right are mean April–September temperatures in Paris during the period of instrumental records. Data smoothed by a 3-yr running mean. Harvest data shaded when *later* than 100-yr running mean (continuous, slightly rising line) (LeRoy Ladurie and Baulant, 1981).

11.3 REGIONAL STUDIES BASED ON HISTORICAL RECORDS

The large number of studies in East Asia and Europe enable a regional synthesis of paleoclimatic conditions to be made from historical data. Elsewhere, either there are insufficient documentary sources (e.g., in Africa or Australia) or detailed studies have yet to be carried out. One neglected resource is the vast archive of marine records, which contain valuable information about wind direction, wind speed, cloud cover, water temperature, and other parameters (Frich and Fryendahl, 1994). At higher latitudes, such records can provide much information about the sea-ice conditions that greatly affected sailing conditions in northern waters (Catchpole, 1992).

11.3.1 East Asia

Dynastic records, local histories, and diaries from China, Korea, and Japan are valuable sources of information about past weather events and their impacts on these largely agricultural societies (Table 11.6). Numerous investigators have sifted through these archives to extract weather-related information (see, for example, the volumes edited by Zhang, 1988 and Mikami, 1992a).

In China, most studies have been carried out on records from the lower Yangtze River basin and regions to the north (R. Wang and S. Wang, 1989; S. Wang, 1991a, b; S. Wang and R. Wang, 1990). A few studies have developed methods that use observations of extreme or intermittent events to estimate monthly or seasonal temperature or rainfall anomalies (R. Wang et al., 1991). This may involve calibrating the data with instrumentally recorded temperatures at representative sites, such as Beijing or Shanghai (W.C. Wang et al., 1992). More often this key procedure is poorly documented, making it difficult to judge how reliable the reconstructions really are. There is a need for much more work on rigorously calibrating these valuable records; bearing this caveat in mind, the following conclusions can be derived from the published literature. The coldest periods in the last 600 yr (or those with the highest frequency of unusually cold events) were in the mid- to late-seventeenth century and in the early to mid-nineteenth century (Fig. 11.11). The 1650s were exceptionally cold throughout eastern China and Korea (Kim, 1984, 1987). Indeed, Kim and Choi (1987) believe that summers in the period 1631–1740 were the coldest of the past 1000 yr. During this interval, snow occasionally fell in southern China and there were even killing frosts that severely damaged vegetation as far south as 20° N (Li, 1992). Temperatures increased during the eighteenth century, reaching levels comparable with the early twentieth century for brief periods in some regions. Cool conditions were again common in the early to mid-nineteenth century, but warmer conditions set in abruptly at the start of the twentieth century, with temperatures reaching the highest level of the last 500 yr in the period 1920–1940 with cooling thereafter. During both of the cold periods, in the seventeenth and nineteenth centuries, the frequency of floods and droughts increased, pointing to greater instability in climate during those times (Zheng and Feng, 1986).

In Japan, many documents dating from the Edo era (late seventeenth to early nineteenth century) contain records of daily weather conditions (Fig. 11.12). This has enabled monthly climatological maps to be constructed for 1700–1870, using daily

TABLE 11.6 Paleoclimate Series Constructed for East Asia Based on Historical Records

	Region	Sp	Su	Fa	Wi	Ann	Record of:	Interval	Ref. period	Period	Source
1	East China	x	x	x	x	x	Temperature anomalies	10	1470–1980	1470–1980	Wang S. and Wang (1990, 1991); Wang S. (1991b)
2	North China	x	x	x	x	x	Temperature anomalies	10	1470–1980	1470–1980	Wang S. and Wang, R. (1990, 1991); Wang (1991a)
3	South China (5 regions)[a]				x		Temperature anomalies	10	1470s–1970s		Zhang (1980); Zhang (ed.) (1988)
4	China				x		Number of cold winters	10		1500s–1970s	Zhang and Gong (1979)
5	Lower Yellow River					x	Number of frosts	10		1440s–1940s	Zhang and Gong (1979)
6	Shandong province				x		Cold winter index	10		1500s–1970s	Zheng and Zheng (1992)
7	Southeast China (Shanghai)	x	x	x	x	x	Temperature anomalies	10	1950–1979	1470s–1970s	Wang et al. (1991)
8	Mid and South China (S. of 35°N)				x		Temperature anomalies	10	1951–1980	1470s–1970s	Wang and Wang (1989)
9	China		x				Thunder events	30		190 BC–1920	Wang (1980)
10	Middle and Lower Yangtse				x		Temperature indices	10		1470s–1970s	Zheng and Feng (1986)
11	Eastern China					x	"Dust Rain" events	1		<300 BC–1933	Zhang (1983)
12	Beijing	x	x				Mean temperatures	1		1724–1986	Wang et al. (1992)
13	S. China				x		Frost, snow (irregular)	1		1488–1900	Li (1992)
14	China				x		Winter monsoon index	10		1390–1980	Guo (1992)
15	Korea					x	Number of cold events	50		1392–1900	Kim (1984)
16	Hirosaki, Japan	x	x	x	x	x	Maximum temperature	20–40	1661–1870	1661–1870	Maejima and Tagami (1983)
17	Japan		x		x		Temperature index	10		601–1900	Maejima and Tagami (1986)
18	Japan		x				Temperature	1	1950–1970	1771–1840	Mikami (1992)
19	C. Japan		x		x		Temperature, precipitation	1		1801–1870	Mizukoshi (1992)
20	C. and S. Japan		x		x		Snow/rain ratio; summer rainfall	1		1670–1860	Tagami and Fukaishi (1992)
21	C. and S. Japan				x		Synoptic type frequency	1		1720–1869	Fukaishi and Tagami (1992)

[a]Region I: Eastern part of Changjiang River Basin; Region II: Central part of Changjiang River Basin; Region III: Hunan and Jiangxi Provinces; Region IV: Southeastern Provinces; and Region V: Guangdong and Guangxi Provinces.

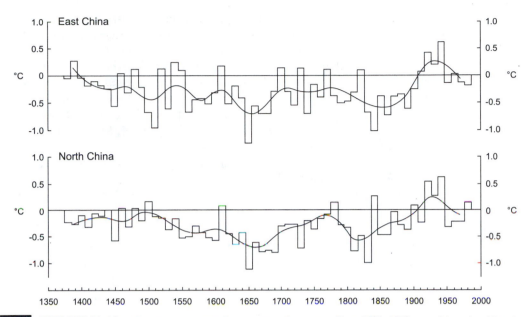

FIGURE 11.11   Ten-year mean annual temperature departures (from 1880–1979 means) based on historical documents from East China (above) and North China (below). East China comprises five provinces: Shanghai, Jiangsu, Jiangxi, Anhui, and Zhejiang. North China covers Beijing, Hebei, Henan, Shandong, and Shanxi provinces (Wang, 1991b).

synoptic information provided by the network of early observers (Yoshimura, 1996). All of the information has been coded and digitized for computer-based analysis (Yoshimura, 1992). These historical records have provided a wealth of information about the climate of Japan during the last 250 yr (Murata, 1992; Mizukoshi, 1992; Tagami and Fukaishi, 1992). For example, Fukaishi and Tagami (1992) classified the observations into those patterns that are typical of characteristic "winter-like" pressure patterns, with high pressure in the west and low pressure in the east. Figure 11.13 shows the number of days with this pattern during the November–March period of each year, from 1720–1869. Unusually cold conditions with heavy snow were common in the years when this pressure pattern prevailed (e.g., 1726–1733, 1777–1785, 1808–1819, and 1826–1836).

11.3.2 Europe

Evidence for significant changes in the climate of Europe over the last few centuries is abundant. In particular, there are many wonderful paintings of alpine glaciers (far in advance of their current positions) that vividly illustrate the environmental consequences of recent climatic changes in the Alps (Zumbühl, 1980; Grove, 1988). These pictorial images make it easy to accept the notion of a "Little Ice Age" (LIA) gripping Europe over the last few hundred yr. In fact, there is enough similar glacial evidence from virtually all mountainous parts of the world to indicate that some large-scale forcing was involved in bringing the Little Ice Age about. Unfortunately, the term is often used without clarity; some authors consider the LIA began in the fourteenth or fifteenth centuries, others date it as starting in the 1600s.

In reality, this deterioration was one of several late Holocene cool episodes (neoglacials) that led Matthes (1940) to introduce the term. He wrote: "We are living in an epoch of renewed but moderate glaciation — a "little ice age" that already has lasted about 4,000 years . . . glacier oscillations of the last few centuries have been among the greatest that have occurred during the 4,000 year period . . . the greatest since the end of the Pleistocene ice age."

It is this latest and most dramatic episode of neoglaciation to which the term "Little Ice Age" is now generally applied; there were a series of post-Medieval cool events, varying in intensity from one region to another, but there seems to have

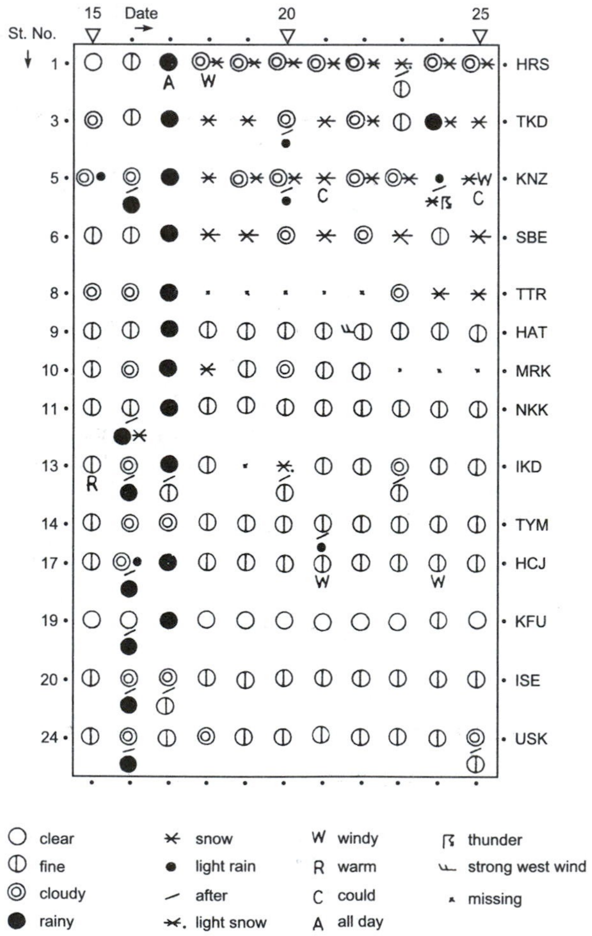

FIGURE 11.12 An example of coded weather data for January 15–25, 1754, derived from diaries kept by feudal clans in 14 different locations throughout Japan (Fukaishi and Tagami, 1992).

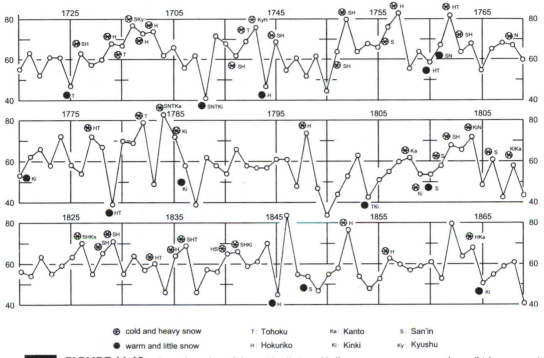

FIGURE 11.13 Annual number of days with a "winter-like" pressure pattern across Japan (high pressure in the west, low pressure in the east) for the period November 1–March 31 each year from 1720–1869, based on an interpretation of mapped weather information obtained from diaries (Fukaishi and Tagami, 1992).

been more widespread climatic deterioration after ~1510 ± 50 (Bradley and Jones, 1992b). There is general agreement that the LIA came to an abrupt end in the mid-nineteenth century (~1850). If we therefore consider the overall "Little Ice Age" as having lasted from ~1510–1850, we find that conditions were not continuously cold, nor was it uniformly cold in all regions. Even within Europe there were temporal and geographical differences in temperature variations (Brázdil, 1996). At some times, in some areas, decadal mean temperatures were comparable to twentieth century values (particularly during the eighteenth century). Particular insight into European climate during this period has been obtained from the comprehensive paleoclimatic reconstructions produced by Pfister for Switzerland (Pfister, 1984, 1985, 1992). Pfister used phenological data, lake freeze/thaw dates, snow cover duration, and tithe auction dates (related to the time of maturation of rye) with other information, to construct monthly indices of temperature for the last 470 yr (Table 11.7). The data were calibrated with instrumentally recorded temperature data from Basel, enabling estimates of seasonal temperature anomalies to be made. Data on floods and low-water levels also enabled estimates of precipitation anomalies to be made (Fig. 11.14). These studies (like those reported by Brázdil) indicate that the record of past anomalies varied between seasons. Winters

TABLE 11.7 Indicators Used by Pfister to Determine the Thermal Character of Individual Months

Month	Cold		Warm
Dec–Jan	Uninterrupted snow cover Freezing of lakes		Scarcity of snow cover Signs of vegetation
March	Long snow cover High snow frequency		Sweet cherry first flower [± 1.3 °C]
April	Snow cover and snow frequency Beech tree leaf emergence		Beech tree leaf emergence Tithe auction date
May		Tithe auction dates [± 0.6 °C] Vine first flower [± 1.2 °C] Barley first beginning	
June		Tithe auction dates [± 0.6 °C] Vine full flower [± 1.2 °C] Vine last flower Coloration of first grapes	
July		Vine yields [± 0.6 °C] Coloration/maturity of first grapes	
April–July		Wine harvest dates [± 0.6 °C]	
August		Wine yields [± 0.6 °C] Tree ring density [± 0.8 °C]	
September		Vine quality Tree ring density [± 0.8 °C]	
October	Snow cover Snow frequency		Reappearance of spring vegetation (cherry flowering etc.)
November	Long snow cover High snow frequency Freezing of lakes		No snowfall Cattle in pastures

Pfister (1992).
Figures in brackets give the standard error of the estimates.

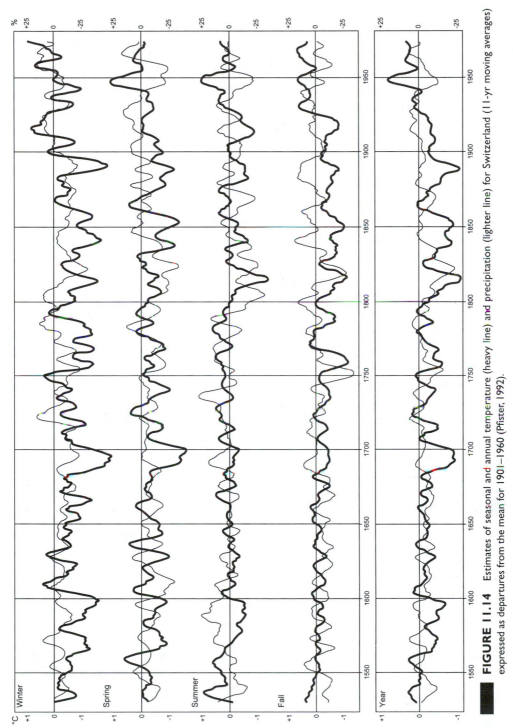

FIGURE 11.14 Estimates of seasonal and annual temperature (heavy line) and precipitation (lighter line) for Switzerland (11-yr moving averages) expressed as departures from the mean for 1901–1960 (Pfister, 1992).

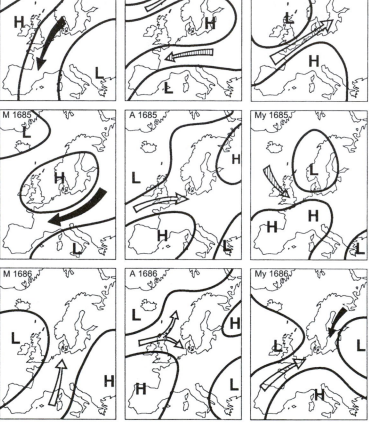

FIGURE 11.15 Monthly circulation patterns over Europe for March–May, 1684–1686 reconstructed from historical evidence of weather conditions across the region (Wanner *et al.*, 1994).

were especially cold in the late seventeenth century and throughout most of the nineteenth century, whereas the coldest summers were in the early 1800s and the late 1800s/early 1900s. Indeed summer temperatures from ~1600–1800 appear to have been quite similar to the twentieth century average. It is also of interest that for much of the last few hundred years conditions were drier than in the twentieth century. Pfister's studies in Switzerland indicate that overall, the coldest conditions of the last 500 yr were in the late seventeenth and nineteenth centuries, especially the early nineteenth century, which can be considered as the "Climatic Pessimum" of the last 1000 yr. Since most of our longest instrumental temperature records began during this time, perhaps the coldest period of the last millennium, much of the "global warming" registered since then represents a recovery from that low point in the early to mid-nineteenth century.

Historical climatology in Europe has greatly benefited from the activities of the EURO-CLIMHIST project led by C. Pfister (see the volumes edited by Frenzel *et al.*, 1992b, 1994). In order to be able to compare diverse historical observations in many different European languages, uniform procedures for coding information have been developed (Schüle and Pfister, 1992; Schwarz-Zanetti *et al.*, 1992). The EURO-CLIMHIST project has helped to coordinate historical climatological studies across Europe, so that daily iconographic maps of weather conditions over Europe can now be constructed, for extended periods of time, as has been carried out in Japan. These have been interpreted by meteorologists familiar with the atmospheric circulation of Europe to produce estimates of the prevailing pressure regimes over the region, consistent with the recorded historical observations (Fig. 11.15). Such maps provide scenarios that can be updated and revised as additional historical data become available. Eventually, it should be possible to establish in great detail how atmospheric circulation in Europe during the Little Ice Age differed from that in the twentieth century, providing clues about the factors causing these changes.

11.4 RECORDS OF CLIMATE FORCING FACTORS

Historical observations have been very important in documenting two factors outside the climate system that may be important in causing climate to change (forcing factors). These are major explosive volcanic eruptions and solar variability. We will also consider here records of El Niño (ENSO) events,[34] which are not strictly external forcings, but involve large-scale reorganizations of the ocean-atmosphere system, with global consequences.

Unusual post-sunset sky colors were commonly recorded by astute observers of the heavens, and many such observations can be linked to large explosive volcanic eruptions that lofted sulfur-rich gases and particulate matter high into the stratosphere. Scattering of solar radiation by these particles can reduce direct radiation and cause dramatic early morning or late evening sky colors (Meinel and Meinel, 1983). For example, Chinese chroniclers reported that, in the reign of Emperor Ling Ti (A.D. 168–189) "several times the sun rose in the east red as blood and lacking light . . . only when it had risen to an elevation of more than two zhàng [24°] was there any brightness. . . .". At the same time, Roman observers noted, "before the war of the deserter [A.D. 186] the heavens were ablaze. . . . stars were seen all the

[34] El Niños (EN) are quasi-periodic changes in oceanographic conditions characterized by unusually warm water off the coasts of Ecuador and northern Peru, especially in December. They are associated with atmospheric anomalies that involved the redistribution of atmospheric mass across the South Pacific, a phenomenon termed the Southern Oscillation (SO). Together, oscillations of the coupled ocean-atmosphere system in the Pacific are termed ENSO events.

day long . . . hanging in the air which was a token of a cloud. . . ." (Wilson *et al.*, 1980). These are typical descriptions of the sky following major explosive volcanic eruptions; it seems likely that this particular event was related to eruptions in Alaska (White River ash), which have been radiocarbon-dated to around that time. Lamb (1970, 1977, 1983) used these kinds of observations as the basis for constructing a chronology of explosive volcanism over the past ~500 yr, which he termed the Dust Veil Index (DVI) (Fig. 11.16). This proved to be invaluable in interpreting the acidity record in ice cores, which is also a register of explosive volcanic events (see Section 5.4.4). Many studies have used the DVI to assess the impact of explosive volcanism on temperature variations (Sear *et al.*, 1987). Historical records have also been useful in documenting the global effects of one of the largest eruptions in the late Holocene (that of Tambora, Indonesia, in April, 1815). This event led to cold conditions in many areas the following year, which became known as "the year without a summer" (Harington, 1992).

Early astronomical observations of the sun noted dark spots on the photosphere, and records of these sunspots extend back to the early seventeenth century (Hoyt and Schatten, 1997). Long-term sunspot observations demonstrated a periodic variation in solar activity with a mean cycle length of ~11 yr. Satellite observations have now shown that these variations involve changes in the solar constant of ~0.1%. There is also documentary evidence that a prolonged episode of little or no

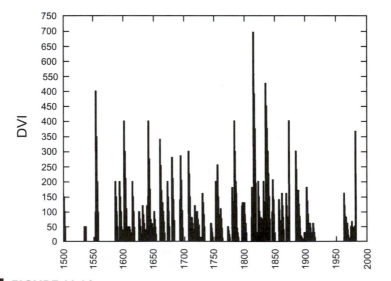

FIGURE 11.16 A dust veil index (DVI) for the northern hemisphere, assuming dust from an individual eruption is apportioned over four yr, with 40% of each DVI assigned to year 1, 30% to year 2, 20% to year 3, and 10% to year 4. Thus, the 1883 eruption of Krakatau (DVI = 1000) results in values of 400 in 1883, declining to 100 in 1886. It is further assumed that all dust from eruptions poleward of 20° N remained in the northern hemisphere. For eruptions equatorward of 15° N, the dust was assigned equally between the two hemispheres and for eruptions between 15° and 20° N and 15° and 20° S, it was assumed that two-thirds of the material remained in the hemisphere of the eruption, and one-third was dispersed to the other hemisphere (Bradley and Jones, 1992b; DVI values from Lamb, 1970, 1977, 1983).

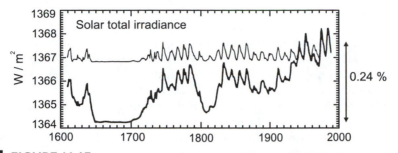

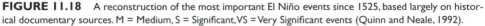

FIGURE 11.17 Reconstruction of solar total irradiance from 1610 to the present. The thin line is the irradiance variability of the Schwabe cycle, and the thick line is the Schwabe cycle plus a longer-term component that accounts for the amplitude of irradiance reductions since the Maunder Minimum (1645–1715) estimated independently from observations of sun-like stars (Lean et al., 1995).

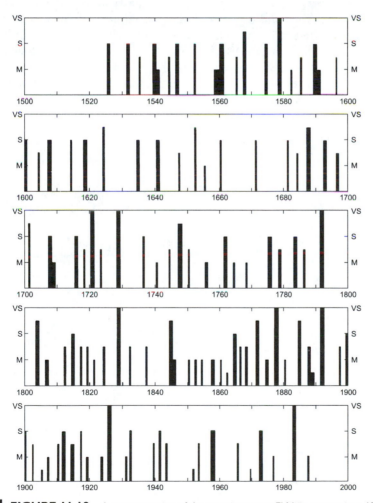

FIGURE 11.18 A reconstruction of the most important El Niño events since 1525, based largely on historical documentary sources. M = Medium, S = Significant, VS = Very Significant events (Quinn and Neale, 1992).

solar activity (the Maunder Minimum) occurred from ~1675–1715 (Eddy, 1976). Lean *et al.* (1992, 1995) estimate that the long-term variability in solar activity from the Maunder Minimum to the present represents an overall increase in solar output of ~0.24% (Fig. 11.17) that is significant enough to have had an effect on global climate (Rind and Overpeck, 1993; Lean and Rind, 1994). Observations of solar variability have also been valuable in showing that both ^{14}C and ^{10}Be records (from tree rings and ice cores, respectively) contain a solar-modulated signal. This may allow these isotopes to be used to reconstruct the history of solar variability, and its influence on climate, long before the beginning of historical sunspot observations (Beer *et al.*, 1996; Stuiver and Braziunas, 1991).

Historical records of unusual weather events in key parts of the world have also been used to reconstruct the history of ENSO events (Quinn, 1993a, b; Quinn and Neal, 1992; Mabres *et al.*, 1993). The ENSO events result in worldwide disruptions of the climate system, with certain regions being particularly affected. Thus, heavy rains in coastal Ecuador and northern Peru are common, often causing floods and landslides. Across the Pacific, ENSOs are associated with droughts in Indonesia and northeastern Australia, where brush fires are also likely. Historical documents have been used to piece together these events in the past and to rank them in terms of their overall magnitude (Fig. 11.18). Once again, this has proven useful in interpreting and verifying other proxy records of ENSO events, such as those derived from corals (see Section 6.8).

12
PALEOCLIMATE MODELS

12.1 INTRODUCTION

Models are simplifications of reality, designed to provide insight into how a system responds to change. Much of the impetus for developing models of the climate system has come from concerns over the increase in trace (greenhouse) gases in the atmosphere due to human activity, and the consequences such increases may have for society. Commonly, general circulation models (GCMs) of the atmosphere are first run with preindustrial levels of CO_2 and then again with 2x CO_2 levels to examine the projected changes ("equilibrium" simulations). Alternatively, models are run sequentially, with CO_2 levels slowly increasing until 2x CO_2 levels are reached (transient simulations). Such experiments, performed by many different modeling research groups, have formed the basis of the Intergovernmental Project on Climatic Change (IPCC) estimates of the probable consequences of anthropogenic increases in greenhouse gases (e.g., IPCC, 1996) However, such projections leave many with a sense of unease. How reliable are such models? Although such models may be able to simulate modern climatic conditions quite well, how do we know that they can reliably simulate a future climate state different from that of today? (Trenberth, 1997). One approach that has been widely adopted to allay such fears is to use the same models to simulate climates of the past. If models can reproduce climatic conditions that are *known* to have occurred (i.e., reconstructed from paleo-data) then confidence in their ability to simulate future (*unknown*) climates will be enhanced. This line of reasoning has driven paleoclimate modeling over the last decade or so, but the resulting experiments have had many collateral benefits. Models have provided considerable insight into the forcing mechanisms that may have been responsible for some of the dramatic changes observed in the paleoclimate record and they have demonstrated how different subsystems of the climate system (atmosphere, surface and deep ocean, ice sheets, biosphere) may have interacted via both positive and negative feedback at different times in the past. Models have also pointed to potentially questionable paleoclimatic data, most notably the apparent inability of some models to generate glaciations with relatively warm tropical sea surface temperatures, as reconstructed by CLIMAP (1981) (see discussion in Section 6.4). Thus, models are now an important part of the field of paleoclimatology. They are tools for understanding how the climate system may have operated in the past, and are used interactively with conventional paleoclimatic data to improve that understanding. Just as paleoclimatic data are constantly being updated, reinterpreted, and extended in time and space, so models are becoming more sophisticated, the methodology for model-data comparisons is improving, and opportunities for complex "total climate system" simulations are increasing as computers become faster. All of these efforts have the dual benefits of improving the ability of models to simulate future climate and broadening our appreciation of the range and complexity of climatic conditions in the past.

12.2 TYPES OF MODELS

There are many types of models used in paleoclimatology and the GCMs referred to in the introduction are most complex. Hydrological balance models, used to understand the paleorecord of lake sediments and lake level changes, have already been discussed in Chapter 7. Models of the changes in forest growth that might be expected with climatic changes of different duration and magnitude were described in Chapter 9. One could also argue that the various approaches to calibrating faunal data from the oceans, and pollen and tree-ring data from the continents, constitute statistical models, which may produce different reconstructions as methodologies are refined and improved. Models are thus embedded in a conceptual framework of how systems work, and as our understanding of such systems improves so we can expect models to be revised. In this sense, model experiments produce results that are moving targets. Some results may be robust and demonstrably correct; others may turn out to be erroneous. By the constant interaction between data-generators (empiricists) and modelers (theorists) the results eventually will converge on a consensus of how (and why) climates varied in the past.

Various types of atmospheric models have been developed with varying levels of complexity. Coupled ocean-atmosphere general circulation models are at the high end of this range but consequently require immense amounts of computer time, making the number of runs and potential applications somewhat limited. Less complex models have the advantage of being much faster to run (allowing many more options to be examined) but the trade-off is in limiting the processes, feedbacks, and/or dimensions being examined.

A very brief summary is provided of the types of models that have been used in climate and paleoclimatic applications, with an emphasis on GCMs. For a much more comprehensive introduction to the theory of paleoclimate modeling, see Saltzman (1985). An introduction to climate models in general, with CD-based examples of different models, is provided by McGuffie and Henderson-Sellers (1997). The volumes edited by Schlesinger (1987) and Trenberth (1992) also provide in-depth discussion of climate models and their components. All of the relatively simple models described here have a role to play in understanding how the climate system works and indeed complex general circulation models have been constructed using insights obtained from the development of simpler, faster models (Schneider, 1992).

12.2.1 Energy Balance Models and Statistical Dynamical Models

Energy balance models (EBMs) consider only surface temperature as a consequence of energy exchange. The simplest are zero-dimensional (i.e., consider the energy balance of the earth as a whole; Budyko, 1969; Sellers, 1969, 1973). One-dimensional EBMs consider the earth in terms of zonal bands, with energy exchanged latitudinally from one zone to the next by diffusive horizontal heat transfer. Two-dimensional EBMs add additional complexity by considering latitude/altitude or latitude/longitude differences (e.g., land vs ocean). Further detail may be added, for example, by considering the earth as a series of linked boxes with different properties (atmosphere-land-mixed layer ocean-deep ocean), and with energy transfers taking place between them by advection and diffusion (Harvey and Schneider, 1985a). Such models enable changes in feedbacks to be investigated in a computationally efficient way, allowing many slightly different simulations (sensitivity experiments) to be carried out. The effects of changes in albedo, solar input to the earth, greenhouse gas increases, and net poleward heat transport have also been investigated with EBMs (North *et al.*, 1981; Harvey and Schneider, 1985b; Wigley, 1991; Wigley and Kelley, 1990).

Statistical-dynamical models (SDMs) are designed to capture the observed record of climate without explicitly creating that record, *ab initio*, from basic physical laws at a network of points, as occurs in general circulation models. Unlike GCMs (considered in what follows), they do not consider explicitly all the synoptic scale variability and atmospheric changes that take place at high frequencies (30-min time steps). Instead, they use equations that describe changes over long time periods, requiring parameterization of many of the phenomena that are dealt with explicitly in GCMs. Consequently, SDMs require less computer time and can be used to examine the long-term evolution of climate, taking into account some of the more slowly responding parts of the climate system (Saltzman, 1985). For example, Gallée *et al.* (1991) developed a quite complex two-dimensional (latitude/altitude) time-dependent climate model of the northern hemisphere, with the surface subdivided into up to 7 different categories, each with distinct properties (ocean, sea ice, snow-covered or snow-free land, and three areas of land ice, representing the Laurentide, Fennoscandian, and Greenland ice sheets). Atmospheric dynamics in the model were zonally averaged, and meridional fluxes, oceanic heat transport, mixed layer dynamics, and land surface hydrology were all parameterized. The atmosphere-land-ocean model was coupled asynchronously to an ice-sheet-bedrock model; this allowed the different response times of these systems to be taken into account. The atmosphere-land-ocean model was run every day for a 20-yr simulation, and the resulting climate was the input to the ice-sheet-bedrock model, which ran at 1-yr time steps for 1000 yr. The procedure was then repeated, taking into account the altered surface boundary conditions, changes in orbital forcing, and CO_2 levels (as deduced from the Vostok ice core) (Gallée *et al.*, 1992; Berger *et al.*, 1993). Figure 12.1 shows the resulting simulation of northern hemisphere continental ice volume for the last 200,000 yr, compared to $\delta^{18}O$ variations in benthic forams, according to SPECMAP (Martinson *et al.*, 1987). Although the model is not correct in absolute terms (e.g., it reconstructs zero ice in the northern hemisphere at the last interglacial, at ~100 ka and ~70–84 ka B.P., which is not supported by the paleorecord) it does well at reproducing the general time evolution of the SPECMAP record. Considering SPECMAP $\delta^{18}O$ represents a *global* ice volume signal, and the model is based only on conditions in the northern hemisphere, the comparison is very favorable. Further improvements in the parameterizations, and incorporation of a southern hemisphere model should make the comparison even better (Berger and Loutre, 1997a,b).

12.2.2 Radiative Convective Models

These models examine radiation processes in a vertical column of the atmosphere; vertical temperature profiles (lapse rates) are maintained within a reasonable range

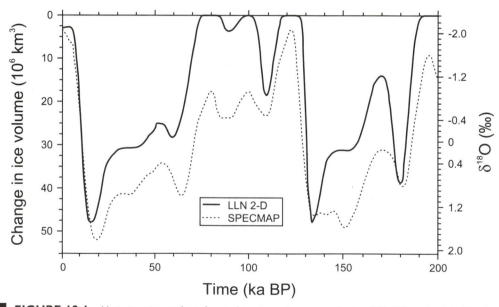

FIGURE 12.1   Variations in northern hemisphere ice volume over the last 200,000 yr simulated by the Louvain-la-Neuve 2D climate model (Gallée *et al.*, 1991) forced by insolation and CO_2 variations (solid line) compared to the SPECMAP $\delta^{18}O$ variations, representing ice volume changes over the entire earth (dashed line) (Berger and Loutre, 1997a).

by convective adjustments (movement of air vertically). Such models have been used to investigate the effects of atmospheric aerosols and clouds on temperature and the effects of changing greenhouse gases (CO_2, CH_4, O_3) (Manabe and Wetherald, 1967; Reck, 1974; Hansen *et al.*, 1978). Radiative convective models (RCMs) have been used extensively to perform sensitivity tests in which the value of one parameter (e.g., greenhouse gas amount) is varied in different model simulations to examine how such changes might lead to feedbacks with other parts of the system (Ramanathan and Coakley, 1978; Kiehl, 1992).

12.2.3 General Circulation Models

Atmospheric general circulation models (AGCMs) simulate atmospheric processes in three dimensions, explicitly taking into account dynamical processes. Basic equations solved in GCMs involve the conservation of energy, mass, and momentum. The earth's surface is divided into a series of grid boxes, extending vertically into the atmosphere (Fig. 12.2). The atmospheric column is also divided into a series of levels (commonly 10–20) with more levels near the earth's surface. Equations are solved at each grid point and at each vertical level at a preset time interval (typically 20–30 min) with vertical and horizontal exchanges of energy, mass, and momentum computed for all points at each time interval. Clearly, this procedure requires immense amounts of computer time and so the most sophisticated GCMs have to run on the fastest computers available, and even then it may take weeks of dedicated computer time to carry out a simulation. For example, the National Center for Atmospheric Research (NCAR) Climate System Model, comprising a 3.75°× 3.75° grid, 18-level atmospheric model, coupled to a 3°× 3° mixed layer ocean, with 30-min time steps requires ~15 h of dedicated time on a CRAY J90 computer (one of the fastest supercomputers currently available) to simulate one yr. The problem is even more acute in ocean general circulation models (OGCMs) where finer spatial resolution is required to simulate effectively the important scale of motion (oceanic eddies) that are < 50 km across (more than an order of magnitude smaller than the typical atmospheric eddies). Massively parallel computers (employing >100 processors simultaneously) are now being used to solve such problems (Chervin and Semtner, 1991).

These GCMs vary considerably in their resolution. Many paleoclimatic simulations have been carried out at a scale of 8° (latitude) × 10° (longitude). Such models have relatively crude geography (for example, small details like the United Kingdom or New Zealand would not be represented!) and poor topographical representation. In such a model, a single grid box in midlatitudes is equivalent in size to the states of Colorado and Utah, or of France and Germany, together. More complex models may have spatial resolution up to 2° × 2° with more highly resolved surface relief. However, such models inevitably trade computational speed for the added sophistication provided by a denser grid network, so the analyst must determine if the nature of the problem being investigated warrants the added time and expense of a high-resolution GCM. Even in the highest resolution AGCMs, many atmospheric processes cannot be represented as they are below the grid scale of the model. Convective thunderstorms, for example, play a critical role on a global scale in latent and sensible heat transfer from the surface to the atmosphere, but individually they

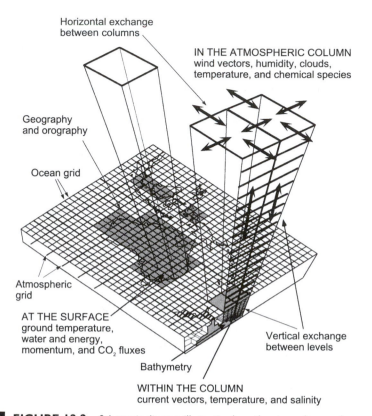

Horizontal exchange
between columns

IN THE ATMOSPHERIC COLUMN
wind vectors, humidity, clouds,
temperature, and chemical species

Geography
and orography

Ocean grid

Atmospheric
grid

AT THE SURFACE
ground temperature,
water and energy,
momentum, and CO_2 fluxes

Vertical exchange
between levels

Bathymetry

WITHIN THE COLUMN
current vectors, temperature, and salinity

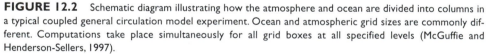

FIGURE 12.2   Schematic diagram illustrating how the atmosphere and ocean are divided into columns in a typical coupled general circulation model experiment. Ocean and atmospheric grid sizes are commonly different. Computations take place simultaneously for all grid boxes at all specified levels (McGuffie and Henderson-Sellers, 1997).

are too small to be represented by even a 2° × 2° grid spacing. In such cases, the process is represented in a simplified manner as a function of other variables, a procedure known as parameterization (parametric representation). This may be based on observed statistical relationships between, for example, temperature and humidity profiles and cloudiness, or on some other simplified model of the process in question. In fact, parameterization of all forms of cloud is one of the most difficult problems in atmospheric GCMs and is the focus of much research at present (Cess *et al.*, 1995).

Atmospheric general circulation models may be coupled to the ocean realm in a variety of ways. At the simplest level, the surface temperature of the ocean is prescribed (predetermined) and the ocean region of the model interacts with the atmosphere only in terms of moisture exchange. This is often termed a "swamp ocean" (Fig. 12.3). At the next level a "slab ocean" is specified as a layer of fixed depth (50–100 m); heat and moisture exchange with the atmosphere occurs, enabling SSTs to vary as the model run progresses. However, in such models there is no

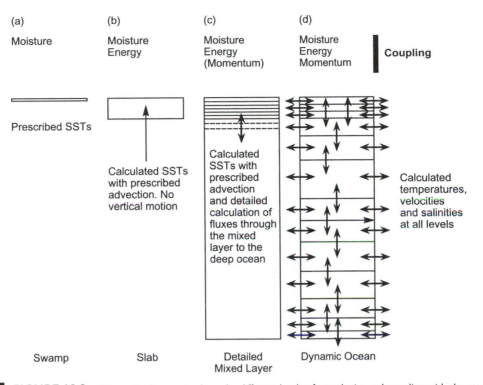

(a)

Moisture

(b)

Moisture
Energy

(c)

Moisture
Energy
(Momentum)

(d)

Moisture
Energy
Momentum

Coupling

Prescribed SSTs

Calculated SSTs
with prescribed
advection. No
vertical motion

Calculated
SSTs with
prescribed
advection
and detailed
calculation of
fluxes through
the mixed
layer to the
deep ocean

Calculated
temperatures,
velocities
and salinities
at all levels

Swamp

Slab

Detailed
Mixed Layer

Dynamic Ocean

FIGURE 12.3 Schematic diagram to show the different levels of complexity and coupling with the atmosphere in various types of ocean model (McGuffie and Henderson-Sellers, 1997).

mechanism for heat exchange with the deep ocean and only a crude representation of horizontal energy fluxes. The mixed layer ocean is a further improvement, also involving prescribed horizontal advection, but with computation of fluxes to and from the deep ocean. The most complex level is a fully coupled ocean-atmosphere GCM (OAGCM) in which the ocean has internal dynamics in three dimensions, and exchanges of energy, moisture, and momentum take place at the ocean-atmosphere interface. It is worth noting that a major problem in coupling atmospheric to oceanic processes is the vastly different response times characteristic of each domain (Fig. 12.4). The slower response time of the deep ocean must be taken into account when the two systems are linked. The problem is further compounded if one is trying to investigate climate system changes involving ice sheets, which have even longer response times. One approach is to couple models of each system "asynchronously," that is, to operate an atmospheric model for a time appropriate for that system, then use the resulting atmospheric conditions as input to a model of the system operating on a different timescale. For example, Schlesinger and Verbitsky (1996) used the climate at 115 ka B.P., generated by an atmospheric GCM (coupled to a mixed layer ocean), to drive an ice sheet/asthenosphere model, in order to investigate the areas most likely to develop ice sheets following the last interglacial

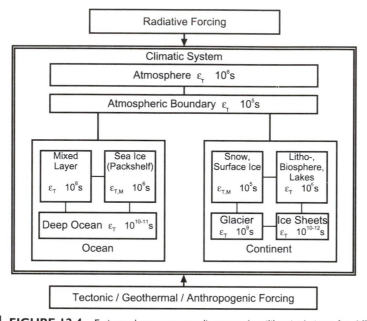

FIGURE 12.4 Estimated response or adjustment (equilibration) times for different components of the climate system. Note that they vary by 6–7 orders of magnitude (Saltzman, 1985).

period (see Section 12.3.1). The ice sheet model was run for 10,000 yr (an appropriate interval of time to examine ice sheet development) but the atmospheric conditions obviously could not be computed over such a long period and so were fixed as those obtained from the 115 ka B.P. simulation. Although this approach clearly has its limitations, it does allow two systems with strikingly different response times to be examined in a somewhat coupled fashion.

GCMs can also be used to trace the pathways of materials within the climate system (Koster *et al.*, 1989). This has been put to good use in paleoclimatic applications (Jouzel, 1991; Jouzel *et al.*, 1993a; Andersen *et al.*, 1998). For example, the long-distance transport of desert dust particles in the atmosphere has been traced using an AGCM for both modern and last glacial maximum (LGM) conditions. Source regions of dust deposited from the atmosphere are identified by "tagging" the dust originating from different areas (Joussaume, 1987, 1990, 1993). Modern-day simulations show a strong seasonality in atmospheric dust production, with atmospheric dust loading in August more than twice that in February. The largest source of dust (by far) is the Sahara /Arabia /central Asia region. Australia is the principal source of dust reaching east and west Antarctica, whereas South America contributes the most dust to central Antarctica. Simulations for the LGM show greater atmospheric dust deposition especially over the tropical Atlantic Ocean and Europe, but the modeled increases significantly underestimate the observed changes (recorded in ice cores). This could be due to many factors, including model resolution (poor representation of source areas) and/or inadequate characterization of dust entrainment, transportation, and depositional processes (wet and dry fallout). In view of the potential climatic significance of dust during glacial periods (Overpeck *et al.*, 1996) these first steps toward fully incorporating the dust cycle into GCMs are important contributions. Further studies with higher resolution GCMs are now needed.

Isotopes (deuterium, ^{18}O) in the hydrological cycle have also been modeled with GCMs, for modern and glacial age conditions (Jouzel *et al.*, 1987c, 1991, 1994; Joussaume and Jouzel, 1993). At each change of phase of water molecules in the model, appropriate fractionation factors are employed to calculate the mass of the isotopes in each reservoir (water vapor, precipitation, ice, groundwater). Isotopic modeling is particularly important in paleoclimatic studies as it allows the direct comparison of model simulations with paleoisotope records (in ice, sediments, biological materials, speleothems, etc.), thus avoiding the need to calibrate the paleo-record in terms of, say, temperature for a comparison with modeled paleotemperature output. Modern simulations reproduce the global pattern of $\delta^{18}O$ and δD very well and the seasonal cycle is well captured at most sites in both high and low latitudes (Jouzel *et al.*, 1987c). The LGM simulations show a similar overall $\delta^{18}O$/temperature relationship to that derived from modern simulations ($\delta^{18}O$ ~0.6°T, where T < −5 °C) and there were large decreases in $\delta^{18}O$ at high latitudes (see Fig. 5.9) (Joussaume and Jouzel, 1993).

Sources of moisture can be traced using GCMs and this is useful in understanding how source regions might have been different in the past; this would be relevant, for example, in the interpretation of ice core geochemistry. Charles *et al.* (1994) used an AGCM to examine how source regions of precipitation reaching Greenland changed from the LGM to the present. The modern (control) simulation showed that 26% of Greenland precipitation was derived from the North Atlantic (30–50° N), 18% from the Norwegian-Greenland Sea, and 13% from the North Pacific. At the LGM, these values changed to 38%, 11%, and 15%, respectively. However, northern Greenland received distinctly more moisture at the LGM from the north Pacific source region, due to displacement of storm tracks around the Laurentide Ice Sheet. Southern Greenland received most of its snowfall from North Atlantic moisture sources. Because of the much longer (and colder) trajectory of the Pacific air masses, snow deposited on Greenland from such sources was much more depleted in $\delta^{18}O$ than snow from North Atlantic sources (~15‰ lower). Charles *et al.* (1994) point out that if there was no change in temperature in Greenland, but only a shift in source region from purely North Atlantic moisture to a 50:50 mix of North Atlantic and North Pacific moisture, changes in $\delta^{18}O$ of snowfall could change by ~7‰, equivalent to the large amplitude oscillations seen during late glacial time in the GISP2/GRIP ice cores. This raises the interesting possibility that abrupt changes in $\delta^{18}O$ seen in the ice cores from Greenland may be partly related to changes in storm tracks rather than large-scale (hemispheric) shifts in temperature. However, it is worth noting that this model did a very poor job of simulating modern precipitation in Greenland (with simulated precipitation exceeding observations by as much as 100%, or ~1 mm day^{-1}, especially in summer); thus this experiment, though interesting, begs the question whether similar results would be found with a more accurate, higher resolution model.

Although the discussion so far has focused on general circulation models of atmospheric and oceanic systems, further development of GCMs will be towards total climate system models (CSMs), which will incorporate in a fully interactive way land surface and cryospheric processes (both terrestrial snow and ice, and sea ice) and biomes. Models of global biomes are now available (Prentice et al., 1992; Haxeltine and Prentice, 1996) and have been used with output from GCM experiments to predict vegetation at times in the past (Harrison et al., 1995; TEMPO, 1996). Models with full biogeochemical cycling of materials in the atmospheric and oceanic realms are also under development (Brasseur and Madronich, 1992; Sarmiento, 1992). Nested models, in which a very detailed grid network for a specific region is used to model detailed geographical variations of the climate, given initial input from a larger scale GCM, will also become more widely available (Hostetler et al., 1994). Such models are especially important in mountainous areas where large-scale GCMs cannot provide the necessary topographic detail to produce meaningful regional simulations.

12.3 SENSITIVITY EXPERIMENTS USING GENERAL CIRCULATION MODELS

Models can be used in sensitivity experiments to examine the effects of changing a boundary condition (e.g., solar radiation; Syktus et al., 1994) or a process (e.g., sea-ice formation; Vavrus, 1995). By comparing climate associated with the specified change with the "no change" (control) simulation the significance of the process or parameter that was altered can be assessed. Such experiments help in understanding how complex feedbacks (amplifying or minimizing the effect of specified changes) operate and interact within the climate system. Generally a control experiment.is run over a number of simulated years "to equilibrium" and a period of years at the end of this run will then be averaged (for a specified month or season) to obtain a reference climate state. This is then "validated" by comparison with modern climate data sets, providing some level of confidence in the model. After the conditions to be investigated have been changed, the simulation is once again run to a new "equilibrium" and a new set of years is averaged. The differences between the control run and the new conditions are considered to be a consequence of those factors that were altered in the sensitivity experiment.

How should the results of sensitivity experiments be evaluated? We know from the outset that each model may have non-trivial limitations in its ability to simulate modern climate. For example, compared to observed climate, one version of the Geophysical Fluid Dynamics Lab (GFDL) atmospheric GCM used in many paleoclimate experiments by Manabe and other researchers (Manabe and Broccoli, 1985a) produces conditions that are too cold in the northern hemisphere, and it generates too much precipitation at high latitudes. It underestimates sea ice in the southern hemisphere and places the Inter-Tropical Convergence Zone (ITCZ) 5°–10° (latitude) too far south. However, the important issue for this and other sensitivity experiments is what *differences* were induced in the model's climate system by the imposed changes. Certainly, model limitations must be kept in mind to ensure the observed differences are not simply artifacts of the model, but in general, sensitivity experiments can provide valuable insights into the role of boundary conditions and other changes. Other models can also be employed to determine which conclusions are robust, regardless of the model used. On the other hand, the simulations may point to discrepancies between modeled paleoclimates and reconstructions based on proxy data, and these may justify a reconsideration of the paleodata and/or calibration method used. Such considerations have led to an international effort (PMIP — Paleoclimate Model Intercomparison Project) to compare paleoclimatic reconstructions for specified times, with all models using exactly the same boundary conditions. This will enable the common climatic characteristics to be identified and problems in particular models to be addressed (Joussaume and Taylor, 1995). In a parallel effort (PMAP — Paleoenvironmental Multiproxy Analysis and Mapping Project) paleodata are being compiled and mapped for selected "time slices" to enable direct comparisons to be made between model simulations and paleoclimatic data (see Section 12.4).

In the next section, we first consider changes in orbital forcing. By examining the range of extremes over the late Quaternary, models can address to what extent orbital changes may have been important in initiating glaciation (at one extreme) and in bringing about post-glacial changes in monsoon climates at the other. Then we examine a set of experiments designed to understand how changes in surface boundary conditions during the last glacial maximum (LGM: 18 ka B.P.)[35] contributed to climatic conditions prevailing at the time. At 18 ka B.P. orbital conditions differed very little from today and so the dramatically different climate state at that time must have been driven largely by orbital changes at some earlier time, with conditions at 18 ka B.P. responding to (but not necessarily maintaining) changes in the boundary conditions. Finally, the significance of changes in the seasonality of incoming radiation is examined, in a series of experiments at intervals from 18 ka B.P to the present.

12.3.1 Orbital Forcing and the Initiation of Continental Glaciation in the Northern Hemisphere

A number of GCM experiments have been carried out to investigate the role of orbital forcing in the initiation of continental glaciation. Paleoclimatic data from marine sediments ($\delta^{18}O$ in benthic forams) and raised corals (recording former sea-level position) indicate that sea level fell by > 50 m from ~115–105 ka B.P. Because of a combination of greater eccentricity, lower obliquity, and perihelion close to the (northern) winter solstice, incoming solar radiation at 115 ka was reduced by ~7% (40 W m^{-2}) during the northern hemisphere summer (and late spring/early summer

---

[35] The dates of all model experiments are given here in calendar years; 18 ka B.P. was originally thought to be the same in both calendar and radiocarbon years. Later experiments recognized that 18 ka B.P. in radiocarbon years corresponds to ~21 ka in calendar years, which alters the relationship between orbital forcing and ^{14}C-dated SSTs, ice-volume estimates, etc. (Kutzbach *et al.*, 1993a). Some experiments have therefore adjusted the orbital parameters accordingly, allowing the large 18 ka B.P. CLIMAP data set (dated in radiocarbon years) to be compared with 21 ka B.P. model simulations (Kutzbach *et al.*, 1997).

in the southern hemisphere) but was higher at other times (Fig. 12.5). In fact, summer insolation in the northern hemisphere was lower at 115 ka B.P. than at any time in the last 200 ka (see Fig. 2.18). This provides an excellent opportunity to investigate whether orbital forcing alone could have brought about changes in the climate system, sufficient to permit the growth of ice in those areas thought to have been critical in the initiation of glaciation (Ives *et al.*, 1975; Clark *et al.*, 1993).

The earliest attempt to address this problem used a coarse-resolution GCM with modern SSTs and CO_2 levels prescribed (Royer *et al.*, 1984). This indicated that temperatures in northeastern North America were reduced and precipitation was increased (more than evaporation), suggesting that conditions at 115 ka B.P. were indeed more favorable for ice-sheet growth, but the model was simply too crude to determine if an ice sheet could be sustained. Rind *et al.* (1989) carried out several experiments using an 8° (latitude) × 10° (longitude) GCM, initially with just modern SSTs and changed orbital parameters, but then with lower CO_2 levels and SSTs as well. They found that with orbital changes alone ice growth could not be achieved in the model, a conclusion also reached by Phillips and Held (1994) using the GFDL GCM coupled to a mixed layer ocean. Lowering CO_2 levels by 70 ppm (to 230 ppm) made little difference. Only when a 10-m thick ice sheet was

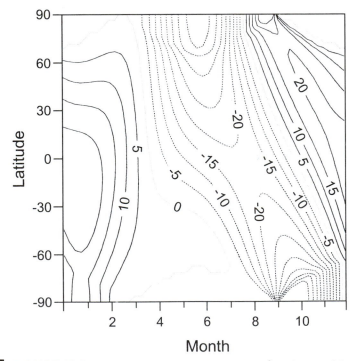

FIGURE 12.5 Difference in solar insolation (in W m⁻²) at the top of the atmosphere at 115 ka B.P. compared to present. Negative anomalies (lower radiation at 115 ka B.P.) are shown as dashed lines (Dong and Valdes, 1995).

specified (to reduce surface albedo), CO_2 levels were lowered to 230 ppm *and* SSTs were reduced to those associated with full glacial conditions (as determined by CLIMAP 1981) could an ice sheet be maintained (Rind *et al.*, 1989). This may be because the model is too coarse and so does not adequately represent topography and upland surfaces where ice sheets probably first formed. It may also not simulate very well the critically important feedbacks between snow, ice, and low clouds. In similar experiments, Gallimore and Kutzbach (1995), using a low-resolution GCM with a mixed layer ocean, concluded that snow cover *could* persist in northwestern and northeastern Canada, for ~11 mo per yr without a reduction in CO_2. If CO_2 levels had been set lower and if cloud cover had not been prescribed, it seems likely that conditions in these areas would have been quite favorable for ice sheet development.

Recently, a higher-resolution GCM has been employed to examine the same problem (Dong and Valdes, 1995). The UGAMP (U.K. Universities Global Atmospheric Modeling Programme) GCM is based on the very successful forecast model of the European Center for Medium Range Weather Forecasts (ECMWF). It has a grid spacing of ~2.8° and includes 19 vertical levels, 5 of which are in the lower atmospheric boundary layer (>850 mb). It includes a 50-m thick mixed layer ocean and an interactive surface hydrology. Topographic resolution is quite good, which may be of particular importance in experiments dealing with the initial stages of glaciation. Several studies have pointed out the critical significance of upland plateaus in Labrador, Baffin Island, and Keewatin in the development of a permanent snow cover (Ives *et al.*, 1975). In the NASA-GISS (Goddard Institute for Space Studies) GCM used by Rind *et al.* (1989), topography is poorly resolved; for example, the Baffin area is represented as <200 m in elevation, whereas in the UGAMP GCM, the value is close to the actual elevation of ~550 m.

The UGAMP GCM simulates modern conditions extremely well. Comparisons between modern conditions and 115 ka B.P. (with modern SSTs prescribed) shows pronounced cooling over all continents except Antarctica. Over North America, surface temperatures were 4–6 °C lower, accompanied by increased cloudiness (+10–15%) and increased soil moisture (+25–30%). However, these changes were not sufficient to maintain snow on the ground throughout the year, except in the Canadian High Arctic Islands and in a small area of Tibet. The CO_2 levels were then reduced slightly (to 280 ppm) and the model was coupled to a shallow mixed layer ocean, allowing SSTs to be determined by the model. This run showed cooling of 8–10 °C in summer over high latitudes, enabling snow to survive all year in many areas (Fig. 12.6). Lower temperatures were brought about by increased sea ice, cooler SSTs in summer, and the presence of permanent snow cover on land, which plays an important role in reinforcing the lower summer insolation effect by sharply increasing surface albedo. This result is quite intriguing because although the model simulated ice growth in northeastern North America and Fennoscandia, where ice sheets are known to have developed, it also generated ice in Siberia and Tibet (Oglesby, 1990). In fact, there has been much debate over whether there really were ice sheets over central Siberia and Tibet during the last glaciation (Velichko *et al.*, 1984; Rutter, 1995; Kuhle, 1991). Although the simulation pertains to an earlier period, and certainly

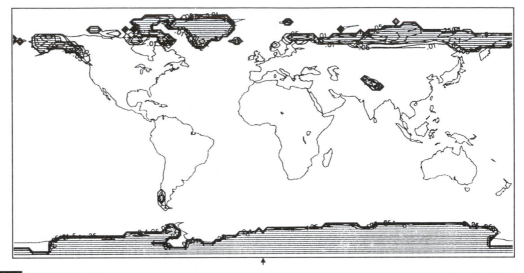

FIGURE 12.6   Modeled geographical distribution of snow depth in August at the end of a 10-yr simulation for 115 ka B.P. with computed SSTs. Shaded areas show snow depth >0.25 m. Many high-latitude regions, and parts of Tibet and Patagonia appear to be "sensitive regions" for the initiation of permanent snow cover (Dong and Valdes, 1995).

does not *prove* that these areas were glaciated during the LGM (or at any other time), it does point to the sensitivity of these regions in the maintenance of a permanent snow cover (and this is also seen in other models) given certain changes in insolation and lower SSTs at high latitudes. On the other hand, the model also indicates that snow would have built up over much of Alaska and there is no evidence that this has occurred for at least several hundred thousand years; more recent glaciations in Alaska have generally been small and confined to mountainous coastal ranges in the south and the Brooks Range in the north. These experiments therefore raise many interesting questions and provide fertile ground for further field studies, as well as additional model simulations and data-model comparisons.

In both of the models already discussed here, lower levels of CO_2 (< 280 ppm) were found to contribute to the cooling needed to maintain a permanent snow cover at high latitudes. Syktus *et al.* (1994), using a GCM coupled to a dynamic upper ocean circulation, set the orbital configuration to 116 ka B.P. and then examined the sensitivity of climate to CO_2 levels from 260–460 ppm. They found that snow cover does not change very much with orbital forcing until CO_2 levels drop to some critical level, in the range of 350–450 ppm. However, with lower levels of CO_2 the climate system is much more sensitive to insolation variations and snow cover can expand rapidly under orbital configurations with low summer insolation. This threshold sensitivity may have played a role in the long-term evolution of glacial-interglacial cycles; if CO_2 levels were at present-day levels (~350 ppm) or higher (as they may have been in the Pliocene), orbital variations may not have produced the same climate response, leading to the growth of large continental ice sheets, as occurred in the late Quaternary (Li *et al.*, 1998).

Schlesinger and Verbitsky (1996) also found that CO_2 levels were critical in generating ice sheets at high latitudes of the northern hemisphere. They used a GCM coupled to a mixed-layer ocean to derive climatic conditions resulting from orbital conditions at 115 ka B.P., with CO_2 levels varying from 326 to 246 ppm. The 246 ppm experiment was carried out to address the combined effects of lower CO_2 and CH_4. The climate so generated was then used as input to a global ice sheet/asthenosphere model. As with the other experiments already discussed here, they found that orbital changes alone were insufficient to produce a significant change in permanent snow cover. However, with an "equivalent CO_2 level" of 246 ppm, ice sheets built up in North America, most of Siberia, and northeastern Europe (but east of northern Fennoscandia). Over a 10,000-yr period (nominally 115–105 ka B.P.) continental ice volume increased, but not as rapidly as the SPECMAP benthic $\delta^{18}O$ record appears to show. However, there were two important limitations in these simulations. First, the model did not allow any precipitation falling as rain on the newly formed ice sheets to accumulate. When this condition was completely changed (all precipitation falling on the newly formed ice sheets was assumed to be snow) the ice sheets grew much larger (volumetrically and geographically) over a 10,000-yr interval, accounting for 86% of the sea-level equivalent of the SPECMAP record (though this record itself may not only represent sea-level change; see Section 6.3.4). The other limitation is that the model climate calculated for 115 ka B.P. was maintained as constant over the ensuing 10,000-yr period during which the ice sheets were allowed to build up. This means that potentially important positive feedbacks, related to snow-albedo changes, were not included, and these would no doubt have played a key role in the process of ice sheet growth.

One additional factor none of the previous simulations of 115 ka B.P. conditions explicitly considered is the role of snow-albedo feedback resulting from changes in vegetation cover. This issue was examined by Gallimore and Kutzbach (1996) using the National Center for Atmospheric Research (NCAR) Community Climate Model Version 1 (CCM1) coupled to a mixed layer ocean with interactive sea ice. With orbital changes alone, tundra vegetation increased in area by 25% relative to today (effectively expanding by 5° latitude to the south, according to a biome model driven by output from the GCM; Harrison *et al.*, 1995). When the *combined* effects of insolation change, reduced CO_2 levels (to 267 ppm) and increased albedo (due to the increased area of tundra) were subsequently considered, summer temperatures over land areas at 60–90° N fell by 8–9 °C, accompanied by more extensive and thicker sea ice. However, other experiments showed that only a slight additional expansion of tundra regions (i.e., > 5° latitudinal expansion) precipitated a dramatic change, with summer temperatures falling by a further 10–15 °C, leading to year-round snow cover at high latitudes. This result points to the potential importance of nonlinear vegetation/snow albedo feedbacks in the initiation of glaciation. Other studies have examined vegetation/albedo feedbacks in different situations (see the next section) and reached similar conclusions: vegetation changes appear to play an important role in the climate system and must be explicitly considered in order to simulate paleoclimatic conditions correctly (i.e., to match the "observed" paleorecord) (Foley *et al.*, 1994; Kutzbach *et al.*, 1996; Crowley and Baum, 1997).

12.3.2 Orbital Forcing and Monsoon Climate Variability

Just as the earth's orbital configuration led to anomalously low solar radiation in the northern hemisphere summer at 115 ka B.P., so the same orbital shifts led to increased seasonality and unusually high summer solar radiation receipts in the early Holocene. At 9 ka B.P., the northern hemisphere received ~7% more radiation than today in July (29–76 Wm^{-2}), mainly as a result of perihelion occurring in July (compared to January 3 today) together with greater eccentricity and an increase in axial tilt (24.23° vs 23.45° today). Several model experiments have examined the effect of the orbital changes on the earth's climate (Kutzbach and Otto-Bliesner, 1982; Kutzbach and Guetter, 1986; Kutzbach and Gallimore, 1988; Mitchell *et al.*, 1988; Hewitt and Mitchell, 1998). The main features of all these studies are similar; here we consider the results of Mitchell *et al.* (1988). They used an 11-layer United Kingdom Meteorological Office GCM (~5° × 7.5° resolution) coupled to a static mixed layer ocean and a simple sea-ice energy balance model. Thus SSTs and sea ice were allowed to change, though the (seasonally varying) heat flux was prescribed and invariant in both the control and 9 ka B.P. experiments. Figure 12.7 (a, b) shows the radiation differences between the two model runs, for incoming solar radiation at the top of the atmosphere and at the surface (after reflection by clouds, scattering and absorption in the atmosphere, etc.) Clearly the main effect is for an increase in the seasonal amplitude of radiation, with higher levels in the northern hemisphere summer, and lower levels in winter (see Fig. 12.5 — the effects are almost the opposite). The excess summer radiation resulted in higher temperatures over the northern hemisphere continents leading to lower pressure and an increase in airflow from the

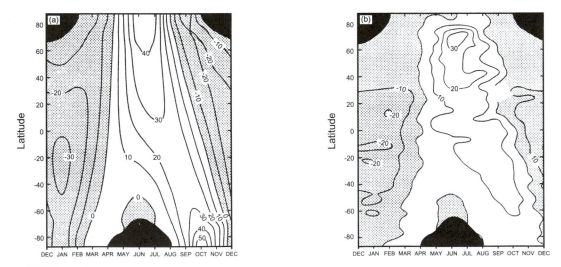

FIGURE 12.7   Difference in solar insolation (in W m^{-2}) at 9 ka B.P. compared to present, **(a)** at the top of the atmosphere and **(b)** at the surface. Negative anomalies (lower radiation at 9 ka B.P.) are shaded (Mitchell *et al.*, 1988).

oceans to the land. Thus, monsoon circulations were enhanced, leading to more precipitation in many parts of the Tropics. Figure 12.8 (a, b) summarizes the temperature changes in terms of zonal averages for land and ocean areas, respectively. The pattern of temperature change clearly reflects the radiation anomaly pattern seen in Fig. 12.7b, taking into account a thermal lag associated with surface heating and various feedbacks operating in the climate system. Cooler temperatures in the northern subtropics in late summer are a consequence of increased cloudiness resulting from enhanced monsoonal airflow. A reduction in sea-ice thickness and prolongation of the ice-free season causes large temperature anomalies over high-latitude ocean areas (Fig. 12.8b), which then maintain somewhat warmer conditions over northern continental regions throughout the winter months in spite of lower radiation amounts (bearing in mind that high-latitude winter radiation totals are small anyway, so that advection dominates temperatures at this time of year). Oceanic temperatures stay warm throughout the year over much of the globe because of the large heat capacity of the ocean and the fact that cloud cover over the oceans was reduced (related to increased subsidence over the oceans, compensating for the enhanced uplift over land areas). Thus, solar radiation increased over oceanic regions relative to the control simulation, enhancing the effect of the radiation anomaly.

Considering the geographical distribution of anomalies, summer temperatures were more than 4 °C warmer over most of Eurasia and 2–4 °C warmer over North America (Fig. 12.9). Sea-level pressure in northern continental interiors was lower by up to 6 mb, leading to enhanced monsoonal airflow and to heavier precipitation in a wide swath from northeastern Africa to southeast Asia, and over much of

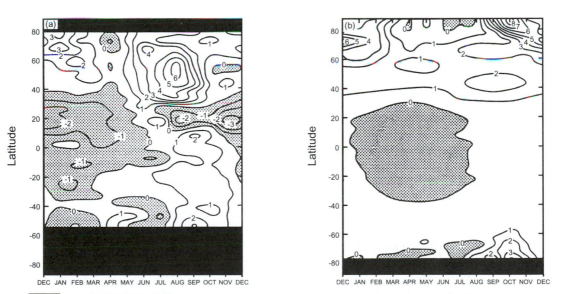

FIGURE 12.8 Modeled zonal mean differences in air temperature: (9 ka B.P. – control) **(a)** over land areas, **(b)** over ocean areas (sea and sea ice). Negative values are shaded (Mitchell *et al.,* 1988).

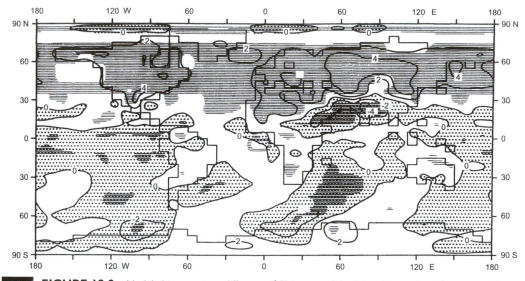

FIGURE 12.9   Modeled temperature difference (°C) between 9 ka B.P. and control run for summer (June–August). Warming was extensive across most of the northern hemisphere at 9 ka B.P. (Mitchell *et al.*, 1988).

northern South America. Increased cloudiness in these areas limited the temperature change so that, on balance, soil moisture levels increased significantly. However, over land areas at higher latitudes, increased levels of evaporation resulting from the large change in temperature were not compensated for by higher rainfall amounts, so soil moisture levels fell over most of North America and northern Eurasia (Fig. 12.10).

These experiments were carried out without taking into account the presence of a substantial ice cap over northeastern North America at 9 ka B.P. When another simulation was run with an ice sheet inserted, temperatures were lowered over the ice sheet and areas downstream (North Atlantic and northwestern Europe), but the principal changes in monsoonal circulation remained, albeit somewhat reduced in intensity.

In many low-latitude regions, orbital conditions at 9 ka B.P. were clearly more favorable than today for the growth of vegetation and the development of lakes in basins of inland drainage. Paleoclimatic evidence supports this scenario (Street-Perrott *et al.*, 1990); the ensuing few thousand years (when radiation anomalies were similar) saw dramatic changes in the environment of arid and semiarid areas from tropical west Africa to India (see Sections 7.6.3 and 9.9.4). At 125 ka B.P. the same orbital configuration led to even greater radiation anomalies, and GCM experiments show that the effects on monsoon climate regimes and high-latitude continental interiors must have been even more intense (Kutzbach *et al.*, 1991; Harrison *et al.*, 1991, 1995). Although limited, paleoclimatic data also point to higher levels of lakes at low latitudes during the last interglacial and drastically reduced sea ice around northern Eurasia at that time (Petit-Maire *et al.*, 1991; Frenzel *et al.*, 1992a).

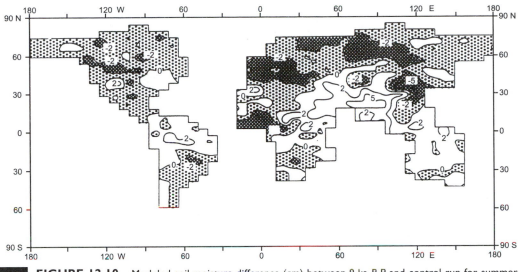

FIGURE 12.10 Modeled soil moisture difference (cm) between 9 ka B.P. and control run for summer (June–August). Shaded areas indicate drier conditions at 9 ka B.P. (Mitchell *et al.*, 1988).

A careful comparison of the paleoclimatic evidence and the orbitally forced model run shows that soil moisture did not change enough in the 9 ka (or other 6 ka) B.P. experiments to account for the dramatic environmental changes recorded in sediments from now dry Saharan lakes that were much more extensive in the early- to mid-Holocene. Kutzbach *et al.* (1996) examined this question by comparing results from the NCAR CCM2 model with 6 ka B.P. orbital changes alone, to a model run with the same orbital configuration, plus a change in the land surface of North Africa, from primarily desert (90–100% bare ground) to mainly grassland (80% grass cover). A further experiment also replaced the desert soil with a more loamy organic-rich soil, more typical of a grassland. These changes in surface boundary conditions brought about an additional increase in precipitation, from 12% more than the control run (at 15°–22° N) due to orbital forcing alone, to a 28% increase with changes in radiation, vegetation, and soil type (Fig. 12.11). Although this experiment involved an unrealistically large change in vegetation cover (extending completely across the Sahara from 15°–30° N) it does at least indicate the potential significance of vegetation/soil feedbacks on the climate system and points to the importance of ensuring that such feedbacks are explicitly considered in studies of future climatic changes. Similar conclusions were reached by Foley *et al.* (1994) and TEMPO[36] (1996) using more sophisticated atmosphere/mixed layer ocean models. They focused on the northern treeline at 6 ka B.P. and compared experiments with orbital forcing alone to experiments in which the area of boreal forest was also greatly expanded at the expense of tundra. The lower albedo of forest

[36] TEMPO (Testing Earth System Models with Paleo-Observations) is the acronym of an interdisciplinary project led by Kutzbach.

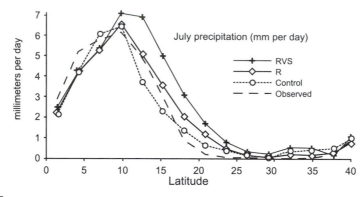

FIGURE 12.11 July rainfall (mm day[-1]) as a function of latitude, averaged over 0°–50° E in North Africa for modern observations and several GCM simulations. R = orbital configuration of 6 ka B.P.; RVS = the same orbital configuration, but with desert vegetation replaced by grassland and desert soil converted to a loamy grassland soil. Control run = modern boundary conditions. Positive vegetation feedbacks enhance the orbital effect of higher rainfall in this region (Kutzbach et al., 1996).

and associated changes in snow cover resulted in additional warming, especially in Spring, again indicating the role of vegetation feedbacks in long-term climate changes.

These modeling studies clearly demonstrate the significance of orbital forcing in generating dramatic changes in the earth's climate. Orbital changes are a necessary, but perhaps not entirely sufficient, condition for the growth and decay of ice sheets at high latitudes and the waxing and waning of the monsoons at lower latitudes. However, many other factors are known to have changed and these must certainly have played important additional roles: changes in trace gases (CO_2, CH_4, N_2O), continental and volcanic aerosols, vegetation cover and surface albedo, topographic effects of the ice sheets, changes in sea level, sea-ice extent, and the oceanic (thermohaline) circulation regime. Many of these factors can be viewed as additional feedbacks induced by orbital forcing, but there may have been many non-linear internal interactions that became the dominant forcing once some critical threshold was crossed (see Section 6.12). Additional sensitivity experiments can be designed to address these forcings explicitly, alone or in combination, so that the complexity of the climate system can eventually be deciphered.

12.3.3 The Influence of Continental Ice Sheets

The most dramatic change in boundary conditions during the last glacial maximum (LGM) was the presence of large ice sheets over North America and Fennoscandia. One can surmise that the very presence of these ice sheets must have had a significant effect on albedo, airflow, and energy exchange not only over the ice sheets themselves but perhaps regionally as well. A general circulation model simulation with topographically realistic ice sheets[37] compared to a similar simulation with only the modern distribution of continental ice enables a quantitative assessment of the effect of large ice sheets to be made, and the role of feedbacks with other parts of the climate system to be examined explicitly. To this end, Manabe and Broccoli (1985a, b) used an atmospheric general circulation model coupled to a mixed layer ocean to investigate climate system conditions with large and extensive (though probably unrealistic!) LGM ice sheets specified according to the "maximum" reconstruction of Hughes *et al.* (1981) (Fig. 12.12).

Comparison of the ice sheet simulation with the control run indicates that very low air temperatures occur downstream (east) of the Laurentide ice sheet (up to 32 °C colder in winter and 8 °C lower in summer, south of Greenland). Temperatures are also considerably depressed in northeastern Asia. By contrast, little or no

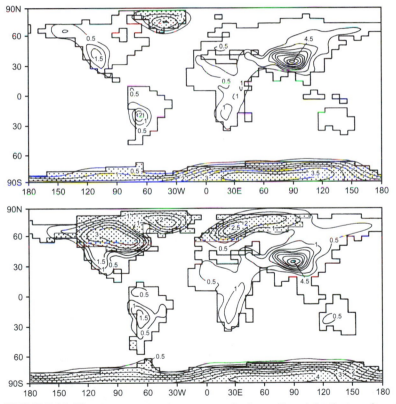

FIGURE 12.12 Continental outlines, topography (km a.s.l.), and distribution of continental ice sheet extent and height prescribed for an experiment to assess the role of a thick ice sheet on the atmospheric circulation. Top = present (control run); bottom = LGM experiment (Manabe and Broccoli, 1985a).

[37] Recent estimates by Peltier (1994) suggest that the height of the ice sheets, especially over North America, may have been overestimated by CLIMAP (1981) for the LGM.

change is observed in the southern hemisphere, which contrasts with marine sediment-based paleoclimatic reconstructions of SSTs, which have SSTs, lower by up to 6 °C in parts of the southern ocean. This result suggests that ice sheets in the northern hemisphere have little impact on southern hemisphere conditions. Although less energy enters the system (because of the higher albedo of the ice sheet) this is compensated for by a reduction in outgoing longwave radiation, so the net effect is small, requiring minimal interhemispheric heat transfer. If the presence of large ice sheets in the northern hemisphere did not cause the southern hemisphere to cool, some other factor (or factors) must be invoked to explain the observed SSTs in this region. To investigate this further, Manabe and Broccoli (1985b) re-ran the LGM ice sheet simulation with CO_2 levels lowered by 100 ppm (to 200 ppm). This change was sufficient to lower SSTs in the southern hemisphere, bringing the model more in line with paleoclimatic reconstructions (Figure 12.13).

The Laurentide ice sheet (~3 km in height at its center in this simulation) has a significant effect on tropospheric circulation, causing a split airflow around it and

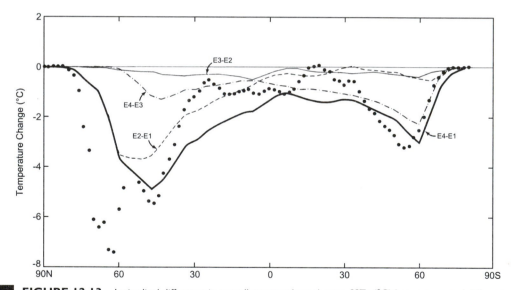

FIGURE 12.13   Latitudinal difference in annually averaged zonal mean SSTs (°C) between several different GCM experiments. E1 used modern continental land and sea-ice distribution, modern land albedo for snow-free areas, and a CO_2 level of 300 ppm. E2 changed only the continental ice and sea-ice distribution to that of the LGM. E3 was the same as E2, but also changed the land albedo for snow-free regions to glacial values. E4 was like E3, but in addition changed CO_2 levels to 200 ppm. E2–E1 thus shows the overall effect of the ice sheets on zonal mean SSTs, indicating little effect on the southern hemisphere. Changing the albedo of unglacierized areas (E3–E2) only further lowers the temperatures slightly. Dropping CO_2 levels to glacial values (E4–E3) produces a more symmetrical response in both hemispheres but has the major effect on temperatures in the southern hemisphere. The combined effects (E4–E1) show broad similarities to differences between CLIMAP LGM SST reconstructions and present-day conditions (shown as black dots) but the model results indicate colder conditions in low latitudes and much warmer temperatures at high latitudes due to reduced sea-ice formation (Manabe and Broccoli, 1985b).

the development of a strong jet stream on its southeastern margin (Shinn and Barron, 1989). Cold air flow around the ice sheet caused extensive, thick sea ice to develop in the Labrador Sea and North Atlantic Ocean. Over continental interiors, air temperature differences between the two model simulations are smaller than in most paleoclimatic reconstructions, perhaps because surface albedos (specified in both models) were too low. Finally, maps of soil moisture change show that large areas of the continents south of the ice sheets were significantly drier during the LGM compared to today (due mainly to lower precipitation amounts) and these regions broadly correspond to known loess regions.

Rind (1987) carried out four experiments with a NASA-GISS GCM with $8° \times 10°$ resolution. Starting with a simulation of modern conditions (the control), he added, sequentially, ice-age SSTs (as determined by CLIMAP 1981), a 10-m thick ice sheet over the area occupied by continental ice sheets at the last glacial maximum, and full ice sheets with appropriate elevations. In this way, he was able to investigate the importance of each of these factors to the atmospheric circulation. As in Manabe and Broccoli's experiment, Rind's simulations showed that thick ice sheets have both thermal and topographic effects; because of their elevation and high albedo, surface temperatures are low and atmospheric water vapor above the ice is limited. Consequently infra red (IR) radiation losses are high due to the (effectively) reduced optical thickness of the atmosphere above the ice sheets. Over North America the jet stream bifurcates around the topographic barrier created by the Laurentide ice sheet (in Europe the ice sheet is thinner so this effect is less significant). High pressure at the surface over the ice sheets sets up strong disturbances to zonal (west-east) flow, creating a pronounced meridional circulation regime. However, Rind found that stationary waves established in the northern hemisphere were accompanied by similar changes in the southern hemisphere, implying some sort of dynamic link between the two hemispheres, a feature not seen in Manabe and Broccoli's simulations. Of particular interest in Rind's experiment is the fact that the mass balance of the ice sheet was strongly negative under the prescribed conditions and, had the "ice sheet" been 10-m thick, it would have melted away completely in just a few years. This suggests that the large continental ice sheets at 18 ka B.P., though clearly having a major influence on atmospheric circulation, were not in balance with prevailing climatic conditions. Furthermore, land temperatures and precipitation amounts at low latitudes were higher than indicated by paleoclimatic data. Rind re-ran his simulation with CO_2 levels 70 ppm lower but this only cooled the atmosphere slightly, not enough to alter the ice mass balance significantly or to reduce low-latitude temperatures and precipitation amounts enough. Only when CLIMAP SSTs were reduced by at least 2 °C at low latitudes were the model simulations brought into alignment with the observational data (Rind and Peteet, 1985). In fact, given the CLIMAP SST boundary conditions specified, the model simulation is not in radiative equilibrium; however, if SSTs are reduced by ~2 °C it brings about an approximate radiative balance.

A further perspective on this problem is provided by Hall *et al.* (1996). They used the high-resolution UGAMP GCM to simulate LGM climate; ice-sheet elevation, land surface albedo, and CLIMAP SSTs were specified, CO_2 levels were reduced to 190 ppm, and the orbital configuration was set to that of 21ka B.P. Temperature differences between the control and LGM runs are shown in Fig. 12.14. Winters were dramatically colder over much of the northern hemisphere extra-tropics (up to 50 °C colder over Iceland as a result of the advection of cold Arctic air). In summer, maximum cooling was associated with the Laurentide (−39 °C) and Fennoscandian (−30 °C) ice sheets. Unlike earlier studies, bifurcation of the jet stream around the Laurentide ice sheet was not observed. Storm tracks were concentrated along the southern margins of the ice sheets and sea-ice boundaries, leading to heavier precipitation (much of it as snow) from coastal Alaska and northern British Columbia, across northeastern North America to central Europe (Fig. 12.15). Snowfall was especially heavy over northwestern North America in winter. By considering both accumulation of snow and ablation, Hall *et al.* estimated that the Laurentide and Fennoscandian ice sheets would have had an overall positive balance, with net accumulation over much of the ice sheets and net ablation confined to a narrow strip along the southern margins. However, they did not use these results as input to an ice sheet model, with ice sheet dynamics, so what the actual mass balance of the ice sheets would have been is not apparent from these results.

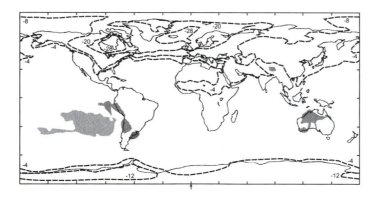

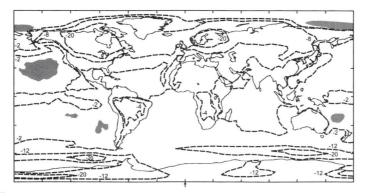

 FIGURE 12.14 Summer (JJA) and winter (DJF) temperature differences between 21 ka B.P. and present, based on the UGAMP GCM with CLIMAP-based ice-sheet elevations, SSTs, and sea-ice extent prescribed (Hall *et al.*, 1996).

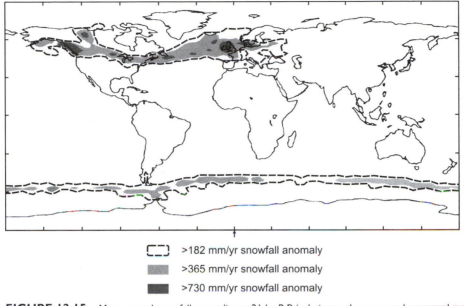

☐ ☐ >182 mm/yr snowfall anomaly

▓▓▓ >365 mm/yr snowfall anomaly

██ >730 mm/yr snowfall anomaly

FIGURE 12.15 Mean annual snowfall anomalies at 21 ka B.P. (relative to the present-day control run) according to the UGAMP GCM, with CLIMAP-based ice-sheet elevations, SSTs, and sea-ice extent prescribed (Hall *et al.*, 1996).

12.4 MODEL SIMULATIONS: 18 KA B.P. TO THE PRESENT

A series of experiments designed to examine the climatic response to orbital changes and surface boundary conditions has been carried out by Kutzbach and associates, using the NCAR Community Climate Model (CCM) (Kutzbach and Guetter, 1986; COHMAP 1988; Kutzbach *et al.*, 1993b). The GCM simulations were carried out at 3000-yr intervals, from 18 ka B.P. to the present, for January and July conditions. Radiation anomalies due to orbital changes over this interval are shown in Fig. 12.16. Potentially important changes in boundary conditions are summarized in Fig. 12.17 (Kutzbach and Ruddiman, 1993). As noted earlier, radiation conditions at 18 ka B.P. are similar to those of today (i.e., the control experiment) so the 18 ka simulation is principally a test of altered boundary conditions, like those of Rind (1987) and Manabe and Broccoli (1985b). Simulations for the later periods, especially from 9–3 ka B.P., mainly examine the way in which the atmosphere responds to changes in seasonality, because boundary conditions are similar to those of today (except for a small ice sheet over North America at 9 ka B.P.) but there are pronounced differences in radiation over the annual cycle (Section 12.3.2).

At 18 ka B.P. global temperatures were lower by ~3.4 °C (average of January and July means), assuming prescribed CLIMAP SSTs are correct. Other estimates for temperature changes at the LGM are given in Table 12.1. As in other LGM simulations, the hydrological cycle was less intense (i.e., both precipitation and evaporation were lower) in spite of generally higher wind speeds. The NCAR CCM

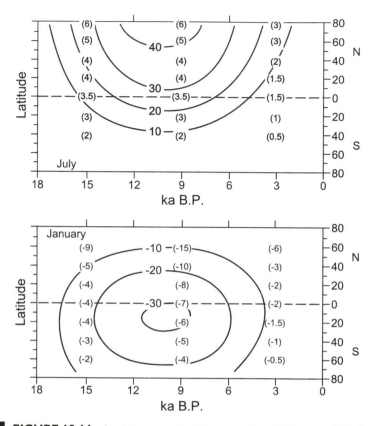

FIGURE 12.16 Insolation anomalies (departures from 1950 values in W m⁻²) at the top of the atmosphere from 18 ka B.P. to present (Kutzbach and Guetter, 1986).

shows a pronounced split in the upper level jet-stream flow around the North American ice sheet and to a lesser extent around the Scandinavian ice sheet (Fig. 12.18), a feature also seen in some other models, as noted earlier. At lower levels, high pressure with anticyclonic flow prevailed. Kutzbach and Wright (1985) compared these results with LGM geological evidence for North America and found strong support for this simulation, particularly in the alignment of sand dunes in the Great Plains, which formed under the prevailing winds. Pollen data also support the model simulation of drier conditions south of the Laurentide ice sheet at this time.

Temperatures increased rapidly from 18–9 ka B.P. in the northern hemisphere as the ice sheets melted and summer insolation increased (Fig. 12.19). This was accompanied by higher precipitation amounts, especially over northern hemisphere, low-latitude, continental regions in summer (Fig. 12.20). Evaporation also increased, but the net effect (P-E) resulted in much higher effective moisture at low latitudes, leading to the higher lake levels and more prolific vegetation in semiarid and arid environments (Kutzbach and Street-Perrott, 1985). Conditions in North Africa (especially the western Sahel/Sahara) were markedly wetter in the early- to mid-Holocene, in keeping with the observed geological record (Hall and Valdes, 1997). These

TABLE 12.1   Model Estimates of the Global Mean Surface Air Temperature Difference Between the Last Glacial Maximum and Today

Source	Model	ΔT (°C) LGM-Control	Month	Comments
Gates (1976a)	GCM	−4.8	July	CLIMAP SSTs prescribed
Manabe and Hahn (1977)	GFDL GCM	−5.4	July	CLIMAP SSTs prescribed
Manabe and Broccoli (1985a, b)	GFDL GCM	−3.6	Annual	70m mixed layer ocean
Kutzbach and Guetter (1986)	NCAR-CCM GCM	−3.9	July	CLIMAP SSTs prescribed
Rind (1987)	NASA-GISS GCM	−3.5 −3.4	July January	CLIMAP SSTs prescribed
Lautenschlager and Herterich (1990)		−4.7	Annual	CLIMAP SSTs prescribed
Hall et al. (1996) Petersen et al. (1979)	UGAMP GCM 1979	−3.8	Annual	CLIMAP SSTs prescribed

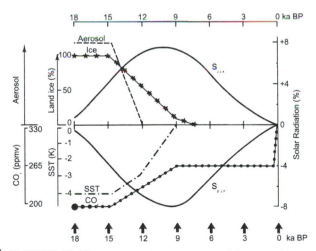

FIGURE 12.17   Principal changes in external forcing and internal climate system boundary conditions, from 18 ka B.P. to the present. Insolation anomalies are shown for the northern hemisphere, for summer (JJA) and winter (DJF) as a percentage difference from present. Global mean annual sea-surface temperatures (SSTs) are as calculated by CLIMAP (1981) for open ocean areas and sea-ice regions. Ice is the percent of 18 ka ice volume, according to Denton and Hughes (1981). Glacial age aerosols are depicted on an arbitrary scale. The arrows along the bottom indicate GCM simulations for each time period, using the changes in forcing and boundary conditions shown (Kutzbach and Street-Perrott, 1985).

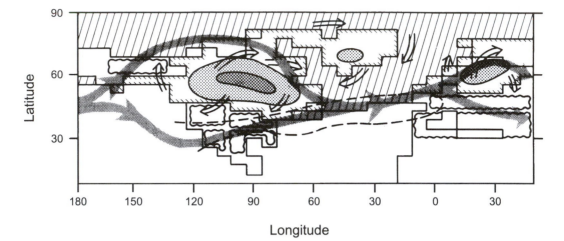

FIGURE 12.18 Schematic diagram illustrating the circulation pattern for 18 ka B.P. in January, based on a GCM experiment with thick Laurentide and European ice sheets (shaded). The main jet stream is split around the ice sheets; the dashed lines along the southern jet stream show the zone of increased precipitation. Surface wind systems are shown by open arrows. This result is from the NCAR GCM; not all GCMs show this bifurcation of upper air winds in LGM simulations (Kutzbach and Wright, 1985).

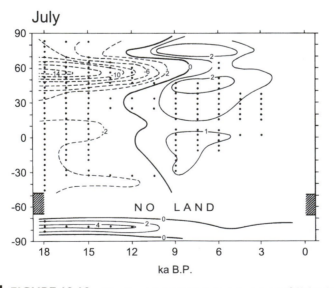

FIGURE 12.19 Zonally averaged temperature departures (°C) for the land surface for July, for the last 18 ka B.P. according to GCM estimates (linearly interpolated between model runs carried out at 3 ka intervals). Dots indicate the value is statistically significant (2-sided t-test) at or above the 90% confidence level based on the model's natural variability (Kutzbach and Guetter, 1986).

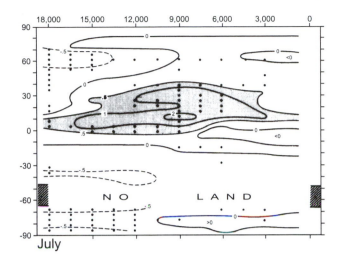

FIGURE 12.20 Zonally averaged precipitation departures (mm day^{-1}) for the land surface for July, for the last 18 ka B.P. according to GCM estimates (linearly interpolated and smoothed, between model runs carried out at 3 ka intervals). Dots indicate the value is statistically significant (2-sided *t*-test) at or above the 90% confidence level based on the model's natural variability. Positive departures are shaded (Kutzbach and Guetter, 1986).

changes were in the opposite direction in the southern hemisphere where January (summer) rainfall amounts fell and P-E was reduced from 12–3 ka B.P. compared to present (Fig. 12.21). The changes are strongly correlated to solar radiation anomalies over the last 18 ka; thus over land areas 0–30° N in July, a 1% increase in solar radiation more or less corresponds to a 3.5% increase in precipitation. This can be thought of as due to the non-linear increase in saturation vapor pressure with temperature, which amplifies (through water vapor feedback) the radiation effect at the surface. A 7% increase in solar radiation outside the atmosphere at 9 ka B.P. is associated with 11% higher net radiation at the surface due to a *decrease* in outgoing longwave radiation (because of increased evaporation and higher water vapor levels in the atmosphere, which absorb long-wave radiation). However, this amplification of solar radiation effects is not as important at higher latitudes where precipitation amounts are much less dependent on solar radiation anomalies and show more sensitivity to surface boundary conditions (Kutzbach and Guetter, 1986).

12.5 COUPLED OCEAN-ATMOSPHERE MODEL EXPERIMENTS

A recurring theme in Chapters 5 and 6 was the evidence for rapid fluctuations in climate during the last glacial period, particularly around the North Atlantic. Many investigators have suggested that these shifts are related to the North Atlantic thermohaline circulation, which may have oscillated between states involving high salinity surface water and deepwater formation ("conveyor on") and low salinity surface water and little or no deepwater formation ("conveyor off"). Both ocean GCMs

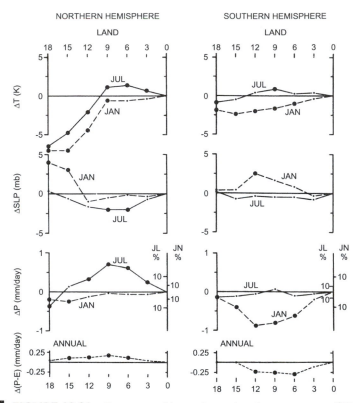

FIGURE 12.21   Departures of (top to bottom): surface temperature (°C) sea-level pressure (mb), precipitation (mm day⁻¹), and precipitation minus evaporation (P–E, mm day⁻¹) for January and July, for northern hemisphere (left) and southern hemisphere (right) land areas for the period 18 ka B.P. to the present at 3000-yr intervals. Departures are expressed as the paleoclimatic experiment minus the control (0 years B.P.). Large dots indicate the departure is statistically significant (2-sided t-test) at or above the 95% confidence level, based on the model's natural variability. For precipitation, the ordinate on the right indicates ±10% departures for January (JN) and July (JL) (Kutzbach and Guetter, 1986).

and coupled ocean-atmosphere GCMs have been used to examine this issue and several studies show that the oceanic circulation in the Atlantic can switch between these modes and the transitions can be rapid (Manabe and Stouffer, 1988; Weaver and Hughes, 1994; Rahmstorf, 1994, 1995). The system is strongly non-linear so that at some critical threshold an abrupt switch from one mode to another can occur, with dramatic consequences for climate in the northern hemisphere, especially in western Europe. However, several authors have argued that the notion of a conveyor system either on or off is not consistent with geological evidence and it is more likely that there were many "modes" or situations in which some regions of deepwater formation ceased to operate, and the main centers shifted geographically. Indeed, there may have been times when deepwater was replaced by intermediate water (see discussion in Sections 6.9 and 6.10.2) (Boyle and Keigwin, 1987; Lehman and Keigwin, 1992a). Experiments with the GFDL ocean circulation model, coupled to an atmospheric GCM show that there can be multiple equilibria in which the main convection sites in the North Atlantic shift in response to varying inputs of meltwater or atmospheric circulation changes, or both (Rahmstorf, 1994). Sea-surface temperatures can change by up to 5 °C during these shifts with Intermediate Water replacing NADW.

Manabe and Stouffer (1997) used a coupled ocean-atmosphere model to examine the response of the thermohaline circulation (and global climate in general) to the effect of a sustained freshwater input across the North Atlantic (50–70° N) compared to a similar input across the western subtropical Atlantic (20–29° N, 52–90° W). This experiment was designed to distinguish between the effects of meltwater drainage from the Laurentide ice sheet southward into the Gulf of Mexico, and thence to the Atlantic, vs drainage through the Gulf of St. Lawrence. Broecker *et al.* (1989) argued that this switch may have been responsible for the Younger Dryas cold episode, which has been noted in many parts of the world, but especially in western Europe (see Section 6.10.2). In Manabe and Stouffer's two experiments, they added 0.1Sv (10^6 m^3 s^{-1}) of freshwater for 500 yr to each region, then ran the model for another 750 yr to determine the long-term response of the overall climate system. Such long simulations are very computer-intensive and so few studies have been carried out to examine questions of this sort. Figure 12.22 shows changes resulting from the northern freshwater input, in sea-surface salinity (SSS), sea-surface temperature (SST), and sea-ice thickness in the Denmark Strait (30° N, 65.3° W) off the coast of East Greenland, where the response to changes in thermohaline circulation is large. Also shown is a measure of the overall thermohaline circulation, and air temperatures over Summit, Greenland (site of the GISP2/GRIP ice cores). Salinity fell by ~3% over the 500-yr period of freshwater input, accompanied by a reduction in thermohaline circulation strength, from ~18 Sv to ~4 Sv. The SSTs also declined rapidly, accompanied by more extensive sea-ice growth in the Denmark Strait. Superimposed on this overall decline are large amplitude decade-to-century scale oscillations of SSS, SSTs, and air temperatures, as shown here for Greenland. These are reminiscent of the high amplitude changes of $\delta^{18}O$ seen in the late glacial section of the GISP2/GRIP ice cores (see Fig. 5.22). In a similar experiment, Manabe and Stouffer (1995) added large amounts of freshwater to the North Atlantic much more rapidly (over 10 yr) and found similar oscillations, but larger in amplitude, associated with a rapid cessation of the thermohaline circulation. Such abrupt changes therefore seem to be characteristic of the change in ocean circulation mode. Once the freshwater input ceased, SSS and the thermohaline circulation recovered their original values within 250 yr, with SSTs returning to control period values more slowly, delayed by the sea-ice cover, which constrained the warming. Figure 12.23 shows the geographical pattern of surface air temperature anomalies (from the control period) for four 100-yr intervals, over the course of the experiment. The largest anomalies at the end of the freshwater influx (years 401–500) are in the North Atlantic (as much as –7 °C over southeast Greenland), with cooler conditions extending throughout the high latitudes of the northern hemisphere. However, low- and mid-latitude experienced little change in temperature, and, indeed, the overall global mean temperature change in years 401–500 was close to zero. However, a notable anomaly is observed in the Southern Circumpolar Ocean, resulting from the reduction in thermohaline circulation and its consequences in the Southern Ocean. This anomaly continued long after the

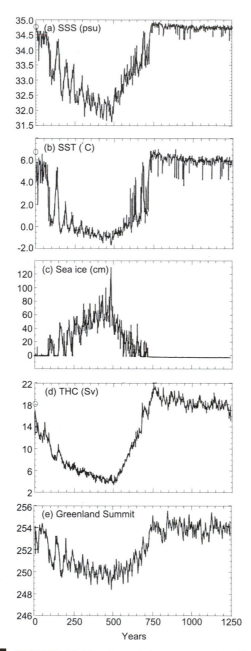

FIGURE 12.22   A coupled ocean-atmosphere GCM experiment was carried out, in which freshwater was introduced into the North Atlantic for a period of 500 yr, followed by a 750-yr period with conditions like those at the start of the experiment. The resulting mean annual time series are shown of: **(a)** sea-surface salinity; **(b)** sea-surface temperature; **(c)** sea-ice thickness (all in Denmark Strait, 30° W, 65.3° N, off the coast of east Greenland); **(d)** shows the strength of the thermohaline circulation (in Sverdrups: units = 10^6 m³ s⁻¹); and **(e)** gives the air temperature over Summit Greenland (degrees Kelvin). Note the decline in thermohaline circulation as the freshwater input continues, accompanied by large amplitude shifts in SSTs and air temperatures around Greenland. Recovery to pre-experiment (control) conditions after the freshwater input ceased (in year 500) was completed within 250 years (Manabe and Stouffer, 1997).

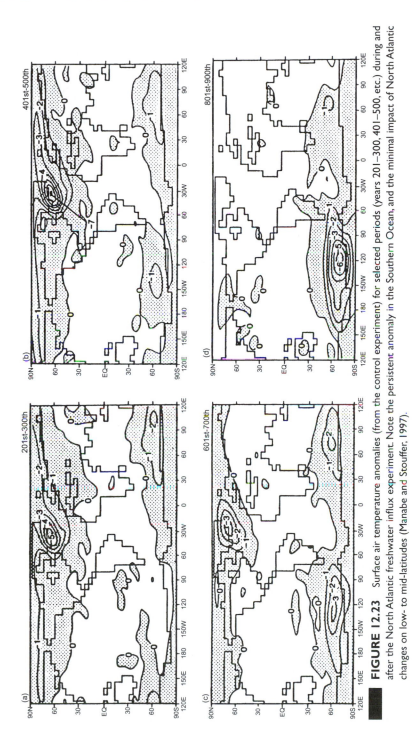

FIGURE 12.23 Surface air temperature anomalies (from the control experiment) for selected periods (years 201–300, 401–500, etc.) during and after the North Atlantic freshwater influx experiment. Note the persistent anomaly in the Southern Ocean, and the minimal impact of North Atlantic changes on low- to mid-latitudes (Manabe and Stouffer, 1997).

freshwater input to the North Atlantic ceased — a persistent cool episode that was especially pronounced west of the Antarctic Peninsula. A comparison with the observed paleoclimatic data shows that the pattern of anomalies is similar to that of the Younger Dryas, though displaced somewhat poleward. This may be because the model did not include an ice cover representative of late glacial conditions; had it done so, the temperature anomaly field would probably have looked even more like the Younger Dryas oscillation, as mapped, for example, by Peteet (1995). It is also of interest that the response to North Atlantic freshwater input is delayed cooling (by up to several centuries in some regions) and this may help to account for the apparent diachronous evidence for late glacial cooling that has often confounded attempts to define a "Younger Dryas" chron.

How do these changes differ from the experiment in which freshwater was injected farther south? Figure 12.24 shows SST anomalies in years 401–500 of the two experiments. The two patterns are remarkably similar, although the magnitude of anomalies associated with the southern freshwater injection is considerably smaller. Determining which forcing was more important will therefore be difficult, as there may have been a wide spectrum of conditions, spanning the range of these two exper-

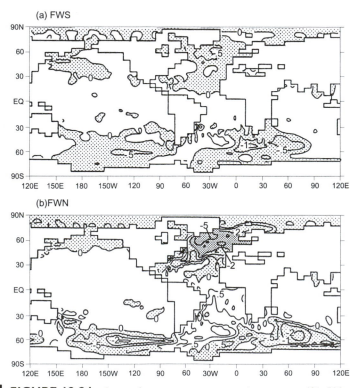

FIGURE 12.24   Sea-surface temperature anomalies in years 401–500 resulting from a 500-yr influx of freshwater **(a)** into the subtropical Atlantic and **(b)** into the North Atlantic. The patterns are similar, but the magnitude of anomalies is larger with the North Atlantic freshwater influx (Manabe and Stouffer, 1997).

iments, at various times in the past. Fanning and Weaver (1997), for example, found in their ocean-atmosphere GCM experiments that North Atlantic freshwater discharge alone (via the St Lawrence) could not shut down the thermohaline circulation, but it could when the Atlantic was first "preconditioned" by freshwater discharge from the Mississippi drainage. These experiments help to explore further the dramatic changes in glacial and late glacial time recorded by paleoclimatic records. They must now be reconciled with arguments over the magnitude and timing of Laurentide (and Scandinavian) discharge as well as the role of ice discharged from the Arctic Ocean. Data from marine sediments, interpreted in terms of deep and intermediate water formation, and the shifting sources of these water masses must also be fitted into the puzzle. Although these lines of evidence do not all fit together perfectly at this time, it is clear that changes in thermohaline circulation and freshwater influx to the North Atlantic are critical factors in resolving these questions, with important implications for understanding the sensitivity of the thermohaline circulation to greenhouse gas-induced changes in the hydrological balance of the North Atlantic basin.

12.6 GENERAL CIRCULATION MODEL PALEOCLIMATE SIMULATIONS AND THE PALEORECORD

It was noted at the start of this chapter that paleodata provides an important test of the ability of GCMs to simulate climates for a set of boundary conditions different from those of today. In discussing paleoclimate model experiments, frequent reference has been made to data-model comparisons. However, several studies have focused explicitly on this important topic. Two approaches have been followed. In the first (termed "forward modeling"), GCM output for a particular time period (e.g., temperature and precipitation at 6 ka B.P.) is used in response function equations[38] (established empirically — see Section 9.6) to produce maps of simulated pollen distribution for a region. These are then compared to maps of pollen distribution based on actual pollen data (derived from radiocarbon-dated lakes and bogs for that time period) (Webb *et al.*, 1987, 1997; Bartlein *et al.*, 1997). A similar approach can be used to produce maps of biomes at selected time intervals in the past, by using the GCM output to drive a biome model. The model-generated biomes can then be compared with the distribution of biomes inferred directly from paleodata (Prentice *et al.*, 1997). Another approach ("inverse modeling") derives temperature and precipitation estimates from pollen data. These can then be compared directly to standard GCM output (Webb *et al.*, 1993b, 1997). Both approaches have proven useful in identifying areas of agreement and differences between model simulations and observational data, forcing a reconsideration of both to determine, on the one hand, where model improvements may be needed and, on the other, where new data or improved data calibrations may be needed. It is this sort of iterative, interactive exchange between data and models that has led to significant improvements in our understanding of Quaternary paleoclimatology, and no doubt it will play an important future role in reconstructing climates of the past.

---

[38] Response functions describe the relationship between pollen abundances (for a particular taxon) in terms of two or more climate variables (i.e., in "climate space") (Bartlein *et al.*, 1986).

APPENDIX

APPENDIX A: FURTHER CONSIDERATIONS ON RADIOCARBON DATING

A. 1 Calculation of Radiocarbon Age and Standardization Procedure

Although it is not necessary for a user of radiocarbon dates to know, in detail, how the actual value is arrived at, some understanding of the procedure is enlightening, particularly when considering adjustments for ^{14}C fractionation effects (see Section 3.2.1.4c). The whole subject is a complex sequence of calibrations, adjustments, and corrections. The following brief explanation is offered for the adventurous.

In order to make dates from different laboratories comparable, a standard material is used by all laboratories for the measurement of "modern" carbon isotope concentrations. This standard used to be "U.S. National Bureau of Standards oxalic acid I" (Ox I) prepared from West Indian sugar cane grown in 1955. Ninety-five percent of the ^{14}C activity of this material is equivalent to the ^{14}C activity of wood grown in 1890, so by this devious means, all laboratories standardized their results to a material that had not been contaminated by "atomic bomb" ^{14}C. Because the original standard has now been exhausted, another oxalic acid standard (Ox II) is used with the old standard equivalent to (0.7459 Ox II). By convention, all dates are given in "years before 1950" (years B.P. or "before physics") so dates are adjusted to this temporal standard, rather than the possibly more logical time of 1890. A second adjustment is needed to correct for the fact that oxalic acid undergoes variable ^{13}C and ^{14}C

fractionation effects during analysis. To make interlaboratory comparisons of samples possible, standardization is necessary to take these fractionation effects into account. Fortunately, the fractionation effect of ^{14}C is extremely close to twice that of ^{13}C, which is far more abundant and can be measured easily in a mass spectrometer.[39] The necessary standardization is thus achieved by measuring ^{13}C rather than ^{14}C. Following the detailed analysis of oxalic acid samples by Craig (1961a), it was agreed that all standard samples should be adjusted to a ^{13}C value of -19.3%, where

$$\delta^{13}C = \frac{(^{13}C/^{12}C)_{ox} - (^{13}C/^{12}C)_{PDB}}{(^{13}C/^{12}C)_{PDB}} \times 10^3 \tag{A.1}$$

C_{OX} refers to the oxalic acid standard, and C_{PDB} refers to another reference standard, a Cretaceous belemnite (*Belemnitella americana*) from the Peedee Formation of South Carolina (Craig, 1957). Updates to the reference standards are discussed by Coplen (1996).

Standardization of the ^{14}C activity in the oxalic acid reference sample is achieved thus:

$$0.95A_{ox} = 0.95A_{ox}^1 \left[1 - \frac{(2\delta^{13}C_{ox} + 19)}{1000} \right] \tag{A.2}$$

where A^1_{ox} is the ^{14}C activity of the reference oxalic acid and $^{13}C_{ox}$ is the ^{13}C of the reference oxalic acid (Eq. A.1). The value of $0.95A_{ox}$ then becomes the universal ^{14}C standard activity from which all dates are calculated.

The procedure normally adopted for calculating the radiocarbon date of a sample can be summarized in three equations.

(a) The activity of the sample is expressed as a departure from the reference standard:

$$\delta^{14}C = \frac{A_{sample} - 0.95A_{ox}}{0.95A_{ox}} \times 10^3$$

where A_{sample} is the ^{14}C activity of the sample, corrected for background radiation, and A_{OX} is the 1950 activity of NBS oxalic acid I (or equivalent), corrected for background and isotope fractionation.

(b) The activity of the sample is corrected for fractionation by normalizing to $\delta^{13}C = -25\%$ PDB, which is the average for wood (see the discussion that follows):

$$\Delta = \delta^{14}C - (2[\delta^{13}C + 25]) \left(1 + \frac{\delta^{14}C}{1000} \right)$$

where

$$\delta^{13}C = \frac{(^{13}C/^{12}C)_{sample} - (^{13}C/^{12}C)_{PDB}}{(^{13}C/^{12}C)_{PDB}} \times 10^3$$

(c) Age (T) is then calculated using the "Libby" half-life of 5570 yr:

$$T = 8033 \log_e [1 + (\Delta/1000)]^{-1}$$

[39] $\delta^{14}C = 2\delta^{13}C + 10^{-3}(\delta^{13}C)^2$.

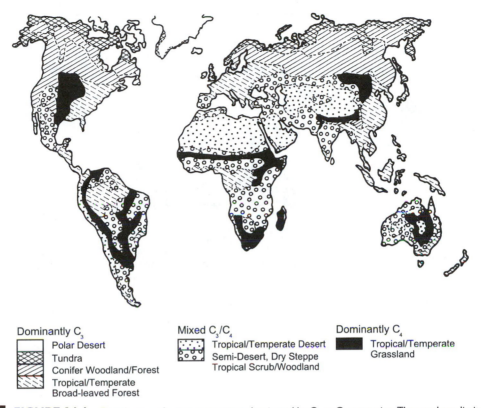

Dominantly C$_3$
- [ ] Polar Desert
- [▨] Tundra
- [▧] Conifer Woodland/Forest
- [▨] Tropical/Temperate Broad-leaved Forest

Mixed C$_3$/C$_4$
- [∴] Tropical/Temperate Desert
- [∘] Semi-Desert, Dry Steppe Tropical Scrub/Woodland

Dominantly C$_4$
- [■] Tropical/Temperate Grassland

FIGURE A1.1   Distribution of major ecosystems dominated by C$_3$ or C$_4$ vegetation. The northern limits of temperate grasslands in North America and Asia include a significant proportion of C$_3$ plants due to the cool growing season (from Cerling and Quade, 1993).

A.2 Fractionation Effects

Because isotopic fractionation occurs to differing extents during plant photosynthesis and during shell carbonate deposition, it is necessary to know the magnitude of the effect so that dates on different materials can be compared. Lerman (1972) and Troughton (1972) have studied fractionation effects in modern plants and found a trimodal distribution; the magnitude of fractionation seems to be related to the particular biochemical pathway evolved by different plant species for photosynthesis. Thus, highest ^{13}C depletion corresponds to so-called C$_3$ plants, which utilize the Calvin photosynthetic cycle (CAL) and lowest depletion occurs in C$_4$ plants utilizing the Hatch-Slack (HS) cycle. Succulents utilize a third metabolic pathway (crassulacean acid metabolism, CAM) and may fix carbon by either the HS or CAL pathways depending on temperature and photoperiod; they thus form a third group. Figure A1.1 shows the large-scale distribution of ecosystems dominated by C$_3$ or C$_4$ plants. The implications for dating are that samples containing, say, a high proportion of HS plants would show ^{14}C/^{12}C ratios characteristic of these less depleted

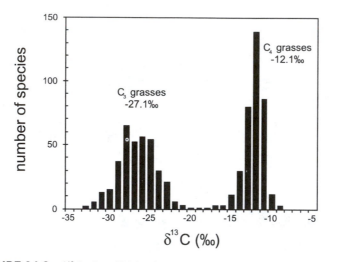

FIGURE A1.2 $\delta^{13}C$ values (‰) for C_3 and C_4 grasses (from Cerling and Quade, 1993).

TABLE A1.1   Radiocarbon Fractionation Errors

	Dating material	Photosynthetic pathway[a]	$\delta^{13}C$ ‰	Years to add to uncorrected date
1.	Wood and wood charcoal	(C_3) CAL	−25 ± 5	0 ± 80
2.	Tree leaves	(C_3) CAL	−27 ± 5	−30 ± 80
3.	Peat, humus, soil[b]	(C_3) CAL	−27 ± 7	−30 ± 110
4.	Grains[c]	(C_3) CAL	−23 ± 4	+30 ± 60
5.	Leaves and straw of grasses and sedges (e.g., totora)[c]	(C_3) CAL	−27 ± 4	−30 ± 60
6.	Bones(European)		−20 ± 4	+80 ± 60
7.	Grains[d]	(C_4) HS	−10 ± 3	+240 ± 50
8.	Leaves and straw of grasses and sedges (e.g., papyrus)[d]	(C_4) HS	−13 ± 4	+200 ± 60
9.	Succulents[e]	CAM	−17 ± 8	+130 ± 120
10.	Aquatic plants (freshwater)		−8 to −24	
	Aquatic plants (marine)		−8 to −17	
11.	Fresh and brackish water shells		0 to −12	
12.	Marine shells		+3 to −2	

[a] CAL = Calvin metabolic pathway; HS = Hatch-Slack metabolic pathway; CAM = crassulacean acid metabolism.

[b] Most nordic and temperate peats, humus, and soils are of C-type. Others must be considered as unidentified.

[c] Wheat, oats, barley, rice, rye, etc. . . . and related grasses.

[d] Maize, sorghum, millet, panic, etc. . . . and related grasses.

[e] Cactus, agave, pineapple, *Tillandsia*, etc. . . .

After Lerman (1972).

plants (Fig. A1.2). An uncorrected date on such material would thus appear younger than a stratigraphically equivalent sample composed of predominantly CAL-type plants. Similar problems presumably exist in dating the remains of animals that fed predominantly on plants using one or another photosynthetic pathway, though this question has not been studied in detail.

By international agreement, all ^{14}C dates are corrected for fractionation effects by standardization to a $\delta^{13}C$ value of $-25‰$, the average value for wood. This results in relatively small "corrections" for CAL plants but larger corrections for HS and CAM plants, freshwater and marine shells, and aquatic plants. In particular, corrections of up to 450 yr may be necessary in the case of marine shells because they produce carbonate in equilibrium with ocean water, which is relatively enriched in ^{13}C. Table A1.1 indicates the magnitude of corrections necessary to standardize ^{14}C dates for fractionation effects. Where no such adjustment was made in a dated sample, these values may act as a guide to the appropriate correction needed.

APPENDIX B: WORLDWIDE WEB-BASED RESOURCES IN PALEOCLIMATOLOGY

The Worldwide Web offers enormous resources for those interested in paleoclimatology. These include paleoclimatic data bases, results from GCM paleoclimate simulations, useful software, information about organizations, and newsletters. Many of these are well-maintained and kept up-to-date; others may be transient. Rather than providing an exhaustive list of all currently relevant Web sites, I list here a few of the more important ones (which are, in turn, linked to many more). Readers can also consult the Paleoclimate Resources Web Page at the University of Massachusetts Climate System Research Center:
(http://www.geo.umass.edu/climate/paleo.html)
where a much more comprehensive and up-to-date inventory will be maintained.

Organizations

American Quaternary Association (AMQUA)
http://www4.nau.edu/amqua/
International Geosphere-Biosphere Programme: Past Global Changes (IGBP-PAGES)
http://www.pages.unibe.ch/
International Union for Quaternary Research (INQUA)
http://inqua.nlh.no/
Quaternary Research Association (QRA)
http://gra.org.uk
Australasian Quaternary Association (AQUA)
http://www.aqua.org.au/
Canadian Quaternary Association (CANQUA)
http://www.mun.ca/canqua/

Data Resources

NOAA Paleoclimatology Program (World Data Center A for Paleoclimatology)
http://www.ngdc.noaa.gov/paleo/
PANGAEA—Publishing Network for Geoscientific and Environmental Data
http://www.pangaea.de/

REFERENCES

Abrahamson, D.E. (1990). *The Challenge of Global Warming*. Washington, DC: Island Press.

Adelseck, C.G. and Berger, W.H. (1977). On the dissolution of planktonic foraminifera and associated microfossils during settling and on the sea floor. In: *Dissolution of Deep-Sea Carbonates* (W.V. Sliter, A.W.H. Bé, and W.H. Berger, eds.). Special Publication No. 13, Cushman Foundation for Foraminiferal Research, Washington, DC, 70–81.

Aharon, P. and Chappell, J. (1986). Oxygen isotopes, sea level changes and the temperature history of a coral reef environment in New Guinea over the last 10^5 years. *Palaeogeography, Palaeoclimatology, Palaeoecology*, **56**, 337–379.

Aitken, M.J. (1974). *Physics and Archaeology,* 2nd edition, Oxford: Clarendon Press.

Aitken, M.J. (1985). *Thermoluminescence Dating*. London: Academic Press.

Aldaz, L. and Deutsch, S. (1967). On a relationship between air temperature and oxygen isotope ratio of snow and firn in the South Pole region. *Earth and Planetary Science Letters*, **3**, 267–274.

Alessio M., Allegri, L., Antonioli, F., Belluomini, G., Improta, S., Manfra, L., and Preite, M. 1998. The Tyrrhenian sea level curve during the last 43 ka, obtained by means of dating submerged speleothems and archaeological data. *Memoria descrittiva della Carta Geologica d'Italia*, LII, 261–276 (in Italian).

Alexandre, P. (1977). Les variations climatiques au Moyen Age (Belgique, Rhenanie, Nord de la France). *Annales: Economies, Sociétés, Civilizations*, **32**, 183–197.

Alley, R.B., Gow, A.J., Johnsen, S.J., Kipfstuhl, J., Meese, D.A., and Thorsteinsson, Th., (1995). Comparison of deep ice cores. *Nature*, **373**, 393–394.

Alley, R.B., Gow, A.J., Meese, D.A., Fitzpatrick, J.J., Waddington, E.D., and Bolzan, J.F. (1997b). Grain-scale processes, folding and stratigraphic disturbance in the GISP2 ice core. *Jour. Geophys. Res.*, **102C**, 26819–26830.

Alley, R., and MacAyeal, D. (1994). Ice-rafted debris associated with the binge-purge oscillations of the Laurentide Ice Sheet. *Paleoceanography,* **9,** 503–511.

Alley, R.B., Mayewski, P.A., Sowers, T., Stuiver, M., Taylor, K.C., and Clark, P.U. (1997c).
Holocene climatic instability: a prominent, widespread event 8200 yr ago. *Geology,* **25,**
483–486.

Alley, R.B., Meese, D.A., Shuman, C.A., Gow, A.J., Taylor, K.C., Grootes, P.M., White,
J.W.C., Ram, M., Waddington, E.D., Mayewski, P.A., and Zielinski, G.A. (1993).
Abrupt increase in Greenland snow accumulation at the end of the Younger Dryas event.
Nature, **362,** 527–529.

Alley, R.B., Perepezko, J.H., and Bentley, C.R. (1988). Long-term climate changes from crystal
growth. *Nature,* **332,** 592–593 (and reply by J.P. Petit, P. Duval, and C. Lorius, p. 593).

Alley, R.B., Shuman, C.A., Meese, D.A., Gow, A.J., Taylor, K.C., Cuffey, K.M., Fitzpatrick,
J.J., Grootes, P.M., Zielinski, G.A., Ram, M., Spinelli, G., and Elder B. (1997a). Visual-
stratigraphic dating of the GISP2 ice core: basis, reproducibility and application. *Jour.
Geophys. Res.,* **102C,** 26367–26381.

Allison, I. and Kruss, P. (1977). Estimation of recent climatic change in Irian Jaya by numerical modelling of its tropical glaciers. *Arctic and Alpine Research,* **9,** 49–60.

Altabet, M.A., Francois, R., Murray, D.W., and Prell, W.L. (1995). Climate-related variations
in denitrification in the Arabian Sea from sediment $^{15}N/^{14}N$ ratios. *Nature,* **373,**
506–509.

Ambach, W., Dansgaard, W., Eisner, H., and Moller, J. (1968). The altitude effect on the isotopic composition of precipitation and glacier ice in the Alps. *Tellus,* **20,** 595–600.

An, Z., Kukla, G., Porter, S.C., and Xiao, J.L. (1991). Late Quaternary dust flow on the Chinese Loess Plateau. *Catena,* **18,** 125–132.

An, Z., Liu, T., Lu, Y., Porter, S.C., Kukla, G., Wu, X., and Hua, Y. (1990). The long term
paleomonsoon variation recorded by the loess paleosol sequence in central China. *Quaternary International,* **7/8,** 91–95.

An, Z., Liu, T., Zhou, Y., Sun, F., and Ding, Z. (1987). The paleosol complex S_5 in the China
Loess Plateau–a record of climatic optimum during the last 1–2 Ma. *Geojournal,* **15,**
141–143.

An, Z., Porter, S.C., Zhou, W., Lu, Y., Donahue, D.J., Head, M.J., Wu, X., Ren, J., and
Zheng, H. (1993). Episode of strengthened summer monsoon climate of Younger Dryas
age on the Loess Plateau of central China. *Quaternary Research,* **39,** 45–54.

Andersen, K. K., Armengaud, A., and Genthon, C. (1998). Atmospheric dust under glacial
and interglacial conditions. *Geophysical Research Letters,* **25,** 2281–2284.

Andersen, T.F. and Steinmetz, J.C. (1981). Isotopic and biostratigraphical records of calcareous nannofossils in a Pleistocene core. *Nature,* **294,** 741–744.

Anderson, D.M., Prell, W. L., and Barratt, N.J. (1989). Estimates of sea surface temperature
in the Coral Sea at the last glacial maximum. *Paleoceanography,* **4,** 615–627.

Anderson, P.M., Bartlein, P.J., Brubaker, L.B., Gajewski, K., and Ritchie, J.C. (1991). Vegetation-
pollen-climate relationships for the arcto-boreal regions of North America and Greenland.
Jour. Biogeography, **18,** 565–582.

Anderson, R.Y., Bradbury, J.P., Dean, W.E., and Stuiver, M. (1993). Chronology of Elk Lake
sediments: coring, sampling and time series construction. In: *Elk Lake, Minnesota: Evidence for Rapid Climatic Change in the North-central United States* (J.P. Bradbury and
W.E. Dean, eds.). Special Paper 276, Geological Society of America, Boulder, 37–43.

Andrée, M., Beer, J., Loetscher, H.P., Moor, E., Oeschger, H., Bonani, G., Hoffman, H.J.,
Morenzoni, E., Ness, M., Suter, M., and Wölfli, W. (1986). Dating polar ice by ^{14}C accelerator mass spectrometry. *Radiocarbon,* **28,** 417–423.

Andrews, J.T., Barry, R.G., Bradley, R.S., Miller, G.H., and Williams, L.D. (1972). Past and
present glaciological responses to climate in eastern Baffin Island. *Quaternary Research,*
2, 303–314.

Andrews, J.T., Davis, P.T., and Wright, C. (1976). Little Ice Age permanent snowcover in the
eastern Canadian Arctic: extent mapped from Landsat-1 satellite imagery. *Geografiska
Annaler,* **58A,** 71–81.

Andrews, J.T., Erlenkeuser, H., Tedesco, K., Aksu, A., and Jull, A.J.T. (1994). Late Quaternary (Stage 2 and 3) meltwater and Heinrich Events, northwest Labrador Sea. *Quaternary Research*, **41**, 26–34.

Andriessen, P.A.M, Helmens, K.F., Hooghiemstra, H., Riezobos, P.A., and Van der Hammen, T. (1993). Absolute chronology of the Pliocene-Quaternary sediment sequence of the Bogotá area, Colombia. *Quaternary Science Reviews*, **12**, 483–501.

Antevs, E. (1948). Climatic changes and pre-white man. *University of Utah Bulletin*, **36**, 168–191.

Appleby, P.G. and Oldfield, F. (1978). The calculation of ^{210}Pb dates assuming a constant rate of supply of unsupported ^{210}Pb to the sediment. *Catena*, **5**, 1–8.

Appleby, P.G. and Oldfield, F. (1983). The assessment of ^{210}Pb data from sites with varying sediment accumulation rates. *Hydrobiology*, **103**, 29–35.

Arakawa, H. (1954). Fujiwhara on five centuries of freezing dates of Lake Suwa in central Japan. *Archive für Meteorologie, Geophysik und Bioklimatologie*, **B6**, 152–166.

Arakawa, H. (1956a). Dates of first or earliest snow covering for Tokyo since 1632. *Quart. Jour. Royal Meteorological Society*, **82**, 222–226.

Arakawa, H. (1956b). Climatic change as revealed by the blooming dates of the cherry blossoms at Kyoto. *Journ. Meteorology*, **13**, 599–600.

Arakawa, H. (1957). Climatic change as revealed by the data from the Far East. *Weather*, **12**, 46–51.

Arnason, B. (1969). The exchange of hydrogen isotopes between ice and water in temperate glaciers. *Earth and Planetary Science Letters*, **6**, 423–430.

Arnow, T. (1980). Water budget and water-surface fluctuations of Great Salt Lake. *Utah Geological and Mineral Survey Bulletin*, **116**, 255–263.

Ashworth, A.C. (1980). Environmental implications of a beetle assemblage from the Gervais formation (Early Wisconsinan?), Minnesota. *Quaternary Research*, **13**, 200–212.

Atkinson, T.C., Harmon, R.S., Smart, P.L., and Waltham, A.C. (1978). Palaeoclimatic and geomorphic implications of ^{230}Th/^{234}U dates on speleothems from Britain. *Nature*, **272**, 24–28.

Atkinson, T.C., Lawson, T.J., Smart, P.L., Harmon, R.S., and Hess, J.W. (1986a). New data on speleothem deposition and palaeoclimate in Britain over the last forty thousand years. *Jour. Quaternary Science*, **1**, 67–72.

Atkinson, T.C., Briffa, K.R., Coope, G.R., Joachim, M., and Perry, D.W. (1986b). Climatic calibration of coleopteran data. In: *Handbook of Holocene Palaeoecology and Palaeohydrology* (B. Berglund, ed.) New York: Wiley, 851–858.

Atkinson, T.C., Briffa, K.R., and Coope, G.R. (1987). Seasonal temperatures in Britain during the last 22,000 years, reconstructed using beetle remains. *Nature*, **325**, 587–592.

Austin, W.E.N., Bard, E., Hunt, J.B., Kroon, D., and Peacock, J.D. (1995). The ^{14}C age of the Icelandic Vedde ash: implications for Younger Dryas marine reservoir age corrections. *Radiocarbon*, **37**, 53–62.

Bada, J.L., (1985). Amino acid racemization dating of fossil bones. *Annual Review of Earth and Planetary Science*, **13**, 241–268.

Bada, J.L., Protsch, R., and Schroeder, R.A. (1973). The racemization reaction of isoleucine used as a palaeotemperature indicator. *Nature*, **241**, 394–395.

Bada, J.L. and Schroeder, R.A. (1975). Amino acid racemization reactions and their geochemical implications. *Naturwissenschaften*, **62**, 71–79.

Baes, C.F., Jr. (1982). Effects of ocean chemistry and biology on atmospheric carbon dioxide. In: *Carbon Dioxide Review: 1982* (W.C. Clark, ed.). New York: Oxford University Press, 187–204.

Baker, A., Smart, P.L., and Edwards, R.L. (1995). Paleoclimate implications of a mass spectrometric dating of a British Flowerstone. *Geology*, **23**, 309–312.

Baker, A., Smart, P.L., Edwards, R.L., and Richards, D.A. (1993). Annual growth banding in a cave stalagmite . *Nature*, **364**, 518–520.

Baker, J.N.L. (1932). The climate of England in the 17th century. *Quart. Jour. Royal Meteorological Society*, **58**, 421–436.

Baksi, A.K., Hsu, V., McWilliams, M.O., and Farrar, E. (1992). ^{40}Ar/^{39}Ar dating of the Brunhes-Matuyama geomagnetic field reversal. *Science,* **256,** 356–357.

Balescu, S. and Lamothe, M. (1994). Comparison of TL and IRSL age estimates of feldspar coarse grains from waterlain sediments. *Quaternary Geochronology (Quaternary Science Reviews),* **13,** 437–444.

Ball, T. (1992). Historical and instrumental evidence of climate: western Hudson Bay, Canada 1714–1850. In: *Climate Since A.D. 1500* (R.S. Bradley and P.D. Jones, eds.). London: Routledge, 40–73.

Bar-Matthews, M., Wasserburg, G.J., and Chen, J.H. (1993). Diagenesis of fossil coral skeletons: correlation between trace elements, textures and ^{234}U/^{238}U. *Geochimica et Cosmochimica Acta,* **57,** 257–276.

Bard, E. (1988). Correction of accelerator mass spectrometry ^{14}C ages measured in planktonic foraminifera: paleoceanographic implications. *Paleoceanography,* **3,** 635–645.

Bard, E., Arnold, M., Fairbanks, R., and Hamelin, B. (1993). ^{230}Th-^{234}U and ^{14}C age, obtained by mass/spectrometry on corals. *Radiocarbon,* **35,** 191–199.

Bard, E., Arnold, M., Mangerud, J., Paterne, M., Labeyrie, L., Duprat, J., Mélieres, M.A., Sønstegaard, E., and Duplessy, J.C. (1994). The North Atlantic atmosphere-sea surface ^{14}C gradient during the Younger Dryas climatic event. *Earth and Planetary Science Letters,* **126,** 275–287.

Bard, E., Fairbanks, R.G., Arnold, M., Maurice, P., Duprat, J., Moyes, J., and Duplessy, J.C. (1989). Sea-level estimates during the last deglaciation based on δ^{18}O and accelerator mass spectrometry ^{14}C ages measured in *Globigerina bulloides. Quaternary Research,* **31,** 381–391.

Bard, E., Hamelin, B., Arnold, M., Montaggioni, L., Cabioch, G., Faure, G., and Rougerie, F. (1996). Deglacial sea-level record from Tahiti corals and the timing of global meltwater discharge. *Nature,* **382,** 241–244.

Bard, E., Hamelin, B., Fairbanks, R.G., and Zindler, A. (1990). Calibration of the ^{14}C time scale over the last 30,000 years using mass spectrometric U-Th ages from Barbados corals. *Nature,* **345,** 405–410.

Bard, E., Rostek, F., and Sonzogni, C. (1997). Interhemispheric synchrony of the last deglaciation inferred from alkenone palaeothermometry. *Nature,* **385,** 707–710.

Barnola, J-M., Pimienta, P., Raynaud, D., and Korotkevich, Y.S. (1991). CO_2-climate relationship as deduced from the Vostok ice core: a re-examination based on new measurements and on a re-evaluation of the air dating. *Tellus,* **B43,** 83–90.

Barnola, J-M., Raynaud, R., Korotkevich, Y.S., and Lorius, C. (1987). Vostok ice core provides 160,000 year record of atmospheric δ^{18}O. *Nature,* **329,** 408–414.

Baron, W.R. (1982). The reconstruction of eighteenth century temperature records through the use of content analysis. *Climatic Change,* **4,** 385–398.

Barriendos, M. (1997). Climatic variations in the Iberian Peninsula during the late Maunder Minimum (A.D. 1675–1715): an analysis of data from rogation ceremonies. *The Holocene,* **7,** 105–111.

Barry, R.G. (1982). Approaches to reconstructing the climate of the steppe-tundra biome. In: *Paleoecology of Beringia.* (D.M. Hopkins, J.V. Matthews, C.E. Schweger, and S.B. Young). New York: Academic Press, 195–204.

Barry, R.G., Andrews, J.T., and Mahaffy, M.A. (1975). Continental ice sheets: conditions for growth. *Science,* **190,** 979–981.

Bartlein, P.J., Anderson, K.H., Edwards, M.E., Thompson, R.S., Webb, R.S., Webb III, T., and Whitlock, C. (1997). Paleoclimate simulations for North America over the past 21,000 years: features of the simulated climate and comparisons with paleoenvironmental data. *Quaternary Science Reviews,* **17,** 549–586.

Bartlein, P.J., Edwards, M.E., Shafer, S.L., and Barker, Jr., E.D. (1995). Calibration of radiocarbon ages and the interpretation of paleoenvironmental records. *Quaternary Research,* **44,** 417–424.

Bartlein, P.J., Prentice, I.C., and Webb, III, T. (1986). Climatic response surfaces from pollen data for some eastern North American taxa. *Jour. Biogeography,* **13,** 35–57.

Bartlein, P.J. and Webb, III, T. (1985). Mean July temperature at 6000 B.P. in eastern North America: regression equations for estimates from fossil-pollen data. In: *Climatic Change in Canada 5* (C.R. Harington, ed.). Syllogeus No. 55, National Museum of Canada, Ottawa, 301–342.

Bartlein, P.J., Webb, III, T., and Fleri, E. (1984). Holocene climatic change in the northern midwest: pollen-derived estimates. *Quaternary Research,* **22,** 361–374.

Bartlein, P.J. and Whitlock, C. (1993). Paleoclimatic interpretation of the Elk Lake pollen record. In: *Elk Lake, Minnesota: Evidence for Rapid Climate Change in the North–Central United States* (J.P. Bradbury and W.E. Dean, eds.). Special Paper 276, Geological Society of America, Boulder, 275–293.

Bassett, I.J. and Terasmae, J. (1962). Ragweeds, *Ambrosia* species in Canada and their history in postglacial time. *Canadian Jour. Botany,* **40,** 141–50.

Bassinot, F.C., Labeyrie, L.D., Vincent, E., Quidelleur, X., Shackleton, N.J., and Lancelot, Y. (1994). The astronomical theory of climate and the age of the Brunhes-Matuyama magnetic reversal. *Earth and Planetary Science Letters,* **126,** 91–108.

Baumgartner, A. (1979). Climatic variability and forestry. In: *Proceedings of the World Climate Conference.* WMO Publication No. 537. World Meteorological Organization, Geneva, 581–607.

Baumgartner, S., Beer, J., Masarik, J., Wagner, G., Meynadier, L., and Synal, H-A. (1998). Geomagnetic modulation of the ^{36}Cl flux in the GRIP ice core, Greenland. *Science,* **279,** 1330–1332.

Baumgartner, T.R., Michaelsen, J., Thompson, L.G., Shen, G.T., Soutar, A., and Casey, R.E. (1989). The recording of inter-annual climatic change by high resolution natural systems: tree rings, coral bands, glacial ice layers and marine varves. In: *Climatic Change in the Eastern Pacific and Western Americas* (D. Peterson, ed.). Washington D.C.: American Geophysical Union, 1–14.

Bé, A.W.H., Damuth, J.E., Lott, L., and Free, R. (1976). Late Quaternary climatic record in western equatorial Atlantic sediment. In: *Investigation of late Quaternary Paleo-oceanography and Paleoclimatology.* (R.M. Cline and J.D. Hays, eds.). Memoir No. 145. Boulder: Geological Society of America, 165–200.

Bé, A.W.H. (1977). An ecological, zoogeographic and taxonomic review of recent planktonic foraminifera. In: *Oceanic Micropalaeontology,* (A.T.S. Ramsay, ed.). London: Academic Press, 1–88.

Beadle, L.C. (1974). *The Inland Waters of Tropical Africa.* London: Longmans.

Beaulieu, J. de and Reille, M. (1992). The last climatic cycle at La Grande Pile (Vosges, France). A new pollen profile. *Quaternary Science Reviews,* **11,** 431–438.

Beck, J.W., Edwards, R.L., Ito, E., Taylor, F.W., Récy, J., Rougerie, F., Joannot, P., and Henin, C. (1992). Sea surface temperature from coral skeletal strontium/calcium ratios. *Science,* **257,** 644–647.

Beck, J.W., Récy, J., Taylor, F., Edwards, R.L., and Cabioch, G. (1997). Abrupt changes in Holocene tropical sea surface temperature derived from coral records. *Nature,* **385,** 705–707.

Becker, B. (1993). An 11,000 year German oak and pine dendrochronology for radiocarbon calibration. *Radiocarbon,* **35,** 201–213.

Becker, B., Kromer, B., and Trimlorn, P. (1991). A stable isotope tree-ring timescale of the late glacial/Holocene boundary. *Nature,* **353,** 647–649.

Beer, J., Johnsen, S.J., Bonani, G., Finkel, R.C., Langway, C.C., Oeschger, H., Stauffer, B., Suter, M., and Woelfli, W. (1992). ^{10}Be peaks as time markers in polar ice cores. In: *The Last Deglaciation: Absolute and Radiocarbon Chronologies* (E. Bard and W. Broecker, eds.). Berlin: Springer-Verlag, 141–153.

Beer, J., Joos, F., Lukasczyk, Ch., Mende, W., Rodriguez, J., Siegenthaler, U., and Stellmacher, R. (1994). ^{10}Be as an indicator of solar variability and climate. In: *The Solar Engine and its Influence on Terrestrial Atmosphere and Climate* (E. Nesmé-Ribes, ed.). Berlin: Springer-Verlag, 221–233.

Beer, J., Mende, W., Stellmacher, R., and White, O.R. (1996). Intercomparisons of proxies for past solar variability. In: *Climatic Variations and Forcing Mechanisms of the Last 2000 Years* (P.D. Jones, R.S. Bradley, and J. Jouzel, eds.). Berlin: Springer-Verlag, 501–517.

Beget, J.E. (1994). Tephochronology, lichenometry and radiocarbon dating at Gulkana glacier, central Alaska Range, USA. *The Holocene, 4,* 307–313.

Beget, J.E., Machida, H., and Lowe, D. (eds.). (1996). *Climatic Impact of Explosive Volcanism: Recommendations for Research,* IGBP PAGES Workshop Report 96–1; Bern, 11.

Beget, J.E., Reger, R.D., Pinney, D., Gillespie, T,. and Campbell, K. (1991). Correlation of the Holocene Jarvis Creek, Tangle Lakes, Cantwell and Hayes tephras in south central and central Alaska. *Quaternary Research, 42,* 301–306.

Bell, B. (1970). The oldest records of the Nile floods. *Geographical Jour.* **136,** 569–573.

Bell, W.T., and Ogilvie, A.E.J. (1978). Weather compilations as a source of data for the reconstruction of European climate during the Medieval Period. *Climatic Change 1,* 331–348.

Bender, M., Labeyrie, L.D., Raynaud, D., and Lorius, C. (1985). Isotopic composition of atmospheric oxygen in ice linked with deglaciation and global primary productivity. *Nature,* **318,** 349–352.

Bender, M., Sowers, T., Dickson, M-L., Orchado, J., Grootes, P., Mayewski, P.A., and Meese, D.A. (1994). Climate correlations between Greenland and Antarctica during the past 100,000 years. *Nature,* **327,** 663–666.

Benedict, J.B. (1967). Recent glacial history of an alpine area in the Colorado Front Range, USA. I. Establishing a lichen growth curve. *Jour. Glaciology,* **6,** 817–832.

Benedict, J.B. (1993). A 2000-year lichen snow-kill chronology for the Colorado Front Range, USA. *The Holocene, 3,* 27–33.

Benson, C.S. (1961). Stratigraphic studies in the snow and firn of the Greenland Ice Sheet. *Folia Geographica Danica, 9,* 13–37.

Benson, L.V. (1981). Paleoclimatic significance of lake-level fluctuations in the Lahontan Basin. *Quaternary Research,* **16,** 390–403.

Benson, W.W. (1982). Alternative models for infrageneric diversification in the humid Tropics: tests with Passion Vine Butterflies. In: *Biological Diversification in the Tropics* (G.T. Prance, ed.). New York: Columbia University Press, 608–640.

Berger, A. (1977a). Long-term variations of the Earth's orbital elements. *Celestial Mech.,* **15,** 53–74.

Berger, A. (1977b). Support for the astronomical theory of climatic change. *Nature,* **269,** 44–45.

Berger, A. (1978). Long-term variations of caloric insolation resulting from the Earth's orbital elements. *Quaternary Research, 9,* 139–167.

Berger, A. (1979). Insolation signatures of Quaternary climatic changes. *Il Nuovo Cimento,* 2(c), 63–87.

Berger, A. (1980). The Milankovitch astronomical theory of paleoclimates. A modern review. *Vistas in Astronomy, 24,* 103–122.

Berger, A. (1988). Milankovitch theory and climate. *Reviews of Geophysics,* **26,** 624–657.

Berger, A. (1990). Testing the astronomical theory with a coupled climate-ice-sheet model. *Global and Planetary Change, 3,* 25–141.

Berger, A., Gallée, H., and Loutre, M.F. (1991). The earth's future climate at the astronomical timescale. In: *Future Climate Change and Radioactive Waste Disposal* (C.M. Goodess and J.P. Palutikof eds.) Climatic Research Center, University of East Anglia, Norwich, 148–165.

Berger, A. and Loutre, M.F. (1991). Insolation values for the climate of the last 10 million years. *Quaternary Science Reviews,* **10,** 297–318.

Berger, A. and Loutre, M.F. (1997a). Palaeoclimate sensitivity to CO_2 and insolation. *Ambio,* **26,** 32–37.

Berger, A. and Loutre, M.F. (1997b). Long-term variations in insolation and their effects on climate: the LLN experiments. *Surveys in Geophysics,* **18,** 147–161.

Berger, A., Loutre, M.F., and Laskar, J. (1992). Stability of the astronomical frequencies over the Earth's history for paleoclimatic studies. *Science*, **255**, 560–566.

Berger, A., Tricot, C., Gallee, H., and Loutre, M.F. (1993). Water vapour, CO_2 and insolation over the last glacial-interglacial cycles. *Phil. Trans. Royal Society of London*, **B341**, 253–261.

Berger, G.W., Pillans, B.J., and Palmer, A.S. (1992). Dating loess up to 800 ka by thermoluminescence. *Geology*, **20**, 403–406 (see also comment by Wintle *et al.* (1993), *Geology*, **21**, 568.

Berger, R., Homey, A.G., Libby, W.F. (1964). Radiocarbon dating of bone and shell from their organic components. *Science*, **144**, 999–1001.

Berger, W.H. (1968). Planktonic foraminifera: selective solution and paleoclimatic interpretation. *Deep-Sea Research*, **15**, 31–43.

Berger, W.H. (1970). Planktonic foraminifera: selective solution and lysocline. *Marine Geology*, **8**, 111–138.

Berger, W.H. (1971). Sedimentation of planktonic foraminifera. *Marine Geology*, **11**, 325–358.

Berger, W.H. (1973a). Deep-Sea carbonates: evidence for a coccolith lysocline. *Deep-Sea Research*, **20**, 917–921.

Berger, W.H. (1973b). Deep-Sea carbonates: Pleistocene dissolution cycles. *Jour. Foraminiferal Research*, **3**, 187–195.

Berger, W.H. (1975). Deep-Sea carbonates: dissolution profiles from foraminiferal preservation. In: *Dissolution of Deep-Sea Carbonates* (W.V. Sliter, A.W.H. Bé, and W.H. Berger, eds.). Special Publication No. 13, Cushman Foundation for Foraminiferal Research, Washington, DC, 82–86.

Berger, W.H. (1977). Deep-Sea carbonate and the deglaciation preservation spike in pteropods and foraminifera. *Nature*, **269**, 301–304.

Berger, W.H. (1990). The Younger Dryas cold spell—a quest for causes. *Palaeogeography, Palaeoclimatology, Palaeocology*, **89**, 219–237.

Berger, W.H. and Gardner, J.V. (1975). On the determination of Pleistocene temperatures from planktonic foraminifera. *Jour. Foraminiferal Research*, **5**, 102–113.

Berger, W.H., Johnson, R.F., and Killingley, J.S., (1977). "Unmixing" of the deep-sea record and the deglacial meltwater spike. *Nature*, **269**, 661–663.

Berger, W.H. and Killingley, J.S. (1977). Glacial–Holocene transition in deep–sea carbonates: selective dissolution and the stable isotope signal. *Science*, **197**, 563–566.

Berger, W.H. and Wefer, G. (1991). Productivity of the glacial ocean: discussion of the iron hypothesis. *Limnology and Oceanography*, **36**, 1899–1918.

Berger, W.H. and Winterer, E.L. (1974). Plate stratigraphy and the fluctuating carbonate line. In: *Pelagic Sediments on Land and in the Ocean* (K.J. Hsu and H. Jenkins, eds.). Special Publication No. 1, International Association of Sedimentologists, Oxford: Blackwell Scientific, 11–48.

Berggren, W.A., Burckle, L.H., Cita, M.B., Cooke, H.B.S., Funnell, B.M., Gartner, S., Hays, J.D., Kennett, J.P., Opdyke, N.D., Pastouret, L., Shackleton, N.J., and Takayanagi, I.Y. (1980). Towards a Quaternary time scale. *Quaternary Research*, **13**, 277–302.

Berggren, W.A., Kent, D.V., Swisher, III, C.C., and Aubry, M.P. (1995). A revised Cenozoic geochronology and chronostratigraphy. In: *Geochronology Time Scales and Global Stratigraphic Correlation* (W.A. Beggren, D.V. Kent, M.P. Aubry and J. Hardenbol, eds.). Special Publication No. 54, Society for Sedimentary Geology, Tulsa, Oklahoma, 129–212.

Bergthorsson, P. (1969). An estimate of drift ice and temperature in Iceland in 1000 years. *Jøkull* **19**, 94–101.

Bernabo, J.C. and Webb, III, T. (1977). Changing patterns in the Holocene pollen record of northeastern North America: a mapped summary. *Quaternary Research*, **8**, 69–96.

Berner, R.A. (1977). Sedimentation and dissolution of pteropods in the ocean. In: *The Fate of Fossil Fuel CO_2 in the Ocean*, (N.R. Anderson and A. Malahoff, eds.). New York: Plenum Press, 243–260.

Beschel, R. (1961). Dating rock surfaces by lichen growth and its application in glaciology and physiography (lichenometry). In: *Geology of the Arctic,* II (G.O. Raasch, ed.). Toronto: University of Toronto Press, 1044–1062.

Betancourt, J.L. (1990). Late Quaternary biogeography of the Colorado Plateau. In: *Packrat Middens: The Last 40,000 Years of Biotic Change* (J.L. Betancourt, T.R. Van Devender, and P.S. Martin eds.). Tucson: University of Arizona Press, 259–292.

Betancourt, J.L., Van Devender, T.R., and Martin, P.S. (1990). *Packrat Middens: The Last 40,000 Years of Biotic Change* (J.L. Betancourt, T.R. Van Devender, and P.S. Martin, eds.). Tucson: University of Arizona Press.

Bickerton, R. and Matthews, J.A. (1992). On the accuracy of lichenometric dates: an assessment based on the 'Little Ice Age' moraine sequence of Nigardsbreen, southern Norway. *The Holocene,* **2,** 227–237.

Birks, H.H., Gulliksen, S., Haflidason, H., Mangerud, J., and Possnert, G. (1996). New radiocarbon dates for the Vedde ash and the Saksunarvatn ash from western Norway. *Quaternary Research,* **45,** 119–127.

Birks, H.J.B. (1978). Numerical methods for the zonation and correlation of biostratigraphical data. In: *Palaeohydrological Changes in the Temperate Zone in the Last 15,000 years* (B.E. Berglund, ed.). **1,** International Geological Correlation Program, Project 158B, Subproject B. University of Lund, Sweden, 99–119.

Birks, H.J.B. and Berglund, B.E. (1979). Holocene pollen stratigraphy of southern Sweden: a reappraisal using numerical methods. *Boreas,* **8,** 257–279.

Birks, H.J.B. and Birks, H.H. (1980). *Quaternary Palaeoecology.* London: E. Arnold.

Birks, H.J.B. and Gordon, A.D. (1985). *Numerical Methods in Quaternary Pollen Analysis.* London: Academic Press.

Blackwell, B. and Schwarcz, H.P. (1995). The uranium series disequilibrium dating methods. In: *Dating Methods for Quaternary Deposits* (N.W. Rutter and N.R. Catto, eds.). Geological Association of Canada, St. John's, 167–208.

Blasing, T.J. and Fritts, H.C. (1975). Past climate of Alaska and northwestern Canada as reconstructed from tree-rings. In: *Climate of the Arctic.* (G. Weller and S.A. Bowling, eds.). Fairbanks: University of Alaska Press, 48–58.

Blinman, E., Mehringer, P.J., and Sheppard, J.C. (1979). Pollen influx and the deposition of Mazama and Glacier Peak tephras. In: *Volcanic Activity and Human Ecology* (P.D. Sheets and D.K. Grayson, eds.). New York: Academic Press, 393–425.

Blunier, T., Chappellaz, J., Schwander, J., Stauffer, B., and Raynaud, D. (1995). Variations in atmospheric methane concentration during the Holocene epoch. *Nature,* **374,** 46–49.

Boersma, A. (1978). Foraminifera. In: *Introduction to Marine Micropaleontology* (B.U. Haq and A. Boersma, eds.). New York: Elsevier/North Holland, 19–77.

Bolin, B. (1981). The carbon cycle. In: *Climatic Variations and Variability: Facts and Theories* (A. Berger, ed.). Dordrecht: D. Riedel, 623–639.

Bolin, B. (1992). The carbon cycle and global change: a focus on CO_2. In: *Trace Gases and the Biosphere* (B. Moore III and D. Schimel, eds.). Boulder: University Corporation for Atmospheric Research, 129–149.

Bond, G., Broecker, W., Johnsen, S., McManus, J., Labeyrie, L., Jouzel, J., and Bonani, G. (1993). Correlations between climate records from North Atlantic sediment and Greenland ice. *Nature,* **365,** 143–147.

Bond, G., Heinrick, H., Broecker, W., Labeyrie, L., McManus, J., Andrews, J., Huon, S., Jantschik, R., Clausen, S., Simet, C., Tedesco, K., Klas, M., Bonani, G., and Ivy, S. (1992). Evidence for massive discharges of icebergs into the North Atlantic ocean during the last glacial period. *Nature,* **360,** 245–249.

Bond, G.C. and Lotti, R. (1995). Iceberg discharges into the North Atlantic on millennial time scales during the Last Glaciation. *Science,* **267,** 1005–1010.

Bonnefille, R., Chalié, F., Guiot, J., and Vincens, A. (1992). Quantitative estimates of full glacial temperatures in equatorial Africa from palynological data. *Climate Dynamics,* **6,** 251–257.

Bonny, A.P. (1972). A method for determining absolute pollen frequencies in lake sediments. *New Phytologist,* **71,** 391–403.

Borisenkov, Y.P., Tsvetkov, A.V., and Agaponov, S.V. (1983). On some characteristics of insolation changes in the past and future. *Climatic Change,* **5,** 237–244.

Borisenkov, Y.P., Tsvetkov, A.V., and Eddy, J.A. (1985). Combined effect of earth orbit perturbations and solar activity on terrestrial insolation. Part I. Sample days and annual mean values. *Jour. Atmos. Science,* **42,** 933–940.

Boto, K. and Isdale, P. (1985). Fluorescent bands in massive corals result from terrestrial fulvic acid inputs to nearshore zone. *Nature,* **315,** 396–397.

Bowen, D.Q., Hughes, S., Sykes, G.A., and Miller, G.H. (1989). Land-sea correlations in the Pleistocene based on isoleucine epimerization in non-marine molluscs. *Nature,* **340,** 49–51.

Bowen, D.Q., Richmond, G.M., Fullerton, D.S., Sibrava, V., Fulton, R.J., and Velichko, A.A. (1986). Correlation of Quaternary glaciations in the Northern Hemisphere. *Quaternary Science Reviews,* **5,** 509–510 (plus chart).

Bowler, J.M. (1976). Aridity in Australia: age origins and expression in aeolian landforms and sediments. *Earth Science Reviews,* **12,** 279–310.

Boyle, E.A. (1988). Cadmium: chemical tracer of deepwater paleoceanography. *Paleoceanography,* **3,** 471–490.

Boyle, E.A. (1992). Cadmium and $\delta^{13}C$ paleochemical ocean distributions during the stage 2 glacial maximum. *Annual Reviews of Earth and Planetary Science,* **20,** 245–287.

Boyle, E.A. (1997). Cool tropical temperatures shift the global $\delta^{18}O$-T relationship: an explanation for the ice core $\delta^{18}O$-borehole temperature conflict? *Geophysical Research Letters,* **24,** 273–276.

Boyle, E.A. and Keigwin, L.D. (1982). Deep circulation of the North Atlantic over the past 200,000 years: geochemical evidence. *Science,* **218,** 784–787.

Boyle, E.A. and Keigwin, L.D. (1985). Comparison of Atlantic and Pacific paleochemical records for the last 215,000 years: changes in deep ocean circulation and chemical inventories. *Earth and Planetary Science Letters,* **76,** 135–150.

Boyle, E.A. and Keigwin, L.D. (1987). North Atlantic thermohaline circulation during the past 20,000 years linked to high-latitude surface temperature. *Nature,* **330,** 35–40.

Boyle, E.A. and Rosener, P. (1990). Further evidence for a link between Late Pleistocene North Atlantic surface temperatures and North Atlantic Deep-Water production. *Palaeogeography, Palaeoclimatology, Palaeoecology (Global and Planetary Change Section),* **89,** 113–124.

Bradbury, J.P., Leyden, B., Salgado-Labouriau, M., Lewis, W.M., Schubert, C., Binford, M.W., Frey, D.G., Whitehead, D.R., and Weibezahn, F.H. (1981). Late Quaternary environmental history of Lake Valencia, Venezuela. *Science,* **214,** 1299–1305.

Bradley, R.S. (1976). *Precipitation History of the Rocky Mountain States.* Boulder, Colo.: Westview Press.

Bradley, R.S. (1988). The explosive volcanic eruption signal in northern hemisphere continental temperature records. *Climatic Change,* **12,** 221–243.

Bradley, R.S. (1990). Holocene paleoclimatology of the Queen Elizabeth Islands, Canadian High Arctic. *Quaternary Science Reviews,* **9,** 365–384.

Bradley, R.S. (1991). Instrumental records of past global change: lessons for the analysis of non-instrumental data. In: *Global Changes of the Past* (R.S. Bradley, ed.). Boulder: University Corporation for Atmospheric Research, 103–116.

Bradley, R.S., Bard, E., Farquhar, G., Joussaume, S., Lautenschlager, M., Molfino, B., Raschke, E., Shackleton, N.J., Sirocko, F., Stauffer, B., and White, J. (1993). Evaluating strategies for reconstructing past global changes—what and where are the gaps? In: *Global Changes in the Perspective of the Past* (J.A. Eddy and H. Oeschger, eds.). Chichester: Wiley, 145–171.

Bradley, R.S. and Eddy, J.A. (1991). Records of past global changes. In: *Global Changes of the Past* (R.S. Bradley, ed.). Boulder: University Corporation for Atmospheric Research, 5–9.

Bradley, R.S. and Jones, P.D. (eds.). (1995). *Climate Since A.D. 1500.* (Revised edition) London: Routledge.

Bradley, R.S. and Jones, P.D. (eds.) (1992a). Introduction. In: *Climate Since A.D. 1500* London: Routledge, 1–16.

Bradley, R.S. and Jones, P. D. (1992b). When was the "Little Ice Age"? In: *Proceedings of the International Symposium on the Little Ice Age Climate*. (T. Mikami, ed.). Dept. of Geography, Tokyo Metropolitan University, Tokyo, 1–4.

Bradley, R.S. and Jones, P.D. (eds.) (1992c). Records of explosive volcanic eruptions over the last 500 years. In: *Climate Since A.D. 1500* London: Routledge, 606–622.

Bradley, R.S. and Jones, P.D. (1993). "Little Ice Age" summer temperature variations: their nature and relevance to recent global warming trends. *The Holocene,* **3,** 367–376.

Bradley, R.S. and Miller, G.H. (1972). Recent climatic change and increased glacierization in the eastern Canadian Arctic. *Nature,* **237,** 385–387.

Bradshaw, R.H.W. (1994). Quaternary terrestrial sediments and spatial scale: the limits to interpretation. In: *Sedimentation of Organic Particles* (A. Traverse, ed.). Cambridge: Cambridge University Press, 239–252.

Brakenridge, G.R. (1978). Evidence for a cold, dry full-glacial climate in the American Southwest. *Quaternary Research,* **9,** 22–40.

Brassell, S.C., Eglinton, G., Marlowe, I.T., Pflaumann, U., and Sarnthein, M. (1986). Molecular stratigraphy: a new tool for climatic assessment. *Nature,* **320,** 129–133.

Brasseur, G.P. and Madronich, S. (1992). Chemistry-transport models. In: *Climate System Modeling* (K.E. Trenberth ed.). Cambridge: Cambridge University Press, 491–517.

Bray, J.R. (1974). Glacial advance relative to volcanic activity since AD 1500. *Nature,* **248,** 42–43.

Bray, J.R. (1979). Surface albedo increase following massive Pleistocene explosive eruptions in western North America. *Quaternary Research,* **12,** 204–211.

Bray, J.R. (1982). Alpine glacier advance in relation to a proxy summer temperature index based mainly on wine harvest dates, AD 1453–1973. *Boreas,* **11,** 1–10.

Brázdil, R. (1996). Reconstructions of past climate from historical sources in Czech lands. In: *Climatic Variations and Forcing Mechanisms of the Last 2000 Year*s (P.D. Jones, R.S. Bradley, and J. Jouzel, eds.). Berlin: Springer-Verlag, 409–431.

Briat, M., Royer, A., Petit, J.R., and Lorius, C. (1982). Late glacial input of eolian continental dust in the Dome C ice core: additional evidence from individual microparticle analysis. *Ann. Glaciology,* **3,** 27–32.

Briffa, K.R. and Schweingruber, F.H. (1992). Recent dendroclimatic evidence of northern and central European summer temperatures. In: *Climate Since A.D. 1500* (R.S. Bradley and P.D. Jones, eds.). London: Routledge, 366–392.

Briffa, K.R., Bartholin, T.S., Eckstein, D., Jones, P.D., Karlén, W., Schweingruber, F.H., and Zetterberg, P. (1990). A 1400-year tree-ring record of summer temperatures in Fennoscandia. *Nature,* **346,** 434–439.

Briffa, K.R., Jones, P.D., Bartholin, T.S., Eckstein, D., Schweingruber, F.H., Karlén, W., Zetterberg, P., and Eronen, M. (1992a). Fennoscandian summers from A.D. 500: temperature changes on short and long timescales. *Climate Dynamics,* **7,** 111–119.

Briffa, K.R., Jones, P.D., Pilcher, J.R., and Hughes, M.K. (1988). Reconstructing summer temperatures in northern Fennoscandia back to A.D. 1700 using tree-ring data from Scots Pine. *Arctic and Alpine Research,* **20,** 385–394.

Briffa, K.R., Jones, P.D., and Schweingruber, F.H. (1992b). Tree-ring density reconstructions of summer temperature patterns across western North America since 1600. *Jour. Climate,* **5,** 735–754.

Briffa, K.R., Jones, P.D., and Schweingruber, F.H. (1994). Summer temperatures across northern North America: regional reconstructions from 1760 using tree-ring indices. *Jour. Geophys. Research,* **99D,** 25835–25844.

Briffa, K.R., Jones, P.D., Schweingruber, F.H., Karlén, W., and Shiyatov, S.G. (1996). Tree-ring variables as proxy climate indicators: problems with low-frequency signals. In: *Climate Variations and Forcing Mechanisms of the Last 2000 Years* (P.D. Jones, R.S. Bradley, and J. Jouzel, eds.). Berlin: Springer-Verlag, 9–41.

Briffa, K.R., Jones, P.D., Schweingruber, F.H., Shiyatov, S.G., and Cook, E.R. (1995). Unusual twentieth century warmth in a 1,000-year temperature record from Siberia. *Nature*, **376**, 156–159.

Broccoli, A.J. and Marciniak, E.P. (1996). Comparing simulated glacial climate and paleodata: a re-examination. *Paleoceanography*, **11**, 3–14.

Broecker, W. (1971). Calcite accumulation rates and glacial to interglacial changes in oceanic mixing. In: *The Late Cenozoic Glacial Ages* (K.K. Turekian, ed.). New Haven: Yale University Press, 239–265.

Broecker, W.S. (1982). Ocean chemistry during glacial time. *Geochimica et Cosmochimica Acta*, **46**, 1689–1705.

Broecker, W.S. (1986). Oxygen isotope constraints on surface ocean temperatures. *Quaternary Research*, **26**, 121–134.

Broecker, W.S. (1987). Unpleasant surprises in the greenhouse? *Nature*, **328**, 123–126.

Broecker, W.S. (1989). The salinity contrast between the Atlantic and Pacific Oceans during glacial time. *Paleoceanography*, **4**, 207–212.

Broecker, W.S. (1990). Salinity history of the northern Atlantic during the last deglaciation. *Paleoceanography*, **5**, 459–467.

Broecker, W.S. (1991). The great ocean conveyor. *Oceanography*, **4**, 79–89.

Broecker, W.S. (1994). Massive iceberg discharges as triggers for global climate change. *Nature*, **372**, 421–424.

Broecker, W.S., Andree, M., Bonani, G., Wolfli, W., Oeschger, H., Klas, M., Mix, A., and Curry, W. (1988a). Preliminary estimates for the radiocarbon age of deep water in the glacial ocean. *Paleoceanography*, **3**, 659–669.

Broecker, W.S., Andree, M., Wolfli, W., Oeschger, H., Bonani, G., Kennett, J., and Peteet, D. (1988b). The chronology of the last deglaciation: implications to the cause of the Younger Dryas event. *Paleoceanography*, **3**, 1–20.

Broecker, W.S. and Bender, M.L., (1972). Age determinations on marine strandlines. In: *Calibration of Hominoid Evolution* (W.W. Bishop and J.A. Miller, eds.). Edinburgh: Scottish Academic Press, 19–38.

Broecker, W.S., Bond, G., Klas, M., Bonani, G., and Wolfli, W. (1990a). A salt oscillator in the glacial Atlantic? 1. The concept. *Paleoceanography*, **5**, 469–477.

Broecker, W.S., Bond, G., Klas, M., Clark, E., and McManus, J. (1992). Origin of the northern Atlantic's Heinrich events, *Climate Dynamics*, **6**, 265–273.

Broecker, W.S. and Broecker, S. (1974). Carbonate dissolution on the eastern flank of the East Pacific Rise. In: *Studies in Paleo-oceanography* (W.W. Hay, ed.). Special Publication No. 20, Society of Economic Paleontologists and Mineralogists, Tulsa, 44–58.

Broecker, W.S., and Denton, G.H. (1989). The role of ocean-atmosphere reorganizations in glacial cycles. *Geochimica et Cosmochimica Acta*, **53**, 2465–2501.

Broecker, W.S., Kennett, J.P., Flower, B.P., Teller, J.T., Trumbore, S., Bonani, G., and Wolfli, W. (1989). Routing of meltwater from the Laurentide Ice Sheet during the Younger Dryas cold episode. *Nature*, **341**, 318–320.

Broecker, W.S., Peng, T-H., Jouzel, J., and Russell, G. (1990b). The magnitude of global fresh-water transports of importance to ocean circulation. *Climate Dynamics*, **4**, 73–79.

Broecker, W.S., Peng, T-H., Ostlund, G., and Stuiver, M. (1985a). The distribution of bomb radiocarbon in the ocean. *Jour. Geophys. Research*, **90**, 6953–6970.

Broecker, W.S., Peteet, D., and Rind, D. (1985b). Does the ocean-atmosphere have more than one stable mode of operation? *Nature*, **315**, 21–25.

Broecker, W.S., Thurber, D.L., Goddard, J., Ku, T-L., Matthews, R.K., and Mesolella, K.J. (1968). Milankovitch hypothesis supported by precise dating of coral reefs and deep-sea sediments. *Science*, **159**, 297–300.

Broecker, W.S. and van Donk, J. (1970). Insolation changes, ice volumes and the ^{18}O record in deep-sea cores. *Reviews of Geophysics and Space Physics*, **8**, 169–198.

Bromwich, D.H. and Weaver, C.J. (1983). Latitudinal displacement of the main moisture source controls $\delta^{18}O$ of snow in coastal Antarctica. *Nature*, **301**, 145–147.

Brook, E.J., Sowers, T., and Orchado, J. (1996). Rapid variations in atmospheric methane concentration during the past 110,000 years. *Science*, **273**, 1087–1091.

Brooks, S.J., Mayle, F.E., and Lowe, J.J. (1997). Chironomid-based Lateglacial climatic reconstruction for southeast Scotland. *Jour. Quaternary Science*, **12**, 161–167.

Brown, I.M. (1990). Quaternary glaciations of New Guinea. *Quaternary Science Reviews*, **9**, 273–280.

Brown, J.S., Colling, A., Park, D., Phillips, J., Rothery, D., and Wright, J. (1989). *Ocean Circulation*. Milton Keynes: The Open University, 238 pp.

Brown, K.S. (1982). Paleoecology and regional patterns of evolution in Neotropical forest butterflies. In: *Biological Diversification in the Tropics* (G.T. Prance, ed.). New York: Columbia University Press, 255–308.

Brown, K.S. and Ab'Saber, A.N. (1979). Ice-age forest refuges and evolution in the Neotropics: correlation of paleoclimatological, geomorphological and pedological data with modern biological endemism. *Paleoclimas*, **5**, 1–30.

Brown, K.S., Sheppard, P.M., and Turner, J.R.G. (1974). Quaternary refugia in tropical America: evidence from race formation in Heliconius butterflies. *Proc. Royal Soc. London, B*, **187**, 369–378.

Brown, T.A., Farwell, G.W., Grootes, P.M., and Schmidt, F.H. (1992). Radiocarbon AMS dating of pollen extracted from peat samples. *Radiocarbon*, **34**, 550–556.

Brunnberg, L. (1995). The Baltic Ice Lake. *Quaternary International*, **28**, 177–178.

Bryson, R.A. (1966). Air masses, streamlines, and the boreal forest. *Geographical Bulletin*, **8**, 228–269.

Bryson, R.A. (1974). A perspective on climatic change. *Science*, **184**, 753–760.

Bryson, R.A., Baerreis, D.A., and Wendland, W.M. (1970). The character of late-glacial and post-glacial climatic changes. In: *Pleistocene and Recent Environments of the Central Great Plains* (W. Dort, Jr. and J.K. Jones, eds.). Lawrence: University of Kansas Press, 53–74.

Bryson, R.A., Irving, W.N., and Larsen, J.A. (1965). Radiocarbon and soil evidence of former forest in the southern Canadian forest. *Science*, **147**, 46–48.

Bryson, R.A. and Murray, T.J. (1977). *Climates of Hunger*. Madison, Wisc: University of Wisconsin Press.

Bryson, R.A. and Wendland, W.M. (1967). Tentative climatic patterns for some late-glacial and post-glacial episodes in central North America. In: *Life, Land and Water* (W.J. Mayer-Oakes, ed.). Winnipeg: University of Manitoba Press, 271–298.

Bucha, V. (1970). Influence of the Earth's magnetic field on radiocarbon dating. In: *Radiocarbon Variations and Absolute Chronology* (I.U. Olsson, ed.). New York: Wiley, 501–511.

Budd, W.F. and Smith, I.N. (1981). The growth and retreat of ice sheets in response to orbital radiation changes. In: *Sea level, Ice and Climatic Change*, Publication No. 131. International Association of Scientific Hydrology, Washington, DC, 369–409.

Budyko, M.I. (1969). The effect of solar radiation variations on the climate of the earth. *Tellus*, **21**, 611–619.

Budyko, M.I. (1978). The heat balance of the earth. In: *Climatic Change* (J. Gribbin, ed.) Cambridge: Cambridge University Press, 85–113.

Burbank, D.W. (1981). A chronology of Late Holocene glacier fluctuations on Mount Rainier, Washington. *Arctic and Alpine Research*, **13**, 369–381.

Burga, C. (1988). Swiss vegetation history during the past 18,000 years. *New Phytologist*, **110**, 581–602.

Burk, R.L. and Stuiver, M. (1981). Oxygen isotope ratios in trees reflect mean annual temperature and humidity. *Science*, **211**, 1417–1419.

Burney, D.A., Brook, G.A., and Cowart, J.B. (1994). A Holocene pollen record for the Kalahari Desert of Botswana from a U-series dated speleothem. *The Holocene*, **4**, 225–232.

Burrows, C.J. and Burrows, V.L. (1976). Procedures for the study of snow avalanche chronology using growth layers of woody plants. University of Colorado INSTAAR Occasional Paper No. 23.

Bush, M.B. and Colinvaux, P.A. (1990). A pollen record of a complete glacial cycle from lowland Panama. *Jour. Vegetation Science,* **1,** 105–118.

Bush, M.B., Colinvaux, P.A., Weimann, M.C., Piperno, D.R., and Liu, K-B. (1990). Late Pleistocene temperature depression and vegetation change in Ecuadorian Amazonia. *Quaternary Research,* **34,** 330–345.

Butzer, K.W., Isaac, G.L., Richardson, J.L., and Washbourn-Kamau, C. (1972). Radiocarbon dating of East African lake levels. *Science,* **175,** 1069–1076.

Calkin, P.E. and Ellis, J.M. (1980). A lichenometric dating curve and its application to Holocene glacier studies in the Central Brooks Range, Alaska. *Arctic and Alpine Research,* **12,** 245–264.

Cande, S.C. and Kent, D.V. (1992). A new geomagnetic polarity time scale for the late Cretaceous and Cenozoic. *Jour. Geophys. Research,* **97B,** 13917–13951.

Cande, S.C., and Kent, D.V. (1995). Revised calibration of the geomagnetic polarity time scale for the late Cretaceous and Cenozoic. *Jour. Geophys. Research,* **100,** 6093–6095.

Carrara, P.E. (1979). The determination of snow avalanche frequency through tree-ring analysis and historical records at Ophir, Colorado. *Geological Society of America Bulletin,* **1,** 90, 773–780.

Carrara, P.E., Trimble, D.A., and Rubin, M. (1991). Holocene treeline fluctuations in the northern San Juan mountains, Colorado, U.S.A., as indicated by radiocarbon-dated conifer wood. *Arctic and Alpine Research,* **23,** 233–246.

Carriquiry, J.D., Risk, M.J., and Schwarcz, H.P. (1994). Stable isotope geochemistry of corals from Costa Rica as proxy indicator of El Niño-Southern Oscillation (ENSO). *Geochimica et Cosmochimica Acta,* **58,** 335–352.

Catchpole, A.J.W. (1992). Hudson's Bay Company ships' log-books as sources of sea ice data, 1751–1870. In: *Climate Since A.D. 1500* (R.S. Bradley and P.D. Jones, eds.). London: Routledge, 17–39.

Catchpole, A.J.W., Moodie, D.W., and Milton, D. (1976). Freeze-up and break-up of estuaries on Hudson Bay in the 18th and 19th centuries. *Canadian Geographer,* **20,** 279–297.

Cerling, T.E. and Quade, J. (1993). Stable carbon and oxygen isotopes in soil carbonates. In: *Climate Change in Continental Isotopic Records* (P.K. Swart, K.C. Lohmann, J. McKenzie and S. Savin, eds.). Washington, DC: American Geophysical Union, 217–231.

Cess, R.D. and others, (1995). Absorption of solar radiation by clouds: observations versus models. *Science,* **267,** 496–499.

Chapman, M.R., Shackleton, N.J., Zhao, M., and Eglinton, G. (1996). Faunal and alkenone reconstructions of sub-tropical North Atlantic surface hydrography and paleotemperature over the last 28 kyrs. *Paleoceanography,* **11,** 343–358.

Chappell, J.M.A. and Polach, H.A. (1972). Some effects of partial recrystallisation on ^{14}C dating of late Pleistocene corals and molluscs. *Quaternary Research,* **2,** 244–252.

Chappell, J.M.A. and Shackleton, N.J. (1986). Oxygen isotopes and sea level. *Nature,* **324,** 137–138.

Chappellaz, J., Barnola, J.M., Raynaud, D., Korotkevich, Y.S., and Lorius, C. (1990). Ice core record of atmospheric methane over the last 160,000 years. *Nature,* **345,** 127–131.

Chappellaz, J., Blunier, T., Raynaud, D., Barnola, J.M., Schwander, J., and Stauffer, B. (1993). Synchronous changes in atmospheric CH_4 and Greenland climate between 40 and 8 kyr B.P. *Nature,* **366,** 443–445.

Chappellaz, J., Brook, E., Blunier, T., and Malaizé, B. (1997). CH_4 and $\delta^{18}O$ of O_2 records from Antarctica and Greenland ice: a clue for stratigraphic disturbance in the bottom part of the GRIP and GISP2 ice cores. *Jour. Geophys. Research,* **102C,** 26547–26557.

Charles, C.D. and Fairbanks, R.G. (1992). Evidence from Southern Ocean sediments for the effect of North Atlantic deepwater flux on climate. *Nature,* **355,** 416–419.

Charles, C.D., Rind, D., Jouzel, J., Koster, R.D., and Fairbanks, R.G. (1994). Glacial-interglacial changes in moisture sources for Greenland: influences on the ice core record of climate. *Science,* **263,** 508–511.

Charlson, R.J., Lovelock, J.E., Andreae, M.O., and Warren, S.G. (1987). Oceanic phytoplankton, atmospheric sulfur, cloud albedo and climate. *Nature, 326,* 655–661.

Charney, J.G., Stone, P.H., and Quirk, W.J. (1975). Drought in the Sahara: a biogeophysical feedback mechanism. *Science, 187,* 434–435.

Chen, C. (1968). Pleistocene pteropods in pelagic sediments. *Nature, 219,* 1145–1147.

Chen, J., Farrell, J.W., Murray, D.W., and Prell, W.L. (1995). Timescale and paleoceanographic implications of a 3.6 m.y. oxygen isotope record from the northeast Indian Ocean (Ocean Drilling Program site 758). *Paleoceanography, 10,* 21–47.

Chervin, R.M. and Semtner, A.J. (1991). Modeling the ocean with supercomputers: the key to simulating and understanding past and future climates. In: *Global Changes of the Past* (R.S. Bradley, ed.). Boulder: University Corporation for Atmospheric Research, 477–488.

Chester, R. and Johnson, L.R. (1971). Atmospheric dusts collected off the Atlantic coasts of North Africa and the Iberian Peninsula. *Marine Geology, 11,* 251–260.

Chinn, T.J.H. (1981). Use of rock weathering-rind thickness for Holocene absolute age–dating in New Zealand. *Arctic and Alpine Research, 13,* 33–45.

Chivas, A.R., De Decker, P., Cali, J.A., Chapman, A., Kiss, E., and Shelley, J.M.G. (1993). Coupled stable-isotope and trace-element measurements of lacustrine carbonates as paleoclimatic indicators. In: *Climatic Change in Continental Isotopic Records* (P.K. Swart, K.C. Lohmann, J. McKenzie, and S. Savin, eds.). Washington, DC: American Geophysical Union, 113–121.

Chu, K'o-chen (1926). Climate pulsations during historical times in China. *Geographical Review, 16,* 274–282.

Chu, K'o-chen (1973). A preliminary study on the climatic fluctuations during the last 5000 years in China. *Scientia Sinica, 16,* 226–256.

Ciais, P., Petit, J.R., Jouzel, J., Lorius, C., Barkov, N.I., Lipenkov, V., and Nicolaïev, V. (1992). Evidence for an early Holocene climatic optimum in the Antarctic deep ice-core record. *Climate Dynamics, 6,* 169–177.

Ciais, P., Tans, P.P., Trolier, M., White, J.W.C., and Francey, R.J. (1995). A large northern hemisphere terrestrial CO_2 sink indicated by $^{13}C/^{12}C$ of atmospheric CO_2. *Science,* 1098–1102.

Clague, J.J. and Mathewes, R.W. (1989). Early Holocene thermal maximum in western North America: new evidence from Castle Peak, British Columbia. *Geology, 17,* 277–280.

Clague, J.J., Mathewes, R.W., Buhay, W.M., and Edwards, T.W.D. (1992). Early Holocene climate at Castle Peak, British Columbia, Canada. *Palaeogeography, Palaeoclimatology, Palaeoecology, 95,* 153–167.

Clapperton, C.M. (1990). Quaternary glaciations in the southern hemisphere: an overview. *Quaternary Science Reviews, 9,* 299–304 (plus chart).

Clapperton, C.M. (1993a). *Quaternary Geology and Geomorphology of South America.* Amsterdam: Elsevier.

Clapperton, C.M. (1993b). Nature of environmental changes in South America at the Last Glacial Maximum. *Palaeogeography, Palaeoclimatology, Palaeoecology, 101,* 189–208.

Clark, D. (1975). *Understanding Canonical Correlation Analysis. Concepts and Techniques in Modern Geography No. 3.* Norwich: University of East Anglia.

Clark, P.U., Alley, R.B., Keigwin, L.D., Licciardi, J.M., Johnsen, S.J., and Wang, H. (1996). Origin of the first global meltwater pulse following the last glacial maximum. *Paleoceanography, 11,* 563–577.

Clark, P.U., Clague, J.J., Curry, B.B., Dreimanis, A., Hicock, S.R., Miller, G.H., Berger, G.W., Eyles, N., Lamothe, M., Miller, B.B., Mott, R.J., Oldale, R.N., Stea, R.R., Szabo, J.P., Thorleifson, L.H., and Vincent, J.-S. (1993). Initiation and development of the Laurentide and Cordilleran ice sheets following the last interglaciation. *Quaternary Science Reviews, 12,* 79–114.

Clarke, M.L., Wintle, A.G., and Lancaster, N. (1996). Infra-red stimulated luminescence dating of sands from the Cronese Basins, Mojave Desert. *Geomorphology, 17,* 199–205.

Clausen, H.B., Gundestrup, N.S., and Johnsen, S.J. (1988). Glaciological investigations in the Crête area, central Greenland: a search for a new deep-drilling site. *Annals of Glaciology,* **10**, 10–15.

Clausen, H.B. and Hammer, C.U. (1988). The Laki and Tambora eruptions as revealed in Greenland ice cores from 11 locations. *Annals of Glaciology*, **10**, 16–22.

Cleaveland, M.K., Cook, E.R., and Stahle, D.W. (1992). Secular variability of the Southern Oscillation detected in tree-ring data from Mexico and the southern United States. In: *El Niño: Historical and Paleoclimatic Aspects of the Southern Oscillation* (H.F. Diaz and V. Markgraf, eds.).Cambridge: Cambridge University Press, 271–291.

Cleaveland, M.K. and Duvick, D.N. (1992). Iowa climate reconstructed from tree-rings, 1640–1982. *Water Resources Research,* **28**, 2607–2615.

CLIMAP Project Members (1976). The surface of the ice-age Earth. *Science,* **191**, 1131–1144.

CLIMAP Project Members (1981). Seasonal reconstructions of the Earth's surface at the last glacial maximum. *Geological Society of America, Map Chart Series*, MC-36.

CLIMAP Project Members (1984). The last interglacial ocean. *Quaternary Research*, **21**, 123–224.

Coale, K.H., Johnson, K.S., Fitzwater, S.E., Gordon, R.M., Tanner, S., Chavez, F.P., Ferioli, L., Sakamoto, C., Rogers, P., Millero, F., Steinberg, P., Nightingale, P., Cooper, D., Cochlan, W.P., Landry, M.R., Constantinou, J., Rollwagen, G., Trasvina, A., and Kudela, R. (1996). A massive phytoplankton bloom induced by an ecosystem-scale iron fertilization experiment in the equatorial Pacific Ocean. *Nature*, **383**, 495–501.

Coe, R.S. and Liddicoat, J.C., (1994). Overprinting of natural magnetic remanence in lake sediments by a subsequent high-intensity field. *Nature*, **367**, 57–59.

COHMAP Members (1988). Climatic changes of the last 18,000 years: observations and model simulations. *Science,* **241**, 1043–1052.

Cole, J.E. and Fairbanks, R.G. (1990). The Southern Oscillation recorded in the oxygen isotopes of corals from Tarawa Atoll. *Paleoceanography,* **5**, 669–683.

Cole, J.E., Fairbanks R.G., and Shen, G.T. (1993). Recent variability in the Southern Oscillation: isotopic results from a Tarawa atoll coral. *Science*, **260**, 1790–1793.

Cole, J.E., Shen, G.T., Fairbanks, R.G., and Moore, M. (1992). Coral monitors of El Niño/Southern Oscillation dynamics across the equatorial Pacific. In: *El Niño: Historical & Paleoclimatic Aspects of the Southern Oscilllation.* (H.F. Diaz and V.K. Markgraf, eds.). Cambridge: Cambridge University Press, 349–376.

Cole, K.L. (1990). Late Quaternary vegetation gradients through the Grand Canyon. In: *Packrat Middens: The Last 40,000 Years of Biotic Change* (J.L. Betancourt, T.R. Van Devender, and P. S. Martin, eds.). Tucson: University of Arizona Press, 240–258.

Cole-Dai, J., Thompson, L.G., and Mosley-Thompson, E. (1995). A 485 year record of atmospheric chloride, nitrate and sulfate: results of chemical analysis of ice cores from Dyer Plateau, Antarctic Peninsula. *Annals of Glaciology,* **21**, 182–188.

Colinvaux, P.A. (1996). Quaternary environmental history and forest diversity in the Neotropics. In: *Environmental and Biological Change in Neogene and Quaternary Tropical America* (J. Jackson and A. Coates, eds.). Chicago: Chicago University Press.

Colinvaux, P.A., De Oliveira, P.E., Moreno, J.E., Miller, M.C., and Bush, M.B. (1996a). A long pollen record from lowland Amazonia: forest and cooling in Glacial times. *Science,* **274**, 85–88.

Colinvaux, P.A., Liu, K-B., De Oliveira, P., Bush, M.B., Miller, M.C., and Steinitz-Kannon, M. (1996b). Temperature depression in the lowland Tropics in Glacial time. *Climatic Change,* **32**, 19–33.

Colman, S.M. and Dethier, D.P. (eds.) (1986). *Rates of Chemical Weathering of Rocks and Minerals.* Orlando: Academic Press.

Colman, S.M. and Pierce, K.L. (1981). Weathering rinds on andesitic and basaltic stones as a Quaternary age indicator, western United States. *U.S. Geological Survey Prof. Paper 1210.*

Colman, S.M., Pierce, K.L., and Birkeland, P.W. (1987). Suggested terminology for Quaternary dating methods. *Quaternary Research,* **28**, 314–319.

Condomes, M., Moraud, P., Camus, G., and Duthon, L. (1982). Chronological and geochemical study of lavas from the Chaine des Puys, Massif Central, France: evidence for crustal contamination. *Contributions to Mineralogy and Petrology*, 81, 296–303.

Cook, E.R., Bird, T., Peterson, M., Barbetti, M., Buckley, B.M., D'Arrigo, R. and Francey, R. (1992a). Climatic change over the last millennium in Tasmania reconstructed from tree-rings. *The Holocene*, 2, 205–217.

Cook, E.R., Briffa, K.R., and Jones, P.D. (1994). Spatial regression methods in dendroclimatology: a review and comparison of two techniques. *International Jour. of Climatology*, 14, 379–402.

Cook, E.R., Briffa, K.R., Meko, D.M., Graybill, D.A., and Funkhauser, G. (1995). The 'segment length curse' in long tree-ring chronology development for paleoclimatic studies. *The Holocene*, 5, 229–237.

Cook, E.R., Briffa, K., Shiyatov, S., and Mazepa, V. (1990). Tree-ring standardization and growth-trend estimation. In: *Methods of Dendrochronology: Applications in the Environmental Sciences* (E.R. Cook and L.A. Kariukstis, eds.). Dordrecht: Kluwer, 104–123.

Cook, E.R. and Jacoby, G.C. (1979). Evidence for the quasi-periodical July drought in the Hudson Valley, New York. *Nature*, 282, 390–392.

Cook, E.R. and Jacoby, G.C. (1983). Potomac River streamflow since 1730 as reconstructed by tree rings. *Jour. Climate and Applied Meteorology*, 22, 1659–1672.

Cook, E.R. and Peters, K. (1981). The smoothing spline: a new approach to standardizing forest-interior tree-ring width series for denroclimatic studies. *Tree Ring Bulletin*, 41, 45–53.

Cook, E.R., Stahle, D.W., and Cleaveland, M.K. (1992b). Dendroclimatic evidence from eastern North America. In: *Climate Since A.D. 1500* (R.S. Bradley and P.D. Jones, eds.). London: Routledge, 331–348.

Cooke, R.U. and Warren, A. (1977). *Geomorphology in Deserts*. Berkeley: University of California Press.

Coope, G.R. (1959). A late Pleistocene insect fauna from Chelford, Cheshire. *Proc. Royal Soc. London, B*, 151, 70–86.

Coope, G.R. (1967). The value of Quaternary insect faunas in the interpretation of ancient ecology and climate. In: *Quaternary Paleoecology* (E.J. Cushing and H.E. Wright, eds.). New Haven: Yale University Press, 359–380.

Coope, G.R. (1974). Interglacial coleoptera from Bobbitshole, Ipswich. *Jour. Geological Society of London*, 130, 333–340.

Coope, G.R. (1975a). Climatic fluctuations in north-west Europe since the last Interglacial, indicated by fossil assemblages of coleoptera. In: *Ice Ages Ancient and Modern,* (A.E. Wright and F. Moseley, eds). *Geological Journal Special Issue No. 6*. Liverpool: Liverpool University Press, 153–168.

Coope, G.R. (1975b). Mid-Weichselian climatic changes in western Europe, reinterpreted from coleopteran assemblages. *Bulletin of the Royal Society of New Zealand*, 13, 101–110.

Coope, G.R. (1977a). Quaternary coleoptera as aids in the interpretation of environmental history. In: *British Quaternary Studies,* (F.W. Shotton, ed.). Oxford: Clarendon Press, 55–68.

Coope, G.R. (1977b). Fossil coleopteran assemblages as sensitive indicators of climatic changes during the Devensian (Last) cold stage. *Proc. Royal Soc. London, B*, 280, 313–337.

Coope, G.R. (1994). The response of insect faunas to glacial-interglacial climatic fluctuations. *Phil. Trans. Royal Society of London*, 344B, 19–26.

Coope, G.R. and Brophy, J.A. (1972). Late glacial environmental changes indicated by a coleopteran succession from North Wales. *Boreas*, 1, 97–142.

Coope, G.R. and Pennington, W. (1977). The Windermere interstadial of the late Devensian. *Phil. Trans. Royal Society of London*, B208, 337–339.

Coplen, T.B. (1996). More uncertainty than necessary. *Paleoceanography*, 11, 369–370.

Cortijo, E., Duplessy, J.C., Labeyrie, L., Leclaire, H., Duprat, J., and van Weering, T.C.E. (1994). Eemian cooling in the Norwegian Sea and North Atlantic Ocean preceding continental ice sheet growth. *Nature*, **372**, 446–449.

Cox, A. (1969). Geomagnetic reversals. *Science*, *163*, 237–245.

Craig, H. (1953). The geochemistry of the stable carbon isotopes. *Geochimica et Cosmochimica Acta*, **3**, 53–92.

Craig, H. (1957). Isotopic standards for carbon and oxygen and correction factors for mass spectrometric analysis of CO_2. *Geochimica et Cosmochimica Acta*, *12*, 133–149.

Craig, H. (1961a). Mass spectrometer analysis of radiocarbon standards. *Radiocarbon*, *3*, 1–3.

Craig, H. (1961b). Standard for reporting concentrations of deuterium and oxygen-18 in natural waters. *Science*, *133*, 1833–1834.

Craig, H. (1961c). Isotopic variations in meteoric waters. *Science*, *133*, 1702–1703.

Craig, H. (1965). The measurement of oxygen isotope paleotemperature. In: *Proceedings of the Spoleto Conference on Stable Isotopes in Oceanographic Studies and Paleotemperatures*. Consiglio Nazionale delle Ricerche Laboratoriodi Geologia Nucleare, Pisa, 3–24.

Croll, J. (1867a). On the eccentricity of the Earth's orbit, and its physical relations to the glacial epoch. *Philosophical Magazine*, *33*, 119–131.

Croll, J. (1867b). On the change in the obliquity of the ecliptic, its influence on the climate of the polar regions and on the level of the sea. *Philosophical Magazine*, *33*, 426–445.

Croll, J. (1875). *Climate and Time*. New York: Appleton and Co.

Crowley, T.J. (1994). Pleistocene temperature changes. *Nature*, **371**, 664.

Crowley, T.J. and Baum, S.K. (1997). Effect of vegetation on an ice-age climate model simulation. *Jour. Geophys. Research*, **102D**, 16463–16480.

Crozaz, G., Picciotto, E., and De Breuck, W. (1964). Antarctic snow chronology with Pb-210. *Jour. Geophys. Research*, **69**, 2597–2604.

Crozaz, G., Langway, Jr., C.C., and Picciotto, E. (1966). Artificial radioactivity reference horizons in Greenland firn. *Earth and Planetary Science Letters*, **1**, 42–48.

Cuffey, K.M., Alley, R.B., Grootes, P.M., Bolzan, J.M., and Anandakrishnan, S. (1994). Calibration of the $\delta^{18}O$ isotopic paleothermometer for central Greenland, using borehole temperatures. *Jour. Glaciology*, **40**, 341–350.

Cuffey, K.M. and Clow, G.D. (1997). Temperature, accumulation and ice sheet elevation in central Greenland through the last deglacial transition. *Jour. Geophys. Research*, **102C**, 26383–26396.

Cuffey, K.M., Clow, G.D., Alley, R.B., Stuiver, M., Waddington, E.D., and Saltus, R.W. (1995). Large Arctic temperature change at the Wisconsin-Holocene transition. *Science*, **270**, 455–458.

Cunningham, J. and Waddington, E.D. (1990). Boudinage: a source of stratigraphic disturbance in glacial ice in central Greenland. *Jour. Glaciology*, **36**, 269–272.

Curry, R.R. (1969). Holocene climatic and glacial history of the Sierra Nevada, California. *Geological Society of America Special Paper*, No. 123, 1–47.

Curry, W.B., Duplessy, J-C., Labeyrie, L.D., and Shackelton, N.J. (1988). Quaternary deep water circulation changes in the distribution of $\delta^{13}C$ of Deep water ΣCO_2 between the last glaciation and the Holocene. *Paleoceanography*, **3**, 317–342.

Curtis, G.H. (1975). Improvements in potassium-argon dating, 1962–1975. *World Archaeology*, **7**, 198–207.

Cushing, E.J. (1967). Evidence for differential pollen preservation in late Quaternary sediments in Minnesota. *Reviews of Palaeobotany and Palynology*, **4**, 87–101.

Cwynar, L.C. and Levesque, A.J. (1995). Chironomid evidence for late-glacial climatic reversals in Maine. *Quaternary Research*, **43**, 405–413.

Dahl, S.O. and Nesje, A. (1996). A new approach to calculating Holocene winter precipitation by combining glacier equilibrium line altitudes and pine-tree limits: a case study from Hardangerjøkulen, central south Norway. *The Holocene*, **6**, 381–398.

Dahl-Jensen, D., Johnsen, S.J., Hammer, C.U., Clausen, H.B., and Jouzel, J. (1993). Past accumulation rates derived from observed annual layers in the GRIP ice core from Summit, Greenland. In: *Ice in the Climate System* (W. Peltier, ed.). Berlin: Springer-Verlag, 517–532.

Dalrymple, G.B. and Lanphere, M.A. (1969). *Potassium-Argon Dating: Principles, Techniques and Applications to Geochronology.* San Francisco: W.H. Freeman.

Damon, P.E. (1970). Radiocarbon as an example of the unity of science. In: *Radiocarbon Variations and Absolute Chronology* (I.U. Olsson, ed.). New York: Wiley, 641–644.

Damon, P.E., Lerman, J.C., and Long, A. (1978). Temporal fluctuations of atmospheric ^{14}C: causal factors and implications. *Annual Reviews of Earth and Planetary Science,* **6,** 457–494.

Dansgaard, W. (1961). The isotopic composition of natural waters with special reference to the Greenland Ice Cap. *Meddelelser øm Grønland,* **165,** 1–120.

Dansgaard, W. (1964). Stable isotopes in precipitation. *Tellus,* **16,** 436–468.

Dansgaard, W. and Johnsen, S.J. (1969). A flow model and a time scale for the ice core from Camp Century. *Jour. Glaciology,* **8,** 215–223.

Dansgaard, W., Clausen, H.B., Gundestrup, N., Hammer, C.U., Johnsen, S.F., Kristinsdottir, P.M., and Reeh, N. (1982). A new Greenland deep ice core. *Science,* **218,** 1273–1277.

Dansgaard, W., Johnsen, S.J., Clausen, H.B., Dahl-Jensen, D., Gundestrup, N.S., Hammer, C.U., Hvidberg, C.S., Steffensen, J.P., Sveinbjornsdottir, A.E., Jouzel, J., and Bond, G. (1993). Evidence for general instability of past climate from a 250-kyr ice-core record. *Nature,* **364,** 218–220.

Dansgaard, W., Johnsen, S.J., Clausen, H.B., Dahl-Jensen, D., Gundestrup, N., Hammer, C.U., and Oeschger, H. (1984). North Atlantic climatic oscillations revealed by deep Greenland ice cores. In: *Climate Processes and Climate Sensitivity* (J.E. Hansen and T. Takahashi, eds.). Washington, DC: American Geophysical Union, 288–298.

Dansgaard, W., Johnsen, S.J., Clausen, H.B., and Gundestrup, N. (1973). Stable isotope glaciology. *Meddelelser øm Grønland,* **197,** 1–53.

Dansgaard, W., Johnsen, S.J., Clausen, H.B., and Langway, C.C. (1971). Climatic record revealed by the Camp Century ice core. In: *The Late Cenozoic Ice Ages* (K.K. Turekian, ed.). New Haven: Yale University Press, 37–56.

Dansgaard, W., Johnsen, S.J., Moller, J., and Langway, Jr., C.C. (1969). One thousand centuries of climatic record from Camp Century on the Greenland ice sheet. *Science,* **166,** 377–381.

Dansgaard, W. and Tauber, H. (1969). Glacier oxygen-18 content and Pleistocene ocean temperatures. *Science,* **166,** 499–502.

Dansgaard, W., White, J.W.C., and Johnsen, S.J. (1989). The abrupt termination of the Younger Dryas climate event. *Nature,* **339,** 532–534.

D'Arrigo, R.D. and Jacoby, G.C. (1992). Dendroclimatic evidence from northern North America. In: *Climate Since A.D. 1500* (R.S. Bradley and P.D. Jones, eds.). London: Routledge, 296–311.

D'Arrigo, R.D., Jacoby, G.C., and Free, R.M. (1992). Tree-ring width and maximum latewood density at the North American treeline: parameters of climatic change. *Canadian Jour. Forest Research,* **22,** 1290–1296.

D'Arrigo, R.D., Jacoby, G.C., and Krusic, P.J. (1994). Progress in dendroclimatic studies in Indonesia. *Terrestrial, Atmospheric and Oceanic Sciences,* **5,** 349–363.

Daultrey, S. (1976). *Principal Components Analysis. Concepts and Techniques in Modern Geography No. 8.* Norwich: University of East Anglia.

Davis, M.B. (1963). On the theory of pollen analysis. *American Journal of Science,* **261,** 899–912.

Davis, M.B. (1973). Redeposition of pollen grains in lake sediment. *Limnology and Oceanography,* **18,** 44–52.

Davis, M.B. (1991). Research questions posed by the paleoecological record of global change. In: *Global Changes of the Past* (R.S. Bradley, ed.). Boulder: University Corporation for Atmospheric Research, 385–396.

Davis, M.B. and Botkin, D.B. (1985). Sensitivity of cool-temperate forests and their fossil pollen record to rapid temperature change. *Quaternary Research, 23,* 327–340.

Davis, M.B., Brubaker, L.B., and Webb, III, T. (1973). Calibration of absolute pollen influx. In: *Quaternary Plant Ecology* (H.J.B. Birks and R.G. West, eds.). Oxford: Blackwell Scientific Publications, 9–25.

Davis, M.B., Woods, K.D., Webb, S.L., and Futyua, R.B. (1986). Dispersal versus climate: expansion of *Fagus* and *Tsuga* into the upper Great Lakes region. *Vegetatio, 69,* 93–103.

Davis, R.B. (1974). Stratigraphic effects of tubificids in profundal lake sediments. *Limnology and Oceanography, 19,* 466–488.

Davis, R.B. and Webb, III, T. (1975). The contemporary distribution of pollen in eastern North America: a comparison with the vegetation. *Quaternary Research, 5,* 395–434.

De Angelis, M., Barkov, N.I., and Petrov, V.N. (1987). Aerosol concentrations over the last climatic cycle (160 kyr) from an Antarctic ice core. *Nature, 325,* 318–321.

De Decker, P. and Forester, R.M. (1988). The use of ostracods to reconstruct continental palaeoenvironmental records. In: *Ostracoda in the Earth Sciences* (P. De Decker, J-P. Colin, and J-P Peypouquet, eds.). Amsterdam: Elsevier, 175–199.

de Jong, A.F.M., Mook, W.G., and Becker, B. (1980). Confirmation of the Suess wiggles, 3200–3700 BP. *Nature, 280,* 48–49.

de Martonne, E. and Aufrere, L. (1928). L'extension des regions privees d'ecoulement vers l'ocean. *Annales de Géographie, 38,* 1–24.

de Vernal, A., Hillaire-Marcel, C., and Bilodeau, G. (1996). Reduced meltwater outflow from the Laurentide ice margin during the Younger Dryas. *Nature, 381,* 774–777.

de Vernal, A., Londeix, L., Mudie, P.J., Harland, R., Morzadec-Kerfourn, M.T., Turon, J-L., and Wrenn, J.H. (1992). Quaternary organic-walled dinoflagellate cysts of the North Atlantic Ocean and adjacent Seas: ecostratigraphy and biostratigraphy. In: *Neogene and Quaternary Dinoflagellate Cysts and Acritarchs* (M.J. Head and J.H. Wrenn, eds.). Dallas: American Association of Stratigraphic Palynologists Foundation, 289–328.

de Vernal, A., Rochon, A., Hillaire-Marcel, C., Thuron, J-L., and Guiot, J. (1993). Quantitative reconstruction of sea-surface conditions, seasonal extent of sea-ice cover and meltwater discharges in high latitude marine environments from dinoflagellate cyst assemblages. In: *Ice in the Climate System* (W.R. Peltier, ed.). Berlin: Springer-Verlag, 611–621.

de Vernal, A., Rochon, A., Thuron, J-L., and Matthiessen, J. (1998). Organic-walled dinoflagellate cysts: palynological tracers of sea-surface conditions in middle to high latitiude marine environments. *GEOBIOS, 30,* 905–920.

de Vernal, A., Thuron, J-L., and Guiot, J. (1994). Dinoflagellate cyst distribution in high latitude marine environments and quantitative reconstruction of sea-surface salinity, temperature and seasonality. *Canadian Jour. Earth Sciences, 31,* 48–62.

de Villiers, S., Nelson, B.K., and Chivas, A.R. (1995). Biological controls on coral Sr/Ca and $\delta^{18}O$ reconstructions of sea surface temperatures. *Science, 269,* 1247–1249.

de Vries, H.L. (1958). Variation in concentration of radiocarbon with time and location on earth. *Proc. Konikl Nederlandse Akademie Wetensch, B61,* 94–102.

de Vries, J. (1977). Histoire du climat et economie: des faits nouveaux, une interpretation differente. *Annales: Economie, Sociétés, Civilizations, 32,* 198–228.

de Vries, J. (1981). Measuring the impact of climate on history: the search for appropriate methodologies. In: *Climate and History: Studies in Interdisciplinary History* (R.I. Rotberg and T.K. Rabb, eds.). Princeton: Princeton University Press, 19–50.

Dean, J.S., Meko, D.M., and Swetnam, T.W. (eds.). (1996). *Tree Rings, Climate and Humanity.* Tucson: Department of Geosciences, University of Arizona.

Deevey, E.S. and Flint, R.F. (1957). Postglacial Hypsithermal interval. *Science, 125,* 1824.

Defant, A. (1961). *Physical Oceanography, 1.* New York: Pergamon/Macmillan.

Delano Smith, C. and Parry, M. (eds.). (1981). *Consequences of Climatic Change.* Nottingham: Department of Geography, University of Nottingham.

Delcourt, P.A., Delcourt, H.R., and Webb, III, T. (1984). Atlas of mapped distributions of dominance and modern pollen percentages for important tree taxa of Eastern North America. *American Association of Stratigraphy Palynologists,* Contribution Series 14. Dallas.

Delmas, R.J. (1992). Environmental information from ice cores. *Reviews of Geophysics,* **30,** 1–21.

Delmas, R.J., Kirchner, S., Palais, J.M., and Petit, J.R. (1992). 1000 years of explosive volcanism recorded at the South Pole. *Tellus,* **44B,** 335–350.

Delmas, R.J., Legrand, M., Aristarain, A.J., and Zanolini, F. (1985). Volcanic deposits in Antarctic snow and ice. *Jour. Geophys. Research,* **90D,** 12901–12920.

Delmas, R.J. and Petit, J.R. (1994). Present Antarctic aerosol composition: a memory of ice age atmospheric dust. *Geophysical Research Letters,* **21,** 879–882.

Denton, G.H. and Hughes, T.J. (eds.). (1981). *The Last Great Ice Sheets.* New York: Wiley.

Denton, G.H. and Karlén, W. (1973a). Holocene climatic variations—their pattern and possible cause. *Quaternary Research,* **3,** 155–205.

Denton, G.H. and Karlén, W. (1973b). Lichenometry: its application to Holocene moraine studies in southern Alaska and Swedish Lapland. *Arctic and Alpine Research,* **5,** 347–372.

Denton, G.H. and Karlén, W. (1977). Holocene glacial and tree-line variations in the White River Valley and Skolai Pass, Alaska and Yukon Territory. *Quaternary Research,* **7,** 63–111.

Deuser, W.G. and Ross, E.H. (1989). Seasonally abundant planktonic foraminifera of the Sargasso Sea: succession, deep-water fluxes, isotopic compositions, and paleoceanographic implications, *Jour. Foraminiferal Research,* **19,** 268–293.

Devine, J.D., Sigurdsson, H., Davis, A.N., and Self, S. (1984). Estimates of sulfur and chlorine yield to the atmosphere from volcanic eruptions and potential climatic effects. *Jour. Geophys. Research,* **89B,** 6309–6325.

Dial, N.P. and Czaplewski, N.J. (1990). Do woodrat middens accurately represent the animals' environments and diets? The Woodhouse Mesa Study. In: *Packrat Middens: The Last 40,000 Years of Biotic Change* (J.L. Betancourt, T.R. Van Devender, and P.S. Martin, eds.). Tucson: University of Arizona Press, 43–58.

Diaz, H.F. and Kiladis, G. (1992). Atmospheric teleconnections associated with the extreme phase of the Southern Oscillation. In: *El Niño: Historical and Paleoclimatic Aspects of the Southern Oscillation* (H.F. Diaz and V. Markgraf, eds.). Cambridge: Cambridge University Press, 7–28.

Dickinson, R.E., Meleshko, V., Randall, D., Sarachik, E., Silva-Dias, P., and Slingo, A. (1996). Climate Processes. In: *Climate Change 1995: The Science of Climate Change* (J.T. Houghton, L.G. Meirha Filho, B.A. Callendar, A. Kattenberg, and K. Maskell, eds.). Cambridge: Cambridge University Press, 193–227.

Dickson, R.R. and Brown, J. (1994). The production of North Atlantic Deep Water: sources rates, and pathways. *Jour. Geophys. Research,* **99C,** 12319–12341.

Dibb, J.E. and Clausen, H.B. (1997). A 200-year ^{210}Pb record from Greenland. *Jour. Geophys. Research,* **102D,** 4325–4332.

Dimbleby, G.W. (1985). *The palynology of archeological sites.* London: Academic Press.

Ding, Z., Rutter, N., Han, J., and Liu, T. (1992). A coupled environmental system formed at about 2.5 Ma in East Asia. *Palaeogeography, Palaeoclimatology, Palaeoecology,* **94,** 223–242.

Ding, Z., Rutter, N.W., and Liu, T.S. (1993). Pedostratigraphy of Chinese loess deposits and climatic cycles in the last 2.5 Ma. *Catena,* **20,** 73–91.

Ding, Z., Yu, N., Rutter, N.W., and Liu, T. (1994). Towards an orbital time scale for Chinese loess deposits. *Quaternary Science Reviews,* **13,** 39–70.

Dong, B. and Valdes, P.J. (1995). Sensitivity studies of northern hemisphere glaciation using an atmospheric general circulation model. *Jour. Climate,* **8,** 2471–2496.

Dorale, J.A., Gonzales, L.A., Reagan, M.K., Pickett, D.A., Murrell, M.T., and Baker, R.G. (1992). A high-resolution record of Holocene climate change in speleothem calcite from Cold Water Cave, northeast Iowa. *Science,* **258,** 1626–1630.

Douglass, A.E. (1914). A method of estimating rainfall by the growth of trees. In: *The Climatic Factor* (E. Huntingdon, ed.). Washington DC: Publication No. 192, Carnegie Institution of Washington, 101–122.

Douglass, A.E. (1919). *Climatic Cycles and Tree Growth*, **1**, Washington, DC: Carnegie Institution of Washington, Publication No. 289.

Dowdeswell, J.A., Maslin, M.A., Andrews, J.T., and McCave, I.N. (1995). Iceberg production, debris rafting and the extent and thickness of Heinrich layers (H-1, H-2) in North Atlantic sediments, *Geology*, **23**, 301–304.

Dowsett, H.J. and Poore, R.Z. (1990). A new planktic foraminifer transfer function for estimating Plio-Holocene paleoceanographic conditions in the North Atlantic. *Marine Micropaleontology*, **16**, 1–23.

Druffel, E.R.M. and Griffin, S. (1993). Large variations of surface ocean radiocarbon: evidence of circulation changes in the southwestern Pacific. *Jour. Geophys. Research*, **98C**, 20,249–20,259.

Dudley, W.C. and Goodney, D.E. (1979). Stable isotope analysis of calcareous nannoplankton: a paleo-oceanographic indicator of surface water conditions. In: *Evolution des Atmospheres Planetaires et Climatologie de la Terre*. Centre National d'Etudes Spatiales, Toulouse, 133–148.

Dugdale, R.C. and Wilkerson, F.P. (1990). Iron addition experiments in Antartica: a re-analysis. *Global Biogeochemical Cycles*, **4**, 13–19.

Duller, G.A.T. (1995). Luminescence dating using single aliquots: methods and applications. *Radiation Measurements*, **24**, 217–226.

Duller, G.A.T. (1996). Recent developments in luminescence dating of Quaternary sediments. *Progress in Physical Geography*, **20**, 127–145.

Dunbar, R.B. and Cole, J.E. (1993). *Coral Records of Ocean-Atmosphere Variability*. Special Report No. 10, NOAA Climate and Global Change Program, Washington, DC.

Dunbar, R.B., Linsley, B.K., and Wellington, G.M. (1996). Eastern Pacific corals monitor El Niño/Southern Oscillation, precipitation and sea surface temperature variability over the past 3 centuries. In: *Climate Variations and Forcing Mechanisms of the Last 2,000 years*. (P.D. Jones, R.S. Bradley, and J. Jouzel, eds.). Berlin: Springer-Verlag, 373–405.

Dunbar, R.B. and Wellington, G.M. (1981). Stable isotopes in a branching coral monitor seasonal temperature variation. *Nature*, **293**, 453–455.

Dunbar, R.B., Wellington, G.M., Colgan, M.W., and Glynn, P.W. (1994). Eastern Pacific sea surface temperature since 1600 A.D.: the ^{18}O record of climate variability in Galapagos corals. *Paleoceanography*, **9**, 291–315.

Duplessy, J-C. (1978). Isotope studies. In: *Climatic Change* (J. Gribbin, ed.). Cambridge: Cambridge University Press, 46–67.

Duplessy, J-C., Arnold, M., Bard, E., Juillet-Leclerc, A., Kallel, N., and Labeyrie, L. (1989). AMS ^{14}C study of transient events and of the ventilation rate of the Pacific Intermediate Water during the last deglaciation. *Radiocarbon*, **31**, 493–502.

Duplessy, J-C., Arnold, M., Maurice, P., Bard, E., Duprat, J., and Moyes, J. (1986). Direct dating of the oxygen-isotope record of the last deglaciation by ^{14}C accelerator mass spectrometry. *Nature*, **320**, 350–352.

Duplessy, J-C., Bard, E., Labeyrie, L., Duprat, J., and Moyes, J. (1993). Oxygen isotope records and salinity changes in the northeastern Atlantic Ocean during the last 18,000 years. *Paleoceanography*, **8**, 341–350.

Duplessy, J-C., Blanc, P-L., and Bé, A.W.H. (1981). Oxygen-18 enrichment of planktonic foraminifera due to gametogenic calcification below the euphotic zone. *Science*, **213**, 1247–1250.

Duplessy, J-C., Labeyrie, L., Arnold, M., Paterne, M., Duprat, D., and van Weering, T.C.E. (1992). Changes in surface salinity of the North Atlantic Ocean during the last deglaciation. *Nature*, **358**, 485–488.

Duplessy, J-C., Labeyrie, L., Juillet-LeClerc, A., Maitre, F., Duprat, J., and Sarnthein, M. (1991). Surface salinity reconstruction of the North Atlantic Ocean during the last glacial maximum. *Oceanologica Acta*, **14**, 311–324.

Duplessy, J-C., Labeyrie, J., Lalou, C., and Nguyen, H.V. (1970b). Continental climatic variations between 130,000 and 90,000 years BP. *Nature*, **226**, 631–632.

Duplessy, J-C., Labeyrie, J., Lalou, C., and Nguyen, H.V. (1971). La mésure des variations climatique continentales: application a la période comprise entre 130 000 et 90000 ans BP. *Quaternary Research,* **1,** 162–174.

Duplessy, J-C., Lalou, C., and Vinot, A.C. (1970a). Differential isotopic fractionation in benthic foraminifera and paleotemperatures re-assessed. *Science,* **168,** 250–251.

Duplessy, J-C. and Maier-Reimer, E. (1993). Global ocean circulation changes. In: *Global Changes in the Perspective of the Past* (J. A. Eddy and H. Oeschger, eds.). Chichester: Wiley, 199–220.

Duplessy, J-C., Moyes, J., and Pujol, C. (1980). Deep water formation in the North Atlantic Ocean during the last ice age. *Nature,* **286,** 479–482.

Duplessy, J-C. and Shackleton, N.J. (1985). Response of global deep-water circulation to Earth's climatic change 135,000–107,000 years ago. *Nature,* **316,** 500–507.

Duplessy, J-C., Shackleton, N.J., Fairbanks, R.G., Labeyrie, L., Oppo, D., and Kallel, N. (1988). Deep-water source variations during the last climatic cycle and their impact on the global deep water circulation. *Paleoceanography,* **3,** 343–360.

Dyakowska, J. (1936). Researches on the rapidity of the falling dowh of pollen of some trees. *Bulletin International de l'Academie Polonaise des Sciences et des Lettres,* **B1** 155–168.

Dylik, J. (1975). The glacial complex in the notion of the late Cenozoic cold ages. *Biuletyn Peryglacjalny,* **24,** 219–231.

Eckstein, D., Wazny, T., Blauch, J., and Klein, P. (1986). New evidence for the dendrochronological dating of Netherlandish paintings. *Nature,* **320,** 465–466.

Eddy, J.A. (1976). The Maunder Minimum. *Science,* **192,** 1189–1202.

Eddy, J.A. (1977). Climate and the changing sun. *Climatic Change,* **1,** 173–190.

Edwards, R.L., Beck, J.W., Burr, G.S., Donahue, D.J., Chappell, J.M.A., Bloom, A.L., Druffel, E.R.M., and Taylor, F.W. (1993). A large drop in atmospheric $^{14}C/^{12}C$ and reduced melting in Younger Dryas, documented with ^{230}Th ages of corals. *Science,* **260,** 962–967.

Edwards, R.L., Chen, J.H., and Wasserburg, G.J. (1987a). ^{238}U - ^{234}U - ^{230}Th - ^{232}Th systematics and the precise measurement of time over the past 500,000 years. *Earth and Planetary Science Letters,* **81,** 175–192.

Edwards, R.L., Chen, J.H., Ku, T-L., and Wasserburg, G.J. (1987b). Precise timing of the last interglacial period from mass spectrometric determination of ^{230}Th in corals. *Science,* **236,** 1547–1553.

Edwards, R.L. and Gallup, C.D. (1993). Dating of the Devils Hole calcite vein. *Science,* **259,** 1626 (see also reply by K.R. Ludwig *et al.,* p. 1626–1627).

Edwards, T.W.D. (1993). Interpreting past climate from stable isotopes in continental matter. In: *Climatic Change in Continental Isotopic Records* (P.K. Swart, K.C. Lohmann, J. McKenzie, and S. Savin, eds.). Washington, DC: American Geophysical Union, 333–341.

Edwards, T.W.D. and Fritz, P. (1986). Assessing meteoric water composition and relative humidity from ^{18}O and ^{2}H in wood cellulose: paleoclimatic implications for southern Ontario, Canada. *Applied Geochemistry,* **1,** 715–723.

Eglinton, G., Stuart, A.B., Rosell, A., Sarnthein, M., Pflaumann, U., and Tiedeman, R. (1992). Molecular record of secular sea surface temperature changes on 100-year timescales for glacial terminations I, II and IV. *Nature,* **356,** 423–426.

Ehhalt, D.H. (1988). How has the atmospheric concentration of CH_4 changed? In: *The Changing Atmosphere* (F.S. Rowland and I.S.A. Isaksen, eds.). Chichester: Wiley, 25–32.

Eicher, U. (1980). Pollen- und Sauerstoffisotopenanalysen an spatglazialen Profilen vom Gerzensee, Faulenseemoos und vom Regenmoos ob Boltigen. *Mitt. Naturforsch. Ges. Bern,* **37,** 65–80.

Eicher, U. and Siegenthaler, U. (1976). Palynological and oxygen isotopic investigations on late-Glacial sediment cores from Swiss lakes. *Boreas,* **5,** 109–117.

Elias, S. (1994). *Quaternary Insects and Their Environments.* Washington, DC: Smithsonian Institution Press.

Elias, S. (1996). Late Pleistocene and Holocene seasonal temperatures reconstructed from fossil beetle assemblages in the Rocky Mountains. *Quaternary Research*, **46**, 311–318.

Elias, S., Anderson K.H., and Andrews, J.T. (1996a). Late Wisconsin climate in the northeastern U.S.A. and southeastern Canada reconstructed from fossil beetle assemblages. *Jour. Quaternary Science*, **11**, 417–421.

Elias, S., Short, S.K., Nelson, C.H., and Birks, H.H. (1996b). The life and times of the Bering Land Bridge. *Nature, 382*, 60–63.

Elmore, S. and Phillips, F.M. (1987). Accelerator mass spectrometry for measurement of long-lived radioisotopes. *Science*, **236**, 543–550.

Emeis, K.C., Anderson, D.M., Doose, H., Kroon, D., and Schulz-Bull, D. (1995). Sea-surface temperatures and the history of monsoon upwelling in the Northwest Arabian Sea during the last 500,000 years. *Quaternary Research, 43*, 355–361.

Emiliani, C. (1954). Depth habitats of some species of pelagic foraminifera as indicated by oxygen isotope ratios. *American Jour. Science, 252*, 149–158.

Emiliani, C. (1955). Pleistocene temperatures. *Jour. Geology, 63*, 538–178.

Emiliani, C. (1966). Paleotemperature analysis of Caribbean cores, P6304–8 and P6304–9 and a generalized temperature curve for the past 425,000 years. *Jour. Geology, 74*, 109–126.

Emiliani, C. (1969). A new paleontology. *Micropaleontology, 15*, 265–300.

Emiliani, C. (1971). Depth habitats of growth stages of pelagic foraminifera. *Science, 173*, 1122–1124.

Emiliani, C. (1972). Quaternary paleotemperatures and the duration of the high temperature intervals. *Science, 178*, 398–401.

Emiliani, C. (1977). Oxygen isotopic analysis of the size fraction between 62 and 250 micrometers in Caribbean cores P6304–8 and P6304–9. *Science, 198*, 1255–1256.

Emiliani, C. and Ericson, D.B. (1991). The glacial/interglacial temperature range of the surface water of the oceans at low latitudes. In: *Stable Isotope Geochemistry: A Tribute to Samuel Epstein* (H.P. Taylor, Jr., J.R. O'Neil, and I.R. Kaplan, eds.). Special Publication No. 3., The Geochemical Society, 223–228.

Endler, J.A. (1982). Pleistocene forest refuges: fact or fancy. In: *Biological diversification in the Tropics* (G.T. Prance, ed.). New York: Columbia University Press, 641–657.

England, J. (1992). Postglacial emergence in the Canadian High Arctic: integrating glacio-isostasy, eustasy and late glaciation. *Canadian Jour. Earth Sciences, 29*, 984–999.

Ennever, F.K. and McElroy, M.B. (1985). Changes in CO_2: factors regulating the glacial to interglacial transition. In: *The Carbon Cycle and Atmospheric CO_2: Natural Variations, Archean to Present* (E.T. Sundquist and W.S. Broecker, eds). Washington, DC: American Geophysical Union, 154–162.

Epstein, S., Buchsbaum, R., Lowenstam, H.A., and Urey, H.C. (1953). Revised carbonate-water isotopic temperature scale. *Bulletin of the Geological Society of America*, **64**, 1315–1326.

Epstein, S. and Mayeda, T. (1953). Variation of ^{18}O content of waters from natural sources. *Geochimica et Cosmochimica Acta, 4*, 213–224.

Epstein, S. and Sharp, R.P. (1959). Oxygen isotope variations in the Malaspina and Saskatchewan glaciers. *Jour. Geology, 67*, 88–102.

Epstein, S. and Yapp, C.J. (1976). Climatic implications of the D/H ratio of hydrogen in C/H groups in tree cellulose. *Earth and Planetary Science Letters, 30*, 252–266.

Epstein, S. and Yapp, C.J. (1977). Isotope tree thermometers (comment). *Nature, 266*, 477–478.

Epstein, S., Yapp, C.J., and Hall, J.H. (1976). The determination of the D/H ratio of nonexchangeable hydrogen in cellulose extracted from aquatic and land plants. *Earth and Planetary Science Letters, 30*, 241–251.

Eronen, M. and Huttunen, P. (1987). Radiocarbon dated sub-fossil pine from Finnish Lapland. *Geografiska Annaler, 69A*, 297–304.

Etheridge, D.M., Steele, L.P., Langenfelds, R.L., Francey, R.J., Barnola J-M., and Morgan, V.I. (1996). Natural and anthropogenic changes in atmospheric CO_2 over the last 1000 years from air in Antarctic ice and firn. *Jour. Geophys. Research*, **101D**, 4115–4128.

Fabre, J. and Petit-Maire, N. (1988). Holocene climatic evolution from two paleolakes near Taoudenni, Mali (22°–23°N). *Palaeogeography, Palaeoclimatology, Palaeoecology*, **65**, 133–148.

Faegri, K. and Iversen, J. (1975). *Textbook of Pollen Analysis,* 3rd edition, New York: Hafner Press.

Faegri, K., Kaland, P.E., and Krzywinski, K. (1989). *Textbook of Pollen Analysis,* 4th edition, Chichester: Wiley.

Fairbanks, R.G. (1989). A 17,000 year glacio-eustatic sea level record: influence of glacial melting rates on the Younger Dryas event and deep ocean circulation. *Nature,* **342**, 637–642.

Fairbanks, R.G. (1990). The age and origin of the "Younger Dryas Climate Event" in Greenland ice cores. *Paleoceanography,* **5**, 937–948.

Fairbanks, R.G. and Dodge, R.E. (1979). Annual periodicity of the 0–18/0–16 and C-13/C-12 ratios in the coral *Montastrea annularis. Geochimica et Cosmochimica Acta,* **43**, 1009–1020.

Fanning, A.F. and Weaver, A.J. (1997). Temporal-geographical meltwater influences on the North Atlantic conveyor: implications for the Younger Dryas event. *Paleoceanography,* **12**, 307–320.

Faul, H. and Wagner, G.A. (1971). Fission track dating. In: *Dating Techniques for the Archaeologist* (H.N. Michael and E.K. Ralph, eds.). Cambridge: MIT Press, 152–156.

Ferguson, R. (1977). *Linear Regression in Geography. Concepts and Techniques in Modern Geography No. 15.* Norwich: University of East Anglia.

Field, W.O. (ed.). (1975). *Mountain glaciers of the Northern Hemisphere.* **1** and **2**. Hanover, Ontario: Cold Regions Research, and Engineering Laboratory.

Fisher, D.A. (1991). Remarks on the deuterium excess in precipitation in cold regions. *Tellus,* **43B**, 401–407.

Fisher, D.A. (1992). Stable isotope simulations using a regional stable isotope model coupled to a zonally averaged global model. *Cold Regions Science and Technology,* **21**, 61–77.

Fisher, D.A., Koerner, R.M., and Reeh, N. (1995). Holocene climatic records from the Agassiz Ice Cap, Ellesmere Island, N.W.T., Canada. *The Holocene,* **5**, 19–24.

Fitch, J.R. (1972). Selection of suitable material for dating and the assessment of geological error in potassium-argon age determination. In: *Calibration of Hominoid Evolution,* (W.W. Bishop and J.A. Miller, eds.). Edinburgh: Scottish Academic Press, 77–91.

Fleischer, R.L. (1975). Advances in fission track dating. *World Archaeology,* **7**, 136–150.

Fleischer, R.L. and Hart, H.R. (1972). Fission track dating techniques and problems. In: *Calibration of Hominoid Evolution* (W.W. Bishop and J.A. Miller, eds.). Edinburgh: Scottish Academic Press, 135–170.

Fleming, S. (1976). *Dating in Archaeology: a Guide to Scientific Techniques.* London: J.M. Dent.

Flint, R.F. (1971). *Glacial and Quaternary Geology.* New York: Wiley.

Flint, R.F. (1976). Physical evidence of Quaternary climatic change. *Quaternary Research,* **6**, 519–528.

Flock, J.W. (1978). Lichen-bryophyte distribution along a snow-cover/soil-moisture gradient, Niwot Ridge, Colorado. *Arctic and Alpine Research,* **10**, 31–47.

Flohn, H. (1949). Klima und Witterungsablauf in Zurich im 16 Jahrhundert. *Vierteljahresheft der Naturforschenden Gesellschaft in Zurich* 94, 28–41.

Flohn, H. (1975). Tropische Zirkulationsformen im Lichte der Satellitenaufnahmen. *Bonner Meteorologische Abhandlungen,* 21.

Flohn, H. (1978). Comparison of Antarctic and Arctic climate and its relevance to climate evolution. In: *Antarctic Glacial History and World Palaeoenvironments* (E.M. Van Zinderen Bakker, ed.). Rotterdam: A.A. Balkema, 3–13.

Foley, J.A., Kutzbach, J.E., Coe, M.T., and Levis, S. (1994). Feedbacks between climate and boreal forests during the Holocene epoch. *Nature,* **371**, 52–54.

Forman, S.L. (1991). Late Pleistocene chronology of loess deposition near Luochuan, China. *Quaternary Research*, **36**, 19–28.

Forman, S.L., Lepper, K., and Pierson, J. (1994). Limitations of infra-red stimulated luminescence in dating High Arctic marine sediments. *Quaternary Geochronology (Quaternary Science Reviews)*, **13**, 545–550.

Forman, S.L., Oglesby, R., Markgraf, V., and Stafford, T. (1995). Paleoclimatic significance of Late Quaternary eolian deposition on the Piedmont and High Plains, central United States. *Global and Planetary Change*, **11**, 35–55.

Fox, A.N. (1991). A quantitative model of Alpine snowline variations in the central Andes. *Boletim IG-USP* (Instituto de Geociências, Universidad de São Paulo, Brazil). *Publicação Especial*, No. 8, 75–88.

Frakes, L.A., Francis, J.E., and Syktus, J.I. (1992). *Climate Modes of the Phanerozoic*. Cambridge: Cambridge University Press.

Frenzel, B., Pésci, M., and Velichko, A.A. (1992a). *Atlas of Paleoclimates and Paleoenvironments of the Northern Hemisphere: Late Pleistocene-Holocene*. Geographical Research Institute, Hungarian Academy of Sciences, Budapest.

Frenzel, B., Pfister, C., and Gläser, B. (eds.). (1992b). *European Climate Reconstructed from Documentary Data: Methods and Results*. Stuttgart: Gustav Fischer Verlag.

Frenzel, B., Pfister, C., and Gläser, B. (eds.). (1994). *Climatic Trends and Anomalies in Europe 1675–1715*. Stuttgart: Gustav Fischer Verlag.

Frich, P. and Freyendahl, K. (1994). The summer climate in the Øresund region of Denmark. In: *Climatic Trends and Anomalies in Europe 1675–1715* (B. Frenzel, C. Pfister, and B. Glaser, eds.). Stuttgart: Gustav Fischer Verlag, 33–41.

Fritts, H.C. (1962). An approach to dendroclimatology screening by means of multiple regression techniques. *Jour. Geophys. Research*, **67**, 1413–1420.

Fritts, H.C. (1965). Tree-ring evidence for climatic changes in western North America. *Monthly Weather Review*, **93**, 421–443.

Fritts, H.C. (1971). Dendroclimatology and dendroecology. *Quaternary Research*, **1**, 419–449.

Fritts, H.C. (1976). *Tree Rings and Climate*. London: Academic Press.

Fritts, H.C. (1991). *Reconstructing Large Scale Climatic Patterns from Tree-Ring Data*. Tucson: University of Arizona Press.

Fritts, H.C., Blasing, T.J., Hayden, B.P., and Kutzbach, J.E. (1971). Multivariate techniques for specifying tree-growth and climate relationships and for reconstructing anomalies in paleoclimate. *Jour. Appl. Meteorology*, **10**, 845–864.

Fritts, H.C., Guiot, J., Gordon, G.A., and Schweingruber, F. (1990). Methods of calibration, verification and reconstruction. In: *Methods of Dendrochronology: Applications in the Environmental Sciences* (E.R. Cook and L.A. Kariukstis, eds.). Dordrecht: Kluwer, 163–217.

Fritts, H.C., Lofgren, G.R., and Gordon, G.A. (1979). Variations in climate since 1602 as reconstructed from tree rings. *Quaternary Research*, **12**, 18–46.

Fritts, H.C. and Shao, X.M. (1992). Mapping climate using tree-rings from western North America. In: *Climate Since A.D. 1500* (R.S. Bradley and P.D. Jones, eds.). London: Routledge, 269–295.

Fritz, S., Juggins, S., Batterbee, R.W., and Engstrom, D.R. (1991). Reconstruction of past changes in salinity and climate using a diatom-based transfer function. *Nature*, **352**, 706–708.

Fritz, S., Juggins, S., and Batterbee, R.W. (1993). Diatom assemblages and ionic characterization of lakes of the northern Great Plains, North America: a tool for reconstructing past salinity and climate fluctuations. *Canadian Jour. Fisheries and Aquatic Sciences*, **50**, 1844–1856.

Fronval, T., Jansen, E., Bloemendal, J., and Johnsen, S. (1995). Oceanic evidence for coherent fluctuations in Fennoscandian and Laurentide ice sheets on millennium timescales. *Nature*, **374**, 443–445.

Fukaishi, K. and Tagami, Y. (1992). An attempt of reconstructing the winter weather situations from 1720 to 1869 by the use of historical documents. In: *Proc. International Symposium on the Little Ice Age Climate* (T. Mikami, ed.). Department of Geography, Tokyo Metropolitan University, 194–201.

Fullerton, D.S. and Richmond, G.M. (1986). Comparison of the marine oxygen isotope record, the eustatic sea level record, and the chronology of glaciation in the United States of America. *Quaternary Science Reviews*, **5**, 197–200.

Gäggeler, H.W., Von Gunten, H.R., Rössler, E., Oeschger, H., and Schotterer, U. (1983). ^{210}Pb-dating of cold alpine firn/ice cores from Colle Gnifetti, Switzerland. *Jour. Glaciology*, **29**, 165–177.

Gajewski, K. and Garralla, S. (1992). Holocene vegetation histories from three sites in the tundra of northwestern Quebec, Canada. *Arctic and Alpine Research*, **24**, 329–336.

Gallée, H., Van Ypersele, J.P., Fichefet, T., Marsiat, I., Tricot, C., and Berger, A. (1992). Simulation of the Last Glacial Cycle by a coupled, sectorially averaged climate-ice sheet model. II. Response to insolation and CO_2 variation. *Jour. Geophys. Research*, **D97**, 15713–15740.

Gallée, H., Van Ypersele, J.P., Fichefet, Th., Tricot, C, and Berger, A. (1991). Simulation of the Last Glacial Cycle by a coupled, sectorially averaged climate-ice sheet model. I. The climate model. *Jour. Geophys. Research*, **D96**, 13139–13161.

Gallimore, R.G. and Kutzbach, J.E. (1995). Snow cover and sea ice sensitivity to generic changes in Earth orbital parameters. *Jour. Geophys. Research*, **100D**, 1103–1120.

Gallimore, R.G. and Kutzbach, J.E. (1996). Role of orbitally induced changes in tundra area in the onset of glaciation. *Nature*, **381**, 503–505.

Galloway, R.W. (1970). The full glacial climate in the southwestern United States. *Annals of the Association of American Geographers*, **60**, 245–256.

Gallup, C.D., Edwards, R.L., and Johnson, R.G. (1994). The timing of sea levels over the past 200,000 years. *Science*, **263**, 796–800.

Ganeshram, R.S., Pedersen, T.F., Calvert, S.E., and Murray, J.W. (1995). Large changes in oceanic nutrient inventories from glacial to interglacial periods. *Nature*, **376**, 755–758.

Gardner, J.V. (1975). Late Pleistocene carbonate dissolution cycles in the eastern Equatorial Atlantic. In: *Dissolution of Deep-sea Carbonates* (W.V. Sliter, A.W.H. Bé, and W.H. Berger, eds.). Special Publication No. 13, Cushman Foundation for Foraminiferal Research, Washington, DC, 129–141.

Gardner, J.V. and Hays, J.D. (1976). Responses of sea-surface temperature and circulation to global climatic change during the past 200,000 years in the eastern Equatorial Atlantic Ocean. In: *Investigation of late Quaternary Paleooceanography and Paleoclimatology*. (R.M. Cline and J.D. Hays, eds.). Geological Society of America Memoir No. 145. Boulder: Geological Society of America, 221–246.

Garnier, M. (1955). Contribution de la phenologie a l'etude des variations climatiques. *La Meteorologie*, **40**, 291–300.

Gascoyne, M. (1992). Paleoclimate determination from cave calcite deposits. *Quaternary Science Reviews*, **11**, 609–632.

Gascoyne, M., Schwarcz, H.P., and Ford, D.C. (1983). Uranium-series ages of speleothem from northwest England in correlation with Quaternary climate. *Phil.Trans. Royal Society of London*, **B301**, 143–164.

Gasse, F., Barker, P., Gell, P.A., Fritz, S.C., and Chalié, F. (1997). Diatom-inferred salinity in palaeolakes: an indirect tracer of climate change. *Quaternary Science Reviews*, **16**, 547–563.

Gasse, F. and Van Campo, E. (1994). Abrupt post-glacial climate events in West Asia and North Africa monsoon domains. *Earth and Planetary Science Letters*, **126**, 435–456.

Gates, W.L (1976a). Modelling the ice age climate. *Science*, **191**, 1138–1144.

Gates, W.L. (1976b). The numerical simulation of ice-age climate with a global general circulation model. *Jour. Atmos. Sciences*, **33**, 1844–1873.

Gaudreau, D.C., Jackson, S.T., and Webb III, T. (1989). Spatial scale and sampling strategy in paleoecological studies of vegetation patterns in mountainous terrain. *Acta Botanica Neerlandica*, **38**, 369–390.

Geitzenauer, K.R., Roche, M.B., and McIntyre, A. (1976). Modern Pacific coccolith assemblages: derivation and application to late Pleistocene paleotemperature analysis. In:

Investigation of Late Quaternary Paleooceanography and Paleoclimatology (R.M. Cline and J.D. Hays, eds.). Memoir **145**, Geological Society of America, Boulder, 423–448.

Genthon, C., Barnola, J.M., Raynaud, D., Lorius, C., Jouzel, J., Barkov, N.I., Korotkevich, Y.S., and Kotlyakov, V.M. (1987). Vostok ice core: climatic response to CO_2 and orbital forcing changes over the last climatic cycle. *Nature, 329,* 414–418.

Geyh, M.A. and Schleicher, H. (1990). *Absolute Age Determination: Physical and Chemical Dating Methods and Their Application.* Berlin: Springer-Verlag.

Gillen, K.P. and Evans, M.E. (1989). New geomagnetic paleosecular variation results from the Old Crow Basin, Yukon Territory, and their use in stratigraphic correlation. *Canad. Jour. Earth Sciences,* **26,** 2507–2511.

Gillot, P.Y., Labeyrie, J., Laj, C., Valladas, G., Guerin, G., Poupeau, G., and Delibrias, G. (1979). Age of the Laschamp paleomagnetic excursion revisited. *Earth and Planetary Science Letters,* **42,** 444–450.

Giresse, P., Maley, J., and Brenac, P. (1994). Late Quaternary palaeoenvironments in the Lake Barombi Mbo (West Cameroon) deduced from pollen and carbon isotopes of organic matter. *Palaeogeography, Palaeoclimatology, Palaeoecology,* **107,** 65–78.

Glock, W. (1937). *Principles and Methods of Tree-Ring Analysis.* Washington DC: Carnegie Institution.

Godfrey-Smith, D.I., Huntley, D.J., and Chen, W.H. (1988). Optical dating studies of quartz and feldspar sediment extracts. *Quaternary Science Reviews,* **7,** 373–380.

Godwin, H. (1956). *The History of the British Flora.* Cambridge: Cambridge University Press.

Godwin, H. (1962). Half-life of radiocarbon. *Nature,* **195,** 984.

Goede, A. (1994). Continuous early Last Glacial palaeoenvironmental record from a Tasmanian speleothem based on stable isotope and minor element variations. *Quaternary Science Reviews,* **13,** 283–291.

Goede, A., McDermott, F., Hawkesworth, C., Webb, J., and Finlayson, B. (1996). Evidence of Younger Dryas and Neoglacial cooling in a Late Quaternary paleotemperature record from a speleothem in eastern Victoria, Australia. *Jour. Quaternary Science,* **11,** 1–8.

Goodfriend, G.A. (1991). Patterns of racemization and epimerization of amino acids in land snail shells over the course of the Holocene. *Geochimica et Cosmochimica Acta,* **55,** 293–302.

Goodfriend, G.A. (1992). Rapid racemization of aspartic acid in molluscan shells and potential for dating over recent centuries. *Nature,* **357,** 399–401.

Goodfriend, G.A., Brigham-Grette, J., and Miller, G.H. (1996). Enhanced age resolution of the marine Quaternary record in the Arctic using aspartic acid racemization dating of bivalve shells. *Quaternary Research,* **45,** 176–187.

Goodfriend, G.A., Hare, P.E., and Druffel, E.R.M. (1992). Aspartic acid racemization and protein diagenesis in corals over the last 350 years. *Geochimica et Cosmochimica Acta,* **56,** 3847–3850.

Goodfriend, G.A. and Meyer, V.R. (1991). A comparative study of the kinetics of amino acid racemization/epimerization in fossil and modern mollusk shells. *Geochimica et Cosmochimica Acta,* **55,** 3355–3367.

Gordon, A.D. and Birks, H.J.B. (1974). Numerical methods in Quaternary paleoecology II. Comparison of pollen diagrams. *New Phytologist,* **73,** 221–249.

Gordon, D., Smart, P.L., Ford, D.C., Andrews, J.N., Atkinson, T.C., Rowe, P.J., and Christopher, N.S.J. (1989). Dating of late Pleistocene interglacial and interstadial periods in the United Kingdom from speleothem growth frequency. *Quaternary Research,* **31,** 14–26.

Gordon, G.A. (1982). Verification of dendroclimatic reconstructions. In: *Climate from Tree Rings* (M.K. Hughes, P.M. Kelley, J.R. Pilcher, and V.C. LaMarche, eds.). Cambridge: Cambridge University Press.

Goslar, T., Arnold, M., Bard, E., Kuc, T., Pazdur, M., Ralska-Jasiewiczowa, M., Rózanski, K., Tisnerat, N., Walanus, A., Wicik, B., and Wieckowski, K. (1995). High concentration of atmospheric ^{14}C during the Younger Dryas cold episode. *Nature,* **377,** 414–417.

Gow, A.J., Epstein, S., and Sheehy, W. (1979). On the origin of stratified debris in ice cores from the bottom of the Antarctic Ice Sheet. *Journal of Glaciology*, **23**, 185–192.

Gow, A.J., Meese, D.A., Alley, R.B., Fitzpatrick, J.J., Anandakrishnan, S., Woods, G.A., and Elder, B.C. (1997). Physical and structural properties of the GISP2 ice core: a review. *Jour. Geophys. Research*, **102C**, 26559–26575.

Grant-Taylor, T.L. (1972). Conditions for the use of calcium carbonate as a dating material. In: *Proc. 8th International Conference on Radiocarbon Dating*, **2**. Royal Society of New Zealand, Wellington, 592–596.

Gray, B.M. (1974). Early Japanese winter temperatures. *Weather*, **29**, 103–107.

Gray, J. and Thompson, P. (1978). Climatic interpretation of $\delta^{18}O$ and δD in tree rings. *Nature*, **271**, 93–94.

Graybill, D.A. and Shiyatov, S.G. (1992). Dendroclimatic evidence from the northern Soviet Union. In: *Climate Since A.D. 1500* (R.S. Bradley and P.D. Jones, eds.). London: Routledge, 393–414.

Grichuk, V.P. (1969). An attempt to reconstruct certain elements of the climate of the northern hemisphere in the Atlantic period of the Holocene. In: *Golotsen* (M.I. Neishtadt, ed.). 8th INQUA Congress, Izd-vo Nauka, Moscow, 41–57 (in Russian). [Translated by G.M. Peterson, Center for Climatic Research, University of Wisconsin, Madison.]

Griffey, N.J. and Matthews, J.A. (1978). Major neoglacial glacier expansion episodes in southern Norway: evidence from moraine ridge stratigraphy with ^{14}C dates on buried palaeosols and moss layers. *Geografiska Annaler*, **60A**, 73–90.

Griggs, R.F. (1938). Timberlines in the northern Rocky Mountains. *Ecology* **19**, 548–564.

Grimm, E.C., Jacobson, Jr., G.L., Watts, W.A., Hansen, B.C.S., and Maasch, K.A. (1993). A 50,000-year record of climate oscillations from Florida and its temporal correlation with the Heinrich events. *Science*, **261**, 198–200.

Grimmer, M. (1963). The space-filtering of monthly surface temperature data in terms of pattern, using empirical orthogonal functions. *Quart. Jour. Royal Meteorological Society*, **39**, 395–408.

Groisman, P.Ya. (1992). Possible regional climatic consequences of Pinatubo eruption. *Geophysical Research Letters*, **19**, 1603–1606.

Groisman, P.Ya., Karl, T.R., and Wright, R.W. (1994a). Observed impact of snow cover on the heat balance and the rise of continental spring temperatures. *Science*, **263**, 198–200.

Groisman, P.Ya., Karl, T.R., Wright, R.W., and Stenchikov, G.L. (1994b). Changes of snow cover, temperature and the radiative heat balance over the northern hemisphere. *Jour. Climate*, **7**, 1633–1656.

Grönvald, K., Óskarsson, N., Johnsen, S.J., Clausen, H.B., Hammer, C.U., Bond, G., and Bard, E. (1995). Ash layers from Iceland in the Greenland GRIP ice core correlated with oceanic and land sediments. *Earth and Planetary Science Letters*, **135**, 149–155.

Grootes, P.M., Stuiver, M., Farwell, G.W., Schaad, T.P., and Schmidt, F.H. (1980). Enrichment of ^{14}C and sample preparation for beta and ion counting. *Radiocarbon*, **22**, 487–500.

Grootes, P.M., Stuiver, M., Thompson, L.G., and Mosley-Thompson, E. (1989). Oxygen isotope changes in tropical ice, Quelccaya, Peru. *Jour. Geophys. Research*, **94D**, 1187–1194.

Grootes, P.M., Stuiver, M., White, J.W.C., Johnsen, S., and Jouzel, J. (1993). Comparison of oxygen isotope records from the GISP2 and GRIP Greenland ice cores. *Nature*, **366**, 552.

Grossman, E.L. (1987). Stable isotopes in benthic foraminifera: a study of vital effect. *Jour. Foraminiferal Research*, **17**, 48–61.

Grotzfeld, H. (1991). Klimageschichte des Vorderen Orients 800–1800 A.D. nach arabischen Quellen. *Würzburger Geographische Arbeiten*, **80**, 21–43.

Grousset, F.E., Biscaye, P.E., Revel, M., Petit, J.R., Pye, K., Joussaume, S., and Jouzel, J. (1992). Antarctic (Dome C) ice core dust at 18 k.y. B.P.: isotopic constraints on origins. *Earth and Planetary Science Letters*, **111**, 175–182.

Grousset, F.E., Labeyrie, L., Sinko, J.A., Cremer, M., Bond, G., Duprat, J., Cortijo E., and Huon, S. (1993). Patterns of ice-rafted detritus in the glacial North Atlantic (40–55° N). *Paleoceanography*, **8**, 175–192.

Grove, A.T. and Warren, A. (1968). Quaternary landforms and climate on the south side of the Sahara. *Geographical Jour., 134*, 194–208.

Grove, J.M. (1979). The glacial history of the Holocene. *Progress in Physical Geography, 3*, 1–54.

Grove, J.M. (1988). *The Little Ice Age*. London: Methuen.

Guilderson, T.P., Fairbanks, R.G., and Rubenstone, J.L.(1994). Tropical temperature variations since 20,000 years ago: modulating inter-hemispheric climate change. *Science, 263*, 663–665.

Guiot, J. (1987). Late Quaternary climatic change in France estimated from multivariate pollen time series. *Quaternary Research, 28*, 100–118.

Guiot, J. (1990). Methodology of paleoclimatic reconstruction from pollen in France. *Palaeogeography, Palaeoclimatology, Palaeoecology, 80*, 49–69.

Guiot, J., de Beaulieu, J.L., Cheddadi, R., David, F., Ponel, P., and Reille, M. (1993). The climate in western Europe during the last Glacial/Interglacial cycle derived from pollen and insect remains. *Palaeogeography, Palaeoclimatology, Palaeoecology, 103*, 73–93.

Guiot, J., de Beaulieu, J.L., Reille, M., and Pons, A. (1992). Calibration of the climatic signal in a new pollen sequence from La Grande Pile. *Climate Dynamics, 6*, 259–264.

Guiot, J., Pons, A., de Beaulieu, J.L., and Reille, M. (1989). A 140,000 year climatic reconstruction from two European pollen records. *Nature, 338*, 309–313.

Guo, Q. (1992). Winter monsoon over East Asia during the Little Ice Age. In: *Proc. International Symposium on the Little Ice Age Climate* (T. Mikami, ed.). Department of Geography, Tokyo Metropolitan University, 227–232.

Gwiazda, R.H., Hemming, S.R., and Broecker, W.S. (1996a). Tracking the sources of icebergs with lead isotopes: the provenance of ice-rafted debris in Henrich layer 2. *Paleoceanography, 11*, 77–93.

Gwiazda, R. H., Hemming, S.R., and Broecker, W.S. (1996b). Provenance of icebergs during Heinrich event 3 and the contrast to their sources during other Heinrich episodes. *Paleoceanography, 11*, 371–378.

Haberle, S. (1997). Late Quaternary vegetation and climate history of the Amazon Basin: correlating marine and terrestrial pollen records. In: *Proc. Ocean Drilling Program, Scientific Results* (R.D. Flood, D.J.W. Piper, A. Klaus, and L.C. Peterson, eds.), *155*, 381–396.

Haffer, J. (1969). Speciation in Amazonian forest birds. *Science, 165*, 131–137.

Haffer, J. (1974). *Avian speciation in tropical South America*. Publication No. 14, Cambridge, Mass: Nuttall Ornithological Club.

Haffer, J. (1982). General aspects of the refuge theory. In: *Biological Diversification in the Tropics* (G.T. Prance, ed.). New York: Columbia University Press, 6–24.

Hage, K.D., Gray, J., and Linton, J.C. (1975). Isotopes in precipitation in western North America. *Monthly Weather Review, 103*, 958–966.

Hajdas, I., Ivy-Ochs, S.D., and Bonari, G. (1995a). Problems in the extension of the radiocarbon calibration curve (10–13 kyr. B.P.). *Radiocarbon, 37*, 75–79.

Hajdas, I., Ivy-Ochs, S.D., Bonani, G., Lotter, A.F., Zolitschka, B., and Schluchter, C. (1995b). Radiocarbon age of the Laacher See tephra: 11,230 ±40 B.P. *Radiocarbon, 37*, 149–154.

Hall, N.M.J. and Valdes, P.J. (1997). A GCM simulation of the climate 6000 years ago. *Jour. Climate, 10*, 3–17.

Hall, N.M.J., Valdes, P.J., and Dong, B. (1996). The maintenance of the last great ice sheets: a UGAMP GCM study. *Jour. Climate, 9*, 1004–1019.

Halme, E. (1952). On the influence of climatic variation on fish and fishery. *Fennia, 75*, 89–96.

Hamelin, B., Bard, E., Zindler, A., and Fairbanks, R.G. (1991). $^{234}U/^{238}U$ mass spectrometry of corals: how accurate is the U-Th age of the last interglacial period? *Earth and Planetary Science Letters, 106*, 169–180.

Hamilton, A.C. (1976). The significance of patterns of distribution shown by forest plants and animals in tropical Africa for the reconstruction of upper Pleistocene palaeoenvironments: a review. *Palaeoecology of Africa, 9*, 63–97.

Hamilton, A.C. and Perrott, R.A. (1979). Aspects of the glaciation of Mt Elgon, East Africa. *Palaeoecology of Africa,* **11,** 153–162.

Hamilton, A.C. and Perrott, R.A. (1980). Modern pollen deposition on a tropical African mountain. *Pollen et Spores,* **22,** 437–468.

Hammer, C.U. (1977). Past volcanism revealed by Greenland ice sheet impurities. *Nature,* **270,** 482–486.

Hammer, C.U. (1980). Acidity of polar ice cores in relation to absolute dating, past volcanism and radio echoes. *Jour. Glaciology,* **25,** 359–372.

Hammer, C.U. (1984). Traces of Icelandic eruptions in the Greenland ice sheet. *Jøkull,* **34,** 51–65.

Hammer, C.U. (1989). Dating by physical and chemical seasonal variations and reference horizons. In: *The Environmental Record in Glaciers and Ice Sheets* (H. Oeschger and C.C. Langway, Jr., eds.). Chichester: Wiley, 99–121.

Hammer, C.U., Clausen, H.B., Dansgaard, W., Gundestrup, N., Johnsen, S.J., and Reeh, N. (1978). Dating of Greenland ice cores by flow models, isotopes, volcanic debris and continental dust. *Jour. Glaciology,* **20,** 3–26.

Hammer, C.U., Clausen, H.B., and Dansgaard, W. (1980). Greenland ice sheet evidence of post-glacial volcanism and its climatic impact. *Nature,* **288,** 230–255.

Hammer, C.U., Clausen, H.B., and Langway, Jr., C.C. (1997). 50,000 years of recorded global volcanism. *Climatic Change,* **35,** 1–15.

Hann, B.J., Walker, B.G., and Warwick, W.F. (1992). Aquatic invertebrates and climate change: a comment on Walker *et al.* (1991). *Canadian Jour. of Fisheries and Aquatic Sciences,* **49,** 1274–1276 (see also, Reply, p.1276–1280).

Hannon, G.E. and Gaillard, M.J. (1997). The plant-macrofossil record of past lake-level changes. *Jour. Paleolimnology,* **18,** 15–28.

Hansen, J.E. and 30 others, (1996). A Pinatubo climate modelling investigation. In: *The Mount Pinatubo Eruption: Effects on the Atmosphere and Climate* (G. Fiocco, D. Fina and G. Visconti, eds.). New York: Springer-Verlag.

Hansen, J.E., Lacis, A., Rind, D., Russell, G., Stone, P., Fung, I., Ruedy, R., and Lerner, J. (1984). Climate sensitivity: analysis of feedback mechanisms. In: *Climate Processes and Sensitivity* (J.E. Hansen and T. Takahashi, eds.). Washington DC: American Geophysical Union, 130–163.

Hansen, J.E., Wang, W.C., and Lacis, A.A. (1978). Mt Agung provides test of a global climate perturbation. *Science,* **199,** 1065–1068.

Haq, B.U. (1978). Calcareous nannoplankton. In: *Introduction to Marine Micropaleontology* (B.U. Haq and A. Boersma, eds.). New York: Elsevier/North Holland, 79–107.

Haq, B.U. and Boersma, A. (eds.). (1978). *Introduction to Marine Micropaleontology.* New York: Elsevier/North Holland.

Hardy, D.R. (1996). Climatic influences on streamflow and sediment flux into Lake C2, northern Ellesmere Island, Canada. *Jour. Paleolimnology,* **16,** 133–149.

Hardy, D.R., Bradley, R.S., and Zolitschka, B. (1996). The climatic signal in varved sediments from Lake C2, northern Ellesmere Island, Canada. *Jour. Paleolimnology,* **16,** 227–238.

Hare, F.K. (1979). Climatic variation and variability: empirical evidence from meteorological and other sources. In: *Proc. World Climate Conference.* Publication No. 537, World Meteorological Organization, Geneva, 51–87.

Hare, P.E. and Mitterer, R.M. (1968). Laboratory simulation of amino acid diagenesis in fossils. *Carnegie Institution of Washington Yearbook,* **67,** 205–208.

Harington, C.R. (ed.). (1992). *1816. The Year Without a Summer?* Canadian Museum of Nature, Ottawa.

Harland, W.B., Armstrong, R.L., Cox, A.V., Craig, L.E., Smith, A.G., and Smith, D.G. (1990). *A Geologic Time Scale 1989.* Cambridge: Cambridge University Press.

Harmon, R.S. (1976). Late Pleistocene glacial chronology of the South Nahanni River Region, Northwest Territories, Canada. *Michigan Academician,* **9,** 147–156.

Harmon, R.S., Schwarcz, H.P., and Ford, D.C. (1978b). Late Pleistocene sea level history of Bermuda. *Quaternary Research,* **9,** 205–218.

Harmon, R.S., Schwarcz, H.P., and O'Neil, J.R., (1979). D/H ratios in speleothem fluid inclusions: a guide to variations in the isotopic composition of meteoric precipitation? *Earth and Planetary Science Letters,* **42**, 254–266.

Harmon, R.S., Ford, D.C., and Schwarcz, H.P. (1977). Interglacial chronology of the Rocky and Mackenzie mountains based on ^{230}Th and ^{234}U dating of calcite speleothems. *Canadian Jour. Earth Sciences,* **14**, 2543–2552.

Harmon, R.S., Thompson, P., Schwarcz, H.P., and Ford, D.C. (1975). Uranium-series dating of speleothems. *National Speleological Society Bulletin,* **37**, 21–33.

Harmon, R.S., Thompson, P., Schwarcz, H.P., and Ford, D.C. (1978a). Late Pleistocene paleoclimates of North America as inferred from stable isotope studies of speleothems. *Quaternary Research,* **9**, 54–70.

Harper, F. (1961). Changes in climate, faunal distribution and life zones in the Ungava Peninsula. *Polar Notes, Dartmouth College, N.H.* No. III, 20–41.

Harrison, S.P. (1989). Lake-levels and climatic changes in eastern North America. *Climate Dynamics,* **3**, 157–167.

Harrison, S.P. (1993). Late Quaternary lake-level changes and climates of Australia. *Quaternary Science Reviews,* **12**, 211–231.

Harrison, S.P. and Digerfeldt, G. (1993). European lakes as palaeohydrological and palaeoclimatic indicators. *Quaternary Science Reviews,* **12**, 233–248.

Harrison, S.P. and Tarasov, P.E. (1996). Late Quaternary lake-level records from northern Eurasia. *Quaternary Research,* **45**, 138–159.

Harrison, S.P., Kutzbach, J.E., and Behling, P.J. (1991). General circulation models, paleoclimatic data and last interglacial climates. *Quaternary International,* **10–12**, 231–242.

Harrison, S.P., Kutzbach, J.E., Prentice, I.C., Behling, P.J., and Sykes, M.J. (1995). The response of northern hemisphere extratropical climate and vegetation to orbitally induced changes in insolation during the last interglaciation. *Quaternary Research,* **43**, 174–184.

Harrison, S.P., Yu, Ge, Tarasov, P.E. (1996). Late Quaternary lake-level record from northern Eurasia. *Quaternary Research,* **45**, 138–159.

Harvey, L.D. and Schneider, S.H. (1985a). Transient climate response to external forcing on 10^0–10^4 year time scales, Part 1: experiments with globally averaged, coupled atmosphere and ocean energy balance models. *Jour. Geophys. Research,* **90D**, 2191–2205.

Harvey, L.D. and Schneider, S.H. (1985b). Transient climate response to external forcing on 10^0–10^4 year time scales, Part 2: experiments with a seasonal, hemispherically averaged, coupled atmosphere, land and ocean energy balance model. *Jour. Geophys. Research,* **90D**, 2207–2222.

Hastenrath, S. (1967). Observations on the snow line in the Peruvian Andes. *Jour. Glaciology,* **6**, 541–550.

Hastenrath, S. (1971). On the Pleistocene snow line depression in the arid regions of the South American Andes. *Jour. Glaciology,* **10**, 255–267.

Hastenrath, S. and Kruss, P.D. (1992). The dramatic retreat of Mount Kenya's glaciers 1963–87: greenhouse forcing. *Annals of Glaciology,* **16**, 127–133.

Haxeltine, A. and Prentice, I.C. (1996). BIOME3: an equilibrium biosphere model based on ecophysiological constraints, resource availability and competition among plant functional types. *Global Biogeochemical Cycles,* **10**, 693–709.

Hay, W.H. (1993). The role of polar deep water formation in global climate change. *Annual Reviews of Earth and Planetary Sciences,* **21**, 227–254.

Hay, W.W. (1974). Introduction. In: *Studies in Paleo-oceanography* (W.W. Hay, ed.). Special Publication No. 20, Society of Economic Paleontologists and Mineralogists. Tulsa, 1–5.

Hays, J.D. (1978). A review of the late Quaternary history of Antarctic Seas. In: *Antarctic Glacial History and World Palaeoenvironrnents.* (E.M. Van Zinderen Bakker ed.). Rotterdam: A.A. Balkema, 57–71.

Hays, J.D., Imbrie, J., and Shackleton, N.J. (1976). Variations in the earth's orbit: pacemaker of the ice ages. *Science,* **194**, 1121–1132 (see also *Science,* **198**, 528–530).

Hecht, A. (1973). A model for determining Pleistocene paleotemperatures from planktonic foraminiferal assemblages. *Micropalaeontology,* **19**, 68–77.

Hecht, A. (1976). The oxygen isotope record of foraminifera in deep sea sediments. In: *Foraminifera*, **2**, (R.H. Hedley and C.G. Adams, eds.). New York: Academic Press, 1–43.

Hecht, A., Barry, R.G., Fritts, H.C., Imbrie, J., Kutzbach, J., Mitchell, Jr., J.M., and Savin, S.M. (1979). Paleoclimatic research: status and opportunities. *Quaternary Research*, **12**, 6–17.

Hecht, A.D. and Savin, S.M. (1970). Oxygen-18 studies of recent planktonic foraminifera: comparisons of phenotypes and of test parts. *Science*, **170**, 69–71 (see also *Science*, **173**, 167–169).

Hecht, A.D. and Savin, S.M. (1972). Phenotypic variation and oxygen isotope ratios in recent planktonic foraminifera. *Jour. Foraminiferal Research*, **2**, 55–67.

Heinrich, M. (1988). Origin and consequences of cyclic ice rafting in the northeast Atlantic Ocean during the past 130,000 years. *Quaternary Research*, **29**, 143–152.

Heller, F. and Liu, T. (1984). Magnetism of Chinese loess deposits. *Geophys. Jour. Royal Astronomical Society*, 77, 125–141.

Heller, F., Shen, C-D., Beer, J., Liu, X-M., Liu, T-S., Bronger, A., Suter, M., and Bonani, G. (1993). Quantitative estimates of pedogenic ferromagnetic mineral formation in Chinese loess and paleoclimatic implications. *Earth and Planetary Science Letters*, **114**, 385–390.

Hendy, C.H. (1970). The use of C-14 in the study of cave processes. In: *Radiocarbon Variations and Absolute Chronology* (I.U. Olsson, ed.). New York: Wiley, 419–442.

Hendy, C.H. and Wilson, A.T. (1968). Paleoclimatic data from speleothems. *Nature*, **219**, 48–51.

Henfling, E. and Pflaubaum, H. (1991). Neue Aspekte zur klimatischen Interpretation der hohen pharaonischen Nilflutmarken am 2. Katarakt aus ägyptologischer und geomorphologischer Sicht. *Würzburger Geographische Arbeiten*, **80**, 87–109.

Hennig, G.J., Grün, R., and Brunnacker, K. (1983). Speleothems, travertines and paleoclimates. *Quaternary Research*, **20**, 1–29.

Herron, M.M. and Langway, C.C. (1979). Dating of Ross Ice Shelf cores by chemical analysis. *Jour. Glaciology*, **24**, 345–357.

Herron, M.M. and Langway, C.C. (1980). Firn densification: an empirical model. *Jour. Glaciology*, **25**, 373–386.

Hesse, P.P. (1994). The record of continental dust from Australia in Tasman Sea sediments. *Quaternary Science Reviews*, **13**, 257–272.

Hewitt, C.D. and Mitchell, J.F.B. (1998). A fully coupled GCM simulation of the climate of the mid-Holocene. *Geophysical Research Letters*, **25**, 361–364.

Hilgen, F.J. (1991). Astronomical calibration of Gauss to Matuyama sapropels in the Mediterranean and implications for the geomagnetic polarity time scale. *Earth and Planetary Science Letters*, **104**, 226–244.

Hoffsummer, P. (1996). Dendrochronology and the study of roof-framing in Belgium. In: *Tree Rings, Environment and Humanity* (J.S. Dean, D.M. Meko, and T.W. Swetnam, eds.) Tucson: University of Arizona, 525–531.

Hofmann, W. (1986). Chironomid analysis. In: *Handbook of Holocene Palaeoecology and Palaeohydrology* (B.E. Berglund, ed.). Chichester: Wiley, 715–727.

Hoganson, J.W., and Ashworth, A.C. (1992). Fossil beetle evidence for climatic change, 18,000–10,000 years B. P. in south-central Chile. *Quaternary Research*, **37**, 101–116.

Hoinkes, H.C. (1968). Glacier variation and weather. *Jour. Glaciology*, **7**, 3–19.

Holdsworth, G. and Peake, E. (1985). Acid content of snow from a mid-troposphere sampling site on Mount Logan, Yukon Territory, Canada. *Annals of Glaciology*, **7**, 153–160.

Hollin, J.T. and Schilling, D.H. (1981). Late Wisconsin-Weichselian mountain glaciers and small ice caps. In: *The Last Great Ice Sheets* (G.H. Denton and T.J. Hughes, ed.). New York: Wiley, 179–206.

Holmes, P.L. (1994). The sorting of spores and pollen by water: experimental and field evidence. In: *Sedimentation of Organic Particles* (A. Traverse, ed.). Cambridge: Cambridge University Press, 9–32.

Holmes, J.A. (1996). Trace-element and stable isotope geochemistry of non-marine ostracod shells in Quaternary paleoenvironmental reconstruction. *Jour. Paleolimnology*, **15**, 223–235.

Holmes, R.L. (1983). Computer-assisted quality control in tree-ring dating and measurement. *Tree-Ring Bulletin*, **44**, 69–75.

Holtmeier, F-K. (1994). Ecological aspects of climatically-caused timberline fluctuations. In: *Mountain Environments in Changing Climates* (M. Beniston, ed.). London: Routledge, 220–233.

Hooghiemstra, H. (1984). Vegetational and climatic history of the high plain of Bogotá, Colombia: a continuous record of the last 3.5 million years. *Dissertationes Botanicae* No. 79, Vaduz: J. Cramer.

Hooghiemstra, H., Melice, J.L., Berger, A., and Shackleton, N.J. (1993). Frequency spectra and paleoclimatic variability of the high resolution 30–1450 kyr Funza I pollen record (eastern Cordillera, Colombia). *Quaternary Science Reviews*, **12**, 141–156.

Hooghiemstra, H., and Ran, E.T.H. (1994). Late Pliocene-Pleistocene high resolution pollen sequence of Colombia: an overview of climatic change. *Quaternary International*, **21**, 63–80.

Hooghiemstra, H., Stalling, H., Agwu, C.O.C., and Dupont, L.M. (1992). Vegetational and climatic changes at the northern fringe of the Sahara 250,000–5,000 years B.P.: evidence from 4 marine pollen records located between Portugal and the Canary Islands. *Review of Palaeobotany and Palynology*, **74**, 1–53.

Hostetler, S.W., Giorgi, F., Bates, G.T., and Bartlein, P.J. (1994). Lake-atmosphere feedbacks associated with paleolakes Bonneville and Lahontan. *Science*, **263**, 665–668.

Hovan, S.A., Rea, D.K., Pisias, N.G., and Shackleton, N.J. (1989). A direct link between the China loess and marine $\delta^{18}O$ records: aeolian flux to the north Pacific. *Nature*, **340**, 296–298.

Hovan, S.A., Rea, D.K., and Pisias, N.G. (1991). Late Pleistocene continental climate and oceanic variability recorded in Northwest Pacific sediments. *Paleoceanography*, **6**, 349–370.

Howe, S.E. and Webb, III, T. (1983). Calibrating pollen data in climatic terms: improving the methods. *Quaternary Science Reviews*, **2**, 17–51.

Hoyt, D.V. and Schatten, K.H. (1997). *The Role of the Sun in Climate Change*. Oxford: Oxford University Press.

Hughen, K.A., Overpeck, J.T., Lehman, S.J., Kashgarian, M., Southon, J., Peterson, L.C., Alley, R., and Sigman, D.M. (1998). Deglacial changes in ocean circulation from an extended radiocarbon calibration. *Nature*, **391**, 65–68.

Hughen, K.A., Overpeck, J.T., Peterson, L.C., and Anderson, R.F. (1996a). The nature of varved sedimentation in the Cariaco Basin, Venezuela, and its paleoclimatic significance. In: *Palaeoclimatology and Palaeoceanography from Laminated Sediments* (A.E.S. Kemp, ed.). Special Publication No. 116, The Geological Society, London, 171–183.

Hughen, K.A., Overpeck, J.T., Peterson, L.C., and Trumbore, S. (1996b). Rapid climate changes in the tropical Atlantic region during the last deglaciation. *Nature*, **380**, 51–54.

Hughes, M.K. and Diaz, H.F. (1994). Was there a "Medieval Warm Period" and if so, where and when? *Climatic Change*, **26**, 109–142.

Hughes, M.K., Kelley, P.M., Pilcher, J.R., and LaMarche, V.C. (1982). *Climate from tree rings*. Cambridge: Cambridge University Press.

Hughes, T.J., Denton, G.H., Anderson, B.G., Schilling, D.H., Fastook, J.L., and Lingle, C.S. (1981). The last great ice sheets: a global view. In: *The Last Great Ice Sheets* (G.H. Denton and T.J. Hughes, ed.). New York: Wiley, 275–318.

Hummel, J. and Reck, R. (1979). A global surface albedo model. *Jour. Appl. Meteorology*, **18**, 239–253.

Hunt, J.B. and Hill, P.G. (1993). Tephra geochemistry: a discussion of some persistent analytical problems. *The Holocene*, **3**, 271–278. (See also discussion in *The Holocene*, **4**, 435–438.)

Huntley, B. (1990a). European vegetation history: palaeovegetation maps from pollen data—13,000 B.P. to present. *Jour. Quaternary Science,* **5**, 103–122.

Huntley, B. (1990b). Dissimilarity mapping between fossil and contemporary pollen spectra in Europe for the past 13,000 years. *Quaternary Research,* **33**, 360–376.

Huntley, B. and Birks, H.J.B. (1983). *An Atlas of Past and Present Pollen Maps for Europe, 0–13,000 Years Ago.* Cambridge: Cambridge University Press.

Huntley, B. and Prentice, I.C. (1988). July temperatures in Europe from pollen data, 6000 years before present. *Science,* **241**, 687–690.

Huntley, B. and Prentice, I.C. (1993). Holocene vegetation and climates in Europe. In: *Global Climates Since the Last Glacial Maximum* (H.E. Wright, Jr., J.E. Kutzbach, T. Webb III, W.F. Ruddiman, F.A. Street-Perrott, and P.J. Bartlein, eds.). Minneapolis: University of Minnesota Press, 136–168.

Huntley, B. and Webb III, T. (1988). *Vegetation History.* Kluwer, Dordrecht.

Huntley, B. and Webb III, T. (1989). Migration: species' response to climatic variations caused by changes in the earth's orbit. *Jour. Biogeography,* **16**, 5–19.

Huntley, D.J., Godfrey-Smith, D.I., and Thewalt, M.L.W. (1985). Optical dating of sediments. *Nature,* **313**, 105–107.

Hurford, A.J. and Green, P.F. (1982). A user's guide to fission track dating calibration. *Earth and Planetary Science Letters,* **59**, 343–354.

Hutson, W.H. (1977). Transfer functions under no-analog conditions: experiments with Indian Ocean planktonic foraminifera. *Quaternary Research,* **3**, 355–367.

Hutson, W.H. (1978). Application of transfer functions to Indian Ocean planktonic foraminifera. *Quaternary Research,* **3**, 87–112.

Hütt, G., Jaek, H.I., and Tchonka, J. (1988). Optical dating: K-feldspars optical response stimulation spectra. *Quaternary Science Reviews,* **7**, 381–385.

Huxtable, J., Aitken, M.J., and Bonhommet, N. (1978). Thermoluminescence dating of sediment baked by lava flows of the Chaine des Puys. *Nature,* **275**, 207–209.

IMAGES Planning Committee (1994). *International Marine Global Change Study: Science and Implementaiton Plan.* Bern: PAGES Report 94–3.

Imbrie, J. (1985). A theoretical framework for the Pleistocene ice ages. *Jour. Geological Society of London,* **142**, 417–432.

Imbrie, J., Berger, A., Boyle, E.A., Clemens, S.C., Duffy, A., Howard, W.R., Kukla, G., Kutzbach, J., Martinson, D.G., McIntyre, A., Mix, A.C., Molfino, B., Morley, J.J., Peterson, L.C., Pisias, N.G., Prell, W.L., Raymo, M.E., Shackleton, N.J., and Toggweiler, J.R. (1992). On the structure and origin of major glaciation cycles, 1. Linear responses to Milankovitch forcing. *Paleoceanography,* **7**, 701–738.

Imbrie, J., Berger, A., Boyle, E.A., Clemens, S.C., Duffy, A., Howard, W.R., Kukla, G., Kutzbach, J., Martinson, D.G., McIntyre, A., Mix, A.C., Molfino, B., Morley, J.J., Peterson, L.C., Pisias, N.G., Prell, W.L., Raymo, M.E., Shackleton, N.J., and Toggweiler, J.R. (1993a). On the structure and origin of major glaciation cycles, 2. The 100,000-year cycle. *Paleoceanography,* **8**, 699–735.

Imbrie, J., Berger, A., and Shackleton, N.J. (1993b). Role of orbital forcing: a two million year perspective. In: *Global Changes in the Perspective of the Past.* (J.A. Eddy and H. Oeschger, eds.). Chichester: Wiley, 263–277.

Imbrie, J., Hays, J.D., Martinson, D.G., McIntyre, A., Mix, A., Morley, J.J., Pisias, N.G., Prell, W., and Shackleton, N.J. (1984). The orbital theory of Pleistocene climate: support from a revised chronology of the marine $\delta^{18}O$ record. In: *Milankovitch and Climate.* (A. Berger, J. Hays, G. Kukla, and B. Salzman, eds.). Dordrecht: Reidel, 269–305.

Imbrie, J. and Imbrie, K.P. (1979). *Ice Ages: Solving the Mystery.* London: Macmillan.

Imbrie, J. and Imbrie, J.Z. (1980). Modeling the climatic response to orbital variations. *Science,* **207**, 943–953.

Imbrie, J. and Kipp, N.G. (1971). A new micropalaeontological method for quantitative paleoclimatology: application to late Pleistocene Caribbean core V28–238. In: *The Late Cenozoic Glacial Ages* (K.K. Turekian, ed.). New Haven: Yale University Press, 77–181.

Imbrie, J., Mix, A.C., and Martinson, D.G. (1993c). Milankovitch theory viewed from Devil's Hole. *Nature,* 363, 531–533 (see also reply by I.C. Winograd and J.M. Landwehr [1993]).

Imbrie, J., van Donk, J., and Kipp, N.G. (1973). Paleoclimatic investigation of a late Pleistocene Caribbean deep-sea core: comparison of isotopic and faunal methods. *Quaternary Research,* 3, 10–38.

Ingram, M.J., Farmer, G., and Wigley, T.M.L. (1981a). Past climates and their impact on Man: a review. In: *Climate and History* (T.M.L. Wigley, M.J. Ingram, and G. Farmer, eds.). Cambridge: Cambridge University Press, 3–50.

Ingram, M.J., Underhill, D.J., and Farmer, G. (1981b). The use of documentary sources for the study of past climates. In: *Climate and History* (T.M.L. Wigley, M.J. Ingram, and G. Farmer, eds.). Cambridge: Cambridge University Press, 180–213.

Ingram, M.J., Underhill, D.J, and Wigley, T.M.L. (1978). Historical climatology. *Nature,* 276, 329–334.

Innes, J.L. (1982). Lichenometric use of an aggregated *Rhizocarpon* "species." *Boreas,* 11, 53–58.

IPCC (1996). *Climate Change 1995: The Science of Climate Change.* Cambridge: Cambridge University Press.

Isdale, P. (1984). Fluorescent bands in massive corals record centuries of coastal rainfall. *Nature,* 310, 378–379.

Isdale, P.J., Stewart, B.J., Tickle, K.S., and Lough, J.M. (1998). Paleohydrological variation in a tropical river catchment: a reconstruction using fluorescent bands in corals of the Great Barrier Reef, Australia. *The Holocene,* 8, 1–8.

Ivanovich, M. and Harmon, R.S. (1982). *Uranium Series Disequilibrium: Applications to Environmental Problems.* Oxford: Clarendon Press.

Ives, J.D. (1974). Permafrost. In: *Arctic and Alpine Environments* (J.D. Ives and R.G. Barry, eds.). London: Methuen, 159–194.

Ives, J.D. (1978). Remarks on the stability of timberline. In: *Geoecological Relations between the Southern Temperate Zone and the Tropical High Mountains* (C. Troll and W. Lauer, eds.). Wiesbaden: Franz Steiner Verlag, 313–317.

Ives, J.D., Andrews, J.T., and Barry, R.G. (1975). Growth and decay of the Laurentide Ice Sheet and comparisons with Fenno-Scandinavia. *Naturwissenschaften,* 62, 118–125.

Izett, G.A. and Obradovich, J.D. (1994). ^{40}Ar/^{39}Ar age constraints for the Jaramillo Normal Subchron and the Matuyama-Brunhes geomagnetic reversal. *Jour. Geophys. Research,* 99B, 2925–2934.

Jackson, S.T. (1994). Pollen and spores in Quaternary lake sediments as sensors of vegetation composition: theoretical models and empirical evidence. In: *Sedimentation of Organic Particles* (A. Traverse, ed.). Cambridge: Cambridge University Press, 253–286.

Jackson, S.T., Overpeck, J.T., Webb III, T., Keattch, S.E., and Anderson, K.H. (1997). Mapped plant macrofossil and pollen records of late Quaternary vegetation change in eastern North America. *Quaternary Science Reviews,* 16, 1–70.

Jacobson, G.L. and Bradshaw, R. (1981). The selection of sites for paleovegetational studies. *Quaternary Research,* 16, 80–96.

Jacobsen, G.L., Webb III, T., and Grimm, E.C. (1987). Patterns and rates of vegetation change during the deglaciation of eastern North America. In: *North America and Adjacent Oceans During the Last Deglaciation. The Geology of North America,* K-3 (W.F. Ruddiman and H.E. Wright, eds.). Boulder: Geological Society of America, 277–288.

Jacoby, G.C., and D'Arrigo, R.D. (1989). Reconstructed northern hemisphere annual temperature since 1671 based on high latitude tree-ring data from North America. *Climatic Change,* 14, 39–49.

Jacoby, G.C., and D'Arrigo, R.D. (1993). Secular trends in high northern latitude temperature reconstructions based on tree-rings. *Climatic Change,* 15, 163–177.

Jacoby, G.C., and D'Arrigo, R.D. (1995). Tree-ring width and density evidence of climatic and potential forest change in Alaska. *Global Biogeochemical Cycles,* 9, 227–234.

Jacoby, G.C. and Ulan, L.D. (1982). Reconstruction of past ice conditions in a Hudson Bay estuary using tree rings. *Nature, 248*, 637–639.

Jaenicke, R. (1981). Atmospheric aerosols and climate. In: *Climatic Variations and Variability: Facts and Theories* (A. Berger, ed.). Dordrecht: D. Riedel, 577–579.

Jagannathan, P., Arlery, R., Ten Kate, H., and Zavarina, M.V. (1967). *A Note on Climatological Normals*. Technical Note 84, WMO No. 208, TP 108. World Meteorological Organization, Geneva.

Janssen, C.R. (1966). Recent pollen spectra from the deciduous and coniferous-deciduous forests and northeastern Minnesota: a study in pollen dispersal. *Ecology, 47*, 804–825.

Jensen, E. and Veum, T. (1990). Evidence for a two-step deglaciation and its impact on North Atlantic deep-water circulation. *Nature, 343*, 612–616.

Jochimsen, M. (1973). Does the size of lichen thalli really constitute a valid measure for dating glacial deposits? *Arctic and Alpine Research, 5*, 417–424.

Johnsen, S.J., Clausen, H.B., Dansgaard, W., Fuhrer, K., Gundestrup, N., Hammer, C.U., Iversen, P., Jouzel, J., Stauffer, B., and Steffensen, J.P. (1992). Irregular glacial interstadials recorded in a new Greenland ice core. *Nature, 359*, 311–313.

Johnsen, S.J., Clausen, H.B., Dansgaard, W., Gundestrup, N., Hammer, C.U., and Tauber, H. (1995). The Eem stable isotope record along the GRIP ice core and its interpretation. *Quaternary Research, 43*, 117–124.

Johnsen, S.J. and Dansgaard, W. (1992). On flow model dating of stable isotope records from Greenland ice cores. In: *The Last Deglaciation: Absolute and Radiocarbon Chronologies* (E. Bard and W. Broecker, eds.). Berlin: Springer-Verlag, 13–24.

Johnsen, S.J., Dansgaard, W., Clausen, H.B., and Langway, Jr., C.C. (1972). Oxygen isotope profiles through the Antarctic and Greenland ice sheets. *Nature, 235*, 429–434 (see also *Nature, 236*, 249).

Johnsen, S.J., Dansgaard, W., and White, J.W.C. (1989). The origin of Arctic precipitation under glacial and interglacial conditions. *Tellus, 41B*, 452–468.

Johnson, R.G. (1982). Brunhes-Matuyama reversal dated at 790,000 yr B.P. by marine and astronomical correlation. *Quaternary Research, 17*, 135–147.

Johnson, R.G. and Wright, Jr., H.E. (1989). Great Basin calcite vein and the Pleistocene time scale: comment. *Science, 246*, 262 (see also reply by I.C. Winograd and T.B. Coplen, 262–263).

Jolly, D. and Haxeltine, A. (1997). Effect of low glacial atmospheric CO_2 on tropical African montane vegetation. *Science, 276*, 786–788.

Jones, P.D. (1994). Recent warming in global temperature series. *Geophysical Research Letters, 21*, 1149–1152.

Jones, P.D. and Bradley, R.S. (1992). Climatic variations over the last 500 years. In: *Climate Since A.D. 1500* (R.S. Bradley and P.D. Jones, eds.), London: Routledge, 649–665.

Jones, P.D., Bradley, R.S., and Jouzel, J. (eds.), (1996). *Climate Variations and Forcing Mechanisms of the Last 2000 years*. Berlin: Springer-Verlag.

Jones, P.D., Briffa, K.R., and Pilcher, J.R. (1984). Riverflow reconstruction from tree rings in southern Britain. *Jour. Climatology, 4*, 461–472.

Jones, P.D., Briffa, K.R., and Schweingruber, F.H. (1995). Tree-ring evidence of the widespread effects of explosive volcanic eruptions. *Geophysical Research Letters, 22*, 1333–1336.

Joussaume, S. (1987). Desert dust and climate: an investigation using an atmospheric general circulation model. In: *Paleoclimatology and Paleometeorology: Modern and Past Patterns of Global Atmospheric Transport* (M. Leinen and M. Sarnthein, eds.). Dordrecht: Kluwer, 253–263.

Joussaume, S. (1990). Three-dimensional simulations of the atmospheric cycle of desert dust particles using a general circulation model. *Jour. Geophys. Research, 95D*, 1909–1941.

Joussaume, S. (1993). Paleoclimatic tracers: an investigation using an atmospheric general circulation model under ice age conditions. 1. Desert dust. *Jour. Geophys. Research, 98D*, 2767–2805.

Joussaume, S. and Jouzel, J. (1993). Paleoclimatic tracers: an investigation using an atmospheric general circulation model under ice age conditions. 2. Water isotopes. *Jour. Geophys. Research*, **98D**, 2807–2830.

Joussaume, S., Jouzel, J., and Sadorny, R. (1984). A general circulation model of water isotope cycles in the atmosphere. *Nature*, **311**, 24–29.

Joussaume, S. and Taylor, K.E. (1995). Status of the paleoclimate modeling intercomparison project (PMIP). In: *Proc. 1st International AMIP Scientific Conference. WCRP-92, WMO TD-732*, World Meteorological Organization, Geneva, 425–430.

Jouzel, J. (1991). Paleoclimatic tracers. In: *Global Changes of the Past* (R.S. Bradley, ed.). Boulder: University Corporation for Atmospheric Research, 449–476.

Jouzel, J., Barkov, N.I., Barnola, J.M., Genthon, C., Korotkevich, Y.S., Kotlyakov, V.M., Legrand, M., Lorius, C., Petit, J.P., Petrov, V.N., Raisbeck, G., Raynaud, D., Ritz, C., and Yiou, F. (1989b). Global changes over the last climatic cycle from the Vostok ice core record (Antarctica). *Quaternary International*, **2**, 15–24.

Jouzel, J., Barkov, N.I., Barnola, J.M., Bender, M., Chappellaz, J., Genthon, C., Kotlyakov, V.M., Lipenkov, V., Lorius, C., Petit, J.R., Raynaud, D., Raisbeck, G., Ritz, C., Sowers, T., Stievenard, M., Yiou, F., and Yiou, P. (1993b). Extending the Vostok ice-core record of palaeoclimate to the penultimate glacial period. *Nature*, **364**, 407–412.

Jouzel, J., Joussaume, S. and Koster, R.D. (1993a). Use of general circulation models to follow climatic tracers on a global scale. In: *Global Changes in the Perspective of the Past* (J.A. Eddy and H. Oeschger, ed.). Chichester: Wiley, 133–142.

Jouzel, J., Koster, R.D., Suozzo R.J., and Russell, G.L. (1994). Stable water isotope behavior during the last glacial maximum: a general circulation model analysis. *Jour. Geophys. Research*, **99D**, 25791–25801.

Jouzel, J., Koster, R.D., Suozzo, R.J., Russell, G.L., White, J.W.C., and Broecker, W.S. (1991). Simulations of the HDO and $H_2^{18}O$ atmospheric cycles using the NASA GISS general circulation model: sensitivity experiments for present day conditions. *Jour. Geophys. Research*, **96D**, 7495–7507.

Jouzel, J., Lorius, C., Merlivat, L., and Petit, J.R. (1987a). Abrupt climatic changes: the Antarctic ice record during the late Pleistocene. In: *Abrupt Climatic Change: Evidence and Implications* (W.H. Berger and L.D. Labeyrie, eds.). Dordrecht: D. Reidel, 235–245.

Jouzel, J., Lorius, C., Petit, J.R., Genthon, C., Barkov, N.I., Kotlyakov, V.M., and Petrov, V.N. (1987b). Vostok ice core: a continuous isotope temperature record over the last climatic cycle (160,000 years). *Nature*, **329**, 403–408.

Jouzel, J. and Merlivat, L. (1984). Deuterium and oxygen 18 in precipitation: modeling of the isotopic effects during snow formation. *Jour. Geophysical Research*, **89D**, 11749–11757.

Jouzel, J., Merlivat, L., and Lorius, C. (1982). Deuterium excess in an East Antarctic ice core suggests higher relative humidity at the oceanic surface during the last glacial maximum. *Nature*, **299**, 688–691.

Jouzel, J., Merlivat, L., Petit, J.R., and Lorius, C. (1983). Climatic information over the last century deduced from a detailed isotopic record in the South Pole snow. *Jour. Geophys. Research*, **88C**, 2693–2703.

Jouzel, J., Petit, J.R., Barkov, N.I., Barnola, J.M., Chappellaz, J., Ciais, P., Kotlyakov, V.M., Lorius, C., Petrov, V.N., Raynaud, D., and Ritz, C. (1992). The last deglaciation in Antarctica: further evidence of a "Younger Dryas" type climatic event. In: *The Last Deglaciation: Absolute and Radiocarbon Chronologies* (E. Bard and W. Broecker, eds.). Berlin: Springer-Verlag, 229–266.

Jouzel, J., Raisbeck, G., Benoist, J.P., Yiou, F., Lorius, C., Raynaud, D., Petit, J.R., Barkov, N.I., Korotkevich, Y.S., and Kotlyakov, V.M. (1989a). A comparison of deep Antarctic ice cores and their implications for climate between 65,000 and 15,000 years ago. *Quaternary Research*, **31**, 135–150.

Jouzel, J., Russell, G.L., Suozzo, R.J., Koster, R.D., White, J.W.C., and Broecker, W.S. (1987c). Simulations of the HDO and $H_2^{18}O$ atmospheric cycles using the NASA GISS general circulation model: the seasonal cycle for present-day conditions. *Jour. Geophys. Research*, **92D**, 14739–14760.

Juillet-LeClerc, A. and Labeyrie, L.D. (1987). Temperature-dependence of the oxygen isotope fractionation between diatom silica and water. *Earth and Planetary Science Letters*, **84**, 69–74.

Junge, C. (1972). The cycle of atmospheric gases—natural and man-made. *Quart. Jour. Royal Meteorological Society*, **98**, 711–729.

Kalela, O. (1952). Changes in the geographic distribution of Finnish birds and mammals in relation to recent changes in climate. *Fennia*, **75**, 38–51.

Kapsner, W.R., Alley, R.B., Shuman, C.A., Anandakrishnan, S., and Grootes, P.M. (1995). Dominant influence of atmospheric circulation on snow accumulation in Greenland over the past 18,000 years. *Nature*, **373**, 52–54.

Karl, T. (1993). Missing pieces of the puzzle. *National Geographic Research and Exploration*, **12**, 234–249.

Karlén, W. (1976). Lacustrine sediments and tree-limit variations as indicators of Holocene climatic fluctuations in Lappland, northern Sweden. *Geografiska Annaler*, **58A**, 1–34.

Karlén, W. (1979). Glacier variations in the Svartisen area, northern Norway. *Geografiska Annaler*, **61A**, 11–28.

Karlén, W. (1980). Reconstruction of past climatic conditions from studies of glacier-front variations. *World Meteorological Organisation Bulletin*, **29**, 100–104.

Karlén, W. (1981). Lacustrine sediment studies. *Geografiska Annaler*, **63A**, 273–281.

Karlén, W. (1993). Glaciological, sedimentological and palaeobotanical data indicating Holocene climatic change in Northern Fennoscandia. In: *Oscillations of the Alpine and Polar Tree Limits in the Holocene* (B. Frenzel, M. Eronen, K.D. Vorren, and B. Gläser, ed.) Stuttgart: Gustav Fischer Verlag, 69–83.

Karrow, P.F. and Anderson, T.W. (1975). Palynological studies of lake sediment profiles from SW New Brunswick: discussion. *Canadian Jour. Earth Sciences*, **12**, 1808–1812.

Karte, J. and Liedtke, H. (1981). The theoretical and practical definition of the term "periglacial" in its geographical and geological meaning. *Biuletyn Periglacjalny*, **28**, 123–135.

Kato, K. (1978). Factors controlling oxygen isotope composition of fallen snow in Antarctica. *Nature*, **272**, 46–48.

Kaufman, A., Broecker, W.S., Ku, T.L., and Thurber, D.L. (1971). The status of U-series methods of mollusc dating. *Geochimica et Cosmochimica Acta*, **35**, 1155–1183.

Kaufman, D.S. and Sejrup, H.P. (1995). Isoleucine epimerization in the high-molecular-weight fraction of Pleistocene Arctica. *Quaternary Science Reviews (Quaternary Geochronology)*, **14**, 337–350.

Kawamura, T. (1992). Estimation of climate in the Little Ice Age using phenological data in Japan. In: *Proc. International Symposium on the Little Ice Age Climate* (T. Mikami, ed.). Department of Geography, Tokyo Metropolitan University, 52–57.

Keigwin, L.D. (1996). The Little Ice Age and Medieval Warm Period in the Sargasso Sea. *Science*, **274**, 1504–1508.

Keigwin, L.D., Curry, W.B., Lehman, S.J., and Johnsen, S. (1994). The role of the deep ocean in North Atlantic climate change between 70 and 130 kyr ago. *Nature*, **371**, 323–326.

Keigwin, L.D. and Jones, G.A. (1994). Western North Atlantic evidence for millennial-scale changes in ocean circulation and climate. *Jour. Geophys. Research*, **C99**, 12397–12410.

Keigwin, L.D. and Jones, G.A. (1995). The marine record of deglaciation from the continental margin off Nova Scotia. *Paleoceanography*, **10**, 973–985.

Keigwin, L.D., Jones, G.A., Lehman, S.J., and Boyle, E.A. (1991). Deglacial meltwater discharge, North Atlantic deep circulation and abrupt climate change. *Jour. Geophys. Research*, **96C**, 16811–16826.

Kellogg, T.B. (1980). and paleo-oceanography of the Norwegian and Greenland seas: glacial-interglacial contrasts *Boreas*, **9**, 115–37.

Kellogg, T.B. (1987). Glacial-interglacial changes in global deepwater production. *Paleoceanography*, **2**, 259–272.

Kellogg, W.W. (1975). Climatic feedback mechanisms involving the polar region. In: *Climate of the Arctic*. (G. Weller and S.A. Bowling, eds.). Fairbanks: Univ. of Alaska Press, 111–116.

Kennedy, J.A. and Brassell, S.C. (1992). Molecular records of twentieth century El Niño events in laminated sediments from the Santa Barbara basin. *Nature, 357,* 62–64.

Kennett, J.P. (1976). Phenotypic variation in some recent and late Cenozoic planktonic foraminifera. In: *Foraminifera,* **2** (R.H. Hedley and C.G. Adams, eds.). New York: Academic Press, 111–170.

Kiehl, J.T. (1992). Atmospheric general circulation modeling. In: *Climate System Modeling* (K.E. Trenberth, ed.). Cambridge: Cambridge University Press, 319–370.

Kiehl, J.T. and Trenberth, K.E. (1997). Earth's annual global mean energy budget. *Bull. American Meteorological Society*, **78**, 197–208.

Kim, G.S. and Choi, I.S. (1987). A preliminary study on long-term variation of unusual climate phenomena during the past 1000 years in Korea. In: *The Climate of China and Global Climate* (D. Ye, C. Fu, J. Chao, and M. Yoshino, eds.). Beijing: China Ocean Press, 30–37.

Kim, Y.O. (1984). The Little Ice Age in Korea: an approach to historical climatology. *Geography-Education*, **14**, 1–16 (in Korean with English abstract). Department of Geography, College of Education, Seoul National University.

Kim, Y.O. (1987). Climatic environment of Chosun Dynasty (1392–1910) based on historical records. *Jour. Geography (Seoul)*, **14**, 411–423. (in Korean with English abstract). Department of Geography, Seoul National University.

King, J.E. and Van Devender, T.R. (1977). Pollen analysis of fossil packrat middens from the Sonoran Desert. *Quaternary Research, 8,* 191–204.

King, K. and Neville, C. (1977). Isoleucine epimerization for dating marine sediments: the importance of analyzing monospecific samples. *Science, 195,* 1333–1335.

King, L. and Lehmann, R. (1973). Beobachtung zur oekologie und morphologie von *Rhizocarpon Geographicurn* (L) D.C. und *Rhizocarpon Alpicola* (Hepp.) Rabenh. in gletschervorfeld des steingletschers. *Berichte der Schweizerischen Botanischen Gesellschaft,* **83,** 139–146.

Kipp, N.G. (1976). New transfer function for estimating past sea-surface conditions from sea-bed distribution of planktonic foraminiferal assemblages in the North Atlantic. In: *Investigation of late Quaternary Paleooceanography and Paleoclimatology*. (R.M. Cline and J.D. Hays, eds.). Memoir No. 145. Boulder: Geological Society of America, 3–42.

Kittleman, L.R. (1979). Geologic methods in studies of Quaternary tephra. In: *Volcanic Activity and Human Ecology* (D.D. Sheets and D.K. Grayson, eds.). New York: Academic Press, 49–82.

Koç Karpuz, N. and Jansen, E. (1992). A high resolution diatom record of the last deglaciation from the S.E. Norwegian Sea: documentation of rapid climatic changes. *Paleoceanography, 7,* 499–520.

Koç Karpuz, N. and Schrader, H. (1990). Surface sediment diatom distribution and Holocene palotemperature variations in the Greenland, Iceland and Norwegian Sea. *Paleoceanography, 5,* 557–580.

Koerner, R.M. (1977). Devon Island Ice Cap: core stratigraphy and paleoclimate. *Science, 196,* 15–18.

Koerner, R.M. (1979). Accumulation, ablation and oxygen isotope variations on the Queen Elizabeth Islands Ice Caps, Canada. *Jour. Glaciology, 22,* 25–41.

Koerner, R.M. (1980). The problem of lichen-free zones in Arctic Canada. *Arctic and Alpine Research, 12,* 87–94.

Koerner, R.M. (1989). Ice core evidence for extensive melting of the Greenland Ice Sheet in the last interglacial. *Science, 244,* 964–968.

Koerner, R.M. and Fisher, D.A. (1979). Discontinuous flow, ice texture and dirt content in the basal layers of the Devon Island Ice Cap. *Jour. Glaciology, 23,* 209–222.

Koerner, R.M. and Fisher, D.A. (1990). A record of Holocene summer climate from a Canadian high-Arctic ice core. *Nature*, 343, 630–631.

Koerner, R.M. and Russell, R.P. (1979). $\delta^{18}O$ variations in snow on the Devon Island Ice Cap, North West Territories, Canada. *Canadian Jour. Earth Sciences*, 16, 1419–1427.

Koerner, R.M. and Taniguchi, H. (1976). Artificial radioactivity layers in the Devon Island Ice Cap, North West Territories. *Canadian Jour. Earth Sciences*, 13, 1251–1255.

Kohler, M.A., Nordenson, T.J., and Baker, D.R. (1966). *Evaporation maps for the United States*, US Weather Bureau Technical Paper No. 37. Washington, DC: US Department of Commerce.

Koide, M. and Goldberg, E.D. (1985). The historical record of artificial radioactive fallout from the atmosphere in polar glaciers. In: *Greenland Ice Core: Geophysics, Geochemistry and the Environment* (C.C. Langway, Jr., H. Oeschger, and W. Dansgaard, eds.). American Geophysical Union, Washington DC: 95–100.

Kolla, V., Biscaye, P.E., and Hanley, A.F. (1979). Distribution of quartz in late Quaternary Atlantic sediments in relation to climate. *Quaternary Research*, 11, 261–277.

Kominz, M.A., Heath, G.R., Ku, T-L., and Pisias, N.G. (1979). Brunhes time scales and the interpretation of climatic change. *Earth and Planetary Science Letters*, 45, 394–410.

Korff, H.Cl. and Flohn, H. (1969). Zusammenhang zwischen dem Temperaturgefälle Äquator-Pol und den planetarischen Luftdruckgurteln. *Annaler Meteorologisch*, 4, 163–164.

Koster, R.D., Broecker, W.S., Jouzel, J., Suozzo, R., Russell, G., Rind, D., and White, J.W.C. (1989). The global geochemistry of bomb-produced tritium: general circulation model compared to available observations and traditional interpretations. *Jour. Geophys. Research*, 94D, 18305–18326.

Koster, R.D., Jouzel, J., Souzzo, R.J., and Russell, G.L. (1992). Origin of July precipitation and its influence on deuterium content: a GCM analysis. *Climate Dynamics*, 7, 195–203.

Krebs, J.S. and Barry, R.G. (1970). The arctic front and the tundra-taiga boundary in Eurasia. *Geographical Review*, 60, 548–554.

Kromer, B. and Becker, B. (1993). German oak and pine ^{14}C calibration, 7200–9439 B.C. *Radiocarbon*, 35, 125–135.

Kroopnick, P.M. (1985). The distribution of ^{13}C in the world oceans. *Deep Sea Research*, 32, 57–84.

Ku, T-L. (1976). The uranium series method of age determination. *Annual Reviews of Earth and Planetary Science*, 4, 347–380.

Ku, T-L. and Oba, T.(1978). A method for quantitative evaluation of carbonate dissolution in deep-sea sediments and its application to paleooceanographic reconstruction. *Quaternary Research*, 10, 112–129.

Kuhle, M. (1991). Observations supporting the Pleistocene inland glaciation of High Asia. *Geojournal*, 25, 133–232.

Kukla, G.J. (1975a). Missing link between Milankovitch and climate. *Nature*, 253, 600–603.

Kukla, G.J. (1975b). Loess stratigraphy of central Europe. In: *After the Australopithecenes* (K.W. Butzer and G. L. Isaac, eds.). The Hague: Mouton, 99–188.

Kukla, G.J. (1977). Pleistocene land-sea correlations. I. Europe. *Earth Science Reviews*, 13, 307–374.

Kukla, G.J. (1978). Recent changes in snow and ice. In: *Climatic Change* (J. Gribbin, ed.). Cambridge: Cambridge University Press, 114–130.

Kukla, G. (1979). Climatic role of snow covers. In: *Sea level, Ice and Climatic Change*, Publication No. 131, Washington DC: International Association of Scientific Hydrology, 79–107.

Kukla, G.J. (1987a). Loess stratigraphy in central China. *Quaternary Science Reviews*, 6, 191–220.

Kukla, G.J. (1987b). Pleistocene climates in China and Europe compared to oxygen isotope record. *Paleoecology of Africa*, 18, 37–45.

Kukla, G.J. and An, Z. (1989). Loess stratigraphy in central China. *Palaeogeography, Palaeoclimatology, Palaeocology*, 72, 203–225.

Kukla, G., An, Z. Melice, J.L., Gavin, J., and Xiao, J.L. (1990). Magnetic susceptibilty record of Chinese loess. *Trans. Royal Society of Edinburgh: Earth Sciences*, **81**, 263–288.

Kukla, G. and Robinson, D. (1980). Annual cycle of surface albedo. *Monthly Weather Review*, **108**, 56–68.

Kullman, L. (1987). Long-term dynamics of high-altitude populations of *Pinus sylvestris* in the Swedish Scandes. *Jour. Biogeography*, **14**, 1–8.

Kullman, L. (1988). Holocene history of the forest-alpine tundra ecotone in the Scandes Mountains (central Sweden). *New Phytologist*, **108**, 101–110.

Kullman, L. (1989). Tree-limit history during the Holocene in the Scandes Mountains, Sweden inferred from sub-fossil wood. *Rev. Paleobotany and Palynology*, **58**, 163–171.

Kullman, L. (1993). Dynamism of the altitudinal margin of the boreal forest in Sweden. In: *Oscillations of the Alpine and Polar Tree Limits in the Holocene* (B. Frenzel, M. Eronen, K.D. Vorren, and B. Glaser, eds.) Stuttgart: Gustav Fischer Verlag, 41–55.

Kumar, N., Anderson, R.F., Mortlock, R.A., Froelich, P.N., Kubik, P., Dittrich-Hannen, D., and Suter, M. (1995). Increased biological productivity and export production in the glacial Southern Ocean. *Nature*, **378**, 675–680.

Kuniholm, P.I., Kromer, B., Manning, S.W., Newton, M., Latini, C.E., and Bruce, M.J. (1996). Anatolian tree rings and the absolute chronology of the eastern Mediterranean, 2220–718 B.C. *Nature*, **381**, 780–783.

Kutzbach, J.E. (1974). Fluctuations of climate–monitoring and modelling. *World Meteorological Organisation Bulletin*, **23**, 155–163.

Kutzbach, J.E. (1976). The nature of climate and climatic variations. *Quaternary Research*, **6**, 471–480.

Kutzbach, J.E. (1980). Estimates of past climate at Paleolake Chad, North Africa, based on a hydrological and energy balance model. *Quaternary Research*, **14**, 210–223.

Kutzbach, J.E. (1983). Monsoon rains of the late Pleistocene and early Holocene: patterns, intensity and possible causes of changes. In: *Variations in the Global Water Budget* (F.A. Street-Perrott, M. Beran and R. Ratcliffe, eds.). Dordrecht: D. Reidel, 371–389.

Kutzbach, J.E., Bartlein, P.J., Prentice, I.C., Ruddiman, W.F., Street-Perrott, F.A., Webb III, T., and Wright, Jr., H.E. (1993a). Epilogue. In: *Global Climate Since the Last Glacial Maximum* (H.E. Wright, J.E. Kutzbach, T. Webb, III, W.F. Ruddiman, F.A. Street-Perrott, and P.J. Bartlein, eds.), Minneapolis: University of Minnesota Press, 536–542.

Kutzbach, J.E., Bonan, G., Foley, J., and Harrison, S.P. (1996). Vegetation and soil feedbacks on the response of the African monsoon to orbital forcing in the early to middle Holocene. *Nature*, **384**, 623–626.

Kutzbach, J.E. and Gallimore, R.G. (1988). Sensitivity of a coupled atmosphere/mixed layer ocean model to changes in orbital forcing at 9000 years B.P. *Jour. Geophys. Research*, **93D**, 803–821.

Kutzbach, J.E., Gallimore, R.G., and Guetter, P.J. (1991). Sensitivity experiments on the effect of orbitally-caused insolation changes on the interglacial climate of high northern latitudes. *Quaternary International*, **10–12**, 223–229.

Kutzbach, J.E., Gallimore, R., Harrison, S.P., Behling, P., Selin, R., and Laarif, F. (1997). Climate and biome simulations for the past 21,000 years. *Quaternary Science Reviews*, **17**, 473–506.

Kutzbach, J.E. and Guetter, P.J. (1986). The influence of changing orbital parameters and surface boundary conditions on climate simulations for the past 18,000 years. *Jour. Atmos. Sciences*, **43**, 1726–1759.

Kutzbach, J.E., Guetter, P.J., Behling, P.J., and Selin, R. (1993b). Simulated climatic changes: results of the COHMAP climate-model experiments. In: *Global Climate Since the Last Glacial Maximum* (H.E. Wright, J.E. Kutzbach, T. Webb, III, W.F. Ruddiman, F.A. Street-Perrott, and P.J. Bartlein, eds.). Minneapolis: University of Minnesota Press, 24–93.

Kutzbach, J.E. and Otto-Bliesner, B. (1982). The sensitivity of the African-Asian monsoonal climate to orbital parameter changes for 9000 years B.P. in a low-resolution general circulation model. *Jour. Atmos. Sciences*, **39**, 1177–1188.

Kutzbach, J.E. and Ruddiman, W.F. (1993). Model description, external forcing and surface boundary conditions. In: *Global Climate Since the Last Glacial Maximum* (H.E. Wright, J.E. Kutzbach, T. Webb III, W.F. Ruddiman, F.A. Street-Perrott, and P.J. Bartlein, eds.), Minneapolis: University of Minnesota Press, 12–23.

Kutzbach, J.E. and Street-Perrott, A.F. (1985). Milankovitch forcing of fluctuations in the level of tropical lakes from 18 to 0 kyr B.P. *Nature*, **317**, 130–134.

Kutzbach, J.E. and Wright, H.E. (1985). Simulation of the climate of 18,000 years B.P.: results for the North American/North Atlantic/European sector and comparison with the geological record of North America. *Quaternary Science Reviews*, **4**, 147–187.

Kvamme, M. (1993). Holocene forest limit fluctuations and glacier development in the mountains of southern Norway, and their relevance to climate history. In: *Oscillations of the Alpine and Polar Tree Limits in the Holocene* (B. Frenzel, M. Eronen, K.D. Vorren, and B. Gläser, eds.). Stuttgart: Gustav Fischer Verlag, 99–113.

Kvamme, T., Mangerud, J., Furnes, H., and Ruddiman, W.F. (1989). Geochemistry of Pleistocene ash zones in cores from the North Atlantic. *Norsk Geologisk Tidsskrift*, **69**, 251–272.

Labeyrie, L., Duplessy, J-C., and Blanc, P.L. (1987). Variations in mode of formation and temperature of oceanic waters over the past 125,000 years. *Nature*, **327**, 477–482.

Labrijn, A. (1945). Het Klimaat van Nederland gedurende de laatste twee en een halve eeuw. *Mededelingen en Verhandlingen*, **49**, 11–105 (Koninklijk Nederlands Met. Inst. No. 102).

Lajoux, J-D. (1963). *The Rock Paintings of the Tassili*. London: Thames and Hudson.

LaMarche, V.C. (1973). Holocene climatic variations inferred from treeline fluctuations in the White Mountains, California. *Quaternary Research*, **3**, 632–660.

LaMarche, V.C. (1982). Sampling strategies. In: *Climate from Tree Rings* (M.K. Hughes, P.M. Kelly, J.R. Pilcher, and V.C. LaMarche, eds.). Cambridge: Cambridge University Press, 2–6.

LaMarche, V.C. and Fritts, H.C. (1971). Tree rings, glacial advance and climate in the Alps. *Zeitschrift für Gletscherkunde und Glazialgeologie*, **7**, 125–131.

LaMarche, V.C. and Mooney, H.A. (1967). Altithermal timberline advance in western United States. *Nature*, **213**, 980–982.

LaMarche, V.C. and Mooney, H.A. (1972). Recent climatic change and development of the bristlecone pine *(P. longaeva* Bailey) Krummholz zone, Mt. Washington, Nevada. *Arctic and Alpine Research*, **4**, 61–72.

Lamb, H.H. (1961). Climatic change within historical time as seen in circulation maps and diagrams. *Annals of the New York Academy of Sciences*, **95**, 124–161.

Lamb, H.H. (1963). On the nature of certain climatic epochs which differed from the modern (1900–1939) normal. In: *Changes of Climate* (Proceedings of the WMO-UNESCO Rome 1961 Symposium on Changes of Climate), UNESCO Arid Zone Research Series XX. UNESCO, Paris, 125–150.

Lamb, H.H. (1965). The early Medieval warm epoch and its sequel. *Palaeogeography, Palaeoclimatology, Palaeoecology*, **1**, 13–37.

Lamb, H.H. (1970). Volcanic dust in the atmosphere; with a chronology and assessment of its meteorological significance. *Phil. Trans. Royal Society of London*, **A266**, 425–533.

Lamb, H.H. (1977). *Climate, Present, Past and Future*, **2**. London: Methuen.

Lamb, H.H. (1983). Update of the chronology of assessments of the volcanic dust veil index. *Climate Monitor*, **12**, 79–90.

Lamb, H.H. (1988). Climate and life during the Middle Ages, studied especially in the mountains of Europe. In: *Weather, Climate and Human Affairs*. London: Routledge, 40–74.

Lambert, G-N., Bernard, V., Doncerain, C., Girardclos, O., Lavier, C., Szepertisky, B., and Trenard, Y. (1996). French regional oak chronologies spanning more than 1000 years. In: *Tree Rings, Environment and Humanity* (J.S. Dean, D.M. Meko, and T.W. Swetnam, eds.). Tucson: University of Arizona, 821–832.

Lamothe, M., Balescu, S., and Auclair, M. (1994). Natural IRSL intensities and apparent luminescence ages of single feldspar grains extracted from partially bleached sediments. *Radiation Measurements*, **23**, 555–561.

Langbein, W.B. (1961). *Salinity and Hydrology of Enclosed Lakes,* US Geological Survey Professional Paper 412, Washington, DC: US Geological Survey.

Langbein, W.B., *et al.* (1949). *Annual runoff in the United States,* US Geological Survey Circular 52. Washington, DC: US Geological Survey.

Langway, Jr., C.C., Clausen, H.B., and Hammer, C.U. (1988). An inter-hemispheric volcanic time-marker in ice cores from Greenland and Antarctica. *Annals of Glaciology,* **10,** 102–108.

Larsen, J.A. (1974). Ecology of the northern continental forest border. In: *Arctic and Alpine Environments* (J.D. Ives and R.G. Barry, eds.). London: Methuen, 341–369.

Lauritzen, S-E. (1996). Calibration of speleothem stable isotopes against historical records: a Holocene temperature curve for north Norway? In: *Climatic Change: the Karst Record* (S-E. Lauritzen, ed.). Charleston, West Virginia: Karst Waters Institute Special Publication 2, 78–80.

Lauritzen, S.E., Haugen, J.E., Løvlie, R., and Gilje-Nielson, H. (1994). Geochronological potential of isoleucine epimerization in calcite speleothems. *Quaternary Research,* **41,** 52–58.

Lauritzen, S.E., and Lundberg, J. (1998). Rapid temperature variations and volcanic events during the Holocene from a Norwegian speleothem record. In: *Past Global Changes and their Significance for the Future.* Bern: IGBP-PAGES, 88.

Lautenschlager, M. and Herterich, K. (1990). Atmospheric response to ice age conditions: climatology near the earth's surface. *Jour. Geophys. Research,* **95D,** 22547–22557.

Lavier, C. and Lambert, G-N. (1996). Dendrochronology and works of art. In: *Tree Rings, Environment and Humanity* (J.S. Dean, D.M. Meko, and T.W. Swetnam, eds.). Tucson: University of Arizona, 543–556.

Lavoie, C. and Payette, S. (1992). Black spruce growth forms as a record of a changing winter environment at treeline, Quebec, Canada. *Arctic and Alpine Research,* **24,** 40–49.

Lawrence, J.R. and White, J.W.C. (1984). Precipitation amounts during the growing season from the D/H ratios of eastern white pine. *Nature,* **311,** 558–560.

Lawrence, J.R. and White, J.W.C. (1991). The elusive climate signal in the isotopic composition of precipitation. In: *Stable Isotope Geochemistry: a Tribute to Samuel Epstein* (H.P Taylor, Jr., J.R. O'Neil, and I.R. Kaplan, eds.). Special Publication No.3, The Geochemical Society, 169–185.

Le Roy Ladurie, E. (1971). *Times of Feast, Times of Famine.* New York: Doubleday.

Le Roy Ladurie, E. and Baulant, M. (1981). Grape harvests from the fifteenth through the nineteenth centuries. In: *Climate and History: Studies in Interdisciplinary History,* (R.I. Rotberg and T.K. Rabb, eds.). Princeton: Princeton University Press, 259–269.

Lea, D.W., Shen, G.T,. and Boyle, E.A. (1989). Coralline barium records temporal variability in equatorial Pacific upwelling. *Nature,* **340,** 373–376.

Lean, J. (1994). Solar forcing of global change. In: *The Solar Engine and its Influence on Terrestrial Atmosphere and Climate,* (ed. E. Nesmé-Ribes). New York: Springer-Verlag, 163–184.

Lean, J., Beer, J., and Bradley, R.S. (1995). Reconstruction of solar irradiance since A.D. 1600 and implications for climate change. *Geophysical Research Letters,* **22,** 3195–3198.

Lean, J. and Rind, D. (1994). Solar variability: implications for global change. *Eos, Transactions, American Geophysical Union,* **75,** 1 and 5–7.

Lean, J., Skumanich, A., and White, O.R. (1992). Estimating the Sun's radiative output during the Maunder Minimum. *Geophysical Research Letters,* **19,** 1591–1594.

Leemann, A. and Niessen, F. (1994a). Varve formation and the climatic record in an Alpine proglacial lake: calibrating annually laminated sediments against hydrological and meteorological data. *The Holocene,* **4,** 1–8.

Leemann, A. and Niessen, F. (1994b). Holocene glacial activity and climatic variations in the Swiss Alps: reconstructing a continuous record from proglacial lake sediments. *The Holocene,* **4,** 259–268.

Legrand, M. and Delmas, R.J. (1987). A 220-year continuous record of volcanic H2SO4 in the Antarctic Ice Sheet. *Nature,* **327,** 671–676.

Legrand, M.R., Delmas, R.J., and Charlson, R.J. (1988). Climate forcing implications from Vostok in ice core sulfate data. *Nature*, **354**, 418–420.

Legrand, M., Feniet-Saigne, C., Saltzman, E.S., Germain, C., Barkov, N.I., and Petrov, V.N. (1991). Ice core record of oceanic emissions of dimethyl-sulphide during the last glacial cycle. *Nature*, **350**, 144–146.

Lehman, S.J., Jones, G.A., Keigwin, L.D., Andersen, E.S., Butenko, G., and Østmo, S.R. (1991). Initiation of the Fennoscandian ice-sheet retreat during the last deglaciation. *Nature*, **349**, 513–516.

Lehman, S.J. and Keigwin, L.D. (1992a). Sudden changes in North Atlantic circulation during the last deglaciation. *Nature*, **356**, 757–762.

Lehman, S.J. and Keigwin, L.D. (1992b). Deep circulation revisited. *Nature*, **358**, 197–198.

Leonard, E.M. (1997). The relationship between glacial activity and sediment production: evidence from a 4450-year varve record of neoglacial sedimentation in Hector Lake, Alberta, Canada. *Jour. Paleolimnology*, **17**, 319–330.

Leopold, L.B. (1951). Pleistocene climate in New Mexico. *American Jour. Science*, **249**, 152–168.

Lerbemko, J.F., Westgate, J.A., Smith, D.G.W., and Denton, G.H. (1975). New data on the character and history of the White River volcanic eruption, Alaska. In: *Quaternary Studies* (R.P. Suggate and M.M. Cresswell, eds.). Bulletin No. 13, Wellington: Royal Society of New Zealand, 203–209.

Lerman, J.C. (1972). Carbon-14 dating: origin and correction of isotope fractionation errors in terrestrial living matter. In: *Proc. 8th International Conference on Radiocarbon Dating*, **2**. Wellington: Royal Society of New Zealand, 613–624.

Letréguilly, A., Reeh, N., and Huybrechts, P. (1991). The Greenland Ice Sheet through the last glacial-interglacial cycle. *Global and Planetary Change*, **90**, 385–394.

Lettau, H. (1969). Evapotranspiration climatonomy, I. A new approach to numerical prediction of monthly evapotranspiration, runoff and soil moisture storage. *Monthly Weather Review*, **97**, 691–699.

Leuenberger, M., and Siegenthaler, U. (1992). Ice-age atmospheric concentration of nitrous oxide from an Antarctic ice core. *Nature*, **360**, 449–451.

Levi, S., Gudmunsson, H., Duncan, R.A., Kristjansson, L., Gillot, P.V., and Jacobsson, S.P. (1990). Late Pleistocene geomagnetic excursion in Icelandic lavas: confirmation of the Laschamp excursion. *Earth and Planetary Science Letters*, **96**, 443–457.

Levi, S. and Karlin, R. (1989). A sixty thousand year paleomagnetic record from Gulf of California sediments: secular variation, late Quaternary excursions and geomagnetic implications. *Earth and Planetary Science Letters*, **92**, 219–233.

Levitus, S. (1982). *Climatological Atlas of the World Ocean*. Rockville, MD: NOAA Professional Paper 13.

Levitus, S. (1989). Interpentadal variability of temperature and salinity at intermediate depths of the North Atlantic Ocean, 1970–1974 versus 1955–1959. *Jour. Geophys. Research*, **94C**, 6091–6131.

Lézine, A.-M. (1989). Late Quaternary vegetation and climate of the Sahel. *Quaternary Research*, **32**, 317–334.

Lhoté, H. (1959). *The Search for the Tassili Frescoes*. London: Hutchinson.

Li, P. (1992). South China climate change during the Little Ice Age (16th–19th century). In: *Proc. International Symposium on the Little Ice Age Climate* (T. Mikami, ed.). Department of Geography, Tokyo Metropolitan University, 138–142.

Li, W-X., Lundberg, J., Dickin, A.P, Ford, D.C., Schwarcz, H.P., McNutt, R., and Williams, D. (1989). High precision mass spectrometric uranium-series dating of cave deposits and implications for paleoclimate studies. *Nature*, **339**, 534–536.

Li, X.S., Berger, A., Loutre, M.F., Maslin, M.A., Haug, G.H., and Tiedemann, R. (1998). Simulating late Pliocene northern hemisphere climate with the LLN 2-D model. *Geophysical Research Letters*, **25**, 915–918 (see also, correction in **25**, 2719).

Libby, W.F. (1955). *Radiocarbon Dating*. Chicago: University of Chicago Press.

Libby, W.F. (1970). Radiocarbon dating. *Phil. Trans. Royal Society of London*, **A269**, 1–10.

Lieth, H. (1975). Primary productivity in ecosystems: comparative analysis of global patterns. In: *Unifying Concepts in Ecology: Report of Plenary Sessions, 1st International Congress on Ecology* (W.H. van Dobben and R.H. Lowe-McConnell, eds.). The Hague: Dr W. Junk BV, 67–88.

Lin, P-N., Thompson, L.G., Davis, M.E., and Mosley-Thompson, E. (1995). 1000 years of climatic change in China: ice-core δ¹⁸O evidence. *Annals of Glaciology*, **21**, 189–195.

Lindsey, A.A. and Newman, J.E. (1956). Use of official data in spring time temperature analysis of Indiana phenological record. *Ecology*, **37**, 812–823.

Lindzen, R., (1993). Absence of scientific basis. *Research and Exploration*, **9**, 191–200.

Linsley, B.K. (1996). Oxygen-isotope record of sea level and climate variations in the Sulu Sea over the past 150,000 years. *Nature*, **380**, 234–237.

Linsley, B.K., Dunbar, R.B., Wellington, G.M., and Mucciarone, D.A. (1994). A coral-based reconstruction of Inter-Tropical Convergence Zone variability over central America since 1707. *Jour. Geophysical Research*, **99C**, 9977–9994.

Lipp, J., Trimborn, P., Fritz, P., Moser, H., Becker, B., and Frenzel, B. (1991). Stable isotopes in tree ring cellulose and climatic change. *Tellus*, **43B**, 322–330.

Litherland, A.E. and Beukens, R.P. (1995). Radiocarbon dating by atom counting. In: *Dating Methods for Quaternary Deposits* (N.W Rutter and N.R. Catto, eds.). Geological Association of Canada, St. John's, 117–123.

Liu, K., Yao, Z. and Thompson, L.G. (1998). A pollen record of Holocene climatic changes from the Dunde ice cap, Qinghai-Tibetan Plateau. *Geology*, **26**, 135–138.

Liu, K-B., and Colinvaux, P.A. (1985). Forest changes in the Amazon Basin during the last glacial maximum. *Nature*, **318**, 556–557.

Liu, T., *et al.* (un-named), (1985). *Loess and The Environment*. Beijing: China Ocean Press.

Liu, T., Ding, Z., Yu, Z., and Rutter, N. (1993). Susceptibility time series of the Baoji section and the bearings on paleoclimatic periodicities in the last 2.5 Ma. *Quaternary International*, **17**, 33–38.

Liu, X, Rolph, T., Bloemendal, J., Shaw, J., and Liu, T. (1994). Remanence characteristics of different magnetic grain size categories at Xifeng, central Chinese Loess Plateau. *Quaternary Research*, **42**, 162–165.

Liu, X, Rolph, T., Bloemendal, J., Shaw, J., and Liu, T., (1995). Quantitative estimates of palaeoprecipitation at Xifeng, in the Loess Plateau of China. *Palaeogeography, Palaeoclimatology, Palaeoecology*, **113**, 243–248.

Livingstone, D.A. (1982). Quaternary geography and Africa and the refuge theory. In: *Biological Diversification in the Tropics* (G.T. Prance, ed.). New York: Columbia University Press, 523–536.

Locke, C.W. and Locke, W.W. (1977). Little ice age snow-cover extent and paleoglaciation thresholds: north-central Barfin Island, NWT, Canada. *Arctic and Alpine Research*, **9**, 291–300.

Locke, W.W., Andrews, J.T., and Webber, P.J. (1979). *A Manual for Lichenometry,* British Geomorphological Research, Group, Technical Bulletin No. 26. Norwich: University of East Anglia.

Löffler, E. (1976). Potassium-argon dates and pre-Wurm glaciations of Mount Giluwe volcano, Papua, New Guinea. *Zeitschrift fur Gletscherkunde und Glazialgeologie*, **12**, 55–62.

Lorenz, E.N. (1968). Climatic determinism. *Meteorological Monographs*, **8**, 1–3.

Lorenz, E.N. (1970). Climatic change as a mathematical problem. *Jour. Applied Meteorology*, **9**, 235–239.

Lorenz, E.N. (1976). Non-deterministic theories of climatic change. *Quaternary Research*, **6**, 495–507.

Lorius, C. (1991). Polar Ice cores: a record of climatic and environmental changes. In: *Global Changes of the Past* (R.S. Bradley, ed.). Boulder: University Corporation for Atmospheric Research, 261–294.

Lorius, C., Barkov, N.I., Jouzel, J., Korotkevich, Y.S., Kotlyakov, V.M., and Raynaud, D. (1988). Antarctic ice core: CO_2 and climatic change over the last climatic cycle. *EOS* **69**, 681 and 683–684.

Lorius, C., Jouzel, J., Raynaud, D., Hansen, J., and Le Treut, H. (1990). The ice core record: climate sensitivity and future greenhouse warming. *Nature, 347,* 139–145.

Lorius, C., Jouzel, J., Ritz, C., Merlivat, L., Barkov, N.I., Korotkevich, Y.S., and Kotlyakov, V.M. (1985). A 150,000 year climatic record from Antarctic ice. *Nature, 316,* 591–596.

Lorius, C., Raynaud, D., Petit, J.R., Jouzel, J., and Merlivat, L. (1984). Late Glacial Maximum-Holocene atmospheric and ice thickness changes from Antarctic ice core studies. *Annals of Glaciology, 5,* 88–94.

Lotter, A.F. (1991). Absolute dating of the late-Glacial period in Switzerland using annually laminated sediments. *Quaternary Research, 35,* 321–330.

Lotter, A.F., Ammann, B., and Sturm, M. (1992). Rates of change and chronological problems during the late-glacial period. *Climate Dynamics, 6,* 233–239.

Lough, J.M. (1992). An index of the Southern Oscillation reconstructed from western North America tree-ring chronologies. In: *El Niño: Historical and Paleoclimatic Aspects of the Southern Oscillation* (H.F. Diaz and V. Markgraf, eds.). Cambridge: Cambridge University Press, 215–226.

Lough, J.M. and Barnes, D.J. (1990). Intra-annual timing of density band formation of *Porites* coral from the central Great Barrier Reef. *Jour. Experimental Marine Biology and Ecology, 135,* 35–57.

Lough, J.M., Barnes, D.J., and Taylor, R.B. (1996). The potential of massive corals for the study of high resolution climate variation in the past millennium. In: *Climatic Variations and Forcing Mechanisms of the Last 2000 Years* (P.D. Jones, R.S. Bradley, and J. Jouzel, eds.). Berlin: Springer-Verlag, 355–371.

Lough, J.M. and Fritts, H.C. (1985). The Southern Oscillation and tree rings: 1600-1961. *Jour. Climate and Appl. Meteorology, 24,* 952–966.

Loutre, M-F., Berger, A., Bretagnon, P., and Blanc, P-L. (1992). Astronomical frequencies for climatic research at the decadal to century time scale. *Climate Dynamics, 7,* 181–194.

Lozano, J.A. and Hays, J.D. (1976). Relationship of radiolarian assemblages to sediment types and physical oceanography in the Atlantic and western Indian Ocean sectors of the Antarctic Ocean. In: *Investigation of Late Quaternary Paleooceanography and Paleoclimatology* (R.M. CLine and J.D. Hays, eds.). Memoir 145. Boulder, Geological Society of America, 303–336.

Luckman, B.H. (1988). Dating the moraines and recession of Athabasca and Dome glaciers, Alberta, Canada. *Arctic and Alpine Research, 20,* 40–54.

Luckman, B.H. (1994). Evidence for climatic conditions between 900–1300 A.D. in the southern Canadian Rockies. *Climatic Change, 26,* 171–182.

Luckman, B.H. (1995). Calendar-dated, early 'Little Ice Age' glacier advance at Robson glacier, British Columbia, Canada. *The Holocene, 5,* 149–159.

Luckman, B.H. (1996). Reconciling the glacial and dendrochronological records for the last millennium in the Canadian Rockies. In: *Climatic Variations and Forcing Mechanisms of the Last 2000 Years* (P.D. Jones, R.S. Bradley, and J. Jouzel, eds.). Berlin: Springer-Verlag, 85–108.

Luckman, B.H. and Kearney, M.S. (1986). Reconstruction of Holocene changes in alpine vegetation and climate in the Maligne Range, Jasper National Park, Alberta. *Quaternary Research, 26,* 244–261.

Ludwig, K.R., Simmons, K.R., Szabo, B., Winograd, I.J., Landwehr, J.M., Riggs, A.C., and Hoffman, R.J. (1992). Mass spectrometric ^{230}Th - ^{234}U -^{238}U dating of the Devil's Hole calcite vein. *Science, 258,* 284–287.

Lund, S.P. and Banderjee, S.K. (1979). Paleosecular geomagnetic variations from lake sediments. *Reviews of Geophysics and Space Physics, 17,* 244–249.

Lundberg, J. and Ford, D.C. (1994). Late Pleistocene sea level change in the Bahamas from mass spectrometric U-series dating of submerged speleothem. *Quaternary Science Reviews, 13,* 1–14.

Luz, B. (1977). Paleoclimates of the South Pacific based on statistical analysis of planktonic foraminifers. *Palaeogeography, Palaeoclimatology, Palaeoecology, 22,* 61–78.

Luz, B. and Shackleton, N.J. (1975). CaCO$_3$ solution in the tropical East Pacific during the past 130,000 years. In: *Dissolution of Deep-sea Carbonates* (W.V. Sliter, A.W.H. Bé, and W.H. Berger, eds.). Special Publication No. 13. Washington, DC: Cushman Foundation for Foraminiferal Research, 142–150.

Lynch, E.A. (1996). The ability of pollen from small lakes and ponds to sense fine-scale vegetation patterns in the central Rocky Mountains, U.S.A. *Reviews of Palaeobotany and Palynology,* **94**, 197–210.

Lynch, T.F. and Stevenson, C.M. (1992). Obsidian hydration dating and temperature controls in the Punta Negra region of northern Chile. *Quaternary Research,* **37**, 117–124.

Lyons, W.B., Mayewski, P.A., Spencer, M.J., Twickler, M.S., and Graedel, T.E. (1990). A northern hemisphere volcanic chemistry (1869–1984) and climatic implications using a South Greenland ice core. *Annals of Glaciology,* **14**, 176–182.

Maarleveld, G.C. (1976). Periglacial phenomena and the mean annual temperature during the last glacial time in the Netherlands. *Biuletyn Peryglacjalny,* **26**, 57–78.

Mabres, A., Woodman, R., and Zeta, R. (1993). Algunos puntos históricos adicionales sobre la cronología de El Niño. *Bulletin de l'Institut Français d'Etudes Andines,* **22**, 395–406.

MacAyeal, D. (1993). Binge/purge oscillations of the Laurentide Ice Sheet as a cause of the North Atlantic's Heinrich events. *Paleoceanography,* **8**, 775–784.

MacCracken, M.C., Budyko, M.I., Hecht, A.D., and Izrael, Y.A. (1990). *Prospects for Future Climate.* Chelsea, Michigan: Lewis Publishers.

Mackereth, F.H. (1971). On the variation in direction of the horizontal component of remanent magnetization in lake sediments. *Earth and Planetary Science Letters,* **12**, 332–338.

Madureira, L.A.S., van Kreveld, S.A., Eglinton, G., Conte, M.H., Ganssen, G., van Hinte, J.E., and Ottens, J.J. (1997). Late Quaternary high-resolution biomarker and other sedimentary climate proxies in a northeast Atlantic core. *Paleoceanography,* **12**, 255–269.

Maejima, I. and Tagami, Y. (1983). Climate of Little Ice Age in Japan. *Geographical Reports of Tokyo Metropolitan University,* **21**, 157–171.

Maejima, I. and Tagami, Y. (1986). Climatic change during historical times in Japan: reconstruction from climatic hazards records. *Geographical Reports of Tokyo Metropolitan University,* **21**, 157–171.

Maher, B.A. and Thompson, R. (1992). Paleoclimatic significance of the mineral magnetic record of the Chinese loess and paleosols. *Quaternary Research,* **37**, 155–170.

Maher, B.A. and Thompson, R. (1995). Paleorainfall reconstructions from pedogenic magnetic susceptibility variations in the Chinese loess. *Quaternary Research,* **44**, 383–391.

Maher, B.A., Thompson, R., and Zhou, L-P. (1994). Spatial and temporal reconstructions of changes in the Asian paleomonsoon: a new mineral magnetic approach. *Earth and Planetary Science Letters,* **44**, 383–391.

Maher, L.J. (1963). Pollen analyses of surface materials from the southern San Juan Mountains, Colorado. *Geological Society of America Bulletin,* **74**, 1485–1504.

Maley, J. (1996). The Africal rainforest—main characteristics of changes in vegetation and climate from the Upper Cretaceous to the Quaternary. *Proc. Royal Society of Edinburgh,* **104B**, 31–73.

Maley, J. (1989a). Late Quaternary climatic changes in the African rain forest: forest refugia and the major role of sea-surface temperature variations. In: *Paleoclimatology and Paleometeorology: Modern and Past Patterns of Global Atmospheric Transport* (M. Leinen and M. Sarnthein, eds.). Dordrecht: Kluwer Academic, 585–616.

Maley, J. (1989b). L'importance de la tradition orale et des donées historiques pour la reconstruction paléoclimatique du dernier millénaire sur l'Afrique nord-tropicale. In: *Sud Sahara, Sahel Nord.* Le Centre Culturel Français d'Abidjan, 53–57.

Maley, J. (1991). The African rain forest and palaeoenvironments during Late Quaternary. *Climatic Change,* **19**, 79–98.

Maley, J. (1997). Middle to Late Holocene changes in Tropical Africa and other continents. Paleomonsoon and sea surface temperature variations. In: *Third Millennium B.C. Climate Change and Old World Collapse* (H.N. Dalfes, G. Kukla, and H. Weiss, eds.). Dordrecht: Kluwer Academic, 611–640.

Manabe, S. and Broccoli, A.J. (1985a). The influence of continental ice sheets on the climate of an Ice Age. *Jour. Geophys. Research*, **90D**, 2167–2190.

Manabe, S. and Broccoli, A.J. (1985b). A comparison of climate model sensitivity with data from the last glacial maximum. *Jour. Atmos. Sciences*, **42**, 2643–2651.

Manabe, S. and Hahn, D.G. (1977). Simulation of the tropical climate of an ice age. *Jour. Geophys. Research*, **82**, 3889–3911.

Manabe, S. and Stouffer, R. (1988). Two stable equilibria of a coupled ocean-atmosphere model. *Jour. Climate*, **1**, 841–866.

Manabe, S. and Stouffer, R. (1995). Simulation of abrupt climate change induced by freshwater input to the North Atlantic Ocean. *Nature*, **378**, 165–167.

Manabe, S. and Stouffer, R. (1997). Coupled ocean-atmosphere model response to freshwater input: comparison to Younger Dryas event *Paleoceanography*, **12**, 321–336.

Manabe, S. and Wetherald, R.T. (1967). Thermal equilibrium of the atmosphere with a given distribution of relative humidity. *Jour. the Atmos. Sciences*, **24**, 241–259.

Mangerud, J. (1972). Radiocarbon dating of marine shells including a discussion of apparent age of recent shells from Norway. *Boreas*, **1**, 143–172.

Mangerud, J., Andersen, S.T., Berglund, B.E., and Donner, J.J. (1974). Quaternary stratigraphy of Norden, a proposal for terminology and classification. *Boreas*, **3**, 109–128.

Mangerud, J., Lie, S.E., Furnes, H., Kristiansen, I.L., and Lomo, L. (1984). A Younger Dryas ash bed in western Norway and its possible correlations with tephra in cores from the Norwegian Sea and the North Atlantic. *Quaternary Research*, **21**, 85–104.

Manley, G. (1969). Snowfall in Britain over the past 300 years. *Weather*, **24**, 428–437.

Manley, G. (1974). Central England temperatures: monthly means 1659 to 1973. *Quart. Jour. Royal Meteorological Society*, **100**, 389–405.

Mann, M.E., Park, J., and Bradley, R.S. (1996). Global inter-decadal and century-scale climate oscillations during the past five centuries. *Nature*, **378**, 266–270.

Mann, M.E., Bradley, R.S., and Hughes, M.K. (1998). Global scale temperature patterns and climate forcing over the past six centuries. *Nature*, **392**, 779–788 (also: *Science*, **280**, 2029–2030).

Margolis, S.V., Kroopnick, P.M., Goodney, D.E., Dudley, W.C., and Mahoney, M.E. (1975). Oxygen and carbon isotopes from calcareous nannofossils as paleo-oceanographic indicators. *Science*, **189**, 555–557.

Markgraf, V. (1980). Pollen dispersal in a mountain area. *Grana*, **19**, 127–146.

Markgraf, V. (1989). Palaeoclimates in Central and South America since 18,000 B.P. based on pollen and lake-level records. *Quaternary Science Reviews*, **8**, 1–24.

Martin, J.H. (1990). Glacial-interglacial CO_2 change: the iron hypothesis. *Paleoceanography*, **5**, 1–13.

Martin, P.S., Sabel, B.E., and Shutler, Jr., D. (1961). Rampart cave coprolite and ecology of the Shasta ground sloth. *American Jour. Science*, **259**, 102–127.

Martinson, D.G., Pisias, N.G., Hays, J.D., Imbrie, J., Moore, T.C., and Shackleton, N.J. (1987). Age dating and the orbital theory of the ice ages: development of a high resolution 0 to 300,000-year chronostratigraphy. *Quaternary Research*, **27**, 1–29.

Matthes, F.E. (1940). Report of the Committee on glaciers. *Trans. American Geophysical Union*, **21**, 396–406.

Matthes, F.E. (1942). Glaciers. In: *Hydrology* (O.E. Meinzer, ed.). New York: Dover/McGraw-Hill, 149–219.

Matthews, J.A. (1980). Some problems and implications of ^{14}C dates from a podzol buried beneath an end moraine at Haugabreen, southern Norway. *Geografiska Annaler*, **62A**, 185–208.

Matthews, J.A. (1993). Deposits indicative of Holocene climatic fluctuations in the timberline areas of northern Europe: some physical proxy data sources and research approaches.

In: *Oscillations of the Alpine and Polar Tree Limits in the Holocene* (B. Frenzel, M. Eronen, K.D. Vorren, and B. Gläser, eds.). Stuttgart: Gustav Fischer Verlag, 85–97.

Matthewson, A.P., Shimmield, G.B., Kroon, D., and Fallick, A.E. (1995). A 300 kyr high resolution aridity record of the North African continent. *Paleoceanography,* **10,** 677–692.

Mayewski, P.A., Lyons, W.B., Spencer, M.J., Twickler, M., Dansgaard, W., Koci, B., Davidson, C.I., and Honrath, R.E. (1986). Sulfate and nitrate concentrations from a South Greenland ice core. *Science* **232,** 975–977.

Mayewski, P.A., Meeker, L.D., Twickler, M.S., Whitlow, S., Yang, Q., and Prentice, M. (1997). Major features and forcing of high latitude northern hemisphere atmospheric circulation using a 110,000 year long glaciochemistry series. *Jour. Geophys. Research,* **102C,** 26345–26366.

Mayewski, P.A., Meeker, L.D., Whitlow, S., Twickler, M.S., Morrison, M.C., Alley, R.B., Bloomfield, P., and Taylor, K. (1993). The atmosphere during the Younger Dryas. *Science,* **261,** 195–197.

Mayewski, P.A., Meeker, L.D., Whitlow, S., Twickler, M.S., Morrison, M.C., Bloomfield, P., Bond, G.C., Alley, R.B., Gow, A.J., Grootes, P.M., Meese, D.A., Ram, M., Taylor, K.C., and Wumkes, W. (1994). *Science,* **263,** 1747–1751.

Mayewski, P.A., Spencer, M.J., and Lyons, W.B. (1992). A review of glaciochemistry with particular emphasis on the recent record of sulfate and nitrate. In: *Trace Gases and the Biosphere* (B. Moore III and D. Schimel, eds.). Boulder: University Corporation for Atmospheric Research, 177–199.

Mayewski, P.A., Twickler, M.S., Whitlow, S.I., Meeker, L.D., Yang, Q., Thomas, J., Kreutz, K., Grootes, P.M., Morse, D.L., Steig, E.J., Waddington, E.D., Saltzman, E.S., Whung, P-Y., and Taylor, K.C. (1996). Climate change during the last deglaciation in Antarctica. *Science,* **272,** 1636–1638.

Mazaud, A., Laj, C., Bard, E., Arnold, M., and Tric, E. (1991). Geomagnetic field control of ^{14}C production over the last 80 ky: implications for the radiocarbon time-scale. *Geophysical Research Letters,* **18,** 1885–1888.

McAndrews, J.H. (1966). Postglacial history of prairie, savanna, and forest in northwestern Minnesota. *Memoirs of the Torrey Botanical Club* **22,** 1–72.

McCarroll, D. (1994). A new approach to lichenometry: dating single-age and diachronous surfaces. *The Holocene,* **4,** 383–396.

McCarthy, D.P. and Luckman, B.H. (1993). Estimating ecesis for tree-ring dating of moraines: a comparative study from the Canadian cordillera. *Arctic and Alpine Research,* **25,** 63–68.

McConnaughey, T.A. (1989). C-13 and O-18 isotopic disequilibria in biological carbonates: I. Patterns. *Geochimica et Cosmochimica Acta,* **53,** 151–162.

McCormac, F.G. and Baillie, M.G.L. (1993). Radiocarbon to calendar date conversion: calendrical bandwidths as a function of radiocarbon precision. *Radiocarbon,* **35,** 311–316.

McCoy, W.D. (1987a). The precision of amino acid geochronology and paleothermometry. *Quaternary Science Reviews,* **6,** 43–54.

McCoy, W.D. (1987b). Quaternary aminostratigraphy of the Bonneville Basin, western United States. *Geological Society of America Bulletin,* **98,** 99–112.

McCulloch, D. and Hopkins, D. (1966). Evidence for an early recent warm interval in northwestern Alaska. *Geological Society of America Bulletin,* **77,** 1089–1108.

McCulloch, M.T., Gagan, M.K., Mortimer, G.E., Chivas, A.R., and Isdale, P.J. (1994). A high resolution Sr/Ca and δ^{18}O coral record from the Great Barrier Reef, Australia, and the 1982–1983 El Niño. *Geochimica and Cosmochimica Acta,* **58,** 2747–2754.

McDougall, I. (1995). Potassium-argon dating in the Pleistocene. In: *Dating Methods for Quaternary Deposits* (N.W. Rutter and N.R. Catto, eds.). St. John's: Geological Association of Canada, 1–14.

McDougall, I. and Harrison, T.M. (1988). *Geochronology and Thermochronology by the $^{40}Ar/^{39}Ar$ Method.* Oxford: Oxford University Press.

McGuffie, K. and Henderson-Sellers, A. (1997). *A Climate Modelling Primer.* Chichester: Wiley.

McIntyre, A. and Molfino, B. (1996). Forcing of Atlantic equatorial and subpolar millennial cycles by precession. *Science,* **274,** 1867–1870.

McIntyre, A., Bé, A.W.H., Hays, J.D., Gardner, J.V., Lozano, J.A., Molfino, B., Prell, W., Thierstein, H.R., Crowley, T., Imbrie, J., Kellogg, T., Kipp, N., and Ruddiman, W.F. (1975). Thermal and oceanic structures of the Atlantic through a glacial-interglacial cycle. In: *Proc. WMO Symposium on Long-term Climatic Fluctuations,* WMO No. 421. World Meteorological Organization, Geneva, 75–80.

McIntyre, A., Kipp, N.G., Bé, A.W.H., Crowley, T., Kellogg, T., Gardner, J.V., Prell, W., and Ruddiman, W.F. (1976). Glacial North Atlantic 18,000 years ago: a CLIMAP reconstruction. In: *Investigation of Late Quaternary Paleooceanography and Paleoclimatology* (R.M. Cline and J.D. Hays, eds.). Memoir 145. Boulder: Geological Society of America, 43–76.

McKenzie, J.A. and Hollander, D.J. (1993). Oxygen-isotope record in recent carbonate sediments from Lake Greifen, Switzerland (1750–1986): application of continental isotopic indicator for evaluation of changes in climate and atmospheric circulation patterns. In: *Climate Change in Continental Isotopic Records* (P.K. Swart, K.C. Lohmann, J. McKenzie, and S. Savin, eds.). Washington DC: American Geophysical Union, 101–111.

McManus, D.A. (1970). Criteria of climatic change in the inorganic components of marine sediments. *Quaternary Research,* 1, 72–102.

McManus, J.F., Bond, G.C., Broecker, W.S., Johnsen, S., Labeyrie, L., and Higgins, S. (1994). High-resolution climate records from the North Atlantic during the last interglacial. *Nature,* 371, 326–329.

Meese, D.A., Gow, A.J., Alley, R.B., Zielinski, G.A., Grootes, P.M., Ram, M., Taylor, K.C., Mayewski, P.A., and Bolzan, J.F., (1997). The GISP2 depth-age scale: methods and results. *Jour. Geophys. Research,* 102C, 26411–26423.

Meese, D.A., Gow, A.J., Grootes, P., Mayewski, P.A., Ram, M., Stuiver, M., Taylor, K.C., Waddington, E.D., and Zielinski, G.A. (1994). The accumulation record from the GISP2 core as an indicator of climate change throughout the Holocene. *Science,* 266, 1680–1682.

Meese, D.A., Gow, A.J., Grootes, P., Mayewski, P.A., Ram, M., Stuiver, M., Taylor, K.C., Waddington, E.D., and Zielinski, G.A. (1995). *Determination of GISP2 Depth-Age Scale: Techniques and Results.* CRREL Report No. 95, Hanover: USA Cold Regions Research and Engineering Laboratory.

Meggers, B.J. (1982). Archeological and ethnographic evidence compatible with the model of forest fragmentation. In: *Biological Diversification in the Tropics* (G.T. Prance, ed.). New York: Columbia University Press, 483–496.

Meinel, A. and Meinel, M. (1983). *Sunsets, Twilights and Evening Skies.* Cambridge: Cambridge University Press.

Meko, D.M. (1992). Spectral properties of tree-ring data in the United States Southwest as related to El Nino/Southern Oscillation. In: *El Niño: Historical and Paleoclimatic Aspects of the Southern Oscillation* (H.F. Diaz and V. Markgraf, ed.). Cambridge: Cambridge University Press, 227–241.

Meko, D.M., Stockton, C.W., and Boggess, W.R. (1980). A tree-ring reconstruction of drought in southern California. *Water Resources Research,* 16, 594–600.

Mesolella, K.J., Matthews, R.K., Broecker, W.S., and Thurber, D.L. (1969). The astronomical theory of climatic change: Barbados data. *Jour. Geology,* 77, 250–274.

Messerli, B., Messerli, P., Pfister, C., and Zumbuhl, H.J. (1978). Fluctuations of climate and glaciers in the Bernese Oberland, Switzerland, and their geological significance, 1600–1975. *Arctic and Alpine Research,* 10, 247–260.

Meyer, C.E., Sarna-Wojcicki, A.M., Hillhouse, J.W., Woodward, M.J., Slate, J.L., and Sorg, D.H. (1991). Fission-track age (400,000 yr) of the Rockland tephra, based on inclusion of zircon grains lacking fossil fission tracks. *Quaternary Research,* 35, 367–382.

Michel, P. (1973). *Les bassins des Fleuves Senegal et Gambie: Etude Geomorphologie,* Memoires no. 63. Paris: Office de la Recherche Scientifique et Technique d'Outre-mer.

Michels, J.W. and Bebrich, C.A. (1971). Obsidian hydration dating. In: *Dating Techniques for the Archaeologist* (H.N. Michael and E.K. Ralph, eds.). Cambridge: MIT Press, 164–221.

Mifflin, M.D. and Wheat, M.M. (1979). *Pluvial lakes and estimated pluvial climates of Nevada*, Bulletin 94, Nevada Bureau of Mines and Geology. Reno: University of Nevada.

Migliazza, E.C. (1982). Linguistic prehistory and the refuge model in Amazonia. In: *Biological Diversification in the Tropics* (G.T. Prance, ed.). New York: Columbia University Press, 497–519.

Mikami, T. (ed.). (1992a). *Proc. International Symposium on the Little Ice Age Climate*. Department of Geography, Tokyo Metropolitan University.

Mikami, T. (ed.). (1992b). Climate variations in Japan during the Little Ice Age. In: *Proc. of the International Symposium on the Little Ice Age Climate*. Department of Geography, Tokyo Metropolitan University, 176–181.

Milankovitch, M.M. (1941). *Canon of insolation and the ice-age problem*. Beograd: Koniglich Serbische Akademie. [English translation by the Israel Program for Scientific Translations, published for the US Department of Commerce, and the National Science, Foundation, Washington, DC (1969).]

Miller, G.H. (1973). Variations in lichen growth from direct measurements: preliminary curves for *Alectoria minuscula* from eastern Baffin Island, NWT, Canada. *Arctic and Alpine Research*, **5**, 333–339.

Miller, G.H. and Andrews, J.T. (1973). Quaternary history of northern Cumberland Peninsula, east Baffin Island, NWT, Canada, Part VI. Preliminary lichen growth curve for *Rhizocarpon geographicum. Geological Society of America Bulletin*, **83**, 1133–1138.

Miller, G.H., Bradley, R.S., and Andrews, J.T. (1975). Glaciation level and lowest equilibrium line altitude in the High Canadian Arctic: maps and climatic interpretation. *Arctic and Alpine Research*, **7**, 155–168.

Miller, G.H. and Brigham-Grette, J. (1989). Amino acid geochronology: resolution and precision in carbonate fossils. *Quaternary International*, **1**, 111–128.

Miller, G.H. and Hare, P.E. (1975). Use of amino acid reactions in some arctic marine fossils as stratigraphic and geochronologic indicators. *Carnegie Institution of Washington Yearbook*, **74**, 612–617.

Miller, G.H. and Hare, P.E. (1980). Amino acid geochronology: integrity of the carbonate matrix and potential of molluscan fossils. In: *Biogeochemistry of Amino Acids* (P.E. Hare, T.C. Hoering, and K. King, Jr., eds.). New York: Wiley, 415–443.

Miller, G.H., Hollin, J.T., and Andrews, J.T. (1979). Aminostratigraphy of UK Pleistocene deposits. *Nature*, **281**, 539–543.

Miller, G.H., Magee, J.W., and Jull, A.J.T. (1997). Low-latitude glacial cooling in the southern hemisphere from amino-acid racemization in emu eggshells. *Nature*, **385**, 241–244.

Miller, G.H. and Mangerud, J. (1985). Aminostratigraphy of European marine interglacial deposits. *Quaternary Science Reviews*, **4**, 215–278.

Miller, J.A. (1972). Dating Holocene and Pleistocene strata using the potassium argon and argon-40/argon-39 methods. In: *Calibration of Hominoid Evolution* (W.W. Bishop and J.A. Miller, eds.). Edinburgh: Scottish Academic Press, 63–73.

Min, G.R., Edwards, R.L., Taylor, F.W., Récy, J., Gallup, C.D., and Beck, J.W. (1995). Annual cycles of U/Ca in coral skeletons and U/Ca thermometry. *Geochimica et Cosmochimica Acta*, **59**, 2025–2042.

Miroshnikov, L.D. (1958). Ostatki drevney lesnoy rastitel'nosti na Taymyrskom poluostrove. *Priroda, Moskva*, **2**, 106–107.

Mitchell, J.F.B., Grahame, N.S., and Needham, K.H. (1988). Climate simulation for 9000 years before present: seasonal variations and the effect of the Laurentide Ice Sheet. *Jour. Geophys. Research*, **93D**, 8283–8303.

Mitchell, J.M. (1976). An overview of climatic variability and its causal mechanisms. *Quaternary Research*, **6**, 481–493.

Mitchell, J.M., Dzerdzeevski, B., Flohn, H., Hofmeyr, W.L., Lamb, H.H., Rao, K.N., and Wallen, C.C. (1966). *Climate Change*, WMO Technical Note No. 79. Geneva: World Meteorological Organization.

Mitsuguchi, T., Matsumoto, E., Abe, O., Uechida, T., and Isdale, P.J. (1996). Mg/Ca thermometry in coral skeletons. *Science*, **274**, 961–963.

Mix, A.C. (1987). The oxygen-isotope record of glaciation. In: *North America and Adjacent Oceans During the Last Deglaciation, The Geology of North America v. K-3,* (W.F. Ruddiman and H.E. Wright, Jr., eds.). Boulder: Geological Society of America, 111–135.

Mix, A.C. and Ruddiman, W.F. (1984). Oxygen isotope analyses and Pleistocene ice volumes. *Quaternary Research,* **21,** 1–20.

Mizukoshi, M. (1992). Climatic reconstruction in central Japan during the Little Ice Age based on documentary sources. In: *Proc. International Symposium on the Little Ice Age Climate.* (T. Mikami, ed.). Department of Geography, Tokyo Metropolitan University, 182–187.

Molfino, B., Kipp, N.G., and Morley, J.J. (1982). Comparison of foraminiferal, coccolithophorid and Radiolarian paleotemperature equations: assemblage coherency and estimate concordancy. *Quaternary Research,* **17,** 279–313.

Monod, Th. (1963). The late Tertiary and Pleistocene in the Sahara. In: *African Ecology and Human Evolution* (F.C. Howell and F. Bouliere, eds.). New York: Viking Publications in Anthropology No. 36, Wenner-Gren Foundation, 117–229.

Moody, D.W. and Catchpole, A.J.W. (1975). *Environmental Data from Historical Documents by Content Analysis: Freeze-up and Break-up of Estuaries on Hudson Bay, 1714-1871.* Manitoba Geographical Studies No. 5. Winnipeg: University of Winnipeg.

Mook, W.G., Bommerson, J.C., and Stoverman, W.H. (1974). Carbon isotope fractionation between dissolved bicarbonate and gaseous carbon dioxide. *Earth and Planetary Science Letters,* **22,** 169–176.

Moore, J.C., Narita, H., and Maeno, N. (1991). A continuous 770-year record of volcanic acidity from East Antarctica. *Jour. Geophys. Research,* **96,** 17353–17359.

Moore, P.D. and Webb, J.A. (1978). *An Illustrated Guide to Pollen Analysis.* London: Hodder and Stoughton.

Moore, T.C. (1978). The distribution of radiolarian assemblages in the modern and ice-age Pacific. *Marine Micropaleontology,* **4,** 229–266.

Moore, T.C., Burckle, L.H., Geitzenauer, K., Luz, B., Molina-Cruz, A., Robertson, J.H., Sachs, H., Sancetta, C., Thiede, J., Thompson, P., and Wenkam, C. (1980). The reconstruction of sea surface temperatures in the Pacific Ocean of 18,000 BP. *Marine Micropaleontology,* **5,** 215–247.

Morgan, A. (1973). Late Pleistocene environmental changes indicated by fossil insect faunas of the English Midland. *Boreas,* **2,** 173–212.

Morgan, A.V. and Morgan, A. (1979). The fossil coleoptera of the Two Creeks forest bed, Wisconsin. *Quaternary Research,* **12,** 226–240.

Morgan, A.V. and Morgan, A. (1981). Paleoentomological methods of reconstructing paleoclimate with reference to interglacial and interstadial insect faunas of southern Ontario. In: *Quaternary Paleoclimate* (W.C. Mahaney, ed.). Norwich: University of East Anglia, 173–192.

Morgan, V.I. (1982). Antarctic Ice Sheet surface oxygen isotope values. *Jour. Glaciology,* **28,** 315–323.

Morley, J.J. and Hays, J.D. (1979). Comparison of glacial and interglacial oceanographic conditions in the South Atlantic from variations in calcium carbonate and radiolarian distributions. *Quaternary Research,* **12,** 396–408.

Morley, J.J. and Hays, J.D. (1981). Towards a high-resolution, global, deep-sea chronology for the last 750,000 years. *Earth and Planetary Science Letters,* **53,** 279–295.

Morley, J.J. and Shackleton, N.J. (1978). Extension of the radiolarian *Stylatractus universus,* as a biostratigraphic datum to the Atlantic Ocean. *Geology,* **6,** 309–311.

Morrison, R. (1965). Quaternary geology of the Great Basin. In: *The Quaternary of the United States* (H.E. Wright and D.G. Frey, eds.). Princeton: Princeton University Press, 265–286.

Mortlock, R.A., Charles, C.D., Froelich, P.N., Zibello, M.A., Saltzmann, J., Hays, J.D., and Burckle, L.H. (1991). Evidence for lower productivity in the Antarctic Ocean during the last glaciation. *Nature,* **361,** 220–223.

Morton, F.I. (1967). Evaporation from large deep lakes. *Water Resources Research,* **3,** 181–200.

Moser, K.A. and MacDonald, G.M. (1990). Holocene vegetation change at treeline north of Yellowknife, Northwest Territories, *Quaternary Research*, **34**, 227–239.

Moser, K.A., MacDonald, G.M., and Smol, J.P. (1996). Applications of freshwater diatoms to geographical research. *Progress in Physical Geography*, **20**, 21–52.

Mott, R.J. (1975). Palynological studies of lake sediment profiles from southwestern New Brunswick. *Canadian Jour. Earth Sciences*, **12**, 273–288.

Müller, F. (1958). Eight months of glacier and soil research in the Everest region. In: *The Mountain World* (M. Barnes ed.). New York: Harper and Bros., 191–208.

Müller, P.J., Kirst, G., Ruhland, G., von Storch, I., and Rosell-Melé, A. (1998). Calibration of the alkenone paleotemperature index $U^{k'}_{37}$ based on core-tops from the eastern South Atlantic and the global ocean (60° N–60° S). *Geochimica et Cosmochimica Acta*, **62**, 1757–1772.

Muller, R.A. (1977). Radioisotope dating with a cyclotron. *Science*, **196**, 489–494.

Muller, R.A. and MacDonald, G.J. (1997). Glacial cycles and astronomical forcing. *Science*, **277**, 215–218.

Mullineaux, D.R. (1974). *Pumice and other Pyroclastic Deposits in Mount Rainier National Park, Washington.* Bulletin 1326, Washington, DC: US Geological Survey.

Mulvaney, R. and Peel, D.A. (1987). Anions and cations in ice cores from Dolleman Island and the Palmer Land plateau, Antarctic Peninsula. *Annals of Glaciology*, **10**, 121–125.

Murata, A. (1992). Reconstruction of rainfall variation of the Baiu in historical times. In: *Climate Since A.D. 1500* (R.S. Bradley and P.D. Jones, eds.) London: Routledge, 224–245.

Naeser, C.W., Briggs, N.D., Obradovich, J.D., and Izett, G.A. (1981). Geochronology of tephra deposits. In: *Tephra Studies* (S. Self and R.J.S. Sparks eds.). Dordrecht: D. Reidel, 13–47.

Naeser, C.W. and Naeser, N.D. (1988). Fission track dating of Quaternary events. In: *Dating Quaternary Sediments* (D.J. Easterbrook, ed.). Special Paper 227. Boulder Geological Society of America, 1–11.

Neftel, A., Oeschger, H., and Stauffer, B. (1988). CO_2 record in the Byrd ice core 50,000-5,000 years B.P. *Nature*, **331**, 609–611.

Negrini, R.M. and Davis, J.O. (1992). Dating Late Pleistocene pluvial events and tephras by correlating paleomagnetic secular variation records from the western Great Basin. *Quaternary Research*, **38**, 46–59.

Nelson, D.E., Korteling, R.G., and Stott, W.R. (1977). Carbon-14: direct detection at natural concentrations. *Science*, **198**, 507–508.

Nesje, A. and Kvamme, M. (1991). Holocene glacier and climate variations in western Norway: evidence for early Holocene glacier demise and multiple Neoglacial events. *Geology*, **19**, 610–612.

Nesje, A., Kvamme, M., and Løvlie, R. (1991). Holocene glacial and climate history of the Jostedalsbreen region, western Norway: evidence from lake sediments and terrestrial deposits. *Quaternary Science Reviews*, **10**, 87–114.

Neuberger, H. (1970). Climate in art. *Weather*, **25**, 46–56.

Newell, R.E. and Chiu, L.S. (1981). Climatic changes and variations: a geophysical problem. In: *Climate Variations and Variability: Facts and Theories* (A. Berger ed.). Dordrecht: D. Reidel, 21–61.

Newhall, C.G. and Self, S. (1982). The Volcanic Explosivity Index (VEI): an estimate of explosive magnitude for historical volcanism. *Jour. Geophys. Research*, **87C**, 1231–1238.

Nichols, H. (1967). The postglacial history of vegetation and climate at Ennadai Lake, Keewatin and Lynn Lake, Manitoba. *Eiszeitalter und Gegenwart*, **18**, 176–197.

Nicholson, S. and Flohn, H. (1980). African environmental and climatic changes and the general circulation in late Pleistocene and Holocene. *Climatic Change*, **2**, 313–348.

Nix, H.A. and Kalma, J.D. (1972). Climate as a dominant control in the biogeography of northern Australia and New Guinea. In: *Bridge and Barrier: the Natural and Cultural History of Torres Strait* (D. Walker, ed.). Publication No. BG3, Department of Biogeography and Geomorphology, Australian National University, Canberra, 61–91.

North, G.R., Calahan, R.F., and Coakley, J.A. (1981). Energy balance climate models. *Reviews of Geophysics and Space Physics*, **19**, 91–121.

Nye, J.F. (1965). A numerical method of inferring the budget history of a glacier from its advance and retreat. *Jour. Glaciology*, **5**, 589–607.

O'Brien, S.R., Mayewski, P.A., Meeker, L.D., Meese, D.A., Twickler, M.S., and Whitlow, S.I. (1995). Complexity of Holocene climate as reconstructed from a Greenland ice core. *Science*, **270**, 1962–1964.

Oches, E.A. and McCoy, W.D. (1995a). Aminostratigraphic evaluation of conflicting age estimates for the "Young Loess" of Hungary. *Quaternary Research*, **44**, 160–170.

Oches, E.A. and McCoy, W.D. (1995b). Aminostratigraphy of central European loess cycles: introduction and data. *Geolines* (Praha), **2**, 34–86.

Oches, E.A. and McCoy, W.D. (1995c). Amino acid geochronology applied to the correlation and dating of central European loess deposits. *Quaternary Science Reviews*, **14**, 767–782.

Oches, E.A., McCoy, W.D., and Clark, P.U. (1996). Amino acid estimates of latitudinal temperature gradients and geochronology of loess deposition during the last glaciation, Mississippi Valley, United States. *Geological Society of American Bulletin*, **108**, 892–903.

Oerlemans, J. (1989). On the response of valley glaciers to climatic change. In: *Glacier Fluctuations and Climatic Change* (J. Oerlemans, ed.). Dordrecht: Kluwer, 353–371.

Oerlemans, J. and Hoogendorn, N.C. (1989). Mass balance gradients and climatic change. *Jour. Glaciology*, **35**, 399–405.

Oeschger, H., Beer, J., Siegenthaler, U., and Stauffer, B. (1984). Late glacial climate history from ice cores. In: *Climate Processes and Climate Sensitivity* (J.E. Hansen and T. Takahashi, eds.). Washington, DC: American Geophysical Union, 299–306.

Oeschger, H. and Langway, Jr., C.C. (eds.). (1989). *The Environmental Record in Glaciers and Ice Sheets*. Chichester: Wiley.

Oeschger, H. and Siegenthaler, U. (1988). How has the atmospheric concentration of CO_2 changed? In: *The Changing Atmosphere* (F.S. Rowland and I.S.A. Isaksen, eds.). Chichester: Wiley, 5–23.

Ogilvie, A. (1992). Documentary evidence for changes in the climate of Iceland, A.D. 1500–1800. In: *Climate Since A.D. 1500* (R.S. Bradley and P.D. Jones, eds.). London: Routledge, 92–117.

Oglesby, R.J. (1990). Sensitivity of glaciation to initial snowcover, CO_2, snow cover, and ocean roughness in the NCAR CCM. *Climate Dynamics*, **5**, 219–235.

Ohkouchi, N., Kawamura, K., Nakamura, T., and Taira, A. (1994). Small changes in the sea surface temperature during the last 20,000 years: molecular evidence from the western tropical Pacific. *Geophysical Research Letters*, **21**, 2207–2210.

Olausson, E. (1965). Evidence of climatic changes in deep sea cores with remarks on isotopic palaeotemperature analysis. *Progress in Oceanography*, **3**, 221–252.

Olausson, E. (1967). Climatological, geoeconomical and paleooceanographical aspects of carbonate deposition. *Progress in Oceanography*, **4**, 245–265.

Olsson, I.U. (1968). Modern aspects of radiocarbon dating. *Earth Science Reviews*, **4**, 203–218.

Olsson, I.U. (1974). Some problems in connection with the evaluation of ^{14}C dates. *Geologiska Foreningens i Stockholm Forhandlingar*, **96**, 311–320.

Olsson, I.U., El-Daoushy, M.F.A.F., Abd-El-Mageed, A.I., and Klasson, M. (1974). A comparison of different methods for pretreatment of bones I. *Geologiska Foreningens i Stockholm Forhandlingar*, **96**, 171–181.

Olsson, I.U. and Eriksson, K.G. (1972). Fractionation studies of the shells of Foraminifera. In: *Etudes sur le Quaternaire dans le Monde*, Proceedings, Congrès INQUA, Paris, 921–923.

Olsson, I.U. and Osadebe, F.A.N. (1974). Carbon isotope variations and fractionation corrections in ^{14}C dating. *Boreas*, **3**, 139–146.

Opdyke, N.D. (1972). Paleomagnetism of deep-sea cores. *Reviews of Geophysics and Space Physics*, **101**, 213–249.

Opdyke, N.D. and Channell, J.E.T. (1996). *Magnetic Stratigraphy.* San Diego: Academic Press.

Oppo, D.W. and Fairbanks, R.G. (1987). Variability in the deep and intermediate water circulation of the Atlantic Ocean during the past 25,000 years: northern hemisphere modulation of the Southern Ocean. *Earth and Planetary Science Letters, 86,* 1–15.

Oppo, D.W. and Lehman, S.J. (1993). Mid-depth circulation of the subpolar North Atlantic during the last glacial maximum. *Science, 259,* 1148–1152.

Oppo, D.W. and Lehman, S.J. (1995). Sub-orbital timescale variability of North Atlantic deep-water during the past 200,000 years. *Paleoceanography, 10,* 901–910.

Osborne, P.J. (1974). An insect assemblage of early Handrian age from Lea Marston, Warwickshire, and its bearing on the contemporary climate and ecology. *Quaternary Research, 4,* 471–486.

Osborne, P.J. (1980). The late Devensian-Flandrian transition depicted by serial insect faunas from West Bromwich, Staffordshire, England. *Boreas, 9,* 139–147.

Osmaston, H.A. (1975). Models for the estimation of firnlines of present and Pleistocene glaciers. In: *Processes in Physical and Human Geography: Bristol Essays* (R.F. Peel, M. Chisholm, and P. Haggett, eds.). London: Heinemann, 218–245.

Østrem, G. (1974). Present alpine ice cover. In: *Arctic and Alpine Environments* (J.D. Ives and R.G. Barry, eds.). London: Methuen, 225–250.

O'Sullivan, P.E. (1983). Annually-laminated sediments and the study of Quaternary paleoenvironmental changes—a review. *Quaternary Science Reviews, 1,* 245–313.

Overpeck, J.E., (1996). Varved sediment records of recent seasonal to millennial-scale environmental variability. In: *Climate Variations and Forcing Mechanisms of the Last 2000 Years* (P.D. Jones, R.S. Bradley, and J. Jouzel, eds.). Berlin: Springer-Verlag, 479–498.

Overpeck, J.T., Rind, D. and Goldberg, R. (1990). Climate-induced changes in forest disturbance and vegetation. *Nature, 343,* 51–53.

Overpeck, J.T., Rind, D., Lacis, A., and Healey, R. (1996). Possible role of dust-induced regional warming in abrupt climate change during the last glacial period. *Nature, 384,* 447–449.

Overpeck, J.T., Webb III, T., and Prentice, I.C. (1985). Quantitative interpretation of fossil pollen spectra: dissimilarity coefficients and the method of modern analogs. *Quaternary Research, 23,* 87–108.

Oviatt, C.G., McCoy, W.D., Nash, W.P. (1994). Sequence stratigraphy of lacustrine deposits; a Quaternary example from the Bonneville Basin, Utah; with Suppl. Data 9402. *Geological Society of America Bulletin, 106,* 133–144.

Paillard, D. (1998). The timing of Pleistocene glaciations from a simple multiple-state climate model. *Nature, 391,* 378–381.

Paillard, D. and Labeyrie, L. (1994). Role of the thermohaline circulation in the abrupt warming after Heinrich events. *Nature, 372,* 162–164.

Palais, J.M. and Sigurdsson, H. (1989). Petrologic evidence of volatile emissions from major historic and prehistoric volcanic eruptions. In: *Understanding Climatic Change,* (A. Berger, R.E. Dickinson, and J.W. Kidson, eds.). Washington, DC: American Geophysical Union, 31–53.

Palais, J.M., Germani, M.S., and Zielinski, G.A. (1992). Inter-hemispheric transport of volcanic ash from a 1259 A.D. volcanic eruption to the Greenland and Antarctic ice sheets. *Geophysical Research Letters, 19,* 801–804.

Palmer, W.C. (1965). *Meteorological Drought.* Research Paper No. 45, US Weather Bureau, Washington, DC.

Parker, M.L. (1971). *Dendrochronological Techniques used by the Geological Survey of Canada,* Paper 71–25. Ottawa: Geological Survey of Canada.

Parker, M.L. and Hennoch, W.E.S. (1971). The use of Engelmann spruce latewood density for dendrochronological purposes. *Canadian Jour. Forest Research, 1,* 90–98.

Parmenter, C. and Folger, D.W. (1974). Eolian biogenic detritus in deep sea sediments: a possible index of Equatorial Ice Age aridity. *Science, 185,* 695–698.

Parry, M.L. (1975). Secular climatic change and marginal agriculture. *Trans. Institute of British Geographers,* No. 64, 1–13.

Parry, M.L. (1981). Climatic change and the agricultural frontier: a research strategy. In: *Climate and History* (T.M.L. Wigley, M.J. Ingram, and G. Farmer, eds.). Cambridge: Cambridge University Press, 319–336.

Paterson, W.S.B. (1994). *The Physics of Glaciers* (3rd edition). Oxford: Pergamon.

Patzelt, G. (1974). Holocene variations of glaciers in the Alps. *Colloques Internationale du Centre Nationale de la Recherche Scientifique*, **219**, 51–59.

Pätzold, J. (1986) Temperature and CO_2 changes in tropical surface waters of the Philippines during the past 120 years: record in the stable isotopes of hermatypic corals. *Berichte 12*, Kiel: Geologisches/Palaontologisches Institut der Universitat Kiel, 1–82.

Pätzold, J. and Wefer, G. (1992). Bermuda coral reef record of the last 1,000 years. *Proc. Fourth International Conf. on Paleoceanography*, Kiel, 224–225.

Pavese, M.P., Banzon, V., Colacino, M., Gregori, G.P., and Pasqua, M. (1992). Three historical data series on floods and anomalous climatic events in Italy. In: *Climate Since A.D. 1500* (R.S. Bradley and P.D. Jones, eds.). London: Routledge, 155–170.

Payette, S., Filion, L., Delwaide, A., and Bégin, C. (1989). Reconstruction of treeline vegetation response to long-term climate change. *Nature*, **341**, 429–432.

Payette, S. and Gagnon, R. (1985). Late Holocene deforestation and tree regeneration in the forest-tundra of Quebec. *Nature*, **313**, 570–572.

Payette, S. and Morneau, C. (1993). Holocene relict woodlands at the eastern Canadian treeline. *Quaternary Research*, **39**, 84–89.

Pearson, S. and Dodson, J.R. (1993). Stick-nest rat middens as sources of paleoecological data in Australian deserts. *Quaternary Research*, **39**, 347–354.

Pearson, G.W. and Stuiver, M., (1993). High-precision bidecadal calibration of the radiocarbon time scale, 500–2500 B.C. *Radiocarbon*, **35**, 25–33.

Pearson, G.W., Becker, B., and Qua, F. (1993). High precision ^{14}C measurement of German and Irish oaks to show the natural ^{14}C variations from 7890 to 500 B.C. *Radiocarbon*, **35**, 93–104.

Pécsi, M. (1992). Loess of the last glaciation. In: *Atlas of Paleoclimates and Paleoenvironments of the Northern Hemisphere* (B. Frenzel, M. Pécsi, and A. Velichko, eds.). Budapest: Geographical Research Institute, Hungarian Academy of Sciences, 110–119.

Pedersen, T.F., Nielsen, B., and Pickering, M. (1991). Timing of late Quaternary productivity pulses in the Panama Basin and implications for atmospheric CO_2. *Paleoceanography*, **6**, 657–678.

Peel, D.A., Mulvaney, R., and Davison, B.M. (1988). Stable-isotope/air-temperature relationships in ice cores from Dolleman Island and the Palmer Land plateau, Antarctic Peninsula. *Annals of Glaciology*, **10**, 130–136.

Peixoto, J.P. and Oort, A. (1992). *The Physics of Climate*. New York: American Institute of Physics.

Peltier, R. (1994). Ice Age paleotopography. *Science*, **265**, 195–201.

Peng, C.H., Guiot, J. and Van Campo, E. (1998). Estimating changes in terrestrial vegetation and carbon storage using palaeoecological data and models. *Quaternary Science Reviews*, **17**, 719–735.

Peng, T-H. and Broecker, W.S. (1995). Reconstruction of radiocarbon distribution in the Glacial Ocean. In: *Radiocarbon After Four Decades* (R.E. Taylor, A. Long, and R.S. Kra, eds.). New York: Springer-Verlag, 75–92.

Pennington, W. (1973). Absolute pollen frequencies in the sediments of lakes of different morphometry. In: *Quaternary Plant Ecology* (H.J.B. Birks and R. West, eds.). Oxford: Blackwell Scientific Publications, 79–104.

Peteet, D. (1992). The palynological expression and timing of the Younger Dryas event—Europe versus eastern North America. In: *The Last Deglaciation: Absolute and Radiocarbon Chronologies* (E. Bard and W.S. Broecker, eds). Berlin: Springer-Verlag, 327–344.

Peteet, D. (1995). Global Younger Dryas? *Quaternary International*, **28**, 93–104.

Petersen, G.M., Webb, T., Kutzbach, J.E., van der Hammen, T., Wijmstra, T.A., and Street, F.A. (1979). The continental record of environmental conditions at 18,000 yr BP: an initial evaluation. *Quaternary Research*, **12**, 47–82.

Peterson, L.C. and Prell, W.L. (1985). Carbonate preservation and rates of climatic change: an 800 kyr record from the Indian Ocean. In: *The Carbon Cycle and Atmospheric CO₂: Natural Variations, Archean to Present* (E.T. Sundquist and W. S. Broecker, eds.). Washington, DC: American Geophysical Union, 251–269.

Petit, J.R., Basile, I., Leruyuet, A., Raynaud, D., Lorius, C., Jouzel, J., Stievenard, M., Lipenkov, Y.Y., Barkov, N.I., Kudryashov, B.B., Davis, M., Saltzman, E., and Kotlyakov, V. (1997). Four climate cycles in Vostok ice core. *Nature*, 387, 359–360.

Petit, J.R., P., Duval, and Lorius, C. (1987). Long-term climatic changes indicated by crystal growth in polar ice. *Nature,* 326, 62–64.

Petit, J.R., Mounier, L., Jouzel, J., Korotkevich, Y.S., Kotlyakov, V.I., and Lorius, C. (1990). Palaeoclimatological and chronological implications of the Vostok ice core dust record. *Nature*, 343, 56–58.

Petit, J.R., White, J.W.C., Young, N.W., Jouzel, J., and Korotkevich, Y.S. (1991). Deuterium excess in recent Antarctic snow. *Jour. Geophys. Research*, 96D, 5113–5122.

Petit-Maire, N., Fontugne, M., and Rouland, C. (1991). Atmospheric methane ratio and environmental changes in the Sahara and Sahel during the last 130 kyr. *Palaeogeography, Palaeoclimatology, Palaeoecology*, 86, 197–204.

Péwé, T.L. and Reger, R.D. (1972). Modern and Wisconsinan snowlines in Alaska. In: *Proceedings No. 24, Section 12, Quaternary Geology,* International Geological Congress, Montreal, 187–197.

Pfister, C. (1978a). Climate and economy in eighteenth century Switzerland. *Jour. Interdisciplinary History*, 9, 223–243.

Pfister, C. (1978b). Fluctuations in the duration of snow-cover in Switzerland since the late seventeenth century. In: *Proc. Nordic Symposium on Climatic Changes and Related Problems* (K. Frydendahl, ed.). Climatological Papers No. 4, Danish Meteorological Institute, Copenhagen, 1–6.

Pfister, C. (1984). *Klimageschichte der Schweiz 1525–1860. Das Klima der Schweiz von 1525–1860 und seine Bedeutung in der Geschichte von Belvölkerung und Landwirschaft*. 2 vol. Bern.

Pfister, C. (1985). Snow cover, snowlines and glaciers in Central Europe since the 16th century. In: *The Climatic Scene* (M.J. Tooley and G.M. Sheail, eds.). London: Allen and Unwin.

Pfister, C. (1992). Monthly temperature and precipitation in central Europe 1525–1979: quantifying documentary evidence on weather and its effects. In: *Climate Since A.D. 1500* (R.S. Bradley and P.D. Jones, eds.). London: Routledge, 118–142.

Pfister, C. and Hächler, S. (1991). Überschwemmungskatastrophen im Schweizer Alpenraum seit dem Spätmittelalter. *Würzburger Geographische Arbeiten*, 80, 127–148.

Pflaumann, U., Duprat, J., Pujol, C., and Labeyrie, L.D. (1996). SIMMAX: a modern analog technique to deduce Atlantic sea surface temperatures from planktonic foraminifera in deep-sea sediments. *Paleoceanography*, 11, 15–35.

Phillips, F.M., Zreda, M.G., Benson, L.V., Plummer, M.A., Elmore, D., and Sharma, P. (1996). Chronology for fluctuations in Late Pleistocene Sierra Nevada glaciers and lakes. *Science*, 274, 749–751.

Phillipps, P.J. and Held, I.M. (1994). The response to orbital perturbations in an atmospheric model coupled to a slab ocean. *Jour. Climate*, 7, 767–782.

Picciotto, E., Crozaz, G., and De Breuck, W. (1971). Accumulation on the South Pole-Queen Maud Land Traverse 1964–1968. In: *Antarctic Snow and Ice Studies II* (A.P. Crary, ed.). Washington, DC: Antarctic Research Series, American Geophysical Union, 257–316.

Picciotto, E., De Maere, X., and Friedman, I. (1960). Isotopic composition and temperature of formation of Antarctic snows. *Nature,* 187, 857–859.

Pichon, J.J., Labeyrie, L.D., Bareille, G., Labracherie, M., Duprat, J., and Jouzel, J. (1992). Surface water temperature changes in the high latitudes of the southern hemisphere over the last glacial-interglacial cycle. *Paleoceanography*, 7, 289–318.

Pienitz, R., Smol, J.P., and Birks, H.J.B. (1995). Assessment of freshwater diatoms as quantitative indicators of past climate change in the Yukon and Northwest Territories, Canada. *Jour. Paleolimnology*, 13, 21–49.

Pierce, K.L., Obradovich, J.D., and Friedman, I. (1976). Obsidian hydration dating and correlation of Bull Lake and Pinedale Glaciations near west Yellowstone, Montana. *Geological Society of America Bulletin*, **87**, 703–710.

Pike, J. and Kemp, A.E.S. (1996). Records of seasonal flux in Holocene laminated sediments, Gulf of California. In: *Palaeoclimatology and Palaeoceanography from Laminated Sediments* (A.E.S. Kemp, ed.). Special Publication No. 116, London: The Geological Society, 157–169.

Pilcher, J.R., Hall, V.A., and McCormac, F.G. (1995). Dates of Holocene Icelandic volcanic eruptions from tephra layers in Irish peats. *The Holocene*, **5**, 103–110.

Pillow, M.Y. (1931). Compression wood records hurricane. *Jour. Forestry*, **29**, 575–578.

Pisias, N.G. and Moore, T.C. (1981). The evolution of Pleistocene climate: a time series approach. *Earth and Planetary Science Letters*, **52**, 450–458.

Pisias, N.G., Martinson, D.G., Moore, Jr., T.C., Shackleton, N.J., Prell, W., Hays, J., and Boden, G. (1984). High resolution stratigraphic correlation of benthic oxygen isotopic records spanning the last 300,000 years. *Marine Geology*, **56**, 119–136.

Pisias, N.G., Roelofs, A., and Weber, M. (1997). Radiolarian-based transfer functions for estimating mean surface ocean temperature and seasonal range. *Paleoceanography*, **12**, 365–379.

Pokras, E.M. and Mix, A.C. (1985). Eolian evidence for spatial variability of late Quaternary climates in tropical Africa. *Quaternary Research*, **24**, 137–149.

Polge, H. (1970). The use of X-ray densitometric methods in dendrochronology. *Tree Ring Bulletin*, **30**, 1–10.

Ponel, P. (1995). Rissian, Eemian and Würmian Coleoptera assemblages from La Grande Pile (Vosges, France). *Palaeogeography, Palaeoclimatology, Palaeoecology*, **114**, 1–41.

Pons, A., Guiot, J., de Beaulieu, J.L., and Reille, M. (1992). Recent contributions to the climatology of the Last Glacial-Interglacial cycle based on French pollen sequences. *Quaternary Science Reviews*, **11**, 439–448.

Porter, S.C., Pierce, K.L., and Hamilton, T.D. (1983). Late Pleistocene glaciation in the western United States. In: *Late Quaternary Environments of the United States* (S.C. Porter, ed.). Minneapolis: University of Minnesota Press, 71–111.

Porter, S.C. (1977). Present and past glaciation thresholds in the Cascade Range, Washington, USA: topographic and climatic controls and paleoclimatic implications. *Jour. Glaciology*, **18**, 101–116.

Porter, S.C. (1979). Hawaiian glacial ages. *Quaternary Research*, **12**, 161–187.

Porter, S.C. (1981a). Glaciological evidence of Holocene climatic change. In: *Climate and History* (T.M.L. Wigley, M.J. Ingram, and G. Farmer, eds.). Cambridge: Cambridge University Press, 82–110.

Porter, S.C. (1981b). Tephrochronology in the Quaternary geology of the United States. In *Tephra Studies* (S. Self and R.J.S. Sparks, eds.). Dordrecht: D. Reidel, 135–160.

Porter, S.C. (1981c). Lichenometric studies in the Cascade Range of Washington: establishment of *Rhizocarpon geographicurn* growth curves at Mount Rainier. *Arctic and Alpine Research*, **13**, 11–23.

Porter, S.C. (1986). Pattern and forcing of Northern Hemisphere glacier variations during the last millennium. *Quaternary Research*, **26**, 27–48.

Porter, S.C. and An, Z. (1995). Correlation between climate events in the North Atlantic and China during the last glaciation. *Nature*, **375**, 305–308.

Porter, S.C. and Denton, G.H. (1967). Chronology of neoglaciation in the North American Cordillera. *American Journal of Science*, **265**, 177–210.

Potter, C.S., Randerson, J.T., Field, C.B., Marson, P.A., Vitousek, P.M., Mooney, H.A., and Klooster, S.A. (1993). Terrestrial ecosystem production: a process model based on global satellite and surface data. *Global Biogeochemical Cycles*, **7**, 811–841.

Potter, N. (1969). Tree-ring dating of snow avalanche tracks and the geomorphic activity of avalanches, northern Absaroka Mountains, Wyoming. In: *US Contributions to Quaternary Research* (S.A. Schumm and W.C. Bradley, eds.). Special Paper 123, Boulder: Geological Society of America, 141–165.

Prahl, F.G., Muehlhausen, L.A., and Zahnle, D.L. (1988). Further evaluation of long-chain alkenones as indicators of paleoceanographic conditions. *Geochimica et Cosmochimica Acta,* **52,** 2303–2310.

Prahl, F.G., Muehlhausen, L.A., Lyle, M. (1989). An organic geochemical assessment of oceanographic conditions at MANOP Site C over the past 26,000 years. *Paleoceanography,* **4,** 495–510.

Prance, G.T. (1974). Phytogeographic support for the theory of Pleistocene forest refuges in the Amazon Basin, based on evidence from distribution patterns in Caryocaraceae, Chrysobalanaceae, Dichapetalaceae and Lecythidaceae. *Acta Amazonica,* 3, 5–26.

Prance, G.T. (1982). Forest refuges: evidence from woody angiosperms. In: *Biological Diversification in the Tropics* (G.T. Prance, ed.). New York: Columbia University Press, 137–158.

Preiss, N., Mélières, M-A., and Pourchet, M. (1996). A compilation of data on lead-210 concentration in surface air and fluxes at the air-surface and water-sediment interfaces. *Jour. Geophys. Research,* **101,** 28847–28862.

Prell, W.L. (1985). The stability of low-latitude sea-surface temperature: an evaluation of the CLIMAP reconstruction with emphasis on the positive SST anomalies. *Dept. Energy Tech Report TR-025,* Washington, DC.

Prell, W.L., Gardner, J.V., Bé, A.W.H., and Hays, J.D. (1976). Equatorial Atlantic and Caribbean foraminiferal assemblages, temperatures and circulation: interglacial and glacial comparisons. In: *Investigation of late Quaternary Paleooceanography and Paleoclimatology.* (R.M. Cline and J.D. Hays, eds.). Memoir No. 145. Boulder: Geological Society of America, 247–266.

Prell, W.L., Hutson, W.H., Williams, D.F., Bé, A.W.H., Keitzenauer, K., and Molfino, B. (1980). Surface circulation of the Indian Ocean during the last glacial maximum approximately 18,000 yr BP. *Quaternary Research,* **14,** 309–336.

Prell, W.L., Imbrie, J., Martinson, D.G., Morley, J.J., Pisias, N.G., Shackleton, N.J., and Streeter, H.F. (1986). Graphic correlation of oxygen isotope stratigraphy: application to the Late Quaternary. *Paleoceanography,* **1,** 137–162.

Prentice, I.C. (1978). Modern pollen spectra from lake sediments in Finland and Finnmark, North Norway. *Boreas,* 7, 131–153.

Prentice, I.C. (1985). Pollen representation, source area and basin size: toward a unified theory of pollen analysis. *Quaternary Research,* **23,** 76–86.

Prentice, I.C. (1986). Vegetation responses to past climate variation: mechanisms and rates. *Vegetatio,* **67,** 131–141.

Prentice, I.C., Bartlein, P.J., and Webb, III, T. (1991). Vegetation and climate change in eastern North America since the last glacial maximum. *Ecology,* **72,** 2038–2056.

Prentice, I.C., Cramer, W., Harrison, S.P., Leemans, R., Monserud, R.A., and Solomon, A.M. (1992). A global biome model based on plant physiology and dominance, soil properties and climate. *Jour. Biogeography,* **19,** 117–134.

Prentice, I.C., Harrison, S.P., Jolly, D., and Guiot, J. (1997). The climate and biomes of Europe at 6000 yr B.P.: comparison of model simulations and pollen-based reconstructions. *Quaternary Science Reviews,* **17,** 659–668

Prentice, I.C. and Sarnthein, M. (1993). Self-regulatory processes in the biosphere in the face of climate change. In: *Global Changes in the Perspective of the Past* (J.A. Eddy and H.Oeschger, eds.). Chichester: Wiley, 29–38.

Prentice, I.C. and Webb III, T. (1986). Pollen percentages, tree abundances and the Fagerlind effect. *Jour. Quaternary Science,* **1,** 35–43.

Price, N.M., Anderson, L.F., and Morel, F.M.M. (1991). Iron and nitrogen nutrition of equatorial Pacific plankton. *Deep-Sea Research,* 38, 1361–1378.

Pye, K. (1984). Loess. *Progress in Physical Geography,* 8, 176–217.

Pye, K. (1987). *Aeolian Dust and Dust Deposits,* London: Academic Press.

Qin Dahe, Petit, J.R., Jouzel, J., and Stievenard, M. (1994). Distribution of stable isotopes in surface snow along the route of the (1990) International Trans-Antarctica Expedition. *Jour. Glaciology,* **40,** 107–118.

Quinn, T.M., Taylor, F.W., and Crowley, T.J. (1993). A 173 year stable isotope record from a tropical South Pacific coral. *Quaternary Science Reviews*, **12**, 407–418.

Quinn, T.M., Taylor, F.W., and Crowley, T.J., and Link, S.M. (1996). Evaluation of sampling resolution in coral stable isotope records: a case study using monthly stable isotope records from New Caledonia and Tarawa. *Paleoceanography*, **11**, 529–542.

Quinn, W.H. (1993a). A study of Southern Oscillation-related climatic activity for A.D. 622–1990 incorporating Nile River flood data. In: *El Niño: Historical and Paleoclimatic Aspects of the Southern Oscillation* (H.F. Diaz and V. Markgraf, eds.) Cambridge: Cambridge University Press, 119–149.

Quinn, W.H. (1993b). The large-scale ENSO event, the El Niño and other important regional features. *Bulletin de l'Institut Français d'Etudes Andines*, **22**, 13–34.

Quinn, W.H. and Neal, V.T. (1992). The historical record of El Nino events. In: *Climate Since A.D. 1500* (R.S. Bradley and P.D. Jones, eds.). London: Routledge, 623–648.

Quinn, W.H., Neal, V., and Antuñez de Mayolo, S. (1987). El Niño occurrences over the past four and a half centuries. *Jour. Geophys. Research,* **92C**, 14449–14461.

Rahmstorf, S. (1994). Rapid climate transitions in a coupled ocean-atmosphere model. *Nature,* **372**, 82–85.

Rahmstorf, S. (1995). Bifurcations of the Atlantic thermohaline circulation in response to changes in the hydrological cycle. *Nature,* **378**, 145–149.

Raisbeck, G.M., Yiou, F., Bourles, D., Lorius, C., Jouzel, J., and Barkov, N.I. (1987). Evidence for two intervals of enhanced [10]Be deposition in Antarctic ice during the last glacial period. *Nature,* **32**, 273–277.

Raisbeck, G.M., Yiou, F., Jouzel, J., Petit, J.R., Barkov, N.I., and Bard, E. (1992). [10]Be deposition at Vostok, Antarctica during the last 50,000 years and its relationship to possible cosmogenic production variations during this period. In: *The Last Deglaciation: Absolute and Radiocarbon Chronologies* (E. Bard and W.S. Broecker, eds.). Berlin: Springer-Verlag, 127–139.

Ram, M. and Illing, M. (1995). Polar ice stratigraphy from laser-light scattering: scattering from meltwater. *Jour. Glaciology,* **40**, 504–508.

Ramanathan, V. and Coakley, J.A. (1978). Climate modeling through radiative convective models. *Reviews of Geophysics and Space Physics*, **16**, 465–489.

Ramesh, R., Battacharya, S.K., and Pant, G.B. (1989). Climatic significance of δD variations in a tropical tree species. *Nature,* **337**, 149–150.

Rampino, M. and Self, S. (1982). Historic eruptions of Tambora (1815), Krakatau (1883), and Agung (1963) their stratospheric aerosols and climatic impact. *Quaternary Research,* **18**, 127–143.

Rampino, M. and Self, S. (1984). Sulfur-rich volcanic eruptions and stratospheric aerosols. *Nature,* **310**, 677–679.

Ramsey, C.N. (1995). Radiocarbon calibration and analysis of stratigraphy: the Oxcal program. *Radiocarbon*, **37**, 425–430.

Rasmussen, T.L., Thomsen, E., van Weering, T.C.E., and Labeyrie, L. (1996). Rapid changes in surface and deepwater conditions at the Faeroe margin during the last 58,000 years. *Paleoceanography*, **11**, 757–771.

Raymo, M. (1998). Glacial puzzles. *Science*, **281**, 1467–1468.

Raymo, M.E. (1992). Global climate change: a three million year perspective. In: *Start of a Glacial* (G.J. Kukla and E. Went, eds.). Berlin: Springer-Verlag, 207–223.

Raymo, M., Ruddiman, W.F., Shackleton, N.J., and Oppo, D. (1990). Evolution of global ice volume and Atlantic-Pacific $\delta^{13}C$ gradients over the last 2.5 m.y. *Earth and Planetary Science Letters*, **97**, 353–368.

Raynaud, D. (1992). The ice record of the atmospheric composition: a summary, chiefly of CO_2, CH_4 and O_2. In: *Trace Gases and the Biosphere* (B. Moore III, and D. Shimel, eds.). Boulder: University Corporation for Atmospheric Research, 165–176.

Raynaud, D. and Barnola, J-M. (1985). CO_2 and climate: information from Antarctica ice core studies. In: *Current Issues in Climate Research* (A. Ghazi and R. Fantechi, eds.). Dordrecht: D. Reidel, 240–246.

Raynaud, D., Chappellaz, J., Barnola, J-M., Korotkevich, Y.S., and Lorius, C. (1988). Climatic and CH_4 cycle implications of glacial-interglacial CH_4 change in the Vostok ice core. *Nature,* **333,** 655–657.

Raynaud, D., Jouzel, J., Barnola, J.M., Chappellaz, J., Delmas, R.J., and Lorius, C. (1993). The ice record of greenhouse gases. *Science,* **259,** 926–934.

Rea, D.K. (1994). The paleoclimatic record provided by eolian deposition in the deep sea: the geologic history of the wind. *Reviews of Geophysics,* **32,** 159–195.

Reck, R.A. (1974). Aerosols in the atmosphere: calculation of the critical absorption/ backscatter ratio. *Science,* **173,** 138–141.

Reeh, N. (1989). Dating by ice flow modeling: a useful tool or an exercise in applied mathematics? In: *The Environmental Record in Glaciers and Ice Sheets* (H. Oeschger and C.C. Langway, Jr., eds.). Chichester: Wiley, 141–159.

Reeh, N. (1991). The last interglacial as recorded in the Greenland Ice Sheet and Canadian Arctic Ice Caps. *Quaternary International,* **10-12,** 123–142.

Reeh, N., Clausen, H.B., Dansgaard, W., Gundestrup, N., Hammer, C.U., and Johnsen, S.J. (1978). Secular trends of accumulation rates at three Greenland stations. *Jour. Glaciology,* **20,** 27–30.

Reeh, N., Thomsen, H.H., and Clausen, H.B. (1987). The Greenland ice sheet margin— a mine of ice for paleo-environmental studies. *Palaeogeography, Palaeoclimatology, Palaeoecology,* **58,** 229–234.

Reeh, N., Oerter, H., Letréguilly, A., Miller, H., and Hubberten, H.W. (1991). A new detailed ice-age oxygen-18 record from the ice-sheet margin in central West Greenland. *Palaeogeography, Palaeoclimatology, Palaeoecology (Global and Planetary Change Section),* **90,** 373–383.

Reeh, N., Oerter, H., and Miller, H. (1993). Correlation of Greenland ice-core and ice-margin $\delta^{18}O$ records. In: *Ice in the Climate System* (W. Peltier, ed.). Berlin: Springer-Verlag, 481–497.

Reeves, C.C. (1965). Pleistocene climate of the Llano Estacado. *Jour. Geology,* **73,** 181–189.

Regnell, J. (1992). Preparing pollen concentrates for AMS dating—a methodological study from a hard-water lake in southern Sweden. *Boreas,* **21,** 373–377.

Rendell, H.M. (1995). Luminescence dating of Quaternary sediments. In: *Non-biostratigraphical Methods of Dating and Correlation* (R.E. Dunay, and E.A. Hailwood, eds.). Special Publication 89. London: The Geological Society, 223–235.

Renne, P.R., Deino, A.L., Walter, R.C., Turrin, B.D., Swisher, C.C., Becker, T.A., Curtis, G.H., Sharp, W.D., and Jaouni, A.R. (1994). Intercalibration of astronomical and radioisotopic time. *Geology,* **22,** 783–786.

Revel, M., Sinko, J.A., Grousset, F.E., and Biscaye, P.E. (1996). Sr and Nd isotopes as tracers of North Atlantic lithic particles: paleoclimatic implications. *Paleoceanography,* **11,** 95–113.

Reynolds-Sautter, L. and Thunell, R.C. (1989). Seasonal succession of planktonic foraminifera: results from a four year time-series sediment trap experiment in the Northeast Pacific. *Jour. Foraminiferal Research,* **19,** 253–267.

Richards, D.A., Smart, P.L., and Edwards, R.L. (1994). Maximum sea levels for the last glacial period from ages of submerged speleothems. *Nature,* **367,** 357–360.

Richman, M.B. (1986). Rotation of principal components. *Jour. Climatology,* **6,** 293–335.

Richmond, G.M. (1965). Glaciation of the Rocky Mountains. In: *The Quaternary of the United States* (H.E. Wright and D.G. Frey, eds.). Princeton: Princeton University Press, 217–230.

Richmond, G.M. (1985). Stratigraphy and correlation of glacial deposits in the Rocky Mountains, the Colorado Plateau and the Ranges of the Great Basin. *Quaternary Science Reviews,* **5,** 99–127.

Ridge, J.C., Brennan, W.J., and Muller, E.H. (1990). The use of paleomagnetic declination to test correlations of late Wisconsinan glaciolacustrine sediments in central New York. *Geological Society of America Bulletin,* **102,** 26–44.

Rind, D. (1987). Components of the ice age circulation. *Jour. Geophys. Research,* **92D,** 4241–4281.

Rind, D. (1993). How will future climate changes differ from those of the past? In: *Global Changes in the Perspective of the Past* (J.A. Eddy and H. Oeschger, eds.). Chichester: Wiley, 39–49.

Rind, D. and Overpeck, J. (1993). Hypothesized causes of decadel-to-century climate variability: climate model results. *Quaternary Science Reviews,* 12, 357–374.

Rind, D., and Peteet, D. (1985). Terrestrial conditions at the last glacial maximum and CLIMAP sea-surface temperature estimates: are they consistent? *Quaternary Research,* 24, 1–22.

Rind, D., Peteet, D., and Kukla, G. (1989). Can Milankovitch orbital variations initiate the growth of ice sheets in a general circulation model? *Jour. Geophys. Research*, **94D**, 12851–12871.

Ritchie, J.C. (1976). The late-Quaternary vegetational history of the western interior of Canada. *Canadian Jour. Botany,* 54, 1793–1818.

Ritchie, J.C. (1986). Climate change and vegetation response. *Vegetatio,* 67, 65–74.

Ritchie, J.C. (1987). *Postglacial Vegetation of Canada.* Cambridge: Cambridge University Press.

Ritchie, J.C. and Hare, F.K. (1971). Late Quaternary vegetation and climate near the Arctic treeline of northwestern North America. *Quaternary Research,* 1, 331–342.

Robertson, I., Field, E.M., Heaton, T.H., Pilcher, J.R., Pollard, M., Switsur, R., and Waterhouse, J.S. (1995). Isotope coherence in oak cellulose. In: *Problems of Stable Isotopes in Tree-rings, Lake Sediments and Peat-bogs as Climatic Evidence for the Holocene* (B. Frenzel, B. Stauffer, and M.M. Weiss, eds.). Stuttgart: Gustav Fischer Verlag, 141–155.

Robertson, S. and Grün, R. (1994). Towards portable radiocarbon dating. *Quaternary Science Reviews (Quaternary Geochronology),* 13, 179–181.

Robin, G. de Q. (1977). Ice cores and climatic change. *Phil. Trans. Royal Society of London, B,* 280, 143–168.

Robinson, W.J. (1976). Tree-ring dating and archaeology in the American Southwest. *Tree Ring Bulletin,* 36, 9–20.

Robinson, W.J., Cook, E., Pilcher, J.R., Eckstein, D., Kariukstis, L., Shiyatov, S., and Norton, D.A. (1990). Some historical background on dendrochronology. In: *Methods of Dendrochronology. Applications on the Environmental Sciences* (E.R. Cook and L.A. Kariukstis, eds.). Dordrecht: Kluwer, 1–21.

Robock, A. (1978). Internal and externally caused climate change. *Jour. Atmos. Sciences,* 35, 1111–1122.

Robock, A. and Mao, J-P. (1995). The volcanic signal in surface temperature observations. *Jour. Climate,* 8, 1086–1103.

Rochefort, R.M., Little, R.L., Woodward, A., and Peterson, D.L. (1994). Changes in subalpine tree distribution in western North America: a review of climatic and other causal factors. *The Holocene,* 4, 89–100.

Rodbell, D.T. (1992). Lichenometric and radiocarbon dating of Holocene glaciation, Cordillera Blanca, Peru. *The Holocene,* 2, 19–29.

Rodbell, D.T. (1993). Subdivision of late Pleistocene moraines in the Cordillera Blanca, Peru based on rock-weathering features, soils and radiocarbon dates. *Quaternary Research,* 39, 133–143.

Rognon, P. (1976). Essai d'interpretation des variations climatiques au Sahara depuis 40,000 ans. *Revue de Géographie Physique et de Géologie Dynamique,* 18, 251–282.

Rognon, P. and Williams, M.A.J. (1977). Late Quaternary climatic changes in Australia and North Africa: a preliminary interpretation. *Palaeogeography, Palaeoclimatology, Palaeoecology,* 21, 285–327.

Rolph, T.C., Shaw, J., Derbyshire, E., and Wang, J.T. (1989). A detailed geomagnetic record from Chinese loess. *Physics of the Earth and Planetary Interiors,* 56, 151–164.

Ropelewski, C.F and Halpert, M.S. (1987). Global and regional scale precipitation patterns associated with El Niño/Southern Oscillation. *Monthly Weather Review,* 115, 1606–1626.

Ropelewski, C.F and Halpert, M.S. (1989). Precipitation patterns associated with high-index phase of the Southern Oscillation. *Jour. Climate* 2, 268–284.

Rosell-Melé, A., Eglinton, G., Pflaumann, U., and Sarnthein, M. (1995). Atlantic core-top calibration of the U^k_{37} index as a sea-surface paleotemperature indicator. *Geochimica et Cosmochimica Acta,* **59**, 3099–3107.

Rostek, F., Ruhland, G., Bassinot, F.C., Müller, P.J., Labeyrie, L.D., Lancelot, Y., and Bard, E. (1993). Reconstructing sea surface temperature and salinity using $\delta^{18}O$ and alkenone records. *Nature,* **364**, 319–321.

Rotberg, R.I. and Rabb, T.K. (eds.). (1981). *Climate and History: Studies in Interdisciplinary History.* Princeton: Princeton University Press.

Rothlisberger, F. (1976). Gletscher- und Klimaschwankungen im Raun Zermatt, Ferpecle und Arolla. *Die Alpen,* **52**, 59–132.

Röthlisberger, F. (1986). *10,000 Jahre Gletschergeschichte der Erde.* Aarau: Verlag Sauerländer.

Royer, A., De Angelis, M., and Petit, J.R. (1983). A 30,000 year record of physical and optical properties of miscroparticles from an East Antarctic core and implications for paleoclimate reconstruction models. *Climate Change,* **5**, 381–412.

Royer, J.F., Deque, M., and Pestiaux, P. (1984). A sensitivity experiment to astronomical forcing with a spectral GCM: simulation of the annual cycle at 125,000 BP and 115,000 BP. In: *Milankovitch and Climate,* **2** (A.L. Berger, J. Imbrie, J. Hays, G. Kukla, and B. Saltzman, eds.). Dordrecht: D. Reidel.

Rozanski, K., Araguás-Araguás, L., and Gonfiantini, R. (1992). Relation between long-term trends of oxygen-18 isotope composition of precipitation and climate. *Science,* **258**, 981–985.

Rozanski, K., Araguás-Araguás, L., and Gonfiantini, R. (1993). Isotopic patterns in modern global precipitation. In: *Climate Change in Continental Isotopic Records* (P.K. Swart, K.C. Lohmann, J. McKenzie, and S. Savin, eds.). Washington, DC: American Geophysical Union, 1–36.

Ruddiman, W.F. (1971). Pleistocene sedimentation in the equatorial Atlantic: stratigraphy and faunal paleoclimatology. *Geological Society of America Bulletin,* **82**, 283–302.

Ruddiman, W.F. (1977a). Investigations of Quaternary climate based on planktonic foraminifera. In: *Oceanic Micropaleontology* (A.T.S. Ramsay). New York: Academic Press, 101–161.

Ruddiman, W.F. (1977b). Late Quaternary deposition of ice-rafted sand in the sub polar North Atlantic (lat. 40-65° N). *Geological Society of America Bulletin,* **88**, 1813–1827.

Ruddiman, W.F. (1987). Synthesis: the ocean ice-sheet record. In: *North American and Adjacent Oceans during the Last Deglaciation* (W.F. Ruddiman and H.E. Wright, eds.). Boulder: Geological Society of America, 463–478.

Ruddiman, W.F. and Esmay, A. (1987). A streamlined foraminiferal transfer function for the subpolar North Atlantic. *Initial Reports of the Deep Sea Drilling Project,* **94**, 1045–1057.

Ruddiman, W.F. and Glover, L.K. (1972). Vertical mixing of ice-rafted volcanic ash in North Atlantic sediments. *Geological Society of America Bulletin,* **83**, 2817–2836.

Ruddiman, W.F. and Heezen, B.C. (1967). Differential solution of planktonic foraminifera. *Deep-Sea Research,* **14**, 801–808.

Ruddiman, W.F. and Kutzbach, J.E. (1989). Forcing of late Cenozoic northern hemisphere climate by plateau uplift in southern Asia and the American West. *Jour. Geophys. Research,* **94**, 18409–18427.

Ruddiman, W.F. and McIntyre, A. (1976). Northeast Atlantic paleoclimatic changes over the past 600,000 years. In: *Investigation of Late Quaternary Paleooceanography and Paleoclimatology* (R.M. Cline and J.D. Hays, eds.). Memoir 145. Boulder: Geological Society of America, 111–146.

Ruddiman, W.F. and McIntyre, A. (1981a). Oceanic mechanisms for amplification of the 23,000 year ice-volume cycle. *Science,* **212**, 617–627.

Ruddiman, W.F. and McIntyre, A. (1981b). The mode and mechanism of the last deglaciation: oceanic evidence. *Quaternary Research,* **16**, 125–134.

Ruddiman, W.F. and McIntyre, A. (1984). Ice-age thermal response and climatic role of the surface Atlantic Ocean, 40° N to 63° N. *Geological Society of America Bulletin,* **95**, 381–396.

Ruddiman, W.F., Raymo, M., and McIntyre, A. (1986). Matuyama 41,000-year cycles: north Atlantic Ocean and northern hemisphere ice sheets. *Earth and Planetary Science Letters,* 80, 117–129.

Rutter, N. (1995). Problematic ice sheets. *Quaternary International, 28,* 19–37.

Rutter, N., Ding, Z., Evans, M.E., and Wang, Y. (1990). Magnetostratigraphy of the Baoji loess-paleosol section in the north-central China loess plateau. *Quaternary International, 7/8,* 97–102.

Rutter, N., Ding, Z., Evans, M.E., and Liu, T. (1991a). Baoji-type pedostratigraphic section, Loess Plateau, north-central China. *Quaternary Science Reviews,* 10, 1–22.

Rutter, N., Ding, Z., and Liu, T. (1991b). Comparison of isotope stages 1–61 with the Baoji-type pedostratigraphic section of north-central China. *Canadian Jour. Earth Sciences,* 28, 985–990.

Rutter, N.W. and Blackwell, B., 1995. Amino acid racemization dating. In: *Dating Methods for Quaternary Deposits* (N.W. Rutter and N.R. Catto, eds.). St John's, Geological Association of Canada, 125–166.

Rutter, N.W., Crawford, R.J., and Hamilton, R.D. (1979). Dating methods of Pleistocene deposits and their problems: IV. Amino acid racemization dating. *Geoscience Canada,* 6, 122–8.

Rybníčková, E. and Rybníček, K. (1993). Late Quaternary forest line oscillations in the West Carpathians. In: *Oscillations of the Alpine and Polar Tree Limits in the Holocene* (B. Frenzel, M. Eronen, K.D. Vorren, and B. Gläser, eds.). Stuttgart: Gustav Fischer Verlag, 187–194.

Saarnisto, M. (1986). Annually laminated lake sediments. In: *Handbook of Holocene Palaeoecology and Palaeohydrology* (B. Berglund, ed.). Chichester: Wiley, 343–370.

Sachs, M.H., Webb, T., and Clark, D.R. (1977). Paleoecological transfer functions. *Annual Reviews of Earth and Planetary Science,* 5, 159–178.

Saigne, C. and Legrand, M. (1987). Measurement of methane-sulphonic acid in Antarctic ice. *Nature,* 330, 240–242.

Salgado-Labouriau, M.L. (1979). Modern pollen deposition in the Venezuelan Andes. *Grana,* 18, 53–68.

Salgado-Labouriau, M.L., Schubert, C., and Valastro Jr., S. (1978). Paleoecologic analysis of a Late-Quaternary terrace from Mucubaji, Venezuelan Andes. *Jour. Biogeography,* 4, 313–325.

Salvesen, H. (1992). The climate as a factor of historical causation. In: *European Climate Reconstructed from Documentary Data: Methods and Results* (B. Frenzel, C. Pfister, and B. Glaser, eds.). Stuttgart: Gustav Fischer Verlag, 219–233.

Saltzman, B. (1985). Paleoclimatic modeling. In: *Paleoclimate Analysis and Modeling.* (A.D. Hecht, ed.). Chichester: Wiley, 341–396.

Sancetta, C. (1979). Oceanography of the North Pacific during the last 18,000 years: evidence from fossil diatoms. *Marine Micropalaeontology,* 4, 103–123.

Sancetta, C. (1983). Fossil diatoms and the oceanography of the Bering Sea during the last glacial event. In: *Siliceous Deposits in the Pacific Region* (A. Iijima, J.R. Hein, and R. Fiever, eds.). Amsterdam: Elsevier, 333–346.

Sancetta, C. (1995). Diatoms in the Gulf of California: seasonal flux patterns and the sediment record for the last 15,000 years. *Paleoceanography,* 10, 67–84.

Sancetta, C., Imbrie, J., and Kipp, N.G. (1973a). Climatic record of the past 130,000 years in the North Atlantic deep-sea core V23–83: correlation with the terrestrial record. *Quaternary Research,* 3, 110–116.

Sancetta, C., Imbrie, J., and Kipp, N.G. (1973b). The climatic record of the past 14,000 years in North Atlantic deep-sea core V23–82: correlation with the terrestrial record. In: *Mapping the Atmospheric and Oceanic Circulations and Other Climatic Parameters at the Time of the Last Glacial Maximum about 17,000 years ago.* Publication No. 2, Norwich: Climatic Research Unit, University of East Anglia, 62–65.

Santer, B.D., Berger, A., Eddy, J.A., Flohn, H., Imbrie, J., Litt, T., Schneider, S.H., Schweingruber, F.H., and Stuiver, M. (1993). How can paleodata be used to evaluate the forcing mechanisms responsible for past climate changes? In: *Global Changes in the Perspective of the Past* (J.A. Eddy and H. Oeschger, eds.). Chichester: Wiley, 343–367.

Santer, B.D., Wigley, T.M.L., Barnett, T.P., and Anyamba, E. (1996). Detection of climate change and attribution of causes. In: *Climate Change 1995: The Science of Climate Change* (J.T. Houghton, L.G. Meirho Filho, B.A. Callendar, A. Kattenberg, and K. Maskell, eds.). Cambridge: Cambridge University Press, 407–443.

Sarmiento, J.L. (1992). Biogeochemical ocean models. In: *Climate System Modeling*. (ed. K.E. Trenberth). Cambridge: Cambridge University Press, 519–551.

Sarnthein, M. (1978). Sand deserts during glacial maximum and climatic optimum. *Nature, 272*, 43–46.

Sarnthein, M., Jansen, E., Arnold, M., Duplessy, J-C., Erlenkeuser, H., Flatøy, A., Veum, T., Vogelsang, E., and Weinelt, M.S. (1992). $\delta^{18}O$ time-slice reconstruction of meltwater anomalies at termination I in the North Atlantic between 50 and 80° N. In: *The Last Deglaciation : Absolute and Radiocarbon Chronologies* (E. Bard and W.S. Broecker, eds.). Berlin: Springer-Verlag, 183–200.

Sarnthein, M., Tetzlaff, G., Koopman, B., Wolter, K., and Pflaumann, U. (1981). Glacial and interglacial wind regimes over the eastern sub-tropical Atlantic and northwest Africa. *Nature, 293*, 193–196.

Sarnthein, M., Winn, K. and Zahn, R. (1987). Palaeoproductivity of oceanic upwelling and the effect on atmospheric CO_2 and climatic change during deglaciation times. In: *Abrupt Climatic Change* (W.H. Berger and L.D. Labeyrie, eds.). Dordrecht: D. Reidel, 311–337.

Sarnthein, M., Winn, K., Duplessy, J-C., and Fortugne, M.R. (1988). Global variations of surface ocean productivity in low and mid latitudes: influence on CO_2 reservoirs of the deep ocean and atmosphere during the last 21,000 years. *Paleoceanography, 3*, 361–379.

Savin, S.M. and Stehli, F.G. (1974). Interpretation of oxygen isotope paleotemperature measurements: effect of the $^{18}O/^{16}O$ ratio of sea water depth stratification of foraminifera, and selective solution. In: *Les Méthodes Quantitatives d'Etude des Variations du Climat au Cours du Pleistocène*. Colloques Internationaux du Centre National de la Recherche Scientifique No. 219, CNRS, Paris, 183–191.

Schimel, D., Alves, D., Enting, I., Heimann, M., Joos, F., Raynaud, D., Wigley, T., Prather, M., Derwent, R., Ehhalt, D., Fraser, P., Sanhueza, E., Zhou, X., Jonas, P., Charlson, R., Rodhe, H., Sadasivan, S., Shine, K.P., Fouquart, Y., Ramaswamy, V., Solomon, S., Srinavasan, J., Albritton, D., Isaksen, I., Lal, M. and Wuebbles, D. (1996). Radiative forcing of climate change. In: *Climate Change 1995: The Science of Climate Change* (eds. J.T. Houghton, L.G. Meirha Filho, B.A. Callendar, A. Kattenberg, and K. Maskell). Cambridge University Press, Cambridge, 65–131.

Schlesinger, M.E. (ed.). (1987). *Physically-Based Climate Models and Climate Modelling* (2 volumes). Dordrecht: Reidel.

Schlesinger, M.E. and Verbitsky, M. (1996). Simulation of glacial onset with a coupled atmosphere general circulation/mixed-layer ocean–ice-sheet/asthenosphere model. *Paleoclimates, 2*, 179–201.

Schmitz, W.J. (1992). On the interbasin-scale thermohaline circulation. *Reviews of Geophysics, 33*, 151–173.

Schneebeli, W. (1976). Untersuchungen yon Gletscherschwankungen in Val de Bagnes. *Die Alpen, 52*, 5–58.

Schneider, S.H. (1992). Introduction to climate modeling. In: *Climate System Modeling* (K.E. Trenberth, ed.). Cambridge: Cambridge University Press, 3–26.

Schneider, S., (1993). Degrees of uncertainty. *Research and Exploration, 9*, 173–190.

Schott, W. (1935). Die Foraminiferen in dem aequatorialen Teil des Atlantischen Ozeans. Deutsche Atlantische Expedition "Meteor" 1925–1927. *Wissenschaftliche Ergebnisse, 3* (3), 43–134.

Schrag, D.P. and DePaulo, D.J. (1993). Determination of $\delta^{18}O$ of sea water in the deep ocean during the last glacial maximum. *Paleoceanography, 8*, 1–6.

Schroeder, R.A. and Bada, J.L. (1973). Glacial-postglacial temperature difference deduced from aspartic acid racemization in fossil bones. *Science, 182*, 479–482.

Schroeder, R.A. and Bada, J.L. (1976). A review of the geochemical applications of the amino acid racemization reaction. *Earth Science Reviews,* **12,** 347–391.

Schubert, C. (1992). The glaciers of the Sierra Nevada de Mérida (Venezuela): a photographic comparison of recent deglaciation. *Erdkunde,* **46,** 58–64.

Schüle, H. and Pfister, C. (1992). Coding climate proxy information for the EURO-CLIMHIST data base. In: *European Climate Reconstructed from Documentary Data: Methods and Results* (B. Frenzel, C. Pfister, and B. Gläser, eds.). Stuttgart: Gustav Fischer Verlag, 235–262.

Schwander, J., and Stauffer, B. (1984). Age difference between polar ice and the air trapped in its bubbles. *Nature,* **311,** 45–47.

Schwarcz, H.P. (1996). Paleoclimate inferences from stable isotope studies of speleothems. In: *Climate Change: the Karst Record* (S-E. Lauritzen, ed.). Charleston, West Virginia: Karst Waters Institute Special Publication **2,** 145–147.

Schwarcz, H.P., Harmon, R.S., Thompson, P., and Ford, D.C. (1976). Stable isotope studies of fluid inclusions in speleothems and their paleoclimatic significance. *Geochimica et Cosmochimica Acta,* **40,** 657–665.

Schwarz-Zanetti, W., Pfister, C., Schwarz-Zanetti, G., and Schüle, H. (1992). The EURO-CLIMHIST data base—a tool for reconstructing the climate of Europe in the pre-instrumental period from high resolution proxy data. In: *European Climate Reconstructed from Documentary Data: Methods and Results* (B. Frenzel, Pfister, C. and B. Gläser). Stuttgart: Gustav Fischer Verlag, 193–210.

Schweingruber, F.H. (1988). *Tree Rings. Basics and Applications of Dendrochronology.* Dordrecht: Kluwer.

Schweingruber, F.H. (1996). *Tree Rings and Environment. Dendroecology.* Berne: Paul Haupt Publishers.

Schweingruber, F.H., Braker, O.U., and Schar, E. (1979). Dendroclimatic studies on conifers from central Europe and Great Britain. *Boreas,* **8,** 427–452.

Schweingruber, F.H. and Briffa, K.R. (1996). Tree-ring density networks for climate reconstruction. In: *Climate Variations and Forcing Mechanisms of the Last 2000 Years* (P.D. Jones, R.S. Bradley, and J. Jouzel, ed.). Berlin: Springer-Verlag, 43–66.

Schweingruber, F.H., Briffa, K.R., and Jones, P.D. (1991). Yearly maps of summer temperatures in western Europe from A.D. 1750 to 1975 and western North America from 1600 to 1982: results of a radiodensitometrical study on tree rings. *Vegetatio,* **92,** 5–71.

Schweingruber, F.H., Briffa, K.R., and Nogler, P., (1993). A tree-ring densitometric transect from Alaska to Labrador. *International Journal of Biometeorology,* **37,** 151–169.

Schweingruber, F.H., Fritts, H.C., Braker, O.U., Drew, L.G., and Schar, E. (1978). The X-ray technique as applied to dendroclimatology. *Tree Ring Bulletin,* **38,** 61–91.

Scott, E.M., Long, Q., and Kra, R. (eds.). (1990). *Proc. International Workshop on Intercomparison of Radiocarbon Laboratories. Radiocarbon,* **32,** 253–397.

Sear, C.B., Kelly, P.M., Jones, P.D., and Godess, C.M. (1987). Global surface air temperature responses to major volcanic eruptions. *Nature,* **330,** 365–367.

Sekiguti, T. (1969). The historical dates of Japanese cherry festivals since the 8th century and their climatic changes. *Geographical Review of Japan,* **35,** 67–76.

Self, S. and Sparks, R.J.S. (eds.). (1981). *Tephra Studies.* Dordrecht: D. Reidel.

Sellers, W.D. (1969). A climate model based on the energy balance of the earth-atmosphere system. *Jour. Appl. Meteorology,* **8,** 392–400.

Sellers, W.D. (1973). A new global climate model. *Jour. Appl. Meteorology,* **12,** 241–254.

Seltzer, G.O. (1990). Recent glacial history and paleoclimate of the Peruvian-Bolivian Andes. *Quaternary Science Reviews,* **9,** 137–152.

Seltzer G.O. (1994). Climatic interpretation of Alpine snowline variations on millennial time scales. *Quaternary Research,* **41,** 154–159.

Seret, G., Guiot, J., Wansard, G., de Beaulieu, J.L., and Reille, M. (1992). Tentative paleoclimatic reconstruction linking pollen and sedimentology in La Grande Pile (Vosges, France). *Quaternary Science Reviews,* **11,** 425–430.

Serre-Bachet, F., Guiot, J., and Tessier, L. (1992). Dendroclimatic evidence from southwestern Europe and northwestern Africa. In: *Climate Since A.D. 1500* (R.S. Bradley and P.D. Jones, eds.). London: Routledge, 349–365.

Seurat, L.G. (1934). *Etudes zoologiques sur le Sahara Central.* Memoires de la Societe d'Histoire Naturelie de l'Afrique du Nord, No. 4, Mission du Hoggar III.

Shackleton, N.J. (1967). Oxygen isotope analyses and Pleistocene temperatures re-assessed. *Nature, 215,* 15–17.

Shackleton, N.J. (1969). The last interglacial in the marine and terrestrial records. *Proc. Royal Society of London,* B174, 135–154.

Shackleton, N.J. (1974). Attainment of isotopic equilibrium between ocean water and the benthonic foraminifera Genus *Uvigerina*: isotopic changes in the ocean during the last glacial. In: *Les Méthodes Quantitatives d'Etude des Variations du Climat au Cours du Pleistocène.* Colloques Internationaux du Centre National de la Recherche Scientifique No. 219, CNRS, Paris, 4–5.

Shackleton, N.J. (1977). The oxygen isotope stratigraphic record of the late Pleistocene. *Phil. Trans. Royal Society of London,* B280, 169–179.

Shackleton, N.J. (1987). Oxygen isotopes, ice volume and sea level. *Quaternary Science Reviews, 6,* 183–190.

Shackleton, N.J. (1993). Last interglacial in Devil's Hole, Nevada. *Nature, 362,* 596 (see also reply by K.R. Ludwig *et al.*, p. 596).

Shackleton, N.J., Backman, J., Zimmerman, H., Kent, D.V., Hall, M.A., Roberts, D.G., Schnitker, D., and Baldauf, J. (1984). Oxygen isotope calibration of the onset of ice rafting and history of glaciation in the North Atlantic region. *Nature,* 307, 620–623.

Shackleton, N.J., Berger, A., and Peltier, W.R. (1990). An alternative astronomical calibration of the lower Pleistocene timescale based on ODP site 677. *Trans. Royal Society of Edinburgh, Earth Science,* 81, 251–261.

Shackleton, N.J., Duplessy, J-C., Arnold, M., Maurice, P., Hall, M.A., and Cartlidge, J. (1988). Radiocarbon age of Last Glacial Pacific deep water. *Nature, 335,* 708–711.

Shackleton, N.J., Hall, M.A., Line, J., and Cang, S. (1983). Carbon isotope data in core V19-30 confirm reduced carbon dioxide concentration of the ice age atmosphere. *Nature,* 306, 319–322.

Shackleton, N.J., Le, J., Mix, A., and Hall, M.A. (1992). Carbon isotope records from Pacific surface waters and atmospheric carbon dioxide. *Quaternary Science Reviews,* 11, 387–400.

Shackleton, N.J. and Matthews, R.K. (1977). Oxygen isotope stratigraphy of late Pleistocene coral terraces in Barbados. *Nature,* 268, 618–620.

Shackleton, N.J. and Opdyke, N.D. (1973). Oxygen isotope and paleomagnetic stratigraphy of equatorial Pacific core V28–238: oxygen isotope temperatures and ice volumes on a 10^5 year and 10^6 year scale. *Quaternary Research, 3,* 39–55.

Shackleton, N.J. and Opdyke, N.D. (1976). Oxygen-isotope and paleomagnetic stratigraphy of Pacific core V28–239. Late Pliocene to latest Pleistocene. In: *Investigation of Late Quaternary Paleooceanography and Paleoclimatology* (R.M. Cline and J.D. Hays, eds.). Memoir 145, Boulder: Geological Society of America, 449–464.

Shackleton, N.J. and Pisias, N.G. (1985). Atmospheric carbon dioxide, orbital forcing and climate. In: *The Carbon Cycle and Atmospheric CO_2: Natural Variations Archean to Present* (E.T. Sundquist and W.S. Broecker, eds.). Washington, DC: American Geophysical Union, 303–317.

Shackleton, N.J., Wiseman, J.D.H., and Buckley, H.A. (1973). Non-equilibrium isotopic fractionation between sea-water and planktonic foraminiferal tests. *Nature, 242,* 177–179.

Shaffer, G. and Bendtsen, J. (1994). Role of the Bering Strait in controlling North Atlantic Ocean circulation and climate. *Nature, 367,* 354–357.

Shane, P.A.R. and Froggatt, P.C. (1994). Discriminant function analysis of glass chemistry of New Zealand and North American tephra deposits. *Quaternary Research,* 41, 70–81.

Sheets, P.D. and Grayson, D.K. (eds.). (1979). *Volcanic Activity and Human Ecology.* New York: Academic Press.

Shemesh, A., Charles, C., and Fairbanks, R.G. (1992). Oxygen isotopes in biogenic silica: global changes in ocean temperature and isotopic composition. *Science,* **256,** 1434–1436.

Shemesh, A. and Peteet, D. (1998). Oxygen isotopes in fresh water biogenic opal—northeastern U.S. Alleröd-Younger Dryas temperature shift. *Geophysical Research Letters,* **25,** 1935–1938.

Shen, G.T., Boyle, E.A., and Lea, D.W. (1987). Cadmium in corals as a tracer of historical upwelling and industrial fallout. *Nature,* **328,** 794–796.

Shen, G.T., Cole, J.E., Lea, D.W., McConnaughey, T.A., and Fairbanks, R.G. (1992a). Surface ocean variability at Galapagos from 1936–1982: calibration of geochemical tracers in corals. *Paleoceanography,* **7,** 563–583.

Shen, G.T., Lim, L.J., Campbell, T.M., Cole, J.E., and Fairbanks, R.G. (1992b). A chemical indicator of trade wind reversal in corals from the eastern tropical Pacific. *Jour. Geophys. Research,* **97,** 12689–12698.

Shen, G.T. and Sanford, C.L. (1990). Trace element indicators of climate variability in reef-building corals. In: *Global Ecological Consequences of the 1981–83 El Niño-Southern Oscillation* (P.W. Glynn, ed.). New York: Elsevier, 255–284.

Shinn, R.A. and Barron, E.J. (1989). Climate sensitivity to continental ice sheet size and configuration. *Jour. Climate,* **2,** 1517–1516.

Shiyatov, S.G. (1993). The upper timberline dynamics during the last 1100 years in the Polar-Ural mountains. In: *Oscillations of the Alpine and Polar Tree Limits in the Holocene* (B. Frenzel, M. Eronen, K.D. Vorren, and B. Gläser, eds.). Stuttgart: Gustav Fischer Verlag, 195–203.

Shiyatov, S.G. (1996). Tree growth decrease between 1800 and 1840 in subarctic and highland regions of Russia. In: *Tree Rings, Environment and Humanity* (J.S. Dean, D.M. Meko, and T.W. Swetnam, eds.). Department of Geosciences, University of Arizona, Tucson, 283–294.

Shopov, Y.Y., Ford, D.C., and Schwarcz, H.P. (1994). Luminescent microbanding in speleothems: high resolution chronology and paleoclimate. *Geology,* **22,** 407–410.

Shotton, F.W. (1972). An example of hard water error in radiocarbon dating of vegetable matter. *Nature,* **240,** 460–461.

Siegenthaler, U. (1991). Glacial-Interglacial atmospheric CO_2 variations. In: *Global Changes of the Past* (R.S. Bradley, ed.) Boulder: University Corporation for Atmospheric Research, 245–260.

Sigafoos, R.S. and Hendricks, E.L. (1961). Botanical evidence of the modern history of Nisqually Glacier, Washington, US Geological Survey Professional Paper 387-A. Washington, DC: US Geological Survey.

Sikes, E.L. and Keigwin, L.D. (1994). Equatorial Atlantic sea surface temperature for the last 30 kyrs: a comparison of U^k_{37}, $\delta^{18}O$ and foraminiferal assemblage temperature estimates. *Paleoceanography,* **9,** 31–45.

Sikes, E.L. and Keigwin, L.D. (1996). A re-examination of northeast Atlantic sea surface temperature and salinity over the last 16 kyr. *Paleoceanography,* **11,** 327–342.

Sikes, E.L., Keigwin, L.D., and Farrington, J.W. (1991). Use of the alkenone unsaturation ration U^{k37} to determine past sea surface temperatures: core-top SST calibration and methodology considerations. *Earth and Planetary Science Letters,* **104,** 36–47.

Simpson, I.M. and West, R.G. (1958). On the stratigraphy and palaeobotany of a late-Pleistocene organic deposit at Chelford, Cheshire. *New Phytologist,* **57,** 239–250.

Sirocko, F. and Sarnthein, M. (1989). Wind-borne deposits in the northwestern Indian Ocean: record of Holocene sediments versus modern satellite data. In: *Paleoclimatology and Paleometeorology: Modern and Past Patterns of Global Atmospheric Transport* (M. Leinen and M. Sarnthein, eds.). Dordrecht: Kluwer Academic, 401–433.

Sirocko, F., Sarnthein, M., Lange, H., and Erlenkeuser, H. (1991). Atmospheric summer circulation and coastal upwelling in the Arabian Sea during the Holocene and the last glaciation. *Quaternary Research,* **36,** 72–93.

Slowey, N.C., Henderson, G.M., and Curry, W.B. (1996). Direct U-Th dating of marine sediments from the two most recent interglacial periods. *Nature*, **383**, 242–244.

Snyder, C.T. and Langbein, W.B. (1962). The Pleistocene lake in Spring Valley, Nevada, and its climatic implications. *Jour. Geophys. Research*, **67**, 2385–2394.

Solomon, A. and Webb III, T. (1985). Computer-aided reconstruction of Late Quaternary landscape dynamics. *Annual Review of Ecology and Systematics*, **16**, 63–84.

Sonnett, C.P. and Finney, S.A. (1990). The spectrum of radiocarbon. *Phil. Trans. Royal Society*, London, **A330**, 413–426.

Sorenson, C.J. (1977). Holocene bioclimates. *Annals of the Association of American Geographers*, **67**, 214–222.

Sorenson, C.J. and Knox, J.C. (1974). Paleosols and paleoclimate related to late Holocene forest-tundra border migrations: Mackenzie and Keewatin, NWT. In: *International Conference on Prehistory and Paleoecology of Western North American Arctic and Subarctic* (S. Raymond and P. Schledermann, eds.). Calgary: Archaeological Association, University of Calgary, 187–203.

Sosef, M.S.M. (1991). New species of Begonia in Africa and their relevance to the study of glacial rain forest refugia. *Wageningen Agricultural University Papers*, **91**, 120–151.

Sowers, T. and Bender, M. (1995). Climate records covering the last deglaciation. *Science*, **269**, 210–214.

Sowers, T., Bender, M., Labeyrie, L., Martinson, D., Jouzel, J., Raynaud, D., Pichon, J-J., and Korotkevich, Y.S. (1993). A 135,000 year Vostok-SPECMAP common temporal framework. *Paleoceanography* **8**, 737–766.

Sowers, T., Bender, M., Raynaud, D., Korotkevich, Y.S., and Orchado, J. (1991). The ^{18}O of atmospheric O_2 from air inclusions in the Vostok ice core: timing of CO_2 and ice volume changes during the penultimate deglaciation. *Paleoceanography* **6**, 679–696.

Sowers, T., Bender, M., Raynaud, D., and Korotkevich, Y.S. (1992). δ^{15}N of N_2 in air trapped in polar ice: a tracer of gas transport in the firn and a possible constraint on ice age-gas-age difference. *J. Geophysical Research* **97D**, 15683–15697.

Spaulding, W.G. (1990). Vegetational and climatic development of the Mojave Desert: the Last Glacial Maximum to the Present. In: *Packrat Middens: The Last 40,000 Years of Biotic Change* (J.L. Betancourt, T.R. Van Devender, and P.S. Martin, eds.). Tucson: University of Arizona Press, 166–199.

Spaulding, W.G. (1991). A middle Holocene vegetation record from the Mojave Desert of North America and its paleoclimatic significance. *Quaternary Research*, **35**, 427–437.

Spaulding, W.G., Betancourt, J.L., Croft, L.K., and Cole, K.L. (1990). Packrat middens: their composition and methods of analysis. In: *Packrat Middens: The Last 40,000 Years of Biotic Change* (J.L. Betancourt, T.R. Van Devender and P.S. Martin, eds.), Tucson: University of Arizona Press, 59–84.

Spell, T.L. and McDougall, I. (1992). Revision to the age of the Brunhes-Matuyama boundary and the Pleistocene geomagnetic polarity time scale. *Geophysical Research Letters*, **19**, 1181–1184.

Staffelbach, T., Stauffer, B., and Oeschger, H. (1988). A detailed analysis of the rapid changes in ice core parameters during the last ice age. *Annals of Glaciology*, **10**, 167–170.

Stager, J.C. and Mayewski, P.A. (1997). Abrupt early to mid-Holocene climatic transition registered at the Equator and the Poles. *Science*, **276**, 1834–1836.

Stahle, D.W. (1996). The hydroclimatic application of tree-ring chronologies. In: *Tree Rings, Environment and Humanity* (J.S. Dean, D.M. Meko, and T.W. Swetnam, eds.) Tucson: Department of Geosciences, University of Arizona, 119–126.

Stahle, D.W. and Cleaveland, M.K. (1988). Texas drought history reconstructed and analyzed from 1698 to 1980. *Jour. Climate*, **1**, 59–74.

Stahle, D.W. and Cleaveland, M.K. (1992). Reconstruction and analysis of spring rainfall over the southeastern United States for the past 1000 years. *Bulletin of the American Meteorological Society*, **73**, 1947–1961.

Stahle, D.W. and Cleaveland, M.K. (1993). Southern Oscillation extremes reconstructed from tree rings of the Sierra Madre Occidental and southern Great Plains. *Jour. Climate,* **6**, 129–140.

Stahle, D.W., Cleaveland, M.K., and Hehr, J.G. (1985). A 450-year drought reconstruction for Arkansas, United States. *Nature,* **316**, 530–532.

Stahle, D.W., Cleaveland, M.K., and Hehr, J.G. (1988). North Carolina climate changes reconstructed from tree-rings: A.D. 372-1985. *Science,* **240**, 1517–1519.

State Meteorological Administration (1981). *Annals of 510 years of precipitation record in China.* Beijing: Cartographic Publishers (in Chinese with English summary).

Stauffer, B.R. (1989). Dating of ice by radioactive isotopes. In: *The Environmental Record in Glaciers and Ice Sheets* (H. Oeschger and C.C. Langway, Jr., eds.). Chichester: Wiley, 123–129.

Stauffer, B.R. and Neftel, A. (1988). What have we learned from the ice cores about atmospheric changes in the concentrations of nitrous oxide, hydrogen peroxide and other trace species? In: *The Changing Atmosphere* (F.S. Rowland and I.S.A. Isaksen, eds.). Chichester: Wiley, 63–77.

Steffensen, J.P. (1985). Microparticles in snow from the South Greenland ice sheet. *Tellus,* **37B**, 286–295.

Steffensen, J.P. (1988). Analysis of the seasonal variations in dust, Cl^-, NO_3^- and SO_4^{-} in two central Greenland firn cores. *Annals of Glaciology,* **10**, 171–177.

Steig, E.J., Grootes, P.M. and Stuiver, M. (1994). Seasonal precipitation and ice core records. *Science,* **266**, 1885–1886.

Sternberg, R.S. (1995). Radiocarbon fluctuations and the geomagnetic field. In: *Radiocarbon After Four Decades* (R.E. Taylor, A. Long, and R.S. Kra, eds.). New York: Springer-Verlag, 93–116.

Stidd, C.K. (1967). The use of eigenvectors for climatic estimates. *Jour. Appl. Meteorology,* **6**, 255–264.

Stirling, C.H., Esat, T.M., McCulloch, M.T., and Lambeck, K. (1995). High-precision U-series dating of corals from western Australia and implications for the timing and duration of the Last Interglacial. *Earth and Planetary Science Letters,* **135**, 115–130.

Stockmarr, J. (1971). Tablets with spores used in pollen analysis. *Pollen et Spores,* **13**, 615–621.

Stockton, C.W. (1975). *Long term Streamflow Records Reconstructed from Tree Rings.* Paper 5, Laboratory for Tree Ring Research, Tucson: University of Arizona Press.

Stockton, C.W. and Boggess, W.R. (1980). Augmentation of hydrologic records using tree rings. In: *Improved Hydrologic Forecasting.* New York: American Society of Civil Engineers, 239–265.

Stockton, C.W. and Fritts, H.C. (1973). Long-term reconstruction of water level changes for Lake Athabasca by analysis of tree rings. *Water Resources Bulletin,* **9**, 1006–1027.

Stokes, M.A. and Smiley, T.L. (1968). *An Introduction to Tree-ring Dating.* Chicago: University of Chicago Press.

Stothers, R.B. (1984). The great Tambora eruption in 1815 and its aftermath. *Science,* **224**, 1191–1198.

Stott, L.D. and Tang, C.M. (1996). Reassessment of foraminiferal-based tropical sea surface $\delta^{18}O$ paleotemperatures. *Paleoceanography,* **11**, 37–56.

Street, F.A. and Grove, A.T. (1976). Environmental and climatic implications of late Quaternary lake-level fluctuations in Africa. *Nature,* **261**, 285–390.

Street, F.A. and Grove, A.T. (1979). Global maps of lake-level fluctuations since 30,000 yr BP. *Quaternary Research,* **12**, 83–118.

Street-Perrott, F.A. (1994). Palaeo-perspectives: changes in terrestrial ecosystems. *Ambio,* **23**, 37–43.

Street-Perrott, F.A. and Harrison, S.P. (1985a). Lake levels and climate reconstruction. In: *Paleoclimate Analysis and Modeling* (A.D. Hecht, ed.). Chichester: Wiley, 291–340.

Street-Perrott, F.A. and Harrison, S.P. (1985b). Temporal variations in lake levels since 30,000 yr B.P.—an index of the global hydrological cycle. In: *Climate Processes and*

Sensitivity (J.E. Hansen and T. Takahashi, eds.). Washington DC: American Geophysical Union, 118–129.

Street-Perrott, F.A., Huang, Y., Perrott, R.A., and Eglinton, G. (1998). Carbon isotopes in lake sediments and peats of last glacial age: implications for the global carbon cycle. In: *Stable Isotopes* (H. Griffith, ed.). Oxford: BIOS Scientific Publishers, 381–396.

Street-Perrott, F.A., Huang, Y., Perrott, R.A., Eglinton, G., Barker, P., Khelifa, L.B., Harkness, D.D., and Olago, D.O. (1997). Impact of lower atmospheric carbon dioxide on tropical mountain ecosystems. *Science,* **278**, 1422–1426.

Street-Perrott, F.A., Marchand, D.S., Roberts, N., and Harrison, S.P. (1983). *Global Lake-level Variations from 18,000 to 0 years ago: a Paleoclimatic Analysis.* Technical Report 046, Department of Energy, Washington, DC.

Street-Perrott, F.A., Mitchell, J.F.B., Marchand, D.S., and Brunner, J.S. (1990). Milan-kovitch and albedo forcing of the tropical monsoons: a comparison of geological evidence and numerical simulations for 9000 yr B.P. *Trans. Royal Society of Edinburgh,* **81**, 407–427.

Street-Perrott, F.A., and Roberts, N., (1983). Fluctuations in closed-basin lakes as an indicator of past atmospheric circulation patterns. In: *Variations in the Global Water Budget,* (Street-Perrott, F.A., Beran, M. and Ratcliffe, R., eds.). Dordrecht: D. Reidel, 331-345.

Street-Perrott, F.A. and Perrott, R.A. (1990). Abrupt climate fluctuations in the tropics: the influence of Atlantic Ocean circulation. *Nature,* **343**, 607–612.

Street-Perrott, F.A. and Perrott, R.A. (1993). Holocene vegetation, lake levels and climate of Africa. In: *Global Climates Since the Last Glacial Maximum* (H.E. Wright, J.E. Kutzbach, T. Webb III, W.F. Ruddiman, F.A. Street-Perrott, and P.J. Bartlein, eds.). Minneapolis: University of Minnesota Press, 318–356.

Studhalter, R.A. (1955). Tree growth: I. Some historical chapters. *Botanical Review,* **21**, 1–72.

Stuiver, M. (1978a). Carbon-14 dating: a comparison of beta and ion counting. *Science,* **202**, 881–883.

Stuiver, M. (1978b). Radiocarbon timescale tested against magnetic and other dating methods. *Nature,* **273**, 271–274.

Stuiver, M. (1993). A note on single-year calibration of the radiocarbon time scale, A.D. 151–954. *Radiocarbon,* **35**, 67–72.

Stuiver, M. (1994). Atmospheric ^{14}C as a proxy of solar and climatic change. In: *The Solar Engine and its Influence on Terrestrial Atmosphere and Climate* (E. Nesmé-Ribes, ed.). Berlin: Springer-Verlag, 203–220.

Stuiver, M. and Braziunas, T.F. (1987). Tree cellulose $^{13}C/^{12}C$ isotope ratios and climatic change. *Nature,* **328**, 58–60.

Stuiver, M. and Braziunas, T.F. (1991). Isotopic and solar records. In: *Global Changes of the Past* (R.S. Bradley, ed.). Boulder: University Corporation for Atmospheric Research, 225–244.

Stuiver, M. and Braziunas, T.F. (1992). Evidence of solar activity variations. In: *Climate Since A.D. 1500* (R.S. Bradley and P.D. Jones, eds.). London: Routledge, 593–605.

Stuiver, M. and Braziunas, T.F. (1993). Modeling atmospheric ^{14}C influence and ^{14}C ages of marine samples to 10,000 B.C. *Radiocarbon,* **35**, 137–189.

Stuiver, M., Braziunas, T.F., Becker, B., and Kromer, B. (1991). Climatic, solar, oceanic and geomagnetic influences on Late-Glacial and Holocene atmospheric $^{14}C/^{12}C$ change. *Quaternary Research,* **35**, 1–24.

Stuiver, M., Grootes, P.M., and Braziunas, T.F. (1995). The GISP2 $\delta^{18}O$ record of the past 16,500 years and the role of the sun, ocean and volcanoes. *Quaternary Research,* **44**, 341–354.

Stuiver, M., Heusser, C.J., and Yang, I.C. (1978). North American glacial history extended to 75,000 years ago. *Science,* **200**, 16–21.

Stuiver, M. and Pearson, G.W. (1993). High-precision bidecadal calibration of the radiocarbon time scale, AD 1950–500 B.C. and 2500–6000 B.C. *Radiocarbon,* **35**, 1–23.

Stuiver, M., Pearson, G.W., and Braziunas, T. (1986). Radiocarbon age calibration of marine samples back to 9000 cal. yr B.P. *Radiocarbon,* **28**, 980–1021.

Stuiver, M. and Quay, P.D. (1980). Changes in atmospheric carbon-14 attributed to a variable sun. *Science,* **207,** 11–19.

Stuiver, M. and Reimer, P.J. (1993). Extended ^{14}C data base and revised Calib 3.0 ^{14}C age calibration program. *Radiocarbon,* **35,** 215–230.

Stute, M., Forster, M., Frischkorn, H., Serejo, A., Clark, J.F., Schlosser, P., Broecker, W.S., and Bonani, G. (1995). Cooling of tropical Brazil (5 °C) during the last glacial maximum. *Science,* **269,** 379–383.

Suess, H.E. (1965). Secular variations of the cosmic-ray produced carbon-14 in the atmosphere and their interpretations. *Jour. Geophys. Research,* **70,** 5937–5952.

Suess, H.E. (1980). The radiocarbon record in tree rings of the last 8000 years. *Radiocarbon,* **22,** 200–209.

Sunda, W.G., Swift, D.G., and Huntsman, S.A. (1991). Low iron requirement for growth in oceanic phytoplankton. *Nature,* **351,** 55–57.

Sundquist, E.T. (1985). Geological perspectives on carbon dioxide and the carbon cycle. In: *The Carbon Cycle and Atmospheric CO_2: Natural Variations Archean to Present* (E.T. Sundquist and W.S. Broecker, eds.) Washington DC: American Geophysical Union, 5–59.

Swain, A.M. (1978). Environmental changes during the last 2000 years in north-central Wisconsin: analysis of pollen, charcoal and seeds from varved lake sediments. *Quaternary Research,* **10,** 55–68.

Swart, P.K., Lohmann, K.C., McKenzie, J., and Savin, S. (eds.). (1993). *Climate Change in Continental Isotopic Records.* Washington DC: American Geophysical Union.

Swetnam, T.W. (1993). Fire history and climate change in giant Sequoia groves. *Science,* **262,** 885–889.

Swetnam, T.W. and Betancourt, J.L. (1990). Fire-Southern Oscillation relations in the southwestern United States. *Science* **249,** 1017–1020.

Switsur, R., Waterhouse, J.S., Field, E.M., Carter, T., and Loader, N. (1995). Stable isotope studies in tree rings from oak–techniques and some preliminary results. In: *Problems of Stable Isotopes in Tree-rings, Lake Sediments and Peat-bogs as Climatic Evidence for the Holocene* (B. Frenzel, B. Stauffer, and M.M. Weiss, eds.). Stuttgart: Gustav Fischer Verlag, 129–140.

Syktus, J., Gordon, H., and Chappell, J. (1994). Sensitivity of a coupled atmosphere-dynamic upper ocean GCM to variations of CO_2, solar constant, and orbital forcing. *Geophysical Research Letters,* **21,** 1599–1602.

Symonds, R.B., Rose, W.I., and Reed, M.H. (1988). Contribution of Cl$^-$ and F$^-$ bearing gases to the atmosphere by volcanoes. *Nature,* **334,** 415–418.

Szabo, B.J. (1979a). Uranium-series age of coral reef growth on Rottnest Island, western Australia. *Marine Geology,* **29,** M11–M15.

Szabo, B.J. (1979b). ^{230}Th, ^{231}Pa and open system dating of fossil corals and shells. *Jour. Geophys. Research,* **84,** 4927–4930.

Szabo, B.J. and Collins, D. (1975). Age of fossil bones from British interglacial sites. *Nature,* **254,** 680–682.

Szabo, B.J., Miller, G.H., Andrews, J.T., and Stuiver, M. (1981). Comparison of uranium series, radiocarbon and amino acid data from marine molluscs, Baffin Island, Arctic Canada. *Geology,* **9,** 451–457.

Szabo, B.J. and Rosholt, J.N. (1969). Uranium-series dating of Pleistocene molluscan shells from southern California—an open system model. *Jour. Geophys. Research,* **74,** 3253–3260.

Szafer, W. (1935). The significance of isopollen lines for the investigation of geographical distribution of trees in the post-glacial period. *Bulletin International de l'Academie Polonaise des Sciences et des Lettres,* **Bl,** 235–239.

Tagami, Y. and Fukaishi, K. (1992). Winter and summer climatic variation in Japan during the Little Ice Age. In: *Proceedings of the International Symposium on the Little Ice Age Climate* (T. Mikami, ed.). Department of Geography, Tokyo Metropolitan University, 188–193.

Talbot, M.R. and Delibrias, G. (1977). Holocene variations in the level of Lake Bosumptwi, Ghana. *Nature,* **268,** 722–724.

Talbot, M.R. and Johannessen, T. (1992). A high resolution palaeoclimate record for the last 27,500 years in tropical West Africa from the carbon and nitrogen isotopic composition of lacustrine organic matter. *Earth and Planetary Science Letters,* **100,** 23–37.

Tarling, D.H. (1975). Archeomagnetism: the dating of archaeological materials by their magnetic properties. *World Archaeology,* **7,** 185–197.

Tarling, D.H. (1978). The geological-geophysical framework of ice ages. In: *Climatic Change* (J. Gribbin, ed.). Cambridge: Cambridge University Press, 3–24.

Tarr, R.S. (1897). Difference in the climate of the Greenland and American side of Davis' and Baffin's Bay. *American Jour. Science,* **3,** 315–320.

Tarusov, P.E., Pushenko, M.Ya., Harrison, S.P., Saarse, L., Andreev, A.A., Aleshinskaya, Z.V., Davydova, N.N., Dorofeyuk, N.I., Efremov, Yu.Y., Elina, G.A., Elovicheva, Ya.K., Filimonova, L.V., Gunova, V.S., Khomutova, V.I., Kvavadze, E.V., Nuestrueva, I.Yu., Pisareva, V.V., Sevastyanov, D.V., Shelekhova, T.S., Subetto, D.A., Uspenskaya, O.N., and Zernitskaya, V.P. (1996). *Lake Status Records from the Former Soviet Union and Mongolia: Documentation of the Second Version of the Database.* Paleoclimatology Publication Series Report No. 3, World Data Center-A for Paleoclimatology, Boulder.

Tauber, H. (1965). Differential pollen dispersion and the interpretation of pollen diagrams. *Danmarks Geologiske Undersogelse,* Series II 89.

Tauxe, L., Deino, A.D., Behrensmeyer, A.K., and Potts, R. (1992). Pinning down the Brunhes-Matuyama and upper Jaramillo boundaries: a reconciliation of orbital and isotopic time scales. *Earth and Planetary Science Letters,* **109,** 561–572.

Tauxe, L., Herbert, T., Shackleton, N.J., and Kok, Y.S. (1996). Astronomical calibration of the Matuyama-Brunhes boundary: consequences for magnetic remanence acquisition in marine carbonates and the Asian loess sequences. *Earth and Planetary Science Letters,* **140,** 133–146.

Taylor, K.C., Hammer, C.U., Alley, R.B., Clausen, H.B., Dahl-Jensen, D., Gow, A.J., Gundestrup, N.S., Kipfstuhl, J., Moore, J.C., and Waddington, E.D. (1993a). Electrical conductivity measurements from the GISP2 and GRIP Greenland ice cores. *Nature,* **366,** 549–552.

Taylor, K.C., Lamorey, G.W, Doyle, G.A., Alley, R.B., Grootes, P.M., Mayewski, P.A., White, J.W.C., and Barlow, L.K. (1993b). The 'flickering switch' of late Pleistocene climate change. *Nature,* **361,** 432–436.

TEMPO (1996). Potential role of vegetation feedback in the climate sensitivity of high latitude regions: a case study at 6000 years B.P. *Global Biogeochemical Cycles,* **10,** 727–736.

Ten Brink, N.W. (1973). Lichen growth rates in west Greenland. *Arctic and Alpine Research,* **5,** 323–331.

Tetzlaff, G., and Adams, L.J. (1983). Present-day and early Holocene evaporation of Lake Chad. In: *Variations in the Global Water Budget* (F.A. Street-Perrott, M. Beran, and R. Ratcliffe, ed.). Dordrecht: D. Reidel, 347–360.

Thierstein, H.R., Geitzenauer, K.R., Molfino, B., and Shackleton, N.J. (1977). Global synchroneity of late Quaternary coccolith datum levels: validation by oxygen isotopes. *Geology,* **5,** 400–404.

Thistlewood, L. and Sun, J. (1991). A paleomagnetic and mineral magnetic study of the loess sequence at Liujiapo, Xian, China. *Jour. Quaternary Science,* **6,** 13–26.

Thompson, L.G. (1991). Ice core records with emphasis on the global record of the last 2000 years. In: *Global Changes of the Past* (R.S. Bradley, ed.). Boulder: University Corporation for Atmospheric Research, 201–224.

Thompson, L.G. (1992). Ice core evidence from Peru and China. In: *Climate Since A.D. 1500* (R.S. Bradley and P.D. Jones, eds.) London: Routledge, 517–548.

Thompson, L.G. and Mosley-Thompson, E. (1987). Evidence of abrupt climatic change during the last 1500 years recorded in ice cores from the tropical Quelccaya Ice Cap, Peru. In: *Abrupt Climatic Change* (W.H. Berger and L.D. Labeyrie, eds.). Dordrecht: Reidel, 99–110.

Thompson, L.G., Davis, M., Mosley-Thompson, E., and Liu, K. (1988a). Pre-Incan agricultural activity recorded in dust layers in two tropical ice cores. *Nature, 336,* 763–765.

Thompson, L.G., Mosley-Thompson, E., and Arnao, B.M. (1984a). Major El Niño/Southern Oscillation events recorded in stratigraphy of the tropical Quelccaya Ice Cap. *Science,* **226,** 50–52.

Thompson, L.G., Mosley-Thompson, E., Bolzan, J.F., and Koci, B.R. (1985). A 1500 year record of tropical precipitation in ice cores from the Quelccaya Ice Cap, Peru. *Science,* **229,** 971–973.

Thompson, L.G., Mosley-Thompson, E., Dansgaard, W., and Grootes, P.M. (1986). The Little Ice Age as recorded in the stratigraphy of the tropical Quelccaya Ice Cap. *Science,* **234,** 361–364.

Thompson, L.G., Mosley-Thompson, E., Davis, M.E., Bolzan, J.F., Dai, J., Gundestrup, N., Yao, T., Wu, X., Klein, L., and Xie, Z. (1990). Glacial stage ice core records from the sub-tropical Dunde Ice Cap, China. *Annals of Glaciology,* **14,** 288–298.

Thompson, L.G., Mosley-Thompson, E., Davis, M.E., Bolzan, J.F., Dai, J., Yao, T., Gundestrup, N., Wu, X., Klein, L., and Xie, Z. (1989). Holocene-late Pleistocene climatic ice core records from Qinghai-Tibetan plateau. *Science,* **246,** 474–477.

Thompson, L.G., Mosley-Thompson, E., Davis, M.E., Lin, P-N., Dai, J. and Bolzan, J.F. (1995a). A 1000 year climate ice-core record from the Guliya ice cap, China: its relationship to global climate variability. *Annals of Glaciology,* **21,** 175–181.

Thompson, L.G., Mosley-Thompson, E., Davis, M.E., Lin, P-N., Henderson, K.A., Cole-Dai, J., Bolzan, J.F., and Liu, K-B. (1995b). Late Glacial Stage and Holocene tropical ice core records from Huascarán, Peru. *Science,* **269,** 46–50.

Thompson, L.G., Mosley-Thompson, E., Davis, M.E., Lin, N., Yao, T., Dyurgerov, M., and Dai, J. (1993). "Recent warming": ice core evidence from tropical ice cores, with emphasis on central Asia. *Global and Planetary Change,* **7,** 145–156.

Thompson, L.G., Mosley-Thompson, E., Grootes, P.M., Pourchet, M., and Hastenrath, S. (1984a). Tropical glaciers: potential for ice core paleoclimatic reconstructions. *Jour. Geophys. Research,* **89D,** 4638–4646.

Thompson, L.G., Yao, T., Davis, M.E., Henderson, K.A., Mosley-Thompson, E., Lin, P-N., Beer, J., Synal, H-A.,Cole-Dai, J., and Bolzan, J.F. (1997). Tropical climate instability: the last glacial cycle from a Qinghai-Tibetan ice core. *Science,* **276,** 1821–1825.

Thompson, L.G., Mosley-Thompson, E., Wu, X., and Xie, Z. (1988b). Wisconsin/Würm glacial stage ice in the sub-tropical Dunde Ice Cap, China. *Geojournal,* **17,** 4, 517–523.

Thompson, P.R. and Saito, T. (1974). Pacific Pleistocene sediments: planktonic foraminifera dissolution cycles and geochronology. *Geology,* **2,** 333–335.

Thompson, P., Schwarcz, H.P., and Ford, D.C. (1976). Stable isotope geochemistry, geothermometry and geochronology of speleothems from West Virginia. *Geological Society of America Bulletin,* **87,** 1730–1738.

Thompson, R. (1977). Stratigraphic consequences of palaeomagnetic studies of Pleistocene and Recent sediments. *Jour. Geological Society of London,* **133,** 51–59.

Thompson, R.S. (1990). Late Quaternary vegetation and climate in the Great Basin. In: *Packrat Middens: The Last 40,000 Years of Biotic Change* (J.L. Betancourt, T.R. Van Devender and P.S. Martin, eds.). Tucson: University of Arizona Press, 200–239.

Thompson, R. and Oldfield, F. (1986). *Environmental Magnetism,* London: Allen and Unwin.

Thorarinsson, S. (1981). The application of tephrochronology in Iceland. In: *Tephra Studies,* (S. Self and R.J.S. Sparks, eds.). Dordrecht: D. Reidel, 109–134.

Thunell, R., Anderson, D., Gellar, D., and Miao, Q. (1994). Sea-surface temperature estimates for the tropical western Pacific during the last Glaciation and their implications for the Pacific Warm Pool. *Quaternary Research,* **41,** 255–264.

Tikhomirov, B.A. (1961). The changes in biogeographical boundaries in the north of USSR as related with climatic fluctuations and activity of man. *Botanisk Tidsskrift,* **56,** 285–292.

Tinner, W., Ammann, B., and Germann, P. (1996). Treeline fluctuations recorded for 12,500 years by soil profiles, pollen and plant macrofossils in the central Swiss Alps. *Arctic and Alpine Research,* **28,** 131–147.

Tranquillini, A. (1993). Climate and physiology of trees in the Alpine timberline regions. In: *Oscillations of the Alpine and Polar Tree Limits in the Holocene* (B. Frenzel, M. Eronen, K.D. Vorren, and B. Gläser, ed.). Stuttgart: Gustav Fischer Verlag, 127–135.

Traverse, A. (ed.). (1994). *Sedimentation of Organic Particles*. Cambridge: Cambridge University Press.

Trenberth, K.E. (ed.) (1992). *Climate System Modeling*. Cambridge: Cambridge University Press.

Trenberth, K.E. (1997). The use and abuse of climate models. *Nature*, **386**, 131–133.

Tric, E., Valet, J-P., Tucholka, P., Paterne, M., and Labeyrie, L.D. (1992). Paleointensity of the geomagnetic field during the last 80,000 years. *Jour. Geophys. Research*, **97B**, 9337–9351.

Tricot, C. and Berger, A. (1988). Sensitivity of present-day climate to astronomical forcing. In: *Long and Short-Term Variability of Climate* (H. Wanner and U. Siegenthaler, ed.). New York: Springer-Verlag, 132–152.

Trimborn, P., Becker, B., Kromer, B., and Lipp, J. (1995). Stable isotopes in tree-rings: a palaeoclimatic tool for studying climatic changes. In: *Problems of Stable Isotopes in Tree-rings, Lake Sediments and Peat-bogs as Climatic Evidence for the Holocene* (B. Frenzel, B. Stauffer, and M.M. Weiss, eds.). Stuttgart: Gustav Fischer Verlag, 163–170.

Troll, C. (1973). The upper timberlines in different climatic zones. *Arctic and Alpine Research*, **5**, A3–A18.

Troughton, J.H. (1972). Carbon isotope fractionation by plants. In: *Proc. 8th International Conference on Radiocarbon Dating*, **2**, (T.A. Rafter and T. Grant-Taylor, eds.). Royal Society of New Zealand, Wellington, 421–438.

Twining, A.C. (1833). On the growth of timber. *American Jour. Science, Arts*, **24**, 391–393.

Tzedakis, P.C. (1994). Hierarchical biostratigraphical classification of long pollen sequences. *Jour. Quaternary Science*, **9**, 257–260.

Urey, H.C. (1947). The thermodynamic properties of isotopic substances. *Jour. Chemical Society*, **152**, 190–219.

Urey, H.C. (1948). Oxygen isotopes in nature and in the laboratory. *Science*, **108**, 489–496.

Vaganov, E.A., Shiyatov, S.G., and Mazepa, V.S. (1996). *Dendroclimatic Study in Ural-Siberian Subarctic*. Novosibirsk: Siberian Branch Russian Academy of Science.

Van Campo, E., Guiot, J., and Peng, C. (1993). A data-based re-appraisal of the terrestrial carbon budget at the last glacial maximum. *Global and Planetary Change*, **8**, 189–201.

van den Dool, H.M., Krijnen, H.J., and Schuurmans, C.J.E. (1978). Average winter temperatures at De Bilt (The Netherlands), 1634–1977. *Climatic Change*, **1**, 319–330.

Van der Hammen, T. (1974). The Pleistocene changes of vegetation and climate in tropical South America. *Jour. Biogeography*, **1**, 3–26.

Van der Hammen, T. and Absy, M.L. (1994). Amazonia during the last glacial. *Palaeogeography, Palaeoclimatology, Palaeoecology*, **109**, 247–261.

Van Devender, T.R. (1990a). Late Quaternary vegetation and climate of the Chihuahuan Desert, United States and Mexico. In: *Packrat Middens: The Last 40,000 Years of Biotic Change* (J.L. Betancourt, T.R. Van Devender, and P.S. Martin, ed.). Tucson: University of Arizona Press, 104–133.

Van Devender, T.R. (1990b). Late Quaternary vegetation and climate of the Sonoran Desert, United States and Mexico. In: *Packrat Middens: The Last 40,000 Years of Biotic Change* (J.L. Betancourt, T.R. Van Devender, and P.S. Martin, eds.). Tucson: University of Arizona Press, 134–165.

Van Devender, T.R., Burgess, T.L., Piper, J.C., and Turner, R.M. (1994). Paleoclimatic implications of Holocene plant remains from the Sierra Bacha, Sonora, Mexico. *Quaternary Research*, **41**, 99–108.

Van Devender, T.R. and Spaulding, W.G. (1979). Development of vegetation and climate in the southwestern United States. *Science*, **204**, 701–710.

Van Loon, H., Taljaard, J.J., Sasamori, T., London, J., Hoyt, D.V., Labitze, K., and Newton, C.W. (1972). Meteorology of the Southern Hemisphere. *Meteorological Monographs*, **13** (35).

Vanzolini, P.E. (1973). Paleoclimates, relief, and species multiplication in Equatorial forests. In: *Tropical Forest Ecosystems in Africa and South America: a Comparative Review* (B.J. Meggers, E.S. Ayensu, and W.D. Duckworth, eds.). Washington DC: Smithsonian Institution, 255–258.

Vanzolini, P.E. and Williams, E.E. (1970). South American anoles: the geographic differentiation and evolution of the *Anolis chrysolepis* species group (Sauria, Iguanidae). *Arquivos de Zoologia*, **19**, 1–298.

Vaughan, T.A. (1990). Ecology of living *packrats*. In: *Packrat Middens: The Last 40,000 Years of Biotic Change* (J.L. Betancourt, T.R. Van Devender, and P.S. Martin, eds.). Tucson: University of Arizona Press, 14–27.

Vavrus, S.J. (1995). Sensitivity of the Arctic climate to leads in a coupled atmosphere-mixed layer ocean model. *Jour. Climate*, **8**, 158–171.

Veeh, H.H. and Chappell, J.M.A. (1970). Astronomical theory of climatic change: support from New Guinea. *Science*, **167**, 862–865.

Velichko, A.A., Isayeva, L.L., Makeyev, V.M., Matishov, G.G., and Faustova, M.A. (1984). Late Pleistocene glaciation of the Arctic shelf and the reconstruction of Eurasian ice sheets. In: *Late Quaternary Environments of the Soviet Union* (A.A. Velichko, H.E. Wright, and C.W. Barnosky, eds.). Minneapolis: University of Minnesota Press, 35–41.

Veron, J.E.N. (1993). *Corals of Australia and the Indo-Pacific*. Honolulu: University of Hawaii Press.

Verosub, K.L. (1975). Paleomagnetic excursions as magnetostratigraphic horizons: a cautionary note. *Science*, **190**, 48–50.

Verosub, K.L. (1977). Depositional and postdepositional processes in the magnetization of sediments. *Reviews of Geophysics and Space Physics*, **15**, 129–143.

Verosub, K.L. (1988). Geomagnetic secular variations and the dating of Quaternary sediments. In: *Dating Quaternary Sediments* (D. Easterbrook, ed.). Special Paper 227, Boulder: Geological Society of America, 123–138.

Verosub, K.L. and Banerjee, S.K. (1977). Geomagnetic excursions and their paleomagnetic record. *Reviews of Geophysics and Space Physics*, **15**, 145–155.

Verosub, K.L., Fine, P., Singer, M.J., and TenPas, J. (1993). Pedogenesis and paleoclimate: interpretation of the magnetic susceptibility record of Chinese loess–paleosol sequences. *Geology*, **21**, 1011–1014.

Veum, T., Jansen, E., Arnold, M., Beyer, J.I., and Duplessy, J-C. (1992). Water mass exchange between the North Atlantic and the Norwegian Sea during the past 28,000 years. *Nature*, **356**, 783–785.

Vibe, C. (1967). Arctic animals in relation to climatic fluctuations. *Meddelelser ømGronland*, **170** (5).

Villanueva, J., Grimalt, J.O., Cortijo, E., Vidal, L., and Labeyrie, L. (1998). Assessment of sea surface temperature variations in the central North Atlantic using the alkenone unsaturation index ($U^{k'}_{37}$). *Geochimica et Cosmochimica Acta*, **62**, 2421–2427.

Vinot-Bertouille, A.C. and Duplessy, J. (1973). Individual isotopic fractionation of carbon and oxygen in benthic foraminifera. *Earth and Planetary Science Letters*, **18**, 247–252.

Volk, T. and Hoffert, M.I. (1985). Ocean carbon pumps: analysis of relative strengths and efficiencies in ocean-driven atmospheric CO_2 strengths. In: *The Carbon Cycle and Atmospheric CO_2: Natural Variations Archean to Present* (E.T. Sundquist and W.S. Broecker, eds.). Washington, DC: American Geophysical Union, 99–110.

Wahl, E.W. and Bryson, R.A. (1975). Recent changes in Atlantic surface temperatures. *Nature*, **254**, 45–46.

Walker, I.R. (1987). Chironomidae (Diptera) in paleoecology. *Quaternary Science Reviews*, **6**, 29–40.

Walker, I.R., Smol, J.P., Engstrom, D.R., and Birks, H.J.B. (1991a). An assessment of Chironomidae as quantitative indicators of past climate change. *Canadian Jour. Fisheries and Aquatic Science*, **48**, 975–987.

Walker, I.R., Mott, R.J., and Smol, J.P. (1991b). Allerød-Younger Dryas lake temperatures from midge fossils in Atlantic Canada. *Science*, **253**, 1010–1012.

Walker, I.R., Wilson, S.E., and Smol, J.P. (1995). Chironomidae (Diptera): quantitative palaeosalinity indicators for lakes of western Canada. *Canadian Jour. Fisheries and Aquatic Science*, **52**, 950–960.

Wang, P.K. (1980). On the possible relationship between winter thunder and climatic changes in China over the past 2200 years. *Climatic Change*, **3**, 37–46.

Wang, P.K. and Zhang, D. (1988). An introduction to some historical government weather records of China. *Bulletin of the American Meteorological Society*, **69**, 753–758.

Wang, P.K. and Zhang, D. (1992). Reconstruction of 18th century summer precipitation of Nanjing, Suzhou and Hangzhou, China based on the Clear and Rain records. In: *Climate Since A.D. 1500* (R.S. Bradley and P.D. Jones, eds.). London: Routledge, 184–209.

Wang, R. and Wang, S. (1989). Reconstruction of winter temperature in China for the last 500 year period. *Acta Meteorologica Sinica*, **3**, No. 3, 279–289.

Wang, R., Wang, S., and Fraedrich, K. (1991). An approach to reconstruction of temperature on a seasonal basis using historical documents from China. *International Jour. Climatology*, **11**, 381–392.

Wang, S. (1991a). Reconstruction of temperature series of North China from 1380s to 1980s. *Science in China, Series B*, **34**, No. 6, 751–759.

Wang, S. (1991b). Reconstruction of paleo-temperature series in China from the 1380s to the 1980s. *Würzburger Geographische Arbeiten*, **80**, 1–19.

Wang, S. and Wang, R. (1990). Seasonal and annual temperature variations since 1470 A.D. in East China. *Acta Meteorologica Sinica*, **4**, 428–439.

Wang, S. and Wang, R. (1991). Little Ice Age in China. *Chinese Science Bulletin*, **36**, No. 3, 217–220.

Wang, S. and Zhao, Z. (1979). The 36 year wetness oscillation in China and its mechanism. *Acta Meteorologica Sinica*, **37**, 64–73.

Wang, S. and Zhao, Z. (1981). Droughts and floods in China, 1470–1979. In: *Climate and History* (T.M.L. Wigley, M.J. Ingram, and G. Farmer, eds.). Cambridge: Cambridge University Press, 271–288.

Wang, S., Zhao, Z., and Chen, Z. (1981). Reconstruction of the summer rainfall regime for the last 500 years in China. *Geojournal*, **5**, 117–122.

Wang, W.C., Portman, D., Gong, G., Zhang, P., and Karl, T. (1992). Beijing summer temperatures since 1724. In: *Climate Since A.D. 1500* (R.S. Bradley and P.D. Jones, eds.). London: Routledge, 210–223.

Wanner, H., Brazdil, R., Frich, P., Freyendahl, K., Jonsson, T., Kington, J., Pfister, C., Rosenorn, S., and Wishman, E. (1994). Synoptic interpretation of monthly weather maps for the late Maunder Minimum (1675–1704). In: *Climatic Trends and Anomalies in Europe 1675–1715* (B. Frenzel, C. Pfister, and B. Glaser, eds.). Stuttgart: Gustav Fischer Verlag, 401–424.

Warburton, J.A. and Young, L.G. (1981). Estimating ratios of snow accumulation in Antarctica by chemical methods. *Jour. Glaciology*, **27**, 347–358.

Wardle, P. (1974). Alpine timberlines. In: *Arctic and Alpine Environments* (J.D. Ives and R.G. Barry, eds.). London: Methuen, 371–402.

Washburn, A.L. (1979a). Permafrost features as evidence of climatic change. *Earth Science Reviews*, **15**, 327–402.

Washburn, A.L. (1979b). *Geocryology*. London: Arnold.

Watts, W.A. and Hansen, B.C.S. (1994). Pre-Holocene and Holocene pollen records of vegetation history from the Florida peninsula and their climatic implications. *Palaeogeography, Palaeoclimatology, Palaeoecology*, **109**, 163–176.

Weaver, A.J. and Hughes, T.M.C. (1994). Rapid interglacial climate fluctuations driven by North Atlantic ocean circulation. *Nature*, **367**, 447–450.

Webb, R.H. and Betancourt, J.L. (1990). The spatial and temporal distribution of radiocarbon ages from packrat middens. In: *Packrat Middens: The Last 40,000 Years of Biotic Change* (J.L. Betancourt, T.R. Van Devender, and P.S. Martin, eds.). Tucson: University of Arizona Press, 85–102.

Webb, R.S., Anderson, K.H., and Webb III, T. (1993). Pollen response-surface estimates of late Quaternary changes in the moisture balance of the northeastern United States. *Quaternary Research,* **40,** 213–227.

Webb III, T. (1974). Corresponding patterns of pollen and vegetation in lower Michigan: a comparison of quantitative data. *Ecology,* **55,** 17–28.

Webb III, T. (1986). Is vegetation in equilibrium with climate? How to interpret late-Quaternary pollen data. *Vegetatio,* **67,** 75–91.

Webb III, T. (1987). The appearance and disappearance of major vegetational assemblages: long-term vegetational dynamics in eastern North America. *Vegetatio,* **69,** 177–187.

Webb III, T. (1988). Eastern North America. In: *Vegetation History* (B. Huntley and T. Webb III, eds.). Dordrecht: Kluwer Academic, 385–414.

Webb III, T. (1991). The spectrum of temporal climate variability: current estimates and the need for global and regional time series. In: *Global Changes of the Past* (R.S. Bradley, ed.). Boulder: University Corporation for Atmospheric Research, 61–82.

Webb III, T., Anderson, K.H., Bartlein P.J., and Webb, R.S. (1997). Late Quaternary climate change in eastern North America: a comparison of pollen-derived estimates with climate model results. *Quaternary Science Reviews,* **17,** 587–606.

Webb III, T., Bartlein, P.J., Harrison, S.P., and Anderson, K.H. (1993b). Vegetational, lake levels and climate in eastern North America for the past 18,000 years. In: *Global Climates Since the Last Glacial Maximum* (H.E. Wright, J.E. Kutzbach, T. Webb III, W.F. Ruddiman, F.A. Street-Perrott, and P.J. Bartlein, eds.). Minneapolis: University of Minnesota Press, 415–467.

Webb III, T., Bartlein, P.J., and Kutzbach, J.E. (1987). Climatic change in eastern North America during the past 18,000 years: comparison of pollen data with model results. In: *North America and Adjacent Oceans during the Last Deglaciation. The Geology of North America,* **K-3.** (W.F. Ruddiman and H.E Wright, eds.). Boulder: Geological Society of America, 447–462.

Webb III, T., Howe, S.E., Bradshaw, R.H.W., and Heide, K.M. (1981). Estimating plant abundances from pollen percentages: the use of regression analysis. *Review of Palaeobotany and Palynology,* **34,** 269–300.

Webb III, T. and Kutzbach, J.E. (1998). An introduction to "Late Quaternary climates: data synthesis and model experiments." *Quaternary Science Reviews,* **17,** 465–472.

Webb III, T. and McAndrews, J.H. (1976). Corresponding patterns of contemporary pollen and vegetation in central North America. In: *Investigation of Late Quaternary Paleoceanography and Paleoclimatology* (R.M. Cline and J.D. Hays, eds.). Memoir 145, Boulder: Geological Society of America, 267–299.

Webb III, T., Ruddiman, W.F., Street-Perrott, F.A., Markgraf, V., Kutzbach, J.E., Bartlein, P.J., Wright, Jr., H.E., and Prell, W.L. (1993a). Climatic changes during the past 18,000 years: regional syntheses, mechanisms and causes. In: *Global Climates Since the Last Glacial Maximum* (H.E. Wright, J.E. Kutzbach, T. Webb III, W.F. Ruddiman, F.A. Street-Perrott and P.J. Bartlein, eds.). Minneapolis: University of Minnesota Press, 514–535.

Webb III, T., Yeracaris, G.Y., and Richard, P. (1978). Mapped patterns in sediment samples of modern pollen from southeastern Canada and northeastern United States. *Geographie Physique et Quaternaire,* **32,** 163–176.

Weber, J.N. and Woodhead, P.M.J. (1972). Temperature dependence of oxygen-18 concentration in reef coral carbonates. *Jour. Geophys. Research,* **77,** 463–473.

Webster, P.J. and Streten, N.A. (1978). Late Quaternary ice age climates of tropical Australasia: interpretations and reconstructions. *Quaternary Research,* **10,** 279–309.

Wehmiller, J.F. (1993). Applications of organic geochemistry for Quaternary research: aminostratigraphy and aminochronology. In: *Organic Geochemistry* (M.H. Engel and S.A. Macko, eds.). New York: Plenum Press, 755–783.

Wells, P.V. (1976). Macrofossil analysis of wood rat *(Neotoma)* middens as a key to the Quaternary vegetational history of arid America. *Quaternary Research,* **6,** 223–248.

Wells, P.V. (1979). An equable glaciopluvial in the West: pleniglacial evidence of increased precipitation on a gradient from the Great Basin to the Sonoran and Chihuahuan Deserts. *Quaternary Research,* **12,** 311–325.

Wells, P.V. and Berger, R. (1967). Late Pleistocene history of coniferous woodland in the Mohave Desert. *Science,* **155,** 1640–1647.

Wells, P.V. and Jorgensen, C.D. (1964). Pleistocene wood rat middens and climatic change in the Mohave desert: a record of Juniper woodlands. *Science,* **143,** 1171–1173.

Wendland, W.M. and Bryson, R.A. (1974). Dating climatic episodes of the Holocene. *Quaternary Research,* **4,** 9–24.

Wenk, T. and Siegenthaler, U. (1985). The high-latitude ocean as a control of atmospheric CO_2. In: *The Carbon Cycle and Atmospheric CO_2: Natural Variations Archean to Present* (E.T. Sundquist and W.S. Broecker, eds.). Washington, DC: American Geophysical Union, 185–194.

Westgate, J.A. and Gorton, M.P. (1981). Correlation techniques in tephra studies. In: *Tephra Studies* (S. Self and R.J.S. Sparks, eds.). Dordrecht: D. Reidel, 73–94.

Westgate, J.A. and Naeser, N.D. (1995). Tephrochronology and fission track dating. In: *Dating Methods for Quaternary Deposits* (N.W. Rutter and N.R. Catto, eds.). St. John's: Geological Association of Canada, 15–28.

Weyl, P.K. (1968). The role of the oceans in climatic change: a theory of the ice ages. *Meteorological Monographs,* **8,** 37–62.

Whitlock, C. and Bartlein, P.J. (1997). Vegetation and climate change in northwest North America during the past 125kyr. *Nature,* **388,** 57–61.

Wigley, T.M.L. (1976). Spectral analysis and the astronomical theory of climatic change. *Nature,* **264,** 629–631.

Wigley, T.M.L. (1978). Climatic change since 1000 AD. In: *Evolution des Atmospheres Planetaires et Climatologie de la Terre.* Toulouse: Centre National d'Etudes Spatiales, 313–324.

Wigley, T.M.L. (1991). Climate variability on the 10–100-year timescale: observations and possible causes. In: *Global Changes of the Past* (R.S. Bradley, ed.). Boulder: University Corporation for Atmospheric Research, 83–101.

Wigley, T.M.L., Ingram, M.J., and Farmer, G. (eds.). (1981). *Climate and History: Studies in Past Climates and their Impact on Man.* Cambridge: Cambridge University Press.

Wigley, T.M.L. and Kelly, P.M. (1990). Holocene climatic change, ^{14}C wiggles and variations in solar irradiance. *Philos. Trans. Royal Society of London,* **A330,** 547–560.

Williams, D.F. and Johnson, W.C. (1975). Diversity of recent planktonic foraminifera in the southern Indian Ocean and late Pleistocene paleotemperatures. *Quaternary Research,* **5,** 237–250.

Williams, J., Barry, R.G., and Washington, W.M. (1974). Simulation of the atmospheric circulation using the NCAR global circulation model with ice age boundary conditions. *Jour. Appl. Meteorology,* **13,** 305–317.

Williams, K.M. and Smith, G.G. (1977). A critical evaluation of the application of amino acid racemization to geochronology and geothermometry. *Origins of Life,* **8,** 91–144.

Williams, L.D. (1975). The variation of come elevation and equilibrium line altitude with aspect in eastern Baffin Island, NWT, Canada. *Arctic and Alpine Research,* **7,** 169–181.

Williams, L.D. (1979). An energy balance model of potential glacierization of northern Canada. *Arctic and Alpine Research,* **11,** 443–456.

Williams, L.D., Wigley, T.M.L., and Kelly, P.M. (1981). Climatic trends at high northern latitudes during the last 4000 years compared with ^{14}C fluctuations. In: *Sun and Climate.* Toulouse: Centre National d'Etudes Spatiales.

Williams, N.E. and Eyles, N. (1995). Sedimentary and paleoclimatic controls on caddisfly (Insecta: Trichoptera) assemblages during the last interglacial-to-glacial transition in southern Ontario. *Quaternary Research,* **43,** 90–105.

Williams, N.E., Westgate, J.A., Williams, D.D., Morgan, A., and Morgan, A.V. (1981). Invertebrate fossils (Insecta: Trichoptera, Diptera, Coloptera) from the Pleistocene Scarborough Formation at Toronto, Ontario and their paleoenvironmental significance. *Quaternary Research,* **16,** 146–166.

Williams, R.B.G. (1975). The British climate during the last glaciation; an interpretation based on periglacial phenomena. In: *Ice ages: Ancient and Modern* (A.E. Wright and F. Moseley, eds.). Geological Jour. Special Issue No. 6. Liverpool: Liverpool University Press, 95–120.

Wilson, A.T. and Hendy, C.H. (1971). Past wind strength from isotope studies. *Nature,* **243,** 344–346.

Wilson, A.T. and Hendy, C.H. (1981). The chemical stratigraphy of polar ice sheets — a method of dating ice cores. *Jour. Glaciology,* **27,** 3–9.

Wilson, C.J.N., Ambraseys, N.N., Bradley, J., and Walker, G.P.L. (1980). A new date for the Taupo eruption, New Zealand. *Nature,* **288,** 252–253.

Windom, H.L. (1975). Eolian contributions to marine sediments. *Jour. Sedimentary Petrology,* **45,** 520–529.

Winograd, I.J., Szabo, B.J., Coplen, T.B., and Riggs, A.C. (1988). A 250,000 year climatic record from Great Basin vein calcite: implications for Milankovitch theory. *Science,* **242,** 1275–1280.

Winograd, I.J., Coplen, T.B., Landwehr, J.M., Riggs, A.C., Ludwig, K.R., Szabo, B., Kolesar, P.T., and Revesz, K.M. (1992). Continuous 500,000 year climate record from vein calcite in Devil's Hole, Nevada. *Science,* **258,** 255–260.

Winograd, I.J. and Landwehr, J.M. (1993). A response to "Milankovitch theory viewed from Devil's Hole" by J. Imbrie, A.C. Mix, and D.G. Martinson. *U.S.G.S. Open File Report,* 93–357.

Winograd, I.J., Landwehr, J.M., Ludwig, K.R., Coplen, T.B., and Rigg, A.C. (1997). Duration and structure of the past four interglaciations. *Quaternary Research,* **48,** 141–154.

Wintle, A.G. (1973). Anomalous fading of thermoluminescence in mineral samples. *Nature,* **244,** 143–144.

Wintle, A.G. (1990). A review of current research on the TL dating of loess. *Quaternary Science Reviews,* **9,** 385–397.

Wintle, A.G. (1993). Luminescence dating of aeolian sands: an overview. In: *The Dynamics and Environmental Context of Aeolian Sedimentary Systems* (K. Pye, ed.). Geological Society Spec. Public. No. 72, 49–58.

Wintle, A.G. and Aitken, M.J. (1977). Thermoluminescence dating of burnt flint: application to a lower palaeolithic site, Terra Amata. *Archaeometry,* **19,** 111–130.

Wintle, A.G. and Huntley, D.J. (1979). Thermoluminescence dating of a deep-sea ocean core. *Nature,* **279,** 710–712.

Wintle, A.G., Lancaster, N., and Edwards, S.R. (1994). Infra-red stimulated luminescence (IRSL) dating of the late-Holocene aeolian sands in the Mohave Desert, California, USA. *The Holocene,* **4,** 74–78.

Wintle, A.G., Li, S.H., and Botha, G.A. (1993). Luminescence dating of colluvial deposits from Natal, South Africa. *South African Jour. Science,* **89,** 77–82.

Wittfogel, M.A. (1940). Meteorological records from the divination inscriptions of Shang. *Geographical Review,* **30,** 110–133.

Wohlfarth, B. (1996). The chronology of the last termination: radiocarbon-dated, high resolution terrestrial stratigraphies. *Quaternary Science Reviews,* **15,** 267–284.

Wohlfarth, B., Björck, S., and Possnert, G. (1995). The Swedish time scale: a potential calibration tool for the radiocarbon time scale during the Late Weichselian. *Radiocarbon,* **37,** 347–359.

Woillard, G.M. (1978). Grande Pile peat bog: a continuous pollen record for the past 140,000 years. *Quaternary Research,* **9,** 1–21.

Woillard, G.M. and Mook, W.G. (1982). Carbon-14 dates at Grande Pile: correlation of land and sea chronologies. *Science,* **215,** 159–161.

Wood, E.M. (1983). *Reef Corals of the World: Biology and Field Guide.* Neptune City, New Jersey, T.F.H. Publications.

Woodwell, G.M., Whittaker, R.H., Reiners, W.A., Likens, G.E., Delwiche, C.C., and Botkin, D.B. (1978). The biota and the world carbon budget. *Science,* **199,** 141–146.

Worthington, L.V. (1968). Genesis and evolution of water masses. *Meteorological Monographs,* **8,** 63–67.

Wright, H.E. (1989). The amphi-Atlantic distribution of the Younger Dryas paleoclimatic oscillation. *Quaternary Science Reviews,* **8,** 295–306.

Wright, H.E. and Patten, H.L. (1963). The pollen sum. *Pollen et Spores,* **5,** 445–450.

Wright, H.E., Kutzbach, J.E., Webb III, T., Ruddiman, W.F., Street-Perrott, F.A., and Bartlein, P.J. (eds.). (1993). *Global Climates Since the Last Glacial Maximum*. Minneapolis: University of Minnesota Press.

Wu, G. and Berger, W.H. (1989). Planktonic foraminifera: differential dissolution of the Quaternary stable isotope record in the west Equatorial Pacific. *Paleoceanography,* **4,** 181–198.

Wu, G., Herguera, J.C., and Berger, W.H. (1990). Differential dissolution: modification of Late Pleistocene oxygen isotope records in the western Equatorial Pacific. *Paleoceanography,* **5,** 581–594.

Xia, J., Haskell, B.J., Engstrom, D.R., and Ito, E. (1997). Holocene climate reconstruction from tandem trace-element and stable isotope composition of ostracodes from Coldwater Lake, North Dakota, U.S.A. *Jour. Paleolimnology,* **17,** 85–100.

Xiao, J., Porter, S.C., An, Z., Kumai, H., and Yoshikawa, S. (1995). Grain size of quartz as an indicator of winter monsoon strength on the loess plateau of central China during the last 130,000 yr. *Quaternary Research,* **43,** 22–29.

Yamamoto, T. (1971). On the nature of the climatic change in Japan since the "Little Ice Age" around 1800 AD. *Jour. Meteorological Society of Japan* 49 (special issue), 798–812.

Yao, S. (1942). The chronological and seasonal distribution of floods and droughts in Chinese history 206 BC-1911 AD. *Harvard Jour. Asiatic Studies,* **6,** 273–312.

Yao, S. (1943). The geographical distribution of floods and droughts in Chinese history 206 BC-AD 1911. *Far East Quarterly,* **2,** 357–378.

Yao, S. (1944). Flood and drought data in the *T'u Shu Chi Ch'eng* and the *Ch'ing Shi Kao. Harvard Jour. Asiatic Studies,* **8,** 214–226.

Yao, T., Thompson, L.G., Jiao, K., Mosley-Thompson, E., and Yang, Z. (1995). Recent warming as recorded in the Qinghai-Tibet cryosphere. *Annals of Glaciology,* **21,** 196–200.

Yapp, C.J. and Epstein, S. (1977). Climatic implications of D/H ratios of meteoric water over North America (9,500–22,000 B.P.) as inferred from ancient wood cellulose C-H hydrology. *Earth and Planetary Science Letters,* **34,** 333–350.

Yapp, C.J. and Epstein, S. (1982). Climatic significance of the hydrogen isotope ratios in tree cellulose. *Nature,* **297,** 636–639.

Yiou, F., Raisbeck, G.M., Bourles, D., Lorius, C., and Barkov, N.I. (1985). ^{10}Be at Vostok Antarctica during the last climatic cycle. *Nature,* **316,** 616–617.

Yoshimura, M. (1992). Historical weather data base system in Japan. In: *Proc. International Symposium on the Little Ice Age Climate*. (T. Mikami, ed.). Department of Geography, Tokyo Metropolitan University, 239–243.

Yoshimura, M. (1996). *Climatic condition in latter half of the Little Ice Age*. Department of Geography, Tamanashi University.

Yoshino, M.M. (1981). Orographically-induced atmospheric circulations. *Progress in Physical Geography,* **5,** 76–98.

Young, M. and Bradley, R.S. (1984). Insolation gradients and the paleoclimatic record. In: *Milankovitch and climate, Part 2* (A.L. Berger, J. Imbrie, J. Hays, G. Kukla, and B. Saltzman, eds.). Dordrecht: D. Reidel, 707–713.

Yu, G. and Harrison, S.P. (1995). *Lake Status Records from Europe: Data Base Documentation*. Paleoclimatology Publication Series Report No. Boulder: 3. World Data Center-A for Paleoclimatology.

Zahn, R., Winn, K., and Sarnthein, M. (1986). Benthic foraminiferal δ^{13}C and accumulation rates of organic carbon: *Uvigerina peregrina* group and *Cibicidoides wuellerstorfi. Paleoceanography,* **1,** 27–42.

Zetterberg, P., Eronen, M., and Briffa, K.R. (1995). A 7500-year pine tree-ring record from Finnish Lapland and its application to paleoclimatic studies. In: *Proc. International Conference on Past, Present and Future Climate* (P. Heilinheimo, ed.). Helsinki: Academy of Finland, 151–154.

Zetterberg, P., Eronen, M., and Lindholm, M. (1996). The mid-Holocene climatic change around 3800 B.C.: tree-ring evidence from northen Fennoscandia. In: *Holocene Treeline Oscillations, Dendrochronology and Palaeoclimate* (B. Frenzel, H.H. Birks, T. Alm, and K-D. Vorren, eds.). Stuttgart: Gustav Fischer Verlag, 135–146.

Zhang, D. (1983). Analysis of dust rain in the historic times of China. *Kexue Tongbao*, **28**, No. 3, 361–366.

Zhang, D. (1980). Winter temperature changes during the last 500 years in South China. *Kexue Tongbao*, **25**, No. 6, 497–500.

Zhang, J. (ed.). (1988). *The Reconstruction of Climate in China for Historical Times*. Beijing: Science Press.

Zhang, P. and Gong, G. (1979). Some characteristics of climatic fluctuations in China since the 16th century. *Acta Meteorologica Sinica*, **34**, 238–247 (in Chinese with English summary: translation available from E.J. Bradley, c/o author).

Zhao, M., Beveridge, N.A.S., Shackleton, N.J., Sarnthein, M., and Eglinton, G. (1995). Molecular stratigraphy of cores off northwest Africa: sea surface temperature history over the last 80 ka. *Paleoceanography*, **10**, 661–675.

Zheng, J. and Zheng, S. (1992). The reconstruction of climate and natural disaster in Shandong province during historical time. In: *Collected Papers in Geography* [publisher unknown].

Zheng, S. and Feng, L. (1986). Historical evidence on climatic instability above normal in cool periods in China. *Scientia Sinica* (Series B), **24**, 441–448.

Zhou, W., Donahue, D.J., Porter, S.C., Jull, T.A., Li, X., Stuiver, M., An, Z., Matsumoto, E., and Dong, G. (1996). Variability of monsoon climate in East Asia at the end of the last glaciation. *Quaternary Research*, **46**, 219–229.

Zielinski, G. (1995). Stratospheric loading and optical depth estimates of explosive volcanism over the last 2100 years as derived from the GISP2 Greenland ice core. *Jour. Geophys. Research*, **100D**, 20937–20955.

Zielinski, G., Mayewski, P.A., Meeker, L.D., Whitlow, S., Twickler, M.S., Morrison, M., Meese, D.A., Gow, A.J., and Alley, R.B. (1994). Record of explosive volcanism since 7000 B.C. from the GISP2 Greenland ice core and implications for the volcano-climate system. *Science*, **264**, 948–952 (see also comments and discussion, (1995): *Science*, **267**, 256–258).

Zielinski, G., Mayewski, P.A., Meeker, L.D., Whitlow, S., Twickler, M.S. (1996). A 110,000-yr record of explosive volcanism from the GISP2 (Greenland) ice core. *Quaternary Research*, **45**, 109–118.

Zielinski, G., Mayewski, P.A., Meeker, L.D., Grönvold, K., Germani, M.S., Whitlow, S., Twickler, M.S., and Taylor, K. (1997). Volcanic aerosol records and tephrochronology of the Summit, Greenland ice cores. *Jour. Geophys. Research*, **102C**, 26625–26640.

Zolitschka, B. (1991). Absolute dating of late Quaternary lacustrine sediments by high resolution varve chronology. *Hydrobiologia*, **214**, 59–61.

Zolitschka, B. (1996a). Recent sedimentation in a high arctic lake, northern Ellesmere Island, Canada. *Jour. Paleolimnology*, **16**, 169–186.

Zolitschka, B. (1996b). High resolution lacustrine sediments and their potential for paleoclimatic reconstruction. In: *Climate Variations and Forcing Mechanisms of the Last 2000 Years* (P.D. Jones, R.S. Bradley, and J. Jouzel, eds.). Berlin: Springer-Verlag, 453–478.

Zöller, L., Oches, E.A., and McCoy, W.D. (1994). Towards a revised chronostratigraphy of loess in Austria, with respect to key sections in the Czech Republic and in Hungary. *Quaternary Science Reviews*, **13**, 465–472.

Zumbühl, H.J. (1980). *Die Schwankungen der Grindelwaldgletscher in den historischen Bild- und Schriftquellen des 12. bis 19. Jahrhunderts*. Basel: Birkhäuser Verlag.

Copyright Acknowledgements

I gratefully acknowledge the following copyright holders who have kindly provided permission to reproduce the figures indicated. Sources of all figures are referenced in each figure caption and the full citations are found in the **References** section (p. 513–594):

Fig. 1.1: Cambridge University Press; Fig. 2.1: World Meteorological Organisation; Fig. 2.5: E.M. Van Zinderen Bakker; Fig. 2.6: G. Kukla and Cambridge University Press; Figs. 2.8, 2.9: American Meteorological Society; Fig. 2.10: G. Kukla; Fig. 2.11, 2.12, 2.13: Cambridge University Press; Fig. 2.15: J. Kutzbach; Fig. 2.16: J. Wiley and Sons; Fig. 2.17: J. Imbrie; Figs. 2.18, 2.19: A. Berger and the Editor, *Il Nuovo Cimento;* Fig. 2.20: © Kluwer Academic Publishers; Figs. 2.21, 2.22: American Geophysical Union; Fig. 3.2: from *Earth Science Reviews,* 4, Olsson, I.U. Modern aspects of radiocarbon dating. 203–18, © 1968, with kind permission from Elsevier Science Ltd, The Boulevard, Langford Lane, Kidlington OX5 1GB, U.K.; Fig. 3.3: J. Mangerud and the Editor, *Boreas;* Fig 3.4: I.U. Olsson; Fig. 3.5: I.U.Olsson and the Editor, *Geologiska Foereningens I Stockholm Foertlandlinger;* Figs. 3.6, 3.7: American Geophysical Union; Fig. 3.8: from Eddy, J.A. 1977. Climate and the changing sun. *Climatic Change* 1, p.178, Fig. 1, © 1977 Kluwer Academic, with kind permission from Kluwer Academic Publishers; Figs. 3.9, 3.10: the Editor, *Radiocarbon;* Figs. 3.11, 3.12: K. Hughen and the publisher, from *Nature,* 391, 65–68, © 1998 Macmillan Magazines Ltd.; Fig. 3.13: reprinted with permission from Stuiver, M. and P.D. Quay 1980. Changes in atmospheric carbon-14 attributed to a variable sun. *Science,* 207, 11–19, © 1980 American Association for the Advancement of Science; Fig. 3.14: *Quaternary Research;* Fig. 3.15: Cambridge University Press; Fig. 3.16: *Quaternary Research;* Fig. 3.17: *World Archeology;* Fig. 3.19: from W.S. Broecker and M. L. Bender in: *Calibration of hominoid evolution* (eds. W.W. Bishop and J.A. Miller) by kind permission of Scottish Academic Press; Fig. 3.20: from Ku, T.-L. 1976. The uranium series method of age determination. *Annual Reviews of Earth and Planetary Science,* 4, 347–80; Fig. 3.21: American Geophysical Union; Fig. 3.22, 3.23: Geological Society of London; Fig. 3.24: *South African Journal of Science;* Fig.3.25: Arnold Publishers; Fig. 4.2: reprinted with permission from Verosub, K.L. 1975. Paleomagnetic excursions as magnetostratigraphic horizons: a cautionary note. *Science,* 190, 48–50, © 1980 American Association for the Advancement of Science; Fig. 4.4: *Quaternary Research;* Fig. 4.6: reprinted from: *Geochimica et Cosmochimica Acta,* 56, Goodfriend, G.A., P.E. Hare and E.R.M. Druffel, Aspartic acid racemization and protein diagenesis in corals over the last 350 years, 3847–3850, © 1992, with kind permission from Elsevier Science; Fig. 4.7: R. Oches; Fig. 4.8: reprinted from: *Quaternary Science Reviews,* 14, Oches, E.A., and W.D. McCoy, Amino acid geochronology applied to the correlation and dating of central European loess deposits, 767–782. © 1995, with kind permission from Elsevier Science; Fig. 4.9: reprinted from: *Quaternary Science Reviews,* 4, Miller, G.H. and J. Mangerud, Aminostratigraphy of European marine interglacial deposits, 215–278, © 1985, with kind permission from Elsevier Science; Fig. 4.10: from G.H. Miller and P.E. Hare, in *Biogeochemistry of Amino Acids* (eds. P.E. Hare et al.) by permission of J. Wiley and Sons; Fig. 4.11: from Wehmiller, J.F., 1993. Applications of organic geochemistry for Quaternary research: aminostratigraphy and aminochronology. In: *Organic Geochemistry* (eds. M.H. Engel and S.A. Macko). Plenum Press, New York, 755–783; Fig. 4.12: *Quaternary Research;* Fig. 4.13: Cambridge University Press; Fig. 4.14a: from *Arctic and Alpine Research,* by permission of the Regents of the Unversity of Colorado; Fig. 414b: Arnold Publishers; Fig. 4.15: J. Matthews; Fig. 4.16: *Quaternary Research;* Fig. 4.17: Luckman, B.H., Evidence for climatic conditions between 900–1300 A.D. in the southern Canadian Rockies. *Climatic Change,* 26, 171–182, © 1994 Kluwer Academic, with kind permission from Kluwer Academic Publishers;.Fig. 5.3: from S. Epstein and R.J. Sharp, *J. Geology,* 67, 88–102 by permission of the publisher, © 1959 by the University of Chicago; Fig. 5.4: from G. de Q. Robin, *Phil. Trans. Royal Society,* Series B, 280, 143–168 by permission of The Royal Society and the author; Fig. 5.5: the Geochemical Society; Fig. 5.6: American Geophysical Union; Fig.5.7: E. Picciotto and publisher, from *Nature* 187, 857–159, © 1960 Macmillan Magazines Ltd.; Fig. 5.8: from *Earth and Planetary Science Letters,* 3, Aldaz, L. and S. Deutsch, On a relationship between air temperature and oxygen isotope ratio of snow and firn in the South Pole region, 267–74, © 1967, with kind permission from Elsevier Science;

Figs. 5.9, 5.10: American Geophysical Union; Fig. 5.11: S. Johnsen and the publisher, from *Nature*, 235, 429–434, © 1972 Macmillan Magazines Ltd.; Fig. 5.12: reprinted from the *Annals of Glaciology* with permission of the International Glaciological Society and J. Steffensen; Fig. 5.13: K. Taylor and the publisher, from *Nature*, 361, 432–436, © 1993 Macmillan Magazines Ltd.; Fig. 5.14: K. Hammer and the publisher, from *Nature*, 288, 230–235, © 1980 Macmillan Magazines Ltd.; Fig. 5.15: reprinted with permission from Meese, D.A., A.J. Gow, P Grootes, P.A. Mayewski, M. Ram, M. Stuiver, K.C. Taylor, E.D. Waddington and G.A. Zielinski, The accumulation record from the GISP2 core as an indicator of climate change throughout the Holocene. *Science* 266, 1680–1682, © 1994 American Association for the Advancement of Science; Fig. 5.16: from *Quaternary International,* 10–12, N. Reeh, The last interglacial as recorded in the Greenland Ice Sheet and Canadian Arctic Ice Caps, 123–142. © 1991, with kind permission from Elsevier Science; Fig. 5.17: C. Lorius, and the publisher, from *Nature* 316, 591–596. © 1985 Macmillan Magazines Ltd; Fig. 5.18: *Quaternary Research*; Fig. 5.19: from Raisbeck, G.M., F. Yiou, J. Jouzel, J.R. Petit, N.I. Barkov and E.Bard 1992. 10Be deposition at Vostok, Antarctica during the last 50,000 years and its relationship to possible cosmogenic production variations during this period, 127–139, and Beer, J., S.J. Johnsen, G. Bonani, R.C. Finkel, C.C. Langway, H. Oeschger, B. Stauffer, M. Suter and W. Woelfli, 10Be peaks as time markers in polar ice cores, 141–153, In: *The Last Deglaciation: Absolute and Radiocarbon Chronologies*, (eds. E. Bard and W. Broecker). © 1992 Springer-Verlag, Berlin; Fig. 5.20: Jouzel, J., J.R. Petit, N.I. Barkov, J.M. Barnola, J. Chappellaz, P. Ciais, V.M. Kotlyakov, C. Lorius, V.N. Petrov, D. Raynaud and C. Ritz 1992. The last deglaciation in Antarctica: further evidence of a "Younger Dryas" type climatic event, 229–266, in: *The Last Deglaciation: Absolute and Radiocarbon Chronologies*, (eds. E. Bard and W. Broecker). © 1992 Springer-Verlag, Berlin; Fig. 5.21: J. Petit and the publisher, from *Nature*, 387, 359–360, © 1997 Macmillan Magazines Ltd.; Fig. 5.22: W. Dansgaard and the publisher, from *Nature*, 364, 218–220, © 1993 Macmillan Magazines Ltd.; Fig. 5.23: S. Johnsen and the publisher, from *Nature*, 359, 311–313, © 1992 Macmillan Magazines Ltd.; Fig. 5.24: W. Dansgaard and the publisher, from *Nature*, 532–534, © 1989 Macmillan Magazines Ltd.; R. Alley and the publisher, from *Nature*, 362, 527–529, © 1993 Macmillan Magazines Ltd.; Fig. 5.26: K. Taylor and the publisher, from *Nature*, 361, 432–436, © 1993 Macmillan Magazines Ltd.; Fig. 5.28: from *Quaternary International*, 10–12, N. Reeh, The last interglacial as recorded in the Greenland Ice Sheet and Canadian Arctic Ice Caps, 123–142. © 1991, with kind permission from Elsevier Science; Fig. 5.29: reprinted with permission from Raynaud, D., J. Jouzel, J.M. Barnola, J. Chappellaz, R.J. Delmas and C. Lorius, The ice record of greenhouse gases. *Science* 259, 926–934. © 1993 American Association for the Advancement of Science; Fig. 5.30: D. Raynaud and University Corporation for Atmospheric Research; Fig. 5.31: Munksgaard International Publishers; Fig. 5.32: J. Jouzel and the publisher, from *Nature*, 364, 407–412, © 1993 Macmillan Magazines Ltd.; Fig. 5.33: T. Blunier and the publisher, from *Nature*, 374, 46–49, © 1995 Macmillan Magazines Ltd.; Fig. 5.34: J. Chappellaz and the publisher, from *Nature*, 366, 443–445, © 1993 Macmillan Magazines Ltd.; Fig. 5.35: M. Leuenberger and the publisher, from *Nature*, 360, 449–451, © 1992 Macmillan Magazines Ltd.; Fig. 5.36: J.R. Petit and the publisher, from *Nature*, 343, 56–58, © 1990 Macmillan Magazines Ltd.; Fig. 5.37: American Geophysical Union; Fig. 5.38: M. Legrand and the publisher, from *Nature*, 327, 671–676, © 1987 Macmillan Magazines Ltd.; Fig. 5.39: American Geophysical Union; Fig. 5.40: M. Bender and the publisher, from *Nature*, 372, 663–666, © 1994 Macmillan Magazines Ltd.; Fig. 5.41: reprinted with permission from Sowers, T. and M. Bender, Climate records covering the last deglaciation. *Science* 269, 210–214, © 1995 American Association for the Advancement of Science; Figs. 5.42, 5.43: American Geophysical Union; Figs. 5.44, 5.45: reprinted with permission from Thompson, L.G., T. Yao, M.E. Davis, K.A. Henderson, E. Mosley-Thompson, P-N. Lin, J. Beer, H-A. Synal, J. Cole-Dai and J.F. Bolzan, Tropical climate instability: the last glacial cycle from a Qinghai-Tibetan ice core. *Science*, 276, 1821–1825.© 1997 American Association for the Advancement of Science; Figs. 5.47: reprinted from the *Annals of Glaciology* with permission of the International Glaciological Society and L. Thompson; Fig. 6.1: W. Hay; Fig. 6.2: W.F. Ruddiman; Fig. 6.3, 6.5: A. De Vernal; Fig. 6.6: A. Boersma, from *Introduction to Marine Micropaleontology* (eds. B.U. Haq and A. Boersma) © 1978 by Elsevier Science; Fig. 6.9: American Geophysical Union;

Fig. 6.10: Fig. 5.32: J-C Duplessy and the publisher, from *Nature*, 358, 485–488, © 1992 Macmillan Magazines Ltd.; Fig. 6.11: American Geophysical Union; Fig. 6.13: B.K. Linsley and the publisher, from *Nature*, 380, 234–237, © 1996 Macmillan Magazines Ltd.; Fig. 6.15: *Quaternary Research*; Fig. 6.16: Raymo, M.E., Global climate change: a three million year perspective. In: *Start of a Glacial* (eds. G. J. Kukla and E. Went), 207–223. © 1992 Springer-Verlag, Berlin; Fig. 6.17: A. Bloom; Fig. 6.20: *Quaternary Research*; Fig. 6.21: from *Late Cenozoic Glacial Ages* (ed. K.K. Turekian) by permission of Yale University Press and J. Imbrie © 1971 Yale University Press; Fig. 6.22: *Quaternary Research*; Figs. 6.23, 6.24, 6.25: from: McIntyre, A., N.G. Kipp, A.W.H. Bé, T. Crowley, T. Kellogg, J.V. Gardner, W. Prell, and W.F. Ruddiman, Glacial North Atlantic 18,000 years ago: a CLIMAP reconstruction. In: *Investigation of Late Quaternary Paleooceanography and Paleoclimatology,* (eds. R.M. Cline and J.D. Hays). Memoir 145, Geological Society of America, Boulder, 43–76. Reproduced with permission of the publisher, the Geological Society of America, Boulder, Colorado, U.S.A. Copyright © 1976 Geological Society of America; Figs. 6.26, 6.27, 6.28: from *Marine Micropaleontology*, 5, 215–247 by permission of T.C. Moore and Elsevier Science; Fig. 6.29: *Quaternary Research*; Fig. 6.30: American Geophysical Union; Fig. 6.31: F. Rostek and the publisher, from *Nature*, 364, 319–321, © 1993 Macmillan Magazines Ltd.; Fig. 6.32: A.W.H. Bé; Fig. 6.33: from Berger, W.H. and E.L. Winterer 1974. Plate stratigraphy and the fluctuating carbonate line. In: *Pelagic Sediments on Land and in the Ocean,* (eds. K.J. Hsu and H. Jenkins). Special Publication No. 1, International Association of Sedimentologists, © 1974 Blackwell Scientific, Oxford; Fig. 6.34: from *Deep Sea Research,* 14, Ruddiman, W.F. and B.C. Heezen, Differential solution of planktonic foraminifera, 801–808, © 1967, with kind permission from Elsevier Science; Figs. 6.35, 6.36, 6.37: American Geophysical Union; Fig. 6.38: *Quaternary Research*; Fig. 6.39: R. Dunbar; Fig. 6.40: reprinted with permission from Mitsuguchi, T., E. Matsumoto, O. Abe, T. Uechida and P.J. Isdale 1996. Mg/Ca thermometry in coral skeletons. *Science*, 274, 961–963, © 1995 American Association for the Advancement of Science; Fig. 6.41: Dunbar, R.B., B.K. Linsley and G.M. Wellington, Eastern Pacific corals monitor El Niño/Southern Oscillation, precipitation and sea surface temperature variability over the past 3 centuries. In: *Climate Variations and Forcing Mechanisms of the Last 2,000 years.* (eds. P.D. Jones, R.S. Bradley and J. Jouzel). © 1996 Springer-Verlag, Berlin; Fig.6.42: American Geophysical Union; Fig. 6.43: The Open University; Fig. 6.44: American Geophysical Union; Fig. 6.45: J. Wiley and Sons; Fig. 6.46: *Quaternary Research*; Fig. 6.47: from *Geology*, 23, 301–304, Dowdeswell, J.A., M.A. Maslin, J.T. Andrews and I.N. McCave, Iceberg production, debris rafting and the extent and thickness of Heinrich layers (H-1, H-2) in North Atlantic sediments. Reproduced with permission of the publisher, the Geological Society of America, Boulder, Colorado, U.S.A. Copyright © 1995 Geological Society of America; Fig. 6.48: G. Bond and the publisher, from *Nature*, 365, 143–147, © 1993 Macmillan Magazines Ltd.; Fig. 6.49: reprinted with permission from Bond, G.C. and R. Lotti, 1995. Iceberg discharges into the North Atlantic on millennial time scales during the Last Glaciation. *Science*, 267, 1005–1010. © 1995 American Association for the Advancement of Science; Fig. 6.50, 6.51, 6.52, 6.53: American Geophysical Union; Fig. 6.54: reprinted from: *Quaternary Science Reviews*, 11, Shackleton, N.J., J. Le, A. Mix and M.A. Hall, 1992. Carbon isotope records from Pacific surface waters and atmospheric carbon dioxide, 387–400. © 1992, with kind permission from Elsevier Science; Fig. 6.55: R.S. Ganeshram and the publisher, from *Nature*, 376, 755–758, © 1995 Macmillan Magazines Ltd.; Fig. 7.1: Academic Press; Fig. 7.3: from *Quaternary International,* 17, Liu, T., Z. Ding, Z. Yu and N. Rutter, 1993. Susceptibility time series of the Baoji section and the bearings on paleoclimatic periodicities in the last 2.5 Ma 33–38. © 1993, with kind permission from Elsevier Science; Fig. 7.4: National Research Council Research Press; Fig. 7.5: reprinted from: *Quaternary Science Reviews,* 13, Ding, Z., N. Yu, N.W. Rutter and T. Liu, 1994. Towards an orbital time scale for Chinese loess deposits, 39–70 © 1994, with kind permission from Elsevier Science; Fig. 7.6: *Quaternary Research*; Fig. 7.7: Arnold Publishers; Fig. 7.8: from *Earth Science Reviews*, 15, Washburn, A.L., Permafrost features as evidence of climatic change, 327–402. © 1979, with kind permission from Elsevier Science; Fig. 7.9: *Biuletyn Periglacjalny;* Fig. 7.10: T. Péwé; Fig. 7.11: reprinted from the *Journal of Glaciology* with permission of the International Glaciological Society and S. Hastenrath; Fig. 7.12: from *Arctic and Alpine Research*, by permission of the Regents of

 **INDEX**

Note: Page numbers followed by the letter f refer to the figure on that page. Page numbers followed by the letter t refer to the table on that page.

International Geophysics Series

EDITED BY

RENATA DMOWSKA
Division of Applied Science
Harvard University
Cambridge, Massachusetts

JAMES R. HOLTON
Department of Atmospheric Sciences
University of Washington
Seattle, Washington

*Out of Print